New Applications of NMR in Drug Discovery and Development

New Developments in NMR

Editor-in-Chief:
Professor William S. Price, *University of Western Sydney, Australia*

Series Editors:
Professor Bruce Balcom, *University of New Brunswick, Canada*
Professor István Furó, *Industrial NMR Centre at KTH, Sweden*
Professor Masatsune Kainosho, *Nagoya University, Japan*
Professor Maili Liu, *Chinese Academy of Sciences, Wuhan, China*

Titles in the Series:
1: Contemporary Computer-Assisted Approaches to Molecular Structure Elucidation
2: New Applications of NMR in Drug Discovery and Development

How to obtain future titles on publication:
A standing order plan is available for this series. A standing order will bring delivery of each new volume immediately on publication.

For further information please contact:
Book Sales Department, Royal Society of Chemistry, Thomas Graham House, Science Park, Milton Road, Cambridge, CB4 0WF, UK
Telephone: +44 (0)1223 420066, Fax: +44 (0)1223 420247
Email: booksales@rsc.org
Visit our website at www.rsc.org/books

New Applications of NMR in Drug Discovery and Development

Edited by

Leoncio Garrido
Consejo Superior de Investigaciones Científicas, Spain
Email: lgarrido@cetef.csic.es

Nicolau Beckmann
Novartis Institutes for BioMedical Research, Switzerland
Email: nicolau.beckmann@novartis.com

RSC Publishing

New Developments in NMR No. 2

ISBN: 978-1-84973-444-8
ISSN: 2044-253X

A catalogue record for this book is available from the British Library

Published by The Royal Society of Chemistry,
Thomas Graham House, Science Park, Milton Road,
Cambridge CB4 0WF, UK

Registered Charity Number 207890

For further information see our website at www.rsc.org

Preface

Nuclear Magnetic Resonance (NMR) is a most informative tool for gaining insight into the inner workings of nature and is useful in many fields of applied science. The gained knowledge allows us to develop advanced medical methods of treating diseases and to develop preventive procedures. Often NMR represents even the ultimate tool of insight in drug discovery and development, for example. The information gained by NMR is in the correct form expected by a medical scientist and it can be applied directly for discovering new potential drugs and for working out prescriptions of their optimum application to treat diseases.

The comprehension of biomedical effects, induced by drugs, requires a molecular view. Molecules are interacting and are leading to beneficial results or to harmful side effects. In this sense, pharmaceutical science is part of applied chemistry where molecules are the objects of central interest. Without a molecular view, it is difficult to understand the functioning of drugs. There is a fortuitous match between the demands of pharmaceutical practice and the outstanding features of NMR as an investigative tool. The results of NMR experiments are also best understood in terms of molecular interactions. This is one of the reasons for the unique importance of NMR in biomedicine. Comprehension is often a question of using the correct language and the proper terms that can be understood and applied directly in the relevant biomedical context.

The match between measurement results and the biomedical needs is implemented rather perfectly in the various chapters of this comprehensive treatise. The *Part I. Small Molecules, Proteins, Cellular Systems*, reflects the fact that many of the most potent drugs are small molecules. Indeed, nature takes advantage of small molecules as agents and transporters with their multivariate functions that adapt well to the needs of the biomedical organism. Small molecules are easier to tailor-make to match the requirements, their

New Developments in NMR No. 2
New Applications of NMR in Drug Discovery and Development
Edited by Leoncio Garrido and Nicolau Beckmann
© The Royal Society of Chemistry 2013
Published by the Royal Society of Chemistry, www.rsc.org

synthesis is manageable in an industrial environment and they are easier to comprehend. Small proteins are the first choices of nature as well as of the drug designer who has to take into account the specific properties of the cellular environment where the drugs must exert their action.

In many, if not in most circumstances, an extreme reductionist approach of analytical science is inappropriate. Physiological action involves the entire biological organism and can not be understood solely in terms of its parts. This view is reflected in *Part II. The Whole Organism in Vivo*, where it is attempted to balance an analytical approach with a holistic view of the entire organism. Again, NMR is astonishingly adaptive to the various degrees of focusing. NMR technology acts like a universal microscope that can be adapted to macroscopic objects, such as a human heart or a brain, or focused onto nano-scale objects, such as cells and tissues, or used to explore truly molecular processes, like the interaction of proteins and nucleic acids. Various tools of magnetic resonance allow the focusing onto the features of actual interest. Many of them are described in this book. They lead to the remarkable adaptability of NMR to the actual practical situation.

In *Part III. Translational drug discovery: From Biological Models to the Clinics*, the possibilities of NMR and MRI are discussed in the context of a number of relevant clinical applications of magnetic resonance techniques, from tissue engineering, to the exploration of psychiatric disorder, to neuro-degenerative diseases, to respiratory diseases, to cardiac MR, and finally to MR applications in cancer.

This volume convincingly demonstrates the enormous breadth of MRI applications in biomedicine. The utility of modern NMR and MRI becomes even more impressive when one realizes that biomedicine is only one of the potential fields of fruitful application of magnetic resonance technology. Other fields are developing equally rapidly today, including magnetic resonance studies of battery materials, NMR in the context of nanotechnology, and materials research in general, to mention just a few. One of the most important fields is and will remain the study of protein and nucleic acid interactions for biomedical understanding and for drug design.

In summary, the volume "New Applications of NMR in Drug Discovery and Development" represents an important addition to the bookshelf of anybody seriously interested in drug discovery and development. It will remain a reliable source of information in this important field for many years to come. I would like to congratulate and thank Nicolau Beckmann and Leoncio Garrido for their efforts towards such a comprehensive survey on this field of remarkable current activity.

Richard R. Ernst
Laboratorium für Physikalische Chemie
ETH Zürich, Switzerland

Contents

New Developments in NMR No. 2
New Applications of NMR in Drug Discovery and Development
Edited by Leoncio Garrido and Nicolau Beckmann
© The Royal Society of Chemistry 2013
Published by the Royal Society of Chemistry, www.rsc.org

Chapter 4 **In-Cell NMR Spectroscopy to Study Protein–Drug Interactions**   **134**
Jacqueline D. Washington, David S. Burz and Alexander Shekhtman

Part II The Whole Organism *In Vivo*

Anatomy, Function, Metabolism and Cellular/Molecular Imaging

Chapter 5 **Increased Sensitivity Using Cryogenic Radiofrequency Coils: Application to *In Vivo* Phenotyping of Mice**   **165**
J. Klohs, A. Seuwen, A. Schröter, D. Marek and M. Rudin

Introduction

LEONCIO GARRIDO AND NICOLAU BECKMANN

The discovery and development of a biological active molecule with therapeutic properties is an ever-increasing complex task, highly unpredictable at the early stages and marked, in the end, by high rates of failure. As a consequence, the overall process leading to the production of a successful drug is very costly. The improvement of the net outcome in drug discovery and development would require, amongst other important factors, a good understanding of the molecular events that characterize the disease or pathology in order to better identify likely targets of interest, to optimize the interaction of an active agent (small molecule or macromolecule of natural or synthetic origin) with those targets and to facilitate the study of the pharmacokinetics, pharmacodynamics and toxicity of an active agent in suitable models and in human subjects. Nuclear magnetic resonance (NMR)-based techniques and methods have proven to be extremely useful and very successful in supporting several steps of this endeavor.

The fruitful application of NMR in pharmaceutical research and development rests on its sensitivity to the chemical structure and molecular dynamics, in addition to the non-invasive and non-destructive nature of the technique. Furthermore, since the discovery of the nuclear magnetic moment of protons by Stern,[1,2] about 75 years ago, the development of a method to measure nuclear magnetic moments by Rabi[3] and the NMR experiments performed by Bloch et al.[4] and Purcell et al.,[5] the technique has been enriched with numerous advances, several of which have been recognized with the Nobel prize (see Table 1). A brief account of the main contributions to NMR advances can be found in ref. 6.

New Developments in NMR No. 2
New Applications of NMR in Drug Discovery and Development
Edited by Leoncio Garrido and Nicolau Beckmann
© The Royal Society of Chemistry 2013
Published by the Royal Society of Chemistry, www.rsc.org

Table 1. List of Nobel prices awarded for scientific contributions in NMR[a].

Recipients	Year	Contribution
Otto Stern	1943	Discovery of magnetic moment of ^{1}H nuclei
Isidor Rabi	1944	First NMR measurement
Felix Bloch and Edward M. Purcell	1952	NMR measurements in solution and solid state
Richard R. Ernst	1991	High resolution NMR spectroscopy
Kurt Wüthrich	2002	NMR for 3D structure elucidation of biological macromolecules
Paul C. Lauterbur and Peter Mansfield	2003	NMR imaging

[a]http://www.nobelprize.org

Concerning drug discovery and development, in addition to the identification of compounds and targets, and the determination of molecular structure, NMR has played an emerging role in screening and hit validation, and more recently in the so-called fragment-based drug-discovery approaches.[7–9]

The versatility and continuous advances of NMR have led to a highly active and fertile field of work with numerous publications. Thus, it represents a challenge to edit a book on the topic, assure novelty and capture the reader's interest. With this in mind and in the framework of the new series entitled *New Developments in NMR*, published by the Royal Society of Chemistry (UK), this book on *New Applications of NMR in Drug Discovery and Development* is produced. The purpose of the book is to summarize recent developments in NMR for applications in pharmaceutical research, focusing not only on consolidated but also on emerging techniques and methods that, at present, are not widely applied, but could contribute to the advancement of this field in the near future.

The increasing availability of high field magnets, cryogenic probes and new methods such as nuclear hyperpolarization are improving the sensitivity of the technique.[10,11] In addition, the development of new and fast data acquisition methods and improved data processing capabilities will undoubtedly facilitate the study *in situ* of the chemistry of living systems. The combination of all these elements is likely to provide new opportunities for drug design and for the *in vivo* investigation of pharmacodynamics, metabolism and toxicity.

The first part of the book consists of four chapters dedicated to the description of NMR as a tool for the analysis of chemicals and their inter-actions with targets. Particular emphasis is put on applications with un-exploited potential for drug discovery and development. The first chapter describes the most important state-of-the-art NMR procedures used in drug discovery, from screening to experiments providing structural information on both ligand and receptor in the bound state. Also, the applicability of advanced NMR methodologies to larger macromolecules and membrane-bound systems is discussed. The second chapter revises multinuclear and multidimensional solid-state NMR spectroscopy applied to the structural characterization of biomolecules that are potential targets for drugs, to the study of pharmaceutical molecules bound to target biomolecules and to the characterization of crystalline and amorphous small-molecule drugs as pure substances and in various complex dosage forms. The third chapter illustrates the application of high-resolution NMR to metabolic profiling in investigations covering distinct phases of drug research and development. A glimpse on how in-cell NMR can provide direct information on protein–drug interactions in the intracellular environment is presented in the fourth chapter.

The second part of the book includes the description of NMR approaches to investigate *in vivo* models of interest in drug discovery and development, with the attention focused to anatomy, function, metabolism and molecular-cellular aspects. Seven chapters cover in detail recent improvements in NMR at high and low fields, alone and combined with other imaging modalities, cryogenic radiofrequency coil design and signal enhancement strategies using hyperpolarization and contrast agents, summarizing efforts to attain the best possible anatomical, biochemical and physiological information in living systems, from cells to small animals to humans.

The third part centers on the application of *in vivo* NMR to the identification and characterization of potential biomarkers with the aim of monitoring the outcome of therapeutic intervention in selected human diseases, including the study of drug metabolism and toxicity. Different aspects of magnetic resonance imaging/spectroscopy (MRI/MRS) in pharmaceutical research are addressed in eight chapters, from opportunities, challenges of translational research and *in vivo* evaluation of drug safety to applications in cartilage therapeutics and tissue engineering, psychiatric disorders, neurodegeneration and respiratory, cardiovascular and cancer diseases.

In the domain of drug discovery, the pharmacological and biomedical questions constitute the center of attention. In this sense, it is fundamental to keep in mind the strengths and limitations of each analytical or imaging technique. Despite its flexibility, NMR will not escape this critical observation. At the end, the judicious application of the technique with the aim of supporting the search for answers to manifold questions arising during a long and painstaking path will continue ensuring NMR a role within the complex area of drug discovery and development.

References

1. R. Frisch and O. Stern, *Zeits. Physik*, 1933, **85**, 4.
2. I. Estermann and O. Stern, *Zeits. Physik*, 1933, **85**, 17.
3. I. Rabi, J. Zacharias, S. Millman and P. A. Kusch, *Phys. Rev.*, 1938, **53**, 318.
4. F. Bloch, W. Hansen and M. E. Packard, *Phys. Rev.*, 1946, **70**, 474.
5. E. M. Purcell, H. Torrey and R. V. Pound, *Phys. Rev.*, 1946, **69**, 37.
6. J. W. Emsley and J. Feeney, *Progr. NMR Spectrosc.*, 2007, **50**, 179.
7. D. A. Middleton, *Ann. Reports NMR Spectrosc.*, 2007, **60**, 39.

8. M. Pellecchia, I. Bertini, D. Cowburn, C. Dalvit, E. Giralt, W. Jahnke, T. L. James, S. W. Homans, H. Kessler, C. Luchinat, B. Meyer, H. Oschkinat, J. Peng, H. Schwalbe and G. Siegal, *Nat. Rev. Drug Discov.*, 2008, **7**, 738.

9. R. Powers, *Expert Opin. Drug. Discov.*, 2009, **4**, 1077.

10. F. A. Gallagher, M. I. Kettunen and K. M. Brindle, *Prog. NMR Spectrosc.*, 2009, **55**, 285.

11. A. Webb, *Anal. Chem.*, 2012, **84**, 9.

PART I
SMALL MOLECULES, PROTEINS,
CELLULAR SYSTEMS

CHAPTER 1

New Applications of High-Resolution NMR in Drug Discovery and Development

MARÍA DEL CARMEN FERNÁNDEZ-ALONSO,
MANUEL ALVARO BERBIS, ÁNGELES CANALES,
ANA ARDÁ, FRANCISCO JAVIER CAÑADA AND
JESÚS JIMÉNEZ-BARBERO*

Centro de Investigaciones Biológicas, Calle Ramiro de Maeztu 9,
28040 Madrid
*Email: jjbarbero@cib.csic.es

1.1 Introduction and Historical Perspective

There is no doubt that NMR is one of the most powerful techniques to determine the structure and characterize properties and interactions of molecules. From the very beginning, as in other chemical disciplines, medicinal chemistry took advantage of the power of NMR for structural elucidation of either natural products or synthetic molecules in the way of developing new drugs. Also very early, NMR revealed itself as a powerful method to study ligand–biomacromolecule transient interactions[1] by observing perturbations of different NMR parameters, including relaxation times,[2,3] chemical shifts,[2,4] inter- and intramolecular nuclear Overhauser effect (NOE)[5–7] and inter-molecular spin diffusion.[8] At those times, in the 1970s, these methodologies allowed to obtain information mainly from the ligand (small molecule) point of view and hardly from the protein itself. Nevertheless, they were immediately

New Developments in NMR No. 2
New Applications of NMR in Drug Discovery and Development
Edited by Leoncio Garrido and Nicolau Beckmann
© The Royal Society of Chemistry 2013
Published by the Royal Society of Chemistry, www.rsc.org

translated to the drug-development field for the characterization of drug–receptor interactions.[4,9] Very remarkably, transferred NOE (tr-NOE) experiments, based on the early work of Bothner-By,[5,10] became a powerful methodology to analyze the conformation of ligands when bound to their macromolecular receptors.[11–14] However, the pioneering work by Akasaka[8] on intermolecular spin diffusion had to wait 20 years to be extensively applied in the field of ligand–receptor studies until the saturation transfer difference (STD) strategy was presented by Meyer and Mayer.[15] Since then, NMR has gained an important role (by itself) with its own personality at different stages of the drug design and development processes.

In the 1980s, further methodological and technological advances in the NMR field and improvements in the procedures for isotopic labeling allowed to obtain structural information of biomacromolecules at atomic resolution. In parallel, the general advances in the techniques for three-dimensional (3D) structure determination of biomacromolecules, including those in the NMR field, encouraged the rational drug design based on the knowledge of the 3D structures of potential pharmacological targets and their complexes with ligands, substrates, inhibitors or any other small or drug-like molecule, in the belief that this information could prime the quest for a new drug.

In this context, significant effort, mainly in academic or basic research institutions, but also in pharmaceutical companies, was put in the development and use of methodologies for structural determination, based on NMR, assuming that these strategies could deliver, on time, full structural information on the biomacromolecules and their complexes with lead compounds.[16–20]

In the 1990s, medicinal chemistry was shaken by the combinatorial chemistry and screening strategies in order to find new lead compounds. Pharmaceutical companies drew renewed interest on NMR, but this time the major efforts moved from the very time-demanding NMR methodologies for obtaining a complete 3D description of the pharmacological targets to more versatile but powerful NMR strategies to study drug–receptor interactions and its applications for screening.[21,22] Then, the more easily available NMR methodologies for measuring NMR parameters directly on large macromolecules (the receptors)[23–28] complemented the previously developed methodologies based on the observation of NMR parameters of small molecules (the drugs) as transferred NOE, relaxation, transferred spin saturation and translational diffusion.

These NMR methodologies have been implemented for screening libraries of compounds,[29] identifying binders in complex mixtures, hit validation, defining binding epitopes of ligands[30] or localizing binding sites on receptor molecules, describing experimentally and theoretically conformations in the bound state,[31,32] and quantifying binding affinities.[33,34] It is worth mentioning that many of these new NMR strategies have been developed inside pharmaceutical companies,[19] such as, for example, structure–activity relationships (SAR) by NMR at Abbot,[35] lead generation (SHAPES) by NMR at Vertex,[25,36] screening methodologies and affinity analysis by means of diffusion-edited spectroscopy or NOE pumping for binder identification at Sandoz-Novartis,[37] magnetization transfer from bulk water[38] or fluorine NMR for high-throughput screening (HTS)[39] by Pharmazia-Pfizer,[18] spin labeling strategies at Novartis,[40] multiplexed NMR screening (RAMPED-UP, Rapid Analysis and Multiplexing of Experimentally Discriminated Uniquely Labeled Proteins) at Eli Lilly[41] or ^{19}F-NMR directed strategies for fragment-based drug discovery (FBDD) at Amgen[42] and others.

In summary, in the last 25 years, an extended portfolio of NMR methodologies has been made available for their use at different stages of the drug-discovery process: from target identification, screening compounds, hit validation and lead-compound characterization, structural and mechanistic characterization of ligand–receptor interactions to the design of new drugs.[43]

1.2 NMR Techniques for Molecular Recognition Studies Based on Monitoring Ligand Resonances

1.2.1 Fundamentals

1.2.1.1 Relaxation

Relaxation processes can be divided into two groups: spin-lattice relaxation (characterized by the longitudinal relaxation time, T_1), which is related to the movement of spin populations back to their Boltzmann distribution values, and spin-spin relaxation (represented by the transverse relaxation time, T_2), which corresponds to the decay of coherences.[24,44]

Dipole–dipole coupling between two nuclear spins in a molecule creates local fields as well as the so-called chemical shift anisotropy (CSA), which is caused by molecular electron currents induced by the external magnetic field. The total field fluctuates slightly in direction and magnitude as the molecule tumbles. Even when very slight fluctuations occur, it is still sufficient to cause spin-lattice relaxation.

1.2.1.1.1 Longitudinal Relaxation. The observed longitudinal relaxation rate $R_{1,obs}$ for a small molecule interacting with its target is given by

$$R_{1,obs} = \frac{[BL]}{[L_{TOT}]} R_{1,bound} + \left(1 - \frac{[BL]}{[L_{TOT}]}\right) R_{1,free} \tag{1.1}$$

where $[L_{TOT}]$ and $[BL]$ are the total ligand concentration and the bound ligand concentration, respectively. $R_{1,bound}$ and $R_{1,free}$ are the longitudinal relaxation rates for the ligand in the bound and free state, respectively.

Longitudinal relaxation experiments can be non-selective $R_{1,ns}$ (performed by inverting all resonances contained in a spectrum) or selective $R_{1,s}$ (inverting only one single resonance). The $R_{1,ns}$ and $R_{1,s}$ for a proton i refer to a sum of uncorrelated pairwise proton dipole–dipole interactions, and the possible contribution by other relaxation mechanisms that are grouped into an extra term, ρ_i^*, as represented by eqn (1.2) and eqn (1.3) respectively

$$R^i_{1,ns} = \sum_{j \neq i} \rho_{ij} + \sum_{j \neq i} \sigma_{ij} + \rho_i^* \tag{1.2}$$

$$R^i_{1,s} = \sum_{j \neq i} \rho_{ij} + \rho_i^* \tag{1.3}$$

where ρ_{ij} and σ_{ij} are the self- and cross-relaxation rates for any H_i–H_j dipole–dipole interaction, respectively, and the sum is extended to all protons that are dipolar coupled to proton i. Re-written in terms of the spectral density, the eqn (1.2) and eqn (1.3) become

$$R^i_{1,ns} = \sum_{j \neq i} \frac{\gamma^4 \hbar^2}{10 r_{ij}^6} \left\{ \frac{3\tau_c}{1 + \omega^2 \tau_c^2} + \frac{12\tau_c}{1 + 4\omega^2 \tau_c^2} \right\} \tag{1.4}$$

$$R^i_{1,s} = \sum_{j \neq i} \frac{\gamma^4 \hbar^2}{10 r_{ij}^6} \left\{ \frac{3\tau_c}{1 + \omega^2 \tau_c^2} + \frac{6\tau_c}{1 + 4\omega^2 \tau_c^2} \tau_c \right\} \tag{1.5}$$

where $\hbar$ is the reduced Planck constant, γ is the proton gyromagnetic ratio, ω is the Larmor frequency, τ_c is the correlation time and r_{ij} is the internuclear distance between the protons i and j.

When a small molecule is bound to a large macromolecule with slow reorientation ($\omega\tau_c \gg 1$), there is a substantial contribution to $R_{1,s}$. It is thus evident that $R_{1,s}$ experiments can be used to monitor binding of small molecules to a receptor. $R_{1,ns}$ lacks a direct τ_c dependence and therefore cannot be used for screening purposes. Measured $R_{1,s}$ changes upon complexation to the target molecule can provide useful information about the binding mode and allow the binding constant to be extracted.

1.2.1.1.2 Transverse Relaxation. In the limit of moderately fast exchange, the transverse relaxation rate, $R_{2,obs}$, for a small molecule interacting with a macromolecule is given by

$$R_{2,obs} = \frac{[BL]}{[L_{TOT}]} R_{2,bound} + \left(1 - \frac{[BL]}{[L_{TOT}]} \right) R_{2,free}$$
$$+ \frac{[BL]}{[L_{TOT}]} \left(1 - \frac{[BL]}{[L_{TOT}]} \right)^2 \frac{4\pi^2 (\delta_{free} - \delta_{bound})^2}{K_{-1}} \tag{1.6}$$

where $R_{2,bound}$ and $R_{2,free}$ are the transverse relaxation rate constants for the ligand in the bound and free states, respectively, δ_{bound} and δ_{free} are the chemical shifts for the resonance of the ligand in the bound and free states, respectively, and $1/K_{-1}$ is the residence time of the ligand bound to the protein. The last term disappears in true fast exchange limit.

In terms of spectral densities, eqn (1.6) can be re-written as

$$R_2^i = \sum_{j \neq i} \frac{\gamma^4 \hbar^2}{10 r_{ij}^6} \left\{ \frac{15\tau_c}{2(1 + \omega^2 \tau_c^2)} + \frac{3\tau_c}{1 + 4\omega^2 \tau_c^2} + \frac{9}{2}\tau_c \right\} \quad (1.7)$$

The third term contains the direct τ_c dependence that can be used for screening. For multiple resonances, dipole–dipole and CSA–dipole cross correlation terms must also be considered. These terms contain spectral densities calculated at zero frequency and therefore for long τ_c they will contribute to the observed differential linewidth of the lines comprising a multiplet. The last term corresponds to the exchange broadening in the intermediate exchange regime, which is a very small contribution and can be neglected for weak binding affinity ligands and when the ligand concentration is in high excess compared to the protein concentration. Compounds interacting with a receptor display an enhanced R_2 value. Since the linewidth is equal to R_2/π, the resonances of a binding molecule display an increased broadening, which is stronger for large receptors. In the case of lower molecular weight receptors, binders can be identified using T_2 or $T_{1\rho}$ (longitudinal relaxation in the rotating frame) filtered experiments.

1.2.1.2 Diffusion

Diffusion in liquids depends on molecular size and viscosity. It is important to take into account the distance range of molecular diffusion on the time scale of the NMR experiment and, therefore, it is possible to distinguish short-range and long-range intermolecular interactions.[24,44] The latter ones are smaller than the first but still lead to significant effects. The effect of diffusion on the spin interaction terms depends on whether these interactions are intramolecular (involving particles on the same molecule) or intermolecular (involving particles on different molecules). In an isotropic liquid, the rotational motion averages all intramolecular spin interactions to their isotropic values. The diffusion motion of the molecules averages the short-range intermolecular interactions to zero. However, the long-range intermolecular interactions are not averaged. In an anisotropic liquid, the rotational motion averages all intramolecular spin interactions to values that are different from their isotropic values. The situation for diffusion motions is the same as for isotropic liquids.

1.2.1.2.1 Diffusion-Editing Experiments. Changes in translational diffusion can probe an interaction of a ligand and its target and be used to detect complex formation. Diffusion-editing experiments are based on the fact that upon binding of the ligand to the target, the hydrodynamic radius r_H of the ligand increases dramatically and hence the diffusion rate decreases. The use of diffusion to identify ligand binding by NMR is based on the fact that the decrease of the signal is proportional to the rate of diffusion and strength/length of the magnetic field gradient pulse.

The translational diffusion coefficient (D) can be determined by using the Stokes–Einstein relation

$$D = \frac{k_B T}{6\pi\eta_W r_H} \tag{1.8}$$

The hydrodynamic radius can be determined experimentally by techniques such as dynamic light scattering or be calculated empirically to estimate a hydration radius compatible with the NMR results. Owing to the 1/r dependence, small molecules that do not aggregate have D values that are about an order of magnitude larger compared to the D value of a large macromolecule. When a small molecule binds to a large macromolecule it will transiently acquire the D value of the receptor and its diffusion will change. In the fast exchange limit, the observed translational diffusion coefficient D_{obs} is then given by

$$D_{obs} = \frac{[BL]}{[L_{TOT}]} D_{bound} + \left(1 - \frac{[BL]}{[L_{TOT}]}\right) D_{free} \tag{1.9}$$

where D_{bound} and D_{free} are the diffusion coefficients of the ligand in the bound and free states, respectively. For high-molecular-weight targets, the D_{bound} contribution to the equation becomes negligible.

Diffusion constants are most conveniently measured with the pulsed field gradient spin echo (PFG-SE)[45] or the pulsed field gradient stimulated echo (PFG-STE)[46] methods. Technical improvements come from the use of bipolar gradients and the use of the longitudinal eddy-current delay, as in the bipolar pulsed field gradient longitudinal eddy-current delay (bpPFGLED) sequence.[47] One limitation of this method is the need to resolve at least one NMR signal per ligand from those of the biological target. If this is not possible, spectra subtraction using spectral differencing methods or isotope labeling may help. In the spectral differencing method, a diffusion-weighted spectrum recorded on the target alone is subtracted from data recorded on the mixture. However, the method can lead to subtraction artifacts. If the biological target is ^{13}C-labeled, its signals may be removed by isotope-filtering NMR techniques and spectra are acquired with different gradient amplitudes. Instead of a quantitative analysis, a qualitative one serves most purposes in the early stages of NMR screening to identify compounds that bind to a receptor from a mixture of non-binding molecules. First, a PFG-STE or PFG-SE spectrum of the chemical mixture in the absence of the protein is recorded at a low gradient strength. Next, the same spectra for the chemical mixture in the presence of the protein are recorded at low and high gradient strengths and subtracted to produce a spectrum that contains only the signals of the compounds not interacting with the protein. The resulted subtracted spectrum is then subtracted from the spectrum of the chemical mixture recorded in the absence of the protein to obtain a spectrum that contains only the signals of the molecule that binds to the receptor.

A useful application of the diffusion coefficient D is in the determination of the dissociation binding constants of weak affinity ligands, knowing the concentration of bound ligand calculated from the equation of D_{obs}.

See more details about diffusion experiments in Section 1.2.3 of this chapter.

1.2.2 Basic Hardware Description

In general, NMR hardware requirements largely depend on the problem to be solved. The NMR experiments described here are meant to detect molecular interactions mainly through the observation of chemical shifts, and relaxation or diffusion changes upon binding. Usually, unless intrinsic limitations (*e.g.* reduced amount of sample) occur, ligand-based methods do not demand the largest sensitivity instrumentation and 500 MHz instruments are sufficient. When the receptor is a large protein, however, improved signal dispersion and higher sensitivities are strongly desirable and higher magnetic fields are employed.

Gradient coils, which allowed the development of enhanced spectroscopy, are a mandatory hardware requirement for the acquisition of diffusion-based methods.

With respect to the probe, most of the heteronuclear methods developed nowadays take advantage of the higher sensitivity offered by proton detection. For this reason *inverse* probes, in which the proton coil sits closer to the sample while the heteronucleus (X) coil sits outermost, is the most used configuration. The X coil can be tuneable for the detection of more than one nucleus (broadband) being usually ^{15}N and ^{13}C. Triple resonance probes allow to pulse and decouple two heteronuclei simultaneously. However, direct observation of heteronuclei other than proton as ^{13}C and ^{19}F is gaining interest in protein studies and drug development.

The more and more affordable isotopic labeling of proteins (see Section 1.5.1 for further details) also facilitates the study of ^{13}C-labeled proteins by direct observation of ^{13}C resonances. New developments for ^{13}C direct observation probes are ongoing, especially at the Centro di Ricerca di Risonanze Magnetiche (CERM) in Florence, Italy (see Section 1.3.1).

For the detection of ^{19}F nuclei, as the proton and fluorine resonance frequencies are relatively close, it would in principle be possible to detect fluorine with a probe designed for proton spectroscopy. However, most spectrometer manufacturers have developed different probes with the inner coil specially dedicated to fluorine detection at high sensitivity and the outer one tuned to ^{1}H, which allows for the manipulation of both ^{1}H and ^{19}F nuclei simultaneously and permits its use for implementing ^{19}F-NMR-based drug-discovery strategies (see Section 1.2.5).

1.2.3 The Basic Experiments: STD, WaterLOGSY and Their Variants

With certainty, the most powerful ligand-detected NMR techniques for probing binding activities are those based on the transfer of magnetization, such as saturation transfer difference (STD) and water-ligand optimized gradient spectroscopy (WaterLOGSY). The common mechanism underlying both techniques is the detection of dipole–dipole interactions between the spins of the ligands and those of the receptor, their main difference being the target of the irradiation: in STD experiments, the magnetization is transferred from the irradiated protein to the binding ligands, while in the WaterLOGSY experiment, the bulk water is selectively excited, and the magnetization is transferred from the water molecules lying at the receptor–ligand interface to the ligand nuclei.

1.2.3.1   Saturation Transfer Difference (STD)

The STD experiment[15] involves the comparison of two ^{1}H-NMR spectra of the same sample, acquired either with (on-resonance spectrum) or without (off-resonance spectrum) saturation of the protein. In the on-resonance experiment, a subset of the protein protons is selectively irradiated by a train of pulses directed towards frequencies of the ^{1}H-NMR spectrum devoid of ligand signals. Typically, the irradiation frequency is set between –1 and 1 ppm, the target of irradiation thus being side chain protons of isoleucine, valine and leucine residues. Alternatively, if the ligand lacks signals in the aromatic region, irradiation can be set at 6–8 ppm. The magnetization then quickly propagates, *via* ^{1}H–^{1}H cross-relaxation pathways, from the irradiated protons to the rest of the protein, and from there to any ligands in direct contact with the protein. This will lead to a spectrum where the intensity of the signals belonging to saturated molecules (both the receptor and binding ligands) is decreased, but leaving signals of non-binding molecules unaffected.

In parallel, an off-resonance spectrum, equivalent to a reference ^{1}H-NMR spectrum of the complex, is obtained by irradiating the sample at spectral regions devoid of both receptor and ligand resonances (*e.g.* 100 ppm). The difference between the reference and the on-resonance spectra yields a saturation spectrum, in which the signals of the saturated nuclei are represented, thus allowing the discrimination of ligands with binding activities in a compound mixture (Figure 1.1).

Typically, the sample is irradiated with a total saturation time between 0.5 and 4 s. During this time, the ligand molecules can enter the binding site, become saturated and leave many times, provided that their binding exchange kinetics is fast enough. Thanks to the slow longitudinal relaxation of the ligands, the saturated state largely persists when ligands dissociate, giving rise to a pool of unbound saturated ligands, which are ultimately responsible for the STD signal. On the basis of this, two considerations can be made. First, the STD experiment is more sensitive to ligands binding strongly enough to become effectively saturated, but weakly enough to be in fast exchange between the bound and the unbound states so that a large population of unbound, saturated molecules is generated. Put into numbers, the optimal range of dissociation constants, K_D is about 10^{-3}–10^{-8} M.[30] Second, the method enables a large excess of ligand to be used, typically 20 : 1 to 100 : 1. Thanks to the efficient saturation of proteins, the concentration of receptor can be kept as low as 1 µM in favourable cases, which means that concentrations between 50 µM and 200 µM of ligands are mostly used in practice. Of course, this is also highly dependent on the water solubility of the screened compounds.

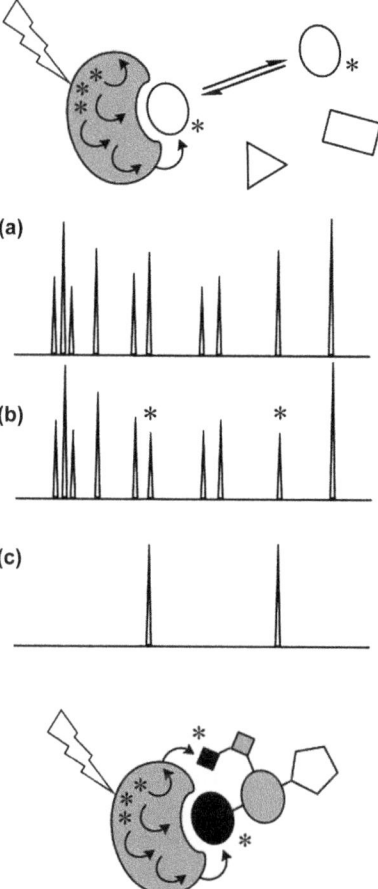

Figure 1.1   Top panel: schematic representation of the STD experiment, with arrows representing the transfer of magnetization from the irradiated protons in the macromolecule to the ligand protons. Middle panel: Simulated STD experiment. (a) Reference (off-resonance) spectrum; (b) on-resonance spectrum; (c) difference spectrum (up-scaled). Bottom panel: illustration of the STD epitope mapping effect.

The small amounts of protein needed for the STD experiment, added to the simplicity of its readout and the possibility to screen several compounds at a time, make it the preferred NMR-based method for screening large quantities of compounds. Moreover, in its more than 10 years of history, the STD has proven versatile and useful for the pharmaceutical industry far beyond its role as a screening technique, being today routinely utilized for the characterization of the ligand epitopes and determination of affinity constants.

The STD experiment was originally designed to screen a library of carbohydrate compounds against a model lectin.[15] In the same study, however, it was proven that this technique was not only useful to detect binding events, but also to map the parts of the ligand making a major contribution to the binding. Indeed, in complex molecules, the ligand epitopes establishing tighter contacts with the protein receive a higher amount of magnetization, which is translated into stronger signals in the STD spectrum. This information facilitates the discrimination between parts of the molecule with major importance for binding (pharmacophore), and those with less relevance, which can be subject to chemical development in order to achieve better affinities or improved absorption, distribution, metabolism and elimination (ADME) properties for the overall drug.

In addition, STD experiments allow binding affinities to be measured, thus enabling the ranking of bioactive compounds. For this purpose, STD-titration curves are built in the presence of variable amounts of the ligand. Special care must be taken to keep the protein concentration constant, as it is an important factor affecting the intensity of the STD signal. Measuring of K_D by STD experiments, however, must be executed carefully, as there are a number of factors that can introduce biases in the estimations, such as rebinding events and the longitudinal relaxation of the ligands. To overcome these sources of error, K_Ds are determined by STD titration curves at zero saturation time, since, at these conditions, neither rebinding nor relaxation occurs. Curves at zero saturation time are extrapolated from titration experiments performed at different saturation times. Affinity data estimated this way have proven to be in agreement with those obtained by other biophysical techniques, such as calorimetric studies.[48]

STD experiments have been successfully applied to whole living cells,[49] enabling the study of ligand binding to membrane proteins, such as G-protein coupled receptors (GPCRs), which are often difficult to isolate and manipulate. A typical problem arising when STD experiments are performed on whole cells is the observation of crowded STD spectra due to the saturated cell components, which can preclude the identification and epitope mapping of the ligand signals. In such cases, a variant technique known as saturation transfer double difference (STDD) is applied. According to this method, the STD spectrum of a control performed in the presence of the cells is subtracted from the STD spectrum performed in the presence of the cells and the ligand, leading to the exclusive observation of signals corresponding to the ligand STD effects in the STDD spectrum.[50]

1.2.3.2   *WaterLOGSY*

In the second magnetization transfer experiment, the magnetization of the bulk water is selectively transferred, *via* the protein–ligand complex, to the ligand in the unbound state. In essence, the WaterLOGSY experiment[38] can be understood as a nuclear Overhauser effect spectroscopy (NOESY) experiment, in which the magnetization of the water molecules is selectively inverted, and during a long mixing time of up to several seconds, the magnetization is transferred, *via* ^1H–^1H cross-relaxation, to the ligand spins at the protein–ligand interface. There are essentially two pathways by which this transfer of the magnetization occurs: (i) *via* direct cross-relaxation from the water molecules tightly bound at the protein–ligand interface, and (ii) *via* chemical exchange between the water protons and the labile protons of amine and hydroxyl groups in the protein, which in turn cross-relax with the protons of the bound ligand.

In either case, an inversion of the NOE sign takes place during the mixing time, since both the protein protons and those of the water molecules bound to the protein adopt the motional and NMR relaxation properties of the receptor, with a slow tumbling rate and a long correlation time. For their part, non-interacting ligands receive the magnetization only in the free state, *via* water molecules involved in their solvation sphere, and the sign of the NOE stays unaltered. As a result, the WaterLOGSY experiment yields a spectrum in which ligands with binding activity are manifested by positive signals, whereas non-binding ligands are represented by negative signals (Figure 1.2).

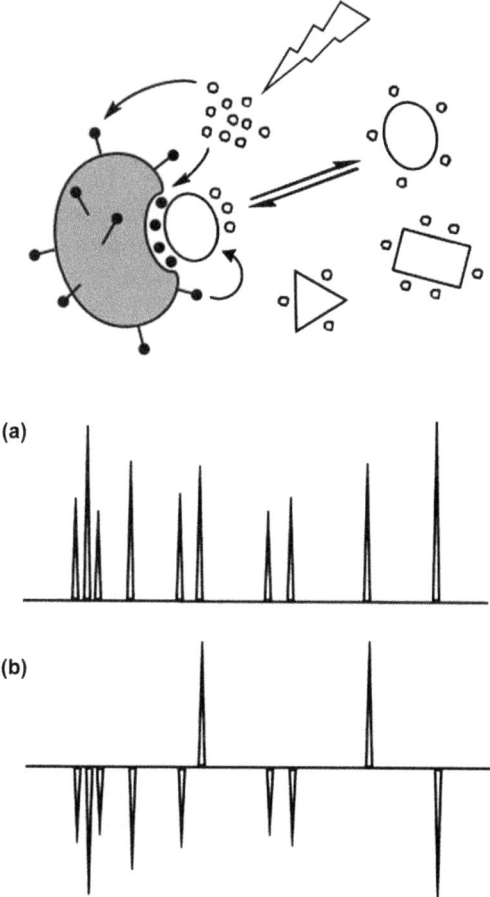

Figure 1.2 Top panel: schematic representation of the WaterLOGSY experiment, with white circles representing water molecules causing negative NOEs, and filled circles representing water molecules causing positive NOEs or labile protons in the protein exchanged from the irradiated water. Bottom panel: Simulated WaterLOGSY experiment. (a) Reference spectrum; (b) WaterLOGSY spectrum.

In practical terms, the same considerations regarding the use of a ligand excess and low protein requirements made earlier with respect to STD experiments can be made here for WaterLOGSY. The sensitivity and range of applicability of both methods is comparable, and WaterLOGSY can also be used to obtain affinity data.[51] For K_D determination by WaterLOGSY, a parallel series of control experiments in the absence of the receptor must be performed. Since each signal in the WaterLOGSY experiment is a weighted combination of the negative NOEs of the hydrated unbound ligand and the positive NOEs of the ligand in the bound state, the WaterLOGSY signals must be normalized by subtracting the signals of the reference spectrum from those of the WaterLOGSY spectrum. Titration curves are then built by plotting the resulting intensities against the respective ligand concentration values.[52] In addition, it has been shown that WaterLOGSY experiments can also provide information on the ligand orientation in the bound state, given the different degrees of bulk water accessibility to parts of the ligand facing the protein (giving rise to positive NOEs) and those protruding from the protein (giving rise to negative NOEs), a concept known as SALMON (Solvent Accessibility, Ligand Binding and Mapping of Ligand Orientation by NMR Spectroscopy).[53] A practical drawback of the WaterLOGSY experiment is that samples must be prepared in protonated H_2O as a solvent, which requires the introduction of an efficient procedure to suppress the water signal. On the other hand, its lesser dependence on spin diffusion makes WaterLOGSY the preferred option when dealing with receptors of low proton densities, such as nucleic acids.[54]

1.2.4   Further Developments

1.2.4.1   Transferred-NOESY (TR-NOESY)

Transferred NOE (tr-NOE) was firstly described by Bothner-By[6,7] and has nowadays become a classical method for studying the binding of ligands to large receptors in fast exchange (weak binding, with K_D usually around the microM range). The experiment consists of acquiring an ordinary NOESY spectrum on a sample containing both the receptor and the ligand. If there is an interaction between them, and the exchange rate for the equilibrium between the bound and free forms is fast, the observed cross-relaxation rate (σ) for the ligand will average between that of its bound state, σ_{bound}, and that of its free state, σ_{free}. Since the cross-relaxation rate for the bound form (σ_{bound}) is much larger than that of the free form (σ_{free}), the observable cross-relaxation rate will be dominated by the bound form. This implies that the NOE observed for the ligand in the spectrum will keep the information of its bound state. Therefore the observed NOEs (tr-NOEs) will be negative (large complex, large correlation time), and will reflect the conformation of the ligand in the binding site. Moreover, if the ligand is small, characterized by a short correlation time, it will show positive NOEs, and will then experience a NOE sign change in the presence of the receptor. Because of the weak interaction prerequisite, usually large excess of ligand is used. The technique has been widely used for defining the binding of a large range of ligand/protein systems, for instance the interaction of blood group antigens with virus like particles,[55] peptide inhibitors with beta-amyloid fibrils,[56] for demonstrating conformational selection,[57,58] for detecting cation–π interactions, among many others.[59]

1.2.4.2   Editing/Filtering Methods

Intermolecular NOEs between protons of a receptor and protons of a ligand have been typically used for the structural characterization of binding complexes, and represent a valuable piece of information for atomic resolution. They are usually obtained through NOESY-type experiments providing information about the interproton distances within the complex. However, this option is only applicable to relatively small receptors (see Figure 1.3 as an example).[60,61] Even in such cases, achieving intermolecular NOE information is significantly hampered by three common situations: severe signal overlapping, the need of resonance assignment (time consuming) and the prerequisite of a relatively slow exchange regime. To overcome the first situation, isotope-labeled, filtered and edited experiments have been designed.[62] They imply the labeling of the receptor (or the ligand) with $^{13}C/^{15}N$ and selecting the signals in the spectrum from protons that are bound either to $^{13}C/^{15}N$ or to $^{12}C/^{14}N$. Examples of such a strategy are common in literature (see more details in Section 1.3.1). These methods can be used to determine the bound conformation of a drug in the presence of a labeled receptor and are applied in those cases where tr-NOEs cannot be detected due to tight binding. In addition, filtered experiments can be used to detect intermolecular NOEs. Originally, these methods were developed by Otting and Wüthrich, as two-dimensional X-double-half-filtered experiments, where X stands for the "NMR active" nuclei (for instance ^{13}C).[63]

The classical scheme to measure NOEs is to study complexes in which one component is isotopically labeled and another one is unlabeled. The presence of ^{13}C in one of the molecules of the complex (usually a protein) allows the use of isotope-filtered experiments that rely on the presence of a scalar coupling between a proton and the attached "NMR active" nuclei (^{13}C) and the absence of a scalar coupling between a proton and an "NMR inactive" nucleus (^{12}C). The scalar coupling can be exploited to discriminate between these two different classes of protons.

Therefore, a ^1H–^1H NOE experiment that is isotope-filtered in one proton dimension and isotope-edited in the other proton dimension can be obtained. This spectrum contains exclusively intermolecular NOEs provided that the scalar coupling is similar for all proton-^{13}C pairs. In addition, double-filtered spectra showing only intramolecular NOEs can also be obtained. The latter are especially useful to detect the intramolecular NOEs of a small drug bound to a protein receptor, since the crowded spectrum of the receptor precludes the identification of the small molecule NOEs. A limitation is that the additional delays in the filtered experiments lead to lower sensitivity in the NOESY spectra. Also, a method that implies ^{13}C edited, $^{13}C/^{15}N$-filtered heteronuclear single quantum correlation NOESY (HSQC-NOESY) spectra acquisition was proposed, that avoids protein

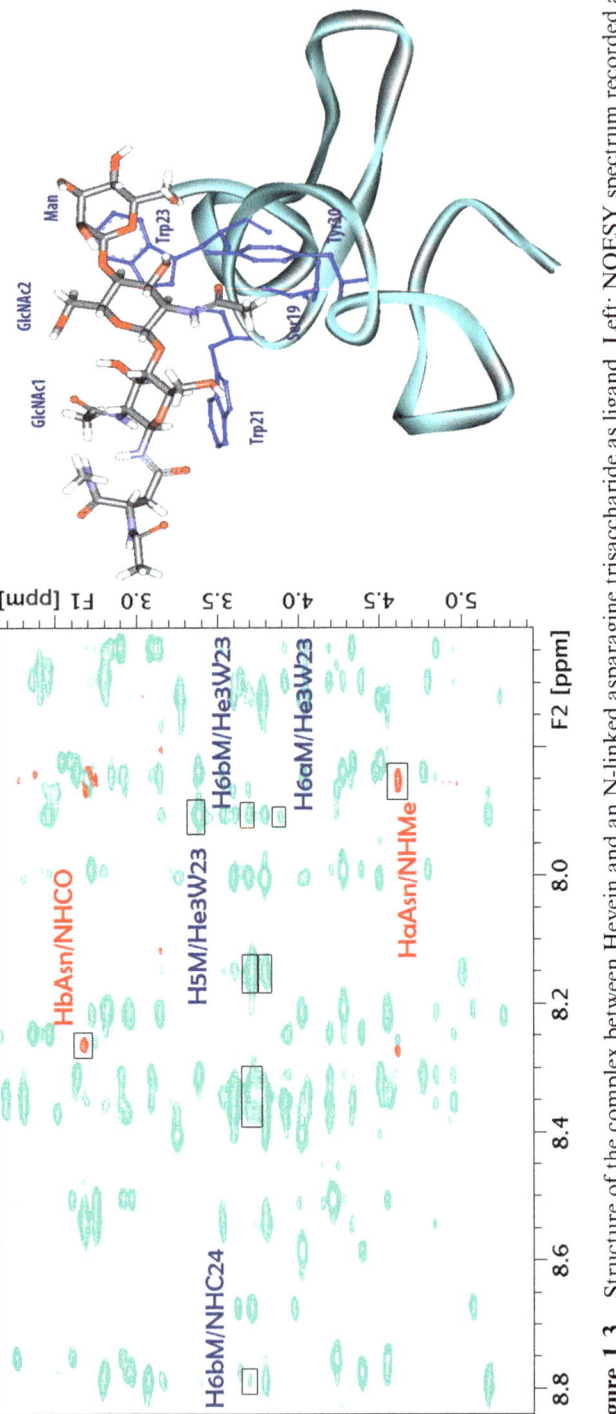

Figure 1.3   Structure of the complex between Hevein and an N-linked asparagine trisaccharide as ligand. Left: NOESY spectrum recorded at 900 MHz, 500 ms mixing time. Some intermolecular NOE cross-peaks are marked. M = Man. Right: 3D structure of the complex. Important residues for the interaction are highlighted (Ser19, Trp21, Trp23 and Tyr30). See Hernández-Gay *et al.*[61] for details.

signal assignment and allows the binding pose to be determined, with the support of molecular docking and scoring procedures.[64] On the other hand, intermediate exchange precludes the detection of intermolecular NOEs by causing signal line broadening. Methods for the identification of such NOE contacts for interactions in the intermediate exchange limit have been proposed. For instance, selective protonation of deuterated proteins allows discriminatory irradiation on certain proton–protein resonances, which can cause NOE build-up on the ligand resonances. The method was applied to study the interaction between a small chemical and the anti-apoptotic protein Bcl-xL (B-cell lymphoma-extra large), allowing to place the binding site for the ligand.[65]

1.2.4.3 DIRECTION

An interesting alternative method, called DIRECTION (Difference of Inversion Recovery Rate with and without Target Irradiation),[66] has been proposed in order to detect binding of a ligand to a large receptor and allowing to asses ligand epitope mapping without the bias of the distinct longitudinal relaxation rates of the different ligand protons.[67] As a matter of fact, the method exploits the different longitudinal relaxation rates of the different ligand protons with and without saturation of the protein protons. The NMR experiment is based on a conventional inversion-recovery sequence with some minor modifications. It was applied to a series of systems yielding consistent conclusions, and compared to the STD-based approach. The authors state that the applicability in terms of molecular size and exchange rates are similar to STD, but with the advantage of more accurate and precise results.

1.2.4.4 TINS (Target Immobilized NMR Screening)

This technique, proposed in 2005, implies attaching the target to a solid support compatible with NMR (sepharose for instance) and acquiring a conventional ^{1}H NMR spectrum.[68] By subtracting it to a control spectrum (acquired under the same conditions) in which the target is missing, only those ligands binding to the target, and thus experiencing line broadening, will show up in the subtracted spectrum, named TINS. The methodology presents the advantage that the target can be reused, so with a single target sample a large number of fragments can be screened. The affinity range for its applicability is larger than other ligand-based NMR methods. It represents an interesting strategy for ligand screening over membrane proteins that are difficult to study because of problems with their solubilization.[69]

1.2.4.5 INPHARMA (Inter-Ligand NOE for Pharmacophore Mapping)

It was proposed as a technique for defining the relative orientation of two ligands competing for the same binding site of a large receptor.[70,71] Technically, it implies the acquisition of a NOESY experiment on a sample containing the receptor and the two ligands in excess, and arises from intermolecular NOE between the protons of the two competitive ligands in the receptor's binding pocket mediated by spin diffusion occurring throughout the protein/ligand complexes. Thus, it results in NOE correlations between protons of the two competing ligands. It applies for systems in which the affinities of both ligands lay in the low μM-mM range.[72] It was for instance applied to study the binding of epothilone A (EpoA) to tubulin in solution[73] and has been demonstrated to be useful in order to discriminate between many different binding poses.[57,74]

1.2.4.6 Paramagnetic Tags: Spin Labels

Paramagnetic effects have been applied in NMR in order to approach different structural issues by taking advantage of mainly two different effects:

(1) the paramagnetic relaxation enhancement (PRE),[75] which is the difference in the ^{1}H relaxation rate between a paramagnetic and diamagnetic state of a molecule. It started to be used as a structural parameter for the study of paramagnetic metal ion-containing proteins (metalloproteins, iron-sulfur proteins), where the paramagnetic source is the metal ion.[76,77] The introduction of a paramagnetic source to non-metal binding proteins was achieved by chemical modification of engineered cysteins and coupling with a paramagnetic tag like nitroxyl radicals (*i.e.* TEMPOL), Mn^{+2} chelate, or Gd(DTPA-BMA) (Omniscan).[78] This allowed to extend the use of PRE to the study of protein–protein and DNA–protein interactions,[79,80] the characterization of minor conformational protein states,[81] the determination of the solution-bound conformation of a protein[82] or to study protein folding. The idea was also applied to the interaction of peptides to micelles containing a paramagnetic tag, allowing to establish the orientation and immersion depth of the peptide in contact with a membrane-like environment.[83,84]

(2) pseudocontact shift ($\Delta\delta^{PCS}$, PCS), which is the change in chemical shift caused by a through-space interaction with a paramagnetic center. It has long been used for ^{15}N-labeled proteins by acquiring ^{1}H–^{15}N HSQC spectra in the presence of a paramagnetic tag and in the presence of a similar diamagnetic ion,[85] in order to study 3D structure, protein–protein complexes and ligand recognition.[86] Lanthanide ions have been found to be suitable paramagnetic probes for creating PCS.[87] For non-natural metal binders, the lanthanide ion is introduced in the system chelated with a lanthanide chelating tag based on EDTA, DTPA or DOTA, mostly introduced in the protein,[88–90] although other biomolecules, like carbohydrates, are starting to be used as well.[91,92] PCS has an r^{-3} distance dependence (r being the distance between the spin and the paramagnetic centers), opposite to the r^{-6} dependency of PRE and thus spins far from the paramagnetic center and not strongly affected by PRE can be measured. PCS has been much less used in order to study interactions

from the ligand point of view. It was first employed in the study of the interaction of a small ligand on fast exchange with a protein target containing a lanthanide ion. The measurement of transferred para-magnetic shifts caused in the ligand permitted the location and orientation of the bound ligand to be defined.[93] Also the interaction between Galectin-3 and lactose was studied by introducing a lanthanide binding peptide into the C-terminal domain of the galectin.[94] More recently, the introduction of a lanthanide ion into a target protein was proposed as a probe to be used as a fragment-based ligand screening method.[95]

1.2.4.7   Diffusion-Based Methods

Diffusion-based methods rely on the different diffusion rate of a ligand when it is free and when bound to a large molecule. Diffusion ordered spectroscopy (DOSY) consists of the acquisition of spin echos in the presence of different gradient strengths (PFG-SE). As the gradient strength increases the signal decays according to the Stokes–Einstein equation and, through a fitting procedure, the corresponding diffusion coefficient can be obtained, as described above.[96,97]

Diffusion experiments can also be applied to obtain epitope information. The idea was proposed by Shapiro, exploiting the fact that during long diffusion times in a stimulated spin-echo, the intermolecular NOE between the protein protons and ligand protons builds up. This fact causes interferences for accurate diffusion rate measurements, but can also be used in order to map those protons in closer contact with the protein, which will present a larger effect than those not exposed to the protein–ligand interface.[67]

1.2.5   Employing Other Nuclei (^{19}F)

The virtue of the proton is, in contrast with a majority of the other elements, its ubiquitous presence in organic compounds. Phosphorous-31 has also a spin $I = \frac{1}{2}$ and 100% isotopic abundance. Depending on its oxidation state, phosphorous nuclei can adopt a broad range of chemical shifts, the typical ^{31}P chemical shift dispersion in organic compounds being of about 100 ppm. In addition to the many important enzymatic reactions involving phosphorous-containing reactants, phosphorous plays a role in many protease inhibitors by mimicking the tetrahedral intermediate of a peptide bond hydrolysis. With these considerations, ^{31}P-NMR has been regarded as a tool for compound screening,[98] but the otherwise limited presence of phosphorous in drug candidates has restricted its use mainly to *in vivo* applications, *e.g.* to assess metabolites such as phospholipids, ATP and inorganic phosphate.

Fluorine-19, on the other side, exhibits a number of properties that make the NMR spectroscopy of this nucleus very attractive for drug research, while at the same time being suitable for a wide range of compounds of pharmaceutical interest. Indeed, around 20% of the currently marketed drugs contain one or more fluorine atoms, a number that is steadily increasing.[99] Fluorinated analogues have long been regarded in the pharmaceutical industry for their high metabolic stability and increased membrane permeability. At the same time, fluorinated drugs are very amenable to trace in order to determine their ADME properties, given the virtual absence of endogenous fluorine in living organisms.

From the NMR viewpoint, ^{19}F is a $I=\frac{1}{2}$ nucleus with a 100% isotopic abundance and a gyromagnetic ratio comparable to that of a proton, the relative sensitivity of ^{19}F with respect to ^{1}H being 83%. Moreover, ^{19}F displays a much broader span of chemical shifts than ^{1}H does, ranging from -272 ppm in the CH_3F molecule to over $+85$ ppm in pentafluorosulfanyl (SF_5) substituents,[100] which minimizes the probability of signal overlapping and, due to the absence of fluorine in biological molecules and in most solvents and buffer components, accounts for very clean ^{19}F-NMR spectra.

Furthermore, and owing to its large chemical shift anisotropy, ^{19}F signals are highly perturbed by local changes, such as binding to a receptor. Indeed, thanks to the exquisite sensitivity of the fluorine nuclei to changes in their micro-environment, and to the potentially large difference between the ^{19}F chemical shifts in the free and bound states, even weak binding between fluorine-containing compounds and a target of interest can be easily detected in simple 1D experiments, by monitoring chemical shift perturbation of the ^{19}F signals after the addition of the receptor. In addition, appreciable changes in transverse relaxation times (T_2) upon binding to a macromolecule enable binding of fluorinated compounds to be assessed by substantial broadening of the ^{19}F signals or by changes in their intensity when a relaxation filter is applied.

The possibility straightforwardly to detect protein–ligand interactions in ^{19}F-NMR 1D experiments makes it a very efficient tool *vis-à-vis* ^{1}H-NMR for the HTS of chemical libraries. In addition, other usual screening procedures well established for ^{1}H-NMR, such as STD, are also feasible by detecting ^{19}F.[101] Of course, the use of NMR spectrometry of ^{19}F also enables structural characterization of fluorochemicals by a wide span of multidimensional experiments, both homonuclear (^{19}F–^{19}F) and heteronuclear (^{19}F–^{1}H and ^{19}F–X), such as correlation spectroscopy (COSY), total correlation spectroscopy (TOCSY), NOESY, heteronuclear Overhauser effect spectroscopy (HOESY), heteronuclear multiple-bond correlation (HMBC) or three-dimensional high constant flow (3D HCF) experiments.[102]

Obviously, a requisite is that the studied compounds contain at least one fluorine atom. While the increasing interest towards ^{19}F-NMR over recent years has prompted the release of chemical libraries composed entirely of fluorine-containing compounds, there are methodologies that allow exploiting the advantages of fluorine NMR for the screening and binding assessment of non-fluorinated compounds, *e.g.* through the use of fluorinated spy molecules.[103] According to this strategy, a reference experiment is first performed in the presence of the spy molecule and the target receptor. Next, the spy molecule is made to compete with a compound or mixture of compounds, and changes in the spectral properties of the spy molecule (*i.e.* ^{19}F chemical shift or transverse relaxation) are monitored. If the affinity data for the spy molecule are known, which can be reliably determined through ^{19}F-NMR titration experiments, competition methods also allow measuring binding constants for the screened compounds.

1.3   NMR Techniques and Applications Based on Monitoring the Receptor Resonances

Receptor-based methods have been applied mainly to proteins. These strategies are based on the direct observation of protein resonances and provide detailed information about the receptor residues involved in drug recognition. Therefore, this method allows the detection of non-specific binding. This site-specific characterization can be used for fragment-based lead generation in which lower affinity molecular fragments that bind to different sites can be linked to yield higher affinity compounds.

In addition, receptor-based methods do not rely on fast exchange and therefore tight binding is not a problem, both higher and lower affinity hits can be identified.

The main disadvantage of these methods is that low-molecular-weight receptors (< 35 kDa) and previous knowledge of the 3D structure of the protein are required.

1.3.1   Fundamentals: A Brief Overview of Biomolecular Structure Elucidation by NMR

Three-dimensional structure determination of biomolecules by NMR gives access to information of paramount importance for drug design. This approach is especially useful in the case of small or flexible receptors that are usually difficult to crystallize.

Regarding small peptides, isotopic labeling is not required and 2D-NMR (NOESY) experiments are sufficient for obtaining the NMR restraints required for structure calculation. In contrast, when medium-sized proteins are considered (10–35 kDa) double-labeled (^{15}N, ^{13}C) or triple-labeled samples (^{15}N, ^{13}C and ^{2}H) and stable proteins, with a lifetime long enough to perform 3D experiments, are required. For labeling purposes, the protein should be over-expressed in large quantities in a proper host. *E. coli* is the most common expression system but it is often not a good option for mammalian proteins, for which alternative expression and labeling techniques have been developed, as for example production in yeast, insect cells and directly on mammalian cells. Labeling and new expression techniques will be addressed in more detail in Section 1.5.1 within this chapter. In addition, the protein should not aggregate at moderate concentration (400 µM–1 mM).

The most common experiments used for the protein backbone assignment are summarized in Table 1.1. Once the HN resonances are assigned, 3D ^{15}N-edited TOCSY-HSQC and 3D HCCH TOCSY experiments can be used for the assignment of the side chain aliphatic protons in the same residue.

Table 1.1 Typical NMR experiments used for protein backbone assignment.

Experiment	Correlations observed
HNCO	$^{13}C'_{i-1}-^1H_i-^{15}N_i$
HN(CA)CO	$^{13}C'_{i-1}-^{13}C'_i-^1H_i-^{15}N_i$
HNCACB	$^{13}C_{\alpha-1}-^{13}C_\alpha-^{13}C_{\beta-1}-^{13}C_\beta-^1H_i-^{15}N_i$
CBCA(CO)NH	$^{13}C_{\alpha-1}-^{13}C_{\beta-1}-^1H_i-^{15}N_i$
HNCA	$^{13}C_{\alpha-1}-^{13}C_\alpha-^1H_i-^{15}N_i$

The next step is to assign and measure the intensities of $^1H-^1H$ NOE cross-correlations in order to derive interatomic distance constraints, what is usually carried out by analysis of ^{15}N-edited-NOESY HSQC and ^{13}C-edited-NOESY-HSQC spectra. It is worth mentioning that the full assignment and the NOE identification is a very time-consuming task.

The strategy described above is based on detection of the HN amide resonances and therefore can not be applied to systems in which proton information is lost due to exchange with the solvent or fast relaxation. In this context, the recent developments of ^{13}C direct detection experiments opened the field of NMR to new systems such as paramagnetic proteins, very large proteins or intrinsically disordered proteins. A number of exclusively hetero-nuclear NMR experiments have been developed and constitute a complete set for studying macromolecules. This includes experiments for spin system identification and sequence-specific assignment as well as for the determination of the most informative NMR observables providing useful data for studying protein structure, dynamics and interactions.[104]

1.3.2 HSQC and Their Variants. SAR by NMR

In order to make the screening process more efficient, only protein backbone assignment was usually carried out and $^1H-^{15}N$ HSQC titration experiments were implemented in the screening programs. The SAR (structure-activity relationship) method has been extensively used in pharmaceutical companies and is based on the detection of secondary binding sites near the main one and in the synthesis of linked inhibitors of both sites with a greatly enhanced affinity (from $K_D \sim$ mM to $K_D \sim$ nM).[35]

The implementation of the transverse relaxation optimized spectroscopy (TROSY)[105] HSQC variant allowed to extend the application of this method to high-molecular-weight receptors, permitting to get high resolution spectra of receptors up to 100 kDa.

As a simple and practical alternative, individual ^{15}N-labeled amino acids can also be incorporated into a protein. To map completely a binding site on the protein surface this approach requires the preparation of several differently ^{15}N-labeled proteins.

$^1H-^{13}C$ HSQC experiments are less used, due to the higher complexity of the spectra. However, these experiments are very useful combined with selective labeling of the methyl groups of valine, leucine and isoleucine or when non-natural methylated amino acids are available (mainly lysines). The favourable relaxation properties of the methyl groups also allow the application of HSQC-based screening to larger proteins.

1.4 Mixed Protocols

1.4.1 Screening. Fragment-Based Drug Design

The screening of a battery of compounds against a macromolecular target of biomedical interest is the fuel that feeds the drug-discovery engine. Regardless of the many biochemical and biophysical techniques used for the detection and evaluation of binding hits, two different philosophies of compound screening coexist in the current pharmaceutical landscape: high-throughput screening (HTS) and fragment-based drug discovery (FBDD).

The aim of the former, which has dominated the drug-discovery activity for much of its history, is the evaluation of large numbers of heavy, drug-like molecules. Typically, hundreds of thousands to several millions of compounds are screened for each target, and high potency inhibitors are sought. For its part, FBDD is based on the screening of small libraries of low-molecular-weight compounds, called fragments, in pursuit of hits with low, but efficient, binding activity, which are then combined or developed into more potent inhibitors *via* different strategies, as illustrated in Figure 1.4. Over the last decade or so, FBDD has gained increasing acceptance both in the industry and in academia. The success of this new drug-discovery paradigm is illustrated by the recent approval of the first fragment-based drug (vemurafenib)[106] by the Food and Drug Administration (FDA).

A thorough review of the theory and state-of-the-art of either HTS or FBDD is out of the scope of this chapter. Instead, it has been our intention to focus on the practical aspects of employing NMR methodologies for the drug-discovery process, essentially from the point of view of FBDD, but much of which can be also applied to non-fragment-based approaches.

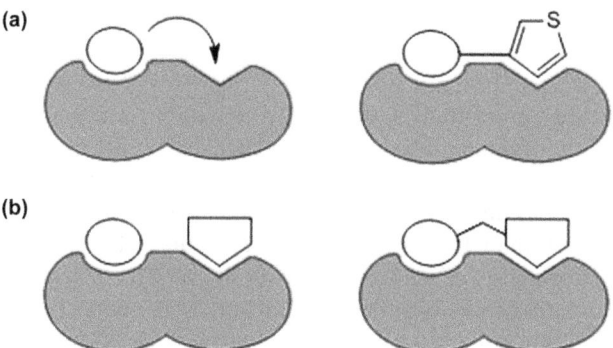

Figure 1.4 Two strategies for lead generation following a fragment-based approach: (a) fragment growth; (b) fragment linking.

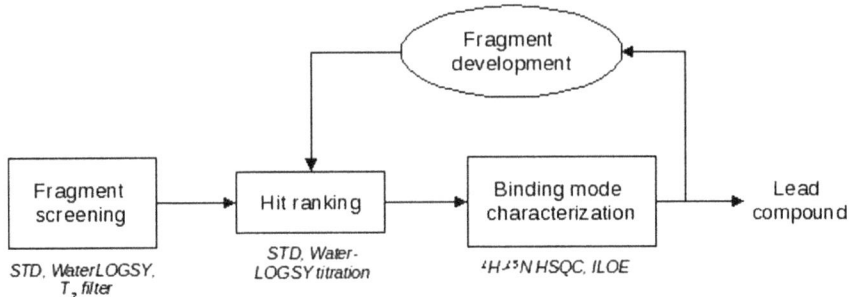

Figure 1.5   Flow chart for an NMR-driven fragment-based screening.

From the beginning, NMR has represented by far the preferred biophysical technique in FBDD programs, due to its reliability and versatility. Indeed, thanks to the wide variety of NMR experiments available, the use of this spectroscopy can guide the whole process (Figure 1.5), from the initial screening of compounds, to the ranking of the hits according to their affinity or efficiency and the characterization of the binding modes, from both the ligand and the receptor viewpoints.[107]

1.4.1.1   Fragment Screening and Ranking. Magnetization Transfer Experiments: STD and WaterLOGSY

Most usually, fragments are screened by using a magnetization transfer-based technique, namely STD or WaterLOGSY. These methods present the advantage that their detection range is suitable to the weak affinities expected for fragment-receptor interactions. To speed up the process and minimize protein consumption, experiments are performed on cocktails of fragments, in a number such that signal overlapping does not preclude the identification of the compounds and their solubility is not compromised. Normally, 6 to 10 fragments are assayed in each experiment. The composition of the mixtures might be entirely random, although it is more advisable to bring together fragments as chemically different as possible, in order easily to identify hits by inspection of the ^{1}H-NMR spectrum without the need to deconvolute the mixture. Alternatively, mixtures can be made up of similar fragments, if one wishes to identify structural features or scaffolds with binding activities towards the target.

Ranking of the fragment hits can be undertaken by determining the dissociation constants with either STD or WaterLOGSY titration experiments, as described in previous units. Sometimes, fragment hits are ranked by their ligand efficiencies, rather than their dissociation constants. The ligand efficiency is defined as the free energy of binding per heavy atom (N), and its measurement is useful for comparing ligands with different potencies and different sizes, as it normalizes the affinity of a given compound to its size, as shown in eqn (1.10):[108]

$$ligand\ efficiency = \frac{-\Delta G_{bind}}{N} = \frac{-RT \cdot \ln K_D}{N} \qquad (1.10)$$

1.4.1.2   Transverse Relaxation Rates

Another popular and fast technique for fragment screening is based on monitoring the changes in T_2 relaxation of the fragment signals upon binding to the target. As we detailed in previous pages, ligands bound to a macromolecular receptor experience an enhancement of their T_2 relaxation, which is translated into a broadening of their signals. In practice, however, rather than directly measuring the linewidth of the ligand signals, a relaxation filter is applied by setting a spin-lock between the 90° excitation pulse and the signal acquisition, and a comparison of spectra recorded in the presence and in the absence of the target will reveal a change in the intensity of signals corresponding to binding compounds. Depending on the duration of the spin-lock, a detection cut-off can be tuned in order to scope for stronger or weaker-binding fragments, with longer spin-lock times enabling to detect weaker interactions. This possibility also introduces a way to qualitatively rank fragment hits, *i.e.* by comparing series of experiments performed with different spin-lock times.

1.4.1.3   Paramagnetic Labeling of Proteins: SLAPSTIC

Paramagnetic tagging of proteins has also been used for compound screening. According to this method, dubbed Spin Labels Attached to Protein Side Chains as Tool to Identify Interacting Compounds (SLAPSTIC),[109] a paramagnetic label is covalently attached to a specific site of the target (*e.g.* a cysteine thiol group), and hits are detected by the transmission of paramagnetic effects (mainly PREs) to the binding fragments, most usually through a transverse relaxation filter. The paramagnetic label should be judiciously placed, so that it lies relatively close to the active site, as paramagnetic effects decrease with the distance, but its attachment does not alter the structure and recognition properties of the protein. Therefore, not only is a previous knowledge of the structure of the target required, but also to perform genetic engineering in the case that a suitable tagging site does not exist in the natural protein.

A modification of this strategy, involving the use of a paramagnetic derivative of a fragment with demonstrated binding activity toward the target, can be used to discover fragments binding to different sites.[40] At saturating concentrations of this fragment, a screening is performed looking for second binders, which will be detected by the transmission of paramagnetic effects from the first fragment, when, and only if it is possible that, both molecules are bound at the same time to the protein. This method presents notorious advantages in comparison to other procedures, which are essentially based on performing any of the mentioned NMR screening experiments (*e.g.* STD) in the presence of saturating concentrations of a first binder, since, for weakly binding compounds, such as fragments, it is often difficult to achieve saturating concentrations before reaching their solubility limit, and false positives can arise.

1.4.1.4   Competition Approaches

Finally, for compound screening, indirect approaches, in the shape of competition experiments involving a spy molecule, can be employed. According to this strategy, the NMR properties of a known ligand for a given target are monitored before and after the addition of the screened compounds, using any of the NMR experiments presented here. If the spy molecule is fluorinated, ^{19}F-NMR experiments can be carried out. Competition approaches present the advantage that, if the binding parameters of the spy molecule are well determined, those of the competitors can be easily derived applying mathematical models. For example, assuming a competitive inhibition model between a fragment hit and a spy molecule for binding to a target, the dissociation constant for the spy molecule (K_D) is shifted to an apparent value ($K_D{}^{ap}$) (eqn (1.11)) in the presence of non-zero concentrations of a fragment hit (F) displaying an inhibition constant K_I

$$K_D{}^{ap} = K_D\left(1 + \frac{[F]}{K_I}\right) \tag{1.11}$$

Therefore, the relation between the observed values for an NMR property of the spy molecule S (*e.g.* the intensity of a signal applying a T_2 relaxation filter) after (*I*) and before (*I_0*) the addition of the competitor fragment can be represented as shown in eqn (1.12):

$$\frac{I}{I_0} = \frac{\dfrac{I_{max} \cdot [S]}{K_D\left(1 + \dfrac{[F]}{K_I}\right) + [S]}}{\dfrac{I_{max} \cdot [S]}{K_D + [S]}} = \frac{K_I \cdot K_D + K_I \cdot [S]}{K_I \cdot K_D + K_D \cdot [F] + K_I \cdot [S]} \tag{1.12}$$

And, for selected experimental conditions where $[S] = K_D$, eqn (1.12) can be re-written as

$$\frac{I}{I_0} = \frac{2 \cdot K_I}{2 \cdot K_I + F} \tag{1.13}$$

The principal drawbacks associated with competition methods are the impossibility to identify the active compounds if screened as mixtures and the occurrence of false-negatives when fragments bind to sites different from that of the spy molecule.

1.4.1.5   Binding Mode Characterization

When it comes to getting a detailed picture on the geometry of the ligand–receptor complex, the versatility of NMR offers a wide variety of experiments, both ligand- and receptor-detected, providing valuable information from different points of view.

1.4.1.5.1   Protein-detected   Techniques: ^{1}H–^{15}N   HSQC   Experiments.
Competition experiments themselves represent a first insight into the

characterization of the binding mode, since, taking allosteric effects aside, if the binding site of the spy molecule is known, so is that of the studied fragment.

In order to delineate in more detail the binding epitope of the protein, changes in the resonances of the latter are monitored upon the addition of the fragment, usually in ^{1}H–^{15}N HSQC experiments. By doing so, the amide groups closer to the ligand are identified by the strong perturbation of their respective cross-peaks in the spectrum. This information is given as input to computational modeling programs in order to get a 3D view of the binding. Understanding the mode of binding of the hits in the three-dimensional context of the protein is of paramount importance to design potentially stronger binding compounds (Figure 1.4a). ^{1}H–^{15}N HSQC experiments demand considerable amounts of uniformly ^{15}N-labeled receptor, since samples normally require protein concentrations above 100 μM for experiments to be acquired within reasonable time scales. Moreover, it requires that the protein resonances, at least for most of their part, have been assigned. The need for the assignment introduces a practical limit in the size of the receptor. Although the availability of increasingly higher magnetic fields and the development of new pulse sequences are continuously pushing the limit to larger sizes, assignment of proteins heavier than 35 kDa becomes a challenge.

1.4.1.5.2 Ligand-Detected Techniques: Inter-Ligand NOEs, STD, TR-NOESY.
When two or more fragments bind to different, but adjacent, sites on the target surface, knowledge on their relative orientation and distance becomes essential in order to attempt the connection of the fragments (Figure 1.4b) *via* a suitable linker to create a larger, more potent compound. In this regard, NOESY experiments can provide valuable information in the shape of inter-ligand NOEs (INOEs) given rise to by pairs of protons belonging to different fragments, but located close in space when docked at the binding site.

It has already been explained how one can take advantage of STD and TR-NOESY experiments to obtain information on the ligand epitopes more exposed to the protein surface and the conformation of the bound ligand, respectively. While for fragments, as relatively simple and small molecules, such experiments might not provide decisive information, they can be useful for the study of the bound structure of lead compounds in more advanced stages.

1.5   Frontiers

1.5.1   Towards Larger and Larger Systems: Labeling Strategies and Instrumentation

The most important limitation in using NMR spectroscopy has traditionally been the molecular weight of the system under study, as imposed by the nuclear spin relaxation (line broadening) and increased spectral complexity associated with macromolecules that are larger than ~35 kDa. Advances on NMR instrumentation and methodology will lead to the improvement of the quality of data obtained for larger systems, and will allow for faster throughput and application of NMR screening to those.[24,44] Furthermore, the screening in living cells (see Chapter 4) will also provide novel information for drug discovery.

Chemical shift mapping by monitoring the perturbation of the amide chemical shift of ^{15}N-labeled proteins upon addition of ligands is a very simple method and has been traditionally used for detecting ligand binding. It allows the characterization of the target binding site as well as to measure dissociation constants and binding kinetics.

In section 1.3.1, the standard ^{15}N and/or ^{13}C labeling procedures have been briefly described. As mentioned, proteins will generally be expressed in *E. coli* in a minimum medium supplemented with [15]NH$_4$Cl and/or ^{13}C-glucose. For human or complex proteins that are subjected to post-translational modifications (such as complicated glycosylation patterns, disulfide bonds, phosphorylation *etc.*), production in bacteria becomes problematic. Alternative expression for such proteins has been developed in the past years and systems such as mammalian cells, yeast or insect cells have gained importance. Strategies for isotopic labeling in these systems have been exploited and recently reviewed for mammalian cells,[110] yeast[111] and insect cells.[112]

Moreover, taking advantage of the possibility of labeling on ^{2}H, spectral complexity is reduced by removing most of the ^{1}H atoms, which improves the relaxation properties of large systems. Protein deuteration[113] can be achieved by growing the bacteria in a medium that uses D$_2$O instead of H$_2$O. In this case about a 70–80% deuteration of the side chains is achieved, as there is a certain amount of contaminating ^{1}H present from the glucose. Higher levels of deuteration can be accomplished by using ^{13}C,[2]H-glucose. As NH groups are exchangeable, they get back to ^{1}H when normal aqueous buffers are used to purify the protein, and most of the standard NH-based experiments for assignment can then be carried out in triple-labeled samples. Pure D$_2$O might create problems for the cells to grow. In such cases, bacteria can be grown in small volumes of minimum medium of increasing percentages of deuterated water to get them adapted.

New developments such as the previously mentioned deuteration, or techniques such as transverse relaxation optimized spectroscopy (TROSY)[105] and cross-relaxation induced polarization transfer (CRIPT)[114] have increased the molecular weight limit of NMR spectroscopy to values greater than 100 kDa. Perdeuteration of a protein eliminates the most important relaxation mechanism of the ^{13}C nuclei, namely the dipolar coupling to the directly bound ^{1}H, and additionally the TROSY method takes advantage of the partial cancellation of dipolar couplings and chemical shift anisotropy, that takes place at very high magnetic fields. The combination of both strategies has resulted in a significant increase in the molecular weight limit of biomolecular NMR. The CRIPT method, on the other hand, is a new approach to polarization transfer in heteronuclear NMR that is much less sensitive to relaxation than the traditionally used insensitive nuclei enhanced by polarization transfer (INEPT).

Moreover, protein perdeuteration combined with selective amino acid labeling, for example methyl ^{13}C/^{1}H selective labeling of residues such as methionine, isoleucine, threonine, valine, leucine, while the protein is uniformly ^{2}H-labeled elsewhere, can be of great help.[115] By using this procedure, the sensitivity is increased with advantages for screening high-molecular-weight protein targets. Perturbations of the observed ^{1}H–^{13}C correlations with known ligands or inhibitors are used to assign active site resonances. New ligands can then be identified using saturation transfer methods. Since the majority of the protein is deuterated, saturation transfer is quite specific for ligand binding near the protonated residues. Identified ligands can then be docked to the protein *via* transfer NOEs to the protonated residues. Methyl labeling of the protein can be carried out by growing the bacteria on minimal medium in deuterated water, using ^{13}C,^{2}H-glucose as the main carbon source. Around one hour prior to induction α-ketobutyrate and α-ketoisovalerate are added to the medium to lead to the desired labeling of isoleucine, valine and leucine.

In addition, only specific aminoacids can be labeled by adding to the growing medium labeled amino acids. In the best case scenario, only those amino acids that are supplied labeled will become labeled in the protein. Unfortunately, *E. coli* metabolism and catabolism cause a certain interconversion between amino acids and provokes scrambling. The situation can be improved using auxotrophic strains but for a complete control over the incorporation of amino acids it is necessary to move onto cell-free methods, which imply the protein to be expressed *in vitro* by adding the DNA or mRNA of the target protein to a cell extract containing the transcription/translation machinery along with the 20 amino acids (which may be unlabeled or labeled), NTPs, several enzymes and buffer, salts, *etc.*[116] Site-specific labeling with unnatural amino acids can be performed by incorporating them *in vivo* into proteins at genetically encoded positions. They might be isotopically labeled, which provides advantages for their use in NMR studies.[117]

Segmental isotopic labeling allows specific segments within a protein to be selectively studied by NMR thus significantly reducing the spectral complexity for large proteins and allowing a variety of solution-based NMR strategies to be applied. This segmental isotopic labeling can be achieved by native chemical ligation, expressed protein ligation or protein trans-splicing.[118] Further developments, such as the Solvent Exposed Amides with TROSY (SEA-TROSY) method,[119] may extend the applicability of ^{15}N-based SAR by NMR to much higher molecular weight proteins. Only backbone amide groups that are in fast exchange with the solvent will be observed. Since most protein–ligand interactions involve surface residues, this subset of amide groups is sufficient to detect binding interactions.

There are some NMR-based approaches to accelerate drug discovery: Nuclear Magnetic Resonance Docking of Compounds (NMR-DOC),[120] which enables the rapid docking of compounds into their binding site once a model of the 3D structure of the protein target is known, and Nuclear Magnetic Resonance Structurally Oriented Library Valency Engineering (NMR-SOLVE),[120] which allows the rapid structural characterization of the ligand binding mode to enzymes with two adjacent binding pockets. Unlike NMR-DOC, NMR-SOLVE works in the complete absence of any structural data on the protein targets.

One additional consideration about the extension of NMR to the analysis of larger proteins are the advances on NMR instrumentation. One of the most important advances comprises the development of high magnetic field strength magnets to improve signal dispersion and sensitivity. Currently, the spectrometer with the highest magnetic field (1000 MHz ^{1}H frequency) is available at the Center for High Field NMR in Lyon, France. Moreover, further technical developments include: (i) cryogenic probes,[121] with the coils working at very low temperature in order to reduce electronic noise and thus improve sensitivity (see also Chapter 5 for applications of cryogenic probes to small animal imaging); (ii) microcoil technology,[122] which is useful and easy to implement for proteins that can be highly concentrated (0.5–1 mM and higher). Due to the excellent mass-based sensitivity of 1 mm microcoil NMR probes, their sensitivity is 8–12 times higher than that provided by regular 5-mm NMR probes,[122] thus speeding up NMR data acquisition; (iii) improved assignment of protein resonances by applying methods as G-matrix Fourier Transform (GFT) spectroscopy,[123] single-scan multidimensional spectroscopy,[124] Hadamard spectroscopy[125] and projection reconstruction techniques[126] could lead to important improvements on the quality of the data recorded by NMR and so extend the applications of the technique.

1.5.2 Membrane Proteins

Integral membrane proteins represent more than 50% of current drug targets. However, due to difficulties in expression and purification, especially on the selection of the proper membrane mimicking system, the study of these structures is still a challenge in structural biology. Despite these problems, great advances have been made recently by using X-ray crystallography[127] and NMR spectroscopy[128] to study membrane proteins. From the NMR viewpoint, three different approaches have been undertaken: full labeling of the protein (^{15}N, ^{13}C and ^{2}H) for 3D structure elucidation, selective labeling of certain residues for interaction studies with drugs and direct observation of ligand binding to membrane proteins in living cells or intact membranes.

The first approach is the most challenging, due to the size of the protein/detergent complexes and the signal overlapping, especially in α-helical proteins. In addition, substantial amounts of the isotopically labeled (^{15}N, ^{13}C and ^{2}H) protein are required.

In most of the membrane proteins the levels of expression are relatively small, requiring large-scale preparations, which for isotopically labeled samples are rather expensive. However, encouraging contributions have recently emerged on the field by using solid and solution NMR spectroscopy (see Chapter 2 for applications of solid-state NMR). In this sense it is worth mentioning the recent report of a high-quality structure of the archaeal protein receptor pSRII (photophobic receptor), showing that the 3D structure of a

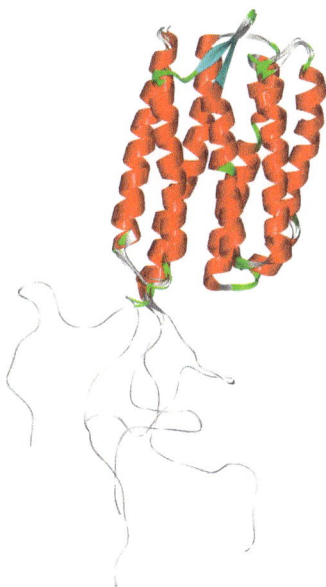

Figure 1.6 Solution NMR structure of the seven-helix transmembrane receptor sensory rhodopsin II by solution NMR spectroscopy (published structure PDB code 2KSY). An ensemble of five of the lowest energy structures are represented and coloured by secondary type.
See Gautier *et al.*[129] for details.

detergent-solubilized seven-transmembrane helix protein can be obtained by solution NMR spectroscopy (Figure 1.6).[129,130]

The second approach exploits the selective labeling of certain residues in the protein receptor. One of the most used methods is the derivatization of the lysine or cysteine residues after the protein purification. In this case, previous information on the systems such as an X-ray crystallography structure is required in order to rationalize the results. Lysine side chains can be selectively labeled with ^{13}C by reductive methylation with ^{13}C-enriched formaldehyde in the presence of cyanoborohydride. This strategy has been successfully used in the study of the binding of small drugs to the β2 adrenergic receptor.[131] As cysteine residues can be labeled with 2,2,2-trifluoroethanethiol (TET), ^{19}F NMR spectroscopy becomes an alternative method for the detection of membrane protein–ligand interactions because no background ^{19}F exists for either proteins or detergents. This method has been used recently to characterize the interaction of β2 adrenergic receptor with various ligands.[132]

The third approach is based on ligand detection and the application of tr-NOE and STD experiments in living cells (see Chapter 4).[49] The disadvantage of this method is that the information obtained is limited to the small molecules interacting with the membrane proteins but the great advantage is that the production of the recombinant receptor and the reconstitution in the detergent are not required. This method has been applied recently to the study of integrin

$\alpha_v\beta_3$ recognition of different ligands in living cancer cells[133] and integrin $\alpha_{IIb}\beta3$ interactions in native platelets.[52] In cells derived from solid tissues there is a strong tendency to aggregate and settle that prevents the applications of good-quality STD experiments. In these cases STD experiments under HR-MAS conditions can be required.[134] Finally, the use of intact membranes rather than reconstituted proteins has also been reported as a good strategy to study interactions with membrane proteins.[77]

With no doubt, in the coming years, advances in hardware and methodologies in all the different contexts will further permit new avenues in the application of NMR methods for drug discovery, employing the different viewpoints mentioned herein and exploiting the synergy among them.

1.6   Conclusions

NMR spectroscopy has shown to be a very powerful tool among other biophysical techniques for its use in drug discovery. The existence of a wide variety of both ligand- and target-detected methodologies exploiting the properties of magnetically active nuclei provides its privileged position in this field.

NMR techniques have been developed for screening large libraries of compounds, identifying hits in compound mixtures, defining binding epitopes of ligands or localizing binding sites on receptor molecules, describing the conformations in both the free and the bound state with the support of computational modeling, quantifying binding affinities and even fully elucidating the structure of target molecules.

Current and prospective advances on NMR instrumentation and methodologies will expand their range of applicability towards larger receptors and to membrane proteins, which represent a majority of the known drug targets.

References

1. O. Jardetzky, N. G. Wade and J. J. Fischer, *Nature*, 1963, **197**, 183.
2. B. D. Sykes, *J. Am. Chem. Soc.*, 1969, **91**, 949.
3. J. J. Fischer and O. Jardetzky, *J. Am. Chem. Soc.*, 1965, **87**, 3237.
4. B. D. Sykes and M. D. Scott, *Annu. Rev. Biophys. Bioeng.*, 1972, **1**, 27.
5. P. Balaram, A. A. Bothner-By and E. Breslow, *Biochemistry*, 1973, **12**, 4695.
6. P. Balaram, A. A. Bothner-By and E. Breslow, *J. Am. Chem. Soc.*, 1972, **94**, 4017.
7. P. Balaram, A. A. Bothner-By and J. Dadok, *J. Am. Chem. Soc.*, 1972, **94**, 4015.
8. K. Akasaka, *J. Magn. Reson.*, 1979, **36**, 135.
9. B. D. Sykes and W. E. Hull, *Ann. NY Acad. Sci.*, 1973, **226**, 60.
10. A. A. Bothner-By and R. Gassend, *Ann. NY Acad. Sci.*, 1973, **222**, 668.
11. J. P. Albrand, B. Birdsall, J. Feeney, G. C. Roberts and A. S. Burgen, *Int. J. Biol. Macromol.*, 1979, **1**, 37.

12. G. M. Clore and A. M. Gronenborn, *J. Magn. Reson.*, 1983, **53**, 423.
13. A. P. Campbell and B. D. Sykes, *Annu. Rev. Biophys. Biomol. Struct.*, 1993, **22**, 99.
14. F. Ni and H. A. Scheraga, *Acc. Chem. Res.*, 1994, **27**, 257.
15. M. Mayer and B. Meyer, *Angew. Chem. Int. Ed.*, 1999, **38**, 1784.
16. S. W. Fesik, E. R. Zuiderweg, E. T. Olejniczak and R. T. Gampe, *Biochem. Pharma.*, 1990, **40**, 161.
17. S. W. Fesik, *J. Med. Chem.*, 1991, **34**, 2937.
18. B. J. Stockman, *Prog. Nucl. Magn. Reson. Spectros.*, 1998, **33**, 109.
19. S. W. Fesik, *J. Biomol. NMR*, 1993, **3**, 261.
20. G. T. Montelione and T. Szyperski, *Curr. Opin. Drug Discov. Devel.*, 2010, **13**, 335.
21. J. M. Moore, *Curr. Opin. Biotechnol.*, 1999, **10**, 54.
22. T. Diercks, M. Coles and H. Kessler, *Curr. Opin. Chem. Biol.*, 2001, **5**, 285.
23. D. C. Fry and S. D. Emerson, *Drug Design Discov.*, 2000, **17**, 13.
24. B. J. Stockman and C. Dalvit, *Prog. Nucl. Magn. Reson. Spectros.*, 2002, **41**, 187.
25. J. W. Peng, J. Moore and N. Abdul-Manan, *Prog. Nucl. Magn. Reson. Spectros.*, 2004, **44**, 225.
26. C. A. Lepre, J. M. Moore and J. W. Peng, *Chem. Rev.*, 2004, **104**, 3641.
27. R. Powers, *Expert Opin. Drug Discov.*, 2009, **4**, 1077.
28. S. W. Homans, *Angew. Chem. Int. Ed.*, 2004, **43**, 290.
29. B. Meyer, T. Weimar and T. Peters, *Eur. J. Biochem.*, 1997, **246**, 705.
30. M. Mayer and B. Meyer, *J. Am. Chem. Soc.*, 2001, **123**, 6108.
31. V. Jayalakshmi and N. R. Krishna, *J. Magn. Reson.*, 2002, **155**, 106.
32. H. N. Moseley, E. V. Curto and N. R. Krishna, *J. Magn. Reson. B*, 1995, **108**, 243.
33. L. Fielding, *Curr. Top. Med. Chem.*, 2003, **3**, 39.
34. L. Fielding, S. Rutherford and D. Fletcher, *Magn. Reson. Chem.*, 2005, **43**, 463.
35. S. B. Shuker, P. J. Hajduk, R. P. Meadows and S. W. Fesik, *Science*, 1996, **274**, 1531.
36. J. Fejzo, C. A. Lepre, J. W. Peng, G. W. Bemis, Ajay, M. A. Murcko and J. M. Moore, *Chem. Biol.*, 1999, **6**, 755.
37. M. F. Lin, M. J. Shapiro and J. R. Wareing, *J. Am. Chem. Soc.*, 1997, **119**, 5249.
38. C. Dalvit, P. Pevarello, M. Tato, M. Veronesi, A. Vulpetti and M. Sundstrom, *J. Biomol. NMR*, 2000, **18**, 65.
39. C. Dalvit, P. E. Fagerness, D. T. Hadden, R. W. Sarver and B. J. Stockman, *J. Am. Chem. Soc.*, 2003, **125**, 7696.
40. W. Jahnke, L. B. Perez, C. G. Paris, A. Strauss, G. Fendrich and C. M. Nalin, *J. Am. Chem. Soc.*, 2000, **122**, 7394.
41. E. R. Zartler, J. Hanson, B. E. Jones, A. D. Kline, G. Martin, H. P. Mo, M. J. Shapiro, R. Wang, H. P. Wu and J. L. Yan, *J. Am. Chem. Soc.*, 2003, **125**, 10941.

42. J. B. Jordan, L. Poppe, X. Y. Xia, A. C. Cheng, Y. Sun, K. Michelsen, H. Eastwood, P. D. Schnier, T. Nixey and W. G. Zhong, *J. Med. Chem.*, 2012, **55**, 678.

43. M. Pellecchia, I. Bertini, D. Cowburn, C. Dalvit, E. Giralt, W. Jahnke, T. L. James, S. W. Homans, H. Kessler, C. Luchinat, B. Meyer, H. Oschkinat, J. Peng, H. Schwalbe and G. Siegal, *Nat. Rev. Drug Discov.*, 2008, **7**, 738.

44. (a) M. H. Levitt, *Spin Dynamics: Basics of Nuclear Magnetic Resonance*, John Wiley & Sons, Ltd, 2001; (b) O. Zerbe, *BioNMR in Drug Research*, Wiley-VCH, 2003.

45. (a) E. O. Stejskal and J. E. Tanner, *J. Chem. Phys.*, 1965, **42**, 288; (b) P. Stilbs, *Prog. NMR Spectrosc.*, 1987, **19**, 1.

46. (a) J. E. Tanner, *J. Chem. Phys.*, 1970, **52**, 2523; (b) C. S. Johnson, *Prog. NMR Spectrosc.*, 1997, **31**, 63.

47. (a) D. Wu, A. Chen and C. S. Johnson, *J. Magn. Reson. A*, 1995, **115**, 260; (b) C. S. Johnson, *Prog. NMR Spectrosc.*, 1999, **34**, 203.

48. J. Angulo, P. M. Enriquez-Navas and P. M. Nieto, *Chemistry*, 2010, **16**, 7803.

49. S. Mari, D. Serrano-Gomez, F. J. Canada, A. L. Corbi and J. Jimenez-Barbero, *Angew. Chem. Int. Ed.*, 2005, **44**, 296.

50. B. Claasen, M. Axmann, R. Meinecke and B. Meyer, *J. Am. Chem. Soc.*, 2005, **127**, 916.

51. C. Dalvit, G. Fogliatto, A. Stewart, M. Veronesi and B. Stockman, *J. Biomol. NMR*, 2001, **21**, 349.

52. S. G. Trevino, N. Zhang, M. P. Elenko, A. Lipták and J. W. Szostak, *Proc. Natl Acad. Sci. USA*, 2011, **33**, 13492.

53. C. Ludwig, P. J. A. Michiels, X. Wu, K. L. Kavanagh, E. Pilka, A. Jansson, U. Oppermann and U. L. Günther, *J. Med. Chem.*, 2008, **51**, 1.

54. C. A. Lepre, J. Peng, J. Fejzo, N. Abdul-Manan, J. Pocas, M. Jacobs, X. Xie and J. M. Moore, *Comb. Chem. High Throughput Screen.*, 2002, **5**, 583.

55. C. Rademacher, J. Guiard, P. I. Kitov, B. Fiege, K. P. Dalton, F. Parra, D. R. Bundle and T. Peters, *Chem. Eur. J.*, 2011, **17**, 7442.

56. Z. Chen, G. Krause and B. Reif, *J. Mol. Biol.*, 2005, **354**, 760.

57. J. Orts, S. Bartoschek, C. Griesinger, P. Monecke and T. Carlomagno, *J. Biomol. NMR*, 2012, **52**, 23.

58. T. Miura, W. Klaus, A. Ross, K. Sakata, M. Masubuchi and H. Senn, *Eur. J. Biochem.*, 2001, **268**, 4833.

59. M. A. Anderson, B. Ogbay, R. Arimoto, W. Sha, O. G. Kisselev, D. P. Cistola and G. R. Marshall, *J. Am. Chem. Soc.*, 2006, **128**, 7531.

60. J. L. Asensio, F. J. Cañada, H.-C. Siebert, J. Laynez, A. Poveda, P. M. Nieto, U. M. Soedjanaamadja, H. J. Gabius and J. Jiménez-Barbero, *Chem. Biol.*, 2000, **7**, 529.

61. J. J. Hernández-Gay, A. Ardá, S. Eller, S. Mezzato, B. R. Leeflang, C. Unverzagt, F. J. Cañada and J. Jiménez-Barbero, *Chem. Eur. J.*, 2010, **16**, 10715.

62. A. L. Breeze, *Prog. Nucl. Magn. Reson. Spectros.*, 2000, **36**, 323.
63. G. Otting and K. Wütrich, *J. Magn. Reson.*, 1989, **85**, 586.
64. K. L. Constantine, M. E. Davis, W. J. Metzler, L. Mueller and B. L. Claus, *J. Am. Chem. Soc.*, 2006, **128**, 7252.
65. M. Reibarkh, T. Malia, B. Hopkins and G. Wagner, *J. Biomol. NMR*, 2006, **36**, 1.
66. Y. Mizukoshi, A. Abe, T. Takizawa, H. Hanzawa, Y. Fukunishi, I. Shimada and H. Takahashi, *Angew. Chem. Int. Ed.*, 2012, **124**, 1391.
67. J. Yan, A. D. Kline, H. Mo, M. J. Shapiro and E. R. Zartler, *J. Magn. Reson.*, 2003, **163**, 270.
68. S. Vanwetswinkel, R. J. Heetebrij, J. van Duynhoven, J. G. Hollander, D. V. Filippov, P. J. Hajduk and G. Siegal, *Chem. Biol.*, 2005, **12**, 207.
69. V. Früh, Y. Zhou, D. Chen, C. Loch, E. Ab, Y. N. Grinkova, H. Verheij, S. G. Sligar, J. H. Bushweller and G. Siegal, *Chem. Biol.*, 2010, **17**, 881.
70. V. M. Sánchez-Pedregal, M. Reese, J. Meiler, M. J. J. Blommers, C. Griesinger and T. Carlomagno, *Angew. Chem. Int. Ed.*, 2005, **44**, 4172.
71. J. Orts, J. Tuma, M. Reese, S. K. Grimm, P. Monecke, S. Bartoschek, A. Schiffer, K. U. Wendt, C. Griesinger and T. Carlomagno, *Angew. Chem. Int. Ed.*, 2008, **47**, 7736.
72. J. Orts, C. Griesinger and T. Carlomagno, *J. Magn. Reson.*, 2009, **200**, 64.
73. M. Reese, V. M. Sánchez-Pedregal, K. Kubicek, J. Meiler, M. J. J. Blommers, C. Griesinger and T. Carlomagno, *Angew. Chem. Int. Ed.*, 2007, **46**, 1864.
74. S. Bartoschek, T. Klabunde, E. Defossa, V. Dietrich, S. Stengelin, C. Griesinger, T. Carlomagno, I. Focken and K. U. Wendt, *Angew. Chem. Int. Ed.*, 2010, **49**, 1426.
75. J. Iwahara, C. Tang and G. M. Clore, *J. Magn. Reson.*, 2007, **184**, 185.
76. F. Arnesano, L. Banci, I. Bertini, I. C. Felli, C. Luchinat and A. R. Thompsett, *J. Am. Chem. Soc.*, 2003, **125**, 7200.
77. D. F. Hansen and J. J. Led, *Proc. Natl Acad. Sci. USA*, 2006, **103**, 1738.
78. G. Pintacuda and G. Otting, *J. Am. Chem. Soc.*, 2001, **124**, 372.
79. J. Iwahara and G. M. Clore, *Nature*, 2006, **440**, 1227.
80. C. Tang, J. Iwahara and G. M. Clore, *Nature*, 2006, **444**, 383.
81. C. Tang, C. D. Schwieters and G. M. Clore, *Nature*, 2007, **449**, 1078.
82. G. A. Bermejo, M.-P. Strub, C. Ho and N. Tjandra, *J. Am. Chem. Soc.*, 2009, **131**, 9532.
83. S. Kosol and K. Zangger, *J. Struct. Biol.*, 2010, **170**, 172.
84. J. Grossauer, S. Kosol, E. Schrank and K. Zangger, *Bioorg. Med. Chem.*, 2010, **18**, 5483.
85. G. Otting, *Annu. Rev. Biophys.*, 2010, **39**, 387.
86. G. Pintacuda, M. John, X.-C. Su and G. Otting, *Acc. Chem. Res.*, 2007, **40**, 206.
87. G. Otting, *J. Biomol. NMR*, 2008, **42**, 1.
88. D. Häussinger, J. R. Huang and S. Grzesiek, *J. Am. Chem. Soc.*, 2009, **131**, 14761.

89. P. H. Keizers, J. F. Desreux, M. Overhand and M. Ubbink, *J. Am. Chem. Soc.*, 2007, **129**, 9292.
90. M. Prudencio, J. Rohovec, J. A. Peters, E. Tocheva, M. J. Boulanger, M. E. P. Murphy, H. J. Hupkes, W. Kosters, A. Impagliazzo and M. Ubbink, *Chem. Eur. J.*, 2004, **10**, 3252.
91. S. Yamamoto, T. Yamaguchi, M. Erdélyi, C. Griesinger and K. Kato, *Chem. Eur. J.*, 2011, **17**, 9280.
92. A. Mallagaray, A. Canales, G. Dominguez, J. Jimenez-Barbero and J. Perez-Castells, *J. Chem. Comm.*, 2011, **47**, 7179.
93. M. John, G. Pintacuda, A. Y. Park, N. E. Dixon and G. Otting, *J. Am. Chem. Soc.*, 2006, **128**, 12910.
94. T. Zhuang, H.-S. Lee, B. Imperiali and J. H. Prestegard, *Protein Sci.*, 2008, **17**, 1220.
95. T. Saio, K. Ogura, K. Shimizu, M. Yokochi, T. Burke and F. Inagaki, *J. Biomol. NMR*, 2011, **51**, 395.
96. K. Bleicher, M. Lin, M. J. Shapiro and J. R. Wareing, *J. Org. Chem.*, 1998, **63**, 8486.
97. P. J. Hajduk, E. T. Olejniczak and S. W. Fesik, *J. Am. Chem. Soc.*, 1997, **119**, 12257.
98. F. Manzenrieder, A. O. Frank and H. Kessler, *Angew. Chem. Int. Ed.*, 2008, **47**, 2608.
99. C. Dalvit and A. Vulpetti, *ChemMedChem*, 2011, **6**, 104.
100. W. R. Dolbier, *Guide to Fluorine NMR for Organic Chemists*, Wiley, Hoboken, NJ, 2009.
101. T. Diercks, J. P. Ribeiro, F. J. Cañada, S. André, J. Jiménez-Barbero and H. J. Gabius, *Chem. Eur. J.*, 2009, **15**, 5666.
102. J. Battiste and R. A. Newmark, *Prog. Nucl. Magn. Reson. Spectros.*, 2006, **48**, 1.
103. C. Dalvit, *Prog. Nucl. Magn. Reson. Spectros.*, 2007, **51**, 243.
104. W. Bermel, I. C. Felli, R. Kümmerle and R. Pierattelli, *Concepts Magn. Reson.*, 2008, **32**, 183.
105. K. Pervushin, R. Riek, G. Wider and K. Wuthrich, *Proc. Natl Acad. Sci. USA*, 1997, **94**, 12366.
106. J. Tsai, J. T. Lee, W. Wang, J. Zhang, H. Cho, S. Mamo, R. Bremer, S. Gillette, J. Kong, N. K. Haass, K. Sproesser, L. Li, K. S. M. Smalley, D. Fong, Y. L. Zhu, A. Marimuthu, H. Nguyen, B. Lam, J. Liu, I. Cheung, J. Rice, Y. Suzuki, C. Luu, C. Settachatgul, R. Shellooe, J. Cantwell, S. H. Kim, J. Schlessinger, K. Y. J. Zhang, B. L. West, B. Powell, G. Habets, C. Zhang, P. N. Ibrahim, P. Hirth, D. R. Artis, M. Herlyn and G. Bollag, *Proc. Natl Acad. Sci. USA*, 2008, **105**, 3041.
107. J. Schultz, in *Fragment-based Drug Discovery, a Practical Approach*, ed. E. R. Zartler and M. J. Shapiro, Wiley, Chippenham, UK, 2008.
108. A. B. Reitz, G. R. Smith, B. A. Tounge and C. H. Reynolds, *Curr. Top. Med. Chem.*, 2009, **9**, 1718.
109. W. Jahnke, S. Rudisser and M. Zurini, *J. Am. Chem. Soc.*, 2001, **123**, 3149.

110. A. Dutta, K. Saxena, H. Schwalbe and J. Klein-Seetharaman, in *Protein NMR Techniques, Methods in Molecular Biology*, ed. A. Shekhtman and D. S. Burz, Springer Science + Business Media, LCC, 2012, vol. 831, p. 55.

111. T. Sugiki, O. Ichikawa, M. Miyazawa-Onami, I. Shimada and H. Takahashi, in *Protein NMR Techniques, Methods in Molecular Biology*, ed. A. Shekhtman and D. S. Burz, Springer Science + Business Media, LCC, 2012, vol. 831, p. 19.

112. K. Saxena, A. Dutta, J. Klein-Seetharaman and H. Schwalbe, in *Protein NMR Techniques, Methods in Molecular Biology*, ed. A. Shekhtman and D. S. Burz, Springer Science + Business Media, LCC, 2012, vol. 831, p. 37.

113. K. H. Gardner and L. E. Kay, *Annu. Rev. Biophys. Biomol. Struct.*, 1998, **27**, 357.

114. J. Fiaux, E. B. Bertelsen, A. L. Horwich and K. Wüthrich, *Nature*, 2002, **418**, 207.

115. V. Tugarinov and L. E. Kay, *J. Am. Chem. Soc.*, 2003, **125**, 13868.

116. (a) G. Zubay, *Ann. Rev. Genet.*, 1973, **7**, 267; (b) T. Kigawa, Y. Muto and S. Yokoyama, *J. Biomol. NMR*, 1995, **6**, 129.

117. D. H. Jones, S. E. Cellitti, X. Hao, Q. Zhang, M. Jahnz, D. Summerer, P. G. Schultz, T. Uno and B. H. Geierstanger, *J. Biomol. NMR*, 2010, **46**, 89.

118. (a) V. Muralidharan and T. W. Muir, *Nat. Methods*, 2006, **3**, 429; (b) L. Dongsheng, R. Xu and D. Cowburn, *Methods Enzymol.*, 2009, **462**, 151.

119. M. Pellecchia, D. Meininger, A. L. Schen, R. Jack, C. B. Kasper and D. S. Sem, *J. Am. Chem. Soc.*, 2001, **123**, 4633.

120. M. Pellecchia, D. Meininger, Q. Dong, E. Chang, R. Jack and D. S. Sem, *J. Biomol. NMR*, 2002, **22**, 165.

121. H. Kovacsa, D. Moskaua and M. Spraul, *Prog. Nucl. Magn. Reson. Spectros.*, 2005, **46**, 131.

122. R. E. Hopson and W. Peti, *Methods Mol. Biol.*, 2008, **4**, 447.

123. H. S. Atreya and T. Szyperski, *Proc. Natl Acad. Sci. USA*, 2004, **26**, 9642.

124. L. Frydman, T. Scherf and A. Lupulescu, *Proc. Natl Acad. Sci. USA*, 2002, **25**, 15858.

125. E. Kupce, T. Nishida and R. Freeman, *Prog. Nucl. Magn. Reson. Spectros.*, 2003, **42**, 95.

126. E. Kupce and R. Freeman, *J. Am. Chem. Soc.*, 2004, **126**, 6429.

127. V. P. Jaakola, M. T. Griffith, M. A. Hanson, V. Cherezov, E. Y. T. Chien, J. R. Lane, A. P. Ijzerman and R. C. Stevens, *Science*, 2008, **322**, 1211.

128. (a) T. Li, D. A. Berthold, R. B. Gennis and C. M. Rienstra, *Protein Sci.*, 2008, **17**, 199; (b) A. McDermott, *Annu. Rev. Biophys.*, 2009, **38**, 385.

129. A. Gautier, H. Mott, M. Bostock, J. Kirkpatrick and D. Nietlispach, *Nat. Struc. Mol. Biol.*, 2010, **17**, 768.

130. T. Warne, M. J. Serrano-Vega, J. G. Baker, R. Moukhametzianov, P. C. Edwards, R. Henderson, A. G. W. Leslie, C. G. Tate and G. F. X. Schertler, *Nature*, 2008, **454**, 486.

131. M. P. Bokoch, Y. Zou, S. G. F. Rasmussen, C. W. Liu, R. Nygaard, D. M. Rosenbaum, J. J. Fung, H. J. Choi, F. S. Thian, T. S. Kobilka, J. D. Puglisi, W. I. Weis, L. Pardo, R. S. Prosser, L. Mueller and B. K. Kobilka, *Nature*, 2010, **463**, 108.
132. J. J. Liu, R. Horst, V. Katritch, R. C. Stevens and K. Wüthrich, *Science*, 2012, **335**, 1106.
133. D. Potenza, F. Vasile, L. Belvisi, M. Civera and E. M. V. Araldi, *Chem. Bio. Chem.*, 2011, **12**, 695.
134. C. Airoldi, S. Giovannardi, B. La Ferla, J. Jiménez-Barbero and F. Nicotra, *Chem. Eur. J.*, 2011, **17**, 13395.

CHAPTER 2

Solid-State NMR in Drug Discovery and Development

Product Development, GlaxoSmithKline plc., 709 Swedeland Rd,
King of Prussia, PA, 19406, USA
Present address: Morgan, Lewis & Bockius LLP, 1701 Market St.,
Philadelphia, PA, 19103-2921, USA
Email: fvogt@morganlewis.com

2.1 Introduction

NMR spectroscopy is a powerful analytical and physical characterization technique that is widely used in pharmaceutical research and development. Solid-state NMR (SSNMR) spectroscopy is an extensive and rapidly growing branch of this field that has increasingly been applied alongside solution-state NMR. Developments in theory and methodology over the last 20 years have allowed for greater use of SSNMR spectroscopy throughout chemistry, biochemistry and materials science, and extensions to different areas of interest in the discovery, development and manufacturing of drugs have quickly followed. This chapter reviews recent applications of SSNMR to drug discovery and development, focusing primarily on techniques that utilize magic-angle spinning (MAS). Key theoretical concepts, instrumentation and experimental methods that are needed to understand the application of SSNMR in these systems are first reviewed. The bulk of this chapter is devoted to two main areas of application of SSNMR: (i) in drug discovery, primarily structural studies of large biomolecules that are drug targets, and (ii) in drug development, primarily in studies of formulations, active pharmaceutical

New Developments in NMR No. 2
New Applications of NMR in Drug Discovery and Development
Edited by Leoncio Garrido and Nicolau Beckmann
© The Royal Society of Chemistry 2013
Published by the Royal Society of Chemistry, www.rsc.org

ingredients (APIs) and manufacturing processes. Particular attention is given to emerging applications such as the use of high static fields, multidimensional experiments and sophisticated pulse sequences designed for the extraction of NMR parameters from complex systems.

A number of recent general reviews covering the broader field of SSNMR are also of interest to drug discovery and development because of the potential for adaptation of techniques applied in these other fields. These include reviews of developments in specific techniques, such as applications of ^{1}H SSNMR, as well as the larger fields of biomaterials, geochemistry, supramolecular chemistry and materials sciences.[1–6] This chapter is focused on analysis of materials that are predominantly crystalline or amorphous solids, and thus the technique of high-resolution magic-angle spinning (HR-MAS) NMR as applied to semi-solid and liquid samples is not discussed (see Chapter 3).[7]

2.2   Background

The theoretical background necessary for an understanding of SSNMR phenomena has been described in numerous monographs.[8–11] Here, a brief review is given of several key aspects of the theory of SSNMR relevant to its use in drug discovery and development. Particular attention is given to aspects of SSNMR that differ from solution-state NMR. First, NMR techniques access nuclear spin states, and require nuclei with a non-zero spin to be present. A number of isotopes covering nearly every element on the periodic table have non-zero spin and are thus NMR-active. The nuclides of interest in most SSNMR studies related to drug discovery and development are summarized in Table 2.1.[12] Some of the nuclides in Table 2.1 have only recently become readily accessible with the advent of higher static field strengths (B_0) and new probe and pulse sequence designs. These emerging nuclides of interest will be discussed in more detail in Sections 2.3 and 2.4. After discussion of the basic theory of SSNMR, instrumentation and experimental methods of interest in drug discovery and development are then reviewed, followed by a brief overview of computational methods used to support SSNMR studies.

2.2.1   Theoretical Background

SSNMR allows access to a number of parameters of interest for analysis of elucidation of structure and dynamics in solids. These parameters arise because of interactions between nuclear spins and their surroundings, including other nuclear spins and electrons. The interactions are normally grouped into categories based on the underlying nuclear spin Hamiltonians that define their energy. These include the chemical shielding interaction, dipolar interaction, J-coupling interaction, quadrupolar interaction and paramagnetic interaction.[8–11] In contrast to solution-state NMR, the lack of rapid molecular tumbling in SSNMR allows access to the anisotropic nature of these interactions.

Because it can provide site-specific resolution, the chemical shift is generally the most useful interaction in studies of many of the most important

Table 2.1   Basic properties of selected NMR nuclides of interest in drug discovery and development applications. Values are from Refs 10 and 12, which contain a complete list of NMR nuclides and their properties. The reference Harris *et al.*[12] contains accurate frequency ratios suitable for ghost referencing.

Isotope	Spin (I)	Frequency in MHz at 2.344 T	Natural Abundance	Relative receptivity[a]
^{1}H	1/2	100.000	99.985%	1.0000
^{2}H	1	15.350	0.0115%	0.00000111
^{11}B	3/2	32.084	80.1%	0.132
^{13}C	1/2	25.145	1.10%	0.000170
^{14}N	1	7.224	99.632%	0.001
^{15}N	1/2	10.137	0.366%	0.00000384
^{17}O	5/2	13.556	0.038%	0.0000111
^{19}F	1/2	94.094	100.0%	0.834
^{23}Na	3/2	26.451	100.0%	0.0927
^{25}Mg	5/2	6.120	10.0%	0.000268
^{27}Al	5/2	26.056	100.0%	0.207
^{29}Si	1/2	19.867	4.67%	0.000367
^{31}P	1/2	40.480	100.0%	0.0665
^{33}S	3/2	7.670	0.76%	0.0000172
^{35}Cl	3/2	9.798	75.78%	0.00358
^{43}Ca	7/2	6.728	0.14%	0.00000868
^{51}V	7/2	26.289	99.75%	0.383
^{67}Zn	5/2	6.254	4.10%	0.000118
^{77}Se	1/2	19.071	7.63%	0.000537

[a]Relative to ^{1}H at 1.0000.

NMR-active nuclides in pharmaceutical and biomolecular solids. The isotropic chemical shift (δ_{iso}), familiar from solution-state NMR, is also the most commonly observed parameter in SSNMR studies and generally provides detailed information about the electronic structure around the observed nucleus. The chemical shift of many isotopes of interest, such as ^{1}H, ^{13}C, ^{15}N, ^{19}F and ^{31}P, is specific to particular molecular functional groups and substituents, and follows established, interpretable trends.[13,14] The chemical shift is caused by the chemical shielding of the nucleus by its surrounding electrons. The isotropic chemical shielding (σ_{iso}) is related to δ_{iso} by

$$\sigma_{iso} \approx -(\sigma_{iso} - \sigma_{ref}) \qquad (2.1)$$

where σ_{ref} is the shielding of a reference nucleus. The chemical shift represents the experimentally accessible quantity measured in SSNMR experiments, while the chemical shielding represents the theoretical quantity that can in many cases be calculated using quantum chemical methods.

SSNMR and solution-state NMR studies both commonly seek to observe δ_{iso} and compute σ_{iso} for comparison, but SSNMR differs from solution-state NMR in that rapid molecular tumbling does not occur, allowing direct access to the full second-rank tensor that describes these interactions.[9] Each tensor has nine independent components that define its magnitude and orientation.

In general, only the three principal components are experimentally accessible in the polycrystalline powders of interest in most pharmaceutical applications of SSNMR. The reported principal component values of the experimental chemical shift tensor typically follow the convention $\delta_{11} \geq \delta_{22} \geq \delta_{33}$, so that the δ_{11} component is designated as that towards lowest frequency and becomes the most deshielded component (typically the leftmost in the spectrum). The orientation of the tensor for a nucleus relative to tensors from other sites in the molecule is commonly obtained through computational methods or can be accessed in certain experiments that utilize another interaction, such as dipolar coupling, to help "align" the orientation of the chemical shift tensor (CST) to a chemical bond. The principal components of the chemical shielding tensor are labeled and ordered following the convention:

$$|\sigma_{zz} - \sigma_{iso}| \geq |\sigma_{xx} - \sigma_{iso}| \geq |\sigma_{yy} - \sigma_{iso}| \quad (2.2)$$

It is also common to report values derived from the principal components, including the asymmetry:

$$\eta_{CSA} = (\sigma_{yy} - \sigma_{xx})/(\sigma_{zz} - \sigma_{iso}) \quad (2.3)$$

and the reduced anisotropy:

$$\sigma_{aniso} = \sigma_{zz} - \sigma_{iso} = (2/3)\sigma_{zz} - (1/3)(\sigma_{xx} + \sigma_{yy}) \quad (2.4)$$

When working with chemical shift or shielding tensors, it is often necessary to rotate the tensors between different frames to predict spectral properties and understand the effects of a range of crystallite orientations in polycrystalline powder, molecular motion within the crystal or mechanical sample spinning. This can be achieved by rotations of tensors written in irreducible form.[9] The CST contains additional information beyond δ_{iso} that can be used to obtain detailed structural information or enhance spectral assignments. For example, typical ^{13}C chemical shift tensor principal component values have been tabulated for organic compounds that can also be extended to drug molecules, and can allow for interpretation of the effects of structural features such as hydrogen bonding.[15]

The process of magic angle spinning (MAS) averages the CST, like other second-rank tensors, over the course of a rotor period (t_R) into a manifold of spinning sidebands. The measurement of CSTs is commonly accomplished by fitting MAS sidebands. An example is shown in Figure 2.1 for the determination of the ^{19}F chemical shift tensors for a crystalline form of the anti-fungal drug voriconazole.[16] This molecule contains three fluorine atoms, designated F1, F2 and F3, and crystalline Form 1 yields a cross-polarization MAS (CP-MAS) spectrum as shown in Figure 2.1. The CP-MAS experiment is discussed in Section 2.2.3. The centerbands, which appear at δ_{iso}, are denoted by colored arrows in the experimental spectrum. The spectrum consists of three sets of overlapping sideband manifolds, which can be fitted to obtain δ_{11}, δ_{22} and δ_{33} for each of the three fluorine sites as shown in Figure 2.1. The sideband manifold encodes these three principal component values across the individual

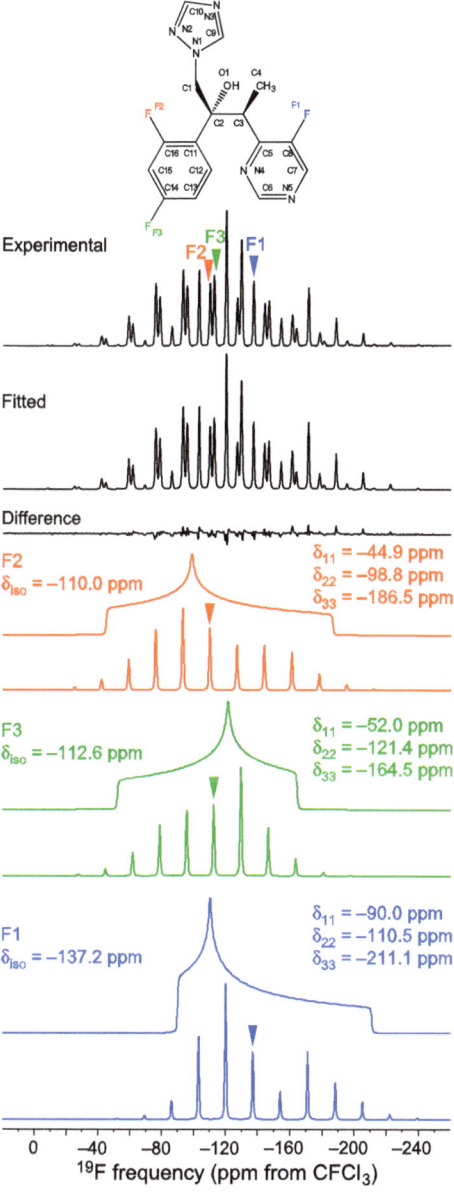

Figure 2.1 ^{19}F CP-MAS spectrum of a crystalline form (see Ravikumar *et al.*[16]) of the antifungal drug voriconazole. A fit of the sideband manifold for the three fluorine sites is shown, along with the three separate sideband manifolds (colored spectra). Arrows denote the MAS centerbands at δ_{iso}. The static lineshapes above each manifold correspond to the fitted chemical shift tensor parameters. The experimental spectrum was obtained using a 4 mm HFX probe at 273 K with ν_r set to 8 kHz and a B_0 field strength of 11.7 T.

sidebands and requires fitting to extract, as discussed in Section 2.2.5. In contrast, the static chemical shift tensors, which were simulated from the fitted MAS patterns as shown in Figure 2.1, contain a direct relationship between δ_{11}, δ_{22} and δ_{33} and the spectral features.

Many drugs, excipients and biomolecules of interest contain quadrupolar nuclei. These nuclei possess a nuclear spin quantum number I that is greater than $\frac{1}{2}$ (see Table 2.1) and thus a quadrupolar moment. The quadrupolar nucleus can couple with the electric field gradient (EFG) at the nuclear site; the EFG results from the surrounding environment, and creates another mechanism by which structural information can be obtained.[8–11] The EFG tensor V is normally reported using the convention

$$|V_{zz}| \geq |V_{yy}| \geq |V_{xx}| \qquad (2.5)$$

and the quadrupolar coupling constant (C_Q) is reported as

$$C_Q = (V_{zz}eQ)/h = (e^2qQ)/h \qquad (2.6)$$

where h is Planck's constant. The quadrupolar asymmetry is reported as

$$\eta_Q = (V_{xx} - V_{yy})/V_{zz} \qquad (2.7)$$

Extremely small values of C_Q, such as those observed commonly in ^{6}Li spectra, only cause line broadening effects on SSNMR spectra. Small but more sizeable C_Q values in the range of 100–200 kHz, as observed for ^{2}H spectra (with I = 1), lead to a first-order quadrupolar lineshape. An example of a first-order quadrupolar lineshape obtained using a quadrupolar echo sequence is shown in Figure 2.2 for the deuterochloroform solvate of the small molecule drug tranilast.[17] Although this ^{2}H lineshape is unaffected by reorientation because the chloroform molecules rotate around the C–H axis in this crystal structure, and can be fitted to a static lineshape as shown in Figure 2.2, ^{2}H lineshapes in pharmaceutical and biomolecular solids are generally affected more significantly by motion.[18] The use of MAS, as shown in Figure 2.2, provides additional resolution between the sidebands and allows for measurement of detailed ^{2}H chemical shift information, which can for example resolve different types of water or different chemically exchanged sites in pharmaceutical hydrates.[18]

Larger values of C_Q are typically encountered for most elements in the periodic table, including many of interest in drug discovery or development. Many of these nuclei also possess half-integral spin (with ^{14}N being a notable exception) and are often studied *via* their central transition spectra.[10,11] The central transition is affected by second order quadrupolar broadening, which is generally observed when the C_Q exceeds 1–2 MHz. MAS incompletely averages the second-order broadening, and other techniques are needed to fully average the lineshape and achieve high resolution. A second-order broadened central transition lineshape for I = 3/2 is observed in the ^{35}Cl MAS spectrum of a crystalline form of raloxifene HCl, a selective estrogen receptor modulator, as shown in Figure 2.3.[19] The spectrum is fitted to a theoretical second-order lineshape to obtain the C_Q and η_Q values given in Figure 2.3. The second-order

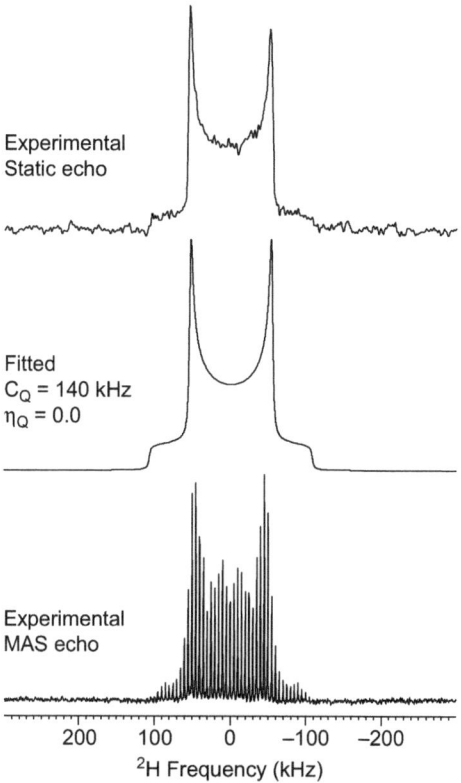

Figure 2.2 ^{2}H first-order quadrupolar lineshape (top spectrum) for the chloroform
solvate of the drug tranilast produced using 99% deuterium-enriched
chloroform, obtained under static conditions (without MAS). A fitted
first-order quadrupolar static lineshape is shown for comparison (middle
spectrum). MAS with v_r set to 5 kHz breaks the first-order experimental
lineshape into a manifold of sidebands as shown (bottom spectrum).
Experimental spectra were obtained using a quadrupolar echo consisting
of two $\pi/2$ pulses spaced by 100 µs (static) or 200 µs (MAS). Spectra were
obtained at 298 K and at a B_0 field strength of 8.4 T.

quadrupolar broadening also affects the observed isotropic chemical shift; this
additional shift can be calculated from knowledge of I, the quadrupolar
coupling and the Larmor frequency.[20] This is taken into account in the fit in
Figure 2.3, showing that the actual ^{35}Cl chemical shift in this form of raloxifene
HCl is 61.6 ppm from solid NaCl.

The dipolar coupling interaction is also of great importance in SSNMR
studies of drugs. The dipolar coupling arises because of through-space
magnetic field effects between two nuclear spins caused by individual nuclear
magnetic moments.[8–11] The dipolar coupling averages in isotropic solution
because of molecular tumbling, so that only its incoherent effects (through
relaxation) are typically observable. In solids, the dipolar coupling is directly
observable. The dipolar coupling between two nuclei depends directly on the

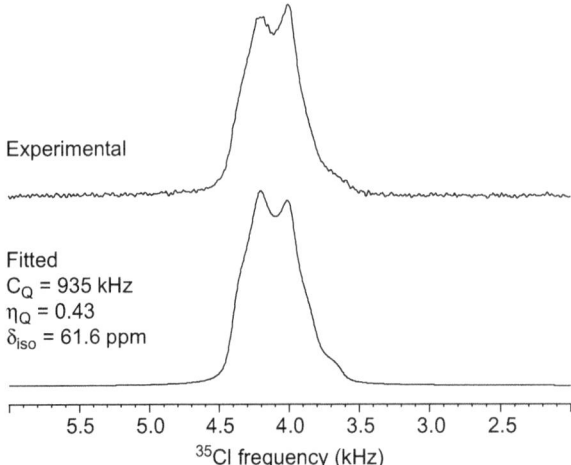

Experimental

Fitted
C_Q = 935 kHz
η_Q = 0.43
δ_{iso} = 61.6 ppm

5.5   5.0   4.5   4.0   3.5   3.0   2.5
^{35}Cl frequency (kHz)

Figure 2.3 ^{35}Cl MAS spectrum (ν_r = 30 kHz) of a crystalline form of raloxifene HCl, showing the effects of second-order quadrupolar broadening on the central transition lineshape for this spin = 3/2 nucleus. The spectrum was obtained at 273 K and at a B_0 field strength of 16.4 T. A fitted spectrum of the central transition lineshape showing the quadrupolar parameters obtained from the fit is also shown. The crystal structure of this phase is reported in Vega *et al.*[19] ^{35}Cl frequency is given relative to crystalline NaCl at 0 kHz.

product of the gyromagnetic ratios of the two nuclei engaged in the coupling and inversely on the cube of the internuclear distance between the nuclei. Because of this inverse cubic dependence, the structural information contained in a measurement of the dipolar coupling can directly yield an internuclear distance. Homonuclear dipolar couplings between ^{1}H nuclei in typical organic solids can reach as high as 30 kHz, while heteronuclear interactions between directly bonded ^{1}H and ^{13}C nuclei are typically about 10 kHz. Longer range interactions that yield dipolar couplings of 200 Hz or less are particularly useful for structural studies and can be measured by various pulse sequences. Systems that exhibit strong homonuclear coupling often engage in a process known as spin diffusion, by which magnetization can be transferred over long distances ranging from a few nm to hundreds of nm. Spin diffusion is normally modeled using phenomenological expressions, but fundamentally involves a series of dipolar coupling interactions mediated by zero- or double-quantum "flip flop" terms in the dipolar interactions.[21,22] Dipolar coupling can also lead to reintroduction of quadrupolar effects from non-observed nuclei.[23] For example, ^{13}C resonances in MAS spectra are often broadened or split by neighboring quadrupolar ^{14}N sites (which have a natural abundance > 99%) experiencing a strong EFG through dipolar coupling.[23] Similar effects can occur for ^{13}C resonances of sites directly bonded to $^{35/37}$Cl, $^{79/81}$Br and ^{127}I. These effects are commonly observed in ^{13}C spectra of API molecules, which often contain a significant number of nitrogen and halogen sites.

Magnetic dipolar coupling information can also be transferred by electrons through the J-coupling interaction, which generally occurs through chemical bonds and is also referred to as indirect dipolar coupling, scalar coupling or electron-mediated dipolar coupling.[8–11] The J-coupling, unlike the dipolar coupling, contains an isotropic component that is observed in solution-state NMR. In solids, the isotropic J-coupling interaction occurs with a similar magnitude to that observed for molecules in solution. The magnitudes of typical heteronuclear J-coupling interactions between directly bonded ^{1}H and ^{13}C nuclei are approximately 150 Hz, and most interactions are much smaller (on the order of a few Hz). The generally larger resonance linewidth in solids often obscures their direct observation except when large J-couplings are present (as often seen in fluorinated drugs because of ^{13}C–^{19}F J-coupling), although a number of pulse sequences are available to detect their effects. Anisotropic J-couplings can be observed in the solid state and contain useful structural information, but have not yet seen significant use in applications to drug discovery or development.[24]

Paramagnetic interactions are currently of particular interest in biomolecular SSNMR studies because of their ability to provide longer-range structural information. The paramagnetic interaction arises through coupling between an unpaired electron spin and a nuclear spin described by the hyperfine coupling tensor (A). Unpaired electron spins can be synthetically inserted into biomolecules as stable organic radicals (typically nitroxides) or as chelated metal ions with an appropriate spin state (typically using selected transition metal ions or lanthanides).[25] The paramagnetic interaction consists of two contributions, the first of which is known as the Fermi contact shift.[26,27] The Fermi contact shift originates because of the presence of unpaired electron spin density at the nucleus. Under MAS, the Fermi contact shift is isotropic and can cause large shifts of 100 ppm or more on resonances from sites within 5–10 Å of the paramagnetic center. The second contribution, known as the pseudocontact shift (PCS) or isotropic dipolar shift, arises from through-space dipolar coupling between the electron and nuclear spins.[26,27] The PCS has a much smaller magnitude than the Fermi contact shift, typically on the order of a few ppm to tens of ppm, but has a range of up to 20–30 Å.[25] In order to observe a PCS, the paramagnetic center must have an anisotropic electron g-tensor, which in turn leads to an anisotropic magnetic susceptibility tensor and then to an observable PCS.[25] Anisotropic electron g-tensors are observed for organic chelates of a subset of paramagnetic transition metal and lanthanide ions.[25]

Knowledge and interpretation of relaxation mechanisms in solids can play an important role in many SSNMR studies in drug discovery and development. Variable temperature studies and studies using different B_0 field strengths may be needed to elucidate which mechanisms occur. In combination with molecular motion and the creation of an appropriate spectral density, many of the interactions described above also contribute to typical relaxation processes in solids.[8,11,28] These processes include spin-lattice relaxation (T_1) and a return to equilibrium population, spin-lattice relaxation in the rotating frame ($T_{1\rho}$), and spin-spin relaxation (T_2) and a loss of phase coherence.[8,11,28–30] Chemical shift anisotropy relaxation is proportional to the square of B_0 and to the magnitude of δ_{aniso}, and requires molecular reorientation to modulate the chemical shift anisotropy.[31] Quadrupolar nuclei can relax *via* the EFG through the quadrupolar interaction, if the EFG at the nuclear site is modulated by molecular motion.[8,11] Dipolar relaxation occurs between spin pairs with a strong dipolar coupling, and in the case of abundant ^1H spins, spin diffusion can lead to an equalization of the relaxation processes across a microcrystal. Relaxation through J-coupling interactions can also occur in rare cases where nuclei are strongly coupled. Spin-rotation relaxation can occur if phonon modes in the solid provide a rotational energy match, and can affect heavy nuclei.[32] Paramagnetic relaxation occurs either because of a paramagnetic center present in the material, or *via* contact with a paramagnetic impurity or purposely inserted paramagnetic center, and is dependent on field and temperature.[25]

In crystalline solids of interest in drug development, the modulation needed for rapid relaxation processes is often enhanced by mobile functional groups such as methyl groups or longer normal or branched aliphatic chains, phenyl groups, CF_3 groups and groups that interact with their through-space neighboring atoms primarily by van der Waals interactions and are in open regions of a crystal structure.[28,33–35] These effects are enhanced particularly if quadrupolar nuclei are present in the moving group or in a less mobile nearby group. These groups are often referred to as relaxation sinks. Enhanced mobility at crystalline defects or at the surface of a crystal, which is itself a defect, can also serve as a relaxation sink and can be enhanced *via* contact with paramagnetic dioxygen in the air. Proton (^1H) T_1 relaxation in crystalline APIs, which are connected to the bulk through spin diffusion and lead to a single ^1H T_1 value across a particle when ν_r is relatively slow (*e.g.* $<12\,kHz$), is often dominated by this type of effect. In amorphous or semi-crystalline solids, relaxation processes are often enhanced because of greater molecular mobility, particularly above the glass transition temperature (T_g).[36] While developments are still needed in the experimental approach to measurement of relaxation rates in complex systems, there are many potential benefits if they can be harnessed in studies of dynamics in biopolymers to applications such as molecular recognition, ligand binding and mechanical properties.[37]

2.2.2 Instrumentation

Modern SSNMR instrumentation is similar to the instrumentation used in solution-state NMR, with several minor but important differences. Amplifiers capable of delivering higher powers, in the range of 300 to 1000 W, are generally used for SSNMR. In addition, preamplifiers and filters capable of handling high powers are necessary. Specialized controllers that supply pressurized air or nitrogen to MAS probes for spinning are also needed. Temperature controllers and chillers capable of high flow rates are typically used, particularly for biomolecular studies, to enable experiments at temperatures of 0 to $-50\,°C$ or lower. These additional needs can boost the cost of an SSNMR installation well beyond that typically encountered for solution-state NMR. The most significant differences between SSNMR and solution-state NMR are encountered in the design of probes. Many probe designs used in SSNMR require a magnet with an 89 mm bore size, which can add significantly to the cost of an instrument. However, other probe designs make use of the conventional 54 mm bore size also used in solution-state NMR magnets, allowing for interchangeable usage. Most probes used in SSNMR studies relevant to drug discovery and development are MAS probes. These probes include a spinning module comprising a stator, coil and bearings to support the spinning rotor. Conventional probe circuit designs for SSNMR probes have been reviewed in detail.[38,39] Currently, transmission line designs are popular because of greater power handling capability. Rotors made of zirconia and other materials with an outer diameter that ranges from 1 mm to more than 7 mm are currently commercially available; these rotors can hold from a few hundred μg up to a few hundred mg of material. Spinning rates for MAS probes in drug applications are typically in the range of 10–15 kHz for mid-sized rotors with outer diameters of 4 to 5 mm. Many laboratories are equipped with probes that can spin from 25 to 40 kHz with smaller rotor sizes, outer diameters being typically in the 2.5 mm to 3.2 mm range. Recently, ultra-fast spinning MAS probes capable of v_r in excess of 70 kHz using rotors with as small as 1 mm outer diameter have become commercially available.[40,41] In applications of fast MAS probes, it should be noted that MAS leads to frictional heating and exerts rapidly increasing pressure on samples with increasing v_r, which can potentially affect the sample.[42] For example, in a recent study of styrene–butadiene rubbers, a 3.2 mm outer diameter rotor with an inner diameter of 2 mm was predicted to exert a pressure of 9.5 MPa with $v_r = 25$ kHz, a much higher value than the pressure of 1.0 MPa exerted with a slower $v_r = 10$ kHz.[43]

Cryogenic cooling of probe electronics to reduce thermal noise is widely employed in solution-state NMR and is used extensively in applications to drug discovery and development. Cooling of the probe coil and electronics to 25 K, or in some cases higher temperatures, while maintaining the sample at room temperature, yields signal-to-noise ratio (SNR) enhancements on the order of 2 to 4 times that of a conventional probe by reducing thermal electronic noise (see also Chapter 5 for *in vivo* applications of a cryogenic coil).[44] A MAS probe with a cooled coil and electronics that allows for the sample to be maintained at room temperature has been developed and used to observe ^{1}H spectra of solid adamantane.[45] A MAS probe that cools the sample to 25 K or less, but maintains the electronics at higher temperatures, has also recently been developed and obtains significant polarization enhancements from the low sample temperature (from population enhancements as predicted by the Boltzmann equation).[46] Initial commercialization of these designs has occurred and new designs are forthcoming. Usage is not yet widespread in drug applications.

Probe designs that allow for simultaneous excitation and detection of ^{19}F, ^{1}H and an X-nucleus (such as ^{13}C or ^{15}N) have become commercially available in the last ten years.[47] These designs have enabled new applications of ^{19}F SSNMR to a variety of problems of interest in drug discovery and development, as discussed below, because of the occurrence of fluorine in many drugs of interest and the possibility of fluorine labeling of biomolecules.

2.2.3   Experimental Methods

A wide variety of experimental methods are available for SSNMR studies in drug discovery and development, and only a brief overview can be given here. Some of the more popular experiments currently utilized for drug discovery and drug development applications are summarized in Tables 2.2 and 2.3, respectively. With limited exceptions, most of these experimental methods make use of MAS, which involves the mechanical spinning of the sample in a gas turbine inclined at an angle of 54.74° relative to the static B_0 magnetic field. Over the course of each rotor period, MAS averages the chemical shift inter-action, first-order quadrupolar interaction and dipolar interaction, and partially averages the second-order quadrupolar lineshape. However, strong multispin dipolar interactions like those prevalent between protons in organic solids are generally only partially averaged by MAS. Cross-polarization (CP) techniques are widely used to enhance the signal of spins with lower gyro-magnetic ratios, such as ^{13}C and ^{15}N nuclei. Modern CP experiments normally utilize a radiofrequency (RF) power ramp on one of the channels, which improves SNR and reproducibility by broadening the Hartmann–Hahn matching condition and improving tolerance for RF inhomogeneity, among other factors.[48] When combined with MAS, the Hartmann–Hahn condition that enables CP splits into a manifold of sidebands, such that actual CP matches in CP-MAS experiments normally are set to occur at the stronger sidebands (normally designated $+/-1$).[49,50]

Heteronuclear dipolar decoupling is commonly used to narrow or simplify spectral resonances by removing the broadening effects of coupling ^1H to nuclei when observing ^{13}C, ^{15}N, ^{19}F or other nuclei.[51] The two-pulse phase modu-lation (TPPM) scheme and the related small phase incremental alteration (SPINAL) decoupling sequence are currently popular methods for hetero-nuclear dipolar decoupling.[51,52] The TPPM sequence offered a significant improvement over earlier continuous-wave (CW) decoupling sequences. In SSNMR applications of interest to pharmaceuticals, where rigid crystalline organic solids are often encountered, it is common to use high RF power levels of 100 kHz or more on the ^1H channel for TPPM or SPINAL decoupling. More recent developments include an improved super-cycled variant of the swept-frequency TPPM sequence,[53] which has also been reported to show superior performance when fast MAS ($v_r = 60$ kHz) is concurrently employed.[54]

The observation of ^{13}C spectra using a combination of CP, MAS and ^1H heteronuclear decoupling is one of the most basic and commonly executed experiments performed in drug discovery and development applications. However, several modifications to this basic CP-MAS experiment are also frequently used.[13] ^{13}C spectra are often complex, and many sites have appreciable anisotropy, leading to spinning sidebands.[15] As a result, sideband suppression sequences are widely used, particularly the total sideband suppression

(TOSS) sequence.[55] An example of the performance of the five-pulse CP-TOSS pulse sequence with a 243-step phase cycle[55] is shown in Figure 2.4 for the small molecule pleuromutilin. No artifacts from the suppressed spinning sidebands are observed, and if any small resonances were observed, they could be confidently assigned to actual signals instead of potential unsuppressed sidebands.

Spectral editing is a useful tool for analysis of more complex spectra. For example, the venerable dipolar dephasing technique can be used to suppress signals from protonated carbons by reintroduction of heteronuclear dephasing to ¹H nuclei through interrupted decoupling (with the exception of methyl carbons, which rotate quickly and partially average dipolar coupling).[56] In

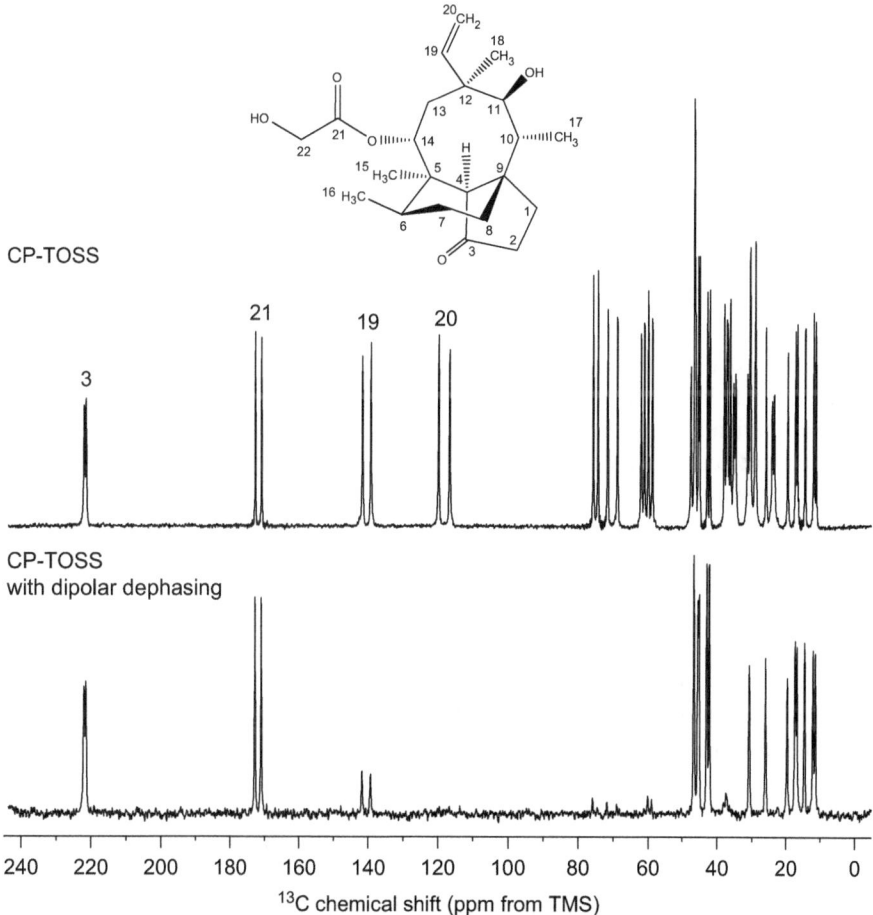

Figure 2.4 ¹³C CP-TOSS spectrum ($v_r = 8$ kHz) of a crystalline form of pleuromutilin with $Z' = 2$, shown in comparison to a spectrum obtained with dipolar dephasing by switching off ¹H decoupling during the TOSS period (which was extended to occupy four rotor periods using a shifted echo). Spectra were obtained at 9.4 T and 273 K.

Figure 2.4, the dipolar dephasing spectrum of pleuromutilin is shown for comparison, displaying only quaternary carbons and methyl carbons. Small residual signals from the 19 position remain despite its protonation state because of rapid motion of the alkene group. Other spectral editing sequences can be used to distinguish CH, CH_2 and CH_3 carbons in ^{13}C spectra using the J-coupling interaction.[57] This experiment can be applied to other nuclei, such as ^{15}N, whose attached proton count in a pharmaceutical molecule can be of significant interest.[58] Because of the variable performance of ^1H homonuclear decoupling, the solid-state attached proton test (APT) experiment is often performed in a pseudo-2D manner using a range of delays. The polarization inversion technique, which is based on the dipolar-driven CP process, can also be used to distinguish CH, CH_2 and CH_3 carbons in ^{13}C spectra although it typically needs to be adjusted for different samples of interest because of its sensitivity to spinning speed and the mobility and the geometry of different spin systems.[59]

Homonuclear dipolar decoupling is primarily used for ^1H nuclei, as this isotope is both abundant in most solids of interest and experiences strong homonuclear dipolar couplings because of its large magnetogyric ratio and typically short internuclear distances. Modern homonuclear decoupling sequences are designed to work at the higher MAS rates currently employed in many studies.[60] These include the original (and still widely used) frequency-switched Lee–Goldburg (FSLG) sequence, the phase-modulated Lee–Goldburg (PMLG) sequence and the "decoupling using mind boggling optimization" (DUMBO) sequence.[60–62] The FSLG sequence typically performs best when the sample is restricted to the center of the MAS rotors to maximize RF homogeneity. The DUMBO class of decoupling sequences has been numerically optimized to achieve better ^1H resolution and also offer superior tolerance to RF inhomogeneity.[60,62] An example of the performance of a 1D windowed-detection DUMBO sequence (with $v_r = 30$ kHz) in comparison to conventional MAS alone (with $v_r = 35$ kHz) is shown for a typical API molecule in Figure 2.5. The spectra have been obtained at a high static field corresponding to a ^1H frequency of 700.13 MHz. This API molecule has been previously studied using 2D FSLG-based techniques.[18] The DUMBO spectrum in Figure 2.5 shows a noticeable improvement in resolution, allowing for several assignments to be made. Homonuclear dipolar decoupling is frequently used in conjunction with 2D experiments whenever a ^1H dimension is included in the experiment.

Heteronuclear dipolar correlation experiments utilize the dipolar coupling between different nuclides to obtain through-space correlations, and hence structural information, usually in a 2D format. A variety of mechanisms are available to recouple the heteronuclear dipolar interaction, which is averaged under MAS, and allow for correlations to be detected. The simplest approach uses conventional flat RF pulses or ramp CP.[48] This is the basis of the popular 2D $^1H–^{13}C$ CP-heteronuclear correlation (CP-HETCOR) experiment, which is typically performed with v_r in the range of 10–25 kHz and with FSLG homonuclear ^1H decoupling during the t_1 evolution period.[63] Typical ramp CP contact times used for this experiment are 200 μs to 500 μs for short-range

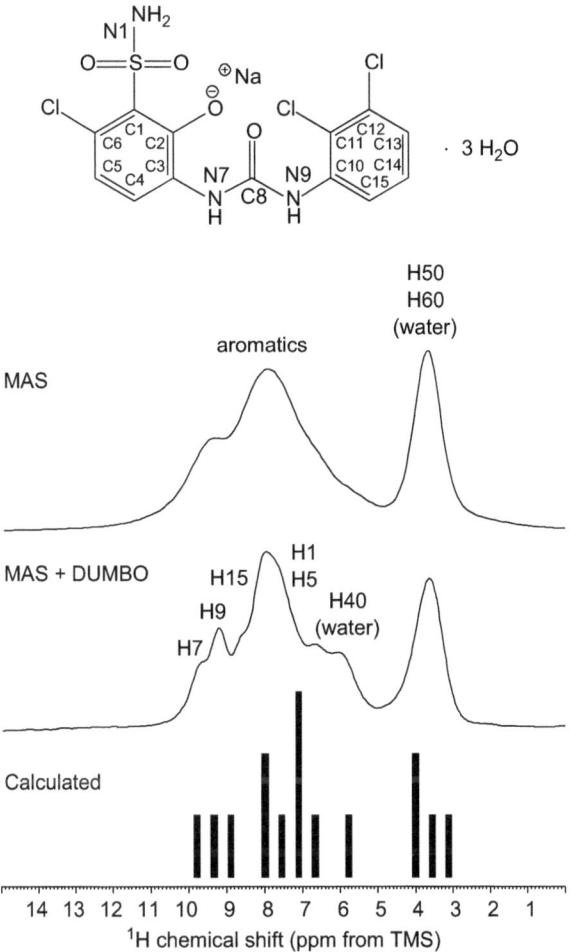

Figure 2.5 ^{1}H SSNMR of a hydrated form of a small-molecule API, showing a comparison between conventional MAS ($v_r = 35$ kHz) and a spectrum obtained using DUMBO homonuclear decoupling ($v_r = 30$ kHz). Spectra were obtained at B$_0$ field strength of 16.4 T (700.13 MHz) and 273 K. A calculated ^{1}H spectrum is shown at the bottom. This spectrum was obtained using the GIPAW method implemented in CASTEP code using the Perdew-Burke-Ernzernhof (PBE) functional and on-the-fly-generated pseudopotentials with a cut-off of 300 eV for hydrogen atoms and ultrasoft pseudopotentials for the other atoms. The crystal structure and SSNMR spectra (at 9.4 T) of this form are discussed in Vogt *et al.*[18]

interactions, and 2 ms or longer when longer range interactions or spin diffusion is observed. To suppress spin diffusion during the CP period, Lee–Goldburg cross polarization (LGCP) can be used in place of ramp CP, in which the ^{1}H spins are spin-locked using a frequency-offset Lee–Goldburg field.[64] An example of heteronuclear dipolar correlation using the 2D ^{1}H–^{19}F LGCP-HETCOR pulse sequence is shown in Figure 2.6 for a crystalline form

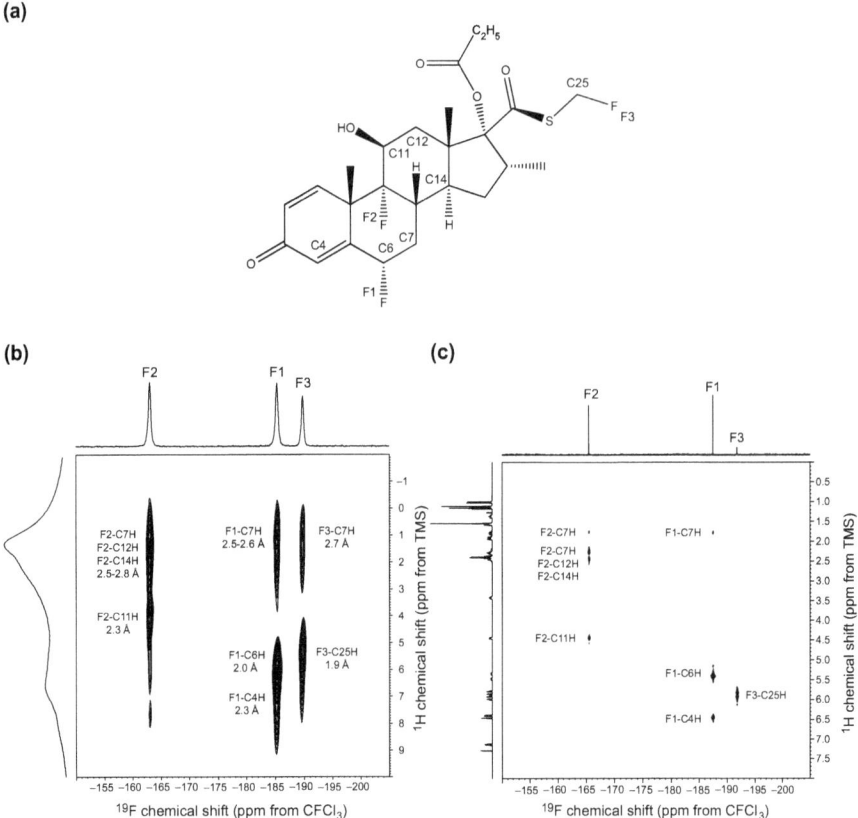

Figure 2.6 Comparison of a typical 2D SSNMR experiment with a 2D solution-state NMR experiment. (a) Chemical structure of the drug fluticasone propionate, showing the atomic numbering scheme. (b) ^1H–^{19}F LGCP-HETCOR spectrum of a crystalline form of fluticasone propionate (see Cejka *et al.*[65]), obtained with a 100 μs contact time using a 2.5 mm HFX probe with v_r set to 25 kHz, at a B_0 field of 11.7 T and at 273 K. ^1H–^{19}F distances are from the crystal structure. (c) Solution-state ^1H–^{19}F HOESY spectrum obtained with 1 s mixing period in 0.8 mL of CDCl$_3$ solution, obtained at a B_0 field of 9.4 T and at 298 K. Both experiments utilized 9 mg of sample and the same approximate acquisition time.

of fluticasone propionate, a drug used in the treatment of respiratory diseases.[65] As shown in Figure 2.6(a), the drug contains three fluorine positions that allow for the use of ^{19}F SSNMR. The ^1H–^{19}F LGCP-HETCOR spectrum in Figure 2.6(b) shows multiple through-space correlations that can be related to short distances in the crystal structure. The intramolecular correlations are comparable to those obtained from a more familiar solution-state NMR experiment, a ^1H–^{19}F heteronuclear Overhauser effect (HOESY) spectrum performed on the same total amount of material in solution as shown in Figure 2.6(c). Although the solution-state HOESY experiment offers much higher resolution, particularly in the ^1H dimension, the solid-state ^1H–^{19}F

Table 2.2   Selected experiments used in SSNMR studies of biomolecules of interest in drug discovery and their typical usage. Basic experiments are not listed here but are summarized in the text. All experiments are performed using MAS unless otherwise noted.

Experiment	Typical Usage	References
fp-RFDR, C7	2D homonuclear dipolar correlation between *e.g.* ^{13}C nuclei. Used to study intraresidue correlations and longer-range distances in uniformly labeled proteins.	76,77
PDSD, DARR	2D proton driven spin diffusion and dipolar correlation between homonuclear spins such as ^{13}C (with labeling). Used to study intraresidue correlations and longer-range interactions.	22,81
XY HETCOR	HETCOR experiments between non-^1H spins, *e.g.* ^{13}C–^{15}N, using ramp CP, adiabatic-passage CP, PAIN-CP or related sequences. Used for assignment and obtaining distance restraints in labeled biomolecules.	66,67
NCA, NCO	2D heteronuclear dipolar correlation *via* selective CP, showing ^{13}C–^{15}N linkages between amide nitrogen (N) and α-carbons (CA) in the NCA experiment and between N and the carbonyl groups (CO) in the NCO experiment.	68
NCACX, NCOCX, NCACB	2D and 3D heteronuclear dipolar ^{13}C–^{15}N CP experiments, including specific CP and homonuclear ^{13}C mixing *via* DARR or PDSD, for assigning backbone and side-chain signals in uniformly labeled proteins. CX refers to any carbon resonance.	69
CANCO	3D heteronuclear dipolar correlation experiment involving ^{13}C and ^{15}N nuclei, designed for sequential residue assignment in uniformly labeled proteins.	70
CTUC-COSY, TOBSY, INADEQUATE	2D homonuclear through-bond J-coupling based experiments, typically used with spin-1/2 nuclei such as ^{13}C for backbone assignments.	84–86
J-based NCA, NCO	2D heteronuclear through-bond J-coupling based experiments, used to detect ^{13}C–^{15}N linkages between N and CA in the NCA experiment and between N and CO in the NCO experiment.	87
J-based NCACO, NCOCA, and CANCO	2D and 3D heteronuclear through-bond J-coupling based experiments, typically used with ^{13}C and ^{15}N nuclei for backbone assignments of uniformly labeled proteins.	88
FS-REDOR	Frequency-selective REDOR measurement of heteronuclear dipolar couplings, typically used for selective distance measurements in uniformly labeled proteins.	75
ZF-TEDOR	3D measurement of heteronuclear dipolar couplings; employed to obtain multiple ^{13}C–^{15}N dipolar distance restraints from a single data set on a uniformly labeled protein.	134
QCPMG	Provides sensitivity-enhanced spectra for quadrupolar central transitions strongly broadened by second-order effects, and can be performed without spinning (static) or using MAS. Used for studies of the environment of the metal site in metalloproteins.	93

LGCP-HETCOR spectrum is sensitive and also shows intermolecular inter-actions not accessible in solution, denoted by distances obtained from the crystal structure and shown on the spectrum in Figure 2.6(b).[65] CP and ramp CP can be used for heteronuclear dipolar correlation between other pairs of nuclei, such as ^{13}C–^{15}N and ^{19}F–^{13}C, in appropriately designed 1D and 2D experiments. Other techniques, such as the adiabatic-passage CP and the proton-assisted insensitive nuclei CP (PAIN-CP) experiments, are also commonly applied for ^{13}C–^{15}N HETCOR experiments particularly in labeled biomolecules.[66,67] A variety of specialized heteronuclear dipolar correlation experiments designed for protein assignments and utilizing ^{13}C–^{15}N CP and other homonuclear correlation methods (described below) are also available and are summarized in Table 2.2.[68–70]

Heteronuclear dipolar distance measurements can also be performed using CP.[64,71] More robust alternatives such as the rotational-echo double resonance (REDOR) and transferred-echo double resonance (TEDOR) experiments are commonly used for heteronuclear distance measurements with isolated, labeled spin pairs.[72] REDOR-type pulse sequences can face challenges in handing multiple spin systems (such as two ^{13}C spins and a single ^{15}N spin),[73] leading to the development of techniques such as θ-REDOR and frequency-selective REDOR (FS-REDOR), which can measure distances in these systems.[74,75] The FS-REDOR experiment is particularly designed for studies of uniformly ^{13}C-labeled biomolecules.[75] Accurate dipolar distance measurements are possible by analyzing the time evolution of the signals from many of these techniques, or by transforming the time domain signal into the frequency domain using a variety of specialized processing methods.[71]

Homonuclear dipolar correlation and distance measurements access the dipolar interaction between like spins, such as ^1H–^1H or ^{13}C–^{13}C spin pairs.[76,77] A range of pulse sequences has been developed to recouple the homonuclear dipolar interaction in various ways and/or measure the homonuclear dipolar coupling.[76,77] The basic rotational resonance MAS method, which involves selection of v_r so that a resonance overlaps with the sideband of another resonance (which causes reintroduction of the homonuclear dipolar coupling averaged by MAS), can be used in a selective manner to study spin pairs in uniformly labeled small molecules, since the rotational resonance condition is narrow enough to allow for pairwise selection of spins with only a minor modification for the analysis of dipolar evolution to account for other passive spins.[78] More sophisticated broadband recoupling sequences are designed to recouple all of the homonuclear couplings in a sample either for correlation or for distance measurement. These include pulse sequences such as finite-pulse RF-driven dipolar recoupling (fp-RFDR), melding of spin-locking and dipolar recovery at the magic angle (MELODRAMA) and the symmetry-derived SPC5 and C7 sequences.[76,77] In addition, multiple-quantum experiments have also been developed for homonuclear spin systems, such as the double-quantum back-to-back (DQ-BABA) pulse sequence for ^1H-^1H correlation.[79] Multiple-quantum dimensions offer resolution improvements, which are particularly valuable for ^1H nuclei, at the expense of sensitivity (because of the need to excite multiple quantum coherence), and are primarily used for abundant nuclei exhibiting strong dipolar couplings. A number of other modifications are possible, including triple-quantum variants, variants that use SPC5 or C7 sequences for excitation and reconversion of multiple-quantum coherence and variants that include ^{1}H homonuclear decoupling.[1,79] Recently, an approach to measure ^{1}H–^{1}H distances using frequency-selective excitation was proposed;[80] if successful, it could open up new directions in studies of small molecules of pharmaceutical interest.

Numerous pulse sequences have also been developed for the detection of dipolar coupling through longer-range spin diffusion.[22,77] Homonuclear 2D correlations in ^{1}H spectra can be observed by simple proton-driven spin diffusion (PDSD) using a longitudinal mixing period.[22] This can be combined with homonuclear ^{1}H decoupling during evolution periods for improved resolution. Correlation between ^{13}C nuclei in uniformly or partially labeled samples can also be achieved by PDSD. However, more efficient spin diffusion between ^{13}C nuclei is obtained using the popular ^{13}C CP-dipolar assisted rotational resonance (DARR) experiments, which employ a weak RF field on the ^{1}H nuclei that causes broadening of the rotational resonance condition, enhancing dipolar recoupling and spin diffusion.[81] ^{13}C CP-DARR experiments, when performed with long mixing times of several hundred ms, can readily detect ^{13}C–^{13}C spin pairs that are 6–8 Å apart.[76,77]

Homonuclear and heteronuclear J-coupling correlation experiments are also used in SSNMR studies of both small molecules and biopolymers.[82,83] The heteronuclear multiple-quantum coherence J-coupling experiment (known as HMQC-J-MAS) and related experiments are popular methods for detecting ^{1}H–^{13}C or ^{1}H–^{15}N J-couplings in both labeled and natural abundance materials, and is often applied to help assign the spectra of small-molecule APIs.[83] Homonuclear J-couplings can also be detected between ^{13}C nuclei in uniformly labeled solids. The CP-incredible natural abundance double quantum transfer (CP-INADEQUATE) experiment is useful for uniformly labeled biomolecules and can be used to detect ^{13}C–^{13}C J-couplings (on the order of 40 Hz) in natural-abundance small molecules in favorable cases with long acquisition times.[84] Other homonuclear J-based sequences more commonly used for biomolecules include total through bond correlation spectroscopy (TOBSY) (which uses isotropic J-mixing) and the constant-time, uniform-sign cross-peak (CTUC) correlation spectroscopy (COSY) experiment.[85,86] A range of 3D heteronuclear J-based sequences designed for proteins have been developed to mimic experiments used in solution-state protein NMR; these include constant-time NCA, NCO, NCACO, NCOCA and CANCO experiments for direct tracing of the backbone of ^{13}C and ^{15}N-labeled proteins.[87,88]

The measurement of the CST for a nucleus provides another important parameter that can be related to structure. While CSTs can be measured by direct observation of static lineshapes or MAS spinning sideband manifolds, these approaches are limited to relatively simple spectra containing only a few resonances. In the more usual case of complex spectra, CST measurement is facilitated by techniques such as the phase-adjusted spinning sidebands (PASS) experiment, five π replicated magic angle turning (FIREMAT) experiment and related methods, which separate the anisotropic portion of the CST in a 2D experiment such that the high-resolution isotropic spectrum is available to help guide the extraction of parameters.[89,90]

Observation of quadrupolar nuclei by SSNMR often requires special techniques to deal with spectral overlap caused by second-order quadrupolar broadening. When the quadrupolar interaction is moderate in size (C_Q of approximately 2 to 8 MHz), a number of methods are available to obtain isotropic spectra. The multiple-quantum MAS (MQMAS) and satellite transition MAS (STMAS) families of pulse sequences are amongst the most popular for observation of half-integral quadrupolar nuclei with moderate C_Q.[91,92] These versatile experiments can be applied to nuclei of interest in both pharmaceutical molecules and target biomolecules, such as ^{23}Na, ^{27}Al and ^{17}O. MQMAS achieves a combination of spatial averaging by MAS and spin space averaging through evolution of different spin coherences during two time evolution periods (t_1 and t_2) in a 2D experiment. Different methods are needed for quadrupolar nuclei with large second-order broadening (*e.g.* $C_Q >$ 10 MHz). One of the most popular methods is the quadrupolar Carr–Purcell–Meiboom–Gill (QCPMG) pulse sequence, performed under both static and MAS conditions, which creates a series of spin echoes that breaks broad powder patterns into a manifold of spinning sidebands for greater sensitivity.[93] Dipolar correlation involving quadrupolar nuclei *via* CP or through dipolar HMQC is also possible. A specialized sequence that is useful for study of spin-1 nuclei with large quadrupolar couplings, such as ^{14}N, is the ^{1}H–^{14}N HMQC experiment, which is best performed using probes capable of $v_r > 60$ kHz.[41,94]

Relaxation time measurement and analysis is commonly performed in drug development applications. Typical measurements involve the observation of T_1 and $T_{1\rho}$, the latter to observe slower molecular motions in the kHz frequency range.[50] In more specialized situations, measurements of T_2 relaxation may be relevant. The measurement of ^{1}H T_1 *via* a heteronucleus plays a special role in studies of phase purity, which can be needed when attempting to assess whether a given sample of a small molecule API consists of one phase or more than one phase.[95] An example of ^{1}H T_1 measurement *via* heteronuclear detection, in this case using ^{31}P, is shown in Figure 2.7 for the phosphorus-containing antiviral API adefovir dipivoxil. The experiment detects the presence of a phase mixture by the observation of different ^{1}H T_1 times for each of the ^{31}P resonances, which correspond to a hydrate and an anhydrate phase. If ^{1}H T_1 values happen to be indistinguishable using this method, this experiment cannot directly prove that a set of resonances arises from the same phase but still can provide some assurance of phase purity if confidence intervals for the T_1 measurements are very narrow (which is often possible if a sensitive heteronucleus like ^{19}F or ^{31}P is employed). Changes in temperature and spinning speed can be used to vary the results and possibly enhance the difference in T_1 measurement in some cases. High spinning speeds

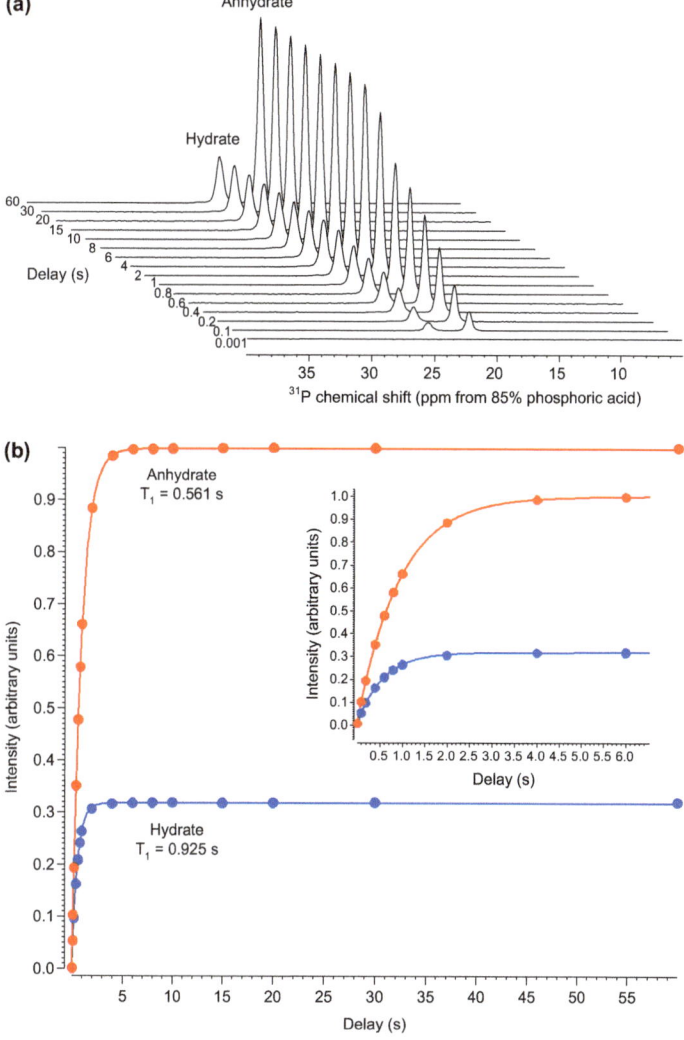

Figure 2.7 (a) ^{31}P-detected ^{1}H T_1 saturation recovery ($\nu_r = 12.5$ kHz) analysis of a sample of adefovir dipivoxil anhydrate contaminated with a hydrated phase impurity. (b) Exponential fits of the peak area of the ^{31}P signals, showing two different ^{1}H T_1 times, which is indicative of phase separation between the anhydrate and hydrate phases. The expanded inset shows the fitting results at shorter delay times.

(typically greater than 15 kHz) should be avoided because of suppression of spin diffusion, which leads to differential ^{1}H T_1 values even within a single phase.

Dynamic nuclear polarization (DNP) methods (see Chapter 9) used in solution-state NMR often make use of a radical species as a polarizing agent, prepared using a separate magnet for a polarizing electron paramagnetic resonance (EPR) spectrometer (such as a 94 GHz spectrometer with a 0.5 GHz sweep), which can increase signal-to-noise up to 10 000 times for some nuclei.[96] Recent advances in DNP methods for solids have occurred with the development of high-frequency microwave sources, enabling EPR experiments to be performed at frequencies in the hundreds of GHz directly in the high-field NMR magnet on frozen matrices containing the paramagnetic polarizing agent.[97] The cross effect involving two electron spins and a nuclear spin generally constitutes the mechanism for the enhancement, and the sample is prepared using a dilute nitroxide radical or biradical as a polarizing agent in a 60% v/v mixture of glycerol in water, which forms a glass as the sample is cooled and disperses the polarizing agent.[97] This approach can be applied to solid nanocrystals, with enhanced polarization transferred from surface contact with the radical by spin diffusion through the bulk crystal. Enhancements in solid frozen protein samples of 25–50 times have been observed.[97] While most experiments have been performed using MAS at cryogenic temperatures (*e.g.* 25–100 K), DNP experiments at temperatures as high as 180 K have recently been successfully performed.[98]

Finally, referencing of SSNMR spectra is an important experimental consideration particularly in drug development, where a large number of samples are analyzed and reproducibility is critical. Most SSNMR spectra are referenced by the external substitution method.[12] For example, ^{13}C spectra are commonly referenced to tetramethylsilane (TMS) *via* susceptibility-corrected external references such as hexamethylbenzene, adamantane and glycine.[99] The ghost referencing process references one isotope to another through frequency ratios and is also a popular approach.[12] Because most SSNMR systems do not use a field-frequency lock, it is important to understand field drift and compensate for it if needed during extended experiments.

2.2.4   Computational Methods

The computational methods used to support SSNMR analyses include a range of techniques that simulate nuclear spin dynamics and calculate electronic structure for prediction of NMR parameters or optimization of molecular structure. Several computational methods with demonstrated applications to drug discovery and development are reviewed briefly in this section. It is important to note the distinction between two major classes of quantum mechanical calculation used in SSNMR: the first calculates electronic structure for the prediction of NMR parameters, such as σ, V and the J-coupling tensor (J), while the second calculates the response of a spin system with a set of NMR parameters under time-dependent propagators such as evolution under internal interactions and external interactions such as RF irradiation and MAS.

Electronic structure calculations are used for prediction of NMR parameters such as σ, J and V, as well as for the optimization of structures prior to these predictions. The prediction of chemical shielding or EFG tensors is an important aid to spectral interpretation. Although *ab initio* methods for solving the electronic wavefunction can be used for this task, density functional theory

(DFT) calculations are the most popular approach because of their combination of speed and accuracy, particularly in systems where electron correlation effects are significant.[100,101] Electron correlation, which is the tendency of the motion of one electron to influence the motion of the other electrons, can affect many pharmaceutical systems of interest.[100] DFT calculations scale more favorably with the number of electrons (N^4) than wavefunction-based methods that incorporate electron correlation, allowing for efficient computations on typical pharmaceutical molecules on small clusters or desktop workstations. A variety of density functionals are used, including hybrid functionals such as B3LYP (Becke, three-parameter, Lee–Yang–Parr), generalized gradient approximation (GGA) functionals such as PBE (Perdew–Burke–Ernzerhof) and PW91 (Perdew–Wang-91), and local density approximation (LDA) functionals, although the latter are generally only employed in EFG calculations.[100,101]

Structural optimization of hydrogen atom positions is a common first step in the preparation of a small-molecule crystal structure for calculations of NMR parameters.[101] In some cases, such as when the crystal structure has been solved using powder X-ray diffraction (PXRD) methods or when a low-temperature structure is compared to high-temperature SSNMR data, it can be beneficial to optimize the entire structure including the heavy atoms. Calculations of NMR parameters carried out on single crystal X-ray diffraction (SCXRD) structures are generally performed as fixed, single-point calculations using the optimized structure and thus do not directly take into account molecular vibrations or unit cell changes between the SCXRD temperature and the SSNMR temperature. The effects of molecular vibrations can often be taken into account by linear scaling of the experimental chemical shifts against the calculated chemical shieldings, particularly if a large number of sites are predicted.[101]

Many calculations of σ, J and V employ the "cluster" method, in which a group of molecules is extracted from the periodic crystal structure and the "central" molecule or molecules is treated with a higher level basis set for the NMR calculation.[101] The surrounding molecules can be treated with a lower level of theory to improve the performance of the calculation. Gaussian or Slater-type basis sets are widely used for these types of calculation in what is often referred to as a "locally dense" approach.[102] When heavy atoms are present, this method allows for inclusion of relativistic effects for the central molecule.[103] Other quantities that may also be of importance in some SSNMR studies, such as electronic g-tensors (for prediction of PCS effects), can be calculated using this method. Alternate levels of theory can be applied to the central molecules and the surroundings *via* the ONIOM (Our own N-layered Integrated molecular Orbital and molecular Mechanics) approach; for example, DFT can be used for central molecules while molecular mechanics can be used to simulate the surroundings.[104] Finally, long-range electrostatic effects can be included *via* point charges or other methods.[101]

Calculations that use periodic boundary conditions and plane waves as a basis set are inherently designed to treat crystalline solids for structural optimization and calculations of σ, J and V. The gauge-including projector augmented wave (GIPAW) method, implemented in the CASTEP package, takes this approach. The GIPAW method utilizes pseudopotentials, which saves computational resource by allowing for removal of the tightly bound core electrons (as these contribute minimally to the chemical shielding) and instead provides a smoothed valence wavefunction near the nucleus.[105,106] To accurately describe chemical shielding, the GIPAW method reconstructs all-electron wavefunctions from their pseudo-equivalents for the smoothed valence wavefunctions near the nucleus. The GIPAW method has been applied to the prediction of chemical shielding using GGA functionals in pharmaceutical small molecules such as flurbiprofen and testosterone, as an aid to assignments or to investigate structural features.[107,108] The GIPAW plane wave pseudopotential method can also be used to predict EFG, and, more recently, calculations of J-coupling parameters have also become possible using this approach.[109] The anisotropy and orientation of the J-tensors and their relationship to the crystal structure can be investigated using this approach.

As an example of the GIPAW approach, the ^{19}F chemical shielding tensors for Form 1 of the antifungal drug voriconazole were calculated for comparison with the experimental chemical shift tensor principal components determined in Figure 2.1. Prior to the GIPAW calculation in CASTEP, the hydrogen atom positions in the Form 1 crystal structure were energy-minimized using the DMol3 package.[110,111] The calculation and comparison of the three principal components of each ^{19}F tensor enables a more confident resonance assignment than use of isotropic values alone, particularly since two of the isotropic values for F2 and F3 are within 3 ppm, which challenges the accuracy of the calculation. The results of the GIPAW calculation are compared to the experimental values in Figure 2.8, where the principal components of the experimental chemical shift tensor (δ_{11}, δ_{22} and δ_{33}) are plotted against the principal components of the chemical shielding tensor (σ_{xx}, σ_{yy} and σ_{zz}). A linear correlation is expected, and with the assignment shown in Figure 2.1 for F1, F2 and F3, the coefficient of correlation (R^2) is 0.9936, indicating excellent agreement.[101] The linear plot allows for assessment of alternative assignments of F1, F2 and F3. If the assignments for F2 and F3 are swapped, for example, the best fit line yields a poorer fit with an R^2 of 0.8958. This result enables confident assignment of the fluorine resonances as originally shown in Figure 2.1. As only the principal components are compared in this analysis, no information about the relative orientation of tensors in Form 1 of voriconazole is experimentally available, although the predicted orientations are provided by the GIPAW calculation.

A second example of a solid-state chemical shielding calculation using the GIPAW method is shown in Figure 2.5. Here, only δ_{iso} is calculated for the ^{1}H nuclei in this API molecule, again using the GIPAW method after optimization of the hydrogen atom positions in the crystal structure. The chemical shielding results from the calculation are presented as a histogram, showing ^{1}H shielding within a 0.5 ppm range after conversion to chemical shift units. Although this simulation does not account for molecular dynamics in this hydrated API, the predicted and experimental spectra are in relatively good agreement and help enhance assignments.

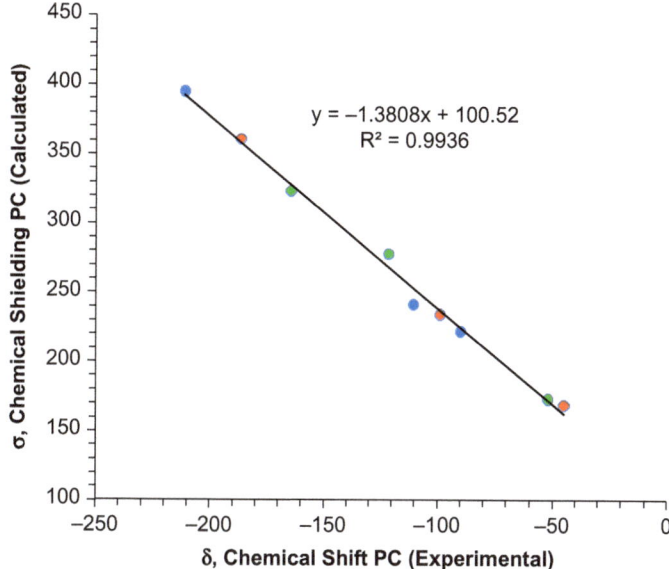

Figure 2.8   Results of a ^{19}F chemical shielding tensor calculation for Form 1 of the antifungal drug voriconazole. The calculated principal components σ_{xx}, σ_{yy} and σ_{zz} are plotted against the experimentally determined principal components δ_{11}, δ_{22} and δ_{33} that were previously shown in Figure 2.1. Blue, red and green circles represent the principal components assigned to fluorine positions F1, F2 and F3, respectively. The GIPAW CASTEP calculation was performed using the PBE GGA density functional, a $2 \times 2 \times 1$ **k**-point set for the Brillioun zone integration and on-the-fly-generated potentials with a 610 eV energy cut-off for ^{19}F atoms. Prior to the GIPAW calculation, the hydrogen atom positions in the Form 1 crystal structure (from Ravikumar *et al.*[16]) were energy-minimized using the DMol3 package with the HCTH/407 functional and the DNP basis set with a $3 \times 3 \times 2$ **k**-point set for the Brillioun zone integration.

Other approaches to solid-state calculations have been developed. For example, calculations of σ and V can be performed in a periodic system using Slater-type basis sets with the ADF-BAND package.[112,113] This package has extensive features for calculations of relativistic effects. The DMol3 package uses highly efficient numerical basis sets that enable EFG calculations on many large API crystal structures of interest, even though σ and J cannot be calculated with this approach.[110,111] However, EFG calculations are highly useful for studies of quadrupolar nuclei even in the absence of other parameters as the quadrupolar interaction often dominates the spectrum.[114] For example, the DMol3 package was used to predict the EFG at the ^2H site in the deuterochloroform solvate of tranilast[17] to compare with the fitted C_Q and η_Q shown in Figure 2.2. The hydrogen atom positions and the positions of the deuterochloroform molecules (which were disordered in the crystal structure) were first optimized at the HCTH/DND level of theory using only the Γ point for the Brillouin zone integration. An EFG calculation was performed at same level of theory. From the EFG calculation, a ^{2}H C_Q of 166 kHz and η_Q of 0.01 were obtained, in good agreement with the experimental results given in Figure 2.2, particularly considering that motional averaging effects on C_Q have not been included. A calculation was performed at the same level of theory to predict the ^{35}Cl EFG for the crystalline form of raloxifene HCl shown in Figure 2.3.[19] Again, good agreement was obtained with a calculated C_Q of –850 kHz and η_Q of 0.38.

Computer simulations of nuclear spin dynamics are also valuable adjuncts to many SSNMR studies. These simulations generally study the propagation of the density matrix for a particular spin system under a set of internal interactions (such as chemical shift tensors or dipolar couplings), MAS and an RF pulse sequence, and can also include dynamic motion.[115] Several laboratories have developed code to perform these types of simulation. The SIMPSON (Simulation Program for Solid-State NMR Spectroscopy) program was among the first widely used, general purpose SSNMR simulation packages.[116] The SIMPSON system can be programmed by defining spin systems and nuclear spin interactions, which can then propagate under RF irradiation (with phase cycling and coherence-order filtering), free precession periods and signal acquisition periods.[116] The Spin Evolution package is another widely used system.[117] These simulations can be of great use in understanding the effects of complex pulse sequences such as DUMBO on complex spin systems exhibiting chemical shift, dipolar and J-coupling interactions.[118] Larger scale simulations of spin diffusion processes have also recently become possible,[119] which should have significant applications particularly to SSNMR studies of API dispersions in polymer matrices. These simulations should find increasing use in studies of complex multispin phenomena in crystalline small-molecule APIs.

2.2.5 Special Processing Methods

Methods for processing of SSNMR data are generally similar or identical to those used in solution-state NMR.[120] However, a number of specialized data processing tools are used in SSNMR studies that are not commonly encountered in solution-state NMR. Specialized lineshape fitting routines are used to extract anisotropic parameters and obtain the principal components of tensors of interest, such as the fits shown in Figures 2.1, 2.2 and 2.3.[121–123] More details about quantitative processing of SSNMR data for drug development applications, including quantitative spectral analysis, deconvolution and non-linear fitting of relaxation data, can be found elsewhere.[123]

2.3 Applications to Drug Discovery

In the context of drug discovery, some of the studies discussed in this section are best described as studies of potential targets and biological mechanisms, as in the case of structural biology. Examples are also given of SSNMR approaches to common drug-discovery activities such as lead identification and studies of drug-target binding. SSNMR has several advantages in comparison to solution-state NMR techniques and X-ray diffraction techniques in studies of biomolecular systems. There is no limitation on the mass, the solubility or the ability to grow high-quality crystals of the biomolecule. Because the molecules are in the solid state, dynamics and exchange rates are generally slowed in comparison to solution. The dipolar interactions observed in many SSNMR experiments scale more favorably with increasing internuclear distance than the nuclear Overhauser effect (NOE) used in many solution-state NMR studies. Finally, SSNMR also enables the study of membrane protein systems in their native solid forms. However, these benefits must be balanced against the disadvantages of SSNMR, such as limitations in the resolution of sites with differing chemical shifts in comparison to solution-state NMR. This section briefly reviews SSNMR studies of biomolecules that are potential targets for drug discovery and special methods used for oriented samples and metalloproteins, drug-receptor interactions and protein–protein interactions.

2.3.1 Labeled Biomolecules

SSNMR has been increasingly applied to structural studies of biomolecules that are of interest as targets in drug discovery. Uniform, segmental or selective labeling of a peptide or protein with ^{13}C, ^{15}N, ^{2}H or other NMR-active isotopes is typically employed in these studies. Labeling is generally achieved using conventional techniques also applied in solution-state NMR, such as the use of fully or selectively labeled ^{13}C-glucose and ^{15}N-ammonia as the carbon and nitrogen sources in protein expression. Advances in stable isotope labeling techniques for solution-state protein NMR spectroscopy, which are also applicable in many cases to SSNMR studies, have recently been reviewed.[124] A review of SSNMR studies of fully labeled proteins has also been published.[125] In general, SSNMR techniques are considered in situations where proteins of interest have not been successfully crystallized, lack solubility or are too large for solution-state NMR. This section focuses on MAS studies of non-aligned samples that are typically polycrystalline powders, although amorphous samples can also be probed but are often less preferable because of resonance broadening. Many SSNMR studies of biomolecules of interest employ aligned samples, which are discussed in Section 2.3.2.

The potential utility of SSNMR in structural studies of uniformly labeled globular and membrane proteins was demonstrated in a number of early studies.[125–127] One of the first examples of a complete protein structure determination by SSNMR was performed on microcrystalline ubiquitin. This study used the ^{13}C DARR experiment to obtain ^{13}C–^{13}C distance restraints in combination with selective labeling, which were combined with torsional angle restraints obtained from ^{13}C chemical shift analyses to yield a structure with a 1 Å resolution.[128] In Figure 2.9, a representative ^{13}C CP-DARR experiment on a uniformly ^{13}C-labeled ubiquitin sample is shown.[128] While this spectrum shows primarily intraresidue correlations, it highlights the complexity observed even for a small protein. Longer-range correlations were detected using mixing times of up to 500 ms and were more easily resolved using site-specific labeling.[128] At long mixing times, DARR transfers magnetization more efficiently than PDSD, allowing for easier detection of interresidue contacts for structure calculations.

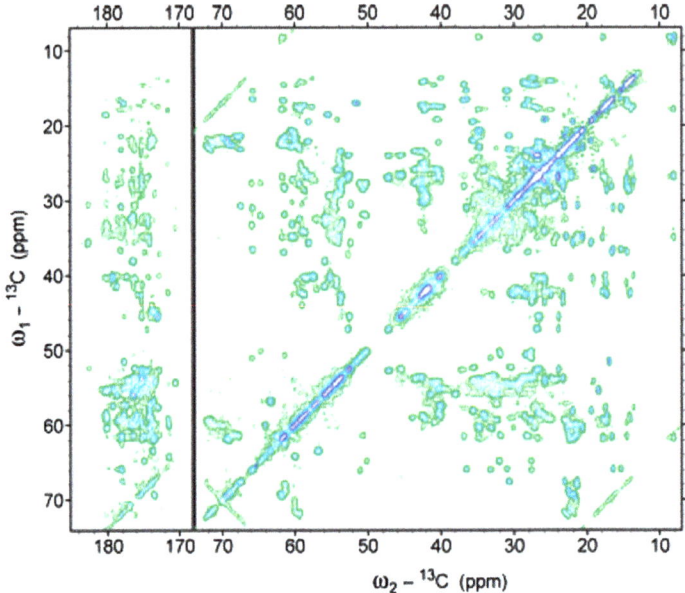

Figure 2.9 $^{13}C-^{13}C$ CP-DARR spectrum of microcrystalline uniformly ^{13}C-labeled ubiquitin obtained with a mixing time of 20 ms. The spectrum shows primarily intraresidue correlations. The experiment was performed using $v_r = 10$ kHz and $B_0 = 17.6$ T.
Reproduced with permission from Zech *et al.*[128] © 2005 American Chemical Society.

REDOR, TEDOR and other related sequences have been applied to many studies of selectively labeled biomolecules.[129] For example, the FS-REDOR experiment has been used to measure $^{13}C-^{15}N$ distances in β-amyloid fibrils in their native solid state.[130–132] This has provided support for a model for the structure of fibrils that includes hydrogen bonds between intermolecular glutamine residues and a salt bridge between an asparagine and a lysine residue related to a bend in the structure. Molecular models for amyloid fibrils, which can contribute to drug discovery, can be developed from SSNMR experiments of this nature, particularly when combined with lower-resolution structural constraints from electron microscopy and other techniques.[133] Efforts have been made to extend the use of REDOR-based techniques to more complex biomolecules and also to allow for simultaneous measurement of many dipolar distance restraints from a single data set. The 3D z-filtered TEDOR (ZF-TEDOR) experiment is an example of such a design.[134] ZF-TEDOR has been used to probe the conformation of an 11-residue amino acid sequence in the amyloidogenic protein transthyretin that was uniformly labeled with both ^{13}C and ^{15}N. The results showed that this peptide fragment has a β-strand secondary structure using 35 distance restraints in the 3 to 6 Å range measured by ZF-TEDOR in conjunction with torsion angles determined from experiments that correlated chemical shift with the dipolar tensor.[134] In Figure 2.10, a section

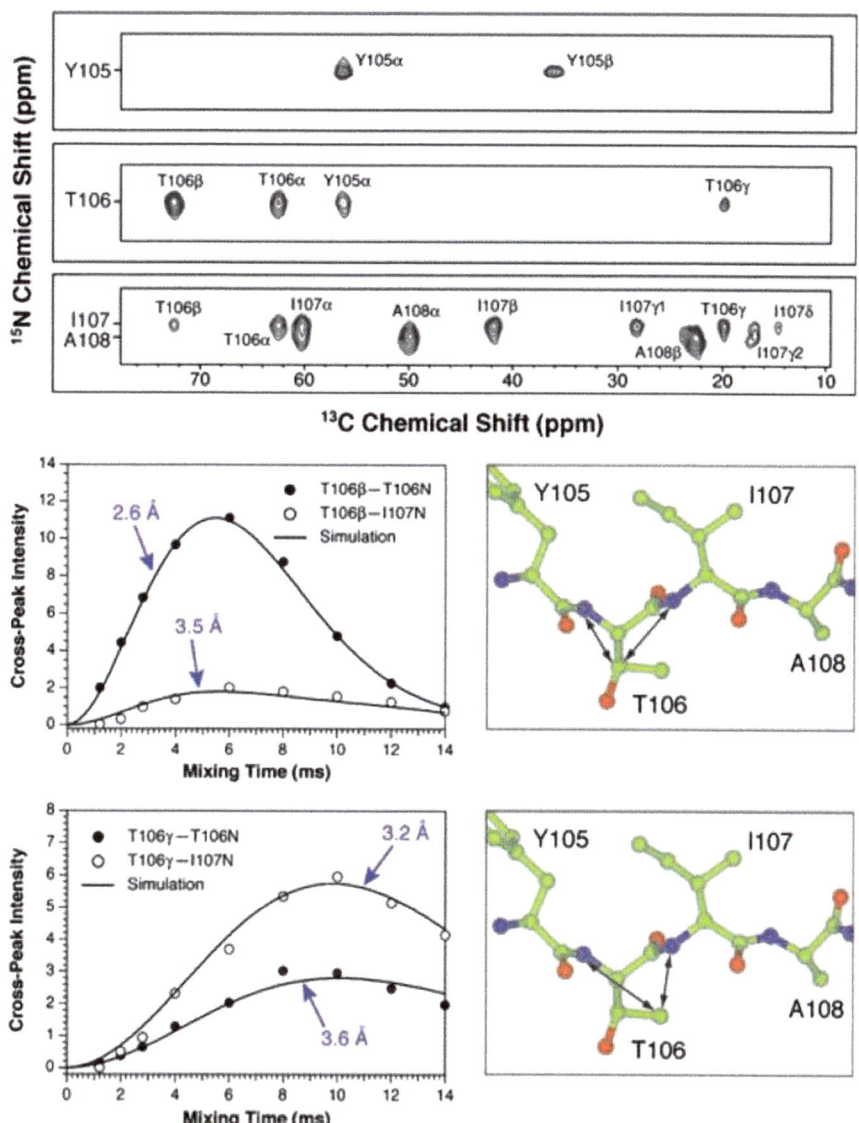

Figure 2.10 ZF-TEDOR ^{13}C–^{15}N distance measurements in an 11-residue peptide fragment of transthyretin amyloid fibrils. Top: sections of the ZF-TEDOR spectrum obtained with a mixing time of 6 ms, showing distances in the 4–6 Å range. Bottom: plots of the time evolution of the TEDOR signal under dipolar mixing, showing the signal evolution (points), fitted signal (solid lines) and the structural element responsible for the signals (right-hand structures).
Reproduced with permission from Jaroniec *et al.*[135] © 2004 National Academy of Sciences of the USA.

of the 3D ZF-TEDOR data set is shown. The intensities of the TEDOR ^{13}C–^{15}N correlations evolve in the third (dipolar) dimension, each of which can be fitted to determine the distances involved in the dipolar interaction as shown.

Efforts to determine the structure of important biomolecules have been hampered by the limited number of longer-range restraints (typically > 5 Å) available from PDSD, DARR, REDOR and other experiments. A relatively new approach that utilizes paramagnetic metal ions to obtain longer-range distance restraints (of 10–20 Å) through PCS has now been demonstrated on a number of systems using labeling schemes that insert paramagnetic centers with significant anisotropy in their g-tensors and observing the shifts using otherwise conventional 2D SSNMR experiments such as ^{13}C DARR.[25] Paramagnetic doping has been used to reduce ^{1}H T_1 and allow for rapid acquisition of uniformly labeled protein 2D SSNMR spectra using a v_r of 40 kHz.[136] It was shown that the experimental time could be reduced by a factor of 5 to 20 without adverse effects on spectral quality. Small amounts of Cu(II)-EDTA were used as the paramagnetic dopant, and typically reduced ^{1}H T_1 by a factor of five or more, allowing for acquisition of 2D RF-driven dipolar recoupling (RFDR) ^{13}C–^{13}C and ^{13}C–^{15}N solid-state NMR spectra for nanomolar quantities of microcrystalline β-amyloid fibrils and ubiquitin within 1 to 2 days. ^{1}H relaxation within the protein microcrystals was enhanced through surface contact with paramagnetic Cu(II), with the bulk of the crystal relaxing *via* spin diffusion to the surface sink.

A 2D CP-HETCOR technique that detects the rate of ^{1}H spin diffusion by way of a X-nucleus was used to detect membrane-embedded segments of a lipid within rigid protein in lipid bilayers.[137] This pulse sequence is also used in several applications in drug development as described in Section 2.4. Using CP-HETCOR, fast spin diffusion can be observed within the rigid protein while slow spin diffusion is observed in the mobile lipids. Membrane-embedded domains in the protein can be identified by the ^{1}H spin-diffusion build-up rates from the lipid chain-end methyl groups to the protein, and simulations of these rates allow for estimation of the insertion depth of protein segments into the membrane. The experiment does not require oriented bilayer preparations.[137]

Finally, in structural studies of biomolecules of pharmaceutical relevance, SSNMR offers the unique possibility of observing protein folding and misfolding in a manner that is not available from solution-state NMR, SCXRD or other biophysical methods because of the disordered nature of the states (*i.e.* a variety of intermediate folded states are simultaneously present).[138] Rapid freezing can be used to trap non-equilibrium, transient structural states on a sub-millisecond time scale for SSNMR studies of these states.[138]

2.3.2 Oriented Biomolecules

Studies of oriented biomolecules have been of interest for many years, as the ability to prepare anisotropic, oriented samples predated many of the pulse sequence developments that have allowed for access to uniformly labeled biomolecules.[139,140] Two approaches to macroscopic orientation are widely used. Mechanical orientation involves the preparation of lipid bilayers on a flat, upright surface, in some cases *via* centrifugation, with the bilayer containing the biomolecule of interest. Magnetic orientation relies on the negative diamagnetic anisotropy of acyl chains within the lipid bilayer, which when placed in a magnetic field align perpendicular to the field.[139] Additives can be used to control the orientation of the bilayer. NMR experiments on oriented samples can be carried out with or without the use of MAS, and allow for the study of the anisotropic chemical shift, dipolar interactions or quadrupolar coupling. The lipid bilayer dynamics and orientation can be studied using ^{2}H and ^{31}P as well as other nuclei. These approaches have allowed for extensive studies of peptide antimicrobials in lipid membranes.[140]

2.3.3 Metal Sites in Proteins and Other Biological Macromolecules

The multinuclear nature of SSNMR, and the availability of many NMR-accessible isotopes for metal nuclei of interest, enables the technique to provide a useful probe of the coordination and oxidation environment of a metal center in a biomolecule such as a metalloprotein.[141] Metalloproteins have emerged as important targets in drug discovery.[142–145] The environment of the metal center can be a critical determinant of the function of a metalloprotein or metal-containing biomolecule.[141–145] For example, many enzymes use metals to catalyze their reactions.[141] SSNMR studies of metal centers in biological macromolecules often probe nuclei that are very challenging to observe, including ^{25}Mg, ^{39}K, ^{51}V, $^{63/65}$Cu and ^{67}Zn. Alkali metal ions also play a role in a variety of biological solids that can be approached using SSNMR.[146] In this role, SSNMR provides another avenue to studies of the structure and function of bioinorganic systems when single crystals are not available for SCXRD studies.[147] SSNMR is complementary to EPR and X-ray absorption fine structure (XAFS) spectroscopy, which are also widely used techniques in studies of non-crystalline or polycrystalline metal-containing biomolecules.[148–150]

^{51}V MAS of central and satellite transitions has been used to study vanadium-containing haloperoxidases, such as a 67.5 kDa vanadium chloro-peroxidase that was examined with direct polarization magic-angle spinning (DP-MAS) methods using a v_r of 15–17 kHz and a static field of 14.4 T.[151] The ^{51}V site exhibited a C_Q of about 10 MHz and a strong chemical shift anisotropy, which were compared with DFT calculations of different active site models to determine that the vanadate cofactor is likely anionic with an axial hydroxo-group and three equatorial hydroxo- and oxo- groups.[151] These enzymes are colorless and diamagnetic in their active state, which limits the ability of other spectroscopic methods such as EPR to characterize the mechanism of their efficient oxidative process. The QCPMG experiment, often performed on static samples at temperatures ranging from room temperature to 10 K, enables studies of extremely challenging isotopes such as ^{55}Mn and ^{67}Zn in model systems for biomolecules of interest.[152,153]

Metalloproteins can be studied in great detail using the PCS approach, as was demonstrated on a 17.6 kDa catalytic domain of matrix metalloproteinase-

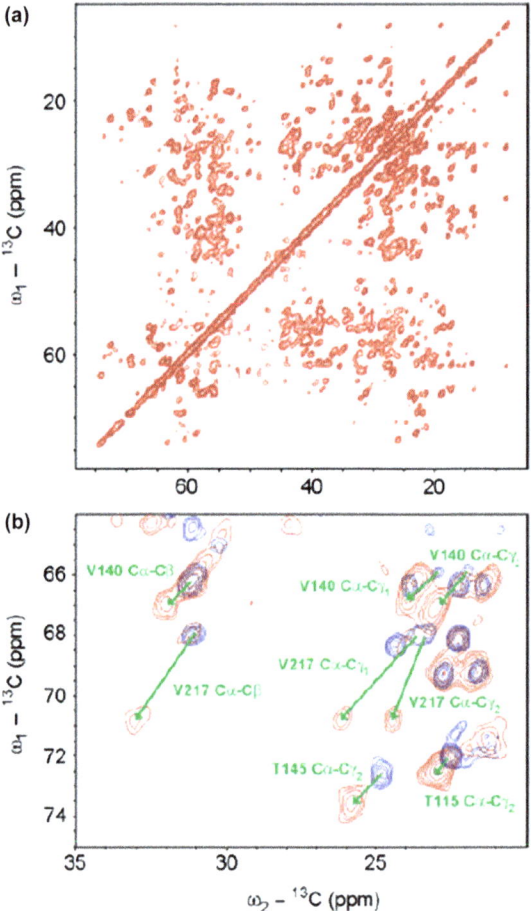

Figure 2.11   (a) $^{13}C-^{13}C$ PDSD spectrum of a microcrystalline sample of cobalt(II)-substituted matrix metalloproteinase-12 (MMP-12) obtained at 16.4 T with $\nu_r = 11.5$ kHz, a mixing time of 60 ms and at 290 K. (b) Superposition of the PDSD spectrum of the diamagnetic zinc MMP-12 (blue) with the spectrum of the paramagnetic cobalt(II) MMP-12 (red). Green arrows indicate the paramagnetic shifts.
Reproduced with permission from Balayssac et al.[154] © 2007 American Chemical Society.

12 (MMP-12) containing 159 amino acid residues (Figure 2.11).[154,155] Unlike the metalloprotein studies described above, which are focused on understanding the metal site, the goal in these studies is to obtain the structure of the protein itself through PCS distance restraints. To achieve this, several hundred PCSs were observed by comparing the 2D $^{13}C-^{13}C$ PDSD spectra of the paramagnetic high-spin cobalt(II) protein with the diamagnetic zinc(II) protein and using the resulting distances to refine a molecular structure, which was completed as a crystal structure after determining the unit cell parameters with

PXRD.[155] The use of these approaches in studies of metalloproteins should also benefit from application of the other SSNMR approaches described as well as other techniques that can provide structural information on paramagnetic centers, such as EPR.[148]

2.3.4 Drug–Receptor Interactions

During drug discovery, it is often of interest to study interactions between a potential drug and a target biomolecule. Solution-state NMR is one of the more common techniques used for this task,[156] and has been of particular interest in the field of fragment-based drug discovery (FBDD) (see Chapter 1).[157] To date, SSNMR has been less utilized in such applications. SSNMR has an advantage in studies of high-affinity drug-receptor binding. In these cases, the dissociation constant K_D, defined as the ratio k_{off}/k_{on} of the off-rate k_{off} to the on-rate k_{on}, is very small (typically less than 10 μM).[158,159] Such systems are consequently difficult to study by solution-state NMR because of broadened signals.[158,159] The basic ^{13}C CP-MAS experiment can be used to detect drug–receptor interactions in some of these situations. When a drug is bound, ^{1}H–^{13}C CP can occur between nuclei on the drug and the macromolecule; CP is not observed when the drug is a free ligand and is highly mobile because of averaging of the dipolar interaction. ^{1}H spin diffusion processes can also mediate and extend the range of the interaction to 10 Å or beyond. When ^{13}C CP-MAS experiments are used, it is generally necessary to synthetically label ^{13}C positions within the small molecule drug. For example, the binding affinities of two transporter protein ligands were studied using ^{13}C labeling and variable contact time CP experiments to obtain binding constants.[160] In some cases, even with ^{13}C labeling of the drug molecule, spectral overlap may require the use of a ^{13}C-depleted biomolecule. Spectral editing approaches can also be used to help identify the ligand in crowded spectra, for example by adding polarization inversion to the CP experiment or by use of a ^{13}C double-quantum filter.[161] In the latter study, the small-molecule transport inhibitor forskolin was both singly and doubly labeled with ^{13}C and was bound to galactose-H^+ symport protein (GalP). A double-quantum filter performed at a low temperature ($-35\,^\circ C$) allowed for detection of forskolin-GalP interactions with effective suppression of natural abundance GalP signals.[161]

Like SCXRD, SSNMR can also offer direct, high-resolution structural information on the bound conformation of a flexible drug molecule. REDOR experiments in combination with selective ^{13}C, ^{15}N and ^{19}F labeling of the anticancer drug paclitaxel have been used to determine its conformation when bound to microtubules.[162,163] This structural determination was not amenable to SCXRD analysis, and electron microscopy was unable to yield a high-resolution structure. The microtubules, with MDa molecular mass, provided a massive natural-abundance background against which the REDOR signals were measured.[162,163] An initial study used two ^{13}C–^{19}F REDOR measurements filtered *via* ^{15}N labels, and a later study used ^{2}H–^{19}F REDOR measurements to avoid natural abundance background issues. The bound conformation identified by SSNMR for paclitaxel influenced the design of next-generation taxoids.[164]

Interactions between candidate drugs and DNA G-quadruplexes have also been probed using REDOR experiments.[165] The DNA G-quadruplex was mimicked using short oligonucleotides labeled with methyl-^{13}C thymidine, and as the drug candidate was a fluoroquinobenzoxazine, ^{19}F–^{31}P and ^{19}F–^{13}C REDOR experiments allowed for construction of a molecular model showing the interaction site.[165] ^{19}F–^{31}P REDOR was also used to study the interaction of the small-molecule antibiotic distamycin with a minor groove in a synthetic oligonucleotide.[166] In these studies, phosphorothioation was used to allow for detection of specific phosphorous signals because of the chemical shift differences in a phosphorous bearing a sulfur from the backbone of the oligonucleotide, where phosphorous is attached only to oxygen.

2.3.5 Protein–Protein Interactions and Interactions Between Biomolecules

Protein–protein interactions and multiprotein complexes or assemblies regulate a variety of critical cell functions, and aberrations can cause a variety of human diseases.[167] As a result, these interactions are currently the target of significant drug-discovery activity. SSNMR studies of protein–protein interactions have shown potential in this challenging field, where protein–protein interactions can often result in formation of insoluble and amorphous protein assemblies that are not amenable to SCXRD studies, solution-state NMR or high-resolution electron microscopy. For example, SSNMR was used to study the RNA viral genome in the human immunodeficiency virus (HIV) enclosed within a cone-shaped core assembly consisting of multiple 26.6 kDa capsid protein molecules.[168] The assembly exhibits different morphologies and undergoes a complex disassembly process as infection occurs. SSNMR experiments, including ^{13}C CP-DARR spectra of uniformly ^{13}C- and ^{15}N-labeled capsid protein within conical assemblies, allowed for assignment of the protein spectrum and also showed no significant protein structural differences between several assembly morphologies.[168] This study highlights the potential of SSNMR to access detail about these challenging biomolecular assemblies that can drive future drug discovery. A set of pulse sequences based on ^{1}H–^{15}N CP-HETCOR with FSLG decoupling, dipolar dephasing and spin diffusion have been adapted for studies of protein–protein interactions in conjunction with differential isotopic labeling of different protein domains.[169]

The binding of a high-affinity biomolecular ligand to a target biomolecule can be detected using 2D SSNMR experiments and related to structural re-arrangements.[170] For example, 2D ^{13}C DARR spectra of a chimeric K$^+$ channel protein complex, handled as a non-oriented powder, exhibited distinct effects upon addition of kaliotoxin (a 38-residue peptide) as shown in Figure 2.12. To simplify the spectra, one of the biomolecules was uniformly labeled for each binding experiment, while the other was not, allowing for assignment of

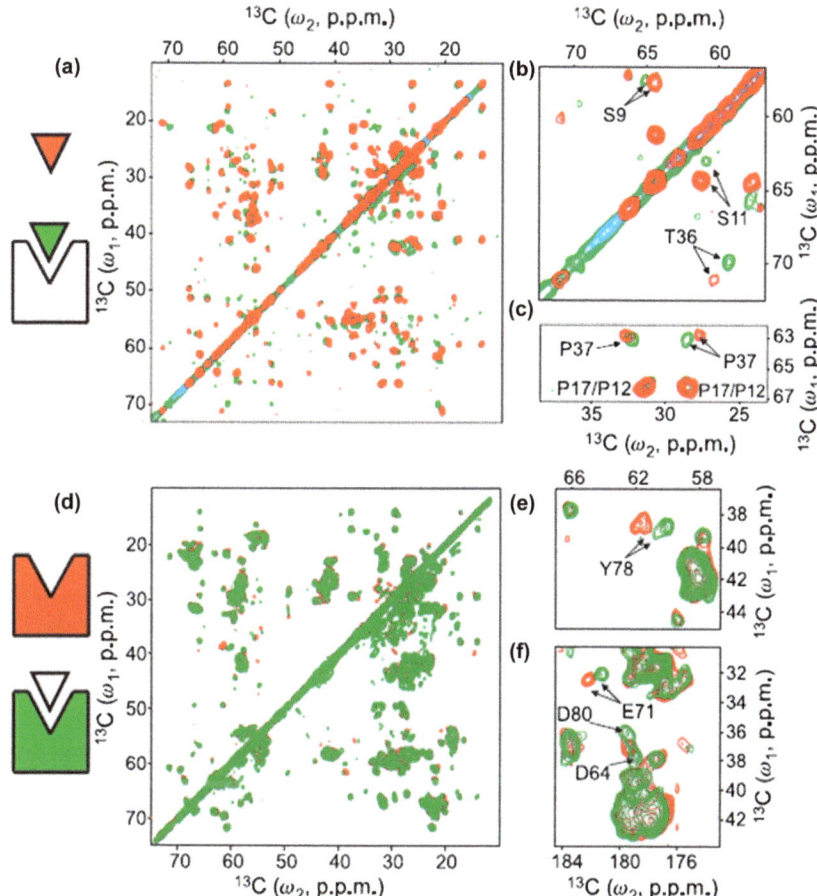

Figure 2.12 Comparison of ^{13}C CP-DARR spectra of (a–c) uniformly $^{13}C,^{15}N$-labeled kaliotoxin free (red) and bound to a chimeric K^+ channel protein complex (green), and (d–f) spectra of uniformly $^{13}C,^{15}N$-labeled chimeric K^+ channel protein complex (green) and the complex with unlabeled kaliotoxin bound (red).
Reproduced with permission from Lange *et al.*[170] © 2006 Nature Publishing Group.

observed spectral changes upon binding to each molecule. These spectral effects can be related to more distant structural changes and to local changes associated with specific binding of kaliotoxin to the active site.[170]

SSNMR studies are not limited to protein–protein interactions. $^{31}P–^{19}F$ REDOR experiments were used to study the interaction between an 11-residue peptide representing an HIV regulatory protein and a 29-mer RNA sequence representing a portion of the HIV transactivation response element of interest for potential therapeutics.[171] A phosphorothioate linkage was inserted to provide the ^{31}P resonance for REDOR, while a ^{19}F label was inserted into the oligonucleotide *via* 2′-fluoro-2′-deoxyuridine. The results showed a major distance change of nearly 4 Å between the two labels upon complexation with the peptide.[171]

2.3.6 Future Directions

Applications of SSNMR in studies that can influence drug discovery are likely to continue to grow, and methods will likely improve as new experiments are designed and sensitivity increases. Multidisciplinary studies involving SSNMR, electron microscopy, X-ray diffraction, EPR[148] or other techniques may enable characterization of previously challenging biomolecular targets, such as protein–protein complexes that are difficult to crystallize for SCXRD. Although only occasionally encountered at present, the ^{19}F isotope has seen increasing use in solution-state NMR studies of fluorinated proteins, which may ultimately translate to increased SSNMR studies of this isotope in proteins.[172] The possibility of site-specific labeling of ^{19}F coupled with uniform or specific labeling of ^{13}C and ^{15}N could lead to further structural insights in many materials of interest in drug discovery, and longer-range distance restraints for structural determinations.

2.4   Applications to Drug Development

SSNMR has been widely employed as an analytical technique during drug development in both industrial and academic settings. Several recent reviews that highlight the role of SSNMR in drug development have been published.[173–175] While SSNMR instrumentation in academia tends to be utilized for the more fundamental studies encountered in drug discovery, as discussed above, SSNMR instrumentation in industry tends to be used for drug development applications, and lower B_0 fields tend to be more frequently employed. Industrial applications include studies of polymorphism, salt and cocrystal formation, hydration and solvation, amorphous content and formulations of small-molecule therapeutics. In many cases, SSNMR studies are focused on the active pharmaceutical ingredient (API), although a growing number of studies are demonstrating the ability of SSNMR to tackle the analysis of complex drug products. Studies of solid drug products, which can be embodied as a number of dosage forms such as oral tablets and capsules, dry powders for inhalation or lyophilized powders for reconstitution and injection, are often of more significance than studies of the API alone because the performance of the drug product is critical in a clinical or commercial setting. Many of the experiments listed in Table 2.3 are commonly applied in the course of these studies. In many cases it is beneficial to co-interpret the results of SSNMR experiments with those of other techniques, including SCXRD, PXRD, vibrational spectroscopy and microscopy, thermal analysis methods, vapor sorption methods, particle size analysis, surface area analysis and many other physical characterization methods.[176–180]

Table 2.3 Selected experiments used in SSNMR studies of small molecules, biopharmaceutical molecules and formulations of interest in drug development. The typical usage of each experiment is given. Several basic experiments are not listed here but are summarized in the text. All experiments are performed with MAS unless otherwise noted.

Experiment	Typical Usage	References
DQ correlation	2D ^{1}H–^{1}H direct dipolar correlation using DQ-CRAMPS, DQ-BABA and other sequences that excite and indirectly detect DQ coherence. Used to probe hydrogen bonding and π-stacking interactions and show molecular association.	79
PDSD	2D proton driven spin diffusion dipolar correlation is used between ^{1}H or other abundant or labeled homonuclei. Used to study molecular association, assess phase purity or perform structural elucidation.	22
DARR	Heteronuclear spin diffusion between high-abundance heteronuclei (*e.g.* ^{19}F, ^{31}P and ^{13}C when labeling is employed), by irradiation of ^{1}H spins to assist dipolar spin diffusion.	81
MAS-J-HMQC	2D heteronuclear through-bond J-coupling based experiment, typically used with spin-1/2 nuclei such as ^{1}H–^{13}C for spectral assignments.	83
CP-HETCOR	2D detection of dipolar connectivity between ^{1}H and various heteronuclei, shows short-range spin diffusion when used with longer contact times. Can be applied using a spin-1/2 heteronucleus or a suitable quadrupolar nucleus.	63
LGCP-HETCOR	2D detection of dipolar connectivity between ^{1}H and heteronuclei as in CP-HETCOR, minimizes spin diffusion, but minimizes spin diffusion and longer-range transfers amongst ^{1}H nuclei.	64
MQ-MAS, ST-MAS	2D multiple-quantum excitation used separate anisotropic effects for nuclei broadened by second-order quadrupolar interactions. Used to observe quadrupolar nuclei such as ^{23}Na in pharmaceutical materials.	91,92
2D PASS, FIREMAT	2D chemical shift anisotropy measurements for spin-1/2 nuclei in complex spectra, useful for structural studies *e.g.* of hydrogen bonding and ionization, and for comparison with calculations.	89,90

2.4.1 Polymorphism

One of the earliest and still most frequent applications of SSNMR in drug development involves studies of polymorphic crystalline phases of the API or excipients. Polymorphism is usually defined as the occurrence of different crystalline solid phases (or "forms") adopted by the same covalent molecular structure.[180] This phenomenon is of great interest in pharmaceutical development because of the propensity of drugs to exhibit polymorphism, and especially because of the observation that polymorphic forms exhibit different properties that can affect the ultimate performance of a finished pharmaceutical product. Properties such as solubility, chemical and physical stability, hygroscopicity, compaction properties and many others are known to differ between polymorphs of drugs and excipients.[176,177,180] As a result, it is generally desirable to identify as many forms as possible for a drug in early development, and then develop API and drug product manufacturing processes that produce and then maintain a single chosen form throughout the shelf-life of the product. To achieve this goal, SSNMR methods are applied for both qualitative purposes, such as for identification of a potential new form, and for quantitative purposes, where an SSNMR method is developed to detect the presence and report the quantity of one or more undesired forms in a form selected for development and manufacture. More details can be found in dedicated reviews of SSNMR in applications to polymorphic organic and inorganic materials.[181,182]

Drug molecules range in size from small molecules to large proteins and antibodies, but with very few exceptions nearly all drug molecules are organic in nature. Techniques that observe the ^{13}C nucleus at natural abundance thus remain the default approach to the study of drug molecules. For example, basic 1D ^{13}C CP-MAS and CP-TOSS experiments have seen extensive application to the study of polymorphism in drugs over the past 15 years.[17,176,183–187] In most applications, the appreciable chemical shift anisotropy (CSA) of many aromatic and carbonyl ^{13}C sites leads to a number of spinning sidebands that can complicate spectral interpretation, and thus either TOSS methods or MAS rates in the 12–15 kHz range are often employed when using B_0 fields in the range of 9.4–11.7 T. Using these techniques, ^{13}C CP-MAS or CP-TOSS spectra of most drug polymorphs offer excellent resolution, particularly at higher fields, and can readily discriminate between polymorphs with reasonable sensitivity using acquisition times of several hours. Well-known examples include prednisolone and its 21-tert-butylacetate ester, enalapril maleate, furosemide, cyclopentathiazide and sulfathiazole.[176,180,182,183] In these studies, ^{13}C spectra are often used to elucidate differences in hydrogen bonding or molecular conformation between polymorphs.[17,176,183–187] Although most typical drug polymorphs of interest have reasonable ^1H T_1 values of 1–5 s, some smaller molecules have been known to exhibit very long ^1H T_1 values, which greatly reduces the sensitivity of ^{13}C CP experiments.

The detection of polymorphic impurities or other crystalline or amorphous phases besides the desired phase is an important part of the drug-development process. Phase purity is generally easy to observe by ^{13}C SSNMR, in comparison to PXRD, because resonances that arise from similar functional groups generally show similar intensity even with the variable effects of CP for different sites. For example, in Figure 2.13, a crystalline phase of a drug molecule known as Form K is shown.[188] Small peaks, denoted by arrows, show a suspected phase impurity. Comparison of the spectrum with ^{13}C CP-TOSS spectra obtained after microwave heating of the sample shows the growth of the phase impurity, and allows for the conclusion that the sample of Form K contains Form A, which can be converted to phase-pure Form A after heating. Other methods for assessing phase purity include heteronuclear-detected ^1H T_1

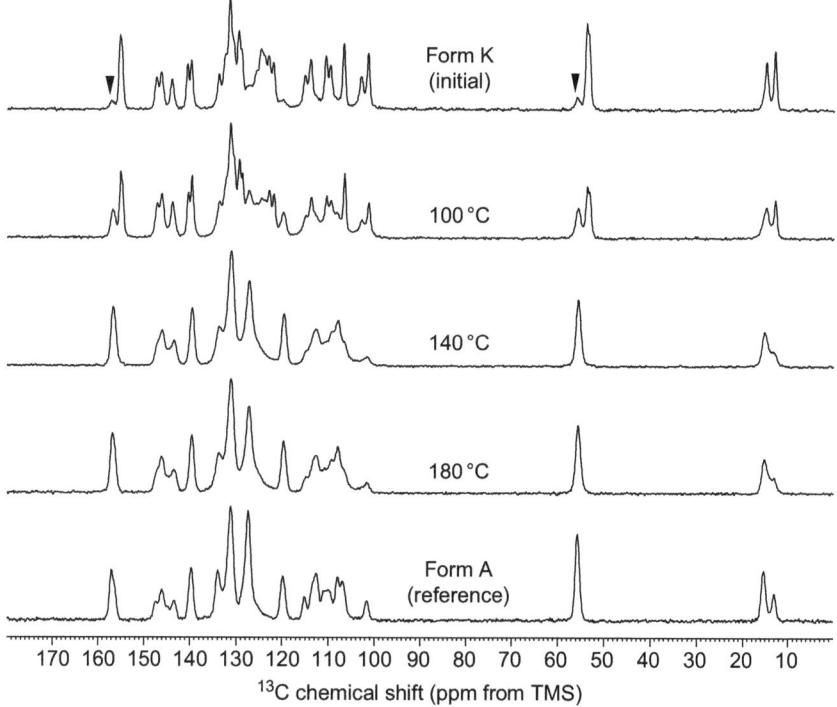

170 160 150 140 130 120 110 100 90 80 70 60 50 40 30 20 10

^{13}C chemical shift (ppm from TMS)

Figure 2.13 ^{13}C CP-TOSS spectra ($\nu_r = 8$ kHz) of a crystalline form of 7-methoxy-1-methyl-5-(4-(trifluoromethyl)phenyl)-[1,2,4]triazolo[4,3-a]quinolin-4-amine known as Form K undergoing conversion to Form A *via* microwave irradiation at different temperatures. The semi-quantitative nature of the spectra is highlighted by the disappearing level of Form K as temperature is increased. The input Form K contains a small amount of Form A. A reference spectrum of Form A is shown at the bottom from a directly-crystallized sample. Arrows denote signals that are specific for the initial Form A phase impurity observed in Form K. Spectra were obtained at 9.4 T and 273 K.

experiments, 2D experiments such as the DARR experiment (when an abundant nucleus such as ^{19}F or ^{31}P is available) or comparison of the ^{13}C spectra of multiple batches (which are likely to show some variation in the level of a phase impurity). The question of phase purity also arises in studies of solvates, inclusion complexes, disordered systems, cocrystals and many other systems of interest in pharmaceutical development that are discussed in later sections. The combined use of ^{13}C and ^{15}N CP-MAS techniques can supply more detailed information on structural differences between polymorphs and also provide greater assurance in questions of phase purity.[17] ^{15}N CP-MAS offers high specificity, readily assignable spectra and unique access to information about protonation state, hydrogen bonding and other structural effects for complex nitrogen-containing pharmaceutical compounds, as shown in Figure 2.14 for a nitrogen-containing drug candidate with a $Z' > 1$. ^{13}C and ^{15}N

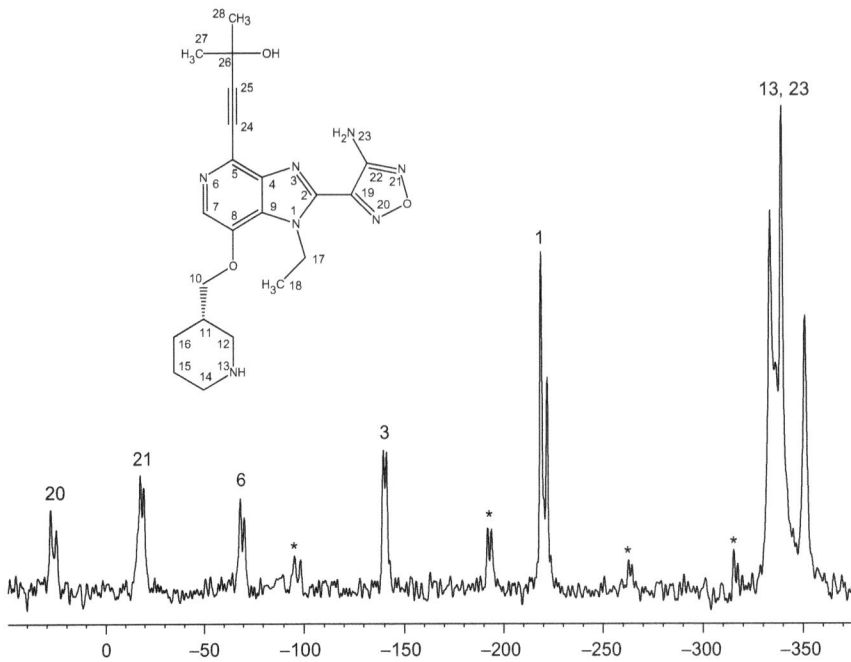

Figure 2.14 ^{15}N CP-MAS ($v_r = 5$ kHz) spectrum of a sample of an API containing multiple nitrogen sites obtained with a 5 ms contact time. Peak assignments are shown in relation to the numbered structure, and spinning sidebands are marked with an asterisk. The spectrum was obtained at 9.4 T and 298 K.

chemical shift tensors have also been used to examine drug polymorphs, because they report on hydrogen bonding and conformation and, as previously discussed, can detect structural effects more readily in some cases than δ_{iso} alone.[189,190]

Basic 1D ^{19}F CP-MAS and DP-MAS experiments have been shown to be specific and extremely sensitive in the analysis of fluorinated drug polymorphs.[191,192] Aryl fluorides generally offer the widest chemical shift range of the different fluorinated functional groups found in drugs.[14] Trifluoromethyl groups rotate in crystalline organic solids and the consequent chemical shift averaging tends to offer less specificity for discriminating between polymorphs,[34] but they are still highly sensitive and useful when present in an API. The presence of fluorine in a drug, which is a frequent occurrence,[193] also allows for other unique experiments to be applied. For example, ^{19}F–^{13}C CP and ^{1}H–^{19}F–^{13}C double CP experiments have been used in the study of polymorphic APIs and can provide information about molecular packing.[190] An example is shown in Figure 2.15 for the previously discussed small molecule API, which contains a trifluoromethyl group, allowing for comparison of the ^{19}F–^{13}C CP-MAS spectra of several polymorphs and a hydrate.[188] The hydrate was previously subjected to a detailed structural analysis by SSNMR, SCXRD and other techniques.[188] The spectra in

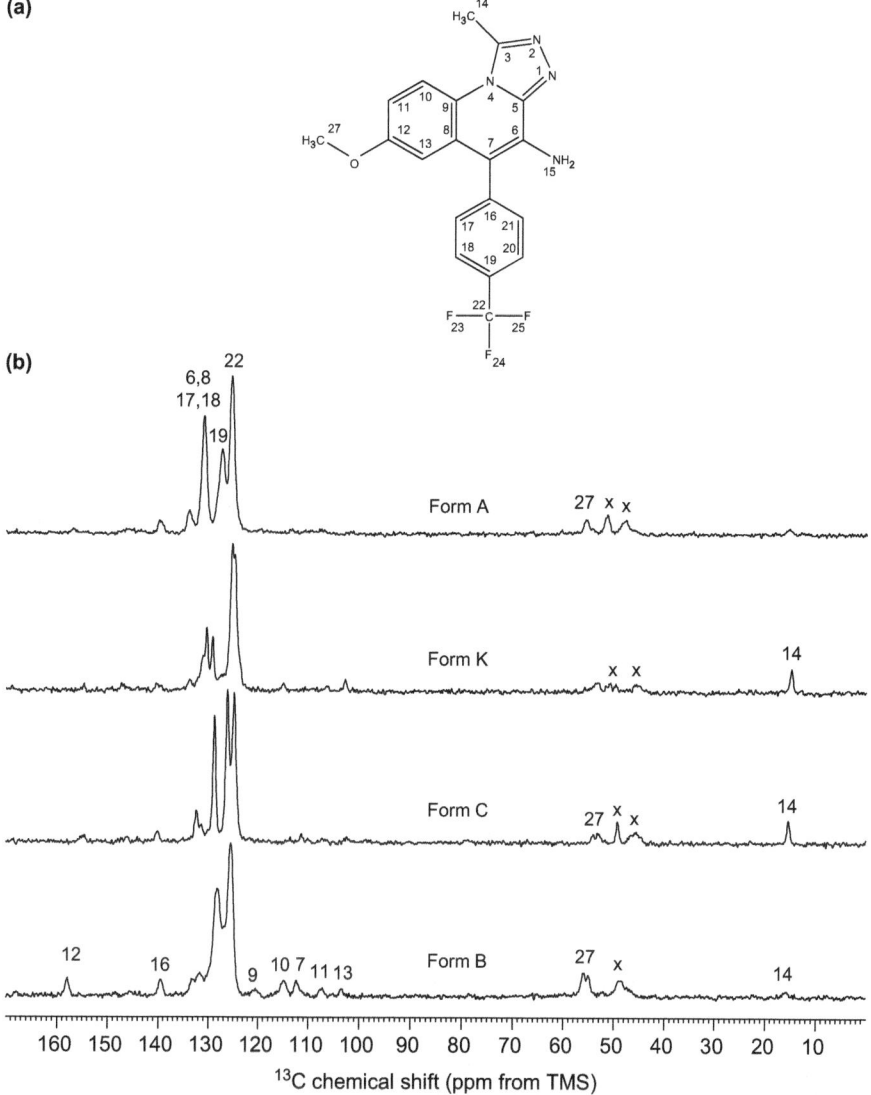

Figure 2.15 (a) Chemical structure and numbering scheme of 7-methoxy-1-methyl-5-(4-(trifluoromethyl)phenyl)-[1,2,4]triazolo[4,3-a]quinolin-4-amine. (b) ^{19}F–^{13}C CP-MAS ($\nu_r = 8\,kHz$) of different crystalline forms of this API obtained with a 5 ms contact time. Differences in intermolecular packing are highlighted by this experiment, which shows short-range interactions between fluorine and carbon sites. Spinning sidebands are marked with an "x". Spectra were obtained at 9.4 T and 273 K. Form B of this compound is discussed in Kang *et al.*[188]

Figure 2.15, obtained with a contact time of 5 ms, show correlations between ^{13}C and ^{19}F nuclei that are within 5–10 Å, and highlight differences in packing between the forms.

In spite of the resolution limitations of ^{1}H SSNMR, direct observation of this sensitive nucleus is increasingly utilized in studies of drug polymorphs. Fast MAS techniques can distinguish polymorphs in some cases,[190] as can ^{1}H MAS experiments performed with DUMBO or windowed PMLG decoupling. The deshielded region of the ^{1}H spectrum is often found to be specific for individual forms because of the sensitivity of proton sites resonating in this region to hydrogen bonding differences in polymorphs. When greater resolution is needed, 2D experiments can be performed such as the ^{1}H double-quantum (DQ) experiments that were applied to the study of two polymorphs of sibenadet hydrochloride.[194] The crystal structure of Form I was known, while that of Form II was not known. ^{1}H DQ build-up was examined and used to determine that Form II had a different packing arrangement from Form I. GIPAW calculations were used to support ^{1}H chemical shift assignments in this study.[194]

Many modern SSNMR studies of polymorphism and of the phenomena discussed in later sections are multinuclear in nature, given the flexibility of modern instrumentation, and seek to take advantage of the availability of any accessible nucleus in a particular drug molecule.[190,195] This is a unique advantage of SSNMR relative to other techniques, such as PXRD, because it both maximizes specificity and allows for selection of sensitive nuclei for quantitative analysis when needed. While the ^{13}C, ^{15}N, ^{1}H and ^{19}F nuclei have all seen use, several emerging nuclei also have potentially wide applicability to studies of polymorphism. For example, the ^{1}H–^{14}N HMQC experiment was recently shown to provide an efficient means to study nitrogen environments in Form A of cimetidine.[58] Study of the ^{14}N nucleus provides a more sensitive way to probe nitrogen environments, while also giving access to the EFG at the nitrogen site in addition to the chemical shift. Nitrogen-containing functional groups are common in many APIs, and their study should benefit from increased application of this experiment. ^{77}Se SSNMR has also been recently demonstrated on a selenium-containing pharmaceutical.[196]

2.4.2 Hydrates, Solvates, Salts and Cocrystals

Solvates are solid-state structures that include at least one type of solvent molecule in addition to a drug molecule. The solvent may be an organic solvent or water, as in the case of a hydrate. Solvates may exist in multiple polymorphic phases and as phases consisting of different stoichiometric ratios between the API and solvent. Because of their close relationship with polymorphism, solvates are in some cases referred to as pseudopolymorphs or solvatomorphs. Organic solvates are typically easily observed by ^{13}C SSNMR because of the distinct carbon resonances detected for the bound solvent.[182] Certain solvated and hydrated phases exhibit a non-stoichiometric relationship between the API and solvent content over a range. For example, many APIs exhibit non-stoichiometric hydration, both as a single phase and in conjunction with one or more phase changes, which can be studied in detail using SSNMR techniques that access conventional nuclei (*e.g.* ^{13}C and ^{15}N) as well as nuclei of specific utility when studying water exchange (*e.g.* ^{2}H and ^{17}O).[18,188,197,198]

Hydrates and solvates represent two well-known classes of multicomponent crystals or inclusion complexes. Cocrystals represent a larger class of multi-component crystal material made up of two or more components, typically in a stoichiometric ratio, where each separate component exists as a solid at room temperature.[199] In comparison to other spectroscopic techniques, SSNMR is a powerful technique for analysis of cocrystals for two main reasons.[200,201] First, SSNMR chemical shifts and chemical shift tensors allow for direct observation of the primary effects used in the design of cocrystals, including protonation state, hydrogen bonding and aromatic π-stacking.[200] Second, SSNMR techniques based on dipolar correlation, particularly the ^1H–^{13}C CP-HETCOR, allow for direct observation of the association between molecules on a molecular scale without the need to grow single crystals, which is important as many cocrystals are first discovered using grinding techniques.[200] Recently, the ^1H–^{14}N HMQC experiment has also been shown to be an effective technique for detection of molecular association in a cocrystal.[202] This experiment offers improved sensitivity when studying nitrogen hydrogen bonding donors and acceptors and also allows simultaneous access to both chemical shift and EFG information. Other newly accessible nuclei, such as ^{35}Cl, may find increasing application to the study of a large class of hydrochloride salts used in many pharmaceutical APIs.[203] The example shown in Figure 2.3 for raloxifene HCl highlights the sensitivity and specificity of ^{35}Cl SSNMR in studies of HCl salts.

2.4.3 Disordered Forms

Structural disorder is often encountered in studies of crystalline APIs. The disorder may arise from dynamic motion of functional groups in the solid structure over a wide range of frequencies, or may be static as in a crystalline solid solution.[204] Both are typically modeled in SCXRD studies by a series of partially occupied sites for each disordered atom. SSNMR can offer particular insight into the disorder model in a solid solution and can also confirm (by 2D methods) that a solid solution has formed between two different compounds.[205] Such a situation may be of interest in API development if a small amount of structurally related impurity forms a solid solution with an API crystal form, or may also occur when the API itself exists as two molecules, as in a racemate, that form a solid solution instead of a racemic crystal or chiral conglomerate.[198]

Composite crystals are another class of disordered material. Composite crystals contain multiple phases with different unit cells within the same crystal, and the ratio of the phases is generally non-stoichiometric.[206,207] SSNMR studies of pharmaceutical composite crystals can identify the presence of mixed salt and cocrystal phases and explain unusual analytical observations from different preparations of API batches.[208]

2.4.4 Amorphous Drugs

Although there is significant motivation to develop a drug molecule as a crystalline form, there are a number of situations where development of an amorphous drug can be beneficial. Because amorphous solids exist in a higher energy state, they can dissolve at a faster rate and help achieve supersaturated conditions, potentially allowing for greater exposure of a poorly soluble drug molecule. SSNMR is a critical technique in studies of amorphous solids because it yields high information content in comparison to other high-resolution spectroscopic methods, such as infrared and Raman spectroscopy, and because conventional PXRD methods are only able to confirm that a material is amorphous, and specialized methods must be used to extract any structural information. In contrast, the chemical shift, dipolar coupling and quadrupolar coupling interactions continue to provide structural information in amorphous materials. Amorphous solids generally yield broad resonances in their SSNMR spectra, which results from a range of chemical environments present because of a range of molecular conformations and intra- and inter-molecular interactions. For example, Figure 2.16 shows the ^{13}C CP-TOSS spectrum of an amorphous oligonucleotide compound that illustrates many typical spectral features of this class of pharmaceutical molecule. The broadened ^{13}C resonances in Figure 2.16 are typical of an amorphous material. However, a number of assignments can still be made and observations of other nuclei (such as ^{15}N and ^{31}P) can provide further insight into such materials. Oligonucleotide and antisense therapeutics, which inhibit gene expression by interfering with translation, are an emerging class of therapeutics.[209] Because of their molecular size (typically 5–20 kDa), and their preparation as solids *via* lyophilization, oligonucleotide therapeutics are likely to be amorphous when

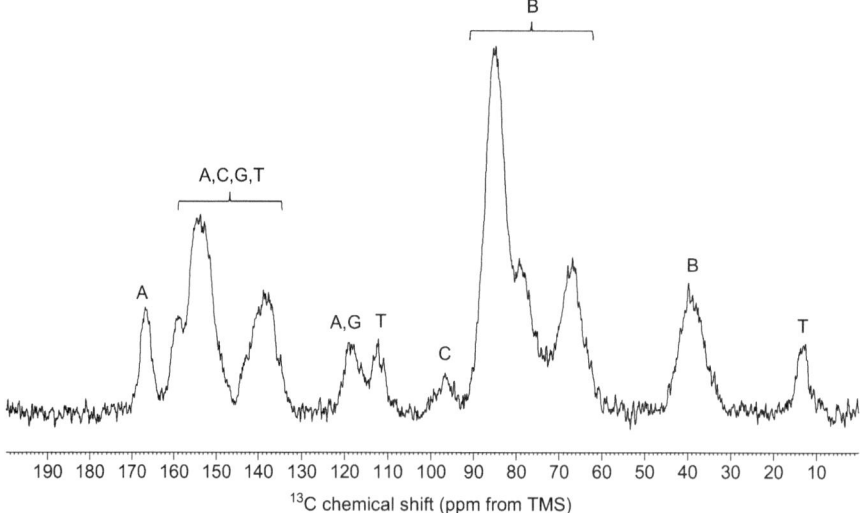

Figure 2.16 ^{13}C CP-TOSS spectrum of an amorphous phosphorothioate oligo-nucleotide with the sequence (5′)TGACTGTGAACGTTCGAGA-TGA(3′). The assignments refer to the nucleobases thymine (T), guanine (G), cytosine (C), adenine (A) and the backbone sugar (B). The spectrum was obtained with $v_r = 8$ kHz, $B_0 = 11.7$ T and at a temperature of 273 K.

delivered as solids, making analysis by conventional PXRD methods intractable, while analysis by SSNMR remains viable. A similar situation occurs for peptide therapeutics and small molecules conjugated to larger molecules. For example, 2D ^{1}H–^{13}C HETCOR and ^{13}C CP-TOSS experiments have been used to probe secondary structure in an amorphous 3.7 kDa peptide API and also to detect the amorphous drug in a formulation.[210]

The measurement of ^{1}H $T_{1\rho}$ relaxation is a sensitive probe of molecular dynamics in the kHz range in amorphous drugs.[211] Differential scanning calorimetry (DSC) and modulated temperature DSC are commonly used to measure the glass transition temperature (T_g) of amorphous drugs, which is used as a predictor of physical stability, because the amorphous glassy solid transitions to a rubbery state above T_g that has greater potential for recrystallization or chemical degradation. Measurement of ^{1}H $T_{1\rho}$ values at a range of temperatures using direct ^{1}H observation after a spin-lock showed evidence that the drug has increased mobility at lower temperatures, well before T_g was observed by DSC methods.[211] This result suggests that molecular motion with frequencies in the kHz range had already begun and could impact chemical or physical stability. Relaxation time measurements can provide information about mobility in amorphous drugs that is not easily obtained with other techniques, and which is critical because higher molecular mobility generally correlates with decreased chemical and/or physical stability upon long-term storage.[212]

2.4.5 Excipients

Excipients serve many functions in modern drug products, ranging from simple fillers to complex functional excipients that control the ultimate performance of the drug product.[213] Excipients can enhance membrane permeability and drug solubility (the latter typically *via* promotion of supersaturation), allowing for better *in vivo* exposure for many APIs.[213] Excipients can also protect drugs against degradation and modify the release properties of drugs.[213,214] A recent review that covers SSNMR studies of excipients in more detail is presented elsewhere.[123] Basic ^{13}C CP-MAS spectra of many common organic and polymeric pharmaceutical excipients have also been compiled.[215] The study of pharmaceutical excipients also offers unique opportunities for multinuclear SSNMR studies. For example, ^{29}Si SSNMR can be used to analyze excipients ranging from silicon dioxide to more complex mesoporous silicates loaded with drugs, ^{31}P SSNMR can be used to study calcium phosphates and ^{23}Na SSNMR can be used to study sodiated excipients such as sodium starch glycolate and croscarmellose sodium.[178]

Many excipients are semi-crystalline or amorphous organic polymers, and can benefit from the ability of ^{13}C SSNMR to observe simultaneously both crystalline and amorphous content. SSNMR is a well-established tool for studies of such materials.[216] In this role, SSNMR is often combined with techniques such as modulated temperature DSC and bulk physical property methods to provide a more complete picture of excipient characteristics.

SSNMR is also sensitive to differences in chemical substitution in random and block polymers, and to differences in mobility caused by variation in water content.[175,217] A study of sodium alginate, an amorphous excipient that performs a functional release-modifying role in many formulations, illustrated the ability of ^{13}C SSNMR in the study of structural and performance characteristics of polymeric excipients by detecting variations in monomer composition, molecular weight and water content in different batches of this material.[218] Because polymers are continuously being developed for pharmaceutical applications and new commercial pharmaceutical products, it is expected that SSNMR will continue to find a role in their analysis. A wide variety of lipids and lipid-like substances are also used in pharmaceutical development. These include the commonly used lubricant magnesium stearate, which is commercially supplied as a mixture of stearate and palmitate groups and typically contains several phases of different hydration states.[219] ^{13}C SSNMR is particularly useful in studies of polymorphism and mesomorphism in these systems.

2.4.6 Formulations and Dispersions of Drugs

Although the characterization of API and excipients is an important task during drug development, the ultimate analytical question usually involves the physical state of these components and any interactions between them in the final drug product dosage form. SSNMR offers unique specificity for the detection of a component of interest, such as the forms of an API, within a complex finished dosage form containing multiple excipients. Conventional ^{13}C SSNMR techniques can be used to study the API form within a formulation when the API is present at sufficiently high levels (typically at least 5% w/w), because ^{13}C spectra usually offer enough resolution to detect at least some of the typical aryl and carbonyl functional groups in APIs without overlap with aliphatic carbons typically found in excipients. This is advantageous in comparison to other techniques, such as PXRD, which may not have sufficient specificity to detect a crystalline drug in the presence of crystalline excipients, or to detect an amorphous drug at all in a formulation. Natural-abundance ^{13}C SSNMR spectra have been successfully used to identify and quantify drug polymorphs in a variety of drug products, including tablets, ointments and inhaled powders.[192,220–224] One of the earliest studies exhibited ^{13}C CP-MAS spectra of aspirin tablets.[221] More recently, both ^{13}C and high-sensitivity ^{19}F SSNMR techniques have been used to characterize the crystalline or amorphous phase of an API present in a formulation. ^{13}C SSNMR techniques can detect the form present at drug loads as low as about 1% w/w in favorable cases, although a more typical level is about 5–10% w/w.[192,220,224] ^{19}F SSNMR is particularly sensitive for formulations containing low drug loads, and can detect undesired forms at levels of 0.1% or lower in dosage forms (*e.g.* punched oral tablets) without difficulty.[191,192]

A [1]H-detected 2D [13]C–[1]H CP-HETCOR experiment, performed at $v_r =$ 40 kHz with a 1.6 mm probe and using the MISSISSIPPI (multiple intense solvent suppression intended for sensitive spectroscopic investigation of protonated proteins instantly) pulse sequence to suppress strong ^{1}H signals, was shown rapidly and specifically to detect ibuprofen and acetaminophen APIs in high drug load (>60%) formulations.[225] A 2D ^{1}H DQ experiment performed with homonuclear ^{1}H decoupling (also referred to as the DQ-CRAMPS experiment, where CRAMPS is an acronym for combined rotation and multiple pulse sequence) also successfully detected the API form in formulations, with additional specificity available from the DQ dimension.[226]

Relaxation time analysis using SSNMR has been shown to be indicative of formulation stability.[227] 2D ^{1}H–^{13}C CP-HETCOR and ^{1}H–^{19}F CP-HETCOR experiments were shown to be capable of demonstrating formation of the desired glass solution in amorphous drug-polymer dispersions.[179,228,229] The approach was demonstrated on a range of drugs and polymers, and is effective in dispersions formed with many polymers including polyvinylpyrrolidone (PVP) and hydroxypropylmethylcellulose acetate succinate (HPMC-AS), both of which are commercially important and are used in processes developed for several marketed drug products. Recently, the ^{1}H–^{14}N HMQC experiment has also been used to characterize hydrogen bonding and show molecular association in a cocrystal and an amorphous dispersion through sensitive access to the nitrogen environment and dipolar correlation to ^{1}H nuclei.[202] Other complex formulations are also tractable using 2D SSNMR methods. For example, inclusion of a drug into a cyclodextrin in an amorphous solid can be detected by the 2D ^{1}H–^{13}C CP-HETCOR and 2D ^{1}H–^{19}F CP-HETCOR experiments.[230,231]

2.4.7 Crystal Structure Determination

The determination of the crystal structure of small pharmaceutical molecules using SSNMR has been of significant interest in recent years. These efforts are relevant because in many cases of interest, single crystals of a suitable size cannot be grown for SCXRD analysis. Typical cases where a crystal structure might necessarily be solved from a powder include polymorphs that are difficult to crystallize, transient solvates and hydrates or cocrystals that can only be produced by solvent-drop grinding methods. While most crystal structure solutions from powders are currently obtained using only PXRD methods,[232] SSNMR can play an important role in supporting PXRD structure solutions, verifying structures and in some cases completely determining the structure without the need for PXRD analysis. The most basic SSNMR experiments, such as a simple 1D ^{13}C CP-TOSS spectrum, can report on symmetry in a crystal phase. For example, the measurement of Z' helps identify the space group for a particular unit cell in conjunction with knowledge of typical density of organic crystals (1.4–1.6 g mL^{-1}). More detailed crystal symmetry and space group information can be obtained from SSNMR spectra in some cases.[233] Basic observation of hydrogen bonding trends (*e.g. via* ^{1}H or ^{13}C NMR chemical shifts or dipolar couplings) or conformational trends (*e.g.* torsion angle) can be used to help verify crystal structures.[190] The chemical shift tensor can be combined with PXRD to help evaluate the quality of crystal structures solved using powder methods.[190,234] Individual distance determinations by REDOR (with appropriate spin labeling) can be used to assist in PXRD structure solutions.[235]

In addition to a supporting role, it is possible to solve entire crystal structures without PXRD or other data solely from SSNMR experiments. Methods based on the modeling of many restraints using ^{1}H spin diffusion have been demonstrated.[236–238] These methods are based on a phenomenological rate matrix analysis approach, which incorporates both direct dipolar transfers and multistep transfers. In these experiments, ^{1}H spin diffusion build-up is typically measured using a series of 2D experiments with a longitudinal mixing period and DUMBO decoupling in both dimensions. A set of distance restraints is obtained and is used to produce a structure for fitting of a unit cell and space group. The method has to date been employed for very small molecules, such as thymol, which are approximately 200 Da or less.

2.4.8   Future Directions

SSNMR is currently an established technique in industrial settings with many applications. However, because of high costs, most pharmaceutical firms have purchased only a limited number of SSNMR instruments in comparison to PXRD instruments and other solid-state analytical tools. The expense of SSNMR instrumentation relative to that of PXRD and vibrational spectroscopic instrumentation provides an impetus for finding additional uses of SSNMR that go beyond providing an orthogonal "fingerprint" using ^{13}C spectral data.[175] For this reason, the future direction of SSNMR in pharmaceutical development is expected to focus increasingly on the use of more detailed structural characterization methods, such as the 2D SSNMR methods described above, which distinguish SSNMR from other competing techniques. This could lead to industrial SSNMR instruments that are equipped with higher-field magnets and a range of specialized probes. These configurations could also approach that needed for many of the aforementioned drug-discovery applications of SSNMR, potentially allowing a greater range of discovery studies to be performed in the industrial setting as well.

2.5   Conclusions

SSNMR is a powerful tool for analysis of the structure and dynamics of systems of interest in drug discovery and development. Continued advancement of new experimental methods and pulse sequences, coupled with increasing availability of static fields exceeding 21 T and MAS rates in excess of 70 kHz, have allowed for more complex systems of interest to be studied. The ability to determine the structures of proteins in the solid state has been demonstrated, and an increasing number of studies have tackled complex interactions between drugs and biomolecules and between multiple biomolecules. In drug development, complex formulations can be analyzed in detail by SSNMR, in many cases allowing for detection of every component of a formulation and observation of interactions between the components. Increased use of ^{1}H and ^{19}F SSNMR techniques is expected, particularly in drug-development applications, because of the sensitivity and availability of these isotopes in many drug candidates. Increased sensitivity during MAS experiments through cryogenic cooling of probe electronics or other approaches, such as DNP, are also expected, and the trend towards higher B_0 will likely continue even as low-field systems continue to provide value. Finally, multidisciplinary studies using SSNMR with other techniques may become more common as acceptance increases and more difficult challenges in drug discovery and development are attempted.

References

1. S. P. Brown, *Solid State Nucl. Magn. Reson.*, 2012, **41**, 1.
2. A. Lesage, *Phys. Chem. Chem. Phys.*, 2009, **11**, 6876.
3. T. W. T. Tsai and J. C. C. Chan, *Ann. Rep. NMR Spectros.*, 2011, **73**, 1.
4. R. Graf, *Solid State Nucl. Magn. Reson.*, 2011, **40**, 127.
5. S. P. Brown, *Macromol. Rapid Commun.*, 2009, **30**, 688.
6. A. E. Aliev and R. V. Law, in *Nuclear Magnetic Resonance*, ed. K. Kamienska-Trela, RSC Publishing, Cambridge, UK, 2011, vol. 40, p. 254.
7. W. P. Power, *Ann. Rep. NMR Spectrosc.*, 2011, **72**, 111.
8. A. Abragam, *The Principles of Nuclear Magnetism*, Clarendon Press, Oxford, 1961.
9. M. Mehring, *Principles of High Resolution NMR in Solids*, Springer-Verlag, New York, 2nd edn, 1983.
10. B. C. Gerstein and C. R. Dybowski, *Transient Techniques in the NMR of Solids*, Academic Press, San Diego, 1985.
11. C. P. Slichter, *Principles of Magnetic Resonance*, Springer-Verlag, Heidelburg, 3rd edn, 1996.
12. R. K. Harris, E. D. Becker, S. M. Cabral de Menezes, R. Goodfellow and P. Granger, *Pure Appl. Chem.*, 2001, **73**, 1795.
13. H. O. Kalinowski, S. Berger and S. Braun, *Carbon-13 NMR Spectroscopy*, Wiley, New York, 1987.
14. S. Berger, S. Braun and H. O. Kalinowski, *NMR Spectroscopy of the Non-metallic Elements*, Wiley, New York, 1997.
15. W. S. Veeman, *Prog. NMR Spectros.*, 1984, **16**, 193.
16. K. Ravikumar, B. Sridhar, K. D. Prasad and A. K. S. B. Rao, *Acta Cryst. E*, 2007, **63**, o565.
17. F. G. Vogt, D. E. Cohen, J. D. Bowman, G. P. Spoors, G. E. Zuber, G. A. Trescher, P. C. Dell'Orco, L. M. Katrincic, C. W. DeBrosse and R. C. Haltiwanger, *J. Pharm. Sci.*, 2005, **94**, 651.
18. F. G. Vogt, J. Brum, L. M. Katrincic, A. Flach, J. M. Socha, R. M. Goodman and R. C. Haltiwanger, *Cryst. Growth Des.*, 2006, **6**, 2333.
19. D. Vega, D. Fernandez and J. A. Ellena, *Acta Cryst. C.*, 2001, **57**, 1092.
20. A. Samoson, *Chem. Phys. Lett.*, 1985, **119**, 29.

21. T. T. P. Cheung, in *The Encyclopedia of NMR*, ed. R. K. Harris and D. M. Grant, John Wiley & Sons, New York, 1996, p. 4518.
22. B. H. Meier, *Adv. Magn. Reson.*, 1994, **18**, 1.
23. R. K. Harris and A. C. Olivieri, *Prog. NMR Spectros.*, 1992, **24**, 435.
24. J. Vaara, J. Jokisaari, R. E. Wasylishen and D. L. Bryce, *Prog. Nucl. Magn. Reson. Spectrosc.*, 2002, **41**, 233.
25. C. P. Jaroniec, *Solid State Nucl. Magn. Reson.*, 2012, **43–44**, 1.
26. H. M. McConnell and R. E. Robertson, *J. Chem. Phys.*, 1958, **29**, 1361.
27. A. Nayeem and J. P. Yesinowski, *J. Chem. Phys.*, 1988, **89**, 4600.
28. P. A. Beckmann, *Phys. Rep.*, 1988, **171**, 85.
29. J. L. Sudmeier, S. E. Anderson and J. S. Frye, *Concepts Magn. Reson.*, 1990, **2**, 197.
30. H. W. Spiess, in *NMR Basic Principles and Progress*, ed. P. Diehl, E. Fluck and R. Kosfeld, Springer-Verlag, Duesseldorf, Germany, 1978, vol. 15, p. 55.
31. H. T. Stokes, D. C. Ailion and T. A. Case, *Phys. Rev. B*, 1984, **30**, 4925.
32. J. B. Grutzner, K. W. Stewart, R. E. Wasylishen, M. D. Lumsden, C. Dybowski and P. A. Beckmann, *J. Am. Chem. Soc.*, 2001, **123**, 7094.
33. X. Wang, A. L. Rheningold, A. G. DiPasquale, F. B. Fallory, C. W. Mallory and P. A Beckmann, *J. Chem. Phys.*, 2008, **128**, 124502/1.
34. P. A. Beckmann, J. Rosenberg, K. Nordstrom, C. W. Mallory and F. B. Mallory, *J. Phys. Chem. A*, 2006, **110**, 3947.
35. A. L. Plofker and P. A. Beckmann, *J. Phys. Chem.*, 1995, **99**, 391.
36. R. Böhmer, G. Diezemann, G. Hinze and E. Rössler, *Prog. NMR Spectros.*, 2001, **39**, 191.
37. J. R. Lewandowski and L. Emsley, in *Solid-state NMR Studies of Biopolymers*, ed. A. E. McDermott and T. Polenova, Wiley, New York, 2010, p. 343.
38. P. L. Gor'kov, W. W. Brey and J. R. Long, in *Solid-state NMR Studies of Biopolymers*, ed. A. E. McDermott and T. Polenova, Wiley, New York, 2010, p. 141.
39. F. D. Doty, in *The Encyclopedia of NMR*, ed. R. K. Harris and D. M. Grant, John Wiley & Sons, New York, 1996, p. 4475.
40. A. Samoson, T. Tuherm, J. Past, A. Reinhold, T. Anupold and N. Heinmaa, *Top. Curr. Chem.*, 2005, **246**, 15.
41. Y. Nishiyama, Y. Endo, T. Nemoto, H. Utsumi, K. Yamauchi, K. Hioka and T. Asakura, *J. Magn. Reson.*, 2011, **208**, 44.
42. I. Kawamura, N. Kihara, M. Ohmine, K. Nishimura, S. Tuzi, H. Saito and A. Naito, *J. Am. Chem. Soc.*, 2007, **129**, 1016.
43. A. Asano, S. Hori, M. Kitamura, C. T. Nakazawa and T. Kurotsu, *Polym. J.*, 2012, **44**, 706.
44. H. Kovacs, D. Moskau and M. Spraul, *Prog. NMR Spectrosc.*, 2005, **46**, 131.
45. T. Mizuno, K. Hioka, K. Fujioka and K. Takegoshi, *Rev. Sci. Instr.*, 2008, **79**, 044706/1.

46. K. R. Thurber and R. Tycko, *J. Magn. Reson.*, 2008, **195**, 179.
47. J. A. Stringer and G. P. Drobny, *J. Chem. Phys.*, 1998, **69**, 3384.
48. G. Metz, X. Wu and S. O. Smith, *J. Magn. Reson. A*, 1994, **110**, 219.
49. D. Rovnyak, *Concepts Magn. Reson.*, 2008, **32A**, 254.
50. E. O. Stejskal and J. D. Memory, *High Resolution NMR in the Solid State*, Oxford University Press, New York, 1994.
51. P. Hodgkinson, *Prog. NMR Spectros.*, 2005, **46**, 197.
52. B. M. Fung, A. K. Khitrin and K. Ermolaev, *J. Magn. Reson.*, 2000, **142**, 97.
53. C. Augustine and N. D. Kurur, *J. Magn. Reson.*, 2011, **209**, 156.
54. V. S. Mithu, S. Paul, N. D. Kurur and P. K. Madhu, *J. Magn. Reson.*, 2011, **209**, 359.
55. O. N. Antzutkin, *Prog NMR Spectros.*, 1999, **35**, 203.
56. S. J. Opella and M. H. Frey, *J. Am. Chem. Soc.*, 1979, **101**, 5854.
57. A. Lesage, S. Steuernagel and L. Emsley, *J. Am. Chem. Soc.*, 1998, **120**, 7095.
58. A. S. Tatton, T. N. Pham, F. G. Vogt, D. Iuga, A. J. Edwards and S. P. Brown, *CrystEngComm*, 2012, **14**, 2654.
59. S. T. Burns, X. Wu and K. W. Zilm, *J. Magn. Reson.*, 2000, **143**, 352.
60. P. K. Madhu, *Solid State Nucl. Magn. Reson.*, 2009, **35**, 2.
61. E. Vinogradov, P. K. Madhu and S. Vega, *Chem. Phys. Lett.*, 1999, **314**, 443.
62. D. Sakellariou, A. Lesage, P. Hodgkinson and L. Emsley, *Chem. Phys. Lett.*, 2000, **319**, 253.
63. B. J. van Rossum, H. Förster and H. J. M. de Groot, *J. Magn. Reson.*, 1997, **124**, 516.
64. B. J. van Rossum, C. P. de Groot, V. Ladizhansky, S. Vega and H. J. M. de Groot, *J. Am. Chem. Soc.*, 2000, **122**, 3465.
65. J. Cejka, B. Kratochvil and A. Jegorov, *Z. Kristallogr. – New Cryst. Struct.*, 2005, **220**, 143.
66. J. R. Lewandowski, G. De Paëpe and R. G. Griffin, *J. Am. Chem. Soc.*, 2007, **129**, 728.
67. M. Baldus, D. G. Geurts, S. Hediger and B. H. Meier, *J. Magn. Reson. A*, 1996, **118**, 140.
68. M. Baldus, A. Petkova, J. Herzfeld and R. G. Griffin, *Mol. Phys.*, 1998, **95**, 1197.
69. J. Pauli, M. Baldus, B. J. van Rossum, H. de Groot and H. Oschkinat, *Chem. Biochem.*, 2001, **2**, 272.
70. Y. Li, D. A. Berthold, H. L. Frericks, R. B. Gennis and C. M. Rienstra, *Chem. Biochem.*, 2007, **8**, 434.
71. F. G. Vogt and K. T. Mueller, in *Encyclopedia of Nuclear Magnetic Resonance*, ed. D. M. Grant and R. K. Harris, John Wiley & Sons, New York, 2002, vol. 9, p. 112.
72. J. Schaefer, in *Encyclopedia of Nuclear Magnetic Resonance*, ed. D. M. Grant and R. K. Harris, 1996, John Wiley & Sons, New York, p. 3977.
73. F. G. Vogt, J. M. Gibson, S. M. Mattingly and K. T. Mueller, *J. Phys. Chem. B*, 2003, **107**, 1272.

74. T. Gullion and C. H. Pennington, *Chem. Phys. Lett.*, 1998, **290**, 88.
75. C. P. Jaroniec, B. A. Tounge, C. M. Rienstra, J. Herzfeld and R. G. Griffin, *J. Am. Chem. Soc.*, 1999, **121**, 10237.
76. R. Tycko, in *Solid-state NMR Studies of Biopolymers*, ed. A. E. McDermott and T. Polenova, Wiley, New York, 2010, pp. 175–188.
77. G. De Paëpe, *Ann. Rev. Phys. Chem.*, 2012, **63**, 661.
78. P. T. F. Williamson, A. Verhoeven, M. Ernst and B. H. Meier, *J. Am. Chem. Soc.*, 2003, **125**, 2718.
79. I. Schnell and H. W. Spiess, *J. Magn. Reson.*, 2001, **151**, 153.
80. G. Mollica, P. K. Madhu, F. Ziarelli, A. Thevand, P. Thureau and S. Viel, *Phys. Chem. Chem. Phys.*, 2012, **14**, 4359.
81. K. Takegoshi, S. Nakamura and T. Terao, *J. Chem. Phys.*, 2003, **118**, 2325.
82. L. J. Mueller and J. J. Titman, in *Solid-state NMR Studies of Biopolymers*, ed. A. E. McDermott and T. Polenova, Wiley, New York, 2010, pp. 297–316.
83. A. Lesage, P. Charmont, S. Steuernagel and L. Emsley, *J. Am. Chem. Soc.*, 2000, **122**, 9739.
84. A. Lesage, M. Bardet and L. Emsley, *J. Am. Chem. Soc.*, 1999, **121**, 10987.
85. E. H. Hardy, R. Verel and B. H. Meier, *J. Magn. Reson.*, 2001, **148**, 459.
86. L. Chen, R. A. Olsen, D. W. Elliott, J. M. Boettcher, D. H. Zhou, C. M. Rienstra and L. J. Mueller, *J. Am. Chem. Soc.*, 2006, **128**, 9992.
87. L. Chen, J. M. Kaiser, J. Lai, T. Polenova, J. Yang, C. M. Rienstra and L. J. Mueller, *Magn. Reson. Chem.*, 2007, **45**, S84.
88. L. Chen, J. M. Kaiser, T. Polenova, J. Yang, C. M. Rienstra and L. J. Mueller, *J. Am. Chem. Soc.*, 2007, **129**, 10650.
89. O. N. Antzutkin, *Prog. NMR Spectros.*, 1999, **35**, 203.
90. D. W. Alderman, G. McGeorge, J. Z. Hu, R. J. Pugmire and D. M. Grant, *Mol. Phys.*, 1998, **95**, 1113.
91. J. Rocha, C. M. Morais and C. Fernandez, *Top. Curr. Chem.*, 2005, **246**, 141.
92. S. E. Ashbrook and S. Wimperis, *Prog. Nucl. Magn. Reson. Spectrosc.*, 2004, **45**, 53.
93. F. H. Larsen, H. J. Jakobsen, P. D. Ellis and N. C. Nielsen, *J. Phys. Chem. A*, 1997, **101**, 8597.
94. Z. H. Gan, J. P. Amoureux and J. Trebosc, *Chem. Phys. Lett.*, 2007, **435**, 163.
95. N. Zumbulyadis, B. Antalek, W. Windig, R. P. Scaringe, A. M. Lanzafame, T. Blanton and M. Helber, *J. Am. Chem. Soc.*, 1999, **121**, 11554.
96. J. H. Adenkjaer, B. Bridlun, A. Gram, G. Hannson, L. Hansson, M. H. Lerche, R. Servin, M. Thaning and K. Golman, *Proc. Natl Acad. Sci. USA*, 2003, **100**, 10158.
97. A. B. Barnes, G. D. Paëpe, P. C. van der Wel, K. N. Hu, C. G. Joo, V. S. Bajaj, M. L. Mak-Jurkauskas, J. R. Sirigiri, J. Herzfeld, R. J. Temkin and R. G. Griffin, *Appl. Magn. Reson.*, 2008, **34**, 237.

98. U. Akbey, A. H. Linden and H. Oschkinat, *Appl. Magn. Reson.*, 2012, **43**, 81.
99. W. L. Earl and D. L. Vanderhart, *J. Magn. Reson.*, 1982, **48**, 35.
100. W. Koch and M. C. Holthausen, *A Chemist's Guide to Density Functional Theory*, Wiley-VCH, Weinheim, 2001.
101. A. M. Orendt and J. C. Facelli, *Ann. Rep. NMR Spectros.*, 2007, **62**, 115.
102. D. B. Chesnut and K. D. Moore, *J. Comp. Chem.*, 1989, **10**, 648.
103. W. Kutzelnigg, in *Calculation of NMR and EPR Parameters*, ed. M. Kaupp, M. Bühl and V. G. Malkin, Wiley-VCH, Weinheim, 2004, p. 43.
104. F. Maseras and K. Morokuma, *J. Comp. Chem.*, 1995, **16**, 1170.
105. C. J. Pickard and F. Mauri, in *Calculation of NMR and EPR Parameters*, ed. M. Kaupp, M. Bühl and V. G. Malkin, Wiley-VCH, Weinheim, 2004, p. 265.
106. R. K. Harris, P. Hodgkinson, C. J. Pickard, J. R. Yates and V. Zorin, *Magn. Reson. Chem.*, 2007, **45**, S174.
107. R. K. Harris, S. Joyce, C. J. Pickard, S. Cadars and L. Emsley, *Phys. Chem. Chem. Phys.*, 2006, **6**, 137.
108. J. R. Yates, S. E. Dobbins, C. J. Pickard, F. Mauri, P. Y. Ghic and R. K. Harris, *Phys. Chem. Chem. Phys.*, 2005, **7**, 1402.
109. J. R. Yates, *Magn. Reson. Chem.*, 2010, **48**, S23.
110. B. Delley, *J. Chem. Phys.*, 1990, **92**, 508.
111. B. Delley, *J. Chem. Phys.*, 2000, **113**, 7756.
112. D. Skachkov, M. Krykunov and T. Ziegler, *Can. J. Chem.*, 2011, **89**, 1150.
113. D. Skachkov, M. Krykunov, E. Kadantsev and T. Ziegler, *J. Chem. Theory Comput.*, 2010, **6**, 1650.
114. P. Schwerdtfeger, M. Pernpointner and W. Nazarewicz, in *Calculation of NMR and EPR Parameters*, ed. M. Kaupp, M. Bühl and V. G. Malkin, Wiley-VCH, Weinheim, 2004, p. 279.
115. M. Mehring and V. A. Weberruss, *Object-Oriented Magnetic Resonance*, Academic Press, San Diego, 2001.
116. M. Bak, J. T. Rasmussen and N. C. Nielsen, *J. Magn. Reson.*, 2000, **147**, 296.
117. M. Veshtort and R. G. Griffin, *J. Magn. Reson.*, 2006, **178**, 248.
118. J. P. Bradley, C. Tripon, C. Filip and S. P. Brown, *Phys. Chem. Chem. Phys.*, 2009, **11**, 6941.
119. J. N. Dumez, M. C. Butler, E. Salager, B. Elena-Herrmann and L. Emsley, *Phys. Chem. Chem. Phys.*, 2010, **12**, 9172.
120. J. C. Hoch and A. S. Stern, *NMR Data Processing*, Wiley, New York, 1996.
121. R. K. Harris and A. C. Olivieri, in *Encyclopedia of Nuclear Magnetic Resonance*, ed. D. M. Grant and R. K. Harris, Wiley, New York, 2002, vol. 9, p. 141.
122. J. Higinbotham and I. Marshall, *Ann. Rep. NMR Spectros.*, 2000, **43**, 60.
123. F. G. Vogt, J. S. Clawson, M. Strohmeier, T. N. Pham, S. A. Watson and A. J. Edwards, in *Pharmaceutical Sciences Encyclopedia: Drug Discovery,*

Development, and Manufacturing, ed. S. C. Gad, John Wiley & Sons, New York, 2011, p. 1, http://dx.doi.org/10.1002/9780470571224.pse418.

124. S. Y. Ohki and M. Kainosho, *Prog. Nucl. Magn. Reson. Spectrosc.*, 2008, **53**, 208.

125. S. K. Straus, *Phil. Trans. R. Soc. Lond. B*, 2004, **359**, 997.

126. F. Castellani, B. van Rossum, A. Diehl, M. Schubert, K. Rehbein and H. Oschkinat, *Nature*, 2002, **420**, 98.

127. A. Lange, S. Becker, K. Seidel, K. Giller, O. Pongs and M. Baldus, *Angew. Chem. Int. Ed.*, 2005, **44**, 2089.

128. S. G. Zech, A. J. Wand and A. E. McDermott, *J. Am. Chem. Soc.*, 2005, **127**, 8618.

129. T. Gullion, *Ann. Rep. NMR Spectrosc.*, 2009, **65**, 111.

130. A. T. Petkova, W.-M. Yau and R. Tycko, *Biochemistry*, 2006, **45**, 498.

131. J. C. C. Chan, N. A. Oyler, W.-M. Yau and R. Tycko, *Biochemistry*, 2005, **44**, 10669.

132. K. L. Sciarretta, D. J. Gordon, A. T. Petkova, R. Tycko and S. C. Meredith, *Biochemistry*, 2005, **44**, 6003.

133. R. Tycko, *Quart. Rev. Biophys.*, 2006, **39**, 1.

134. C. P. Jaroniec, C. Filip and R. G. Griffin, *J. Am. Chem. Soc.*, 2002, **124**, 10728.

135. C. P. Jaroniec, C. E. MacPhee, V. S. Bajaj, M. T. McMahon, C. M. Dobson and R. G. Griffin, *Proc. Natl Acad. Sci. USA*, 2004, **101**, 711.

136. N. P. Wickramasinghe, S. Parthasarathy, C. R. Jones, C. Bhardwaj, F. Long, M. Kotecha, S. Mehboob, L. W. M. Fung, J. Past, A. Samoson and Y. Ishii, *Nature Meth.*, 2009, **6**, 215.

137. D. Huster, X. Yao and M. Hong, *J. Am. Chem. Soc.*, 2002, **124**, 874.

138. K. N. Hu and R. Tycko, *Biophys. Chem.*, 2010, **151**, 10.

139. C. S. Sanders, B. J. Hare, K. P. Howard and J. H. Prestegard, *Prog. NMR Spectrosc.*, 1994, **26**, 421.

140. M. Ouellet and M. Auger, *Ann. Rep. NMR Spectrosc.*, 2008, **63**, 1.

141. A. S. Lipton, T. Polenova and P. D. Ellis, in *Solid-state NMR Studies of Biopolymers*, ed. A. E. McDermott and T. Polenova, Wiley, New York, 2010, p. 491.

142. S. Johnson, E. Barile, B. Farina, A. Purves, J. Wei, L. H. Chen, S. Shiryaev, Z. Zhang, I. Rodionova, A. Agrawal, S. M. Cohen, A. Osterman, A. Strongin and M. Pellecchia, *Chem. Biol. Drug. Des.*, 2011, **78**, 211.

143. S. M. Cohen and M. Rouffet, *Dalton Trans.*, 2011, **40**, 3445.

144. J. L. Mauriz, J. Martín-Renedo, A. García-Palomo, M. J. Tuñón and J. González-Gallego, *Cur. Drug Targets*, 2010, **11**, 1439.

145. M. Lopez, S. Köhler and J. Y. Winum, *J. Inorg. Chem.*, 2012, **111**, 138.

146. G. Wu and J. Zhu, *Prog. NMR Spectrosc.*, 2012, **61**, 1.

147. R. W. Strange and S. S. Hasnain, in *Protein-Ligand Interactions. Methods in Molecular Biology*, ed. G. U. Nienhaus, Springer, Berlin, 2005, vol. 305, p. 167.

148. W. R. Hagen, *Dalton Trans.*, 2006, **37**, 4415.
149. N. Dimakis and G. Bunker, *Biophys. J.*, 2006, **91**, L87.
150. S. S. Hasnain and K. O. Hodgson, *J. Synchrotron Rad.*, 1999, **6**, 852.
151. N. Pooransingh-Margolis, R. Renirie, Z. Hasan, R. Wever, A. J. Vega and T. Polenova, *J. Am. Chem. Soc.*, 2006, **128**, 5190.
152. P. D. Ellis, J. A. Sears, P. Yang, M. Dupuis, T. T. Boron, V. L. Pecararo, T. A. Stich, R. D. Britt and A. S. Lipton, *J. Am. Chem. Soc.*, 2010, **132**, 16727.
153. P. D. Ellis and A. S. Lipton, *J. Am. Chem. Soc.*, 2007, **129**, 9192.
154. S. Balayssac, I. Bertini, M. Lelli, C. Luchinat and M. Maletta, *J. Am. Chem. Soc.*, 2007, **129**, 2218.
155. C. Luchinat, G. Parigi, E. Ravera and M. Rinaldelli, *J. Am. Chem. Soc.*, 2012, **134**, 5006.
156. T. Diercks, M. Coles and H. Kessler, *Cur. Opin. Chem. Biol.*, 2001, **5**, 285.
157. C. W. Murray and D. C. Rees, *Nat. Chem.*, 2009, **1**, 187.
158. D. A. Middleton, *Ann. Rep. NMR Spectrosc.*, 2007, **60**, 39.
159. P. T. F. Williamson, *Concepts Magn. Reson.*, 2009, **34A**, 144.
160. S. G. Patching, G. Psakis, S. A. Baldwin, J. Baldwin, P. J. Henderson and D. A. Middleton, *Mol. Membr. Biol.*, 2008, **25**, 474.
161. A. N. Appleyard, R. B. Herbert, P. J. F. Henderson, A. Watts and P. J. R. Spooner, *Biochim. Biophys. Acta*, 2000, **1509**, 55.
162. Y. Li, B. Poliks, L. Celgelski, M. Poliks, Z. Gryczynski, G. Piszczek, P. G. Jagtap, D. R. Studelska, D. G. I. Kingston, J. Scahefer and S. Bane, *Biochemistry*, 2002, **39**, 281.
163. Y. Paik, C. Yang, B. Metaferia, S. Tang, S. Bane, R. Ravindra, N. Shanker, A. A. Alcaraz, S. A. Johnson, J. Schaier, R. D. O'Connor, L. Celgelski, J. P. Snyder and D. G. I. Kingston, *J. Am. Chem. Soc.*, 2007, **129**, 361.
164. I. Ojima and M. Das, *J. Nat. Prod.*, 2009, **72**, 554.
165. A. K. Mehta, Y. Shayo, H. Vankayalapati, L. H. Hurley and J. Schaefer, *Biochemistry*, 2004, **43**, 11953.
166. G. L. Olsen, E. A. Louie, G. P. Drobny and S. T. Sigurdsson, *Nucl. Acids Res.*, 2003, **31**, 5084.
167. G. Zinzalla and D. E. Thurston, *Future Med. Chem.*, 2009, **1**, 65.
168. Y. Han, J. Ahn, J. Concel, I. L. Byeon, A. M. Gronenborn, J. Yang and T. Polenova, *J. Am. Chem. Soc.*, 2010, **132**, 1976.
169. J. Yang, M. L. Tasayco and T. Polenova, *J. Am. Chem. Soc.*, 2008, **130**, 5798.
170. A. Lange, K. Giller, S. Hornig, M. F. Martin-Eauclaire, O. Pongs, S. Becker and M. Baldus, *Nature*, 2006, **440**, 959.
171. G. L. Olsen, T. E. Edwards, P. Deka, G. Varani, S. T. Sigurdsson and G. P. Drobny, *Nucl. Acids Res.*, 2005, **33**, 3347.
172. J. L. Kitevski-LeBlanc and R. S. Prosser, *Prog. Nucl. Magn. Reson. Spectrosc.*, 2012, **62**, 1.
173. R. T. Berendt, D. M. Sperger, P. K. Isbester and E. J. Munson, *Trends Anal. Chem.*, 2006, **25**, 977.
174. R. K. Harris, *J. Pharm. Pharmacol.*, 2007, **59**, 225.

175. F. G. Vogt, *Fut. Med. Chem.*, 2010, **2**, 915.
176. S. R. Byrn, R. R. Pfeiffer and J. G. Stowell, *Solid-state Chemistry of Drugs*, SSCI, West Lafayette, IN, 1999.
177. T. Threllfall, *Analyst*, 1995, **120**, 2435.
178. F. G. Vogt, *Am. Pharm. Rev.*, 2008, **11**, 50.
179. F. G. Vogt and G. R. Williams, *Am. Pharm. Rev.*, 2010, **13**, 58.
180. J. Bernstein, *Polymorphism in Molecular Crystals*, Oxford University Press, New York, 2002.
181. F. G. Vogt, in *The Encyclopedia of Analytical Chemistry*, ed. R. A. Meyers, Wiley, New York, 2011, p. 1, http://dx.doi.org/10.1002/9780470027318. a9088.
182. R. K. Harris, *Analyst*, 2006, **131**, 351.
183. D. C. Apperley, R. A. Fletton, R. K. Harris, R. W. Lancaster, S. Tavener and T. L. Threllfall, *J. Pharm. Sci.*, 1999, **88**, 1275.
184. R. M. Wenslow, M. W. Baum, R. G. Ball, J. A. McCauley and R. J. Varsolona, *J. Pharm. Sci.*, 2000, **89**, 1271.
185. R. K. Harris, P. Y. Ghi, H. Puschmann, D. C. Apperley, U. J. Griesser, R. B. Hammond, C. Ma, K. J. Roberts, G. J. Pearce, J. R. Yates and C. J. Pickard, *Org. Proc. R&D*, 2005, **9**, 902.
186. S. R. Byrn, P. A. Sutton, B. Tobias, J. Frye and P. Main, *J. Am. Chem. Soc.*, 1988, **110**, 1609.
187. U. J. Griesser, R. K. R. Jetti, M. F. Haddow, T. Brehmer, D. C. Apperley, A. King and R. K. Harris, *Cryst. Growth Des.*, 2008, **8**, 44.
188. F. Kang, F. G. Vogt, J. Brum, R. Forcino, R. C. B. Copley and G. Williams, *Cryst. Growth Des.*, 2012, **12**, 60.
189. J. Smith, E. MacNamara, D. Raftery, T. Borchardt and S. R. Byrn, *J. Am. Chem. Soc.*, 1998, **120**, 11710.
190. F. G. Vogt, L. M. Katrincic, S. T. Long, R. L. Mueller, R. A. Carlton, Y. T. Sun, M. N. Johnson, R. C. B. Copley and M. E. Light, *J. Pharm. Sci.*, 2008, **97**, 4756.
191. R. M. Wenslow, *Drug Dev. Ind. Pharm.*, 2002, **28**, 555.
192. L. M. Katrincic, Y. T. Sun, R. A. Carlton, A. M. Diederich, R. L. Mueller and F. G. Vogt, *Int. J. Pharm.*, 2009, **366**, 1.
193. K. Müller, C. Faeh and F. Diederich, *Science*, 2007, **317**, 1881.
194. J. P. Bradley, C. J. Pickard, J. C. Burley, D. R. Martin, L. P. Hughes, S. D. Cosgrove and S. P. Brown, *J. Pharm. Sci.*, 2012, **101**, 1821.
195. Y. Garro Linck, A. K. Chattah, R. Graf, C. B. Romanuk, M. E. Olivera, R. H. Manzo, G. A. Monti and H. W. Spiess, *Phys. Chem. Chem. Phys.*, 2011, **13**, 6590.
196. F. G. Vogt and G. R. Williams, *Pharm. Res.*, 2012, **29**, 1866.
197. F. G. Vogt, P. C. Dell'Orco, A. M. Diederich, Q. Su, J. L. Wood, G. E. Zuber, L. M. Katrincic, R. L. Mueller, D. J. Busby and C. W. DeBrosse, *J. Pharm. Biomed. Anal.*, 2006, **40**, 1080.
198. F. G. Vogt, R. C. B. Copley, R. L. Mueller, G. P. Spoors, T. N. Cacchio, R. A. Carlton, L. M. Katrincic, J. M. Kennady, S. Parsons and O. V. Chetina, *Cryst. Growth Des.*, 2010, **10**, 2713.

199. N. Schultheiss and A. Newman, *Cryst. Growth Des.*, 2009, **9**, 2950.
200. F. G. Vogt, J. S. Clawson, M. Strohmeier, A. J. Edwards, T. N. Pham and S. A. Watson, *Cryst. Growth Des.*, 2009, **9**, 921.
201. J. R. Patel, R. A. Carlton, T. E. Needham, C. O. Chichester and F. G. Vogt, *Int. J. Pharm.*, 2012, **436**, 685.
202. A. S. Tatton, T. N. Pham, F. G. Vogt, D. Iuga, A. J. Edwards and S. P. Brown, *Mol. Pharmaceutics*, 2013, **10**, 999.
203. H. Hamaed, J. M. Pawlowski, B. F. T. Cooper, R. Fu, S. H. Eichhorn and R. W. Schurko, *J. Am. Chem. Soc.*, 2008, **130**, 11056.
204. S. Datta and D. J. W. Grant, *Nat. Rev. Drug. Disc.*, 2004, **3**, 42.
205. F. G. Vogt, J. A. Vena, M. Chavda, J. S. Clawson, M. Strohmeier and M. E. Barnett, *J. Mol. Struct.*, 2009, **932**, 16.
206. P. Coppens, K. Maly and V. Petricek, *Mol. Cryst. Liq. Cryst.*, 1990, **181**, 81.
207. A. Yamamoto, *Acta Cryst. A*, 1993, **49**, 813.
208. J. S. Clawson, F. G. Vogt, J. Brum, J. Sisko, D. B. Patience, W. Dai, S. Sharpe, A. D. Jones, T. N. Pham, M. N. Johnson and R. C. B. Copley, *Cryst. Growth Des.*, 2008, **8**, 4120.
209. F. Eckstein, *Expert Opin. Biol. Ther.*, 2007, **7**, 1021.
210. W. P. Kelley, S. Chen, P. D. Floyd, P. Hu, S. Kapsi, A. S. Kord, M. Sun and F. G. Vogt, *Anal. Chem.*, 2012, **84**, 4357.
211. A. Forster, D. Apperley, J. Hempenstall, R. Lancaster and T. Rades, *Pharmazie*, 2003, **58**, 761.
212. S. Yoshioka and Y. Aso, *J. Pharm. Sci.*, 2007, **96**, 960.
213. J. Hamman and J. Steenekamp, *Exp. Opin. Drug Deliv.*, 2012, **9**, 219.
214. O. Pillai and R. Panchagnula, *Curr. Opin. Chem. Biol.*, 2001, **5**, 447.
215. D. E. Bugay and W. P. Findlay, *Pharmaceutical Excipients: Characterization by IR, Raman, and NMR spectroscopy*, Marcel Dekker, New York, 1999.
216. K. Schmidt-Rohr and H. W. Spiess, *Multidimensional Solid-State NMR and Polymers*, Academic Press, London, 1994.
217. D. M. Sperger and E. J. Munson, *AAPS PharmSciTech*, 2011, **12**, 821.
218. D. M. Sperger, S. Fu, L. H. Block and E. J. Munson, *J. Pharm. Sci.*, 2011, **100**, 3441.
219. R. Rajala and E. Laine, *Thermochim. Acta*, 1995, **248**, 177.
220. P. Gao, *Pharm. Res.*, 1996, **13**, 1095.
221. C. Chang, L. E. Diaz, F. Morin and D. M. Grant, *Magn. Reson. Chem.*, 1986, **24**, 768.
222. P. J. Saindon, N. S. Cauchon, P. A. Sutton, C. J. Chang, G. E. Peck and S. R. Byrn, *Pharm. Res.*, 1993, **10**, 197.
223. S. Stephanie, Z. Fabio, V. Stephanie, D. Corinne and C. Stafano, *J. Pharm. Biomed. Anal.*, 2008, **47**, 683.
224. R. K Harris, P. Hodgkinson, T. Larsson and A. Muruganantham, *J. Pharm. Biomed. Anal.*, 2005, **38**, 858.
225. D. H. Zhou and C. M. Rienstra, *Angew. Chem. Int. Ed.*, 2008, **47**, 7328.
226. J. M. Griffin, D. R. Martin and S. P. Brown, *Angew. Chem. Int. Ed.*, 2007, **46**, 8036–8038.

227. J. W. Lubach, D. Xu, B. E. Segmuller and E. J. Munson, *J. Pharm. Sci.*, 2007, **96**, 777.
228. T. N. Pham, S. A. Watson, A. J. Edwards, M. Chavda, J. S. Clawson, M. Strohmeier and F. G. Vogt, *Mol. Pharmaceutics*, 2010, **7**, 1667.
229. J. R. Patel, R. A. Carlton, F. Yuniatine, T. E. Needham, L. Wu and F. G. Vogt, *J. Pharm. Sci.*, 2012, **101**, 641.
230. K. Anzai, H. Kono, J. Mizoguchi, T. Yanagi, F. Hirayama, H. Arima and K. Uekama, *Carbohydrate Res.*, 2006, **341**, 499.
231. F. G. Vogt and M. Strohmeier, *Mol. Pharmaceutics*, 2012, **9**, 3357.
232. *Structure Determination from Powder Diffraction Data*, International Union of Crystallography, Monographs on Crystallography, No. 13, ed. W. I. F. David, K. Shankland, L. M. McCusker and C. Baerlocher, Oxford University Press, New York, 2002.
233. F. Taulelle, *Solid State Sci.*, 2004, **6**, 1053.
234. J. K. Harper, D. H. Barich, E. M. Heider, D. M. Grant, R. R. Franke, J. H. Johnson, Y. Zhang, P. L. Lee, R. B. Von Dreele, B. Scott, D. Williams and G. B. Ansell, *Cryst. Growth Des.*, 2005, **5**, 1737.
235. D. A. Middleton, X. Peng, D. Saunders, K. Shankland, W. I. F. David and A. J. Markvardsen, *Chem. Commun.*, 2002, 1976.
236. B. Elena, G. Pintacuda, N. Mifsud and L. Emsley, *J. Am. Chem. Soc.*, 2006, **128**, 9555.
237. C. J. Pickard, E. Salager, G. Pintacuda, B. Elena and L. Emsley, *J. Am. Chem. Soc.*, 2007, **129**, 8932.
238. E. Salager, R. S. Stein, C. J. Pickard, B. Elena and L. Emsley, *Phys. Chem. Chem. Phys.*, 2009, **11**, 2610.

CHAPTER 3

High-Resolution NMR-Based Metabolic Profiling in Drug Discovery and Development

NIGEL J. WATERS

Epizyme, Inc., 400 Technology Square, Cambridge, MA, USA
Email: nwaters@epizyme.com

3.1 Background to Metabolic Profiling by NMR Spectroscopy

The application of NMR spectroscopy in the field of biology and medicine has grown consistently and steadily over the last 40 years or so, with the advancements in high field magnets, computational infrastructure and probe technology all aiding its progress to what is now a fundamental tool in the life sciences. Moreover, it has played an ever-increasing role in a plethora of applications within drug discovery and development including compound structure elucidation and chemical characterization, macromolecule secondary and tertiary structure determination, measurement of molecular dynamics and rates of turnover and magnetic resonance imaging (MRI). One area where NMR spectroscopy has really proved invaluable is that of metabolic profiling. The first reported application of NMR to metabolic profiling was in 1974, when Hoult *et al.* measured the complement of phosphorus-containing metabolites in isolated skeletal muscle by ^{31}P NMR spectroscopy.[1] This early work culminated in the emergence of a whole new field of metabolism science, which has in recent years been termed metabonomics or metabolomics, to describe the

New Developments in NMR No. 2
New Applications of NMR in Drug Discovery and Development
Edited by Leoncio Garrido and Nicolau Beckmann
© The Royal Society of Chemistry 2013
Published by the Royal Society of Chemistry, www.rsc.org

analogy with other 'omics platforms such as the megavariate nature of the approach and the chemometric tools used to process, analyze and interpret these large data sets. Metabonomics was originally defined by Prof. Jeremy Nicholson at Imperial College London and in the seminal paper is classified as "the quantitative measurement of the dynamic multiparametric metabolic response of living systems to pathophysiological stimuli or genetic modification".[2] Metabolomics was a term originally employed by investigators working with *in vitro* cell cultures,[3] and so is distinct from the whole animal systems that are typically studied by metabonomics. However, in recent years, the terminology has become somewhat interchangeable. Much of the early work in metabolic profiling utilized NMR spectroscopy and, even with the advent and advances in high resolution mass spectrometry, NMR spectroscopy continues to play a central role in metabolic science. This is exemplified in the rising number of publications on the topic through time (Figure 3.1).

Proton NMR spectroscopy has been the predominant methodology in metabonomics as it provides a reproducible, linear response with a high dynamic range. It is a universal detection tool (for any compound containing a proton) and is particularly useful in the identification of unknowns due to its non-biased nature with no *a priori* assumptions of what is being measured, unlike other analytical techniques relying on extinction coefficients and adequate ionization for detection. In addition, as a result of compounds being separated spectroscopically, minimal sample preparation is required. This aspect together with no need for upfront separation of analytes, has facilitated the measurement of

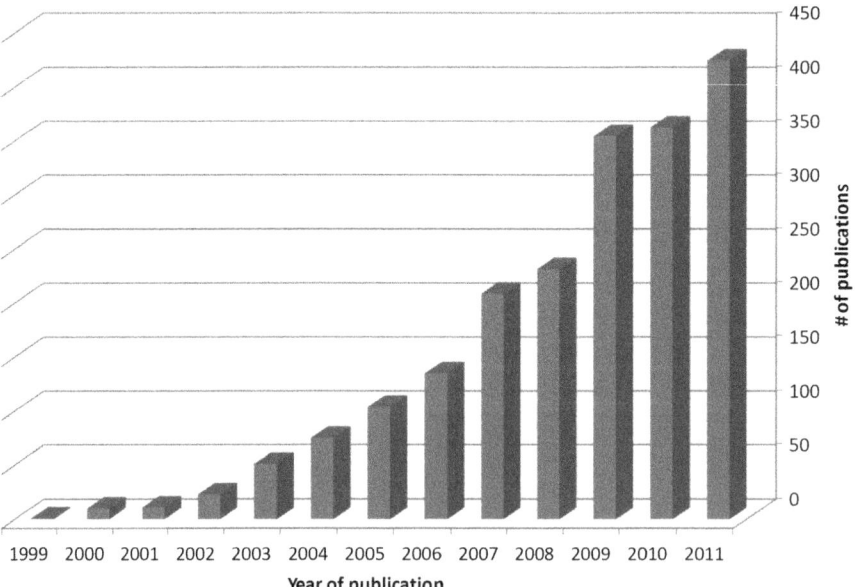

Figure 3.1   Number of publications on NMR-based metabonomics/metabolomics over time. Literature search was performed using terms NMR, metabonomics and metabolomics in Scifinder (web version) in January 2012.

the low-molecular-weight complement of intermediary metabolism in various biofluids, cell suspensions or tissue extracts by NMR spectroscopy. The range of biofluids studied by this approach has included blood plasma, erythrocyte suspensions, urine, seminal fluid, bile, cerebrospinal fluid, synovial fluid as well as others. Solid-state NMR spectroscopy has also provided the opportunity to acquire similar high-resolution spectra on isolated, intact tissue sections. For non-rigid solid materials or highly viscous liquids, high-resolution proton magic angle spinning (MAS) NMR spectroscopy offers an approach whereby some of the major line broadening contributions related to restricted molecular motion are eliminated or considerably reduced. High resolution ^{1}H NMR spectroscopic studies of excised tissues normally require relatively large amounts of tissue (>0.25 g), which must be extracted *via* protein precipitation methods to enable the collection of sharp line spectra. However, MAS-NMR of tissues is non-destructive, requires no sample preparation and additionally allows molecular compartmentation and dynamic interactions to be investigated.[4]

One-dimensional ^{1}H NMR spectra of biofluids acquired at 600 MHz or higher may contain several thousand signals and this inherent complexity would limit the information gain of the approach without further downstream spectral standardization, normalization and integration prior to a statistical analysis. And so in most cases a computer-based multivariate analysis of the spectra is performed. A single NMR spectrum can be thought of as an n-dimensional object with each dimension representing a single discrete segment of the overall spectral intensity distribution, *e.g.* for proton NMR data, spectra are segregated into bins of nominal width (usually 0.04 ppm) before integration. The resulting metabolic fingerprint (or row within a data matrix of spectral integrations) can then be mapped in the multivariate space relative to the positioning of other spectra acquired from control groups or other treatment groups. This allows two distinct levels of interpretation: (i) the first involves the metabolic fingerprints of the sample without necessarily identifying the specific components that are changing. This provides a numerical representation of the metabolic milieu and is amenable to multivariate statistical analysis and the development of classification models of responders and non-responders; (ii) the second, more powerful, strategy is the comprehensive and quantitative elucidation of metabolites changing in response to the stimuli under investigation. Statistical analyses are supervised or unsupervised in nature. Unsupervised techniques such as principal components analysis (PCA) allow visualization of spatial and temporal patterns in the data, looking for clustering of treatment groups or trajectories of response onset and recovery. Supervised techniques such as partial least squares discriminant analysis (PLS-DA) are able to maximize class separation and are supervised in the sense that class membership (*e.g.* vehicle control, treatment 1, treatment 2, *etc.*) is assigned prior to analysis. This approach requires additional validation steps to ensure models are robust. The application of chemometrics also aids xenobiotic metabolite identification by comparison of treated animals with vehicle controls or pre-dose data. More detail on these pattern recognition tools is nicely described in recent review articles.[5,6]

3.2   Advancements in NMR Spectroscopy

The scientific contributions afforded by investigations utilizing NMR-based metabolic profiling have really been borne out of the technological advances that have been made over the last few decades. These have included improvements in field strength, probe technology, pulse sequences and two-dimensional experiments, advances in hyphenated techniques, automation and computing power.

Most recent metabolic profiling studies have made use of NMR spectrometers operating at observation frequencies of 600 MHz or above, although some have been performed at lower observation frequencies such as 400 or 500 MHz and, as such, have a corresponding lower spectral dispersion and sensitivity. The availability of higher field magnets has allowed some investigations utilizing higher observation frequencies including 750, 800 and 900 MHz. However, technology is somewhat limiting in some cases since many important biochemical components at the sub-nanomole level remain undetectable in biofluids *e.g.* hormones, prostaglandins, *etc.*

Sensitivity improvements have also been facilitated with the introduction of cryogenic NMR probes where the NMR detector coil and preamplifier are cooled to approximately 20 K in order to decrease the thermal electronic noise (see Chapter 5 for the use of cryogenic probes in imaging applications). This leads to markedly greater signal-to-noise ratios (as much as 5-fold) or time savings in the order of 25-fold. This time saving is especially pertinent for 2D NMR experiments and the application of natural abundance ^{13}C NMR spectroscopy. A further development has been the commercial availability of miniaturized NMR probes permitting the measurement of samples in the low microliter range.[7] The high natural abundance and the ubiquity of the proton in most biomolecules have put ^{1}H NMR spectroscopy at the forefront of metabolic profiling. That said, other nuclei such as ^{13}C, ^{15}N and ^{31}P have been measured in these types of metabolomic studies, typically as a secondary measurement to the proton. ^{19}F NMR has also shown utility for studies of fluorinated drugs or xenobiotics, as resonances are usually well resolved as a consequence of the large chemical shift range for this nucleus, as well as the low background and high specificity relative to proton NMR spectroscopy. Although the downside with ^{19}F is that little structural information is obtained and thus further characterization with measurement of other spin nuclei is required.

One of the great advantages of NMR spectroscopy is not only the ability to obtain detailed information on molecular structure but also to be able to probe the dynamics, compartmentation and mobility of metabolites of interest through the interpretation of NMR spin relaxation times and measurement of molecular diffusion coefficients. Despite the complexity of high-resolution 1D ^{1}H NMR spectra of biofluids, identification of metabolites is possible based on assignment by chemical shift or signal multiplicity, or by adding an authentic standard and re-acquiring the spectrum. NMR pulse sequence techniques have enabled the suppression of solvent signals present in the sample. In the case of

^{1}H NMR spectroscopy of biofluids, this has meant suppression of the large water resonance that would otherwise dominate the proton NMR spectra. In addition, complex NMR spectra of biofluids can be simplified by attenuating resonances from macromolecules or other species with short spin-spin relaxation times by utilizing a spin-echo editing method such as the Carr–Purcell–Meiboom–Gill (CPMG) pulse sequence. Conversely, resonances from rapidly diffusing species can be attenuated through diffusion edited spectra by application of a longitudinal-eddy-current-delay pulse sequence, for example. In cases where there is much overlap in signal intensities, 2D NMR spectroscopy becomes invaluable. Confirmation of spectral assignments can be facilitated with ^{1}H–^{1}H COSY and the ^{1}H–^{1}H total correlation spectrum (TOCSY) as well as inverse-detected heteronuclear correlation methods such as HMQC and HSQC. The 2D J-resolved pulse sequence allows the coupling pattern to be plotted in the second dimension, and can help deconvolute the multitude of splittings in a standard biofluid 1D NMR spectrum. In addition, macromolecule resonances are attenuated as effectively as with spin-echo experiments. Further simplification of other 2D experiments can be achieved with spin-echo methods, T_1 and/or T_2 relaxation editing, diffusion editing or multiple quantum filtering. The multivariate analysis of 2D J-resolved spectra, retaining both chemical shift and J-coupling information, has been reported recently using parallel factor analysis.[8] The multivariate analysis of proton NMR spectra has also extended to the statistical total correlation spectroscopy (STOCSY) principle, which enhances the information recovery from complex biological mixtures. Based on the multiple colinearity of the range of spectral intensities in a ^{1}H NMR spectrum, a pseudo-2D spectrum is produced that highlights the correlation between various peaks. In this way, identification of peaks arising from the same molecule will show strong correlation and, in addition, connectivities between different molecules in the same metabolic pathway can show positive or negative correlation. The STOCSY spectral analysis can also be combined with supervised chemometric methods in order to identify the molecular basis for any observed metabolic variation. This approach was first demonstrated on a study of insulin resistance in which the metabolite 3-hydroxyphenylpropionic acid was identified,[9] and is illustrated in Figure 3.2.

In light of the complexity of biofluids, metabolite resonances can be attenuated by protein binding or metal binding with a consequential broadening of proton resonances. This can be addressed through sample preparation procedures such as ultrafiltration to remove macromolecular contributions. The presence of macromolecules in some biofluid samples has been used in a diagnostic sense, for example, by allowing the measurement of enzyme activity. Alanine aminotransferase (ALT) is known to be elevated in blood plasma following liver or kidney damage as a result of leakage from injured cells. The addition of deuterated water to such plasma samples (especially if lyophilized and reconstituted in D_2O) leads to incorporation of deuterium at the alpha-methine position of alanine. Due to a loss of coupling, this is observed as the collapse of the alanine methyl doublet to a singlet with a

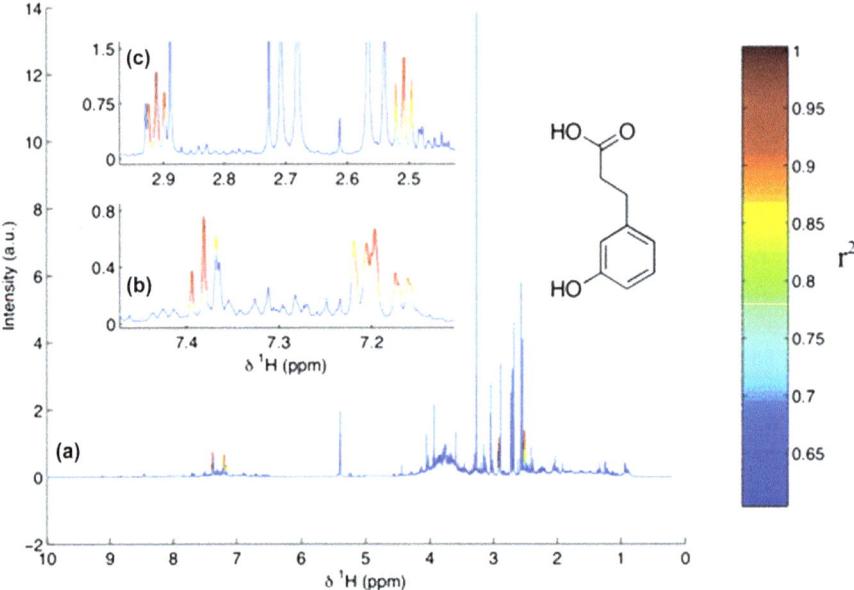

Figure 3.2 One-dimensional STOCSY analysis for the selected variable corresponding to delta 2.512 ppm. The degree of correlation across the
spectrum has been color coded and projected on the spectrum that has the
maximum for this variable. (a) Full spectrum; (b) same spectrum between
7.1 and 7.5 ppm; (c) same spectrum between 2.4 and 3 ppm.
Reproduced with permission from Cloarec *et al.*[9] © 2005 American
Chemical Society.

small deuterium isotope shift.[10] Similar observations have been made in intact
tissues by ^{1}H MAS-NMR spectroscopy, where activity of ALT, lactate dehydrogenase and 3 glycolytic enzymes (fructose bisphosphate aldolase, triose
phosphate isomerase and glyceraldehyde phosphate dehydrogenase) were
implicated.[11] Use of this isotope exchange concept has allowed the
measurement of metabolic flux with one of the first examples being the addition
of stable isotope labeled compounds to suspensions of erythrocytes and
monitoring for the subsequent fate of the label with time.[12] Use of tracer
metabolites enriched in ^{13}C or ^{15}N is essential in mapping specific molecules in
biosynthetic pathways and in studying metabolic flux. This has been demonstrated using [U–[13]C] glucose and [U–[13]C, ^{15}N] glutamine as source tracers in
human lung adenocarcinoma A549 cells cultured *in vitro*.[13] Metabolic reactions
have also been studied in the absence of stable isotope labeling such as the
hydrolysis of phosphoryl choline and nucleotides in seminal fluids.[14]

A number of biofluids contain metabolites capable of metal chelation
including free amino acids (*e.g.* glutamine) and organic acids (*e.g.* citrate). The
main endogenous metal ions, Ca^{2+}, Mg^{2+} and Zn^{2+}, are involved in
complexation reactions, which can be monitored by NMR spectroscopy.
Moreover, metal chelating agents such as ethylenediaminetetraacetic acid

(EDTA) can be added to biofluids with consequent changes in the NMR resonances. EDTA can also lead to a sharpening of NMR signals if it complexes with any paramagnetic ions that are present.

Furthermore, with some biofluids containing cells or intact membrane fractions, the addition of paramagnetic species has enabled the study of membrane transport of small endogenous species. Paramagnetic species that are not membrane permeable, such as ferric desferrioxamine and copper-cyclohexane-diaminetetraacetic acid, selectively broaden resonances from the extracellular population of the metabolite of interest allowing the intracellular resonance to be monitored exclusively over time. The application of MAS-NMR spectroscopy (see discussion below) has also allowed the study of molecular compartmentation *e.g.* observation of intracellular and extracellular water resonances in erythrocyte suspensions.[15,16] The line broadening of polar species brought about by the shortening of the T_2 relaxation time induced by binding of paramagnetic ions has also been used to assist in the attenuation of the water resonance, the broadened water resonance being eliminated by the application of a spin-echo pulse sequence. Another effective method of reducing the spectral contribution of water in biofluid NMR is that of freeze-drying and reconstitution in the desired solvent. Reconstitution in D_2O can lead to selective deuteration of exchangeable protons *e.g.* NH, OH and SH as well as certain methine or methylene protons such as the CH_2 group of acetoacetate, which undergoes keto-enol tautomerism. The use of aprotic solvents such as d_6-DMSO enables the study of protons normally broadened by chemical exchange in aqueous solutions or those overlaying with the water resonance. One important consideration is the choice of solvent given the solubility of the metabolites of interest. Lyophilization also has implications on molecular mobility for biofluids such as bile wherein the micellar compart-mentation is disrupted, and in addition leads to the loss of volatile components, *e.g.* acetone. Certain biofluids, such as urine, require buffering of the sample pH to *ca.* 7 in order to stabilize pH drift over time. However, relatively few metabolites show major chemical shift variations over the normal pH range with the exception of histidine and citrate. In some cases, acidification of samples is utilized to enhance signals arising from species such as histidine, tyrosine and phenylalanine. In addition to pH modification, variable temperature ^1H NMR studies have been applied to mapping the phase diagram of the cholesteryl esters abundant in bile.

Hyphenation of liquid chromatography (LC) with NMR spectroscopy has led to efficient structure elucidation of both xenobiotic and endogenous metabolites. The ability to detect signals from analytes of interest in the presence of excessive ^1H NMR signals originating from the high-performance liquid chromatography (HPLC) solvents has been facilitated with the use of solvent suppression methods or deuterated solvents. This approach is not typically applied to investigations of global metabolic profiles but rather to facilitate the identification of specific metabolites.

The development of high-resolution magic angle spinning (HR-MAS) proton NMR spectroscopy has proved a beneficial adjunct to conventional biofluid ^1H

NMR spectroscopy, enabling the acquisition of high-quality spectra on small sections of untreated intact tissues.[4] The approach has been applied to a whole host of intact biological matrices including bone, brain, cartilage, gut, kidney, liver, muscle, ocular tissue, prostate, skin, testes, tumor sections and vascular tissue as well as various cells and cell lines. An example of the spectral information that can be obtained is illustrated in Figure 3.3 relative to aqueous and lipophilic tissue extracts. The technique requires rapid spinning of the

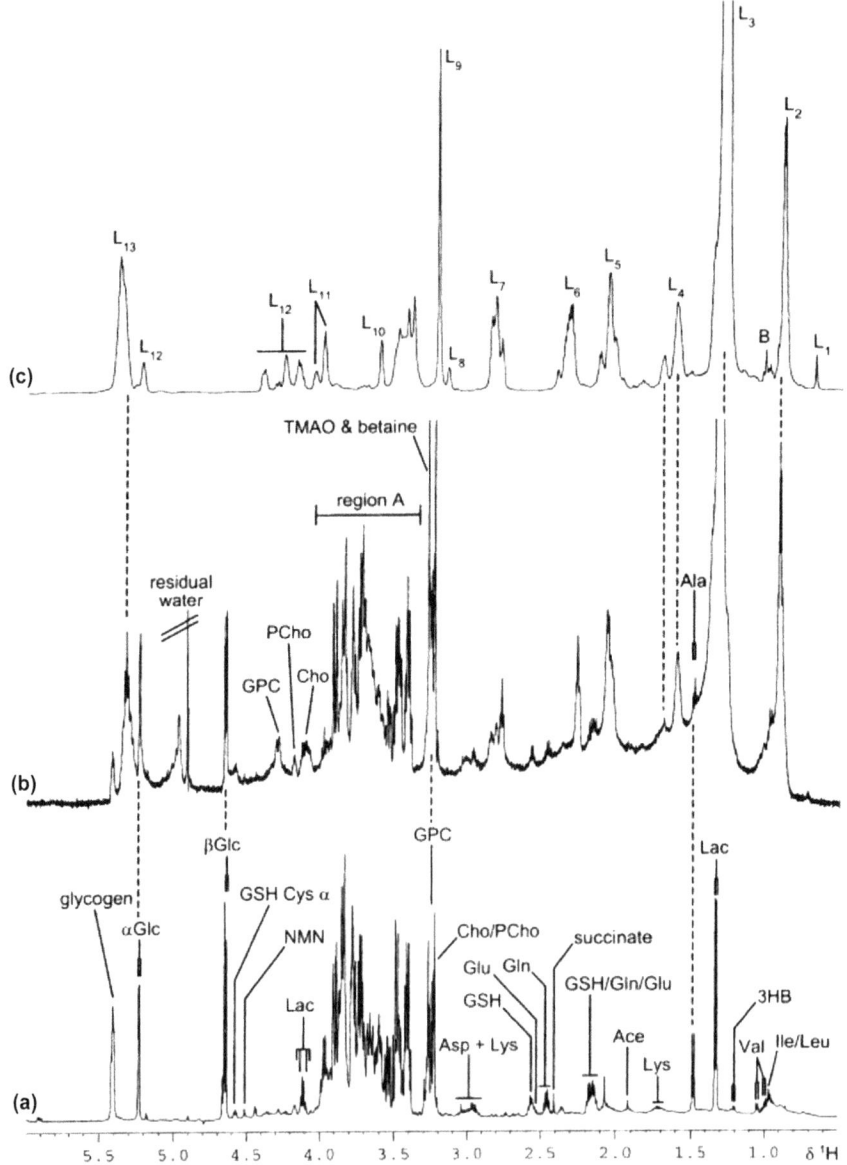

sample at an angle of 54.7° (the so-called magic angle) relative to the magnetic field, which precludes the line broadening effects typically observed for non-liquid samples.[17,18] This line broadening is caused by sample heterogeneity and residual anisotropic behavior that is usually averaged out with the isotropic and rapid tumbling of molecules free in solution. The chemical shift anisotropy is relatively small for protons and so the major causes of line broadening in ^{1}H NMR spectra are a consequence of [1]H–[1]H dipolar couplings and differential magnetic susceptibility effects across a sample. A spinning rate of only a few hundred hertz is sufficient to reduce the line broadening effects but would lead to spinning side bands within the ppm range of interest and so in order to negate this issue rotations that place the spinning side bands outside the spectral window are typically employed, for example, sample spinning at 6000 Hz on a 600 MHz instrument. The stability and integrity of samples spun at these frequencies has been addressed.[11,19,20] The sample preparation is relatively simple with tissue sections being thawed, cut to size, weighed and inserted into a zirconia rotor with a small volume of D_2O to provide the NMR lock signal. Chemical shift references such as TSP are typically not utilized because of sequestration and compartmentation effects and so internal references such as the anomeric proton of alpha-glucose at δ 5.23 or the methyl signal of lactate at δ 1.33 are used. The experimental setup is straightforward with the MAS probe coupled to a pneumatic unit to control sample spinning and with the potential to apply many of the pulse sequences and 2D experiments normally used in solution-state NMR-based metabolic profiling. The enhanced engineering of cryogenic MAS probes could permit the sensitivity improvements that have already been afforded in conventional biofluid NMR spectroscopy. Along similar lines, the sensitivity and resolution capabilities of a novel magic angle coil spinning (MACS) microcoil have been explored recently, offering the opportunity to acquire high-resolution spectra on nanoscale samples such as needle biopsies from clinical studies or small volume tissues or cells.[21] Detection using a microcoil, commonly applied in

---

Figure 3.3   Series of 600 MHz single pulse ^{1}H NMR spectra (δ 0.5–6.0) of control liver taken from one individual: (a) liver tissue aqueous extract, (b) single pulse ^{1}H MAS-NMR spectrum of intact control liver after Lorentzian–Gaussian function and (c) liver tissue chloroform/methanol extract. Key: 3 HB, 3-β-hydroxybutyrate; Ace, acetate; Ala, alanine; Asp, aspartate; B, C21 bile acid methyl; Cho, choline; Glc, glucose; Gln, glutamine; Glu, glutamate; GPC, glycerophosphorylcholine; GSH, glutathione; Ile, isoleucine; L1 C18/C19 cholesteryl methyls; L2, triglyceride terminal methyls; L3, lipid $(CH_2)n$; L4, lipid $CH_2{}^*CH_2CO$; L5, lipid $CH_2C{=}C$; L6, lipid CH_2CO; L7, lipid $C{=}CCH_2C{=}C$; L8, lipid CH_2NH^{3+}; L9, lipid $N^+(CH_3)_3$; L10, lipid $(CH_3)3N+CH_2{}^*$; L11, lipid CH_2OPO^{2-}; L12, lipid CH_2OCOR; L13, lipid $CH{=}CH$; Lac, lactate; Leu, leucine; Lys, lysine; NMN, *N*-methyl nicotinamide; PCho, phosphocholine; glycogen; TMAO, trimethylamine-*N*-oxide; Val, valine. $^*\alpha CH$ resonances of amino acids.
Reproduced with permission from Waters *et al.*, *Biochem. Pharmacol.*, 2002, **64**, 67–77. © 2002 Elsevier Science Publishers.

liquid state NMR spectroscopy but able to rotate simultaneously with the sample at the magic angle, enables the collection of sharp line spectra on solid and semi-solid samples in the nanoliter range.

It is clear that HR-MAS has an important role in metabolic profiling, (i) providing biochemical information on tissue sections and biopsies normally only examined for histopathology, (ii) enabling dynamic measurements on intact tissues whilst maintaining tissue integrity unlike NMR analysis of tissue extracts and (iii) more closely resembling the spectra obtained from *in vivo* magnetic resonance spectroscopy (MRS) with much higher resolution and sensitivity.

Ultimately there are several attributes that greatly impart ^{1}H NMR spectroscopy to the study of metabolic science and they include: (a) minimal and simple sample preparation; (b) no *a priori* bias in metabolite detection such that all metabolites containing non-exchangeable protons (above a threshold concentration) are observed in a single analysis; (c) analysis is high-throughput with no prerequisite for chromatographic separation; (d) measurements have a high degree of reproducibility; and (e) it is non-destructive enabling repeat or alternative measurements on the same sample.

3.3   Applications in Drug Discovery and Development

The ability of NMR spectroscopy to record a diverse range of complex metabolite information with no *a priori* selection or bias to specific sub-classes of metabolites has rendered the technique extremely powerful when coupled with multivariate statistical analysis. For this reason, it has afforded complementary biochemical information to that derived from the allied fields of genomics, transcriptomics and proteomics. This data-rich paradigm has led to many studies that have successfully integrated the various 'omic endpoints into a more complete picture of the biochemical processes under study. Much of the early work in the field of NMR-based metabolic profiling was largely focused on drug-induced toxicities in terms of both identifying and characterizing biomarkers as well as gaining understanding of the underlying mechanisms. This also led to large studies designed to characterize the metabolic variance in control populations. More recently, the approach has received a lot of attention in disease etiology and pathophysiology, both pre-clinically and clinically. It is also increasingly playing a role in drug metabolism and pharmacokinetics, beyond the traditional application of NMR spectroscopy in structure elucidation of drug metabolites. The breadth of application across various drug discovery and development disciplines is summarized in Figure 3.4. The following section is intended to illustrate the breadth of application and hence is not an exhaustive list of all published studies.

3.3.1   Understanding of Disease Etiology in Pre-Clinical Models

3.3.1.1   *Immunology*

There have been a number of metabolic profiling studies aimed at characterizing the metabolic phenotype in animal models of various immune-mediated diseases.

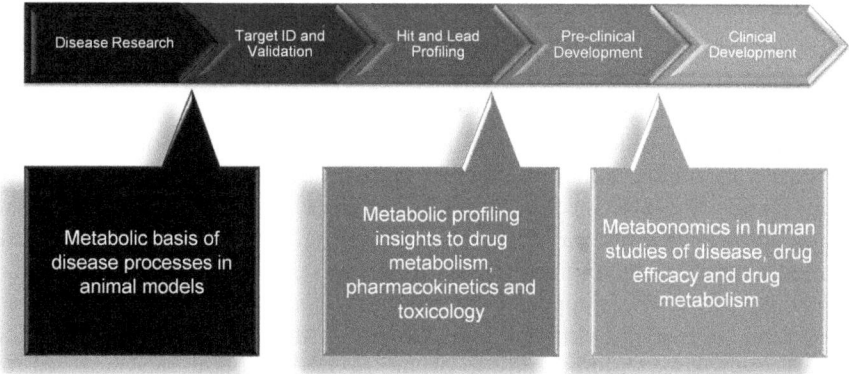

Figure 3.4 Impact of NMR-based metabolic profiling across the drug discovery and development axis.

Smolinska *et al.* used high field ^{1}H NMR with cryoprobe technology to acquire high-resolution spectra on very low volumes of rat cerebrospinal fluid (CSF) (*ca.* 10 µL).[22] The metabolic profiles derived from the two models of either peripheral inflammation (induced by complete Freund adjuvant (CFA) emulsion) or neurological inflammation were compared using Partial Least Squares Discriminant Analysis (PLS-DA) and ANOVA Principal Component Analysis (ANOVA-PCA). Neurological inflammation was manifest in the experimental autoimmune/allergic encephalomyelitis (EAE) model, induced by CFA and myelin basic protein (MBP), which has become an important tool in the understanding of human disease given the resemblance in clinical expression and pathology to that of multiple sclerosis, although the neurodegeneration and demyelination processes are not mimicked in this model. A key metabolite signature differentiating peripheral and neural inflammation was observed and included primarily increases in choline, *N*-acetyl aspartate and creatine, all of which are critical intermediary metabolites in neural function. The model derived from this analysis was then applied prospectively to predict the disease progression in an independent cohort of EAE animals, illustrating the power of coupling NMR-based metabolic profiling with multivariate statistical analysis.

Baur and co-workers used a similar approach using the TNF$^{\Delta ARE}$/WT mouse model of Crohn's disease-like ileitis.[23] The mice developed a similar phenotype to the human disease with ileal histological abnormalities, reduced fat mass and body weight, as well as signs of malabsorption with higher energy wasting. The underlying biochemical processes involved in the onset and progression of inflammatory bowel disease (IBD) are somewhat poorly characterized in terms of both systemic and gastrointestinal metabolism. And therefore, this investigation aimed at better understanding the pathogenesis of IBD using a combination of proton NMR spectroscopy and liquid chromatography-mass spectrometry (LC-MS) approaches to analyze tissue extracts of various discrete sections of the GI tract. Various 2D NMR experiments were performed such as JRES, TOCSY and HSQC to aid in metabolite identification, leading to the elucidation of a metabolic phenotype involving significantly altered metabolism of cholesterol, triglycerides, phospholipids, plasmalogens and sphingomyelins.

Kominsky and colleagues, using both *in vitro* and *in vivo* models, explored the global metabolic signature of mucosal inflammation.[24] A combination of ^{1}H, ^{13}C and ^{31}P NMR spectroscopy was applied to colonic tissue from murine colitis as well as epithelial cell lines. Interestingly, these model systems all showed alterations in specific metabolites associated with cellular methylation reactions, which was supported by microarray analysis showing increased expression of S-adenosylmethionine synthetase and S-adenosylhomocysteine (SAH) hydrolase. Administration of an SAH hydrolase inhibitor led to exacerbation of colitis whilst folate supplementation promoted methylation and partially alleviated the disease severity. These results identified the importance of methylation in inflammatory processes and the protective role it has.

Rheumatoid arthritis has also been studied using the K/BxN transgenic mouse model.[25] A proton NMR spectroscopy approach was developed based on an analysis of sera from both arthritic K/BxN mice and a control group with the same genetic background but lacking the arthritogenic T-cell receptor KRN transgene. Ultrafiltration was employed as a sample preparation step to remove the proteinaceous contributions before quantitative "targeted profiling" of metabolites, which was followed by supervised Orthogonal Partial Least Squares Discriminant Analysis (O-PLS-DA) pattern recognition analysis and metabolic network analysis to aid in the interpretation of results. Spectral features associated with nucleic acid, amino acid and fatty acid metabolism, lipolysis, reactive oxygen species generation and methylation were all observed, indicative of a shift from metabolites involved in numerous reactions (so-called hub metabolites) towards metabolic endpoints associated with arthritis. The authors argue that these downstream variations in metabolites are directly reflective of the organism phenotype relative to genomic, transcriptomic or proteomic endpoints. In addition, metabolic pathways are typically highly conserved across species and so there is likely merit in the translation of disease biomarkers from the pre-clinical model system to humans.

3.3.1.2   Infection

The metabolic effects of the host–parasite interaction in infectious diseases has received much attention by NMR-based metabonomics of various pre-clinical model systems. The identification of biomarkers that have a diagnostic or prognostic application is highly sought, and has included studies of the impact of trematode Schistosoma infection in mouse[26] and Syrian hamster[27] and Trypanosoma infection in mouse.[28]

Utilizing a combination of high-resolution ^{1}H MAS-NMR spectroscopy of excised tissue sections and conventional biofluid NMR, the panorganismal effects of parasitic infection have been explored for the experimental infection of female NMRI mice with the intestinal fluke *Echinostoma caproni*.[29] Tissues including brain, colon, ileum, jejunum, kidney, liver and spleen were analyzed together with urine, blood plasma and fecal water, and metabolic profiles compared to an uninfected control group, using chemometric approaches. NMR spectral methods such as the CPMG pulse sequence and diffusion editing were employed. Metabolic changes in tissues were limited to the liver, renal cortex and intestinal sections, with the primary observation of malabsorption in the small intestine as evidenced by lowered levels of various amino acids in the ileum. Study of the metabolic covariates showed plasma and urine most closely reflected changes occurring in the ileum.

A similar NMR-based metabonomics approach has been applied to the characterization of *Plasmodium berghei* infection in the mouse.[30] Metabolic profiles from urine and plasma were compared with pre-infection samples from the same animals, with a marked progression observed over time. Levels of plasma lactate and pyruvate were concomitant with lowered plasma glucose, creatine and glycerophosphoryl choline relative to pre-infection, indicating a glycolytic upregulation. In the urine, elevations in phenylacetylglycine and dimethylamine were coexistent with decreased taurine and trimethylamine-*N*-oxide, a pattern suggesting perturbations in gut microbiota. Interestingly, pipecolic acid was observed in urine with the structural identity confirmed by 2D techniques (COSY and TOCSY). The origin and role of pipecolic acid in Plasmodium infection remains unclear, but warrants further study and may provide a useful diagnostic for malaria in the future.

The interaction of gut microflora and mammalian host has received much attention recently, particularly with respect to metabolism and health consequences. The impact of oral antibiotic treatment on gut microbiota and the subsequent recolonization following drug withdrawal has been assessed by ^{1}H NMR urinalysis.[31] The gut microbial composition and host metabolic phenotype was assessed in male Han Wistar rats relative to matched controls, with two treatment groups of penicillin and streptomycin sulfate exposure over 8 days or over 4 days. Bacterial suppression, as measured by fluorescence *in situ* hybridization, led to a reduction in the urinary excretion of the mammalian-microbial cometabolites hippurate, phenylpropionic acid, phenylacetylglycine and indoxyl-sulfate, with a rise in urinary taurine, glycine, citrate, 2-oxoglutarate and fumarate. The group that received antibiotic treatment over only 4 days started to become recolonized with two cage-dependent subgroups; *Lactobacillus/Enterococcus* counts were dominant in one of these subgroups. This dichotomy was manifested in the metabolic phenotypes with subgroup differentiation based on tricarboxylic acid cycle intermediates and indoxyl-sulfate excretion.

3.3.1.3   Neurology

In the field of neurological diseases several pre-clinical animal models have been studied by NMR-based metabolomics. For example, Huntington's disease is a hereditary disorder for which the exact mechanisms of pathogenesis remain unclear. Using the presymptomatic Huntington's disease transgenic rat,

Verwaest and colleagues were able to differentiate the transgenic group from the wild-type control group based on the proton NMR spectra of serum and CSF.[32] Even at the presymptomatic stage, mitochondrial dysfunction was implicated as a result of perturbations in serum *N*-acetyl aspartate (NAA), glutamine, succinate, glucose and lactate, with the latter two also increased in CSF. This metabolite profile was indicative of a shutdown in neuronal-glial glutamate-glutamine cycling and impairment of mitochondrial energy production. A comparable approach was applied to the study of traumatic brain injury in rats.[33] Although no discernible changes were seen in blood plasma, there were changes in brain metabolism identified from tissue extracts of brain regions including hippocampus and cortex. Evidence of oxidative stress (ascorbate), excitotoxic damage (glutamate), membrane disruption (phosphorylated choline derivatives) and neuronal injury (NAA) were all observed.

In addition to the study of animal models, the combined application of nanoprobe NMR technology and *in vivo* microdialysis has enabled the study of neurochemistry in defined regions of the brain of freely moving rats.[34,35] This setup facilitates close to real time measurements on low volumes of CSF (20 µL), which equates to detection on the nanogram level. The impact of stimulatory or drug-induced changes in extracellular metabolites was nicely demonstrated using the neurotoxin, tetrodotoxin.

3.3.1.4   Oncology

The search for novel cancer biomarkers has been facilitated by the global molecular profiling approaches of genomics, transcriptomics, proteomics and metabolomics. In the latter case, this has incorporated the use of high-resolution NMR spectroscopy to a variety of *in vitro* and *in vivo* biomatrices.

The metabolic phenotype of the immortalized, non-tumorigenic prostate epithelial cell line, RWPE-1, has been compared with two tumorigenic sub-lines of increasing malignancy, WPE1-NB14 and WPE1-NB11.[36] Employing NMR spectroscopic analysis of cell media and aqueous extracts of cell pellets allowed measurement of the intracellular and extracellular metabolic profiles in the different cell lines. The malignant cell line showed aberrant intracellular levels of various metabolites including choline and branched-chain amino acids. Also changes in intracellular glycine and lactate and extracellular lactate and alanine were concomitant. These pathways have also been shown to be altered in human prostate cancer and warrant further study as a source of valuable biomarkers.

In another *in vitro* investigation, the biomarker profile was monitored in response to drug treatment. The process of O-linked β-*N*-acetylglucosamine glycosylation is important in a number of biological processes including tumor metastasis, where uridine diphospho-*N*-acetylglucosamine (UDP-GlcNAc) is the donor molecule. This GlcNAc donating species was measured by high-resolution ^{1}H MAS-NMR spectroscopy in intact cells from four brain cancer lines treated with and without cisplatin.[37] In this study cisplatin responding cells showed increasing levels of UDP-GlcNAc with increasing exposures of cisplatin prior to the microscopic signs of cell death. Changes in UDP-GlcNAc were not detected in those cells unresponsive to cisplatin. Therefore, a mechanistic link between glycosylation status and cancer cell death following chemotherapeutic treatment was established.

Metabolic profiling by high-resolution MAS-NMR and solution state NMR spectroscopy has also been applied to transgenic animal models of tumorigenesis as well as to xenograft models. Backshall and co-workers characterized the metabolic profile of intestinal tissues from the ApcMin/+ mouse model of early gastrointestinal tumorigenesis.[38] In particular, an increase in the phospho-choline/glycerophosphocholine (PC/GPC) ratio was observed relative to wild-type, consistent with previous reports of malignant transformation in cells and with the role of Apc as a tumor suppressor. Rantalainen and colleagues were able to integrate findings from proteomic and metabonomic approaches in their studies on a human tumor xenograft mouse model.[39] Multiple associations between metabolites and proteins in blood plasma were obtained *e.g.* sero-transferrin precursor and tyrosine and 3-D-hydroxybutyrate. A combination of *in vivo* MRS and *ex vivo* high-resolution MAS-NMR techniques have been utilized in studying transgenic mice over expressing smoothened (SMO) receptor (which exhibit a high incidence of medulloblastomas).[40] The metabolic phenotype reported as perturbations in taurine, glycine and choline-containing metabolites closely mirrored that observed in human medulloblastomas, confirming that SMO mice are a realistic model for investigations into the metabolic basis of this disease. Analogous model validation has been undertaken in mice bearing xenografts of human glioblastoma multiforme cells.[41]

3.3.1.5   Endocrinology

The use of ^{1}H NMR-based metabolic profiling lends itself well to the study of endocrine dysfunction since a large complement of the intermediary metabolism from tissues such as liver and pancreas is observable in systemic biofluids such as blood plasma or excretory matrices such as urine. This has resulted in a number of studies looking at the various metabolic aberrations induced by diabetes or dyslipidemias. Cobbold and colleagues used MAS-NMR spectroscopy to profile the liver metabolism of three mouse strains fed either control or high-fat diets.[42] A spectrum of non-alcoholic fatty liver disease (NAFLD) was observed from normal to steatohepatitis, which was concomitant with increasing total lipid to water ratio, a decrease in poly-unsaturation indices and a decrease in total choline. In a similar study, Dumas and co-workers utilized mouse as a model of dietary-induced impaired glucose homeostasis and NAFLD, with NMR-based metabonomics of plasma and urine.[43] They were able to show that this syndrome was associated with perturbed choline metabolism, such that low circulating plasma phos-phatidylcholine and high urinary excretion of methylamines were observed. This pattern was ascribed to the role of gut microbiota reducing the systemic availability of choline and hence mimicking a choline-deficient diet causing NAFLD. Further insights in this field have been brought about with studies in transgenic animals. Cheng *et al.* studied the low density lipoprotein (LDL) receptor null mouse using NMR-based metabolic profiling of plasma and urine following administration of control or high-fat diets.[44] These animals developed hypercholesterolemia and atherosclerosis when fed a high-fat diet and again choline metabolism was implicated, particularly the choline oxidation pathway. These findings were consistent with those observed in the apolipoprotein E (ApoE) knock-out mouse, and as such provide means to monitor atherosclerosis in various animal models. The metabolic signature of streptozotocin-induced diabetic nephropathy (DN) has been acquired by Zhao and colleagues who characterized the metabolic changes in urine and kidney tissue extracts in 2-week and 8-week DN rats.[45] Lipid and ketone body synthesis pathways were enhanced whilst glycolysis and tricarboxylic acid cycle pathways were decreased in DN rats compared to controls. This animal model was also studied by Maher *et al.*, however in this case the STOCSY concept was utilized to scale the NMR data and enhance latent biomarker recovery.[46] This scaling approach, referred to as STOCSYS (statistical total correlation spectroscopy scaling), takes advantage of the fixed proportionality in a set of NMR spectra to suppress or augment spectral features correlated with a resonance of interest. In the study by Maher *et al.*, the dominant and interfering signals from glucose and lactate were dampened to allow recovery of otherwise obscured markers such as glycine.

In addition to the characterization of disease states and progression in various animal models, another important development in the application of NMR-based metabolic profiling has been in the assessment of drug efficacy. This was nicely demonstrated by Connor and co-workers using ^{1}H NMR urinalysis to monitor non-invasively the efficacy of an orally-active anti-hypoglycaemic agent, BRL49653, in a genetic mouse model of diabetes, C57BL/KsJ db/db.[47] Dosing of BRL49653 at 3 µmol kg^{-1} for 24 days significantly reduced urinary excretion of sugars, acetate, lactate and ketone bodies, whilst reducing blood glucose concentrations to control levels.

3.3.1.6 *Others*

The application of NMR-based metabolic profiling to pre-clinical models of disease has also shown utility in other areas of relevance to drug research, including aging and senescence. The serum metabolic signature of the SAMP8 mouse, a senescence-prone model of age-related learning and memory deficits as well as Alzheimer's disease, was shown to be distinct from control animal cohorts with the nucleoside, inosine, representing an important metabolite in this distinction.[48] Other differential metabolites included polyunsaturated fatty acid (PUFA), choline species, the lipoprotein profile, as well as glucose and ketone bodies. Conversely, the plasma metabolic phenotype has also been explored in three other aging-related murine models of longevity including 30% dietary restriction, insulin receptor substrate 1 null and Ames dwarf.[49] In agreement with work on the SAMP8 mouse model, modulations in the choline species and plasma lipoproteins were observed in all three models. However, the directionality of the lipoprotein profile did differ between the dietary restriction, insulin receptor substrate 1 null and Ames dwarf models suggesting the metabolic basis for longevity is a complex picture. Representative NMR spectra of plasma from these studies are shown in Figure 3.5, nicely illustrating

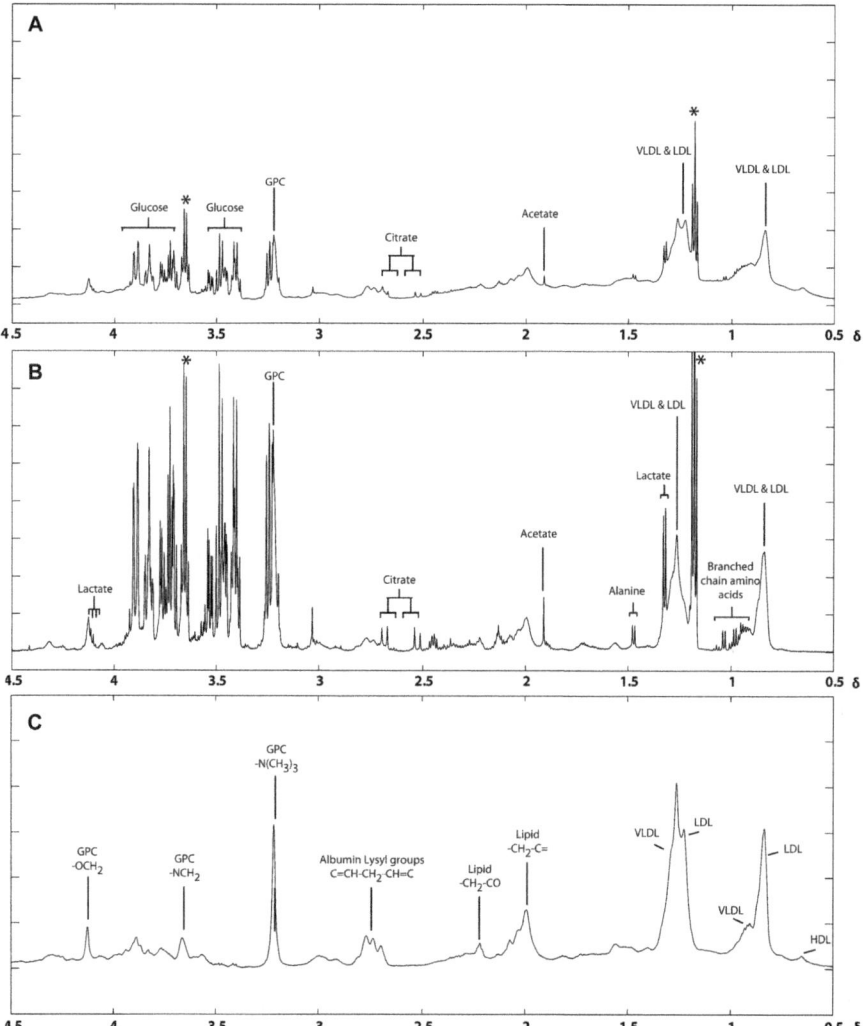

Figure 3.5 Aliphatic region (δ^1H 0.5–4.5) of ^{1}H-NMR plasma spectra. Acquired using (A) standard 1D pulse sequence with pre-saturation suppression of the water peak, (B) the Carr–Purcell–Meiboom–Gill sequence to attenuate broad signals from proteins and lipoproteins that may overlap signals from low-molecular-weight metabolites, also using pre-saturation, and (C) a diffusion edited pulse sequence to analyze high-molecular-weight molecules such as lipids and proteins. *Ethanol contaminant.
Reproduced with permission from Wijeyesekera *et al.*[49] © 2012 American Chemical Society.

the application of the CPMG pulse sequence and diffusion editing to focus on the low- and higher-molecular-weight complement in the samples, respectively.

3.3.2 Toxicology and Safety Assessment

Metabonomics made significant progress in the 1990s and early 2000s in the field of toxicology (see Chapter 13 for applications of *in vivo* NMR techniques to toxicology), building a platform for what has subsequently led to application in the other areas of biomedical research. The fundamental work in the metabonomics area involved metabolic characterization of a number of toxicities induced by a range of prototypical toxicants.[50–54] Many of these agents had previously been part of the groundbreaking work at NIH in the 1950s led by Bernard Brodie, James Gillette and others, in understanding the role of reactive metabolites and drug-induced tissue injury. Biofluid ^{1}H NMR spectroscopy was applied to the study of kidney cortical toxins such as mercury chloride,[53] *p*-aminophenol[55] and ifosfamide,[56] kidney medullary toxins including propylene imine[57] and 2-bromoethanamine,[58] a wide range of liver toxins including hydrazine,[59] allyl alcohol,[54] galactosamine[50] and thioacetamide,[54] together with the testicular toxin, cadmium chloride.[51] This provided the foundation for the study of novel compounds, new chemical entities and marketed drugs such as triazole anti-fungals,[60] bisphosphonates,[61] psychotropics,[62] non-steroidal anti-inflammatory drugs (NSAIDs) (including acetaminophen),[63] immunosuppressants,[64] simvastatin,[65] cisplatin,[66] vancomycin,[67] doxorubicin[68] and the C-C chemokine receptor type 5 (CCR5) antagonist, MrkA (N-[3-{[4-(3-benzyl-1-ethyl-1H-pyrazol-5-yl)piperidin-1-yl]methyl}-4-(3-fluorophenyl) cyclopentyl]-*N*-methylvaline).[69] The approach has even been applied to understanding species differences in toxicity.[70]

The Consortium for Metabonomic Toxicology, formed between five pharmaceutical companies (Bristol Myers Squibb, Eli Lilly & Co., Hoffmann-La Roche, NovoNordisk and Pfizer Inc.) and Imperial College, London, investigated the utility of metabonomics in evaluating xenobiotic toxicity by profiling rodent urine and blood serum using proton NMR spectroscopy.[71–75] Studies on 147 model toxins covering the breadth of liver and kidney toxicities allowed the development of a toxicity-prediction expert system. The probabilistic model based on CLOUDS (Classification of Unknowns by Density Superposition) enabled classification of toxin class (liver or kidney) of each treatment. This technique was validated with a leave-one-out approach to predict the main target organ of toxicity using the three most similar matches in the database. The sensitivity to liver and kidney toxicity was 67 and 41%, respectively, with corresponding specificities of 77 and 100%, and an error rate of 8%. Interestingly, at subtoxic doses, similar metabolic fingerprints were observed to those at higher doses, which caused a histopathologically detectable lesion. This represents an important advantage of this technology, being able to identify metabolic trajectories indicating toxicity but preceding observations of pathological lesions. This Consortium initiative also provided a baseline model of control biofluids in rat and mouse, which aided understanding of the typical biological variation in these populations.[75] Characterization of this "control space" facilitated outlier detection and initial deviations when a full time-course was not possible. The consistency and robustness of NMR measurements between labs and across the multiple studies was also evaluated and shown to have a high degree of reproducibility. The expert system developed on the proton NMR "fingerprints" of biofluids is a powerful approach as it leads to identification of combination biomarkers (*i.e.* perturbation of multiple intermediary metabolites) being unbiased in its assessment of the most important variables, and is likely a stronger prospect for application in early pre-clinical toxicology studies.

The Liver Toxicity Biomarker Study (LTBS) is following a similar protocol to address the issue of unanticipated clinical drug-induced liver injury (DILI) and the negative impact this has on patients, healthcare systems and the pharmaceutical industry.[76] This initiative has been formed between the FDA National Center for Toxicological Research (NCTR) and BG Medicine Inc. and is supported by seven pharmaceutical companies (Mitsubishi Chemical Holdings Corporation, Eisai Co. Ltd, Daiichi Sankyo Co. Ltd, UCB Pharma, Orion Pharma, Johnson & Johnson, Inc. and Pfizer, Inc.) with the goal of testing the ability of 'omics technologies (including LC-MS and NMR-based metabonomics) to detect DILI when conventional indicators of liver toxicity in pre-clinical studies have failed. This hypothesis is based on the notion that biochemical signals or molecular biomarkers in liver or biofluids can be used to distinguish between drug candidates that either do or do not mediate DILI, thus leading to the discovery of candidate biomarkers, which, once validated, can be incorporated into the drug development process. A crucial aspect of this strategy is the *in vivo* assessment in rat of the broad array of 'omics techniques as part of a comprehensive molecular systems analysis and *via* pairwise comparison of closely related compounds that have differential "clean" *vs.* "toxic" phenotypes. In the Phase I report, the catechol-*O*-methyl transferase inhibitors, entacapone (clean) and tolcapone (toxic), were compared; both compounds were reportedly free of liver toxicity in pre-clinical studies, whilst in clinical trials of tolcapone dose-related increases in liver transaminases were observed.[76] Furthermore, post-marketing tolcapone caused several acute hepatotoxicity events, which in some cases were fatal. In the LTBS study, metabonomics was performed on polar, lipid and amino acid LC-MS/MS platforms (liver and plasma) and by proton NMR spectroscopy (urine), together with proteomics, gene expression profiling and conventional measures of pathology and clinical chemistry. Data sets were integrated and analyzed by principal components analysis to facilitate cross-compartment correlations and mechanistic interpretation. Although full data analysis is yet to be reported, the preliminary results showed the promise of metabonomics identifying a number of tolcapone-specific markers. In addition, many of the 'omic platforms were able to detect molecular markers preceding the observation of classical toxicology endpoints.

There has been a multitude of studies of drug-induced tissue injury, using HR-MAS NMR spectroscopic analysis of intact tissues. Early investigations in pre-clinical species were able to identify tissue-specific metabolic perturbations in various organ toxicities, such as bromoethanamine-induced effects in liver and kidney,[77] the testicular toxicity mediated by ethylene glycol monomethyl ether[78] and the tissue injury caused by metals like mercuric II chloride,[79] lanthanum,[80] arsenic[81] and cadmium.[82] This led to broader analyses of drug-mediated toxicities by combining data sets obtained from biofluid NMR spectroscopy on urine, blood plasma and tissue extracts with spectral information from HR-MAS NMR measurements on intact tissues. Such an integrated metabonomic approach has been applied to a range of hepatotoxins with differing mechanisms of action and different pathological loci including ANIT,[83] bromobenzene,[84] ethionine,[85] hydrazine[86] and thioacetamide,[87] together with drugs such as acetaminophen[88] and proprietary drug candidates. The multiple 'omic paradigms of genomics, proteomics and metabonomics have also been applied in parallel to the study of toxicant-induced tissue injury in what has been termed systems toxicology; for example methapyrilene[89] and hydrazine.[90]

Utilizing statistical correlation techniques it has been possible to relate specific xenobiotic-derived metabolites to changes in the endogenous metabolite profile, implicating bioactivation *vs.* detoxification pathways. This concept was demonstrated using the pancreatic toxin, 1-cyano-2-hydroxy-3-butene (CHB), a naturally occurring nitrile. Proton NMR spectroscopy was performed on serially collected urine following subcutaneous administration in SD rats. STOCSY was applied to detect both structural and novel toxicological connectivities between xenobiotic and endogenous metabolite signals.[91] For example, CHB was found at high levels in the urine of high-dose animals at 8 h and interestingly the level was positively correlated to that of urinary glucose, suggesting that the compound was affecting the endocrine pancreas. In addition, the *N*-acetylcysteine conjugate was present in the urine of high-dose animals at 8 and 24 h, and its levels at 8 h were anti-correlated to glucose excretion. This indicates that the *N*-acetylcysteine conjugate of CHB is a major detoxification pathway, being a downstream metabolite of the glutathione conjugate. The metabonomic STOCSY approach allowed the investigation of differential pathway correlations between xenobiotic and endobiotic metabolites to evaluate drug metabolite toxicity and detoxification processes. The STOCSY approach has also shown value in understanding the metabolic signature of L-arginine induced exocrine pancreatitis.[92]

Metabolomic profiling of *in vitro* systems has also shown promise in the field of toxicology. Integrating microtechnology such as cell culture in microfluidic systems with NMR-based metabolic profiling of cell media has been coined "metabolomics on a chip" and was recently demonstrated with the characterization of toxicities of ammonia, DMSO and *N*-acetyl-paraaminophenol in liver, kidney and co-cultures of both tissues.[93] In addition, the metabolic effects of D-galactosamine on liver spheroid cultures have been compared with those observed in intact liver using proton MAS-NMR spectroscopy.[94] The biomarkers present in intact liver were much better represented in the spheroid system relative to isolated hepatocytes, demonstrating the potential of metabolic profiling to enhance the validation of *in vitro* safety profiling assays.

3.3.3 Drug Metabolism and Pharmacokinetics

High-resolution NMR spectroscopy has been pivotal in drug metabolism and pharmacokinetics (DMPK) endeavors to identify and characterize drug and xenobiotic derived metabolites. However, the application of global metabolic profiling to problems in DMPK has been relatively limited to date. More recently, though, the global, unbiased nature of metabonomics has led to some studies where more substantial consideration of endogenous components and the interplay between endobiotic and xenobiotic has been placed, *e.g.* pharmacometabonomics.

Pharmacometabonomics describes the use of metabolic profiling of biofluids, tissues and tissue extracts to predict, prior to dosing, the beneficial and adverse effects of an intervention such as drug administration, and as such is analogous to pharmacogenomics. The concept of pharmacometabonomics in humans has been demonstrated using the analgesic acetaminophen.[95] In this study a clear association was shown between the individual's metabolic phenotype (defined as the pre-dose urinary metabolite profile obtained by ^{1}H NMR spectroscopy) and the metabolic fate of a normal dose of acetaminophen. The pre-dose biomarker of drug fate was found to be a human-gut microbiome cometabolite and individuals with high pre-dose urinary levels of p-cresol sulfate exhibited low post-dose urinary ratios of acetaminophen sulfate to acetaminophen glucuronide. The extent to which a compound undergoes sulfation can be potentially limited by either availability of the sulfate donor, 3-phospho-adenosine-5-phosphosulfate (PAPS), or by the characteristics and expression level of the relevant sulfotransferase. Both p-cresol and acetaminophen are substrates for the same human sulfotransferase isoform, SULT1A1, and can therefore compete for the active site as well as for PAPS. Clearly, high microbial generation of p-cresol can reduce the systemic capacity to sulfate acetaminophen and potentially other drugs that are sulfated, with downstream implications on metabolic pathways, variable drug response and adverse drug reactions. This study, using a seemingly well understood drug, illustrates the tremendous potential of this approach above and beyond pharmacogenomics. One critical advantage is the ability to capture the impact of both genomic and environmental factors affecting metabolism. For example, enzyme induction, which influences drug metabolism and toxicity, is driven by environmental determinants and is not captured in genomic analysis. Pharmacometabonomics, as with any global metabolic phenotyping paradigm, can be applied with relative ease on matrices such as blood plasma and urine, with the potential to discover unexpected biomarker signatures without pre-specification of what those analytes should be. More recently, other clinical studies have corroborated these findings and the enormous potential of pharmacometabonomics to provide advanced warning and facilitate early intervention of drug-induced liver injury.[96] In addition, this strategy proved effective in explaining the response phenotype in a repeat dosing study in rodent with the hepatotoxin, galactosamine.[97] The pre-dose urine and fecal metabolite profiles were able to classify whether resistant or sensitive phenotypes presented on challenge with galactosamine. Furthermore, the concept of longitudinal pharmacometabonomics has also recently been proposed,[98] presenting the opportunity for monitoring or stratifying of patients over a period of time such as during clinical trials or pre- and post-surgery.

Rahmioglu and co-workers performed a metabonomic approach to identify a metabolic signature associated with variation in the activity of induced cytochrome P450 3A4 (CYP3A4).[99] A cohort of approximately 300 female twins was administered the potent CYP3A inducer, St John's Wort, over a two-week period with the activity of CYP3A4 quantified by the urinary ratio of 3-hydroxyquinine to quinine after administration of the probe drug quinine sulfate. The pre- and post-intervention urine samples were analyzed by ^{1}H NMR spectroscopy and, based on multiple linear regression, a combination of seven metabolites (indoxyl sulfate, glycine, N-acetylated unknowns, proline-betaine, guanidinoacetate, scyllo-inositol and alanine) from pre-intervention profiles and seven covariates showed a strong correlation with the 3-hydroxy-quinine:quinine probe measure of CYP3A4 activity. This represents a promising demonstration of the capability for metabolic profiling to predict CYP3A4-mediated activity without the need for dosing of exogenous probe substrates.

In the characterization of an organic anion transporting polypeptide (OATP) 1A1 and 1A4 knock-out mouse model, Gong *et al.* performed an NMR urinalysis to assess the urinary excretion of organic anions alongside determinations of mRNA, protein and transport of prototypic substrates such as estradiol-17beta-D-glucuronide, estrone-3-sulfate and taurocholate.[100] Similarly, NMR-based metabolic profiling has been used to assess the metabolic capacity of *in vitro* preparations of primary hepatocytes. Using proton NMR spectroscopy of aqueous extracts, Ellis and colleagues were able to demonstrate the normal metabolic phenotype in cultures of primary hepatocytes was stabilized in the presence of trichostatin A, a histone deacetylase inhibitor with anti-proliferative and differentiation inducing effects.[101] This has implications in the use of primary hepatocytes for the study of xenobiotic metabolism and endogenous hepatic metabolism.

The STOCSY concept has facilitated the study of flucloxacillin biotransformation taking advantage of the sparse spectral resonances and high resolution offered by ^{19}F NMR to aid extraction of the most relevant ^{1}H NMR resonances from complex proton spectra.[102] In essence, this represents a heteronuclear STOCSY and offers a powerful tool to aid structure elucidation of drug metabolites where a spin active nucleus is present (*e.g.* [19]F, ^{13}C, ^{15}N, ^{31}P).

Molecular chirality is an important component of drug metabolism studies and often requires the use of derivatizing agents or chromatographic separations based on chiral stationary phases. Recently, a simple, rapid and direct alternative to these approaches has been proposed, which enabled the differentiation and identification of chiral drug enantiomers in human urine.[103] Using ibuprofen as an example, the enantiomers of the parent drug as well as the diastereoisomers of one of the main metabolites, the acyl glucuronide, were resolved by proton NMR spectroscopy by the direct addition of the chiral co-solvating agent, beta-cyclodextrin.

The global, unbiased nature of the NMR-based metabonomic approach lends itself nicely to several applications in the field of drug metabolism, including the identification of novel drug metabolites. From a pharmacokinetic perspective, it offers the opportunity for multiple analyte evaluation in an integrated fashion and thus has potential in studies of pharmacokinetic/pharmacodynamic (PK/PD) relationships and the corresponding toxicity concentration-effect relationship. The study of traditional herbal medicines is starting to receive attention in this regard.[104]

3.3.4 Clinical Applications

The utilities of NMR spectroscopic-based metabolic profiling in clinical settings have been many and varied, with studies on biofluids such as urine, serum or blood plasma and CSF, exploring the metabolic basis of a broad range of disorders and diseases including inborn errors of metabolism,[105,106] multiple sclerosis,[107] diabetes mellitus,[108] cardiovascular disease[109] and various cancers.[110–115] The following examples taken from the recent literature exemplify the translation of the NMR-based metabolic profiling platform to clinical investigations of disease progression, and the potential for therapeutic monitoring given the non-invasive nature of most routine biofluid sampling, such as urine and blood serum. In most cases, such studies in the human population require a high level of statistical rigor to ensure confidence in the observed differentiation of patients and healthy volunteers.

NMR spectroscopy coupled with multivariate statistical analysis has been shown to predict the cardiovascular risk in a large cohort of healthy volunteers.[116] The biochemical signature acquired on 864 plasma samples was able to classify low and high risk, in agreement with common clinical markers and cardiovascular risk factors. The metabolic discrimination was based on well-established biochemical indicators such as total cholesterol, triglycerides, LDL and high-density lipoprotein (HDL) as well as a pattern of novel low-molecular-weight compounds including 3-hydroxybutyrate, alpha-ketoglutarate, threonine and dimethylglycine. Similar diagnostic studies have been able to distinguish the presence and severity of coronary heart disease.[109]

This approach has also shown utility in the evaluation of liver and kidney function since NMR spectroscopy is ideally suited to the study of metabolic perturbations. In a study of liver cirrhosis patients (n = 124), serum metabolomics was able to classify the severity of chronic liver failure based on an O-PLS analysis.[117] The basis for the differentiation in disease severity was manifest in lower HDL and phosphocholine and higher lactate, glucose, pyruvate, amino acids and creatinine with increasing levels of chronic liver failure. In an analogous fashion, proton NMR spectroscopic fingerprints from urine of 77 patients with glomerulonephritides (based on histopathology assessment from renal biopsy) were able to predict the presence of renal damage.[118] As shown in the representative NMR spectra of Figure 3.6, patients with mild to severe tubulointerstitial lesions were differentiated from healthy individuals in O-PLS-DA models of NMR-based metabolic signatures. The

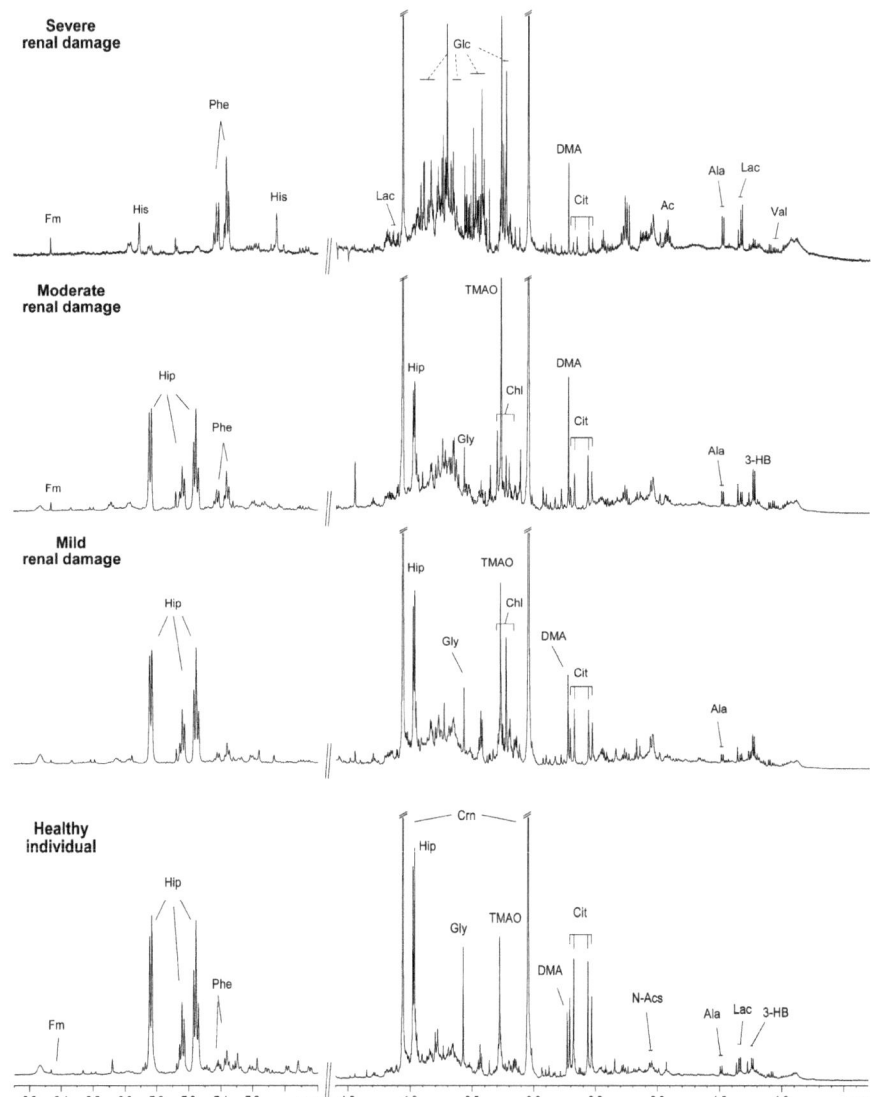

Figure 3.6 ^{1}H NMR 500 MHz spectra of urine (delta 0.3–4.6 and 6.8–8.7 ppm) from one healthy subject and patients with mild, moderate and severe renal damage. Abbreviations: 3-HB, 3-hydroxybutyrate; Ac, acetate; Ala, alanine; Chl, choline headgroup containing metabolites; Cit, citrate; Crn, creatinine; DMA, dimethylamine; Fm, formate; Gly, glycine; Glc, glucose; Hip, hippurate; His, histidine; Lac, lactate; N-Acs, N-acetyl groups from glycoproteins; Phe, phenylalanine; TMAO, trimethylamine-N-oxide; Val, valine.
Reproduced with permission from Psihogios et al.[118] © 2007 American Chemical Society.

lesion onset was characterized by reduced excretion of citrate, hippurate, glycine and creatinine with lesion progression leading to glycosuria, selective aminoaciduria, ablation of citrate and hippurate levels with gradually higher excretion of lactate, acetate and trimethylamine-*N*-oxide. These works provide further support to the role of NMR-based metabonomics in evaluating and monitoring of liver and kidney function in the clinic. Furthermore, the study of kidney disease in Type I diabetic patients has been studied using NMR spectroscopy across three distinct molecular "windows", namely the lipoproteins, low-molecular-weight compounds and lipids.[119] Diverse metabolic phenotypes of diabetic kidney disease were identified; high levels of unsaturated fatty acids, phospholipids, intermediate density lipoprotein (IDL) and LDL lipids were present in the sub-clinical phase, higher saturated fatty acids and low HDL were implicated in accelerated disease progression, and the sphingolipid pathway was also involved.

In the oncology arena, high-resolution ^{1}H MAS-NMR spectroscopy has been applied to explore the metabolic phenotypes of head and neck squamous cell carcinoma *versus* normal adjacent tissue. Somshekar and colleagues analyzed a total of 43 tissues and using multivariate statistics were able to classify normal tissue and tumor tissue based on a metabolic signature comprised of elevated lactate, selective amino acids, choline-containing compounds, creatine, taurine, glutathione and decreased levels of triglycerides.[120] This profile of metabolite changes was associated with upregulated glycolysis, increased amino acid influx into the TCA cycle (anaplerosis), oxidative and osmotic protective mechanisms and membrane phospholipid metabolism providing an alternative energy source *via* lipolysis, all of which support rapid tumor cell proliferation and growth. Napoli and co-workers have studied the urinary metabolome of pancreatic ductal adenocarcinoma patients by proton NMR spectroscopy, with the aim of identifying a useful, disease-specific molecular fingerprint for screening purposes.[121] Discrimination of patient samples (n = 33) from healthy, age-matched controls (n = 54) was statistically significant and showed promise for ultimately moving towards a non-invasive method that is capable of large-scale screening paradigms. In a similar manner, MacIntyre *et al.* applied the technology to study a cohort of chronic lymphocytic leukemia (CLL) patients.[122] The metabolic profiles of serum from CLL patients were differentiated from controls based on elevated resonances from pyruvate and glutamate with decreases in isoleucine. In addition, the metabolic phenotype was also distinguishable between patients with and without mutations in immunoglobulin heavy chain variable regions of CLL cells, a useful prognostic marker for high-risk patients. Patients without the mutation exhibited higher levels of cholesterol, lactate, uridine and fumarate with lower levels of pyridoxine, glycerol, 3-hydroxybutyrate and methionine.

As these case studies exemplify, the application of NMR-based metabolic profiling is showing promise in a number of clinical paradigms and has the potential to have a greater impact in the hospital setting whereby non-invasive monitoring of a patient's metabotype can be achieved before, during and after the course of surgical and medical procedures, as recently proposed by others.[123]

3.4   Summary and Future Directions

Global metabolic profiling afforded by high-resolution NMR spectroscopy has made a number of key contributions to the understanding of disease etiology and mechanisms of drug-induced toxicities in both pre-clinical and clinical settings, and in doing so has provided some sophisticated tools and models enhancing the drug discovery and development process.

The ability to leverage the study of biological fluids and tissues from human subjects (patients or healthy volunteers) or from pre-clinical models induced by administration of exogenous agents or manifest through genetic modification has empowered investigators with the opportunity to further understand the metabolic basis of disease. Interestingly, many of the biological markers have a commonality across disease states; in immune-mediated disorders markers related to lipid signaling and methylation status are often observed, in infectious diseases there is a prevalence of host-parasite co-metabolites such as hippurate, phenylacetylglycine and indoxyl-sulfate, and for the neurological indications, excitatory and mitochondrial metabolism predominates. Future directions will likely see metabolic profiling play a role in the monitoring of disease modification through administration of therapeutic modalities. Pre-clinically this would facilitate target validation and the search for disease-relevant biomarkers. In clinical development, the translation of biomarkers would enable pharmacodynamic assessment and, where appropriate, patient stratification as well as diagnostic capabilities.

It should be acknowledged that high-resolution NMR spectroscopy in isolation would not have advanced the field of metabolic science, without the effective coupling of multivariate statistics and chemometric tools capable of handling the complex spectral data sets and reducing the dimensionality of the data to one that enables robust and insightful interpretation. These techniques have also identified latent metabolite information otherwise hidden in simple viewing of the spectra. This is especially true with the development of the STOCSY concept for analysis of complex mixtures, demonstrated in multiple studies described earlier. Furthermore, the statistical rigor of these pattern recognition analyses has become extremely important in clinical studies of the much more heterogeneous human population.

The vast array of biological matrices that can be analyzed by high-resolution NMR spectroscopy, in both the liquid and solid state, has clearly enabled a diverse series of problems in biomedicine to be addressed. This has included the study of *in vitro*, *ex vivo* and *in vivo* systems, and with appropriate sample pre-treatments or various NMR pulse sequences researchers have been able to extract relevant metabolic information from a complex milieu. Major advancements have been made through the integration of multiple, allied 'omic endpoints such that the interdependencies and correlations between gene expression, protein levels and metabolite profile can be established for the pathophysiological process in question. In addition, a number of techniques such as microdialysis and microfluidics have become useful adjunct approaches in metabonomics. Through technological innovations, high-resolution NMR spectroscopy has evolved to become an enormously powerful tool in the study of metabolic science, and it appears that evolution is set to continue.

Acknowledgements

The author would like to thank Prof. Jeremy K. Nicholson, Imperial College, for insightful discussion, and Dr. Robert A. Copeland, Epizyme, and the book editors for critical review of the manuscript.

References

1. D. I. Hoult, S. J. Busby, D. G. Gadian, G. K. Radda, R. E. Richards and P. J. Seeley, *Nature (London)*, 1974, **252**, 285.
2. J. K. Nicholson, J. C. Lindon and E. Holmes, *Xenobiotica*, 1999, **29**, 1181.
3. O. Fiehn, *Comp. Funct. Genomics*, 2001, **2**, 155.
4. J. C. Lindon, O. P. Beckonert, E. Holmes and J. K. Nicholson, *Prog. Nucl. Mag. Res. Sp.*, 2009, **55**, 79.
5. J. C. Lindon and J. K. Nicholson, *Trends Anal. Chem.*, 2008, **27**, 194.
6. J. C. Lindon and J. K. Nicholson, *Annu. Rev. Anal. Chem.*, 2008, **1**, 45.
7. R. L. Haner, W. Llanos and L. Mueller, *J. Magn. Reson.*, 2000, **143**, 69.
8. A. Yilmaz, N. T. Nyberg and J. W. Jaroszewski, *Anal. Chem.*, 2011, **83**, 8278.
9. O. Cloarec, M. E. Dumas, A. Craig, R. H. Barton, J. Trygg, J. Hudson, C. Blancher, D. Gauguier, J. C. Lindon, E. Holmes and J. Nicholson, *Anal. Chem.*, 2005, **77**, 1282.
10. M. L. Anthony, C. R. Beddell, J. C. Lindon and J. K. Nicholson, *J. Pharmaceut. Biomed.*, 1993, **11**, 897.
11. N. J. Waters, S. Garrod, R. D. Farrant, J. N. Haselden, S. C. Connor, J. Connelly, J. C. Lindon, E. Holmes and J. K. Nicholson, *Anal. Biochem.*, 2000, **282**, 16.
12. R. J. Simpson, K. M. Brindle, F. F. Brown, I. D. Campbell and D. L. Foxall, *Biochem. J.*, 1981, **193**, 401.
13. T. W. M. Fan and A. N. Lane, *J. Biomol. NMR*, 2011, **49**, 267.
14. A. M. Tomlins, P. J. Foxall, M. J. Lynch, J. Parkinson, J. R. Everett and J. K. Nicholson, *Biochim. Biophys. Acta*, 1998, **1379**, 367.
15. J. H. Chen, B. M. Enloe, Y. Xiao, D. G. Cory and S. Singer, *Magn. Reson. Med.*, 2003, **50**, 515.
16. E. Humpfer, M. Spraul, A. W. Nicholls, J. K. Nicholson and J. C. Lindon, *Magn. Reson. Med.*, 1997, **38**, 334.
17. E. R. Andrew, A. Bradbury and R. G. Eades, *Nature (London)*, 1958, **182**, 1659.
18. I. J. Lowe, *Phys. Rev. Lett.*, 1959, **2**, 285.
19. O. Beckonert, M. Coen, H. C. Keun, Y. Wang, T. M. Ebbels, E. Holmes, J. C. Lindon and J. K. Nicholson, *Nat. Protoc.*, 2010, **5**, 1019.
20. K. S. Opstad, B. A. Bell, J. R. Griffiths and F. A. Howe, *NMR Biomed.*, 2008, **21**, 1138.

21. A. Wong, B. Jimenez, X. Li, E. Holmes, J. K. Nicholson, J. C. Lindon and D. Sakellariou, *Anal. Chem.*, 2012, **84**, 3843.
22. A. Smolinska, A. Attali, L. Blanchet, K. Ampt, T. Tuinstra, H. van Aken, E. Suidgeest, A. J. van Gool, T. Luider, S. S. Wijmenga and L. M. Buydens, *J. Proteome Res.*, 2011, **10**, 4428.
23. P. Baur, F.-P. Martin, L. Gruber, N. Bosco, V. Brahmbhatt, S. Collino, P. Guy, I. Montoliu, J. Rozman, M. Klingenspor, I. Tavazzi, A. Thorimbert, S. Rezzi, S. Kochhar, J. Benyacoub, G. Kollias and D. Haller, *J. Proteome Res.*, 2011, **10**, 5523.
24. D. J. Kominsky, S. Keely, C. F. MacManus, L. E. Glover, M. Scully, C. B. Collins, B. E. Bowers, E. L. Campbell and S. P. Colgan, *J. Immunol.*, 2011, **186**, 6505.
25. A. M. Weljie, R. Dowlatabadi, B. J. Miller, H. J. Vogel and F. R. Jirik, *J. Proteome Res.*, 2007, **6**, 3456.
26. Y. Wang, E. Holmes, J. K. Nicholson, O. Cloarec, J. Chollet, M. Tanner, B. H. Singer and J. Utzinger, *Proc. Natl Acad. Sci. USA*, 2004, **101**, 12676.
27. Y. Wang, J. Utzinger, S. H. Xiao, J. Xue, J. K. Nicholson, M. Tanner, B. H. Singer and E. Holmes, *Mol. Biochem. Parasit.*, 2006, **146**, 1.
28. Y. Wang, J. Utzinger, J. Saric, J. V. Li, J. Burckhardt, S. Dirnhofer, J. K. Nicholson, B. H. Singer, R. Brun and E. Holmes, *Proc. Natl Acad. Sci. USA*, 2008, **105**, 6127.
29. J. Saric, J. V. Li, Y. Wang, J. Keiser, K. Veselkov, S. Dirnhofer, I. K. Yap, J. K. Nicholson, E. Holmes and J. Utzinger, *J. Proteome Res.*, 2009, **8**, 3899.
30. J. V. Li, Y. Wang, J. Saric, J. K. Nicholson, S. Dirnhofer, B. H. Singer, M. Tanner, S. Wittlin, E. Holmes and J. Utzinger, *J. Proteome Res.*, 2008, **7**, 3948.
31. J. R. Swann, K. M. Tuohy, P. Lindfors, D. T. Brown, G. R. Gibson, I. D. Wilson, J. Sidaway, J. K. Nicholson and E. Holmes, *J. Proteome Res.*, 2011, **10**, 3590.
32. K. A. Verwaest, T. N. Vu, K. Laukens, L. E. Clemens, H. P. Nguyen, B. Van Gasse, J. C. Martins, A. Van Der Linden and R. Dommisse, *BBA Mol. Basis Dis.*, 2011, **1812**, 1371.
33. M. R. Viant, B. G. Lyeth, M. G. Miller and R. F. Berman, *NMR Biomed.*, 2005, **18**, 507.
34. P. Khandelwal, C. E. Beyer, Q. Lin, P. McGonigle, L. E. Schechter and A. C. Bach, II, *J. Neurosci. Meth.*, 2004, **133**, 181.
35. P. Khandelwal, C. E. Beyer, Q. Lin, L. E. Schechter and A. C. Bach, II, *Anal. Chem.*, 2004, **76**, 4123.
36. O. Teahan, C. L. Bevan, J. Waxman and H. C. Keun, *Int. J. Biochem. Cell B.*, 2011, **43**, 1002.
37. X. Pan, M. Wilson, L. Mirbahai, C. McConville, T. N. Arvanitis, J. L. Griffin, R. A. Kauppinen and A. C. Peet, *J. Proteome Res.*, 2011, **10**, 3493.
38. A. Backshall, D. Alferez, F. Teichert, I. D. Wilson, R. W. Wilkinson, R. A. Goodlad and H. C. Keun, *J. Proteome Res.*, 2009, **8**, 1423.

39. M. Rantalainen, O. Cloarec, O. Beckonert, I. D. Wilson, D. Jackson, R. Tonge, R. Rowlinson, S. Rayner, J. Nickson, R. W. Wilkinson, J. D. Mills, J. Trygg, J. K. Nicholson and E. Holmes, *J. Proteome Res.*, 2006, **5**, 2642.
40. S. K. Hekmatyar, M. Wilson, N. Jerome, R. M. Salek, J. L. Griffin, A. Peet and R. A. Kauppinen, *Brit. J. Cancer*, 2010, **103**, 1297.
41. J. Moroz, J. Turner, C. Slupsky, G. Fallone and A. Syme, *Phys. Med. Biol.*, 2011, **56**, 535.
42. J. F. L. Cobbold, Q. M. Anstee, R. D. Goldin, H. R. T. Williams, H. C. Matthews, B. V. North, N. Absalom, H. C. Thomas, M. R. Thursz, R. D. Cox, S. D. Taylor-Robinson and I. J. Cox, *Clin. Sci.*, 2009, **116**, 403.
43. M. E. Dumas, R. H. Barton, A. Toye, O. Cloarec, C. Blancher, A. Rothwell, J. Fearnside, R. Tatoud, V. Blanc, J. C. Lindon, S. C. Mitchell, E. Holmes, M. I. McCarthy, J. Scott, D. Gauguier and J. K. Nicholson, *Proc. Natl Acad. Sci. USA*, 2006, **103**, 12511.
44. K.-K. Cheng, G. M. Benson, D. C. Grimsditch, D. G. Reid, S. C. Connor and J. L. Griffin, *Physiol. Genomics*, 2010, **41**, 224.
45. L. Zhao, H. Gao, F. Lian, X. Liu, Y. Zhao and D. Lin, *Am. J. Physiol.*, 2011, **300**, F947.
46. A. D. Maher, J. M. Fonville, M. Coen, J. C. Lindon, C. D. Rae and J. K. Nicholson, *Anal. Chem.*, 2012, **84**, 1083.
47. S. C. Connor, M. G. Hughes, G. Moore, C. A. Lister and S. A. Smith, *J. Pharm. Pharmacol.*, 1997, **49**, 336.
48. N. Jiang, X. Yan, W. Zhou, Q. Zhang, H. Chen, Y. Zhang and X. Zhang, *J. Proteome Res.*, 2008, **7**, 3678.
49. A. Wijeyesekera, C. Selman, R. H. Barton, E. Holmes, J. K. Nicholson and D. J. Withers, *J. Proteome Res.*, 2012, **11**, 2224.
50. B. M. Beckwith-Hall, J. K. Nicholson, A. W. Nicholls, P. J. Foxall, J. C. Lindon, S. C. Connor, M. Abdi, J. Connelly and E. Holmes, *Chem. Res. Toxicol.*, 1998, **11**, 260.
51. K. P. Gartland, C. R. Beddell, J. C. Lindon and J. K. Nicholson, *Mol. Pharmacol.*, 1991, **39**, 629.
52. K. P. Gartland, F. W. Bonner and J. K. Nicholson, *Mol. Pharmacol.*, 1989, **35**, 242.
53. E. Holmes, A. W. Nicholls, J. C. Lindon, S. C. Connor, J. C. Connelly, J. N. Haselden, S. J. Damment, M. Spraul, P. Neidig and J. K. Nicholson, *Chem. Res. Toxicol.*, 2000, **13**, 471.
54. E. Holmes, A. W. Nicholls, J. C. Lindon, S. Ramos, M. Spraul, P. Neidig, S. C. Connor, J. Connelly, S. J. Damment, J. Haselden and J. K. Nicholson, *NMR Biomed.*, 1998, **11**, 235.
55. K. P. Gartland, F. W. Bonner, J. A. Timbrell and J. K. Nicholson, *Arch. Toxicol.*, 1989, **63**, 97.
56. P. J. Foxall, E. M. Lenz, J. C. Lindon, G. H. Neild, I. D. Wilson and J. K. Nicholson, *Ther. Drug Monit.*, 1996, **18**, 498.
57. M. L. Anthony, B. C. Sweatman, C. R. Beddell, J. C. Lindon and J. K. Nicholson, *Mol. Pharmacol.*, 1994, **46**, 199.

58. E. Holmes, F. W. Bonner and J. K. Nicholson, *Arch. Toxicol.*, 1995, **70**, 89.
59. A. W. Nicholls, E. Holmes, J. C. Lindon, J. P. Shockcor, R. D. Farrant, J. N. Haselden, S. J. Damment, C. J. Waterfield and J. K. Nicholson, *Chem. Res. Toxicol.*, 2001, **14**, 975.
60. D. R. Ekman, H. C. Keun, C. D. Eads, C. M. Furnish, R. N. Murrell, J. C. Rockett and D. J. Dix, *Metabolomics*, 2006, **2**, 63.
61. F. Dieterle, G. Schlotterbeck, M. Binder, A. Ross, L. Suter and H. Senn, *Chem. Res. Toxicol.*, 2007, **20**, 1291.
62. G. A. McLoughlin, D. Ma, T. M. Tsang, D. N. C. Jones, J. Cilia, M. D. Hill, M. J. Robbins, I. M. Benzel, P. R. Maycox, E. Holmes and S. Bahn, *J. Proteome Res.*, 2009, **8**, 1943.
63. S. Y. Um, M. W. Chung, K.-B. Kim, S. H. Kim, J. S. Oh, H. Y. Oh, H. J. Lee and K. H. Choi, *Anal. Chem.*, 2009, **81**, 4734.
64. V. Schmitz, J. Klawitter, J. Bendrick-Peart, W. Schoening, G. Puhl, M. Haschke, J. Klawitter, J. Consoer, J. Rivard Christopher, L. Chan, V. Tran Zung, D. Leibfritz and U. Christians, *Nephron. Exp. Nephrol.*, 2009, **111**, e80.
65. H.-j. Yang, M.-J. Choi, H. Wen, H. N. Kwon, K. H. Jung, S.-W. Hong, J. M. Kim, S.-S. Hong and S. Park, *PLoS One*, 2011, **6**, e16641.
66. E. Holmes, J. K. Nicholson, A. W. Nicholls, J. C. Lindon, S. C. Connor, S. Polley and J. Connelly, *Chemometr. Intell. Lab.*, 1998, **44**, 245.
67. I. K. S. Yap, J. V. Li, J. Saric, F.-P. Martin, H. Davies, Y. Wang, I. D. Wilson, J. K. Nicholson, J. Utzinger, J. R. Marchesi and E. Holmes, *J. Proteome Res.*, 2008, **7**, 3718.
68. J.-C. Park, Y.-S. Hong, Y. J. Kim, J.-Y. Yang, E.-Y. Kim, S. J. Kwack, D. H. Ryu, G.-S. Hwang and B. M. Lee, *J. Toxicol. Env. Heal. A*, 2009, **72**, 374.
69. R. J. Mortishire-Smith, G. L. Skiles, J. W. Lawrence, S. Spence, A. W. Nicholls, B. A. Johnson and J. K. Nicholson, *Chem. Res. Toxicol.*, 2004, **17**, 165.
70. M. E. Bollard, H. C. Keun, O. Beckonert, T. M. Ebbels, H. Antti, A. W. Nicholls, J. P. Shockcor, G. H. Cantor, G. Stevens, J. C. Lindon, E. Holmes and J. K. Nicholson, *Toxicol. Appl. Pharm.*, 2005, **204**, 135.
71. T. Ebbels, H. Keun, O. Beckonert, H. Antti, M. Bollard, E. Holmes, J. Lindon and J. Nicholson, *Anal. Chim. Acta*, 2003, **490**, 109.
72. O. Beckonert, M. E. Bollard, T. M. D. Ebbels, H. C. Keun, H. Antti, E. Holmes, J. C. Lindon and J. K. Nicholson, *Anal. Chim. Acta*, 2003, **490**, 3.
73. T. M. D. Ebbels, H. C. Keun, O. P. Beckonert, M. E. Bollard, J. C. Lindon, E. Holmes and J. K. Nicholson, *J. Proteome Res.*, 2007, **6**, 4407.
74. J. C. Lindon, H. C. Keun, T. M. D. Ebbels, J. M. T. Pearce, E. Holmes and J. K. Nicholson, *Pharmacogenomics*, 2005, **6**, 691.
75. J. C. Lindon, J. K. Nicholson, E. Holmes, H. Antti, M. E. Bollard, H. Keun, O. Beckonert, T. M. Ebbels, M. D. Reily, D. Robertson, G. J. Stevens, P. Luke, A. P. Breau, G. H. Cantor, R. H. Bible, U. Niederhauser, H. Senn, G. Schlotterbeck, U. G. Sidelmann, S. M. Laursen, A. Tymiak, B. D. Car,

L. Lehman-McKeeman, J.-M. Colet, A. Loukaci and C. Thomas, *Toxicol. App. Pharm.*, 2003, **187**, 137.

76. R. N. McBurney, W. M. Hines, L. S. von Tungeln, L. K. Schnackenberg, R. D. Beger, C. L. Moland, T. Han, J. C. Fuscoe, C.-W. Chang, J. J. Chen, Z. Su, X.-h. Fan, W. Tong, S. A. Booth, R. Balasubramanian, P. L. Courchesne, J. M. Campbell, A. Graber, Y. Guo, P. J. Juhasz, T. Y. Li, M. D. Lynch, N. M. Morel, T. N. Plasterer, E. J. Takach, C. Zeng and F. A. Beland, *Toxicol. Pathol.*, 2009, **37**, 52.

77. S. Garrod, E. Humpher, S. C. Connor, J. C. Connelly, M. Spraul, J. K. Nicholson and E. Holmes, *Magn. Reson. Med.*, 2001, **45**, 781.

78. T. Yamamoto, I. Horii and T. Yoshida, *J. Toxicol. Sci.*, 2007, **32**, 515.

79. Y. Wang, M. E. Bollard, J. K. Nicholson and E. Holmes, *J. Pharmaceut. Biomed.*, 2006, **40**, 375.

80. P.-Q. Liao, H.-F. Wu, X.-Y. Zhang, X.-J. Li, Z.-F. Li, W.-S. Li, Y.-J. Wu and F.-K. Pei, *Chem. J. Chinese U.*, 2006, **27**, 1448.

81. J. L. Griffin, L. Walker, R. F. Shore and J. K. Nicholson, *Xenobiotica*, 2001, **31**, 377.

82. J. L. Griffin, L. A. Walker, R. F. Shore and J. K. Nicholson, *Chem. Res. Toxicol.*, 2001, **14**, 1428.

83. N. J. Waters, E. Holmes, A. Williams, C. J. Waterfield, R. D. Farrant and J. K. Nicholson, *Chem. Res. Toxicol.*, 2001, **14**, 1401.

84. N. J. Waters, C. J. Waterfield, R. D. Farrant, E. Holmes and J. K. Nicholson, *J. Proteome Res.*, 2006, **5**, 1448.

85. E. Skordi, I. K. Yap, S. P. Claus, F. P. Martin, O. Cloarec, J. Lindberg, I. Schuppe-Koistinen, E. Holmes and J. K. Nicholson, *J. Proteome Res.*, 2007, **6**, 4572.

86. S. Garrod, M. E. Bollard, A. W. Nicholls, S. C. Connor, J. Connelly, J. K. Nicholson and E. Holmes, *Chem. Res. Toxicol.*, 2005, **18**, 115.

87. N. J. Waters, C. J. Waterfield, R. D. Farrant, E. Holmes and J. K. Nicholson, *Chem. Res. Toxicol.*, 2005, **18**, 639.

88. M. Coen, E. M. Lenz, J. K. Nicholson, I. D. Wilson, F. Pognan and J. C. Lindon, *Chem. Res. Toxicol.*, 2003, **16**, 295.

89. A. Craig, J. Sidaway, E. Holmes, T. Orton, D. Jackson, R. Rowlinson, J. Nickson, R. Tonge, I. Wilson and J. Nicholson, *J. Proteome Res.*, 2006, **5**, 1586.

90. T. G. Kleno, B. Kiehr, D. Baunsgaard and U. G. Sidelmann, *Biomarkers*, 2004, **9**, 116.

91. E. Bohus, A. Racz, B. Noszal, M. Coen, O. Beckonert, H. C. Keun, T. M. Ebbels, G. H. Cantor, J. A. Wijsman, E. Holmes, J. C. Lindon and J. K. Nicholson, *Magn. Reson. Chem.*, 2009, **47**(1), S26.

92. E. Bohus, M. Coen, H. C. Keun, T. M. Ebbels, O. Beckonert, J. C. Lindon, E. Holmes, B. Noszal and J. K. Nicholson, *J. Proteome Res.*, 2008, **7**, 4435.

93. L. Shintu, R. Baudoin, V. Navratil, J.-M. Prot, C. Pontoizeau, M. Defernez, B. J. Blaise, C. Domange, A. R. Pery, P. Toulhoat, C. Legallais, C. Brochot, E. Leclerc and M.-E. Dumas, *Anal. Chem.*, 2012, **84**, 1840.

94. M. E. Bollard, J. Xu, W. Purcell, J. L. Griffin, C. Quirk, E. Holmes and J. K. Nicholson, *Chem. Res. Toxicol.*, 2002, **15**, 1351.
95. T. A. Clayton, D. Baker, J. C. Lindon, J. R. Everett and J. K. Nicholson, *Proc. Natl Acad. Sci. USA*, 2009, **106**, 14728.
96. J. H. Winnike, Z. Li, F. A. Wright, J. M. Macdonald, T. M. O'Connell and P. B. Watkins, *Clin. Pharmacol. Ther.*, 2010, **88**, 45–51.
97. M. Coen, F. Goldfain-Blanc, G. Rolland-Valognes, B. Walther, D. G. Robertson, E. Holmes, J. C. Lindon and J. K. Nicholson, *J. Proteome Res.*, 2012, **11**, 2427.
98. J. K. Nicholson, J. R. Everett and J. C. Lindon, *Expert Opin. Drug Metab. Toxicol.*, 2012, **8**, 135.
99. N. Rahmioglu, G. Le Gall, J. Heaton, K. L. Kay, N. W. Smith, I. J. Colquhoun, K. R. Ahmadi and E. K. Kemsley, *J. Proteome Res.*, 2011, **10**, 2807.
100. L. Gong, N. Aranibar, Y.-H. Han, Y.-C. Zhang, L. Lecureux, V. Bhaskaran, P. Khandelwal, C. D. Klaassen and L. D. Lehman-McKeeman, *Toxicol. Sci.*, 2011, **122**, 587.
101. J. K. Ellis, P. H. Chan, T. Doktorova, T. J. Athersuch, R. Cavill, T. Vanhaecke, V. Rogiers, M. Vinken, J. K. Nicholson, T. M. D. Ebbels and H. C. Keun, *J. Proteome Res.*, 2010, **9**, 413.
102. H. C. Keun, T. J. Athersuch, O. Beckonert, Y. Wang, J. Saric, J. P. Shockcor, J. C. Lindon, I. D. Wilson, E. Holmes and J. K. Nicholson, *Anal. Chem.*, 2008, **80**, 1073.
103. M. Perez-Trujillo, J. C. Lindon, T. Parella, H. C. Keun, J. K. Nicholson and T. J. Athersuch, *Anal. Chem.*, 2012, **84**, 2868.
104. K. Lan and W. Jia, *Curr. Drug Metab.*, 2010, **11**, 105–114.
105. E. Holmes, P. J. Foxall, J. K. Nicholson, G. H. Neild, S. M. Brown, C. R. Beddell, B. C. Sweatman, E. Rahr, J. C. Lindon, M. Spraul and P. Neidig, *Anal. Biochem.*, 1994, **220**, 284.
106. S. H. Moolenaar, U. F. H. Engelke and R. A. Wevers, *Ann. Clin. Biochem.*, 2003, **40**, 16.
107. F. Koschorek, W. Offermann, J. Stelten, W. E. Braunsdorf, U. Steller, H. Gremmel and D. Leibfritz, *Neurosurg. Rev.*, 1993, **16**, 307.
108. J. D. Bell, J. C. C. Brown, P. J. Sadler, A. F. Macleod, P. H. Sonksen, R. D. Hughes and R. Williams, *Clin. Sci.*, 1987, **72**, 563.
109. J. T. Brindle, H. Antti, E. Holmes, G. Tranter, J. K. Nicholson, H. W. Bethell, S. Clarke, P. M. Schofield, E. McKilligin, D. E. Mosedale and D. J. Grainger, *Nat. Med.*, 2002, **8**, 1439.
110. V. M. Asiago, L. Z. Alvarado, N. Shanaiah, G. A. N. Gowda, K. Owusu-Sarfo, R. A. Ballas and D. Raftery, *Cancer Res.*, 2010, **70**, 8309.
111. I. Barba, C. Sanz, A. Barbera, G. Tapia, J.-L. Mate, D. Garcia-Dorado, J.-M. Ribera and A. Oriol, *Exp. Hematol.*, 2009, **37**, 1259.
112. W. Chen, X. Zhou, D. Huang, F. Chen and X. Du, *Chinese J. Chem.*, 2011, **29**, 2511.

113. V. Righi, C. Durante, M. Cocchi, C. Calabrese, G. Di Febo, F. Lecce, A. Pisi, V. Tugnoli, A. Mucci and L. Schenetti, *J. Proteome Res.*, 2009, **8**, 1859.
114. B. Sitter, T. F. Bathen, T. E. Singstad, H. E. Fjosne, S. Lundgren, J. Halgunset and I. S. Gribbestad, *NMR Biomed.*, 2010, **23**, 424.
115. A. N. Zira, S. E. Theocharis, D. Mitropoulos, V. Migdalis and E. Mikros, *J. Proteome Res.*, 2010, **9**, 4038.
116. P. Bernini, I. Bertini, C. Luchinat, L. Tenori and A. Tognaccini, *J. Proteome Res.*, 2011, **10**, 4983.
117. R. Amathieu, P. Nahon, M. Triba, N. Bouchemal, J.-C. Trinchet, M. Beaugrand, G. Dhonneur and L. Le Moyec, *J. Proteome Res.*, 2011, **10**, 3239.
118. N. G. Psihogios, R. G. Kalaitzidis, S. Dimou, K. I. Seferiadis, K. C. Siamopoulos and E. T. Bairaktari, *J. Proteome Res.*, 2007, **6**, 3760.
119. V.-P. Makinen, T. Tynkkynen, P. Soininen, T. Peltola, A. J. Kangas, C. Forsblom, L. M. Thorn, K. Kaski, R. Laatikainen, M. Ala-Korpela and P.-H. Groop, *J. Proteome Res.*, 2012, **11**, 1782.
120. B. S. Somashekar, P. Kamarajan, T. Danciu, Y. L. Kapila, A. M. Chinnaiyan, T. M. Rajendiran and A. Ramamoorthy, *J. Proteome Res.*, 2011, **10**, 5232.
121. C. Napoli, N. Sperandio, R. T. Lawlor, A. Scarpa, H. Molinari and M. Assfalg, *J. Proteome Res.*, 2012, **11**, 1274.
122. D. A. MacIntyre, B. Jimenez, E. J. Lewintre, C. R. Martin, H. Schaefer, C. G. Ballesteros, J. R. Mayans, M. Spraul, J. Garcia-Conde and A. Pineda-Lucena, *Leukemia*, 2010, **24**, 788.
123. J. M. Kinross, E. Holmes, A. W. Darzi and J. K. Nicholson, *Lancet*, 2011, **377**, 1817.

CHAPTER 4

In-Cell NMR Spectroscopy to Study Protein–Drug Interactions

JACQUELINE D. WASHINGTON, DAVID S. BURZ AND
ALEXANDER SHEKHTMAN*

Department of Chemistry, State University of New York at Albany,
1400 Washington Ave., Albany, NY 12222, USA
*Email: ashekhta@albany.edu

4.1 Introduction

In-cell NMR spectroscopy is valued for its ability to shed light on molecular structures and interactions under physiological conditions. One of its greatest potentials lies in the area of drug discovery where it can reveal molecular interactions at the atomic level at all stages of structure-based drug discovery. Drug therapy is based on the ability specifically to enhance or impede biological activity by binding to a single target protein or protein complex. Until recently, mostly *in vitro* techniques have been used to study macromolecular interactions that govern biological processes under conditions remote from those existing in the cell. With the advent of in-cell nuclear magnetic resonance (NMR) spectroscopy,[1] these processes can now be studied within a cellular environment.

By using in-cell NMR spectroscopy the structure of the target protein and drug candidate can be deduced through uniform and/or selective isotopic labeling, and both can be monitored simultaneously through the use of differential isotopic labeling; in addition, structural changes as a result of protein maturation processes or post-translational modifications (PTMs) can be analyzed. Non-specific and specific target interactions are distinguished

New Developments in NMR No. 2
New Applications of NMR in Drug Discovery and Development
Edited by Leoncio Garrido and Nicolau Beckmann
© The Royal Society of Chemistry 2013
Published by the Royal Society of Chemistry, www.rsc.org

through concentration-dependent chemical shift analyses and signal broadening, which also helps identify the interacting surface of the target, and the affinity of the interaction can be approximated. Techniques for incorporating isotopically labeled proteins into both prokaryotic and eukaryotic cells provide a means for drug delivery as well. Control experiments easily distinguish intracellular signals from extracellular, facilitating the identification of valid drug–protein interactions. Finally, high-throughput techniques are increasingly able to screen larger numbers of drug candidates without the need for extensive *in vitro* experimentation.

In this review, the basics of in-cell NMR spectroscopy are described with emphasis on applications that are relevant to the structure-based drug-discovery process. Applications such as determining three-dimensional (3D) structures *de novo*, mapping interaction surfaces and measuring changes in protein structure due to post-translational biochemical modifications, folding states, metabolic processing and protein maturation under physiological or near physiological conditions are described. The latest methods for incorporating target molecules into the intracellular milieu at sufficient concentrations for NMR spectroscopy are detailed along with the physiological relevance and pitfalls of each technique. Lastly, specific examples of drug–protein interactions identified by using in-cell NMR are presented.

4.2 In-Cell NMR Spectroscopy

In-cell NMR spectroscopy provides atomic level resolution of molecular structures under physiological conditions. NMR-active nuclei in biological macromolecules are extremely sensitive to changes in the chemical environment resulting from specific and non-specific binding interactions as well as changes due to biochemical modifications. This sensitivity can be exploited to screen for the binding of small drug-like molecules to selected targets in a cellular environment. These interactions alter molecular surfaces and can result in tertiary and quaternary conformational changes, all of which are reflected by changes in the chemical shifts of these nuclei.

Molecular interactions form the basis of biological activity. One of the most difficult challenges in studying protein interactions in cells arises from the fact that proteins bind many types of ligands, from protons and other ions to small drug-like molecules and ligands as well as large macromolecules such as nucleic acid polymers, membranes and other proteins. The range of affinity with which these ligands are bound varies over a broad range of concentrations, offering the possibility of unrestricted protein interactions. To dissect the differences between in-cell and *in vitro* studies of protein–protein interactions, the interactions between expressed proteins and other intracellular molecules can be divided into two parts: specific interactions with affinities lower than $10\,\mu M$ and non-specific interactions with affinities much greater than $10\,\mu M$.

To detect specific interactions by using in-cell NMR spectroscopy, interactor molecules have to be present in a stoichiometric ratio with the target protein so that the bulk of the labeled population is bound. Typically, target proteins are over-expressed to concentrations 10 to 100-fold greater than their physiological concentrations, which results in a large population of free protein and no changes in the NMR spectrum. Under these conditions, the contribution of undesired specific interactions between the target protein and intracellular binding partners to the resulting in-cell NMR spectrum can generally be ignored. Interactor molecules are later introduced at concentrations that are comparable to that of the target protein, thereby giving rise to specific interactions inside the cell that are detectable by in-cell NMR.

Non-specific interactions between labeled proteins and intracellular molecules are omnipresent and establish the proper physiological environment for the labeled proteins that uniquely distinguishes in-cell NMR from *in vitro* techniques. This is best evidenced by small differences in the NMR spectra and solution structures of proteins measured in cells and *in vitro*.[2] Because the concentrations of molecules inside eukaryotic and prokaryotic cells are very similar to each other, the in-cell NMR spectra of cytosolic proteins acquired in either cell type will also be very similar.

4.2.1   Cell Types

The choice of cell is a key element in the drug screening process. Methods exist for delivering labeled target molecules to the cytosol of both prokaryotic and eukaryotic cells. However, both cell types offer benefits and drawbacks. The delivery of the candidate drugs to the cell is also a critical feature. Finally, cell viability throughout the selection process must be considered.

4.2.1.1   Prokaryotic Cells

Early work in the field of in-cell NMR utilized prokaryotes as host cells.[1] Prokaryotic cells, specifically *Escherichia coli* (*E. coli*), were exploited because they are easy to handle and grow very rapidly. Proteins can be uniformly labeled with NMR-active isotopes, primarily ^{13}C and ^{15}N, and over-expressed to high enough intracellular levels to yield high-quality heteronuclear single quantum coherence (HSQC) spectra with little or no interfering background. Moreover, the ability to selectively label proteins and grow cells in deuterated medium provides the capability to study high-molecular-weight proteins in their native cellular environment. By appropriately engineering the transcriptional machinery of prokaryotic cells, post-translational modifications can be studied under conditions in which the effects of PTMs on protein structure can be examined without competing reactions, in effect, turning the bacteria into "cellular test tubes". Although *E. coli* are well-suited for simulating an intracellular environment to study eukaryotic proteins, in that they present the lowest potential for intrinsic binding partners, these cells lack the inherent ability to affect post-translational modifications and provide compartmentalization for selective activity.

4.2.1.2 Eukaryotic Cells

In-cell NMR in eukaryotic cells is hampered by difficulties in over-expressing isotopically labeled proteins inside the cell, the inability to regulate post-translational modification activities and the inefficient delivery of labeled proteins to the cytosol. Early in-cell NMR spectroscopy using eukaryotic cells was limited to very large and mechanically manipulable cells, such as *Xenopus laevis* (*X. laevis*) oocytes;[3–5] in these studies, isotope-labeled molecules were microinjected into the cells. Later studies developed methods to introduce isotopically labeled proteins into a range of cell types including human embryonic kidney (HEK293F) and HeLa cells.[6,7] The delivery methods included using active transport of isotope-labeled molecules into the cell by linking them to cell-penetrating peptides,[6] and, more recently, using a pore-forming bacterial toxin, streptolysin O (SLO),[7] to allow target molecules to diffuse passively into the cell interior before the pores are sealed. The primary problem of attaining a sufficiently high intracellular concentration of target to perform in-cell NMR has been largely overcome by these newer methods.

4.2.1.3 Cell Viability

In-cell NMR is an effective tool for studying proteins in their native environment. However, the long experimental times required to collect in-cell spectra can lead to a loss of cell viability. One consequence of reduced cell viability is protein leakage, which can lead to the detection of sharp signals from the protein molecules in the less viscous extracellular medium, thereby masking the broader signals obtained from intracellular proteins.[8] Another important factor that limits the acquisition time of in-cell NMR experiments is cell lysis, which also results in leakage of labeled target from the cells.[9] Stabilizing cells by using known cell protectants, such as glycerol or sucrose,[9] may also extend the in-cell NMR acquisition time. Control experiments are performed to ensure that the NMR signal arises from proteins located inside the cell.

Typically, following data acquisition, the cells are sedimented and a $^1H\{^{15}N\}$-HSQC spectrum of the supernatant is collected. Little or no signal above noise level is a good indication that the NMR signals were intracellular in origin. Li *et al.*[10] collected a spectrum of an in-cell sample supernatant before and immediately after acquiring the in-cell spectrum. A comparison between the spectra obtained using two proteins, the disordered, 14 kDa α-synuclein (αSN), and the 7 kDa globular protein chymotrypsin inhibitor-2 (CI2), showed that ~20% of the CI2 leaks from the cell, while αSN remains in-cell with only small metabolites found in the spectrum of the supernatant. To improve the in-cell spectrum of αSN, Li *et al.*[10] used an alginate encapsulation method[11] to stabilize the cells, which, in turn, yielded a clean $^1H\{^{15}N\}$-HSQC for αSN. These results show that each protein must be separately evaluated to optimize the in-cell signal and minimize protein leakage.

Xie *et al.*[12] added glycerol (to 10%) to samples as a cryoprotectant for prolonged storage of the cells at −80°C. Adding glycerol to the NMR buffer maintains the viability of *E. coli* cells at room temperature for more than four hours at densities sufficient to obtain in-cell NMR spectra and to minimize cell lysis. To minimize cellular degradation during their study of the tau protein, Bodart *et al.*[3] suspended *Xenopus* oocytes in a 20% Ficoll solution, which allowed the cells to remain stable overnight.[13] Control experiments showed no tau in the extracellular medium. These studies demonstrated methods for extending the lifetimes of cells in the NMR tube, a critical prerequisite for collecting data at physiological concentrations.

Another method used to assess bacterial cell viability is colony plating, in which the number of colonies grown on antibiotic selection plates inoculated with the in-cell sample before NMR spectroscopy is compared with the number of colonies grown using cells plated after acquiring experimental data. Colonies are counted by using a molecular imager. The cells are considered to be viable if the number of colonies on the plates are within 10% of each other.[14]

4.2.2   Target Labeling

The use of NMR spectroscopy to study biological macromolecules in living cells requires that the labeled targets must be easily distinguished from all other species present. Specific isotopic labeling schemes are employed to detect and resolve in-cell NMR protein resonance peaks and to yield the lowest background signals.

4.2.2.1   Backbone Group Probes

The most commonly used scheme employs $^{15}NH_4Cl$ as the sole nitrogen source to incorporate NMR-active ^{15}N nuclei into the peptide backbone of proteins over-expressed in bacterial cells. This uniform, [U-^{15}N], labeling strategy results in diminished background resonance peaks because bacterial growth is significantly reduced during protein over-expression. The protein resonance peaks of the resulting $^{1}H\{^{15}N\}$ HSQC spectra are adequately resolved provided that the protein is expressed to a sufficiently high concentration within the cell.[15]

Another backbone labeling scheme uses auxotrophic bacterial strains to incorporate high levels of amino acids that have been specifically enriched with ^{15}N.[16,17] Arginine, histidine and lysine are ideal for this type of labeling since these amino acids lie at the end of their respective biosynthetic pathways.[16] Labeled amino acids are chosen so as to comprise a substantial number of the total residues in the protein being expressed, since only those residues will contribute to the NMR spectrum. As a result of the less extensive labeling, the resulting in-cell spectrum will necessarily exhibit lower resolution than a spectrum obtained for a uniformly labeled target, but will be essentially devoid of background signals. Other amino acids can be used with the caveat that there will be an unavoidably larger background signal due to the presence of multiple metabolic products.

Proteins may also be uniformly labeled in bacteria by using ^{13}C as the sole carbon source during bacterial over-expression. Uniform $[U\text{-}^{13}C]$ labeling results in a high background due to metabolic reactions and the natural abundance of ^{13}C (1.1%), and is generally not used in this capacity.[18] Selective ^{13}C-isotopic labeling of constituent amino acids offers a better opportunity for enhancing the signal and minimizing the background. This isotope is best suited for labeling amino acid side chain residues, particularly methyl and methylene groups.[18]

Other labeling techniques, such as selective labeling of individual amino acids or the use of fluorine probes[19] may also be used in drug screening applications. These schemes will only be effective when the probe is close enough to the binding site to be affected by changes in the chemical environment.

4.2.3   In-Cell Delivery of Protein Target Molecules

Proper preparation of samples for in-cell NMR spectroscopy is critical for acquiring high-quality in-cell spectra. To study drug–protein interactions, it is critical that the concentration of labeled nuclei be high enough to provide well-resolved resonances for unambiguous identification. This often requires concentrations that exceed normal physiological levels by an order of magnitude or more. Such high concentrations can be problematic for assessing the efficacy of a drug–protein interaction, which would occur at much lower target protein concentrations. While the use of tightly regulated promoters for protein over-expression remains the method of choice for prokaryotic cells, new techniques for delivering target molecules to eukaryotic cells at physiological concentrations that provide a good signal-to-noise ratio (SNR) have been developed. These methods have their advantages and disadvantages, but all have proven effective in facilitating the acquisition of reproducible in-cell NMR spectra.

4.2.3.1   Endogenous Over-Expression

Endogenous over-expression has the advantage of creating the target protein within the cellular milieu in which it will be studied; no exogenous trans-locations or extracellular manipulations are involved, making it the simplest and most straightforward method for generating samples for use in in-cell NMR spectroscopy. The most popular and convenient method for producing labeled targets for in-cell NMR spectroscopy utilizes an inducible plasmid in bacterial cells to over-express isotopically labeled protein. The concentration of over-expressed protein can be controlled by using promoters in which the level of transcription is proportional to the concentration of the inducing molecule (*e.g.* the arabinose P_{BAD}[20] and rhamnose P_{RHA}[21]). Other methods for controlling the levels of intracellular concentration of over-expressed protein include varying the induction time and using plasmids with greater or lesser copy number. Polyacrylamide gel electrophoresis (PAGE) or Western blot analyses are typically employed to estimate the intracellular concentration of over-expressed target proteins. Endogenous over-expression in eukaryotic cells does not produce sufficiently high levels of target protein to perform in-cell NMR; in these cells, exogenous delivery is required to attain high levels of target protein.

4.2.3.2 *Microinjection*

Microinjecting labeled protein into cells has advantages over endogenously over-expressing labeled protein: the concentration of labeled protein can be accurately and reproducibly controlled and the background generated by over-expression is eliminated. The only significant contribution to background arises from the natural abundance of ^{13}C (1.1%) when employing such labeling. The primary disadvantages are that only large, easily manipulated cells, such as *X. laevis* oocytes, are amenable to this procedure, the process is tedious and there is an inherent variability in the oocytes.

Selenko *et al.*[5] used a robotic microinjection device to administer precise quantities of the B1 domain of streptococcal protein G (GB1)[22] into *X. laevis* oocytes.[23] GB1 was uniformly $[U-^{15}N]$ labeled in and purified from *E. coli*. As a prelude to in-cell NMR spectroscopy, the initial NMR experiments were performed using *X. laevis* egg extracts as a model system to mimic the intra-cellular milieu. Results showed that intracellular components do not appear to affect the folded state of the protein domain and revealed no intracellular binding partners. The spectrum of ^{15}N-GB1 obtained in extracts was virtually identical to that obtained *in vitro* using purified protein over the concentration range examined.

Next, *X. laevis* oocytes were injected with purified ^{15}N-GB1 to intracellular concentrations ranging from 50–500 µM. Intracellular ^{15}N-GB1 produced high-quality 2D spectra, generally matching peaks seen in spectra acquired *in vitro* by using purified protein. Some peaks displayed a distorted, split profile that was attributed to different intracellular environments. This observation was substantiated by dissecting single oocyte nuclei and demonstrating that GB1 was found in both the cytoplasm and the nucleus.

There were no discernible differences in the quality of the resulting NMR spectra acquired under identical conditions of temperature and data acquisition time; however, oocytes that had been automatically injected were found to be more viable than those that were manually injected. Control experiments showed no leakage of labeled protein from the cells. This work shows the feasibility of performing in-cell NMR spectroscopy in *X. laevis* oocytes and *Xenopus* egg extracts for high-resolution NMR spectroscopy in eukaryotic cells.

Bodart *et al.*[3] studied the neuronal protein, tau, inside *X. laevis* oocytes, an environment in which it is not normally found.[24] Tau interacts with the microtubular network present in oocytes, thus providing an opportunity to perform in-cell NMR spectroscopy on a target protein in the bound state. Tau was uniformly $[U-^{15}N]$ labeled in and purified from bacteria.[25,26] The protein was microinjected into *X. laevis* oocytes to a final concentration of $\sim 5\,\mu M$, which is close to physiological levels. The resulting $^1H\{^{15}N\}$-HSQC spectrum was similar to that of a purified sample obtained *in vitro*,[27] but the cross-peaks were broadened and many, attributed to free tau, were missing. After mechanically lysing the cells and re-acquiring the spectrum, the peak intensity increased but still lacked the peaks associated with free tau. Instead the spectrum resembled that of tau bound to tubulin.[27] Therefore, the peak broadening, which leads to lower resolution spectra, is likely due to tau interacting with tubulin and possibly other proteins present in the oocyte.

To optimize the reproducibility of spectra acquired from different in-cell preparations the same solution was injected into the equivalent site on oocytes selected in a qualitatively reproducible manner. They concluded that while the overall spectral profile is improved, the physiological state of the individual oocytes likely contributes to variations in signal strength, and therefore collecting a series of spectra on different samples will not be straightforward.

The results show that tau can be studied in cells at an intracellular concentration of $\sim 5\,\mu M$, extending the lower limit of concentration that can be studied by using in-cell NMR spectroscopy, but many of the resonances are broadened due to protein–protein interactions, yielding low resolution data.

4.2.3.3 Cell-Penetrating Peptides

Inomata *et al.*[6] utilized a heretofore untested procedure for introducing isotopically labeled proteins into eukaryotic cells by using cell penetrating peptides (CPPs) to transduce human HeLa cells. Uniformly $[U\text{-}^{15}N]$ labeled protein was covalently tagged or conjugated with a cell-penetrating peptide derived from the Tat protein (CPP$_{\text{Tat}}$) of human immunodeficiency virus-1 (HIV-1),[28] and incubated with human HeLa cells and pyrenebutyrate. Pyrenebutyrate mediates the translocation of CPP-tagged proteins into the cytosol.[29]

Experiments were performed using a ubiquitin (Ub-3A) that was mutated at three sites (L8A, I44A, V70A)[30] to preclude binding with ubiquitin-interacting proteins (UIPs). The $[U\text{-}^{15}N]$ labeled Ub-3A contained a C-terminal CPP fusion (Ub-3A-CPP$_{\text{Tat}}$). Following transduction, a well-resolved $^{1}H\{^{15}N\}$-HSQC spectrum was observed. The in-cell spectrum lacked the cross-peak corresponding to the C-terminal CPP residue and showed an intense signal corresponding to the C-terminal glycine (G76) of Ub-3A. Control experiments showed that Ub-3A-CPP$_{\text{Tat}}$ was cleaved between G76 and D77, presumably by endogenous ubiquitin-specific C-terminal proteases (deubiquitylases, DUBs)[31] to yield free Ub-3A.

Cleavage of the peptide tag is not only desirable, but necessary. It is well known that CPPs aggregate with many cytosolic components, including the inner plasma membrane.[32] Proteins bound to CPPs also form aggregates, producing broad, overlapping signals in NMR spectra. Experiments performed using an uncleavable CPP demonstrated that cleavage is essential for a well-resolved spectrum of the target protein. In addition, cleaved proteins exhibited a uniform intracellular distribution, whereas CPP-tagged proteins are heterogeneously dispersed throughout the cytosol. CPP cleavage is therefore essential for uniform protein distribution.

The concentration of transduced Ub-3A-CPP$_{Tat}$ was estimated to be 20–30 μM in the cells, about twice the physiological concentration. Cell leakage was negligible; cell viability and membrane integrity testing indicated that no significant toxicity was associated with pyrenebutyrate treatment. Ub-3A-CPP$_{Tat}$ was also transduced into monkey COS-7 cells, demonstrating the versatility of this method.

The in-cell spectrum of wild-type ubiquitin showed extensive peak broadening relative to the in-cell spectrum of Ub-3A and included the G76 cross-peak observed for cleaved Ub-3A. This indicated that at least some of the wild-type ubiquitin existed in a C-terminally unconjugated state. The differences in peak intensity between the two spectra likely reflect the inter-action of wild-type ubiquitin with endogenous proteins, since the mutated residues prevented binding with UIPs. A similar effect was seen in in-cell NMR experiments performed by using *X. laevis* oocytes.[4]

Other methods of CPP-linked transduction were tested. CPPs linked to cargo proteins by using disulfide bonds are cleaved in the cytosol by autonomous reduction.[33] The ^{1}H{^{15}N}-HSQC spectrum of GB1 conjugated to CPP$_{Tat}$ and transduced into HeLa cells was well resolved and virtually identical to that of a spectrum acquired *in vitro*. Similar results were obtained using Ub-3A.

A final delivery method was used to demonstrate the feasibility of studying drug–protein interactions. A [U-^{15}N] labeled fusion protein consisting of Ub containing an N-terminal CPP$_{Tat}$ and C-terminal FKBP12 domain was transduced into HeLa cells. Cleavage was predicted to be mediated by DUBs, which would release free FKBP12 into the cytosol. The resulting in-cell ^{1}H{^{15}N} correlation spectrum of FKBP12 was identical to the reference *in vitro* spectrum, whereas the contribution of CPP$_{Tat}$-Ub to the spectrum was minimal. It was concluded that DUB-mediated cleavage released free FKBP12 to yield an analyzable in-cell NMR spectrum while CPP$_{Tat}$-Ub underwent CPP-mediated aggregation within the cell.

The interaction of free FKBP12 with two immunosuppressant drugs, FK506 and rapamycin, was also examined. Transduced HeLa cells were incubated with either FK506 or rapamycin. The changes observed in the in-cell ^{1}H-^{15}N correlation spectra obtained for both drugs were distinct from one another but consistent with the reference spectra, acquired *in vitro*, of FKBP12 complexed with each drug. The results showed that exogenously administered drugs entered the cells and formed specific complexes with FKBP12.

This work demonstrated that using cell penetrating peptides to deliver sufficient concentrations of isotopically labeled proteins into eukaryotic cells to perform high-resolution in-cell NMR spectroscopy is effective for studying in-cell protein dynamics, protein–protein and protein–drug interactions.

4.2.3.4 *Membrane Permeabilization*

Ogino *et al.*[7] used the bacterial toxin streptolysin O (SLO) to deliver ^{15}N-labeled actin-sequestering protein, thymosin β4 (Tβ4), into human embryonic kidney (293F) host cells. SLO forms 35 nm diameter pores on cholesterol-containing plasma membranes, allowing molecules up to 150 kDa to diffuse into and out of the cell; the pores are then repaired by adding Ca^{2+} into the cytosol. Tβ4 conjugated with fluorescein isothiocyanate (FITC) and propidium iodide (PI), which stains cells with damaged plasma membranes, were used, in conjunction with flow cytometry, to assess the efficiency of pore formation and resealing, and to estimate the intracellular concentration of Tβ4. In-cell NMR was then used to directly observe the protein in living cells.

Most of the SLO-treated cells were FITC-positive and PI-negative relative to untreated cells, indicating that the formation of the pores and subsequent resealing were successful. These cells provide a good in-cell NMR signal resulting exclusively from intracellular protein. FITC-positive and PI-positive cells were also observed; these cells produce undesired, extracellular NMR signals due to leakage from unrepaired pores. The presence of FITC-negative cells indicates that the formation of pores was unsuccessful, and would result in no intracellular NMR signal.

The concentration of SLO was optimized to achieve a pore-forming efficiency of 50% with resealing efficiency of 70–80%. Higher concentrations of SLO resulted in more FITC-positive cells, but also more PI-positive cells. Confocal fluorescence microscopy and sodium dodecyl sulfate polyacrylamide gel electrophoresis (SDS-PAGE) revealed that the Tβ4 protein was evenly distributed throughout the cytoplasm and the nucleus, the cells remained viable and Tβ4 did not undergo proteolytic degradation. By comparing the fluorescent signals of lysates from cells treated with and without SLO, the intracellular concentration of Tβ4 was estimated to be 50 μM after incubating SLO-permeabilized cells with 1 mM Tβ4. Assuming 2×10^7 cells containing 50 μM Tβ4 in 200 μL of NMR buffer, the authors estimated that the final concentration of Tβ4 in the NMR sample would be 4 μM.

In-cell $^1H\{^{15}N\}$-HSQC spectra were collected on cells suspended in CD293 medium containing 20% D_2O and 30% RediGrad, a colloidal medium used to stabilize the cell suspension.[3] After the experiment, the cells were sedimented and a HSQC spectrum of the supernatant showed that almost all of the in-cell signals were intracellular. The *in vitro* spectrum of Tβ4 differed from the in-cell spectrum; the chemical shifts were consistent with those of chemically synthesized Tβ4 with an N-terminal acetylation modification. Interaction with G-actin was ruled out as no significant line broadening was observed and the affected residues were clustered near the N-terminus of Tβ4. Endogenous Tβ4 is known to be enzymatically acetylated in the cytosol.[34] The results showed that this PTM occurs in 293F cells.

The study demonstrates an alternative way to introduce isotopically labeled protein into mammalian cells using SLO reversible permeabilization and Ca^{2+} resealing. The method is advantageous because it is applicable to many types of cells, and does not require any synthetic PTMs or special equipment.

The methods described for delivering target proteins for in-cell NMR spectroscopy meet the requirements for accurately assessing drug–protein interactions. Intracellular concentrations of ≤50 μM can be typically achieved. Each target protein must be individually evaluated for optimum SNR, leakage, folded state and interactions with other endogenous molecules before being used in the drug screening process.

4.3   Methods and Applications

Identifying a drug–protein interaction is merely the first step of a complex process. The structural details and consequent activity resulting from such an interaction depend on the stage of maturation of the target protein including its folded state and other structural characteristics arising from metabolic processing and PTMs, and finally the structures of the unbound and bound states. This section describes studies that have illuminated the properties of target proteins that are relevant to the drug-discovery process.

4.3.1   STructural INTeractions NMR (STINT-NMR)

The primary criterion for determining drug–protein interactions entails verifying that a specific interaction is taking place. Burz *et al.*[35,36] developed an in-cell NMR-based method for mapping the structural interactions (STINT)-NMR that underline protein complex formation. The method entails sequentially expressing a target protein within a single bacterial cell in a time-controlled manner[37] and monitoring its interaction with another protein or small molecule by using in-cell NMR spectroscopy.[38] The resulting NMR data provide a complete titration of the interaction and define the structural details of the interacting surface of the target protein at atomic resolution. Unlike the case when interacting molecules are simultaneously over-expressed in labeled medium, in STINT-NMR spectral complexity is minimized because only the target protein is labeled with NMR-active nuclei, which leaves the interactor molecule(s) cryptic.

The target protein, whose NMR structure must be known, is first over-expressed on uniformly labeled (U-^{15}N) medium to yield a high-resolution, isotope-edited HSQC (^{1}H{^{15}N}-HSQC) backbone spectrum of the target protein inside the bacterial cells. The growth medium is changed, and the unlabeled interactor is over-expressed or delivered to the cytosol. As the interactor binds to the target, the HSQC spectrum of the target changes to reflect the different chemical environment for residues that have been affected by the binding interaction. The corresponding changes in the peak widths and chemical shifts of the target protein resonances define the interface between the protein and its ligand (interactor), providing atomic resolution information on the interactions.

Changes in chemical shifts and differential broadening of some assigned peaks may be more widespread, however, reflecting rearrangements of secondary structural elements, or a global or allosteric change in the conformation of the target. To assess accurately the changes in the NMR spectrum of a target molecule upon complexation with an interactor molecule, it is imperative that the resonance assignments of the target protein be known beforehand, and that the target is stable and well behaved in the absence of the interactor protein.

The efficacy of this method was demonstrated by over-expressing $[U\text{-}^{15}\text{N}]$ ubiquitin followed by over-expression of either one of two ubiquitin ligands containing the ubiquitin interacting motif (UIM):[39] a 28-amino acid peptide from ataxin-3 (ataxin-3 ubiquitin interacting motif, AUIM; $\sim 4\,\text{kDa}$) or the signal-transducing adapter molecule-2 (STAM2;[40] $\sim 50\,\text{kDa}$) in unlabeled medium. AUIM binds ubiquitin *in vitro* with $\sim 230\,\mu\text{M}$ affinity[41,42] and STAM2, which contains two ubiquitin interacting surfaces, binds with a higher ($\sim 10\,\mu\text{M}$) overall affinity. These two systems simulated a range of protein–protein interaction affinities and molecular weights.

The $^1\text{H}\{^{15}\text{N}\}$-HSQC spectrum of ubiquitin, which was maintained at a single concentration, changed as the concentration of interactor was increased (Figure 4.1). The chemical shift changes were mapped onto the 3D structure of ubiquitin. Peaks that underwent substantial (>0.1 ppm) chemical shifts came exclusively from surface-exposed amides. Control experiments using a mutant AUIM, which does not bind to ubiquitin, demonstrated that the chemical shift changes result from specific interactions and not merely over-expression of the interacting molecule.

The method is limited primarily by the concentration level of interacting protein that can be achieved and can be used to study interacting proteins whose structure is unknown, since only one of the interacting species is labeled. Proteins that are difficult to purify or are proteolytically labile can be studied, since there is no need for purification. A limiting consideration is the integrity

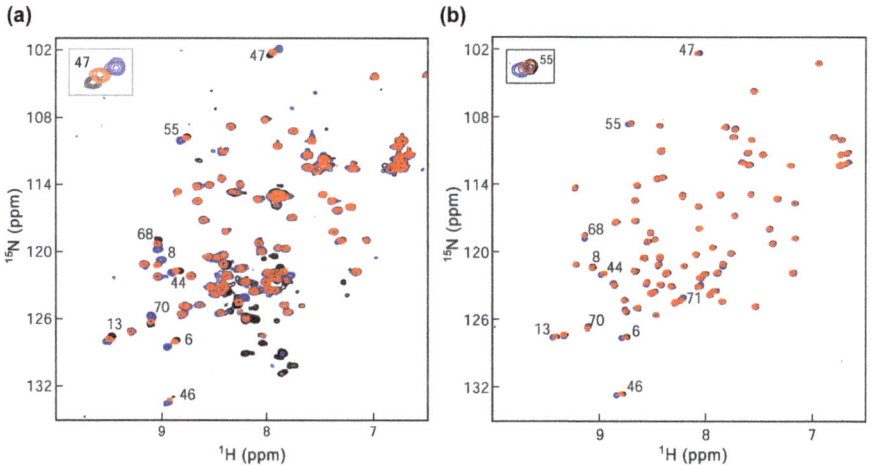

Figure 4.1 Ubiquitin–ligand complexes in *E. coli*. Left: overlay of $^1\text{H}\{^{15}\text{N}\}$-HSQC spectra of *E. coli* after 3-hour over-expression of $[U\text{-}^{15}\text{N}]$-Ubiquitin and 0-h (black), 2-h (red) and 3-h (blue) over-expression of ataxin-3 ubiquitin inter-acting motif (AUIM). Individual peaks exhibiting large chemical shifts are labeled with corresponding assignments. Right: overlay of $^1\text{H}\{^{15}\text{N}\}$-HSQC spectra of free $[U\text{-}^{15}\text{N}]$ ubiquitin (black) and $[U\text{-}^{15}\text{N}]$-ubiquitin-AUIM complexes at a molar ratio of 1:1 (red) and 1:2 (blue).
Reproduced with permission from Burz *et al.*[35,36] © 2006 Nature Publishing Group.

of the interacting molecules. For example, over-expressed proteins may degrade into components that bind non-specifically to each other, thereby presenting multiple and/or incorrect interaction surfaces. For this reason, the target should be stable over the course of the experiment. Should sample stability become questionable, SDS-PAGE and Western blots can be used to assess the extent of degradation.

4.3.2   Protein Maturation

Assessing the state of maturation of a target protein is critical to evaluating its interaction with drug candidates. Different structures may bind small molecules with varying affinities and the resulting structural complexes may exhibit profoundly different activities. Banci *et al.*[43] used in-cell NMR to characterize the maturation of human copper, zinc superoxide dismutase-1 (hSOD1) in *E. coli*. Apo-hSOD1 has been linked to familial amyotrophic lateral sclerosis (ALS), a fatal motor neurodegenerative disease.[43–45] In its immature form, hSOD1 lacking one Zn^{2+} ion and one catalytic Cu^+ ion per subunit and having a misfolded structure is believed to play a critical role in ALS pathology.[46] NMR experiments were performed to determine the folded state of apo-hSOD1 in the cytosol, to characterize zinc binding both in-cell and *in vitro* and to monitor how disulfide bond formation affects protein self-association.

To identify the folded state of apo-hSOD1, uniformly labeled protein over-expression was induced in metal-free ^{15}N-labeled M9 medium and ^{1}H{^{15}N}-selective optimized flip-angle short-transient heteronuclear multiple quantum coherence (SOFAST-HMQC)[47] in-cell spectra were collected. Under favorable relaxation conditions, ^{1}H{^{15}N}-SOFAST-HMQC can lead to up to a ten-fold decrease in the NMR acquisition time, as compared to ^{1}H{^{15}N}-HSQC, with minimal loss in signal sensitivity. The in-cell spectrum contained peaks between 7.5 and 8.5 ppm, implying that the protein was largely unfolded in the cytosol. Following the in-cell NMR experiment, cells were lysed and the spectrum was re-acquired. A few broadened amide peaks in the lysate spectrum indicate the presence of some structure; however, most of the peaks correspond to unfolded protein. The lysate spectrum is similar to that of monomeric apo-hSOD1 (E,E-hSOD1$^{SH\text{-}SH}$), which contains reduced cysteines, and exhibits many of the peaks corresponding to the unstructured regions seen in the in-cell spectrum.

To determine if the protein is completely unfolded or partially unfolded inside the cell, the in-cell spectrum was compared to that of E,E-hSOD1$^{SH\text{-}SH}$ denatured with guanidinium chloride *in vitro*. The *in vitro* spectrum exhibits peaks only in the region expected for unfolded protein, but with different chemical shift values from that of the in-cell spectrum. For example, the cross-peak of the side chain of W32, which is located in a β-strand of the hSOD1 β-barrel, has a chemical shift of ∼10.3 ppm in-cell and ∼10.1 ppm for *in vitro* denatured protein. Furthermore, the in-cell chemical shift of W32 is the same as in the lysate spectrum suggesting that apo-E,E-hSOD1$^{SH\text{-}SH}$ is not completely unfolded in the cytosol and that cross-peaks attributed to folded protein are broadened due to chemical or conformational exchange phenomena.

To rule out that the loss of signal was due to chemical or conformational exchange, ^{1}H{^{15}N}-SOFAST-HMQC spectra were collected at 500 MHz (315 K) and at 800 MHz (288 K and 298 K) to monitor the change in the SNR, which is expected to increase with increasing magnetic field strength and temperature. No increase of the SNR was observed, suggesting that exchange phenomena were not responsible for the loss of signals.

Signal loss or line broadening may also be due to oligomerization or interactions with other cellular components. The formation of soluble apo-hSOD1 oligomers *in vitro* leads to the complete loss of signals in the HSQC spectrum. Therefore the presence of in-cell signals combined with the recovery of missing signals upon cell lysis indicates that no apo-hSOD1 oligomers are forming in the cytosol. Because apo-hSOD1 is present in high intracellular concentrations, it is capable of engaging in a myriad of non-specific interactions. Each interaction places apo-hSOD1 in different chemical environments with correspondingly different chemical shifts. The resulting cellular anisotropy gives rise to inhomogeneous broadening in which cross-peaks from folded regions exhibit a larger chemical shift dispersion that is undetectable by NMR.

Binding of zinc to a native site on the protein is critical for maturation of hOSD1. To understand the metal binding properties of hSOD1, protein over-expression was induced in ^{15}N-labeled M9 medium supplemented with 10 µM to 1 mM ZnSO$_4$. The in-cell ^{1}H{^{15}N}-SOFAST-HMQC spectrum of hSOD1 expressed in zinc medium was different from that of in-cell apo-hSOD1. The appearance of dispersed peaks and the broadening of some signals in the unfolded region of the spectrum indicate that in-cell, hSOD1 binds zinc when zinc is in excess relative to the total amount of protein over-expressed in the cell. The in-cell spectrum is very similar to that of *in vitro* E,Zn-hSOD1^{SH-SH} (the species with one zinc bound to the zinc binding site) and not with that of *in vitro* Zn,Zn-hSOD1^{SH-SH} (the non-physiological species with two zinc ions bound to both metal binding sites). Amide signals from G61 and T135, located near the metal binding sites and indicators of binding site occupancy, have chemical shift values that are close to those of the *in vitro* E,Zn-hSOD1^{SH-SH} spectrum and further from the signals corresponding to the *in vitro* spectrum of Zn,Zn-hSOD1^{SH-SH}. The results demonstrate that intracellular hSOD1 exhibits greater selectivity for metal ion binding than *in vitro* because only the native zinc binding site is occupied in-cell when the concentration of zinc ions is in molar excess. This is unlike what is observed at physiological pH *in vitro* in which sub-stoichiometric amounts of zinc give rise to a mixture of three species, E,E-hSOD1^{SH-SH}, E,Zn-hSOD1^{SH-SH} and Zn,Zn-hSOD1^{SH-SH}, and when two equivalents of zinc are present per protein subunit, only Zn,Zn-hSOD1^{SH-SH} is formed. Following the in-cell experiments, the cells were washed with metal-free medium and NMR spectra were collected on the cleared lysates. Only the Zn,Zn-hSOD1^{SH-SH} species was detected, indicating that there was an excess of zinc still present inside the cells and zinc was bound to the hSOD1 copper binding site.

To determine the redox state of the cysteines, ^{1}H{^{15}N}-SOFAST-HMQC spectra were acquired on both in-cell and *in vitro* samples grown in metal-free media. Monitoring the ^{1}H-^{15}N signals of selectively labeled cysteines showed that in apo-hSOD1, all four are in a reduced state in the cytosol and in cell lysates. Following exposure to air, oxidized C57 and C146 displayed chemical shifts consistent with an oxidized state and identical to those of *in vitro* dimeric E,E-hSOD1^{S-S43}. Upon reduction by adding dithiothreitol (DTT), C57 and C146 exhibit peaks consistent with the reduced state observed in-cell and in cell lysates. When the same analysis was performed for cells and lysates in which hSOD1 was expressed in the presence of Zn(II), the C57 and C146 residues of E,Zn-hSOD1 were found to be reduced in the cytosol.

Disulfide bond formation is tightly linked to dimerization. The spectra of ^{15}N-cysteine labeled protein provide evidence on the assembly state of hSOD1 in the cytosol. *In vitro*, E,E-hSOD1^{SH-SH} is monomeric, while E,E-hSOD1^{S-S} and E,Zn-hSOD1^{SH-SH} are homodimers. The *in vitro* NMR spectra of both E,E-hSOD1 and E,Zn-hSOD1 exhibit a large chemical shift difference for C146 between the oxidized and reduced states, because C146 is directly involved in disulfide formation. C6 is used as a marker to indicate the state of assembly because C6 is located close to the interaction surface of the homodimer. Upon E,E-hSOD1 dimerization and consequent disulfide bond formation, the chemical environment of C6 changes. For E,Zn-hSOD1, which is dimeric regardless of the oxidation state, the chemical shift of C6 remains unchanged. The chemical shift of C6 for both in-cell E,E-hSOD1^{SH-SH} and E,Zn-hSOD1^{SH-SH} species matches that of the corresponding *in vitro* species, which suggests that, in the cytosol, E,E-hSOD1^{SH-SH} is monomeric and E,Zn-hSOD1^{SH-SH} is dimeric.

This study also investigated the maturation of hSOD1. The processes observed included determining the folded and assembly states of the protein, the redox state of key cysteine residues as an indicator of quaternary structure and zinc binding to native and non-native sites on the protein. The work demonstrates that linked functions necessary for protein maturation can be identified and dissected from one another by using in-cell NMR spectroscopy. The results provide insight into the myriad of structures that are available for drug binding interactions that may affect the onset of familial ALS.

4.3.3   Metabolic Processing of Proteins

The effect of metabolic processing on in-cell NMR spectra was examined by Sakai *et al.*[4] Reduced spectral quality resulted from interactions involving the target protein due to endogenous enzymatic activity and binding of metal ions. The results highlight the effect of protein–protein interactions on NMR spectra and the structure and activity of the resulting species. Uniformly [U-^{15}N] labeled ubiquitin and its derivatives were expressed and purified in bacteria.[48] The purified protein was microinjected into *X. laevis* oocytes to a maximum final concentration of $\sim 100\,\mu$M. The magnitude of the signal for the ^{1}H{^{15}N}-HSQC spectrum of wild-type ubiquitin injected into the oocytes was very weak. Control experiments verified that the ubiquitin spectrum originated from an intracellular environment and not from labeled protein that had leaked from the cells into the surrounding medium.

Experiments were performed using a series of mutant ubiquitins, in which residues implicated in binding UIPs, L8, I44 and V70,[30] were changed to alanines. The results showed that the in-cell $^1H\{^{15}N\}$-HSQC spectrum of ubiquitin could be largely recovered when the UIP interface is perturbed. The perturbation disrupts protein–protein interactions, which decreases peak broadening and resolves the spectrum. Ubiquitin molecules carrying single mutations affecting the UIP binding site (L8A-D77, I44A-D77 and V70A-D77) partially restored the in-cell ubiquitin $^1H\{^{15}N\}$-HSQC spectrum and the protein containing all three mutations ((L8A, I44A, V70A)-D77) dramatically improved the spectrum. Thus, the inability to resolve an in-cell spectrum for ubiquitin results from its interaction with endogenous UIPs, preventing a sufficient in-cell concentration of free ubiquitin for analysis by NMR spectroscopy.

The ubiquitin used in this study contained a C-terminal D77 residue (Ub-D77), whereas mature ubiquitin has a G76 in that location.[49] The D77 protein, therefore, mimics a ubiquitin precursor and can act as a substrate for ubiquitin C-terminal hydrolase (UCH). The cross-peaks attributed to the G76 and D77 amide groups are missing from the in-cell spectrum of Ub-D77, and a single peak corresponding to wild-type G76 is present. This suggests that the G76-D77 bond in this mutant is cleaved in the oocyte.

By pre-injecting oocytes with ubiquitin aldehyde, which specifically inhibits UCH,[49] the in-cell spectrum of ^{15}N-Ub-D77 showed G76 and D77 cross-peaks and a G76 signal that was weaker than control cells pre-injected with water that displayed no G76 and D77 cross-peaks. Further experiments showed a dose dependence on residual UCH activity. Thus, by pre-injecting oocytes with ubiquitin aldehyde, the UCH activity was inhibited, thereby reducing and curtailing ubiquitin processing. These observations suggest that in-cell NMR spectroscopic analyses of metabolic processing may be possible under select conditions.

Sakai *et al.*[4] also microinjected [U-^{15}N] labeled calmodulin into oocytes. The in-cell $^1H\{^{15}N\}$-HSQC spectrum of calmodulin, acquired without Ca^{2+} in the buffer, resembled that of apo-calmodulin acquired *in vitro*, except the peaks were broader, indicating that the majority of the in-cell calmodulin was Ca^{2+}-free. When an excess of Ca^{2+} was co-injected with the protein, the overall spectrum changed and the resulting cross-peaks were further broadened, precluding exact assignments of individual resonances. However, more than ten cross-peaks that are consistent with those observed for Ca^{2+}-bound calmodulin were identified. Since Ca^{2+}-bound calmodulin is more likely to interact with downstream effector proteins than apo-calmodulin,[50,51] the reduced spectral quality observed in the presence of Ca^{2+} suggests interactions between injected and endogenous proteins.

4.3.4 Regulation of Post-Translational Modifications (In-Cell Biochemistry)

PTMs can alter the structure, reactivity and activity of proteins. Each resulting conformation has the potential to interact with drug-like molecules differently to produce different outcomes. Burz and Shekhtman[52] developed an in-cell methodology to introduce PTMs onto interactor proteins in bacterial cells and identify the changes in the interaction surface of a target protein when bound to the biochemically modified interactors. Modifying the interactor protein causes structural changes that manifest on the interacting surface of the target protein and these changes are monitored by using STINT-NMR.[35,36] By creating specific populations of post-translationally modified target proteins, the contribution of these modifications to drug–protein interactions can be quantitatively evaluated.

The method was used to phosphorylate tyrosine residues on the STAM2 and hepatocyte growth factor-regulated tyrosine kinase substrate (Hrs), and to identify changes in the interaction surface of ubiquitin resulting from these post-translational modifications. STAM2 and Hrs are components of an endocytic pathway present in eukaryotic cells. Each binds ubiquitin *via* a UIM.[53] In addition, STAM2 has a VHS (Vps-27, Hrs and STAM) domain capable of binding ubiquitin. Evidence suggests that receptor sorting through endocytosis and subsequent degradation is controlled by ubiquitination of both the internalized receptors and components of the endocytic machinery.[54] The work demonstrated that post-translational modification of over-expressed proteins in bacterial cells can be regulated by tight temporal control over protein expression, a process dubbed "in-cell biochemistry".

To study ubiquitin binding to STAM2 and Hrs, [U-^{15}N]-ubiquitin over-expression was induced prior to or following 3 or 4 hours of over-expression of STAM2 or Hrs alone, or co-expression of both. STAM2 and Hrs were phosphorylated by inducing over-expression of the constitutively active Src-family tyrosine-kinase, Fyn, for the final 2 hours of STAM2, Hrs or STAM2-Hrs over-expression. STAM2 and Hrs phosphorylation were confirmed by using Western blots and mass spectroscopic analyses.

The STINT-NMR spectra of ubiquitin revealed no changes in the interaction surface when bound to non-phosphorylated or phosphorylated Hrs (Figure 4.2). The spectrum of ubiquitin interacting with STAM2 revealed that a smaller surface is involved in the interaction with phosphorylated STAM2, corresponding to the loss of the interaction surface attributed to the VHS domain (Figure 4.2). Mutational analysis revealed that two STAM2 tyrosines, Y371 and Y374, located in the conserved ITAM (immunoreceptor tyrosine-based activation motif) domain, were responsible for these changes. The ITAM domain has been identified as necessary for tyrosine phosphorylation of STAM2 by Janus kinase-1 (Jak1).[55]

A similar result was obtained for the interaction between ubiquitin and the STAM2-Hrs heterodimer: ubiquitin interacted with the phosphorylated ternary complex in much the same way that it interacted with phosphorylated STAM2 and phosphorylated Hrs, involving contact with only the UIMs of both interactor proteins (Figure 4.2). The commensurate weakening of the binding due to the loss of the second interaction surface is consistent with the idea that phosphorylation mediates the disassembly of ubiquitin-mediated scaffold complexes during endocytosis.

The introduction of in-cell biochemistry using STINT-NMR facilitates biochemical modification and examination of protein–protein interaction

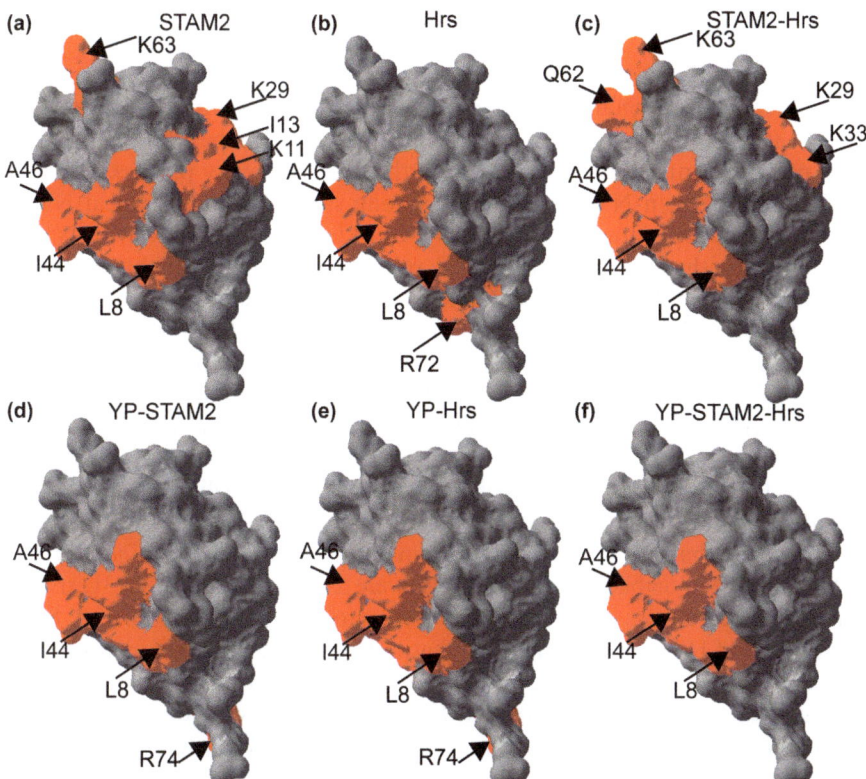

Figure 4.2 Interaction surface maps of ubiquitin–ligand complexes. Interaction surface of ubiquitin mapped onto the 3D structure of ubiquitin (Protein Data Bank code 1D3Z). Individual residues exhibiting either a chemical shift change >0.05 ppm or significant differential broadening are indicated in red. All perturbed residues lie on the ubiquitin surface and, therefore, reflect changes in the interaction surface of the molecule rather than changes in tertiary or quaternary structure. (a) STAM2-ubiquitin interaction; (b) Hrs-ubiquitin interaction; (c) STAM2-Hrs-ubiquitin interaction; (d) phosphorylated STAM2-ubiquitin interaction (YP-STAM2); (e) phosphorylated Hrs-ubiquitin interaction (YP-Hrs); (f) phosphorylated STAM2-Hrs-ubiquitin interaction (YP-STAM2-Hrs). Ubiquitin ligands are indicated in each panel.
Reproduced with permission from Burz and Shekhtman.[52] © 2008 Burz, Shekhtman.

surfaces at the atomic level. The ability to control PTMs in an environment that normally lacks the ability to provide such modifications, *i.e.* bacterial cells, affords an opportunity to examine the effects of PTMs on protein structure without competing reactions, thus tailoring the ability of the target to interact optimally with the desired drug candidate. The methodology can be applied to any stable target molecule and may be extended to include other post-translational modifications.

4.3.5 *De Novo* Structure Determination

The exact conformation of an intracellular protein or protein complex may be quite different from what is observed *in vitro*, and the structural changes may affect the resulting activity. Therefore, determining the structure of a drug–protein complex in-cell is critical to the drug-discovery process. The low sensitivity of protein NMR requires very long sampling times: typical NMR experiments collect data for 1–2 days. Without fresh supplies of nutrients and gases, along with waste removal, *E. coli* cells cannot survive for that long. To minimize the sampling time, Sakakibara *et al.*[2] employed a novel non-linear sampling scheme[3,24,25] to solve *de novo* the structure of TTHA1718, a 66 amino acid, putative heavy-metal binding protein from *Thermus thermophilus* (*T. thermophiles*). By combining this scheme with maximum entropy processing,[56] the time required to collect sufficient multidimensional data was reduced to several hours.

TTHA1718 was over-expressed in uniform isotopic-labeling medium from *E. coli* to an intracellular concentration of 3–4 mM. Control experiments demonstrated that the NMR signal originated from protein within the cells. Sixty-three out of 66 backbone resonances were assigned by using six 3D triple-resonance experiments with fresh samples prepared for each experiment; 86% of Hα, 71% of Hβ and 34% of the aliphatic ^{1}H/^{13}C side chain resonances were identified. NMR spectra collected *in vitro* using purified protein were used to confirm assignments from in-cell NMR. After 6 hours of data acquisition colony plating tests showed that the viability of the bacteria was 85%.

To improve the SNR, each 3D experiment was performed several times and the data sets were combined to enhance the protein signal. The 3D experiments were bracketed by a 2D ^{1}H{^{15}N}-HSQC experiment, which acted as a control to ensure that only data collected from intact cells were included in the combined data set. Each control spectrum, following a 3D experiment, was compared against the reference ^{1}H{^{15}N}-HSQC spectrum collected at the beginning of the run. If the control spectrum was significantly different from the reference spectrum, the 3D spectrum that was collected prior to the control was not added to the composite data.

Side chain methyl groups of Ala, Leu and Val were selectively ^{13}C-labeled allowing 78% of their side chain resonances to be assigned. In addition, out of a possible total of 148 NOEs involving methyl groups 69 of 89 long-range NOEs were assigned. NOE distance restraints, backbone torsion restraints and restraints for hydrogen bonds were incorporated into the calculation of a 3D structure for TTHA1718 using CYANA.[57] The final calculated structure was similar to the structure determined *in vitro* for purified TTHA1718, and had a root mean square deviation (rmsd) of 0.96 Å and a backbone rmsd of 1.16 Å.

This study demonstrates the feasibility of determining high-resolution 3D structures of proteins in living bacterial cells. The work was possible because of innovations that allow rapid data collection and unambiguous identification of long-range nuclear Overhauser effect (NOE) interactions based on selective labeling of methyl groups. Although eukaryotic cell stability is still a limiting factor, in-cell structure determination in eukaryotic cells may be possible since labeled proteins can be introduced into *Xenopus* oocytes at concentrations up to 0.7 mM.[5] In-cell determination of the structure of drug–protein complexes will provide a linchpin to understanding the mechanisms of drug activity.

4.4 Drug Screening

Efficient delivery of drug candidates to the intracellular medium is a necessary component of the drug-discovery process. In addition to monitoring the target protein and identifying potential structures for drug design, the ability to control the concentration of drug candidate species during in-cell screening is key for the evaluation of drug–protein interactions and the effectiveness of potential drug therapies. This section details in-cell NMR-based strategies that apply to the drug-discovery process including the delivery of drugs to intracellular targets, chemical shift mapping of changes in the target or drug upon binding and a high-throughput method to rapidly screen small-molecule libraries for compounds capable of strengthening or weakening protein–protein interactions within a biomolecular complex.

4.4.1 In-Cell Delivery of Drug Molecules

Efficient delivery of drugs to the intracellular medium and the ability to control the concentration of drug candidate species during in-cell screening is key to evaluating drug–protein interactions and potential drug therapies. Arnesano *et al.*[58] used solution and in-cell NMR spectroscopy to explore intracellular drug delivery and the interaction of cisplatin (cis-$[PtCl_2(NH_3)_2]$) with Atox1, a human copper chaperone that mediates Cu(I) delivery to copper-transporting P-type ATPases.[59,60] Cisplatin is a platinum (Pt) based anticancer drug that forms a stable complex with DNA to interfere with DNA replication and transcription processes.[61] Cellular uptake of platinum drugs is tightly connected to Cu transport by virtue of surface-exposed metal binding sites,[58,62] which are used to bind metallo-drug compounds. Atox1 has a $\beta_1\alpha_1\beta_2\beta_3\alpha_2\beta_4$ ferredoxin-like structure and a metal-binding site, CxxC, located in the loop between β_1 and α_1, which is highly conserved among metallochaperones and soluble Cu-transport ATPases.[60,62]

To identify the residues of Atox1 that are involved in platinum coordination $^1H\{^{15}N\}$-HSQC NMR spectra were collected on a purified [*U*-^{15}N]-Atox1 in the absence and presence of unlabeled cisplatin. Chemical shift changes were localized to the CxxC Cu(I)-binding region of Atox1 (Figure 4.3). $^1H\{^{13}C\}$-HSQC NMR spectra, in which cysteine residues were selectivity labeled, revealed fast-to-intermediate exchange between free and coordinated Atox1 in the presence of cisplatin, due to the complete broadening of cross-peaks Cys12 and Cys15. To monitor the fate of the amine ligands released by cisplatin upon interaction with Atox1, $^1H\{^{15}N\}$-HSQC NMR spectra were collected on [*U*-^{15}N]-cisplatin in the absence and presence of unlabeled Atox1.

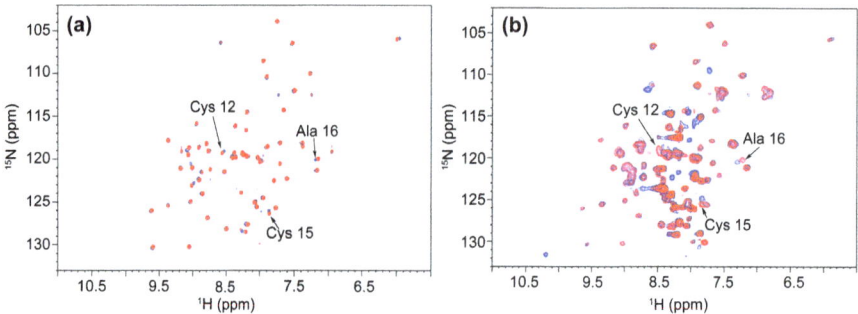

Figure 4.3 Atox1-cisplatin complexes in *E. coli*. (a) Overlay of ^{1}H{^{15}N}-HSQC spectra of free [U-^{15}N]-Atox1 in the absence (red) and in the presence (blue) of 1 mol equivalent of cisplatin (4 h after mixing). (b) Overlay of ^{1}H{^{15}N}-SOFAST-HMQC spectra of *E. coli* after a 4 h induction of [U-^{15}N]-Atox1 in the absence (red) and in the presence (blue) of 10 µM cisplatin. Cys12, Cys15 and Ala16 cross-peaks are indicated with arrows. Reproduced with permission from Arnesano *et al.*[58] © 2011 American Chemical Society.

In the presence of Atox1 the amine signal decreased as the reaction progressed, indicating the release of ligand.

In-cell ^{1}H{^{15}N}-SOFAST-HMQC spectra[47] collected on [U-^{15}N]-Atox1 over-expressed in *E. coli* were similar to the *in vitro* spectrum and consistent with a reduced apo state for Atox1. Cross-peaks for the in-cell spectrum were broadened relative to the *in vitro* spectrum of purified protein. Controls indicate that there was no leakage of protein from the cells. Purified Atox 1 is sensitive to oxidation and must be handled with precautions. Typically Atox1 is handled in the presence of a reducing agent such as DTT or TCEP (*tris*(2-carboxyethyl)phosphine); however, these materials are good ligands for Pt(II); this problem was overcome by observing Atox1 in-cell where the intracellular environment provided suitable conditions for preserving the protein in its active form.

To determine whether platinum drugs could enter the cell and interact with Atox1, [U-^{15}N]-Atox1 was over-expressed in *E. coli* grown in the presence of platinum drugs. Cisplatin concentrations >10 µM inhibited cell growth, thus setting an upper limit to the extracellular concentration used in these experiments. The uptake of cisplatin was estimated by atomic absorption spectroscopy to be ~10%. In-cell, ^{1}H{^{15}N}-SOFAST-HMQC spectra were collected on cells grown with and without cisplatin. The largest spectral changes were observed for residues that comprise the CxxC metal-binding motif, Cys12, Cys15 as well as Ala16. The changes were similar to those observed for the cisplatin-Atox1 interaction *in vitro* suggesting that cisplatin entered the cell and was bound to Atox1 (Figure 4.3) without inducing any major structural re-arrangements.

This work highlights the advantages of in-cell NMR spectroscopy to probe intracellular drug delivery by monitoring a known drug–protein interaction.

The binding-competent form of the protein target was identified and the fate of the ligand upon binding was elucidated. In addition, the intracellular milieu provided a stable environment for the otherwise labile drug, confirming its efficacy *in vivo*.

4.4.2 Screening of Small Molecule Interactor Library NMR (SMILI-NMR)

The ability to rapidly screen large numbers of candidate compounds, *i.e.* high-throughput screening (HTS), presents one of the major challenges to the drug-discovery process. HTS in combination with in-cell NMR requires minimal sample preparation and eliminates the need for extensive protein purification. The technique takes advantage of the fact that drug candidates can pass through the cell membrane to interact with the cytosolic target. Furthermore, SMILI-NMR can be automated by making use of robotic HTS accessories available for modern NMR spectrometers, such as liquid handlers and NMR tube changers.

Xie *et al.*[12] developed an in-cell NMR spectroscopy-based screening procedure, screening of small molecule interactor library (SMILI-NMR), to rapidly screen for compounds capable of disrupting and enhancing specific interactions between two or more components of a biomolecular complex. SMILI-NMR utilizes STINT-NMR[35,36] technology to produce biomolecular complexes inside the cell in which one of the constituent proteins is uniformly [U-^{15}N] labeled with NMR-active nuclei. By monitoring the in-cell NMR spectrum of the labeled protein, the formation of high-affinity ternary complexes is observed. STINT-NMR analyzes changes in structure induced by binding of a small drug-like molecule that disrupts or enhances the stability of the complex and reveals biologically relevant, functional interaction surfaces. In this way, STINT-NMR serves as a direct assay for protein–drug inter-actions, identifying small drug-like molecules that bind to this surface, and thus facilitates HTS.

A system of two interacting proteins, FKBP and FRB, was used as a model to show the effectiveness of SMILI-NMR to screen small molecules that facilitate heterodimerization. The FKBP-FRB interaction constitutes one of the immunomodulatory systems in mammalian cells.[63] In complex with rapamycin, a macrolide antifungal antibiotic currently undergoing clinical trials for a variety of cancer treatments,[64,65] FKBP binds to FRB. When [U-^{15}N]-FKBP is over-expressed in bacterial cells, the ^{1}H{^{15}N}-HSQC spectrum shows no well-resolved peaks, implying that the single species is part of a large complex and therefore invisible to NMR. When unlabeled FRB is then over-expressed in the same cells, the NMR spectrum of FKBP becomes evident but only at the highest FRB concentrations, indicating the formation of a complex. Similar results were obtained when [U-^{15}N]-FRB and unlabeled FKBP were sequentially over-expressed in the same cells. These observations demonstrated that creating a proper protein complex is necessary for high-resolution studies.

Adding rapamycin to the cell suspension results in visible changes in the $^{1}H\{^{15}N\}$-HSQC-spectrum of $[U-^{15}N]$-FKBP indicating the formation of a high-affinity ternary complex between FKBP-rapamycin and FRB (Figure 4.4). Adding rapamycin to cells over-expressing labeled FKBP in the absence of FRB or labeled FRB in the absence of FKBP does not produce an NMR spectrum. In each case, co-expression of the second protein is required to generate an in-cell NMR spectrum.

A dipeptide chemical library[66] composed of 17×17 dipeptides[66,67] was chosen to provide a collection of compounds that are capable of interacting with the target molecule at a detectable level. These compounds provide

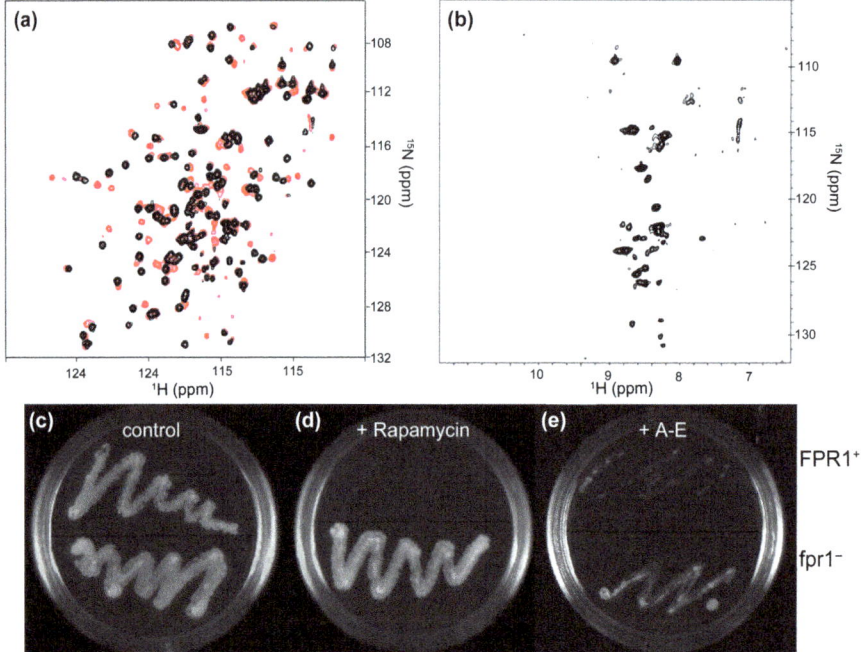

Figure 4.4   Top: ternary FKBP-FRB-ligand complexes in *E. coli*. (a) Overlay of $^{1}H\{^{15}N\}$ HSQC spectra of *E. coli* after 4 h of over-expression of $[U-^{15}N]$-FKBP and 4 h sequential over-expression of FRB in the absence (black) and presence (red) of 150 μM rapamycin. (b) $^{1}H\{^{15}N\}$ HSQC spectrum of *E. coli* after 4 h of over-expression of $[U-^{15}N]$-FKBP and 4 h sequential over-expression of FRB in the presence (red) of 5 mM A-E. Bottom: yeast assay for biological activity of the dipeptide, A-E. Isogenic haploid yeast strains (*S. cerevisiae*) that express (*FPR1*$^{+}$) or lack (*frp1*$^{-}$) the FKBP proline isomerase were grown for 3 days on YPD medium. (c) Control plate. (d) 100 μM rapamycin. Expression of FKBP allows the formation of a toxic FKBP-rapamycin-FRB bio-complex. (e) 5 mM A-E. Results indicate that the dipeptide induces formation of a bio-complex similar to that induced by rapamycin. The reduced growth in the *frp1*$^{-}$ strain likely reflects the weaker affinity of A-E for FKBP.
Reproduced with permission from Xie *et al.*[12] © 2009 American Chemical Society.

suitable starting points for subsequent optimization into credible drug candidates[68] and are considered as potential sources of novel lead structures. Dipeptides have been shown[67,69] to be an excellent starting point for drug design since (1) they can be prepared at low cost, (2) a library containing only 289 compounds can provide a data set that spans a broad spectrum of physicochemical properties, (3) no deconvolution is required to identify the lead structures[68] and (4) dipeptides can pass through the cell membrane and interact with the target protein directly *in vivo*.[70,71]

A standard procedure called the matrix method, in which compounds located in one row or one column of a matrix plate are mixed and tested, is used to screen the library. Individual mixtures are examined for their ability to change the in-cell NMR spectrum of the FKBP-FRB bio-complex. Samples exhibiting similar spectral changes, located at the intersection of rows and columns, are used in the second round of screening to deconvolute and validate the initial findings. In this way, a matrix of 289 (17×17) compounds can be screened by examining 34 (17 + 17) samples.

Most of the dipeptide mixtures showed no interaction with the FKBP target. The mixture of A-X (where X is all possible amino acids) elicited a totally different spectrum from that of the rapamycin-induced ternary complex (Figure 4.4). Formation of the dipeptide-induced complex resulted in extreme broadening and the disappearance of some peaks in the NMR spectrum at the highest concentration used. The mixtures of D-X, T-X, L-X, X-E, X-I, X-A, X-T also caused a similar broadening of the spectrum. Thus, the dipeptides located at the intersection of rows A-X, D-X, T-X, L-X and columns X-E, X-I, X-A, X-T were screened in a second round and titrated into cells individually. Only A-E showed the same interaction with FKBP, suggesting that A-E facilitated heterooligomerization of FKBP and FRB. Competition experiments with rapamycin confirmed that A-E binds specifically to the FKBP-FRB complex. Further confirmation that A-E exhibited biological activity comparable to that of rapamycin was obtained by using a yeast growth assay (Figure 4.4).

SMILI-NMR provides an important means to bridge the gap between biochemical identification of small ligands capable of interfering with target bio-complexes and the biological activity resulting from the inhibition of cellular processes by these ligands. The extension of SMILI-NMR to eukaryotic cells, in which compartmentalization may be important for drug–protein interactions, will rely on the development of new techniques for monitoring protein targets in eukaryotic cells.

4.5 Conclusions

In-cell NMR offers many possibilities for drug development as improvements in instrumentation and new procedures for introducing target proteins and drug candidates into prokaryotic and eukaryotic cells are developed. In-cell NMR reveals how the intracellular environment influences molecular structure and how, in turn, these structures affect protein–drug interactions.

Understanding the details of such interactions can lead to the design of new molecules capable of binding with high affinity to proteins at different stages of spatial and temporal maturation. Indeed, it is at this level of specificity where the efficacy of drug activity is best optimized, and it is through the use of in-cell NMR that these properties are best examined.

References

1. Z. Serber and V. Dotsch, *Biochemistry*, 2001, **40**, 14317.
2. D. Sakakibara, A. Sasaki, T. Ikeya, J. Hamatsu, T. Hanashima, M. Mishima, M. Yoshimasu, N. Hayashi, T. Mikawa, M. Walchli, B. O. Smith, M. Shirakawa, P. Guntert and Y. Ito, *Nature*, 2009, **458**, 102.
3. J. F. Bodart, J. M. Wieruszeski, L. Amniai, A. Leroy, I. Landrieu, A. Rousseau-Lescuyer, J. P. Vilain and G. Lippens, *J. Magn. Reson.*, 2008, **192**, 252.
4. T. Sakai, H. Tochio, T. Tenno, Y. Ito, T. Kokubo, H. Hiroaki and M. Shirakawa, *J. Biomol. NMR*, 2006, **36**, 179.
5. P. Selenko, Z. Serber, B. Gadea, J. Ruderman and G. Wagner, *Proc. Natl Acad. Sci. USA*, 2006, **103**, 11904.
6. K. Inomata, A. Ohno, H. Tochio, S. Isogai, T. Tenno, I. Nakase, T. Takeuchi, S. Futaki, Y. Ito, H. Hiroaki and M. Shirakawa, *Nature*, 2009, **458**, 106.
7. S. Ogino, S. Kubo, R. Umemoto, S. Huang, N. Nishida and I. Shimada, *J. Am. Chem. Soc.*, 2009, **131**, 10834.
8. C. O. Barnes and G. J. Pielak, *Proteins*, 2011, **79**, 347.
9. C. Cruzeiro-Silva, F. P. Albernaz, A. P. Valente and F. C. Almeida, *Cell. Biochem. Biophys.*, 2006, **44**, 497.
10. C. Li, L. M. Charlton, A. Lakkavaram, C. Seagle, G. Wang, G. B. Young, J. M. Macdonald and G. J. Pielak, *J. Am. Chem. Soc.*, 2008, **130**, 6310.
11. W. M. Kühtreiber, R. P. Lanza and W. L. Chick, *Cell Encapsulation Technology and Therapeutics*, Birkhauser, Boston, 1999.
12. J. Xie, R. Thapa, S. Reverdatto, D. S. Burz and A. Shekhtman, *J. Med. Chem.*, 2009, **52**, 3516.
13. H. P. Richter, C. Hoock and B. Neumcke, *Biol. Cell.*, 1995, **84**, 129.
14. T. Maniatis, E. F. Fritsch and J. Sambrook, *Molecular Cloning: A Laboratory Manual*, Cold Spring Harbor Laboratory, Cold Spring Harbor, NY, 1982.
15. M. S. Almeida, W. Peti and K. Wuthrich, *J. Biomol. NMR*, 2004, **29**, 453.
16. D. S. Waugh, *J. Biomol. NMR*, 1996, **8**, 184.
17. L. P. McIntosh and F. W. Dahlquist, *Q. Rev. Biophys.*, 1990, **23**, 1.
18. Z. Serber, W. Straub, L. Corsini, A. M. Nomura, N. Shimba, C. S. Craik, P. Ortiz de Montellano and V. Dotsch, *J. Am. Chem. Soc.*, 2004, **126**, 7119.
19. C. Li, G. F. Wang, Y. Wang, R. Creager-Allen, E. A. Lutz, H. Scronce, K. M. Slade, R. A. Ruf, R. A. Mehl and G. J. Pielak, *J. Am. Chem. Soc.*, 2010, **132**, 321.

20. L. M. Guzman, J. J. Barondess and J. Beckwith, *J. Bacteriol.*, 1992, **174**, 7716.
21. A. Haldimann, L. L. Daniels and B. L. Wanner, *J. Bacteriol.*, 1998, **180**, 1277.
22. A. M. Gronenborn, D. R. Filpula, N. Z. Essig, A. Achari, M. Whitlow, P. T. Wingfield and G. M. Clore, *Science*, 1991, **253**, 657.
23. K. Schnizler, M. Kuster, C. Methfessel and M. Fejtl, *Receptors Channels*, 2003, **9**, 41.
24. D. L. Gard and M. W. Kirschner, *J. Cell. Biol.*, 1987, **105**, 2191.
25. G. Lippens, J. M. Wieruszeski, A. Leroy, C. Smet, A. Sillen, L. Buee and I. Landrieu, *ChemBioChem*, 2004, **5**, 73.
26. C. Smet, A. Leroy, A. Sillen, J. M. Wieruszeski, I. Landrieu and G. Lippens, *ChemBioChem*, 2004, **5**, 1639.
27. A. Sillen, P. Barbier, I. Landrieu, S. Lefebvre, J. M. Wieruszeski, A. Leroy, V. Peyrot and G. Lippens, *Biochemistry*, 2007, **46**, 3055.
28. S. R. Schwarze, A. Ho, A. Vocero-Akbani and S. F. Dowdy, *Science*, 1999, **285**, 1569.
29. T. Takeuchi, M. Kosuge, A. Tadokoro, Y. Sugiura, M. Nishi, M. Kawata, N. Sakai, S. Matile and S. Futaki, *ACS Chem. Biol.*, 2006, **1**, 299.
30. L. Hicke, H. L. Schubert and C. P. Hill, *Nat. Rev. Mol. Cell. Biol.*, 2005, **6**, 610.
31. F. Loison, P. Nizard, T. Sourisseau, P. Le Goff, L. Debure, Y. Le Drean and D. Michel, *Mol. Ther.*, 2005, **11**, 205.
32. P. A. Wender, W. C. Galliher, E. A. Goun, L. R. Jones and T. H. Pillow, *Adv. Drug Deliv. Rev.*, 2008, **60**, 452.
33. I. Giriat and T. W. Muir, *J. Am. Chem. Soc.*, 2003, **125**, 7180.
34. A. Wodnar-Filipowicz, U. Gubler, Y. Furuichi, M. Richardson, E. F. Nowoswiat, M. S. Poonian and B. L. Horecker, *Proc. Natl Acad. Sci. USA*, 1984, **81**, 2295.
35. D. S. Burz, K. Dutta, D. Cowburn and A. Shekhtman, *Nat. Methods*, 2006, **3**, 91.
36. D. S. Burz, K. Dutta, D. Cowburn and A. Shekhtman, *Nat. Protoc.*, 2006, **1**, 146.
37. R. Lutz and H. Bujard, *Nucleic Acids Res.*, 1997, **25**, 1203.
38. Z. Serber, L. Corsini, F. Durst and V. Dotsch, *Methods Enzymol.*, 2005, **394**, 17.
39. P. P. Di Fiore, S. Polo and K. Hofmann, *Nat. Rev. Mol. Cell. Biol.*, 2003, **4**, 491–497.
40. K. G. Bache, C. Raiborg, A. Mehlum and H. Stenmark, *J. Biol. Chem.*, 2003, **278**, 12513.
41. A. Shekhtman and D. Cowburn, *Biochem. Biophys. Res. Commun.*, 2002, **296**, 1222.
42. E. Mizuno, K. Kawahata, M. Kato, N. Kitamura and M. Komada, *Mol. Biol. Cell.*, 2003, **14**, 3675–3689.
43. L. Banci, L. Barbieri, I. Bertini, F. Cantini and E. Luchinat, *PLoS One*, 2011, **6**, e23561.

44. M. J. Lindberg, L. Tibell and M. Oliveberg, *Proc. Natl Acad. Sci. USA*, 2002, **99**, 16607.
45. Y. Furukawa and T. V. O'Halloran, *J. Biol. Chem.*, 2005, **280**, 17266–17274.
46. L. Banci, I. Bertini, M. Boca, S. Girotto, M. Martinelli, J. S. Valentine and M. Vieru, *PLoS One*, 2008, **3**, e1677.
47. P. Schanda, E. Kupce and B. Brutscher, *J. Biomol. NMR*, 2005, **33**, 199.
48. T. Tenno, K. Fujiwara, H. Tochio, K. Iwai, E. H. Morita, H. Hayashi, S. Murata, H. Hiroaki, M. Sato, K. Tanaka and M. Shirakawa, *Genes Cells*, 2004, **9**, 865.
49. S. C. Johnston, S. M. Riddle, R. E. Cohen and C. P. Hill, *EMBO J.*, 1999, **18**, 3877.
50. J. S. Nair, C. J. DaFonseca, A. Tjernberg, W. Sun, J. E. Darnell, Jr., B. T. Chait and J. J. Zhang, *Proc. Natl Acad. Sci. USA*, 2002, **99**, 5971.
51. M. Zhang and T. Yuan, *Biochem. Cell Biol.*, 1998, **76**, 313.
52. D. S. Burz and A. Shekhtman, *PloS One*, 2008, **3**, e2571. doi:10.1371/journal.pone.0002571.
53. S. Polo, S. Sigismund, M. Faretta, M. Guidi, M. R. Capua, G. Bossi, H. Chen, P. De Camilli and P. P. Di Fiore, *Nature*, 2002, **416**, 451.
54. L. Hicke and H. Riezman, *Cell*, 1996, **84**, 277.
55. A. Pandey, M. M. Fernandez, H. Steen, B. Blagoev, M. M. Nielsen, S. Roche, M. Mann and H. F. Lodish, *J. Biol. Chem.*, 2000, **275**, 38633.
56. E. D. Laue, M. R. Mayger, J. Skilling and J. Staunton, *J. Magn. Reson.*, 1986, **68**, 14.
57. P. Guntert, *Methods Mol. Biol.*, 2004, **278**, 353.
58. F. Arnesano, L. Banci, I. Bertini, I. C. Felli, M. Losacco and G. Natile, *J. Am. Chem. Soc.*, 2011, **133**, 18361.
59. R. McRae, B. Lai and C. Fahrni, *J. Biol. Inorg. Chem.*, 2010, **15**, 99.
60. A. K. Boal and A. C. Rosenzweig, *Chem. Rev.*, 2009, **109**, 4760.
61. E. R. Jamieson and S. J. Lippard, *Chem. Rev.*, 1999, **99**, 2467.
62. F. Arnesano, L. Banci, I. Bertini, S. Ciofi-Baffoni, E. Molteni, D. L. Huffman and T. V. O'Halloran, *Genome Res.*, 2002, **12**, 255.
63. M. W. Harding, A. Galat, D. E. Uehling and S. L. Schreiber, *Nature*, 1989, **341**, 758.
64. M. E. Pavel, J. D. Hainsworth, E. Baudin, M. Peeters, D. Hörsch, R. E. Winkler, J. Klimovsky, D. Lebwohl, V. Jehl, E. M. Wolin, K. Oberg, E. van Cutsem, J. C. Yao and RADIANT-2 Study Group, *Lancet*, 2011, **378**, 2005.
65. M. Agulnik, *Cancer*, 2012, **118**, 1486.
66. D. C. Horwell, W. Howson, G. S. Ratcliffe and D. C. Rees, *Bioorg. Med. Chem. Lett.*, 1993, **3**, 799.
67. D. C. Horwell, J. Hughes, J. C. Hunter, M. C. Pritchard, R. S. Richardson, E. Roberts and G. N. Woodruff, *J. Med. Chem.*, 1991, **34**, 404.
68. P. Boden, J. M. Eden, J. Hodgson, D. C. Horwell, J. Hughes, A. T. McKnight, R. A. Lewthwaite, M. C. Pritchard, J. Raphy,

K. Meecham, G. S. Ratcliffe, N. Suman-Chauhan and G. N. Woodruff, *J. Med. Chem.*, 1996, **39**, 1664.

69. M. A. Ondetti, B. Rubin and D. W. Cushman, *Science*, 1977, **196**, 441.
70. E. R. Olson, D. S. Dunyak, L. M. Jurss and R. A. Poorman, *J. Bacteriol.*, 1991, **173**, 234.
71. H. Saito and K. Inui, *Am. J. Physiol.*, 1993, **265**(2 Pt 1), G289.

PART II
THE WHOLE ORGANISM *IN VIVO*

Anatomy, Function, Metabolism and Cellular/Molecular Imaging

CHAPTER 5

Increased Sensitivity Using Cryogenic Radiofrequency Coils: Application to In Vivo Phenotyping of Mice

J. KLOHS,[a] A. SEUWEN,[a] A. SCHRÖTER,[a] D. MAREK[b] AND M. RUDIN*[a,c]

[a] Institute for Biomedical Engineering, ETH & University of Zurich, 8093 Zurich, Switzerland; [b] Bruker BioSpin AG, CH-8117 Fällanden, Switzerland; [c] Institute of Pharmacology and Toxicology, University of Zurich, 8052 Zurich, Switzerland
*Email: rudin@biomed.ee.ethz.ch

5.1 Introduction

5.1.1 Mouse Models in Biomedical Research

Genetically engineered mice have become indispensable tools for investigating the function of genes during development and in human disease *in vivo*. Such transgenic mouse lines have been included in studies of molecular mechanisms and cellular processes underlying pathological states, and in the discovery of novel and optimization of available therapeutic regimes.[1] Magnetic resonance imaging (MRI) and magnetic resonance spectroscopy (MRS) have become increasingly important in this context, motivated by the need to phenotype these engineered mouse lines. Structural MRI scans can reveal

New Developments in NMR No. 2
New Applications of NMR in Drug Discovery and Development
Edited by Leoncio Garrido and Nicolau Beckmann
© The Royal Society of Chemistry 2013
Published by the Royal Society of Chemistry, www.rsc.org

anatomical differences such as altered organ size or morphology, or eventually the occurrence of pathological tissue transformation.[2–5] Quantitative analyses of the physical properties of molecules in tissue, such as water diffusion, relaxation parameters T_1, T_2 and T_2* or magnetization transfer provide valuable information on tissue microstructure.[6,7] Beyond anatomy, functional MRI (fMRI) can measure physiological processes such as tissue perfusion, heart function and brain activity.[8–11] Importantly, the non-invasive nature of the technique enables the repetitive assessment of animals, which is attractive not only from an animal welfare perspective but also in view of the high value of genetically engineered mice. MRI allows conducting longitudinal studies to investigate changes in phenotype with age.

5.1.2   Challenges in MRI Phenotyping of Mice

Many of the diagnostic MRI techniques used in clinical imaging can be applied in the mouse. However, given the small dimensions of mice the major challenge is obtaining MRI and MRS data with comparable signal-to-noise ratio (SNR) and relative resolution as human clinical scans. Voxel dimensions must be decreased by a factor of 10–15 in each dimension to adapt the scans to the mouse scale. Special applications like microstructural imaging and the confection of MRI based digital atlases require even ultrahigh spatial resolution.[3,12] Furthermore, comprehensive phenotyping may involve the acquisition of multiple imaging data sets (multi-parametric imaging) with large groups of animals at repeated time points; hence throughput becomes an essential requirement. Substantial efforts have been made to increase the sensitivity as defined by the SNR for imaging applications in mice. SNR may be improved by increasing either the signal and/or by decreasing the noise of the experiment. Correspondingly, several strategies to improve SNR have been implemented. Increasing the magnetic field strength increases the polarization according to Boltzmann's law. MRI systems with field strengths up to 17.6 T are currently being used for mouse imaging.[13] However, moving to very high fields involves several drawbacks such as increased T_1 relaxation time and decreased T_2 and T_2* relaxation times as well as increased power deposition in the tissue.[14] Susceptibility effects scale with the magnetic field strengths causing imaging artifacts at high fields such as geometrical distortions and intravoxel signal loss. Alternatively, the MR signal can be increased by signal averaging, which is inefficient as the SNR gain scales with the square root of the number of acquisitions. This has been extensively used for example in the MRI examination of fixed mouse brains where high-resolution images have been obtained at the expense of long acquisition times.[3] However, for *in vivo* applications the acquisition times need to be kept as short as possible to avoid lengthy periods of anesthesia, to guarantee physiological stability of the animal under study and to enable reasonable throughput. Alternatively SNR may be increased by reducing the denominator, *i.e.* the noise. As will be outlined in Section 5.2.1, there are two dominant sources for noise in MRI/MRS experiments: the sample itself and the scanner electronics, in particular the receiver coil and the preamplifier. For small sample volumes (typically a few mL) the sample noise becomes sufficiently low such that reducing the electronic noise can lead to significant improvements in the final SNR. Hence, reducing electronic noise by the use of cryogenic RF coils (cryogenic probes, CRP) is an attractive concept that has already been widely and successfully implemented in high-resolution nuclear magnetic resonance spectroscopy (see also Section 1.5.1 of Chapter 1).

5.2 Cryogenic RF Coils as a Tool in Small Animal MRI

5.2.1 Improving Signal-to-Noise Ratio by Lowering the Coil Temperature: Theoretical Background

The SNR is defined as the ratio of the signal divided by noise. Both must be considered when the SNR of a system is to be analyzed. Here a short summary of the relevant contributions is presented. We are referring to Figure 5.1(a), and looking at the terminal A-B of the entire coil, before it is connected to a tuning and matching network and the preamplifier:

1. *Signal*: According to the principle of reciprocity, a magnetic moment $\vec{M}$ at a given voxel position within the sample rotating with a frequency ω around the main magnetic field aligned along the z-axis induces an electromotive force of the amplitude E in a detector coil (Figure 5.1(b)), given by

$$E = -\frac{\partial}{\partial t}\left(\frac{\vec{B}_1 \cdot \vec{M}}{I}\right) = \omega \cdot \left(\frac{\vec{B}_1 \cdot \vec{M}}{I}\right) \quad (5.1)$$

where $\vec{B}_1$ is the RF magnetic field at the voxel position induced by a current I in the coil.[15]

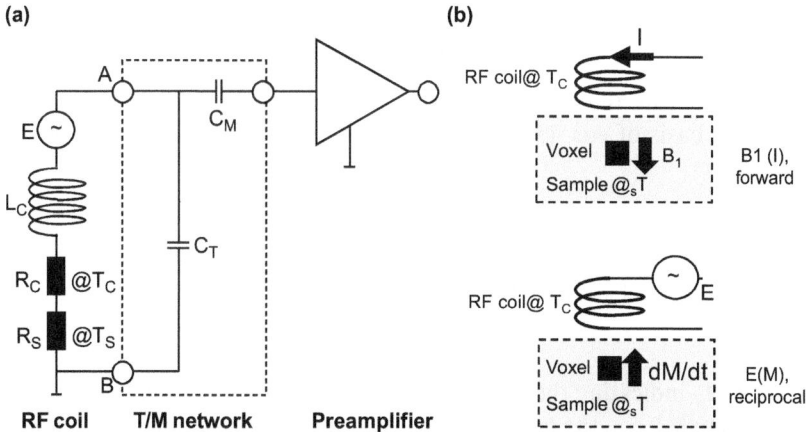

Figure 5.1 (a) Signal and noise analysis in the detector chain sample/coil, tuning and matching (T/M) network and preamplifier. Sample and coil noise are lumped in the terms RS and RC. (b) Principle of reciprocity. A current in a radiofrequency coil induces a magnetic field B1. Analogously, a change in the sample magnetization dM/dt induces an electromotive force in the rf coil.

2. *Noise*: There are a number of noise sources that contribute to the total noise. The most important ones are the coil noise, sample noise, pre-amplifier and spin noise. The spin noise can be neglected with the parameters in a typical MRI setup and for simplicity reasons we also neglect the preamplifier noise in the following analysis. The noise can then be regarded as arising from the thermal noise of the lossy elements in the receiving chain. In the following, we will have a closer look at the coil and sample noise (Figure 5.1(a)).

The noise of the RF coil (inductance L_C) is given by the coil RF resistance R_C, which is at the coil temperature T_C. When the coil is in the proximity of the sample, additional losses occur due to the RF losses of the sample, which at the typical MRI frequencies are predominantly caused by the electrical conductivity of the sample. These losses result in an additional effective damping of the receiver coil. This can be described by an effective resistance R_S, which appears in series with the coil resistance R_C. One important point to note is that for noise analysis, each resistance has to have an associated temperature. Here, the temperature of the sample effective resistance R_S is the sample temperature T_S. This is very relevant for the performance and limitations of cryogenic coils, as the Johnson noise power of a resistor is proportional to its temperature, so the same amount of coil damping (or equivalent resistance) induced from the sample is much more detrimental for the SNR than the same damping from within the coil itself, which is obvious in eqn (5.2).

The total RMS voltage spectral noise density n, which appears in the detection coil (for simplicity, we are looking at the signal and noise at the terminals A-B, before connecting to the preamplifier *via* the T/M network consisting of the capacitors C_T und C_M), is thus given by

$$n = \sqrt{n_S^2 + n_C^2} = \sqrt{4 \cdot k_B \cdot (R_S \cdot T_S + R_C \cdot T_C)} \tag{5.2}$$

with k_B being the Boltzmann constant. The SNR can then be estimated on the basis of eqs. (5.1) (the signal being proportional to E) and (5.2), yielding

$$SNR \propto \cdot \frac{\omega \cdot (B_1/I)}{\sqrt{4 \cdot k_B \cdot (R_S \cdot T_S + R_C \cdot T_C)}} \cdot M_T \cdot V_{voxel} \cdot \sqrt{\frac{N_{FE}}{BW}} \cdot \sqrt{N_{PE}} \cdot \sqrt{N_A}$$

$$\tag{5.3}$$

with M_T indicating the transverse magnetization component giving rise to the signal, V_{voxel} the voxel volume, N_{FE} and N_{PE} the number of frequency and phase encoding steps, BW the acquisition bandwidth and N_A the number of averages collected.[16] Eqn (5.3) comprises terms characteristic for the MR pulse sequence, such as the amplitude of the transverse magnetization generated and the geometric parameters defining the image, which shall not be further considered here. We are rather interested in optimizing SNR for a given pulse sequence and a given set of parameters defining field of view and voxel dimensions, hence we only consider the first term of eqn (5.3).

3. *Quantitative SNR analysis*: In the review paper of Darrasse and Ginefri,[16] analytical expressions can be found for the two noise contributions in eqn (5.2) and (5.3). The setup consisted of a circular normal metal coil of radius r_{coil} with ℓ turns located at a distance d from a lossy semi-infinite sample characterized by an electrical conductivity value σ (Figure 5.2a). The resistances R_S and R_C are given by

$$R_S = \frac{2}{3\pi} \cdot \sigma \cdot \mu_0^2 \cdot \omega^2 \cdot n^2 \cdot r_{coil}^3 \cdot \arctan\left(\frac{\pi \cdot r_{coil}}{8 \cdot d}\right) \qquad (5.4)$$

$$R_C = \sqrt{\frac{1}{2} \cdot \rho_C \cdot \mu_0 \cdot \omega} \cdot \ell^2 \cdot \left[\xi \cdot \frac{r_{coil}}{r_{wire}}\right] \qquad (5.5)$$

In eqn (5.5) the coil resistivity is given by ρ_C, while ξ denotes the influence of the proximity effect given by the RF current distribution around the coil wire. While the sample resistance R_S (and hence the noise power) increases with the third power of the coil radius (defining the sampling volume for surface coils) and the square of the resonance frequency, the coil resistance R_C increases only as the square root of ω and is typically constant with the coil radius (*i.e.* because one would typically scale the wire radius r_{wire} together with the coil radius r_{coil}). Hence it becomes obvious that the relative contribution of coil noise to the total noise increases for small values of r_{coil} (as typically encountered in mouse MRI applications) and also for low values of ω.

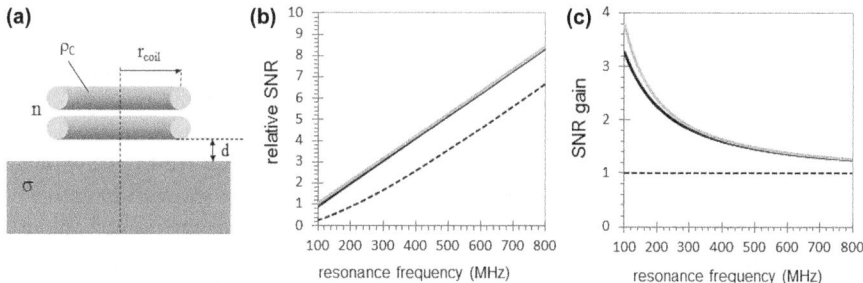

Figure 5.2   (a) Scheme of circular surface copper coil of n turns with a radius r_{coil} positioned at a distance d from a lossy sample characterized by the electrical conductivity σ. (b) Relative SNR as a function of ^{1}H resonance frequency for RF coils operating at 293 K (gray), 30 K (black) and 0 K (dotted). (c) Gain in SNR for a receiver coil operating at 30 K (black) and 0 K (dotted) as compared to RT detection. Parameters used for the simulation were $r_{coil} = 10\,\text{mm}$, $n = 1$, $\sigma = 0.66\,\text{S/m}$, $\xi \cdot r_{coil}/r_{wire} = 40$, $T_S = 310\,\text{K}$ (37 °C), $T_{C,RT} = 293\,\text{K}$, $T_{C,LT} = 0\,\text{K}$, 30 K, 293 K, $d = 2\,\text{mm}$, $\rho_c(293\,\text{K}) = 1.68 \times 10^{-8}\,\Omega\text{m}$, $\rho_c(30\,\text{K}) = 1.73 \times 10^{-9}\,\Omega\text{m}$.

Comparing the SNR for a low-temperature system (SNR_{LT}) to a geometrically identical system at room temperature with the SNR_{RT}, the SNR gain G achieved by cooling the RF coil can then be written as

$$G \equiv \frac{SNR_{LT}}{SNR_{RT}} = \frac{\sqrt{(R_S \cdot T_S + R_{C,RT} \cdot T_{C,RT})}}{\sqrt{(R_S \cdot T_S + R_{C,LT} \cdot T_{C,LT})}} \qquad (5.6)$$

Here, $R_{C,RT}$ denotes the resistance of the room temperature coil and $R_{C,LT}$ that of the cryogenic coil. Similarly, the temperatures of the room temperature coil and of the cryogenic coils are given by $T_{C,RT}$ and $T_{C,LT}$, respectively.

In order to maximize G, the numerator has to be maximized and the denominator minimized. The effective sample resistance R_S is given by the sample constitution (conductivity, shape, *etc.*) together with the coil geometry, which for the sake of this comparison are considered to stay constant. T_S denotes the sample temperature, which is the animal's body temperature, which of course cannot be reduced for *in vivo* work. In the denominator, only the expression $R_{C,LT} \cdot T_{C,\,LT}$ can be varied, *i.e.* reduced by cooling. The cooling has a two-fold effect: while the reduced coil temperature $T_{C,LT}$ by itself already reduces the noise, by cooling a normal metal, the electrical resistance is also lowered at the same time. In case of superconductors, the resistance can be lowered even further.[17]

From eqn (5.6) it becomes obvious that significant SNR gains can be achieved only if the coil noise was dominant to start with, *i.e.* if

$$R_{C,RT} \cdot T_{C,RT} \geq R_S \cdot T_S \qquad (5.7)$$

Figures 5.2(b),(c) show the overall SNR and the SNR gain for circular receiver coils of 10 mm radius as a function of the frequency. It becomes clear that all signals increase with increasing field strength and frequency, which becomes apparent from eqn (5.1). However, the gain in SNR by lowering the coil temperature is most prominent for low frequencies. For example, for the parameters used in the computation we estimate an SNR gain of 3.7 at 100 MHz as compared to 2.2 at 400 MHz. Hence, cryogenic receiver coils are of particular interest when using protons at low magnetic fields B_0 or X-nuclei with a small gyromagnetic ratio. Similarly, it can readily be shown using eqn (5.1) and (5.3)–(5.5) that the SNR gain decreases with increasing coil radius as the sample noise becomes more dominant, as mentioned above.

It should be noted that while eqn (5.4) and (5.5) are approximations to a various extent; *e.g.* eqn (5.4) assumes a sample extending over the entire semi-infinite half space. This is of course not the case in real applications, as the sample is of limited size. This results in reality in a lower sample resistance than assumed on the basis of eqn (5.4) and consequently in the ensuing results: lower sample resistance results in a lower sample noise contribution and thus a potentially higher SNR gain upon lowering the coil temperature. So the results derived above are underestimating the possible SNR gain to a certain degree. Given the approximations taken in the consideration, in particular neglecting the other elements of the detector chain, these relations and predicted gain factors should be taken as guiding principles only. More accurate theoretical predictions would require a detailed analysis of RF signals and losses of a specific detector setup.

5.2.2 Comparison of Signal-to-Noise Ratio Obtained with Room Temperature and Cryogenic RF Coils

Several studies have assessed the practical gain in SNR of CRPs at different field strengths, in comparison to RF coils operating at room temperature.[11,18,19] A phantom study conducted at 4.7 T has demonstrated SNR gains for a CRP in the range from 2.1 ± 0.01 to 2.28 ± 0.01 and from 2.12 ± 0.01 to 2.3 ± 0.01 for spin echo (SE) and gradient echo (GE) sequences, respectively.[18] However, a direct comparison of coil performances is not trivial and SNR may be influenced by factors unrelated to coil temperature. The Biot–Savart law and the analysis in the previous section predict that a small loop provides a great SNR on the surface but has a limited penetration depth and will have a smaller SNR further from the surface compared to a bigger coil; hence the assessment of the SNR gain depends on the size and configuration of the coils used, which may in turn be chosen depending on the location of the ROI (region of interest) under study within the sample. Many other parameters of the coil (details being beyond the scope of this contribution) will also affect the SNR. A study assessing SNR in the mouse brain has shown that the average gains for a quadrature CRP for an SE sequence are 1.88 ± 0.23 and 3.20 ± 0.21 relative to a room-temperature surface coil and a room-temperature volume coil at 4.7 T, respectively.[18] For GE sequences the average gain was found to be 1.91 ± 0.18 and 2.97 ± 0.35. At 9.4 T an average SNR gain of 2.55 for a GE sequence and 2.44 for an SE sequence was reported (Figure 5.3).[19]

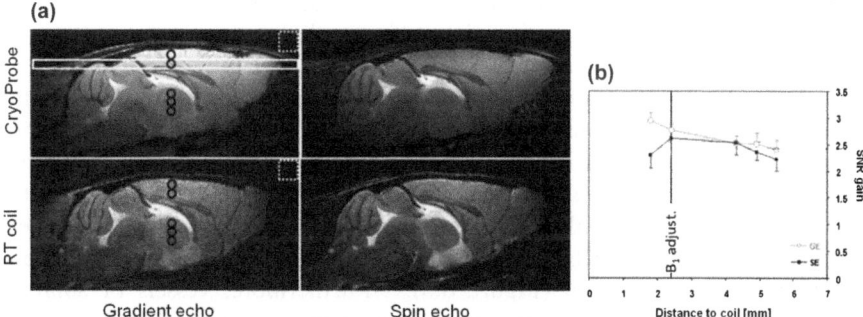

Figure 5.3   (a) Sagittal GE and SE images of the mouse brain obtained with a CRP (top row) and with a similar setup operating at room temperature (bottom row) at 9.4 T. Acquisition times were 2 min. 12 s and 3 min. 22 s, respectively. Modified with permission from Baltes *et al.*[19] © 2009 John Wiley & Sons, Ltd. (b) SNR gains for increasing distances from the coil for both SE (black) and GE (grey) sequences. As the used CRP is a transmit/receive system, excitation pulse angles were adjusted to a distance of 0.9 mm from the brain surface to values $\alpha = 90°$ and $30°$ for SE and GE sequences, respectively. Values represent mean ± standard deviation.

Furthermore, an SNR gain of 3.10 ± 0.66 was found at 9.4 T for a gradient echo-echo planar imaging (GE-EPI) acquisition of the mouse brain with a CRP in comparison to a room-temperature setup.[11]

5.3 Cryogenic RF Coils for Mouse Phenotyping

5.3.1 Application of Cryogenic RF Coils for High-Resolution MR Imaging

High-resolution MR imaging can yield information about the dimensions and morphology of organs/tissues. In brain research, for example, these techniques have been used to study the changes in brain structure during development and the effects of genetic mutations, sexual dimorphism and the effect of learning on brain plasticity.[20–22] MRI techniques such as T_1- or T_2-mapping, diffusion-weighted imaging or quantitative susceptibility mapping can be used quantitatively to assess changes in tissue structure and composition during disease states.[23–25] Images of the mouse brain with very high resolution can be obtained by restraining the mouse head on the MRI support during scanning to reduce motion artifacts. Optimally, these scans are three-dimensional with isotropic voxel dimensions to map accurately small brain structures.

The SNR gain provided by CRPs can be invested into improved spatial resolution and increased contrast-to-noise ratio (CNR), enabling better delineation of structures or reduced measurement time. For instance, GE images of the mouse brain acquired at 9.4 T revealed a distinctly superior image quality compared to the room-temperature coil setup when using the CRP and the same imaging parameters (Figure 5.4). The magnified inserts depict significantly more structural details in the image acquired with the CRP compared to the room-temperature coil due to superior CNR. In a different study the microstructural integrity of the olfactory bulb was investigated *in vivo* with high-resolution scans using a CRP at 9.4 T.[26]

High CNR is also important for MR angiography (MRA). In time-of-flight MRA (TOF-MRA), a commonly used technique to visualize cerebral blood vessels, the CNR depends not only on blood flow velocity but also on vessel dimensions, as vessels smaller than the voxel dimension will be subject to partial volume effects. Evaluation of TOF angiograms revealed that a higher number of individual vessels can be visualized when a CRP is used for detection compared to an RT setup (Figure 5.5).[19] Furthermore, vessels of smaller diameter can be detected in the angiograms obtained with the CRP. The increased CNR when using CRPs translates into an improved quality of angiographic data of the mouse brain as reflected by the reconstruction of the mouse middle cerebral artery (MCA) from a 3D TOF-MRA data set covering the whole mouse brain (Figure 5.5(c)).

High-resolution MRA was used for characterizing mouse models of cerebral amyloidosis.[4,12] TOF-MRA and contrast-enhanced MR microangiography (CE-µMRA) were applied to transgenic arcAβ mice. Recently, the arcAβ mouse was generated, which expresses the human APP695 containing both the

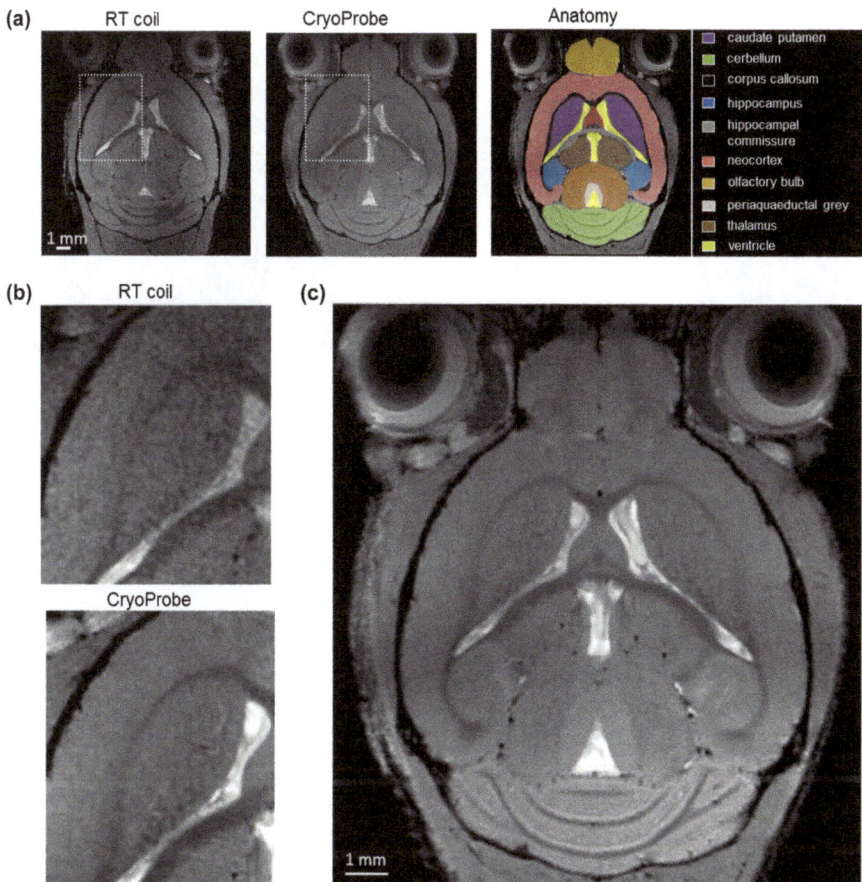

Figure 5.4 (a) Horizontal 2D GE magnitude images of a mouse brain with an in-plane resolution of $52 \times 52 \, \mu m^2$ and slice thickness of $170 \, \mu m$ acquired at 9.4 T with a quadrature receive only room-temperature coil (RT coil) and with a quadrature transmit/receive CRP, respectively (see also ref. 19), using the same sequence parameters (echo time TE = 5.5 ms, repetition time TR = 400 ms, $\alpha = 40°$; 4 averages). A digital brain atlas is shown to depict anatomical structures. Data acquisition time was 10 min. 40 s. (b) Enlarged sections depict the parts of the midbrain (dotted line). (c) High-resolution horizontal cross-section through mouse brain with a spatial resolution of $52 \times 52 \times 170 \, \mu m^3$ acquired at 9.4 T with a CRP. Data acquisition time was 12 min.

Swedish and the Arctic mutation under the control of a prion protein promoter.[27] The arcAβ mouse shows an age-dependent parenchymal and vascular deposition of amyloid-β and cerebrovascular dysfunction. Quantitative analysis of contrast-enhanced microangiograms revealed a reduction of intracortical microvessels in 24-month-old arcAβ mice compared to age-matched controls, while no significant differences were seen between arcAβ mice and wild-type controls at 4 months of age (Figure 5.6(a)).[12] In addition to

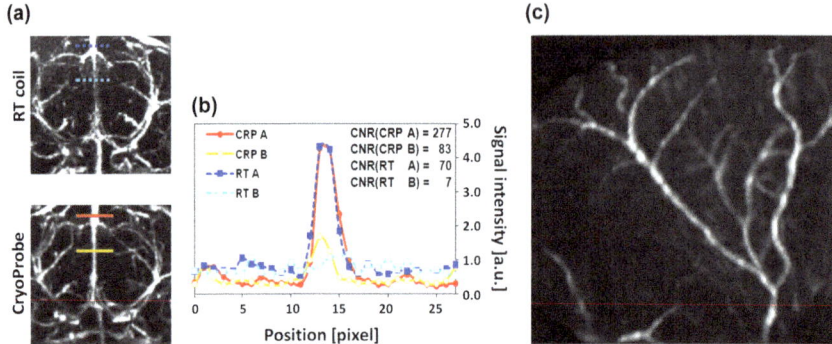

Figure 5.5   (a) Sagittal maximum intensity projections of 3D time-of-flight angiograms of a mouse brain acquired at 9.4 T with an RT coil and with a CRP, respectively, using the same sequence parameters.[19] Data acquisition time was 10 min. 40 s. (b) CNR evaluation of two vessels (A, B) at different distances from the CRP (red, orange) and the room-temperature (RT) coil (blue, light blue), respectively.[19] (c) Sub-region of high-resolution TOF-MRA of mouse brain acquired at 9.4 T with a CRP displaying the right middle cerebral artery. The scale bar indicates 3 mm.

decreased microvascular perfusion the occurrence of cerebral microbleeds was observed at an advanced stage of disease in these mice.[4] T_2*-weighted 3D GE data sets from arcAβ and wild-type mouse brain with high resolution were acquired for quantitative susceptibility mapping. On quantitative susceptibility maps, focal areas of increased magnetic susceptibility were detected in the arcAβ but not in wild-type mice, indicating severe vascular pathology with occurrence of cerebral microbleeds in the transgenic mice (Figure 5.6(b)). Moreover, quantitative susceptibility mapping provided increased detection sensitivity of cerebral microbleeds and improved contrast when compared with the conventional GRE magnitude imaging, and it appeared not to be subject to the blooming effects observed in GRE magnitude and phase images. Another study has demonstrated that inflammatory lesions can be visualized with high resolution at 9.4 T in the brain of mice with experimental autoimmune encephalomyelitis.[28]

5.3.2   The Use of Cryogenic RF Coils for Functional MRI of the Mouse Brain

MRI not only provides anatomical and morphological information, but it also enables functional/physiological studies to be conducted. Of particular interest is functional MRI (fMRI), which has matured to a prominent investigational tool in neuroscience providing information on brain function and connectivity. fMRI measures alterations in the MRI signal intensity arising from changes in hemodynamics, typically in blood oxygenation (so-called blood oxygen-level dependent (BOLD) contrast), elicited by the activity of a large number of neurons.[29] These changes in the BOLD signal intensity or in the cerebral blood

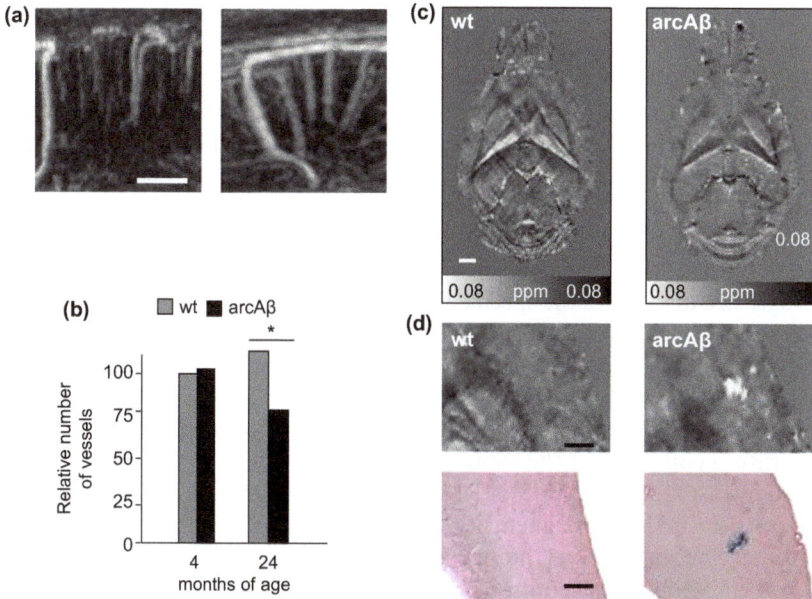

Figure 5.6 (a) Contrast-enhanced MR microangiograms of a wild-type mouse acquired at 9.4 T with a CRP. Pre-contrast images were subtracted from post-contrast images, *i.e.* after administration of superparamagnetic iron oxide nanoparticles. Data acquisition time was 36 min. for each scan. Sagittal maximum intensity projections were derived from the 3D stacks of difference images. Modified from Klohs *et al.*[12] © 2012 The authors. (b) Quantitative analysis of contrast-enhanced MR microangiograms revealed a reduction of intracortical microvessels in 24-month-old arcAβ mice compared to age-matched controls, while no significant differences were seen between arcAβ mice and wild-type controls at 4 months of age. Values represent mean ± standard deviation. (c) Horizontal quantitative susceptibility maps of an 18-month-old wild-type animal and of an age-matched transgenic arcAβ mouse. Data acquisition time was 29 min. 51 s. Modified with permission from Klohs *et al.*[4] © 2011 International Society of Cerebral Blood Flow and Metabolism. (d) Quantitative susceptibility maps with corresponding tissue section after Prussian blue/eosin staining show that focal areas of high susceptibility in the cortex of 18-month-old arcAβ mice correspond to areas of focal iron accumulation, indicating the occurrence of cerebral microbleeds in this mouse strain.

volume (CBV) are small, typically of the order of only a few percent (1–5%). In view of the small sample volumes associated with mouse fMRI it constitutes a major challenge reliably to detect these small activity-related fluctuations in the signal over a period of time. In addition, the quantitative analysis of the fMRI signal time course requires high temporal resolution as the hemodynamic response function is characterized by time constants of the order of seconds.[30,31] Even though susceptibility effects, and, hence, BOLD contrast and perfusion, increase in amplitude at higher field strengths, increases in BOLD signal above 7 T have not been observed in small animal fMRI.[32,33] Instead, the use of higher magnetic field strengths comes at the cost of unwanted side effects. The susceptibility-related effects scale with the magnetic field strength, *i.e.* image artifacts due to susceptibility differences at tissue interfaces (*e.g.* soft tissue–bone, soft tissue–air) become more severe for higher magnetic fields. Particularly when using echo-planar imaging (EPI) sequences, brain regions closest to tissue borders are prone to fMRI signal loss due to intravoxel signal dephasing caused by susceptibility gradients. Alternatively, data averaging or the increase in the dimensions of the voxels being sampled can improve fMRI signal intensity; see eqn (5.3).[32] However, data averaging compromises temporal resolution and increasing the voxel size diminishes the specificity of the measurement.

The application of CRPs has recently been shown to lead to significant sensitivity improvements in fMRI studies.[11] The authors investigated how increased SNR provided by a CRP translates into enhanced BOLD sensitivity in a sensory forepaw stimulation experiment in comparison to a room-temperature coil setup at the same field strength.

In contrast to structural MRI studies, physiological noise has to be considered in fMRI experiments. Physiological noise originates, for example, from the fluctuations in the metabolic rate, blood flow, blood volume, de-/oxyhemoglobin concentrations and cardiac and respiratory motion.[34–37] The total temporal or time-series SNR (tSNR), which is of prime importance in fMRI studies, can be written as

$$tSNR = \frac{S}{\sqrt{n_C^2 + n_S^2 + n_{phys}^2}} = \frac{SNR_{stat}}{\sqrt{1 + \lambda^2 \cdot SNR_{stat}^3}} \qquad (5.8)$$

and includes this physiological noise term in the denominator. SNR_{stat} corresponds to the SNR of an individual image and λ is a measure for the SNR degradation caused by physiological noise ($\lambda = n_{phys}/S$). In order to achieve significant improvements in SNR upon coil cooling in serial measurements (such as fMRI studies) the noise level arising from the receiver electronics has to be comparable to that of the sum of both static and physiological noise. This is the case for fMRI studies in mice: the use of CRPs has significantly improved SNR and tSNR by factors 3.1 ± 0.7 and 1.8 ± 1.0 when comparing the CRP and the room-temperature coil.[11] This illustrates the significant contribution of physiological noise sources. In addition it has been shown that by properly adjusting the temperature of the surface of the coil housing of the CRP in contact with the animal, the BOLD amplitude could be optimized. Increases in BOLD signal were attributed to alterations in baseline conditions, *i.e.* in basal CBV values.[11] The gains in SNR and tSNR not only allow the detection of fMRI signal changes at higher spatial or temporal resolutions, but can also be exploited to measure very weak fMRI responses, which is particularly interesting for studies in mice.

fMRI in genetically engineered mice is of high interest for biomedical research. It allows studying molecular mechanisms underlying the propagation and neural processing of peripheral nerve input. As an example, an fMRI study using a CRP in global and nociceptor-specific cannabinoid-1-receptor (CB1-R) knock-out mice has revealed that CB1-R are involved in the hyperalgesia that was observed following administration of lidocaine at a low dose.[10] Alternatively, mouse fMRI can be used to characterize altered functional processing in mouse models of human disease. fMRI experiments probing alteration in CBV have demonstrated a compromised vascular response in the arcAβ mouse. Following the administration of a vasoactive compound both the vascular reactivity as well as the maximal vascular dilation were found to be reduced in transgenic arcAβ mice as compared to wild-type animals.[38] Finally, fMRI in combination with genetically engineered mice may be used to demonstrate the specificity of pharmacological interventions. While administration of a serotonergic $5HT_{1A}$ receptor agonist prompted a strong response in the brain structures associated with serotonergic signaling, this response was largely suppressed in heterozygous and completely suppressed in homozygous $5HT_{1A}$-R knock-out mice.[39]

5.3.3 Metabolic Phenotyping Using MRS and Chemical Shift Imaging

Apart from specific structural and function parameters a phenotype might be characterized by alterations in tissue metabolism. MRS provides a unique tool to study metabolism non-invasively. Because the concentrations of the metabolites of interest (*e.g.* *N*-acetyl-aspartate, creatine, phosphorylcholine, lactate) are in the millimolar range, as compared to 80 molar of water protons, MRS signals are weak and data averaging is required to identify clearly metabolite signals. MRS studies in mice are even more challenging due to the small volumes of interest (see eqn (5.3)). The use of CRPs providing a gain in SNR of a factor 2 to 3 is attractive as this translates into a reduction of measurement time by a factor of 4 to 9 with regard to data acquisition using a room-temperature coil of equal dimensions (Figure 5.7).[18] Alternatively, the SNR gain can be invested into improved spectral quality for the same acquisition time.

A quadrature transceiver CRP was used for an MRS study evaluating the levels of intramyocellular lipids in tibialis anterior muscle of ob/ob and ob/+ control mice.[40] Ob/ob mice expressing leptin deficiency are used as a model organism for studying obesity and associated metabolic disorders. High-quality spectra were recorded from voxels of 3 mL in the mouse tibialis anterior muscle in typically 12 to 16 min. Intramyocellular lipid levels were found to be significantly higher in ob/ob mice as compared to age-matched ob/+ control animals. CRPs might thus be useful for non-invasive measurement of changes in lipid metabolism in the murine skeletal muscle that accompanies obesity and to assess underlying molecular mechanisms *via* genetic or pharmacological intervention.

Similarly, MRS has become an important tool for assessing brain metabolism. Using proper experimental conditions up to 20 metabolites/neurotransmitter might be measured simultaneously (see also Chapter 8).[41] Again the use of transgenic models is of high value either as a model of human disease or to study molecular mechanisms. A critical aspect in these studies is

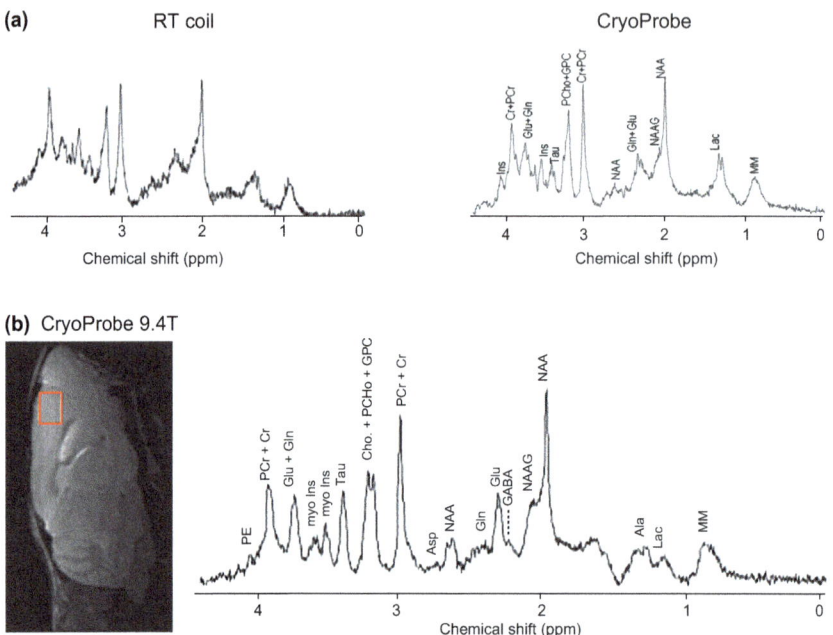

Figure 5.7   (a) MR spectra of a 64 μl volume of the mouse brain acquired at 4.7 T with a room-temperature (RT) coil and with a CRP.[18] (b) MR spectrum of 1 μl volume acquired in the pre-frontal cortex (red box) of the mouse brain at 9.4 T with a phased array CRP in cross-coil mode. The spectrum was acquired with a point resolved spectroscopy (PRESS) sequence with the parameters TE = 12 ms, TR = 2500 ms, 512 averages. Data acquisition time was 21 min. The metabolites are Ala: alanine, Asp: aspartate, Cho: choline, Cr: creatine, g-amino butyric acid (GABA), Gln: glutamine, Glu: glutamate, GPC: glycerolphosphoryl-choline, myo Ins: *myo*-inositol, Lac: lactate, MM: macromolecules, NAA: *N*-acetyl-aspartate, NAAG: *N*-acetylaspartylglutamate, PCho: phosphorylcholine, PCr: phospho-creatine, PE: phosphoethanolamine and Tau: taurine.

accurate quantification. Typically pathologies or genetic/pharmacological interventions will lead to altered metabolite levels. These alterations may be small in the order of a few percent to a few tens percent of the basal levels. In this respect, it is attractive to invest the SNR gain provided by CRP receivers into an improved spectral quality (Figure 5.7(b)) as this immediately translates into better quantification.[18] Alternatively, the sensitivity gain may be used to increase the temporal resolution, which might enable carrying out dynamic studies on tissue metabolism in response to a challenge.

5.3.4   Advantages and Limitations of Cryogenic RF Coils for Mouse Phenotyping

MRI in combination with other imaging modalities can provide a wealth of information about the phenotype of genetically engineered mice, which are being used to address basic research questions or as models of human disease. Due to its inherent high soft-tissue contrast MRI is of particular interest for the structural characterization of tissue. It has been demonstrated that valuable information on tissue morphology can be obtained from *ex vivo* studies of perfusion fixed mice or mouse organs. In these studies the SNR issues associated to the high demands on spatial resolution have been solved by using long acquisition times of typically several hours yielding high-resolution data sets displaying high contrast. Addition of contrast agent to the perfusate was shown to enhance contrast.[42]

The use of coils with improved sensitivity such as CRPs offers a number of substantial advantages:

1) '*In vivo histology*': CRPs can be used to achieve high-resolution images with high CNR necessary for an accurate morphological characterization also under *in vivo* conditions. The increased SNR not only enables the spatial resolution of sequences to be adapted to the dimensions of the mouse, but it also allows depicting structures that are small independently of the scaling between species, like microvessels or cell layers such as the Purkinje cell layer in the living mouse brain.[19] *In vivo* MRI of the mouse brain with isotropic spatial resolution of $(50 \,\mu\text{m})^3$ comes close to histological imaging in revealing such structures. An attractive feature is the administration of contrast agents during the examination, which allows enhancing specific structures, *e.g.* the microvasculature of the cerebral cortex.[12]

2) *Longitudinal studies*: The potential to carry out high-resolution phenotyping studies *in vivo* is of particular advantage as animals can be used repetitively *e.g.* in longitudinal studies, which helps to reduce the number of animals needed for a study. At a time when the general public is questioning the use of animals in research, efforts need to be taken to reduce the number of animals in a study. In addition, the availability of genetically engineered animals may be limited and these animals are expensive.

3) '*Measurement of tissue physiology, function and/or metabolism*': The ability to do *in vivo* imaging with high SNR is not only an advantage for structural MRI, but is also a prerequisite to obtain functional/ physiological and metabolic information. Such investigations can only be performed in the intact living organism where all regulatory processes are intact. The gain in SNR that comes with use of CRPs can be used for example to measure very weak fMRI responses and to detect changes in fMRI signal at high spatial or temporal resolutions. A reduction in the signal noise will lead to a higher statistical power and thus lead to a reduction of animals needed for fMRI studies. As we have discussed, the gain is reduced in such studies as factors like physiological noise become relevant. Hence, stable physiology and positional stability during the acquisition period are a prerequisite for an optimal exploitation of the sensitivity gain by the CRP.

4) '*Reduced acquisition times*': The sensitivity gain provided by the CRP can be invested in shortening the acquisition time when keeping the SNR value in the image unchanged: an SNR gain of 2 compared to a standard probe can be turned into a 4-fold faster measurement. This is particularly useful for MRS and chemical shift imaging, which suffer from low sensitivity and therefore demand signal averaging and lengthy acquisition protocols under anesthesia. The use of a CRP might thus enable high-quality spectra to be acquired at acquisition times that are well tolerable to mice.

The reduction in acquisition time makes phenotyping of mice more efficient, thereby enhancing the throughput, which is relevant when considering for instance drug-discovery/-development studies. Multiparametric MRI protocols may be required unambiguously to stratify groups of animals *e.g.* for comparison between strains (transgenic *versus* wild-type) or for evaluating treatment effects. Such investigations are in general rather time consuming. In addition, the use of unassisted computed readouts usually requires groups of sufficient size in order to detect potentially minute changes in morphology and tissue function. A reduction in acquisition time can help to increase the throughput and thus diminish the cost of data acquisition.

In spite of these important advantages the use of a CRP suffers from limitations:

1) *SNR gain decreases with increasing sampling volume*: As discussed the SNR gain depends on the relative contributions of sample noise, which increases with the volume sampled. Hence, the use of CRPs is only beneficial when sampling small volumes of the order of 1 mL to a few mL.
2) *Sensitivity gain decreases with increasing frequency*: From eqn (5.6) it becomes obvious that the gain in SNR by using CRPs becomes smaller with increasing frequency (or field strength for a given nucleus) as the sample noise power scales with ω^2 and the coil noise power with $\sqrt{\omega}$. Hence the ratio coil noise/sample noise decreases with increasing ω. While high magnetic fields are attractive in small animal imaging as they enhance the polarization of the sample and hence the signal, they are less attractive regarding noise reduction through cooling of the detector device. Yet, the dependence is relatively weak (see Figure 5.2(d)) and for currently used field strengths up to 17.6 T significant SNR benefits can be expected.
3) *The design of a CRP is demanding:* The coil temperature is typically equal to or lower than 77 K (liquid nitrogen). Since the coil has to be as close to the sample as possible for efficient coupling, thermal insulation is a challenge. Nevertheless, this has been solved and the coil designs used to acquire the data reported here are relatively straightforward to handle and can be inserted into and taken out of the magnet bore in the cold state, which facilitates operation.

Today, CRPs are commercially available (Cryoprobe®, Bruker BioSpin) and increasingly used by research groups throughout the world. They constitute an attractive option for increasing the SNR in MRI and MRS studies, either as an additional SNR boost at high fields for the ultimate SNR, or as a cost-effective alternative to increasing the magnetic field strength, which requires expensive magnets. Operating at lower fields with increased sensitivity also offers advantages such as more favorable relaxation times, better efficiency of contrast agents and fewer susceptibility artifacts. It is foreseeable that CRPs will emerge as a standard tool for imaging small animals.

References

1. A. Doyle, M. P. McGarry, N. A. Lee and J. J. Lee, *Transgenic Res.*, 2012, **21**, 327.
2. N. A. Bock, G. Zadeh, L. M. Davidson, B. Qian, J. G. Sled, A. Guha and R. M. Henkelman, *Neoplasia*, 2003, **5**, 546.
3. A. Badea, A. A. Ali-Sharief and G. A. Johnson, *Neuroimage*, 2007, **37**, 683.
4. J. Klohs, A. Deistung, F. Schweser, J. Grandjean, M. Dominietto, C. Waschkies, R. M. Nitsch, I. Knuesel, J. R. Reichenbach and M. Rudin, *J. Cereb. Blood Flow Metab.*, 2011, **31**, 2282.
5. J. P. Lerch, J. G. Sled and R. M. Henkelman, *Methods Mol. Biol.*, 2011, **711**, 349.
6. K. C. Chan, J. S. Cheng, S. Fan, I. Y. Zhou, J. Yang and E. X. Wu, *Neuroimage*, 2012, **59**, 2274.
7. C. R. McCreary, T. A. Bjarnason, V. Skihar, J. R. Mitchell, V. W. Yong and J. F. Dunn, *Neuroimage*, 2009, **45**, 1173.
8. Y. Sun, N. O. Schmidt, K. Schmidt, S. Doshi, J. B. Rubin, R. V. Mulkern, R. Carroll, M. Ziu, K. Erkmen, T. Y. Poussaint, P. Black, M. Albert, D. Burstein and M. W. Kieran, *Magn. Reson. Med.*, 2004, **51**, 893.
9. C. J. Berry, J. D. Miller, K. McGroary, D. R. Thedens, S. G. Young, D. D. Heistad and R. M. Weiss, *J. Cardiovasc. Magn. Reson.*, 2009, **11**, 27.
10. S. C. Bosshard, C. Baltes, M. T. Wyss, T. Mueggler, B. Weber and M. Rudin, *Pain*, 2010, **151**, 655.
11. C. Baltes, S. Bosshard, T. Mueggler, D. Rating and M. Rudin, *NMR Biomed.*, 2010, **24**, 439.
12. J. Klohs, C. Baltes, F. Princz-Kranz, D. Rating, R. M. Nitsch, I. Knuesel and M. Rudin, *J. Neurosci.*, 2012, **32**, 1705.
13. R. Pohmann, G. Shajan and D. Z. Balla, *Magn. Reson. Med.*, 2011, **66**, 1572.
14. R. C. van de Ven, B. Hogers, A. M. van den Maagdenberg, H. J. de Groot, M. D. Ferrari, R. R. Frants, R. E. Poelmann, L. van der Weerd and S. R. Kiihne, *Magn. Reson. Med.*, 2007, **58**, 390.
15. D. I. Hoult and R. E. Richards, *J. Magn. Reson.*, 1969, **24**, 71.
16. L. Darrasse and J. C. Ginefri, *Biochimie*, 2003, **85**, 915.
17. R. D. Black, T. A. Early, P. B. Roemer, O. M. Mueller, A. Mogro-Campero, L. G. Turner and G. A. Johnson, *Science*, 1993, **259**, 793.

18. D. Ratering, C. Baltes, J. Nordmeyer-Masner, D. Marek and M. Rudin, *Magn. Reson. Med.*, 2008, **59**, 1440.
19. C. Baltes, N. Radzwill, S. Bosshard, D. Marek and M. Rudin, *NMR Biomed.*, 2009, **22**, 834.
20. S. J. Sawiak, N. I. Wood, G. B. Williams, A. J. Morton and T. A. Carpenter, *Neurobiol. Dis.*, 2009, **33**, 12.
21. J. P. Lerch, A. P. Yiu, A. Martinez-Canabal, T. Pekar, V. D. Bohbot, P. W. Frankland, R. M. Henkelman, S. A. Josselyn and J. G. Sled, *Neuroimage*, 2011, **54**, 2086.
22. X. Yu, B. J. Nieman, A. Sudarov, K. U. Szulc, D. J. Abdollahian, N. Bhatia, A. K. Lalwani, A. L. Joyner and D. H. Turnbull, *Neuroimage*, 2011, **56**, 1251.
23. P. R. Rau, J. Sellner, S. Heiland, K. Plaschke, P. D. Schellinger, U. K. Meyding-Lamadé and W. R. Lamadé, *Life Sci.*, 2006, **78**, 11750.
24. N. El Tannir El Tayara, B. Delatour, C. Le Cudennec, M. Guégan, A. Volk and M. Dhenain, *Neurobiol. Dis.*, 2006, **22**, 199.
25. F. Schweser, K. Sommer, A. Deistung and J. R. Reichenbach, *Neuroimage*, 2012, **62**, 2083.
26. H. P. Burmeister, T. Bitter, P. M. Heiler, A. Irintchev, R. Fröber, M. Dietzel, P. A. Baltzer, L. R. Schad, J. R. Reichenbach, H. Gudziol, O. Guntinas-Lichius and W. A. Kaiser, *Neuroimage*, 2012, **60**, 1662.
27. M. Knobloch, U. Konietzko, D. C. Krebs and R. M. Nitsch, *Neurobiol. Aging*, 2007, **28**, 1297.
28. H. Waiczies, J. M. Millward, S. Lepore, C. Infante-Duarte, A. Pohlmann, T. Niendorf and S. Waiczies, *PLoS One*, 2012, **7**, e32796.
29. S. Ogawa, T. M. Lee, A. R. Kay and D. W. Tank, *Proc. Natl Acad. Sci. USA*, 1990, **87**, 9868.
30. J. Martindale, J. Mayhew, J. Berwick, M. Jones, C. Martin, D. Johnston, P. Redgrave and Y. Zheng, *J. Cereb. Blood Flow. Metab.*, 2003, **23**, 546.
31. K. E. Stephan, L. M. Harrison, W. D. Penny and K. J. Friston, *Curr. Opin. Neurobiol.*, 2004, **14**, 629.
32. J. S. Gati, R. S. Menon, K. Ugurbil and B. K. Rutt, *Magn. Reson. Med.*, 1997, **38**, 296.
33. J. U. Seehafer, D. Kalthoff, T. D. Farr, D. Wiedermann and M. Hoehn, *J. Neurosci.*, 2010, **30**, 5234.
34. G. Kruger and G. H. Glover, *Magn. Reson. Med.*, 2001, **46**, 631.
35. G. A. Kruger, G. H. Kastrup and G. H. Glover, *Magn. Reson. Med.*, 2001, **45**, 595.
36. C. Triantafyllou, R. D. Hoge, G. Krueger, C. J. Wiggins, A. Potthast, G. C. Wiggins and L. L. Wald, *Neuroimage*, 2005, **26**, 243.
37. D. Kalthoff, J. U. Seehafer, C. Po, D. Wiedermann and M. Hoehn, *Neuroimage*, 2011, **54**, 2828.
38. F. L. Princz-Kranz, T. Mueggler, M. Knobloch, R. M. Nitsch and M. Rudin, *Neurobiol. Dis.*, 2010, **40**, 284.

39. T. Mueggler, F. Razoux, H. Russig, A. Buehler, T. B. Franklin, C. Baltes, I. M. Mansuy and M. Rudin, *Eur. Neuropsychopharmacol.*, 2011, **21**, 344.
40. Q. Ye, C. F. Danzer, A. Fuchs, W. Krek, T. Mueggler, C. Baltes and M. Rudin, *NMR Biomed.*, 2011, **24**, 1295.
41. J. Pfeuffer, I. Tkác, S. W. Provencher and R. Gruetter, *J. Magn. Reson.*, 1999, **141**, 104.
42. G. A. Johnson, G. P. Cofer, S. L. Gewalt and L. W. Hedlund, *Radiology*, 2002, **222**, 789.

CHAPTER 6

Recent Developments of Contrast Agents, CEST and Low Fields

S. AIME* AND D. L. LONGO

Department of Molecular Biotechnologies and Health Sciences, University of Torino, Molecular Imaging Center, Via Nizza 52, Torino, 10126, Italy
*Email: silvio.aime@unito.it

6.1 Introduction

Contrast in an MR image is determined by a complex interplay of factors that are both intrinsic to the tissue examined and of instrumental nature. The most important intrinsic factors are the proton density and the longitudinal (T_1) and transverse (T_2) relaxation times of tissue water protons. The instrumental factors, such as type of sequence used, are exploited to enhance the contrast based on the difference in T_1 or T_2. The T_1-weighted sequence, for example, is used to enhance the contrast based on differences in T_1, where the tissue with the shorter T_1 will show a more intense (usually brighter) signal in the image. On the contrary, in a T_2-weighted sequence, there is a loss of signal in the regions with short T_2 (darker in the image). The success of these techniques has been determined by the fortunate coincidence that the relaxation times are related to the biochemical environment of a given tissue, and that they are modified in the presence of a pathological process.

Contrast in MR images can be further enhanced with the administration of suitable contrast agents (CAs). The presence of the CA causes a great increase in

New Developments in NMR No. 2
New Applications of NMR in Drug Discovery and Development
Edited by Leoncio Garrido and Nicolau Beckmann
© The Royal Society of Chemistry 2013
Published by the Royal Society of Chemistry, www.rsc.org

the water proton relaxation rate, thus adding further physiological information to the already extraordinary anatomical resolution usually obtained without the CA. Therefore, contrast media are routinely used in several protocols and are particularly useful to evaluate organ perfusion, and any abnormality in the blood-brain barrier and in renal clearance. Nowadays, CAs are used in *ca.* 35% of MR diagnostic assays. Unlike probes used in nuclear medicine, MRI CAs are not directly visualized in the image. Only their effects are observed: contrast is affected by the variation that the CA causes on water proton relaxation times and, consequently, on the intensity of the NMR signal. Generally, the purpose is to reduce T_1 in order to obtain an intense signal in shorter times and a better signal-to-noise ratio (SNR) with the acquisition of a higher number of measurements. CAs that reduce either T_1 or T_2 are called positive, whereas those that mainly affect T_2 are called negative. Since unpaired electrons are able to reduce markedly T_1 and T_2, the search for positive CAs is mainly oriented towards paramagnetic compounds, particularly towards paramagnetic metal complexes. The paramagnetic metal ions most extensively studied have been either in the transition metals or in the lanthanide series.

6.2 Gd(III)-Based Contrast Agents

As far as lanthanides are concerned, the attention is essentially focused on Gd(III) ion both for its high paramagnetism (seven unpaired electrons) and for its favorable properties in terms of electronic relaxation.[1] This metal does not possess any physiological function in mammals, and its administration as free ion is strongly toxic even at low doses (LD_{50} 0.4 mmol kg^{-1}).[2] For this reason, it is necessary to use ligands that form very stable chelates.[3,4] The high affinity shown by Gd(III) ions towards some polyaminocarboxylic acids, either cyclic or linear, has been exploited to form complexes endowed with very high stability (up to log $K_{ML} > 20$). The first CA approved for clinical use, Gd-DTPA (Magnevist®, Bayer Healthcare, Germany), in more than 25 years of clinical use has been administered to many millions of patients (Figure 6.1). Other Gd(III)-based CAs similar to Magnevist have been marketed, namely Gd-DOTA (Dotarem®, Guerbet, France), Gd-DTPA-BMA (Omniscan®, GE Healthcare, USA) and Gd-HPDO3A (ProHance®, Bracco Imaging, Italy).[5] These CAs have very similar pharmacokinetic properties because they distribute in the extracellular fluid and are eliminated *via* glomerular filtration. They are particularly useful to delineate lesions in the blood-brain barrier. Other commercial systems include Gd-EOB-DTPA (Eovist®, Bayer Healthcare, Germany) and Gd-BOPTA (MultiHance®, Bracco Imaging, Italy).[6,7] They are Gd-DTPA derivatives endowed with an enhanced lipophilicity owing to the introduction of an aromatic substituent on the ligand surface.

6.2.1 Determinants of the Relaxivity of Gd(III) Complexes

An MRI CA should be endowed with high thermodynamic and kinetic stability, and have at least one water molecule coordinated to the metal ion in fast exchange

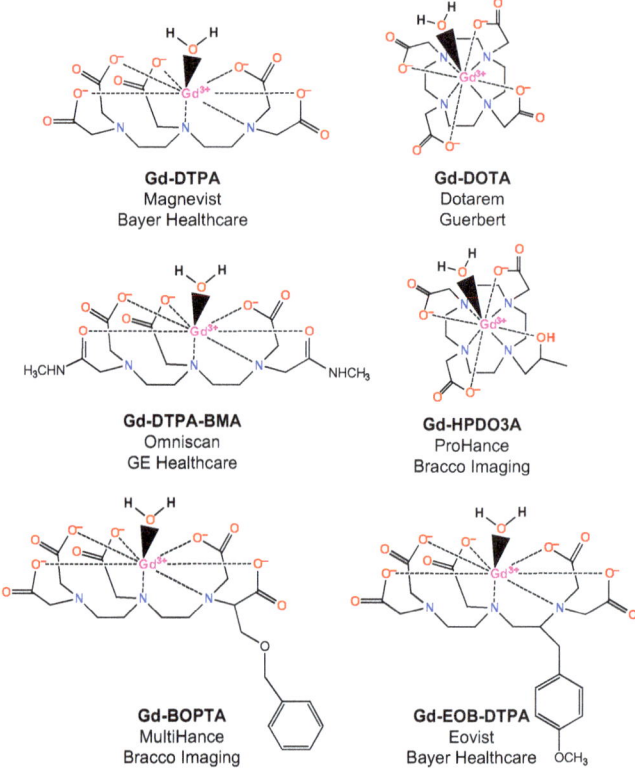

Figure 6.1   Structures of commercial MRI contrast agents used in the clinical practice.

with the bulk water. This would permit to influence strongly the relaxation process of all protons present in the solvent in which the CA is dissolved. The Gd(III) chelate efficiency is commonly estimated *in vitro* through the measure of its relaxivity (r_{1p}), which for CAs such as Magnevist, Dotarem, ProHance and Omniscan is around 3.4–$3.5 \, \mathrm{m \, M^{-1} s^{-1}}$ (at $20 \, \mathrm{MHz}$ and $37 \, ^{\circ}\mathrm{C}$). The observed longitudinal relaxation rate (R_1^{obs}) of the water protons in an aqueous solution containing a paramagnetic complex is the sum of three contributions: (i) a diamagnetic one, whose value corresponds to a proton relaxation rate measured in the presence of a diamagnetic (La, Lu, Y) complex of the same ligand; (ii) a paramagnetic one, relative to the exchange of water molecules from the inner coordination sphere of the metal ion with bulk water (R_{1p}^{is}); and (iii) a paramagnetic one relative to the contribution of water molecules that diffuse in the external coordination sphere of the paramagnetic center (R_{1p}^{os}).[8] Sometimes also a fourth paramagnetic contribution is taken into account that is due to the presence of mobile protons or water molecules (normally bound to the chelate through hydrogen bonds) in the second coordination sphere of the metal ion.[9]

    The inner sphere contribution is directly proportional to the molar concentration of the paramagnetic complex and to the number of water molecules coordinated to the paramagnetic center, q, and is inversely proportional to the sum of the mean residence lifetime, τ_M, of the coordinated water protons and their relaxation time, T_{1M}. This latter parameter is directly correlated to the sixth power of the distance between the metal center and the coordinated water protons and depends on the molecular reorientational time, τ_R, of the chelate, on the electronic relaxation times, T_{iE} ($i = 1,2$), of the unpaired electrons of the metal (which depend on the applied magnetic field strength) and on the observed frequency itself. The outer sphere contribution depends on T_{iE}, on the distance of the maximum approach between the solvent and the paramagnetic solute, on the relative diffusion coefficients and, again, on the magnetic field strength. The dependence of R_{1p}^{is} and R_{1p}^{os} on the magnetic field is very important because, from the analysis of the magnetic field dependence, it is possible to assess the principal parameters characterizing the relaxivity of a Gd(III) chelate. This information can be obtained through an NMR instrument in which the magnetic field is changed (field-cycling relaxometer) to obtain the measure of R_1 over a wide range of frequencies (typically 0.01–80 MHz). At 0.5–1.5 T R_1 is generally determined by the τ_R of the chelate so that high-molecular-weight systems display a higher relaxivity. A quantitative analysis of R_1 dependence on the different structural and dynamic parameters shows that, for systems with long τ_R, the maximum attainable R_1 values can be achieved through the optimization of τ_M and T_{1M}.[8]

On this basis, much attention has been devoted to the design of systems characterized by long τ_R values. In principle, this task can be tackled either by designing high-molecular-weight systems or by pursuing the non-covalent interaction of small-sized, properly functionalized Gd(III) complexes with endogenous macromolecules. In macromolecular systems the relaxation induced by paramagnetic species usually displays remarkable changes, primarily related to the increase of the molecular reorientational time τ_R on going from the free to the bound form, which results in a marked increase of the inner sphere R_{1p}^{is} term. For the supramolecular protein-Gd(III) complexes adduct, from the measurement of the relaxivity enhancement, it is possible to assess the affinity (and the number of binding sites) between the interacting partners.[10]

As anticipated above, high relaxivities can be attained by means of an elongation of the molecular reorientational time τ_R, *i.e.* dealing with slowly moving paramagnetic systems. Over the years, this prompted a number of studies on the interaction of Gd(III) complexes with proteins and other macromolecular substrates. While avoiding the use of covalent conjugates (*i.e.* systems based on Gd(III) chelates covalently bound to macromolecules, such as Gd-DTPA-HSA, whose metabolic fate may be problematic[11]), research activities have been addressed to design Gd(III) chelates bearing on their surface suitable functionalities that promote the reversible binding to a target-protein.[12]

Human serum albumin (HSA) has been by far the most investigated protein for binding Gd(III) chelates. Besides the attainment of high relaxivities, a high

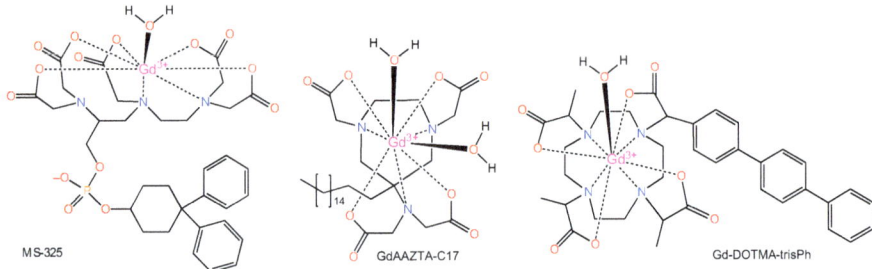

Figure 6.2 Structures of ligands whose Gd(III) complexes have been investigated for binding to HSA.

binding affinity to HSA enables the Gd(III) chelate to have a long intravascular retention time, which is the property required for a good blood-pool agent for MR angiography. Moreover, the presence of a good binding to HSA is useful to carry out dynamic contrast enhanced MRI (DCE-MRI) studies aimed at assessing changes in vascular permeability.[13] In blood, HSA has a concentration of about 0.6 mM and its main physiological role deals with the transport of a huge number of substrates.[14,15] For many of them, the binding region has been identified on the basis of extensive competitive assays. The availability of the solid state X-ray crystal structure of HSA, combined to molecular modeling procedures, allows more insight into the structural details of the binding interaction and the corresponding relaxivity enhancement to be gained.[16–18] The information gained from the studies of the interaction of the various substrates to HSA has been very important to address the design of Gd(III)-based blood-pool agents (Figure 6.2).[16,19,20]

6.2.2   Gd(III)-Based Contrast Agents with q > 1

As underlined by the theory of paramagnetic relaxation, the number of bound water molecules has a strong effect on the relaxivity of Gd(III) complexes. A straightforward approach to increase the inner-sphere relaxivity can be pursued through the increase of the hydration number q, resulting in an increase of this contribution at any field. The use of hepta- or hexadentate ligands would, in principle, result in Gd(III) complexes with 2- and 3-coordinated water molecules, respectively, but the decrease of the ligand coordination number is likely to be accompanied by a decrease of their thermodynamic stability and an increase of their toxicity. Furthermore, systems with $q = 2$ may suffer a "quenching" effect upon interacting with endogenous anions or with proteins, as donor atoms from lactate or Asp or Glu residues may replace the coordinated water molecules.[21] Commercial Gd(III)-based CAs have $q = 1$ but some stable Gd(III) chelates containing two inner sphere water molecules have been identified and are currently under intense scrutiny.

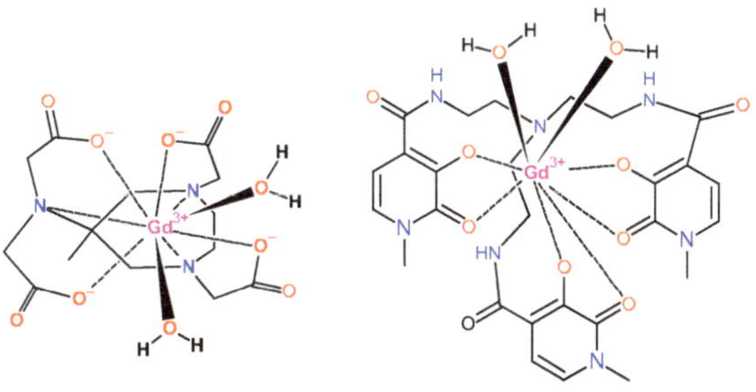

GdAAZTA **Gd-TREN-1-Me-3,2-HOPO**

Figure 6.3 Structures of Gd(III)-complexes with q > 1.

A novel Gd(III) chelate with the heptadentate AAZTA ligand(AAZTA: 6-amino-6-methylperhydro-1,4-diazepinetetraacetic acid; see Figure 6.3) has recently been characterized.[22] AAZTA is readily obtained in high yields and its Gd(III) complex displays interesting properties to be considered the prototype of a new class of enhanced MRI agents. It is characterized by a quite high relaxivity value (7.1 mM^{-1}s^{-1} at 20 MHz and 298 K), a relatively fast exchange rate of the coordinated water molecules ($\tau_M = 90$ ns at 298 K), a high thermodynamic stability in aqueous solution and a nearly complete inertness towards the coordination of bidentate endogenous anions.[23] Another interesting class is represented by Gd-HOPO complexes developed by Raymond and co-workers. HOPO ligands (Figure 6.3) are based on 4-carboxyamido-3,2-hydroxypyridinone chelating units and act as hexadentate ligands towards Gd(III) thus leaving two water molecules in the inner coordination sphere.[24,25] The peculiar coordinating geometry of Gd-HOPO complexes does not allow an easy replacement of the two water molecules by other ligands. Moreover, the exchange rate of the coordinated water molecules is in the range of the optimal values and the electronic relaxation appears to be slow enough to allow the attainment of very high relaxivities.[26]

6.2.3 Nanosized Carriers

In recent years a considerable amount of systems containing a high payload of Gd(III) chelates have been reported. Most of these systems have been endowed with specific cell receptor targeting capabilities. They consist of a different kind of nanocarrier, such as micelles, liposomes, dendrimers, coupled to biological vectors such as peptides, proteins, lipoproteins, monoclonal antibodies and viral capsids.[27,28] In general, these systems share an enhanced relaxivity and longer excretion lifetimes. Gd(III) complexes can be loaded either inside the nanovesicular carriers (like liposomes or apoferritin),[29,30] or they can be loaded onto the outer surface of the system through covalent linkages (*e.g.* dendrimers) or being one of the components of the hydrophobic aggregations (micelles, liposomes, solid lipid nanoparticles).[16,31] When properly designed, these nanosized probes may show several advantages (the level of modification introduced by the loading of the paramagnetic complexes being minimal) such as the absence of immunological reactions, limited uptake by macrophages and optimal biodistribution properties. From the relaxometric point of view, the higher efficiency of these systems is related to the restricted rotational mobility (long τ_R) of the paramagnetic complex upon binding to the nanocarrier, even though for supramolecular adducts the presence of a long spacer may decouple the fast motion of the complex and the slow tumbling of the carrier, resulting in a reduced relaxivity. For nanovesicular carriers, where the Gd(III) complexes are encapsulated in the inner aqueous cavity, one can observe either an increase of the relaxivity (*e.g.* apoferritin, due to the contribution arising from the internal mobile protons of the protein) or a decrease due to the "quenching" effect, as in the case of liposomes, characterized by low water permeability. The latter disadvantage can be exploited for the visualization of drug-delivery/-release processes, upon the "lighting-up" of contrast when the vesicle is destroyed.

Concerning the exploitation of natural nanosized carriers, a nice example is represented by apoferritin, a protein devoted to the storage of iron in cells. Ferritin consists of 24 proteins that self-assemble by means of saline and hydrogen bonds to yield a spherical aggregate containing the iron core, displaying ten channels for communicating between the inner and outer compartments. A method has been reported to replace the iron core in the inner cavity of ferritin with up to 8–10 Gd-HPDO3A molecules per apoferritin. Very interestingly, the relaxivity of Gd-HPDO3A entrapped in the apoferritin cavity shows a relaxivity that, at 20 MHz, is *ca.* 20 times higher than that of the free Gd-HPDO3A complex.[30] The presence of a high number of ferritin transporters on hepatocytes allows the Gd-loaded apoferritin system to be quickly taken up by the liver. The synthesis of a Gd-loaded apoferritin derivative containing biotinylated residues on its surface, coupled with the use of a targeting peptide recognized by avidin, has allowed the MRI visualization of the over-expression of NCAM (neural cell adhesion molecule) epitopes in neo-formed tumor endothelia.[32]

6.3 CEST (Chemical Exchange Saturation Transfer) Agents

The advent of the MR cellular and molecular era has prompted the search for new paradigms in the design of MR imaging reporters. MRI-CEST agents are frequency-encoding probes.[33] They offer the possibility of designing novel experiments such as the multiplex detection in *in vivo* cell tracking or to map a specific physico-chemical parameter of the microenvironment in which they distribute. The CEST contrast arises from the decrease in the intensity of the bulk water signal following the saturation of the exchanging protons of the CEST agent by a selective radiofrequency (RF) pulse. Hence, the basic requisite for a CEST agent is that the exchange rate between its mobile protons and the bulk water protons has to be smaller than the frequency difference between the absorption frequencies of the exchanging spins. In the CEST experiment, the RF pulse is applied at the absorption frequency of the exchanging protons to saturate their longitudinal magnetization. Because such nuclei exchange slowly with the nuclei of the bulk water, the intensity of the longitudinal magnetization of the latter will decrease from its equilibrium value.

The idea of transferring saturated magnetization to water proton resonance by irradiating the absorption of a proton pool of another molecule that is in a slow/intermediate exchange regime with water protons has been extensively exploited over the years in several fields of NMR spectroscopy. Its translation to MRI led to magnetization transfer contrast (MTC), which is the result of selectively observing the interaction of bulk water protons with the semi-solid macromolecular protons of a tissue. Differently from the semi-solid macromolecules, the absorption of the mobile proton of CEST molecules is generally relatively sharp and their frequencies precisely defined.[34] The latter feature associates each CEST agent with its defined frequency-encoded property, thus opening the field to multiplex detection of different contrast agents in the same anatomical region.

Research in this field has progressed along different paths in order to tackle the issue of the overall safety of the MRI-CEST experiment and the sensitivity/reliability of the CEST detection.[35–38] The systems developed so far range from small diamagnetic molecules to nanosized liposomes, passing through paramagnetic and supramolecular adducts. In an attempt to accelerate the entry of these agents into the clinical practice, it has been deemed of interest to consider chemicals already approved for human use and potentially able to generate CEST contrast. On this basis, we have undertaken a systematic study of either currently used X-ray contrast agents containing mobile protons (*e.g.* Iopamidol – Isovue® Bracco SpA, Italy, and Iopromide – Ultravist®, Bayer-Schering AG, Germany, as diamagnetic CEST (DIACEST) agents) or the paramagnetic lanthanide analogues of a clinically approved MRI contrast agent (Gadoteridol –ProHance®, Bracco SpA, Italy, as paramagnetic CEST (PARACEST) agent). The *in vivo* results are very encouraging for a clinical translation of these CEST agents for relevant cellular and molecular imaging applications.

6.3.1 Iopamidol and Iopromide as DIACEST Agents

The currently used iodinated CAs were discovered about 50 years ago and since then they have been important tools in clinical routine for radiographic procedures. Recently, it has been suggested that the presence of amide and alcoholic groups in these molecules may be exploited for their direct visualization in MR images as CEST agents.[39] Iopamidol (Isovue® – Bracco Imaging, Italy) is characterized by the presence of two types of functionalities that may be exploited for the generation of CEST effect (Figure 6.4).[40] The exchange rate of the two amide mobile proton pools is markedly pH-dependent. Thus, a ratiometric method for pH assessment has been set up based on the comparison of the saturation transfer effects induced by selective

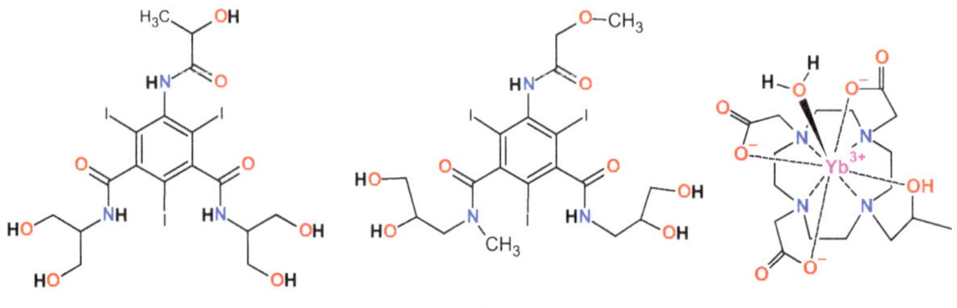

Figure 6.4   Structures of DIACEST (Iopamidol and Iopromide) and PARACEST (Yb-HPDO3A) agents.

irradiation of the two resonances. This ratiometric approach allows the knowledge of the local concentration of the contrast agent to be dispensed with, showing a good pH-dependence relationship in the physiological range 5.5–7.4. The Iopromide molecule, having a similar chemical structure to two non-equivalent amide proton pools, shares an analogous ability as pH-responsive CA for MRI-CEST procedure (see Figure 6.4).

The i.v. administration of Iopamidol into healthy mice at clinical doses ($0.75 \, g \, I \, kg^{-1}$ b.w.) allowed the quantification of the percentile saturation transfer (ST%) effect for as long as 45 min after the injection, exploiting the complete renal extraction of the molecule itself and its accumulation within the kidneys. The ratiometric procedure applied to the ST% maps acquired at 4.2 and 5.5 ppm (corresponding to the absorption frequency of the two non-chemically equivalent amide proton pools) yielded the *in vivo* parametric pH maps (Figure 6.5). Many pathological conditions are associated to pH alterations, such as inflammation processes, ischemia, tumor and kidney diseases. Therefore, pH may be considered as biomarker of kidney injuries and might find considerable clinical relevance. In particular, pathological altered renal physiology resulting from acute kidney injury (AKI) or tubular acidosis is associated with a perturbation of renal pH.[41] Clinical biomarker of kidney damage, such as blood urea nitrogen and serum creatinine apparently report on kidney damage only after a significant loss (50%) of renal function has occurred.[42] We recently demonstrated the use of Iopamidol to monitor the disease evolution *in vivo* by imaging pH variations in a glycerol-induced AKI model, as well as in an ischemia-reperfusion injury model.[43] In contrast with control mice, showing pH values around 6.7 after Iopamidol injection, we observed in the AKI model a constant increase of the pH values peaking after three days (mean pH values 7.1 and 7.3 after 1 and 3 days, respectively) and then from weeks 1 to 3 we observed a recovery to control values (pH = 6.7 after 3 weeks) (Figure 6.5). This evolution closely followed the time evolution of blood urea nitrogen levels as well as the damage scoring obtained by histo-logical methods. These results highlight the role that responsive CAs may acquire as an *in vivo* functional imaging tool to assess damage evolution and therapeutic response.

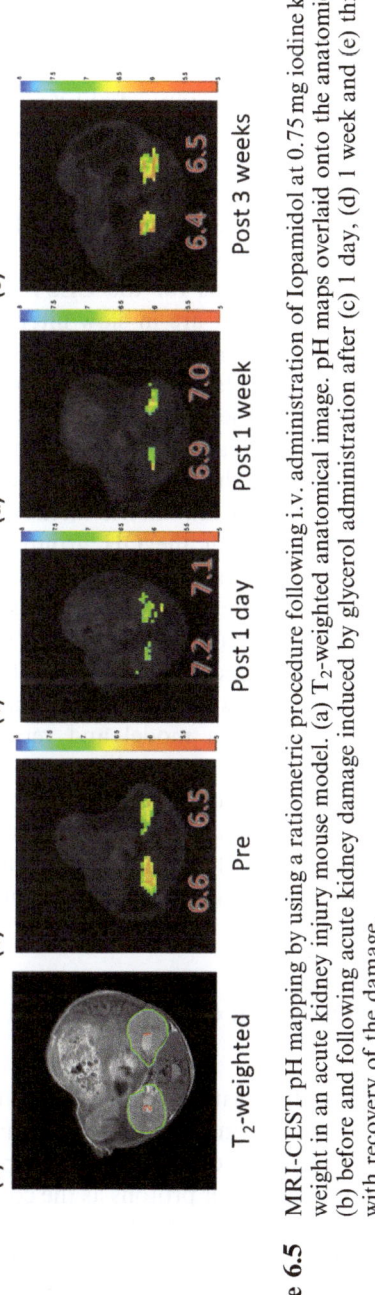

Figure 6.5 MRI-CEST pH mapping by using a ratiometric procedure following i.v. administration of Iopamidol at 0.75 mg iodine kg^{-1} body weight in an acute kidney injury mouse model. (a) T$_2$-weighted anatomical image. (b) T$_2$-weighted anatomical image. pH maps overlaid onto the anatomical image before and following acute kidney damage induced by glycerol administration after (c) 1 day, (d) 1 week and (e) three weeks with recovery of the damage.

6.3.2   Yb-HPDO3A as PARACEST Agent

Yb-HPDO3A is chemically an analog of Gd-HPDO3A (known as Gadoteridol or ProHance®), one of the most used MRI CAs in the clinical setting. The close chemical analogy anticipates for Yb-HPDO3A the same extracellular biodistribution and safety profile of ProHance®, thus making it an optimal candidate for being the first PARACEST agent considered for translation to the clinical practice. Yb-HPDO3A exists in solution as four main isomeric forms that are couples of enantiomers yielding an NMR spectrum in which two sets of signals are clearly detected. Based on their structures, the two observed isomers are indicated as SAP (square anti-prismatic) geometry and TSAP (twisted square anti-prismatic) geometry, respectively.[44]

The exchangeable pools of protons for the CEST experiments are represented by the hydroxyl moieties that are in slow/intermediate exchange with water protons. Since the –OH groups are directly coordinated to the paramagnetic center, their chemical shifts are very large, thus allowing the exploitation of large exchange rates that lead to very good values of ST% upon the application of the proper RF irradiation. The exchange rates of the two OH protons are dependent on pH, but the base catalysis is operating to a different extent, thus allowing the setup of a ratiometric method for the assessment of pH. Moreover, the chemical shift of the OH groups is highly sensitive to temperature, making this system an excellent agent for the simultaneous monitoring of pH and temperature. By acquiring a single Z-spectrum it is possible to obtain information concerning both the chemical shifts of the mobile protons (that report the temperature) and the saturation transfer ability that, analyzed on a ratiometric basis, allows the pH readout. The method has been tested on a murine melanoma model and representative images are reported in Figure 6.6.

The obtained extracellular pH values indicate a relationship between tumor growth and progressive acidification of the diseased region. *In vivo* pH mapping may represent a powerful tool to advance our understanding of the complex biology of cancer and provide an additional aid to monitor the effects of the selected therapeutic treatment.

6.3.3   Multiplexing Detection of Cells Labeled with PARACEST Agents

As mentioned above, one of the great advantages of the CEST methodology is the possibility of detecting more than one agent in the same anatomical region. This task is precluded to agents (such as Gd(III) complexes) that act on the relaxation enhancement of tissue water protons as the co-presence of more than one agent results in the sum of their effects. *Vice versa*, two CEST agents characterized by distinct absorption frequencies of their exchangeable proton pools can be separately "interrogated" by setting the proper irradiation off-set.

We have recently applied this approach to visualize two cell populations injected under the mouse skin.[45] Macrophages have been labeled *ex vivo* with

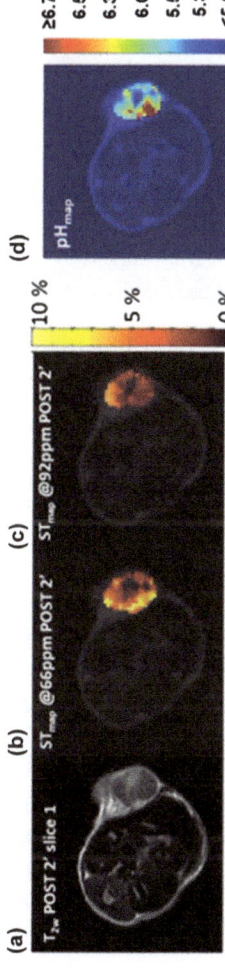

Figure 6.6 MRI-CEST pH maps in tumor using Yb-HPDO3A. (a) T_2-weighted anatomical image. Percentile saturation transfer (ST%) maps superimposed on the anatomical images at (b) 66 ppm and at (c) 92 ppm. (d) The ratio of the two ST% maps allows the corresponding pH map to be calculated.

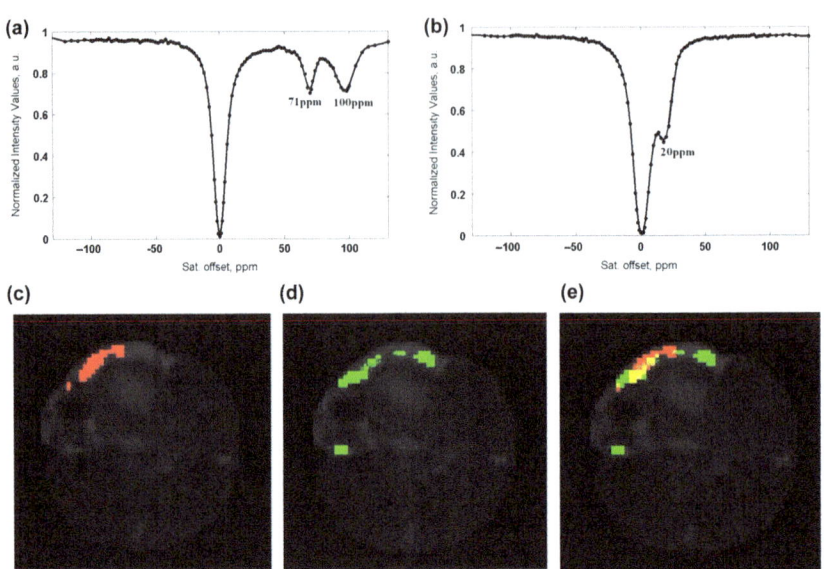

Figure 6.7 Z-spectra of the two PARACEST agents used for cell labeling. (a) Yb-HPDO3A and (b) Eu-HPDO3A in PBS at pH 7.4 and 20 °C. Falsecolors maps overimposed on the anatomical image 24 h after the injection of a mixture of Eu-labeled melanoma cells and Yb-labeled macrophages in mice: (c) T_2-weighted image; (d) ST% map at 66 ppm for Yb-HPDO3A in green; (e) ST% map at 15.6 ppm for Eu-HPDO3A in red; (e) merge of b and c, in yellow. Adapted with permission from Ferrauto *et al.*[45] © 2012 John Wiley & Sons.

Yb-HPDO3A whose Z-spectrum consists of two absorptions, as discussed above. For the study reported herein, irradiation at 71 ppm resulted in being the most efficient. Conversely, melanoma cells have been labeled by pinocytosis with Eu-HPDO3A whose Z-spectrum shows one absorption (for the OH moiety) at 20 ppm. The two sets of labeled cells (*ca.* 1×10^6 suspended in matrigel) were then injected under the mouse skin and, after a couple of hours, were visualized by the CEST procedure. As shown in Figure 6.7, upon irradiating at 20 ppm only the response of Eu-HPDO3A-labeled cells lighted up, whereas, upon moving the RF offset to 71 ppm, the response of Yb-HPDO3A-labeled cells could be detected. By combining the two images, one could clearly assign the regions where the two cell populations were present simultaneously.

6.4 Contrast Agents & Low Fields MRI Scanners

In recent years much attention has been devoted to the design of pre-clinical MRI scanners working at fields <1.5T. These systems appear particularly interesting for a generalized diffusion of MRI studies in biological laboratories. The possibility of dealing with permanent magnets that do not require cryogenic liquids otherwise necessary for superconducting magnets is

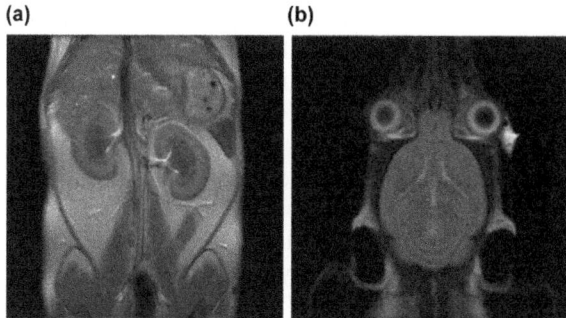

Figure 6.8 Representative anatomical images acquired with a 1.0 T MRI scanner (Icon – Bruker, Germany). (a) Coronal view of a mouse abdomen (FLASH TR/TE/FA/NEX 16 ms/6.7 ms/25/70, FOV 35 mm, matrix 256 × 256, 1 slice, slice thickness 1.25 mm, in plane spatial resolution 134 microns, acquisition time 5 min). (b) Coronal view of a mouse brain (RARE TR/TE/RF/NEX 2500 ms/60 ms/8/8, FOV 25 mm, matrix 192 × 192, slices 5, slice thickness 1 mm, in plane spatial resolution 130 microns, acquisition time 8 min).

particularly compelling. The B_0 homogeneity of the available permanent magnets are definitively good for mice studies with spatial in-plane resolution less than 150 μm, therefore, good enough to obtain high-definition anatomical images (Figure 6.8).

Moreover, we think that an important advance relies on the use of Gd-based CAs at 1.0 T. In fact, the relaxivity field dependence of Gd(III) complexes allows a maximum of efficiency to be exploited at *ca.* 1.0 T when the paramagnetic complex is part of a macromolecular system.[46] According to the theory of paramagnetic relaxation given by the Solomon–Bloembergen–Morgan equations, when the exchange rate of the metal coordinated water is fast enough (20–40 ns), the long reorientational correlation time of the paramagnetic macromolecular adduct dominates the contribution to the relaxivity. The relaxivity peak shown by these systems at 1.0 T makes the observed SNR markedly higher than the one attainable for a higher magnetic field (*e.g.* 7.0 T). By using Gd-containing systems linked to a macromolecular moiety able to recognize specifically a given target in a diseased region, 1.0 T MRI scanners can provide the biologist with the basic information concerning the overexpression of given targets.[47,48] Furthermore, blood-pool agents may yield an enhanced diagnostic information upon discriminating the different vascular properties between healthy and tumor tissues.

6.4.1 Low-density Lipoproteins as Carriers of Amphiphilic Gd(III) Complexes

Among the different classes of lipoproteins, attention has been focused on low-density lipoproteins (LDLs). These systems are particles of *ca.* 25 nm diameter consisting of a high-molecular-weight protein (ApoB-100) that wraps around the lipidic core mainly made of cholesterol ester and stearic acid. The magnetic labeling of these particles has been attained by using amphiphilic paramagnetic complexes formed by a hydrophilic Gd(III) chelate functionalized with one or two long aliphatic chains. Such systems insert their lipophilic moieties into the lipidic core of LDLs and expose the Gd-containing cages to the aqueous phase.[16]

Tumor cells over-expressing LDL receptors have been successfully targeted with the Gd-loaded LDL adducts consisting of *ca.* 300 Gd(III) amphiphilic complexes incorporated in the lipophilic LDL particles. From the NMRD profiles, one may see that the relaxivity peak of such a system is five times higher at 1.0 T than at 7.0 T, because of the lengthening of the reorientational time once the Gd complex is immobilized on the surface of the LDL particle. Upon labeling a B16 melanoma cell line with such particles the signal intensity enhancement in the MR images measured at 1.0 T was markedly higher than the one observed at 7.0 T.[49] The good sensitivity observed at 1.0 T yielded a detection threshold for Gd-labeled cells below 5000 cells μl^{-1}. These *in vitro* results were confirmed *in vivo* on a C57BL/6 mouse model bearing a tumor xenograft obtained by subcutaneous injection of B16 cells. The intravenous administration of the Gd-loaded/LDL particles at a 0.06 mmol Gd kg^{-1} dose resulted in a more than double contrast-to-noise ratio (CNR) value at 1.0 T (10 ± 3) in comparison to that measured at 7.0 T (4 ± 1.5). These results highlight the role that such low-field MRI scanners may have in the molecular imaging scenario when using Gd(III)-based probes possessing long molecular correlational time and properly functionalized with targeting moiety. It is worth noting that the well-known benefits provided by high field scanners (higher SNR, higher spatial resolution and faster sampling rate) may be well counterbalanced by the relaxation enhancement (=CNR) brought by the combination of Gd(III)-based macromolecular adducts with scanners operating at 1.0 T (Figure 6.9).

6.4.2   Blood-Pool Agents for DCE-MRI Applications

DCE-MRI allows the investigation of microvascular structure and function by tracking the pharmacokinetics of injected Gd-based contrast agents as they pass through the tumor vasculature. The obtained enhancement pattern reflects vascular perfusion and permeability of the tumor, showing the potential to monitor changes in the tumor microvasculature following antiangiogenic therapy.[50,51] At the clinical level, small-molecular-weight contrast media (*e.g.* Gd-DTPA) have been used at both low and high fields, allowing DCE-MRI to be proposed as an imaging biomarker of drug efficacy in Phase 1 clinical trials of angiogenesis inhibitors (see also Chapter 19).[52] Drug efficacy has been demonstrated with DCE-MRI in several clinical trials with antiangiogenic and antivascular drugs.[53,54]

We have shown that the DCE-MRI procedure can be efficiently carried out on a 1.0 T MRI scanner provided that the used Gd reporter has "blood-pool agent" characteristics. Gd-complexes able to bind to HSA have been extensively exploited in the 1990s for the setup of MR angiographic procedures.

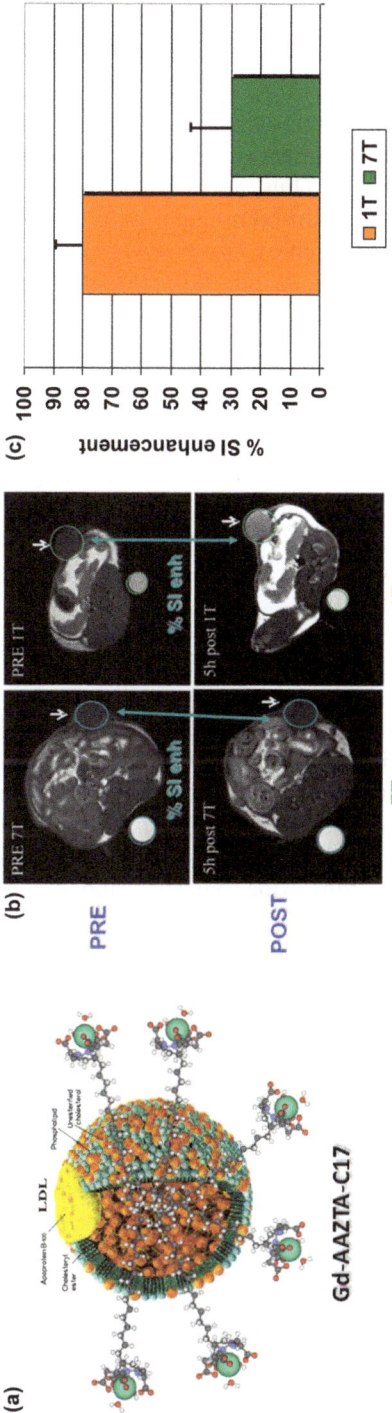

Figure 6.9 Comparison of signal intensity enhancements between a 1.0 T and a 7.0 T MRI scanner when using nanocarrier Gd-based systems. (a) Schematic representation of LDL labeled with the Gd-AAZTAC17 complex. (b) T_1-weighted multislice spin echo MR images of C57BL/6 mice grafted subcutaneously with B16 melanoma cells. Images were obtained before (PRE – top) and 5 hours after (POST – bottom) the administration of Gd-AAZTAC17/LDL 0.06 mmol kg^{-1} on a 7.0 T scanner (left) and on a 1.0 T scanner (right). (c) Calculated signal intensity enhancements of the same tumor regions at the two fields. Adapted with permission from Geninatti-Crich *et al.*[49] © 2011 John Wiley & Sons.

The search for systems with optimal relaxometric properties led to the development of many Gd complexes with an enhanced control of parameters such as the residence lifetime of the coordinated water, the tumbling rate and improved binding affinity to HSA.[17,18] When used in aDCE-MRI protocol, their longer extravasation time associated to their enhanced relaxivity allows the setup of highly efficient protocols for the assessment of changes in vascular permeability following antiangiogenic therapies.[55] Thus, at the pre-clinical level, the use of non-albumin-bound or strong albumin-bound blood-pool Gd-based CAs is emerging as a new tool to correctly report morphologic and functional characteristics of tumor vasculature *in vivo*. Again, the exploitation of the relaxivity hump that these CAs show at 1.0 T combined with their improved pharmacokinetic profiles in comparison to small-molecular-weight CAs yields a better characterization of vascular changes following antiangiogenic therapies.[56] Novel antiangiogenic therapies recently proposed, like those based on DNA vaccination against angiomotin receptors, a specific marker of neo-vessel formation, have been evaluated *in vivo* by combining a blood-pool contrast agent with a 1.0 T scanner. The semi-quantitative approach used to analyze the DCE data allowed a clear assessment of the vessel modification following angiomotin vaccination.[57]

6.5 Conclusions

The field of MRI CAs is still growing and new concepts are continuously forwarded. The limited sensitivity of MRI probes is the major limitation for a widespread diffusion of MRI in molecular imaging studies (see also Chapter 11). The examples reported in this chapter have dealt mainly with paramagnetic metal complexes, but it has to be outlined that important advances have been achieved also in the fields of iron oxide particles and of ^{19}F-containing agents.[58–60] Actually, the real breakthrough in the field of MRI CAs might be represented by the introduction of hyperpolarized molecules in the clinical practice (see Chapter 9). By means of hyperpolarization-based procedures, the sensitivity issue is tackled at its roots and a number of new MR spectroscopy imaging applications can be envisaged.[61]

Finally, new horizons have been opened by conjugating imaging science and nanotechnology. Among others, this merging will lead to the development of new protocols for imaging-guided therapy that might find great interest in the domain of personalized medicine.

Acknowledgements

The support MIUR (FIRB and PRIN projects), Regione Piemonte (PIIMDMT and Nano-IGT-Converging Technologies), EC-FP7-projects ENCITE (FP7-HEALTH-2007-A) and EU-COST TD1004 action are acknowledged.

References

1. P. Caravan, J. J. Ellison, T. J. McMurry and R. B. Lauffer, *Chem Rev.*, 1999, **99**, 2293.
2. M. F. Tweedle, G. T. Gaughan, J. Hagan, P. W. Wedeking, P. Sibley, L. J. Wilson and D. W. Lee, *Int. J. Rad. Appl. Instrum. B*, 1988, **15**, 31.
3. E. Brucher and A. D. Sherry, *The Chemistry of Contrast Agents in Medical Magnetic Resonance Imaging*, Wiley, Chichester, 1st edn, 2001.
4. J.-C. G. Bünzli and G. R. Choppin, *Lanthanide Probes in Life, Chemical, and Earth Sciences: Theory and Practice*, Elsevier, Amsterdam, 1st edn, 1989.
5. H.-J. Weinmann, A. Mühler and B. Radüchel, *Methods in Biomedical Magnetic Resonance Imaging and Spectroscopy*, Wiley, 1st edn, 2000.
6. F. Uggeri, S. Aime, P. L. Anelli, M. Botta, M. Brocchetta, C. D. Haeen, G. Ermondi, M. Grandi and P. Paoli, *Inorg. Chem.*, 1995, **34**, 10.
7. H. Schmitt-Willich, M. Brehm, C. L. Ewers, G. Michl, A. Muller-Fahrnow, O. Petrov, J. Platzek, B. Raduchel and D. Sulzle, *Inorg. Chem.*, 1999, **38**, 1134.
8. E. Toth, L. Helm and A. E. Merbach, *The Chemistry of Contrast Agents in Medical Magnetic Resonance Imaging*, Wiley, Chirchester, 1st edn, 2001.
9. M. Botta, *Eur. J. Inorg. Chem.*, 2000, **3**, 399.
10. R. A. Dwek, *Nuclear Magnetic Resonance in Biohemistry; Applications to Enzyme Systems*, Clarendon Press, Oxford, 1st edn, 1973.
11. U. Schmiedl, M. Ogan, H. Paajanen, M. Marotti, L. E. Crooks, A. C. Brito and R. C. Brasch, *Radiology*, 1987, **162**, 205.
12. S. Aime, M. Chiaussa, G. Digilio, E. Gianolio and E. Terreno, *J. Biol. Inorg. Chem.*, 1999, **4**, 766.
13. P. Farace, F. Merigo, S. Fiorini, E. Nicolato, S. Tambalo, A. Daducci, A. Degrassi, A. Sbarbati, D. Rubello and P. Marzola, *Eur. J. Radiol.*, 2011, **78**, 52.
14. S. Curry, *Drug Metab. Pharmacokinet.*, 2009, **24**, 342.
15. J. Ghuman, P. A. Zunszain, I. Petitpas, A. A. Bhattacharya, M. Otagiri and S. Curry, *J. Mol. Biol.*, 2005, **353**, 38.
16. E. Gianolio, G. B. Giovenzana, D. Longo, I. Longo, I. Menegotto and S. Aime, *Chemistry*, 2007, **13**, 5785.
17. S. Avedano, L. Tei, A. Lombardi, G. B. Giovenzana, S. Aime, D. Longo and M. Botta, *Chem. Commun. (Camb.)*, 2007, **45**, 4726.
18. S. Aime, E. Gianolio, D. Longo, R. Pagliarin, C. Lovazzano and M. Sisti, *ChemBioChem*, 2005, **6**, 818.
19. P. Caravan, N. J. Cloutier, M. T. Greenfield, S. A. McDermid, S. U. Dunham, J. W. Bulte, J. C. Amedio, Jr., R. J. Looby, R. M. Supkowski, W. D. Horrocks, Jr., T. J. McMurry and R. B. Lauffer, *J. Am. Chem. Soc.*, 2002, **124**, 3152.
20. E. K. Yucel and R. B. Lauffer, *Semin. Intervent. Radiol.*, 1998, **15**, 215.
21. S. Aime, E. Gianolio, E. Terreno, G. B. Giovenzana, R. Pagliarin, M. Sisti, G. Palmisano, M. Botta, M. P. Lowe and D. Parker, *J. Biol. Inorg. Chem.*, 2000, **5**, 488.

22. S. Aime, L. Calabi, C. Cavallotti, E. Gianolio, G. B. Giovenzana, P. Losi, A. Maiocchi, G. Palmisano and M. Sisti, *Inorg. Chem.*, 2004, **43**, 7588.
23. Z. Baranyai, F. Uggeri, G. B. Giovenzana, A. Benyei, E. Brucher and S. Aime, *Chemistry*, 2009, **15**, 1696.
24. S. M. Cohen, J. Xu, E. Radkov, K. N. Raymond, M. Botta, A. Barge and S. Aime, *Inorg. Chem.*, 2000, **39**, 5747.
25. E. J. Werner, S. Avedano, M. Botta, B. P. Hay, E. G. Moore, S. Aime and K. N. Raymond, *J. Am. Chem. Soc.*, 2007, **129**, 1870.
26. C. J. Jocher, E. G. Moore, J. Xu, S. Avedano, M. Botta, S. Aime and K. N. Raymond, *Inorg. Chem.*, 2007, **46**, 9182.
27. W. Y. Huang and J. J. Davis, *Dalton Trans.*, 2011, **40**, 6087.
28. A. Louie, *Chem. Rev.*, 2010, **110**, 3146.
29. D. A. Sipkins, D. A. Cheresh, M. R. Kazemi, L. M. Nevin, M. D. Bednarski and K. C. Li, *Nat. Med.*, 1998, **4**, 623.
30. S. Aime, L. Frullano and S. Geninatti Crich, *Angew. Chem. Int. Ed. Engl.*, 2002, **41**, 1017.
31. I. R. Corbin, H. Li, J. Chen, S. Lund-Katz, R. Zhou, J. D. Glickson and G. Zheng, *Neoplasia*, 2006, **8**, 488.
32. S. Geninatti Crich, B. Bussolati, L. Tei, C. Grange, G. Esposito, S. Lanzardo, G. Camussi and S. Aime, *Cancer Res.*, 2006, **66**, 9196.
33. E. Terreno, D. D. Castelli and S. Aime, *Contrast Media Mol. Imaging*, 2010, **5**, 78.
34. A. X. Li, R. H. Hudson, J. W. Barrett, C. K. Jones, S. H. Pasternak and R. Bartha, *Magn. Reson. Med.*, 2008, **60**, 1197.
35. J. Stancanello, E. Terreno, D. D. Castelli, C. Cabella, F. Uggeri and S. Aime, *Contrast Media Mol. Imaging*, 2008, **3**, 136.
36. E. Terreno, J. Stancanello, D. Longo, D. D. Castelli, L. Milone, H. M. Sanders, M. B. Kok, F. Uggeri and S. Aime, *Contrast Media Mol. Imaging*, 2009, **4**, 237.
37. P. Z. Sun, C. T. Farrar and A. G. Sorensen, *Magn. Reson. Med.*, 2007, **58**, 1207.
38. P. Z. Sun, Y. Murata, J. Lu, X. Wang, E. H. Lo and A. G. Sorensen, *Magn. Reson. Med.*, 2008, **59**, 1175.
39. S. Aime, L. Calabi, L. Biondi, M. De Miranda, S. Ghelli, L. Paleari, C. Rebaudengo and E. Terreno, *Magn. Reson. Med.*, 2005, **53**, 830.
40. D. L. Longo, W. Dastru, G. Digilio, J. Keupp, S. Langereis, S. Lanzardo, S. Prestigio, O. Steinbach, E. Terreno, F. Uggeri and S. Aime, *Magn. Reson. Med.*, 2011, **65**, 202.
41. P. C. Pereira, D. M. Miranda, E. A. Oliveira and A. C. Silva, *Curr. Genomics*, 2009, **10**, 51.
42. N. Paragas, A. Qiu, Q. Zhang, B. Samstein, S. X. Deng, K. M. Schmidt-Ott, M. Viltard, W. Yu, C. S. Forster, G. Gong, Y. Liu, R. Kulkarni, K. Mori, A. Kalandadze, A. J. Ratner, P. Devarajan, D. W. Landry, V. D'Agati, C. S. Lin and J. Barasch, *Nat. Med.*, 2011, **17**, 216.
43. D. L. Longo, A. Busato, S. Lanzardo, F. Antico and S. Aime, *Magn. Reson. Med.*, 2012, in press.

44. D. Delli Castelli, E. Terreno and S. Aime, *Angew. Chem. Int. Ed. Engl.*, 2011, **50**, 1798.
45. G. Ferrauto, D. Delli Castelli, E. Terreno and S. Aime, *Magn. Reson. Med.*, 2012, in press.
46. R. B. Lauffer, *Chem. Rev.*, 1987, **87**, 901.
47. N. Protti, F. Ballarini, S. Bortolussi, P. Bruschi, S. Stella, S. Geninatti, D. Alberti, S. Aime and S. Altieri, *Appl. Radiat. Isot.*, 2011, **69**, 1842.
48. K. C. Briley-Saebo, S. Geninatti-Crich, D. P. Cormode, A. Barazza, W. J. Mulder, W. Chen, G. B. Giovenzana, E. A. Fisher, S. Aime and Z. A. Fayad, *J. Phys. Chem. B*, 2009, **113**, 6283.
49. S. Geninatti-Crich, I. Szabo, D. Alberti, D. Longo and S. Aime, *Contrast Media Mol. Imaging*, 2011, **6**, 421.
50. K. Miyazaki, D. J. Collins, S. Walker-Samuel, J. N. Taylor, A. R. Padhani, M. O. Leach and D. M. Koh, *Eur. Radiol.*, 2008, **18**, 1414.
51. D. J. Jonker, L. S. Rosen, M. B. Sawyer, F. de Braud, G. Wilding, C. J. Sweeney, G. C. Jayson, G. A. McArthur, G. Rustin, G. Goss, J. Kantor, L. Velasquez, S. Syed, O. Mokliatchouk, D. M. Feltquate, G. Kollia, D. S. Nuyten and S. Galbraith, *Ann. Oncol.*, 2011, **22**, 1413.
52. N. Hylton, *J. Clin. Oncol.*, 2006, **24**, 3293.
53. K. Mross, J. Drevs, M. Muller, M. Medinger, D. Marme, J. Hennig, B. Morgan, D. Lebwohl, E. Masson, Y. Y. Ho, C. Gunther, D. Laurent and C. Unger, *Eur. J. Cancer*, 2005, **41**, 1291.
54. G. Liu, H. S. Rugo, G. Wilding, T. M. McShane, J. L. Evelhoch, C. Ng, E. Jackson, F. Kelcz, B. M. Yeh, F. T. Lee, Jr., C. Charnsangavej, J. W. Park, E. A. Ashton, H. M. Steinfeldt, Y. K. Pithavala, S. D. Reich and R. S. Herbst, *J. Clin. Oncol.*, 2005, **23**, 5464.
55. T. P. Roberts, K. Turetschek, A. Preda, V. Novikov, M. Moeglich, D. M. Shames, R. C. Brasch and H. J. Weinmann, *Acad. Radiol.*, 2002, **9**(2), S511.
56. K. Turetschek, A. Preda, V. Novikov, R. C. Brasch, H. J. Weinmann, P. Wunderbaldinger and T. P. Roberts, *J. Magn. Reson. Imaging*, 2004, **20**, 138.
57. M. Arigoni, G. Barutello, S. Lanzardo, D. Longo, S. Aime, C. Curcio, M. Iezzi, Y. Zheng, I. Barkefors, L. Holmgren and F. Cavallo, *Angiogenesis*, 2012, **15**, 305.
58. S. Laurent, S. Boutry, I. Mahieu, L. Vander Elst and R. N. Muller, *Curr. Med. Chem.*, 2009, **16**, 4712.
59. F. Liu, S. Laurent, H. Fattahi, L. Vander Elst and R. N. Muller, *Nanomedicine (Lond)*, 2011, **6**, 519–528.
60. J. Chen, G. M. Lanza and S. A. Wickline, *Wiley Interdiscip. Rev. Nanomed. Nanobiotechnol.*, 2010, **2**, 431.
61. A. Viale and S. Aime, *Curr. Opin. Chem. Biol.*, 2010, **14**, 90.

CHAPTER 7

Pharmacological fMRI in Drug Discovery and Development

ALEXANDRE COIMBRA,*[a] RICHARD BAUMGARTNER[b] AND ADAM J. SCHWARZ[c]

[a] Clinical Imaging Group, Early Clinical Development, Genentech, Inc., South San Francisco, CA, USA; [b] Biometrics, Merck Research Laboratories, Merck & Co., Inc., Rahway, NJ, USA; [c] Translational Medicine, Lilly Research Laboratories, Eli Lilly & Co., Indianapolis, IN, USA
*Email: coimbra.alexandre@gene.com

7.1 Introduction

Assessment of brain neuronal activity is critical to understanding normal brain function and the pathology of neurologic diseases and psychiatric disorders. It is also paramount in assessing and understanding drug action *in vivo* in the central nervous system (CNS). A series of related magnetic resonance imaging (MRI) techniques allow for indirect assessment of spatially and temporally localized phenomena associated with neuronal function. Here, these techniques are referred to broadly as functional MRI, or fMRI.

Although powerful and informative, none of these fMRI techniques provides a direct measure of the electric or biochemical activity related to functioning neurons. Instead, fMRI techniques infer neuronal function or changes in functional state by direct or indirect measurement of local hemodynamic properties associated with the microvasculature, including blood flow and volume and local changes in blood oxygenation levels, associated with the delivery of oxygen *via* the bloodstream to support increased oxidative

New Developments in NMR No. 2
New Applications of NMR in Drug Discovery and Development
Edited by Leoncio Garrido and Nicolau Beckmann
© The Royal Society of Chemistry 2013
Published by the Royal Society of Chemistry, www.rsc.org

metabolism taking place in active neurons.[1–3] The relationship between neuronal activity and perfusion and oxygen metabolism factors is referred to as neurovascular coupling.

The development and use of fMRI techniques in the past two-and-a-half decades, and the understanding of neurovascular coupling, have provided a window into normal brain function. This understanding would not have been possible without the development of the appropriate data analysis tools to extract pertinent information associated with brain function from the wealth of data acquired during the functional imaging experiments. Many investigators have used fMRI to study changes or differences in brain function in patients with neurological and psychiatric disorders.[4–9] And more recently, fMRI has been used to infer action of pharmacological agents in the CNS.[10–13]

In order to provide the reader with a perspective on how fMRI has played and may continue to play a role in drug development, this chapter offers an overview of various fMRI techniques available, the type of physiological information they assess and their relationship to neuronal function. Also, a review is given of possible applications in both the pre-clinical and clinical domains as well as applications translating from pre-clinical to clinical areas. A brief discussion of good practices in clinical studies is also provided. Finally, methods for analyses of fMRI experiments and extraction of quantitative features of the data and challenges associated with these techniques are discussed.

7.2 Physiology of fMRI

Functional MRI techniques rely on a phenomenon called neurovascular coupling to infer neuronal function.[3,14] Neurovascular coupling refers to the tight relationship between local neuronal activity and the associated, anatomically concordant, hemodynamic response. The expression "hemodynamic response" is used here to refer to changes in hemodynamic parameters (blood flow and volume, and blood oxygenation levels) associated with neuronal activity (see Figure 7.1).

Neuronal function is accompanied by electrical and biochemical activity involving not only neurons, but glia and vascular cells as well. These activities include restoration of ionic gradients across cell membranes and neurotransmitter recycling, and require energy in the form of adenosine triphosphate (ATP). In the brain, ATP is typically produced by oxidative glucose metabolism, *i.e.* a metabolism that depends on constant supply of glucose and oxygen regulated through blood perfusion. Thus, in the chain of events that ensue, increase in neuronal activity induces increases in oxygen and glucose consumption, which are followed by increases in cerebral blood flow (CBF) and blood volume (CBV) to comply with the energetic demands of the local tissue. However, because oxygen delivered *via* the capillary bed is transported to the cells by a process of passive diffusion, requiring a positive oxygen concentration gradient from the vasculature to the parenchyma to the cells, the fractional increase in CBF, CBV and oxygenated:deoxygenated blood ratio is greater than the increase in oxygen consumption. This net increase in the local concentration of oxygen in blood and tissue, and the fact that oxygenated

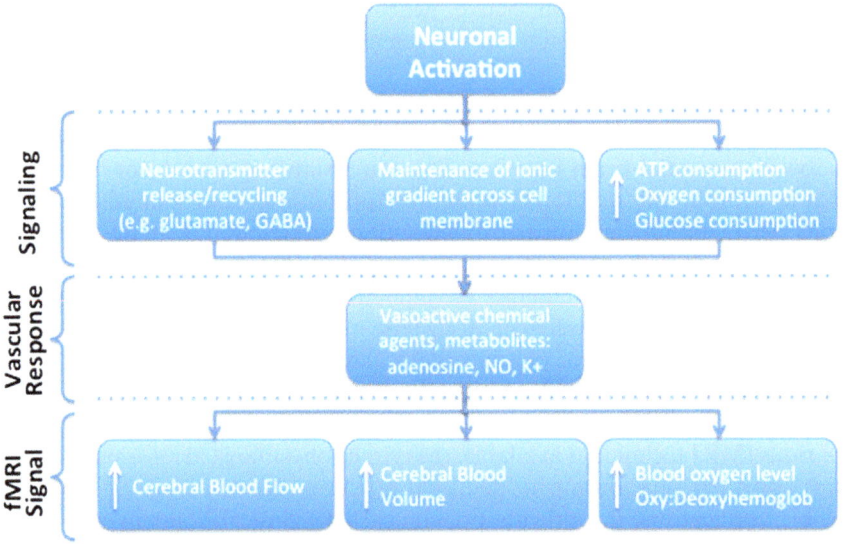

Figure 7.1   Neurovascular coupling: schematic summarizing the link between neuronal activation, vascular response and possible fMRI signal readouts.

hemoglobin is diamagnetic whereas deoxygenated hemoglobin is paramagnetic, gives rise to what has been termed the blood oxygen-level dependent (BOLD) signal, first described by Seigi Ogawa.[15]

7.3 fMRI Techniques

The most common fMRI techniques are those that use MRI methods to detect signals related to hemodynamic parameters of the neurovasculature.

7.3.1 BOLD Imaging

As the name indicates, this refers to the family of MRI techniques that are sensitive to the BOLD signal. This includes techniques that render spin-spin relaxation time (T_2-) and effective spin-spin relaxation time (T_2^*-)weighted images sensitive to paramagnetic deoxyhemoglobin. The most commonly used pulse sequences are T_2^*-weighted fast gradient echo and echo-planar imaging (EPI) sequences.[16,17] With commercially available hardware, the EPI method allows for acquisition of whole-brain data with fair spatial (on the order of millimeters in humans) and temporal ($\sim$2–3 seconds) resolutions. Novel sequences, with specialized hardware and using parallel imaging can improve the time resolution to sub-second scales. The BOLD signal, however, is not an absolute measure and can only be evaluated as a response relative to some reference level that is subject-specific and also specific to an experimental session. Typical response levels obtained for most applications are on the order of 0.25–3% and are dependent on the brain region and the experimental paradigm.[18]

7.3.2 CBF Imaging with Arterial Spin Labeling

Arterial spin labeling (ASL) is used to measure blood flow. It refers to pulse sequences that use preparation pulses to excite (label) water molecule spins in the arterial blood feeding the perfused tissue. After a pre-determined amount of time from the labeling pulse, a label-sensitive image is acquired and compared to a paired, control (non-labeled) image. The difference between control and labeled image is proportional to blood perfusion.[19] ASL data combined with appropriate data perfusion models can be used to compute absolute CBF,[20] which represents an advantage over relative BOLD signal changes. With absolute quantification of CBF, the response signal can be directly measured not only within, but also across, experimental sessions and individuals. Unfortunately, currently available ASL techniques yield relatively low signal-to-noise ratios (SNRs). Sample CBF data are shown in Figure 7.2.

7.3.3 CBV Imaging with Exogenous Contrast

Blood-pool contrast agents have been used to enhance the contrast-to-noise signal in fMRI. The most promising contrast agents are based on small and ultra-small super-paramagnetic iron-oxide particles, SPIO and USPIO, respectively. The blood T_2^* signal decreases in a manner dependent on the

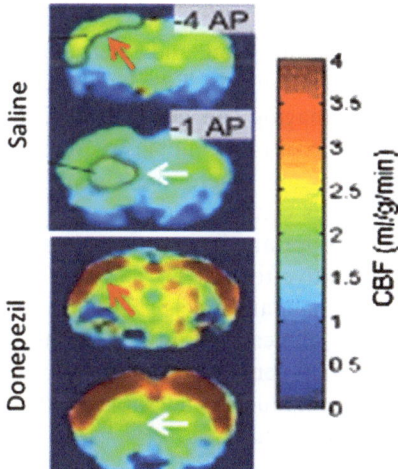

Figure 7.2 Coronal maps of basal cerebral blood flow (CBF) obtained with the arterial spin labeling (ASL) technique in awake rats following challenge with (top) saline solution and (bottom) the acetyl-cholinesterase inhibitor donepezil.[1] The labels −4 AP and −1 AP refer to the approximate anteroposterior position (in mm) of the slices relative to the bregma. The corresponding maps show a marked increase in CBF in the cortex (red arrows), and stable CBF in the striatum (white arrows) suggesting that the increase in cortical CBF induced by donepezil is not due solely to a systemic increase in blood flow.

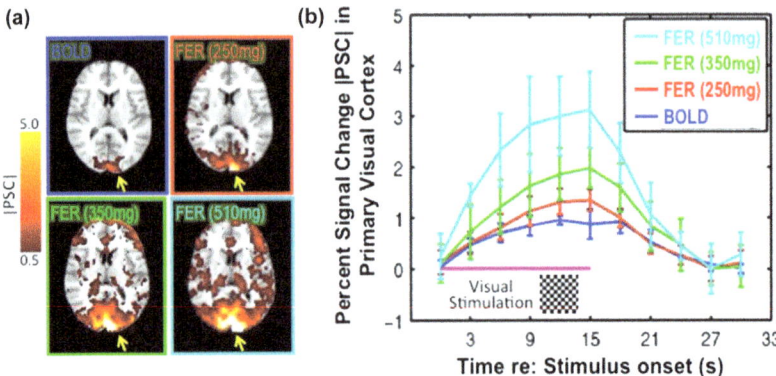

Figure 7.3 Dose-dependent enhancement of the fMRI response to visual stimulation obtained with ferumoxytol, an FDA approved therapeutic compound for iron deficiency that can be used as a USPIO contrast for CBV imaging. Absolute percent signal change (|PSC|) was used as a measure to compare the fMRI responses between BOLD and CBV imaging technique with 250, 350 and 510 mg doses of the contrast agent. a) A dose-related stronger and broader response is measured with CBV imaging as compared to BOLD particularly in the visual cortices (arrows). b) A closer analysis of the |PSC| response in the primary visual cortex indicates an enhancement on the order of $3\times$ with the FDA-approved 510 mg dose of ferumoxytol.[2]

concentration of these particles in blood, and cerebral blood volume (CBV) can be directly inferred from the amount of T_2^* signal change. The same fast EPI sequences used for BOLD imaging can be used for CBV imaging when high temporal resolution (on the order of seconds) is required. For applications such as pharmacological intervention, where the signal changes of interest are slower, T_2-weighted sequences can be used instead, providing reduced image distortion.[21,22] With the appropriate dosing of contrast, SNR can be enhanced by several fold ($2\times$–$5\times$) relative to BOLD.[23,24] Until recently, the use of this technique was restricted to animal models. New iron-oxide particles such as ferumoxytol, approved for clinical use in treatment of anemia, are being investigated as fMRI signal enhancing contrast agents in humans[2,25] (Figure 7.3). Other MRI techniques using sophisticated pulse sequences and endogenous contrasts are emerging.[26]

7.4 fMRI Applications in Drug Research: Principles and Examples

7.4.1 Pre-Clinical Drug Discovery

As with other methods of experimental biology, the application of fMRI methods to pre-clinical drug discovery most commonly involves experiments in the rodent. These imaging experiments have typically been performed on anesthetized rodents in order to allow tight control of physiological parameters and to minimize motion and stress to the animal. More recently, fMRI methods in conscious rodents have been demonstrated and, if stress can be minimized by habituating the animals to the scanner environment (constrained holder, noise), this avoids the potential confound of anesthesia on the pharmacology of interest.[1,27,28] However, the animals are nevertheless tightly restrained within the magnet bore, restricting the range of fMRI paradigms that can be applied. This excludes, for example, rodent analogs of human (or primate) fMRI of cognitive tasks, which require a freely moving behavioral state.

7.4.1.1 Applications of fMRI in Pre-Clinical Drug Discovery

The key applications of fMRI methods to drug discovery are primarily the following.

Screening. This involves comparative testing of different compounds to aid selection of a candidate therapeutic. However, to fit into a screening cascade, relatively rapid experimental turnaround times are required;

Model characterization. Understanding the central neurobiological substrates underlying behavioral effects in pre-clinical models;

Mechanism of action. Understanding pharmacological action of a particular mechanism and verifying with the candidate therapeutic;

Biomarker development. Informing the choice of human biomarker studies for the compound if it progresses into clinical development.

7.4.1.2 Key fMRI Methods in Pre-Clinical Drug Discovery

Several types of fMRI experiments have been successfully applied as tools for CNS pharmacology in pre-clinical drug discovery.

Forepaw and hindpaw stimulation and pain models. In these "task-based" experiments, the paw is electrically stimulated[29,30] or a punctuate, thermal or compressive stimulus is applied in analogy to experimental pain models.[29] Visceral pain models (rectal distension) have also been studied with fMRI.[31] The evoked activation pattern, reflecting the brain's response to the stimulus, can then be modulated by treatment with a pharmacological agent, *e.g.* a known or putative analgesic.

Pharmacological MRI (phMRI). In this paradigm, the direct hemodynamic response to an acute drug challenge is measured to determine the spatial (and, in the case of time-series approaches, temporal) profile of the functional response to a pharmacological, rather than a sensory, stimulus (*e.g.* agonism or antagonism of a particular cell surface receptor). These experiments can be used to map the response to a single compound of interest, or to study the interaction between the functional effect of two compounds, in which one "probe" compound is used to elicit the measured hemodynamic response, and another (*e.g.* administered as a pre-treatment) modulates it.[32–34]

Measurements of basal hemodynamic properties. This approach can be used to probe effects of acute or chronic drug treatment, of behavioral conditioning or interactions between the two, on resting brain function. Most commonly, measures of resting CBV,[35] using blood-pool contrast agents, or of CBF, using ASL,[1,36] have been used.

Resting state connectivity. Recently, experiments measuring resting state "functional connectivity", an approach in which temporally correlated, low-frequency (typically 0.01–0.1 Hz) fMRI signal oscillations are measured in the absence of a task or stimulus, have been successfully developed for the rodent.[37–39] This provides a back-translation of a method widely adopted in human fMRI.[40–42] BOLD has been the most commonly used fMRI signal in resting functional connectivity experiments; however, CBF and CBV have also been investigated as indicators of functional connectivity.[43,44]

7.4.1.3   Anesthetized vs. Awake Animal Experiments

A key question in pre-clinical fMRI experiments is to what extent anesthesia confounds the fMRI data and interpretation of results. There are several reports indicating anesthesia affects not only neuronal activity but also neurovascular coupling.[45–47] It is also generally accepted that these phenomena are differently affected by different anesthetics and that the effects are dependent on the level of anesthetic used.[45,47–49] Finally, it is typically unknown *a priori* whether there are any potentially confounding pharmacological interactions between the anesthetics and the new compounds being tested.[49] On the other hand, data from awake animal experiments may suffer from other confounding effects such as stress, and in some jurisdictions it may be challenging to obtain ethics approval to run conscious animal experiments. It should also be noted that for some pharmacological challenges there is nevertheless a strong experimental concordance between phMRI responses observed in appropriately anesthetized rodents and those observed using other parameters of brain function in the conscious state. For example, functional responses to NMDAR antagonists such as ketamine or PCP are very similar whether measured using phMRI in the anesthetized rat or 2-deoxyglucose autoradiography in the conscious animal[50] and the phMRI response to nicotine (anesthetized)[51] closely matches that of c-fos early gene expression in the conscious rat.[52]

Overall, it remains a matter of continuous debate how exactly these effects should influence interpretation of fMRI results and how the knowledge of these effects should inform strategies on the use of anesthetics or awake animal setups. In any event, experiments with anesthetized animals are much more streamlined compared to experiments with awake animals where additional resources must be put into minimizing stress by habituation of the animals to the MRI environment and/or training to perform tasks. This renders the anesthesia paradigm better suited operationally to drug screening applications. The choice of anesthetic, however, will have to be addressed on a case-by-case basis taking into account animal species, target fMRI method and technique and potential drug–anesthesia interaction. For applications such as investigation of compound mechanism-of-action and biomarker development, where throughput is of lesser importance and translatability to clinic of higher importance, closer consideration should be given to the awake animal paradigm. Investigators interested in performing experiments in awake animals should provide evidence of their ability to scan animals that are not stressed in the MRI environment. In rats, a few groups have successfully adopted the animal habituation procedure proposed by King *et al.*[1,27,48] Following this procedure animals have been shown to normalize physiological indicators of stress (heart and respiratory rates, arterial blood pressure and blood cortisol levels) within 3–5 days following daily 1–2 hour habituation sessions to mechanical restraint in a simulated MRI environment. Similar procedures have been published and adopted for non-human primates.[53] Some reports suggest that in the awake rat paradigm there is better concordance between phMRI endpoints and compound target-binding when compared to anesthetized animals.[54] Similarly, in awake rats, there seems to be better concordance between phMRI endpoints and behavioral outcome.[1] The field would benefit from continued investigation of the cost-to-benefit ratio of the awake animal paradigm in drug development applications.

7.4.1.4   Case Study: Dopamine D_3 Antagonists in Drug Addiction

The dopamine D_3 receptor has a focal distribution that overlaps the mesolimbic dopamine "reward" pathways, including a high density in the nucleus accumbens.[55,56] Highly selective antagonists of the D_3 receptor are efficacious in animal models of drug dependence.[57,58] A series of phMRI experiments in the rat demonstrated that acute pre-treatment with three different selective D_3 antagonist molecules, from different chemical classes, consistently potentiated the rCBV (relative CBV) response to *d*-amphetamine challenge in the nucleus accumbens[59,60] (see Figure 7.4). This included an independent replication of the effect with one of the compounds, the widely used tool compound SB277011A. Given that the dopaminergic response to amphetamine in the nucleus accumbens is blunted in human drug addicts,[61,62] these results provided a compelling mechanistic hypothesis for the efficacy of these compounds in models of substance abuse, and suggested analogous human experiments for translation. Importantly, the replication of functional effect and convergence across different molecules with the same mechanism of action enhances the confidence in these findings.

7.4.2   Clinical Drug Development

In clinical drug development, fMRI biomarker studies may be performed on healthy volunteers or in patient populations and can be integrated into any clinical phase. However, the most typical applications are in Phase Ib/IIa, where the pharmacokinetic (PK) and safety profile, and pharmacodynamic

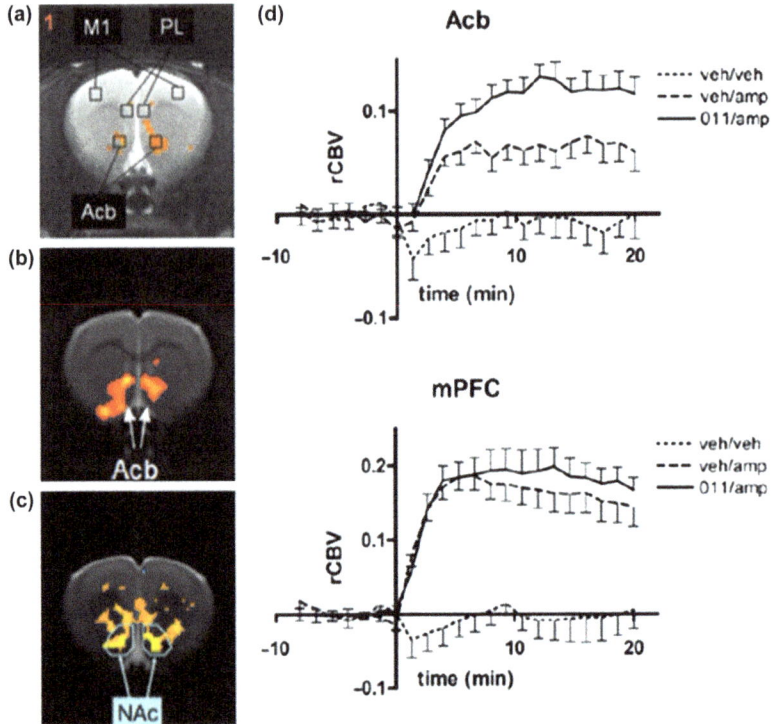

Figure 7.4 A series of phMRI studies in the rat demonstrated a consistent potentiation of the *d*-amphetamine-evoked phMRI signal when animals were pre-treated with selective dopamine D3 antagonists. This effect was anatomically specific, with the effect in the nucleus accumbens being observed consistently across cohorts and test compounds. (a) Potentiation of *d*-amphetamine phMRI signal by SB277011A. (b) Replication of SB277011A-induced potentiation in an independent set of animals. (c) Consistent potentiation observed with a different compound (cpd 35). (d) PhMRI time courses illustrating the anatomical specificity of the D3 antagonist effect, showing increased phMRI response in the accumbens, but not in the medial prefrontal cortex, following pre-treatment with SB277011A *vs.* vehicle. Abbreviations: Acb or NAc = nucleus accumbens; M1 = primary motor cortex; PL = prelimbic cortex; mPFC = medial prefrontal cortex.
Permissions: (a) Reprinted with permission from Schwarz *et al.*[60] (b), (d) Reprinted with permission from Schwarz *et al.*[59] (c) Reprinted with permission from Micheli *et al.*, *J. Med. Chem.*, 2007, **50**, 5076. © 2007 American Chemical Society.

(PD), biological and clinical effects of the developmental compound are first elucidated in humans.

7.4.2.1 Applications of fMRI in Clinical Drug Development

For use as a clinical biomarker, an fMRI study should contribute to some decision relating to the compound's further development. A pharmacological effect observed using an fMRI method is fundamentally a PD measure, providing information on the functional* effect of the compound downstream from engagement of the molecular target. Two ways in which fMRI studies can inform clinical development are the following:

Dose selection. Central PD effects observed *via* fMRI studies can be used to help select doses for subsequent clinical trials. This can be important in allowing doses lower than the maximum tolerated dose (MTD) to be tested if a modulation of the fMRI signal is observed at lower doses, potentially providing a better tradeoff between pharmacological efficacy and side effects.

Proof of mechanism. An appropriately designed fMRI experiment can also be used to understand the compound's functional effect in humans and generate or test hypotheses regarding its mechanism of action at the brain system level – *i.e.* how and in which brain regions the fMRI signal is modified. For example, a putative analgesic might be expected to modulate the fMRI response evoked by a painful stimulus in a core set of brain regions that have been identified as mediating the different factors (*e.g.* sensory, emotional, cognitive) underlying the central response to pain[12,63–67] or a putative antidepressant might be expected to attenuate the exaggerated fMRI response in the amygdala to negative affective stimuli.[13,68,69] Likewise, in the resting brain of healthy individuals receiving a single dose of the antidepressant citalopram, ASL measurements revealed reduction of basal perfusion levels in brain regions known to be hyper-perfused in patients with major depression.[70] Similar findings have been reported in patients, *e.g.* with Parkinson's[71] and Alzheimer's disease,[72] treated with active compounds. Examples also exist where functional connectivity across brain regions known to be implicated in disease mechanisms has been shown to be specifically modulated by analgesics[12,73] and cholinesterase inhibitors,[72,74] among others. Depending on how well characterized the fMRI paradigm being deployed is, and the hypothesis being tested, such results could influence the clinical indication pursued with the compound and impact calculations of its probability of technical success.

For both of these applications, the particular fMRI method being applied should be appropriate to either the compound's pharmacology and/or to the clinical indication being pursued. The method should also be sufficiently well characterized and validated such that basic measurement characteristics are known. These include test-retest reliability (which informs whether a within-subject or between-subject design would be optimal and also provides data for powering subsequent studies) and, ideally, comparator data with reference compounds (see also Section 7.5 of this chapter). The issue of dose selection can be especially important if a positron emission tomography (PET) radioligand is not available to measure receptor occupancy at the compound's target.[75,76] However, data reflecting both receptor occupancy (PET) and PD (fMRI) effects in the brain are likely to be complementary and provide a more complete

---

*Assuming possible confounds, that can mimic or obscure true functionally driven hemodynamic changes, have been adequately accounted for – see Section 7.7.

picture of not only the compound's dose-occupancy profile but also its functional effects on relevant brain circuits.[77]

Whether the fMRI study is performed in healthy volunteers or in a patient population depends on the pharmacological hypothesis being tested. Healthy volunteer studies can be quicker to run and may have less variability across subjects. Patient studies provide data in a specified pathological population, and are necessary if the compound targets a pathological process not present in healthy subjects.

7.4.2.2   Case Study: Combined Receptor Occupancy-PET and fMRI of Opioid Antagonists

A combined receptor occupancy-PET and fMRI study of novel (GSK1521498) and comparator (naltrexone) µ-opioid antagonists illustrates the complementary value of measuring target engagement (receptor occupancy) and downstream functional consequences (PD) in early phase drug development.[77] Attenuation of µ-opioid signaling has potential therapeutic benefit in disorders of compulsive consumption; in a previous study, GSK1521498 reduced pleasurable response to, and consumption of, high fat/high sugar snack items in an experimental model of overeating behavior in overweight volunteers,[78] while naltrexone is moderately effective in similar models.[79,80] In the imaging study, µ-opioid receptor occupancy was measured following single acute doses of both compounds using [^{11}C]-carfentanil and revealed dose-dependent occupancy in the high-µ-opioid receptor density region of the ventral striatum (nucleus accumbens). The fMRI paradigm was designed to test the hypothesis that these compounds attenuate the functional brain response to a palatable gustatory stimulus (juice). GSK1521498 attenuated the BOLD responses in the amygdala, a brain region hypothesized *a priori*, and showed stronger attenuating effects in both the amygdala and nucleus accumbens than did naltrexone. Since both modalities were acquired from the same subjects, the fMRI analysis could take account of within-subject receptor occupancy. This fMRI study confirmed a PD effect of the novel compound, stronger in certain brain regions than the comparator, on central brain responses to a task central to the mechanism. The experiment was key to demonstrate the potential therapeutic benefit of the compound.

7.4.3   Translation

There is a strong interest in the translational potential of functional imaging for drug development.[†] Many methods can be applied in both pre-clinical species and in humans, including measures of basal, task-free, brain function (*e.g.* resting perfusion, functional connectivity), phMRI and somatosensory stimulus paradigms.[11,75,81,82]

[†]A caveat – not specific to imaging – is that some pharmacological effects might not effectively translate, for example due to different receptor affinities or pharmacodynamic variation across species.

For most effective translation the experimental paradigm applied and physiological endpoint measured in both pre-clinical species and in humans should match. However, methods that are most applicable in the rodent might not be as easily available in the clinical context, and *vice versa*. So it is, for example, that phMRI response to ketamine challenge has been investigated in the rat with rCBV techniques using ultra-small superparamagnetic iron oxide (USPIO) contrast agents. A similar anatomical profile of activation to ketamine challenge as well as temporal activation patterns were observed in humans with BOLD technique (see Figure 7.5). With functional imaging techniques, which often rely on on-site expertise and experience with specific methods, this is often dependent on the imaging sites involved and hence provides a practical constraint on effective translation. For example, basal perfusion using ASL is a promising imaging biomarker, both as a probe of central pharmacology and of disease states (*e.g.* Alzheimer's disease).[70,83] While ASL may be implemented in an in-house pre-clinical imaging facility, currently its availability for clinical studies remains limited to sites with a particular

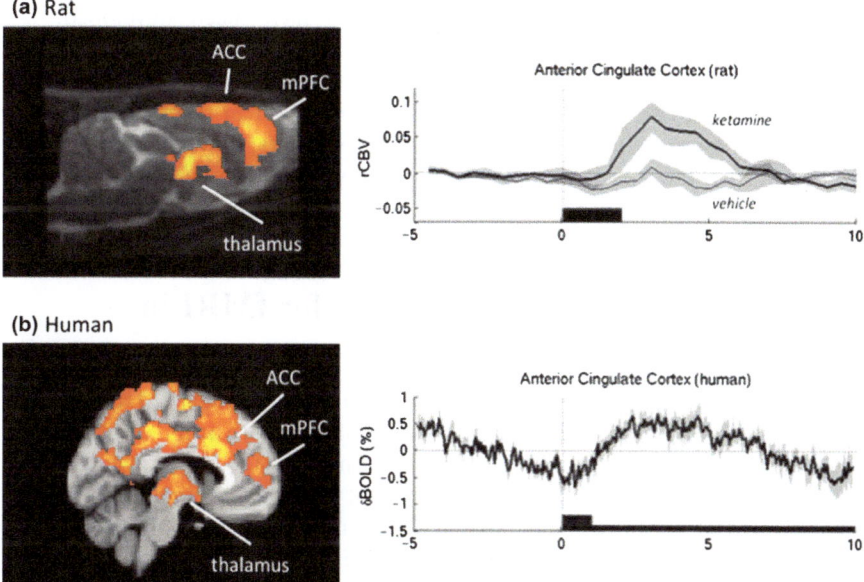

Figure 7.5 Consistent neuroanatomical profile of phMRI signal changes induced by acute ketamine challenge in rats and humans. (a) Activation pattern and temporal rCBV signal changes induced by 1 mg kg^{-1} ketamine (N = 5) *vs.* vehicle (saline; N = 4) in the isoflurane anaesthetized rat. (b) Activation pattern and temporal BOLD signal changes induced in the human brain by the administration of 0.12 ± 0.0026 mg kg^{-1} during the first minute followed by the infusion of 0.31 mg kg^{-1} h^{-1}, designed to achieve a 75 ng mL^{-1} plasma concentration (N = 5). The thick black bars in the time course plots indicate the infusion duration. Data courtesy of Eli Lilly & Company and King's College, London.

research interest in this method. This is acceptable for early-phase, single site PD studies,[70,71,81,84,85] for which an appropriate site can be chosen, but can limit its use in larger, multisite clinical trials.

In contrast to rodent studies, however, a wider range of task-evoked paradigms is available in humans due to the subject's ability to perform the task within the scanner: a vast range of neuropsychological paradigms and even gustatory[77,86] and olfactory[87] stimuli have been adapted for fMRI. These can be used to probe pharmacological effects on specific brain regions and circuits involved in particular mental tasks, providing a high degree of specificity in the experimental design. Although direct back-translation to rodent fMRI is limited, two alternative translational approaches are becoming available. First, alternative measures of hemodynamic variables can be acquired in freely moving rodents, opening the way to cognitive and other paradigms that can be back-translated from human neuropsychology to rodent behavioral assays. For example, localized measures of oxygen tension can be acquired from specific brain regions using voltammetry probes,[88,89] and have been shown to closely mimic human fMRI responses in the nucleus accumbens and medial prefrontal cortex for reward paradigms.[90] Second, methods for performing fMRI in awake, behaving non-human primates (NHP) have been developed,[91–93] allowing cognitive tasks or other paradigms to be performed. These experiments are carried out in specialized vertical bore scanners and/or clinical horizontal bore systems equipped with special mechanical restraint frames to prevent head motion during scanning. Such experiments, however, may only be performed at a few specialized centers currently. NHP studies also provide a phylogenetically closer pre-clinical species than rodents.

7.5 Good Imaging Practice (GIP) for fMRI in Drug Studies

Compared with many procedures employed in clinical trials, fMRI is not a standard diagnostic procedure that can be straightforwardly applied in a comparable way across laboratories provided only that the required equipment is available. That is, it is for the most part not a "turnkey" imaging method. Secondly, fMRI is an extremely flexible technique. This is a strength for its use in academic research, but means that every fMRI experiment comprises many details, pertaining to both acquisition and analysis, that may differ between laboratories, paradigms and experimenters.

For use in a clinical trial, the acquisition details and primary analysis steps need to be specified *a priori*; it is important that the pharmacological intervention be the only variable, and that the methodological parameters have been determined and characterized. In many cases, a pilot study may be necessary to allow appropriate characterization of the method and optimization of the analysis approach.[82]

Clinical studies require a pre-defined analysis plan and in the standard industry framework the statistical analysis takes place within a regulated environment that includes audit trails and independent verification of analysis code. This database and software environment works with numeric data; it is thus good practice where possible to pre-determine numeric endpoints from fMRI studies (for example, region-of-interest summary measures) that can be analyzed within the industry statistics framework and easily combined with pharmacokinetic and behavioral data. It is also critical that the many fMRI analysis steps that are applied in order to generate these endpoints are pre-defined and documented prior to image analysis. These details do not naturally fit into the pharmaceutical industry documentation structure and so additional templates may be required. The analysis documentation should be sufficient to allow an independent analyst to reconstruct the same endpoints from the same input image data. An audit trail, tracing all operations applied to raw image data, is a useful practice.

It is also good practice to agree upon and document the experimental acquisition details (*e.g.* sequence parameters, paradigm timings and content) and data de-identification, transfer and backup processes prior to trial execution. A clear naming convention for the image data can be very helpful for independent data quality control (QC) and both pre-specified and exploratory analyses. A well-thought-out naming specification also aids scripting for automated analyses. It is important that all data and information required to perform or repeat the analyses be transferred and/or archived together. That is, in addition to the raw image data, paradigm input files or timing parameters and subject responses may be critical information without which the analysis cannot proceed.

Another key aspect of good imaging practice is assiduous QC of ancillary equipment (*e.g.* infusion pumps, paradigm presentation equipment, secondary computers or button boxes) as well as the image data itself. Pre-defined QC steps and decision points should be built into the image analysis pipeline and documented as the image processing proceeds. Additional image QC, independent of analysis group, may also be considered.

These considerations are some of the key points in what can be described as "good imaging practice" (GIP) for pharmaceutical fMRI studies, where the challenge is to reconcile a flexible cutting edge neuroscience technique with the rigor and operational processes demanded by the regulated pharmaceutical industry. These considerations are elaborated in greater depth by Schwarz *et al.*[94,95]

7.6 Analytical Tools for Functional Magnetic Resonance Imaging (fMRI) in Pharmacological Experiments

The analysis of pharmacological fMRI experiments reflects both the nature of the data acquired (*e.g.* time series or "snapshot" for each subject/session) and the study design (*e.g.* parallel group or crossover). In addition, two aspects of particular importance for pharmacological studies are (a) consideration of potentially confounding effects of the compound on the signals measured and (b) compound pharmacokinetics. As mentioned in

Section 7.4.1.2, fMRI experiments can be categorized into four groups, namely "task-based" fMRI, phMRI, resting state connectivity experiments and assays of basal hemodynamic properties.

7.6.1 Primary Data Analysis Components

For all pharmacological fMRI studies, the analysis can be considered as comprising three main parts:

1. *Analysis of the data from each individual scanning session*
   This part of the analysis (often referred to as "first level analysis") involves processing the image data from each individual scan or session in order to generate voxel-wise maps of the individual response and, preferably, numeric summary values (*e.g.* the mean response within a defined region of interest).

   This part of the analysis varies between the four types of experiment described above (Section 7.4.1.2), and is discussed in greater depth for each case in subsequent sub-sections.

2. *Group analysis*
   Here, group-level inference (often referred to as "second level analysis") is performed. Most typically, a mixed linear model approach is used. The statistical model should take account of the study design and consider factors likely to explain variability in the data. For example, an analysis of data from a crossover design should normally include session as well as treatment condition as factors to account for any systematic effects due to repeating the experimental paradigm on the same subject. These analyses are likely to be performed on both numeric summary values and voxel-wise (over the whole brain); it is advisable to align as closely as possible the statistical models applied to these two analyses.

   A common approach to deriving numeric summary values from first-level analyses is to pre-specify regions-of-interest (ROIs) and extract the mean response amplitude from voxels within each ROI for each subject and scanning session. These numeric values can then be analyzed within a standard statistical framework.

   When group analysis is performed in a "massively univariate" fashion over the whole brain (voxel-wise), a severe multiplicity problem ensues. This may be addressed either by application of Gaussian random fields (GRF) developed by Worsley[96] or permutation tests with false-discovery rate control as proposed by Nichols.[97] Both approaches are similar, as they attempt to approximate the distribution of maximum statistic that is obtained under null condition, *i.e.* the distribution when there is no activation present. The random field theory (RFT) represents a parametric approach and significance threshold above which a voxel being declared significant is derived from the assumption that the fMRI data follows the GRF model. In permutation testing on the other hand, the distribution of the maximum statistic is obtained by resampling of the fMRI data at hand which may be computationally intensive. For a detailed comparison of RFT and resampling approaches see Nichols and Hayasaka.[98]

3. *Pharmacokinetic/pharmacodynamic (PK/PD) analyses*

For pharmacological studies, the relationship between the PD effect of the compound (here, the fMRI signal change) and the systemic exposure to the compound (usually plasma PK) is of central interest. Most simply, summary statistics of the compound PK and fMRI responses by dose or treatment group provide a summary of PD as a function of PK at the group level. If sufficient dynamic range is present in the data, more explicit analyses of the relationship between the plasma levels of the compound and the induced change in the fMRI response in individual subjects and treatment sessions can be performed.[1,71]

7.6.2 Task-Based fMRI and phMRI Experiments

Both "task-based" fMRI and "compound-driven" phMRI experiments generate image time series.

The main workhorse for analysis of these experiments follows the general linear model (GLM) as historically proposed by Worsley and Friston.[96] The essence of this approach is the formation of a design matrix, containing a number of fixed regressors that are expected to model, in linear combination, the temporal changes (beyond residual noise) in the signal. For task-based fMRI, these regressors include the expected hemodynamic responses, derived from the executed task or the stimulus applied, and confounding variables (*e.g.* head motion or measured confounds such as blood pressure variations). In phMRI, the stimulus that evokes the neuronal response is the administration of a pharmacological agent; the design matrix may include similar confounds as for task-based fMRI but the regressor of interest (signal model for the phMRI response to the compound) will usually be informed by other considerations, most directly by pilot data with the same compound. Estimates of parameters of interest indicating neuronal response are computed by fitting this matrix to the signal time course from each voxel in the brain.

Since the fMRI data have a complex spatio-temporal structure, appropriate pre-processing steps need to be performed in order to condition the data to the GLM approach. Typical analysis steps comprising the "first level" analysis in an fMRI analysis pipeline include:

- *Pre-processing*: This includes slice timing correction, motion correction, spatial co-registration of the images and/or "normalization" to a standard template space for subsequent group analysis, spatial and temporal smoothing.
- *GLM estimation of the time series data*: This involves fitting the design matrix to each voxel time course. The GLM also includes specification of

the residual noise model, *e.g.* to account for temporal correlation in the time series. Usually non-white structure of the noise is assumed.[‡]

In prospective studies where the drug mechanism is known or assumed one is primarily interested in modulation of the fMRI signal at the level of meaningful brain structures or ROIs. Regional summaries of the response are of main interest in these cases.[99–101] Typically, average ROI percent-signal-change derived from the GLM has been proposed as an endpoint of interest.[99,100] This approach has two advantages: first, the numeric summary values so generated can be easily accommodated within the statistical analysis framework of the pharmaceutical company and, second, the specification of a small number of endpoints greatly alleviates the multiplicity problem (for typical sample sizes used in early phase drug development (N ~ 12–20), it is rare for a compound to have such a strong effect to survive family-wise error correction at the whole brain level). Moreover, it is important to define target ROIs prospectively in order to avoid selection bias by "double dipping", *i.e.* defining ROIs with the same data as those being analyzed. This issue has been widely discussed as a general challenge in fMRI analysis, not germane only to pharmaceutical applications. Prospective definition of ROI should be based on either anatomical constraints, *e.g.* using anatomical atlas, or using functional maps obtained from *independent* cohorts.[102]

7.6.3 Resting State Connectivity

The approaches to task-free ("resting state") connectivity analysis fall into two major groups: (1) seed region analysis and (2) sophisticated exploratory methods such as independent component analysis (ICA) and graph theoretical methods.

The seed region analysis explores functional connectivity between a specified brain region (the "seed") and the rest of the brain. It proceeds by selecting a seed voxel, or region, the corresponding time course of which is then correlated against the time courses from all voxels in the brain, most simply using the Pearson correlation coefficient although other similarity metrics have also been used (*e.g.* partial correlation, mutual information). The correlation coefficient is then converted to statistically meaningful z-score using Fisher's trans-formation. Finally, a group analysis (similar to the task-based experiment) is carried out to identify regions systematically correlated across subjects and so by inference "functionally connected" to the seed region. This method can be simplified for prospective drug studies, by pre-defining a set of target seed regions where changes in functional connectivity are expected. Statistical inferences will then be based on correlation matrices with reduced dimension.

ICA is an advanced analysis technique that decomposes the data into spatial and temporal components based on their statistical independence.[40] Applied to resting state fMRI data, a set of independent components (ICs) is obtained.

[‡]Typically, an auto-regressive, AR(1), type of noise structure as described by Worsley and Friston.[96]

Traditionally in the analysis of fMRI resting state connectivity, the independence has been enforced on spatial components (also referred to as spatial maps), since for typical acquisitions the number of voxels in the image greatly exceeds the number of temporal samples – this approach to ICA was referred to as spatial ICA (sICA).[40] Recently, with the development of fast fMRI sampling sequences, temporal ICA (tICA), with independent time components,[103] has been performed as well (see also below).

From the ICA-derived ICs, some reflect data noise or confounds (*e.g.* a brain "edge" pattern associated with motion) while others define reproducible, neuronally driven resting state networks (RSNs).[40,104] The number of relevant RSNs that are identified depends in part on the sample size; on the order of 8–10 components have been systematically reported across a number of publications (*e.g.* ref. 40) whereas analyses of larger groups have revealed a larger number of more finely grained network components.[103,105] With advances in hardware and pulse sequences, larger numbers and more refined RSNs are being discovered.[103,106]

For prospective clinical studies, a low-dimensional endpoint quantification of ICA is desirable. To meet this goal, a goodness-of-fit (GOF) method was proposed by Greicius *et al.*[107] For GOF computation, each of the ICs extracted by the ICA is compared against a set of pre-defined template RSNs, or networks of interest (NOI, *e.g.* default mode network). Indices of similarity between each template NOI and each of the subject's ICs are estimated. The goodness-of-fit is defined as the maximal similarity index across IC comparisons to the template NOI. This process both identifies the IC associated with the NOI, and quantifies the level of connectivity within the NOI. The quantitative GOF measure serves as an endpoint suitable for statistical analysis of prospectively planned clinical trials (*e.g.* placebo *vs.* treatment).[108]

Recently, dual regression was developed as an extension of ICA analysis.[73,109] As for GOF computation, one considers a pre-defined set of NOIs (template maps). (Originally, these template NOIs were obtained from the group ICA applied to the subject cohort under investigation. To avoid selection bias issues, researchers now typically use template NOIs originated from independent cohorts.) First, the template NOIs are regressed (subject-wise) to each fMRI time series using a GLM approach. With this fit by spatial regression, a set of time courses for each subject is obtained. Using these subject-specific time courses as explanatory variables, the procedure follows the traditional GLM as applied for stimulus driven experiments. This way voxel-wise group activation maps are obtained and statistical inferences are made accordingly. For a more thorough description of the method, the reader is referred to Filippini *et al.*[109] and Khalilili-Mahani *et al.*[73] Although originally developed in the context of ICA, dual-regression is a standalone technique and it does not necessarily have to be used in conjunction with ICA.

Graph theory is another prominent approach that has been applied to "resting state" fMRI. This approach considers the brain as a collection of "nodes", some of which are connected by "edges". Most commonly, the brain is parcellated into regions based on some atlas or template and, for each image time series, the average time course from each region is computed. A measure of similarity in the temporal responses (most simply, linear correlation) is then computed between all pairs of time courses. The strength of this measure is then used to determine whether any pair of regions (nodes) is "connected" or not – *i.e.* edges whose similarity measure is above a certain threshold are retained, and the others are suppressed. Networks of functional connectivity constructed in this way, like many other types of biological networks, exhibit "small world" properties.[110] Such binary networks can then be analyzed using measures based on the topology of the connections – either whole-network (*i.e.* whole-brain) or node-wise summary measures can be computed.[111] These measures can then serve as metrics to evaluate functional connectivity.[112] A key issue with the graph theory approach is the need to ensure any finding is not specific to the binarization threshold used in creating the graphs – for this reason, comparisons are usually made for a range of thresholds. More recently, approaches to retaining edge weight information within a graph-theoretic framework have been proposed.[113–115]

7.6.4 Measures of Basal Hemodynamics

Image data reflecting resting brain perfusion (CBF and CBV) generally require less first-level analysis than the time-series-based methods discussed above, since the data are in the form of a single "snapshot" rather than a time series. Depending on the sequence and level of reconstruction on the scanner, the acquired data may need to be transformed into physiological units (*e.g.* mL $100 \ g^{-1} \ min^{-1}$ in the case of CBF). However, spatial pre-processing (*e.g.* smoothing, normalization to a standard brain template space) are likely to be used. Group-level and PK/PD analyses may be performed as described in Section 7.6.1.

7.7 Challenges

Application of fMRI as a biomarker in early drug development faces challenges, determined by the nature and limitations of the technique. As the fMRI signal is an indirect measure of the neural activity, reflecting a hemodynamic response function to the activation paradigm (or a drug in phMRI experiments), there are several potential confounding effects that need to be considered. Important physiological confounds include breathing and cardiac motion, as they impact the global signal across the brain during an fMRI experiment. Some of these confounds may be partially dealt with in the pre-processing stage of the fMRI analysis pipeline. For example, fast breathing artifacts typically alias into lower frequency and high pass filtering of the fMRI data may be beneficial to remove them. However, independent recording of the physiological readouts is strongly recommended, particularly for resting state fMRI,[116–119] as they may be included in the GLM analysis framework as covariates of no interest.

Signal dynamic range is also a challenge. The "activation" signal changes obtained with BOLD fMRI typically range between 0.25 and 3%, which together with physiological and measurement noise limit the sensitivity of fMRI. As a consequence, establishment of a dose–response relationship for compounds may be problematic, especially in cases when there are subtle changes of the fMRI signal induced by the pharmacological intervention under investigation.

Optimization of the experimental protocol for an fMRI study is of paramount importance to maximize the dynamic range and mitigate confounding factors. This includes definition of ROIs and pre-specified quantitative endpoints, where there is expected compound action, optimization of the measurement and processing sequence as well as statistical analysis of the derived endpoints. Understanding of the impact of measurement and physiological noise on particular ROIs in the brain is important for proper statistical analysis, as it determines the standardized effect size of the treatment effect and in turn appropriate sample size and power calculations.

As relates to study design, in pharmacological fMRI studies in early drug development, crossover designs are commonly employed. In this framework, within-subject standard deviation and an assumed effect size are necessary to obtain required sample sizes. A realistic assessment of a range of within-subject standard deviation may be obtained from the confidence interval of a pilot test-retest study, performed before the active treatment study.[18,120]

Furthermore, for the applicability of the fMRI in pre-clinical as well as clinical studies related to drug development, further standardization of the measurement protocols of the imaging procedures applied is necessary. An important step in the right direction was the establishment of the Functional Bioinformatics Research Network (fBIRN), where stability, test-retest repeatability and reproducibility across sites of fMRI endpoints were studied for a battery of sensory paradigms (including visual, finger tapping and auditory stimuli). The guidelines developed by fBIRN are germane for the design of fMRI clinical trials with novel drugs and more complex paradigms.[121,122]

A related important aspect of a successful execution of a clinical fMRI study is qualification of the experimental site. Due to the complex nature of the fMRI experiment, particularly when using specialized paradigms and the fact that, in early development, the studies do not exceed 20–25 subjects and therefore can be carried out in one center, well-trained sites with experience should be preferentially considered. However, the recent emergence of experiments with resting state networks that do not utilize any pre-defined stimulus paradigm alleviate to some extent the necessity of highly specialized sites.

As described above (Section 7.4.1.2), there is a variety of functional MRI techniques to assess neuronal function, each comprising a vast number of paradigms and variations on similar paradigms. Characteristics of the outcome measures (brain regions/networks involved, dynamic range, statistical properties) can differ between different types of paradigm (for example, test-retest repeatability in task-based techniques[18,123]). While some characteristics may be indicative of ranges across similar paradigms, caution must be exercised when applying generalization of these characteristics when decision-making is at stake. In general, it is necessary to define paradigms that are fit-for-purpose (*e.g.* as indicators of PD activity in specific brain pathways) and characterize the paradigm-specific outcome measures in pilot studies.[82]

Further development and validation of fMRI techniques has benefited and is continuing to benefit from concerted efforts in pre-competitive consortia. This approach is both cost-effective and enriched by the diversity of perspectives and resources of the participating members.

References

1. A. Coimbra, D. Welsh, D. Posavec, A. Vanko, R. Baumgartner, C. Regan, A. Danziger, M. Baran, K. Groover, J. J. Cook, J. Lynch, J. Uslaner and D. S. Williams, *Proc. Intl Soc. Mag. Reson. Med.*, 2011, **19**, 1533.
2. R. Baumgartner, W. Cho, A. Coimbra, C. Gargano, R. Iannone, A. Struyk, R. Fox, Z. Wang, F. Zhao, D. S. Williams, T. Reese, B. Henry, E. Petersen, C. Chen, D. Feng, S. Apreleva and J. Evelhoch, *Proc. Intl Soc. Mag. Reson. Med.*, 2012, **20**, 2885.
3. P. J. Magistretti and L. Pellerin, *Philos. Trans. R. Soc. Lond. B Biol. Sci.*, 1999, **354**, 1155.
4. A. Fornito and E. T. Bullmore, *Curr. Opin. Psychiatry*, 2010, **23**, 239.
5. R. E. Gur and R. C. Gur, *Dialogues Clin. Neurosci.*, 2010, **12**, 333.
6. G. G. Brown and W. K. Thompson, *Curr. Top. Behav. Neurosci.*, 2010, **4**, 181.
7. A. Meyer-Lindenberg, *Nature*, 2010, **468**, 194.
8. P. Fusar-Poli, A. Placentino, F. Carletti, P. Landi, P. Allen, S. Surguladze, F. Benedetti, M. Abbamonte, R. Gasparotti, F. Barale, J. Perez, P. McGuire and P. Politi, *J. Psychiatry Neurosci.*, 2009, **34**, 418.
9. R. A. Sperling, B. C. Dickerson, M. Pihlajamaki, P. Vannini, P. S. LaViolette, O. V. Vitolo, T. Hedden, J. A. Becker, D. M. Rentz, D. J. Selkoe and K. A. Johnson, *Neuromolecular Med.*, 2010, **12**, 27.
10. R. G. Wise and I. Tracey, *J. Magn. Reson. Imaging*, 2006, **23**, 862.
11. D. Borsook, L. Becerra and R. Hargreaves, *Nat. Rev. Drug Discov.*, 2006, **5**, 411.
12. J. Upadhyay, J. Anderson, A. J. Schwarz, A. Coimbra, R. Baumgartner, G. Pendse, E. George, L. Nutile, D. Wallin, J. Bishop, S. Neni, G. Maier, S. Iyengar, J. L. Evelhoch, D. Bleakman, R. Hargreaves, L. Becerra and D. Borsook, *Neuropsychopharmacology*, 2011, **36**, 2659.
13. A. Patin and R. Hurlemann, *Neuropsychologia*, 2011, **49**, 706.
14. R. Freeman and B. Pasley, *Scholarpedia*, 2008, **3**, 5340.
15. S. Ogawa, T. M. Lee, A. R. Kay and D. W. Tank, *Proc. Natl Acad. Sci. USA*, 1990, **87**, 9868.
16. P. A. Bandettini, E. C. Wong, A. Jesmanowicz, R. S. Hinks and J. S. Hyde, *NMR Biomed.*, 1994, **7**, 12.
17. J. Hennig, O. Speck, M. A. Koch and C. Weiller, *J. Magn. Reson. Imaging*, 2003, **18**, 1.
18. M. M. Plichta, A. J. Schwarz, O. Grimm, K. Morgen, D. Mier, L. Haddad, A. B. Gerdes, C. Sauer, H. Tost, C. Esslinger, P. Colman, F. Wilson, P. Kirsch and A. Meyer-Lindenberg, *Neuroimage*, 2012, **60**, 1746.

19. D. S. Williams, J. A. Detre, J. S. Leigh and A. P. Koretsky, *Proc. Natl Acad. Sci. USA*, 1992, **89**, 212.
20. R. B. Buxton, *J. Magn. Reson. Imaging*, 2005, **22**, 723.
21. T. Reese, B. Bjelke, R. Porszasz, D. Baumann, D. Bochelen, A. Sauter and M. Rudin, *NMR Biomed.*, 2000, **13**, 43.
22. A. J. Schwarz, T. Reese, A. Gozzi and A. Bifone, *Magn. Reson. Imaging*, 2003, **21**, 1191.
23. F. P. Leite, D. Tsao, W. Vanduffel, D. Fize, Y. Sasaki, L. L. Wald, A. M. Dale, K. K. Kwong, G. A. Orban, B. R. Rosen, R. B. Tootell and J. B. Mandeville, *NeuroImage*, 2002, **16**, 283.
24. J. B. Mandeville, B. G. Jenkins, Y. C. Chen, J. K. Choi, Y. R. Kim, D. Belen, C. Liu, B. E. Kosofsky and J. J. Marota, *Magn. Reson. Med.*, 2004, **52**, 1272.
25. D. Qiu, G. Zaharchuk, T. Christen, W. W. Ni and M. E. Moseley, *Neuroimage*, 2012, **62**, 1726.
26. H. Lu and P. C. van Zijl, *Neuroimage*, 2012, **62**, 736.
27. J. A. King, T. S. Garelick, M. E. Brevard, W. Chen, T. L. Messenger, T. Q. Duong and C. F. Ferris, *J. Neurosci. Methods*, 2005, **148**, 154.
28. C. L. Chin, G. B. Fox, V. P. Hradil, M. A. Osinski, S. P. McGaraughty, P. D. Skoubis, B. F. Cox and Y. Luo, *Neuroimage*, 2006, **33**, 1152.
29. F. Zhao, D. Welsh, M. Williams, A. Coimbra, M. O. Urban, R. Hargreaves, J. Evelhoch and D. S. Williams, *Neuroimage*, 2012, **59**, 1168.
30. F. Zhao, M. Williams, X. Meng, D. C. Welsh, A. Coimbra, E. D. Crown, J. J. Cook, M. O. Urban, R. Hargreaves and D. S. Williams, *Neuroimage*, 2008, **40**, 133.
31. A. C. Johnson, B. Myers, J. Lazovic, R. Towner and B. Greenwood-van Meerveld, *PLoS One*, 2010, **5**, e8573.
32. Y. C. Chen, W. R. Galpern, A. L. Brownell, R. T. Matthews, M. Bogdanov, O. Isacson, J. R. Keltner, M. F. Beal, B. R. Rosen and B. G. Jenkins, *Magn. Reson. Med.*, 1997, **38**, 389.
33. R. A. Leslie and M. F. James, *Trends Pharmacol. Sci.*, 2000, **21**, 314.
34. B. G. Jenkins, *Neuroimage*, 2012, **62**, 1072.
35. A. Gozzi, M. Tessari, L. Dacome, F. Agosta, S. Lepore, A. Lanzoni, P. Cristofori, E. M. Pich, M. Corsi and A. Bifone, *Neuropsychopharmacology*, 2011, **36**, 2431.
36. A. Bruns, B. Kunnecke, C. Risterucci, J. L. Moreau and M. von Kienlin, *Magn. Reson. Med.*, 2009, **61**, 1451.
37. C. P. Pawela, B. B. Biswal, Y. R. Cho, D. S. Kao, R. Li, S. R. Jones, M. L. Schulte, H. S. Matloub, A. G. Hudetz and J. S. Hyde, *Magn. Reson. Med.*, 2008, **59**, 1021.
38. E. Jonckers, J. van Audekerke, G. de Visscher, A. van der Linden and M. Verhoye, *PLoS One*, 2011, **6**, e18876.
39. L. Becerra, G. Pendse, P. C. Chang, J. Bishop and D. Borsook, *PLoS One*, 2011, **6**, e25701.
40. C. F. Beckmann, M. DeLuca, J. T. Devlin and S. M. Smith, *Philos. Trans. R. Soc. Lond. B Biol. Sci.*, 2005, **360**, 1001.

41. M. D. Fox and M. Greicius, *Front. Syst. Neurosci.*, 2010, **4**, 19.
42. M. E. Raichle, *Brain Connect.*, 2011, **1**, 3.
43. M. Magnuson, W. Majeed and S. D. Keilholz, *J. Magn. Reson. Imaging*, 2010, **32**, 584.
44. K. H. Chuang, P. van Gelderen, H. Merkle, J. Bodurka, V. N. Ikonomidou, A. P. Koretsky, J. H. Duyn and S. L. Talagala, *Neuroimage*, 2008, **40**, 1595.
45. H. Takuwa, T. Matsuura, T. Obata, H. Kawaguchi, I. Kanno and H. Ito, *Brain Res.*, 2012, **1472**, 107.
46. K. Masamoto and I. Kanno, *J. Cereb. Blood Flow Metab.*, 2012, **32**, 1233.
47. V. C. Austin, A. M. Blamire, K. A. Allers, T. Sharp, P. Styles, P. M. Matthews and N. R. Sibson, *Neuroimage*, 2005, **24**, 92.
48. T. Tsurugizawa, A. Uematsu, H. Uneyama and K. Torii, *Neuroscience*, 2010, **165**, 244.
49. A. Gozzi, A. Schwarz, V. Crestan and A. Bifone, *Magn. Reson. Imaging*, 2008, **26**, 999.
50. A. Bifone and A. Gozzi, *Expert Opin. Drug Discov.*, 2012, **7**(11), 1071.
51. A. Gozzi, A. Schwarz, T. Reese, S. Bertani, V. Crestan and A. Bifone, *Neuropsychopharmacology*, 2006, **31**, 1690.
52. A. M. Mathieu-Kia, C. Pages and M. J. Besson, *Synapse*, 1998, **29**, 343.
53. A. C. Silva, J. V. Liu, Y. Hirano, R. F. Leoni, H. Merkle, J. B. Mackel, X. F. Zhang, G. C. Nascimento and B. Stefanovic, *Methods Mol. Biol.*, 2011, **711**, 281.
54. C. L. Chin, J. R. Pauly, B. W. Surber, P. D. Skoubis, S. McGaraughty, V. P. Hradil, Y. Luo, B. F. Cox and G. B. Fox, *Synapse*, 2008, **62**, 159.
55. D. Levesque, J. Diaz, C. Pilon, M. P. Martres, B. Giros, E. Souil, D. Schott, J. L. Morgat, J. C. Schwartz and P. Sokoloff, *Proc. Natl Acad. Sci. USA*, 1992, **89**, 8155.
56. A. C. Tziortzi, G. E. Searle, S. Tzimopoulou, C. Salinas, J. D. Beaver, M. Jenkinson, M. Laruelle, E. A. Rabiner and R. N. Gunn, *Neuroimage*, 2011, **54**, 264.
57. C. A. Heidbreder and A. H. Newman, *Ann. NY Acad. Sci.*, 2010, **1187**, 4.
58. C. Heidbreder, *CNS Neurol. Disord. Drug Targets*, 2008, **7**, 410–421.
59. A. J. Schwarz, A. Gozzi, T. Reese, C. A. Heidbreder and A. Bifone, *Magn. Reson. Imaging*, 2007, **25**, 811.
60. A. Schwarz, A. Gozzi, T. Reese, S. Bertani, V. Crestan, J. Hagan, C. Heidbreder and A. Bifone, *Synapse*, 2004, **54**, 1.
61. D. Martinez, R. Gil, M. Slifstein, D. R. Hwang, Y. Huang, A. Perez, L. Kegeles, P. Talbot, S. Evans, J. Krystal, M. Laruelle and A. Abi-Dargham, *Biol. Psychiatry*, 2005, **58**, 779.
62. D. Martinez, R. Narendran, R. W. Foltin, M. Slifstein, D. R. Hwang, A. Broft, Y. Huang, T. B. Cooper, M. W. Fischman, H. D. Kleber and M. Laruelle, *Am. J. Psychiatry*, 2007, **164**, 622.
63. G. D. Iannetti, L. Zambreanu, R. G. Wise, T. J. Buchanan, J. P. Huggins, T. S. Smart, W. Vennart and I. Tracey, *Proc. Natl Acad. Sci. USA*, 2005, **102**, 18195.

64. G. D. Iannetti and A. Mouraux, *Exp. Brain Res.*, 2010, **205**, 1.
65. D. Borsook, G. Pendse, M. Aiello-Lammens, M. Glicksman, J. Gostic, S. Sherman, J. Korn, M. Shaw, K. Stewart, R. Gostic, S. Bazes, R. Hargreaves and L. Becerra, *Drug Develop. Res.*, 2007, **68**, 23.
66. D. Borsook and L. R. Becerra, *Mol. Pain*, 2006, **2**, 30.
67. I. Tracey and P. W. Mantyh, *Neuron*, 2007, **55**, 377.
68. Y. I. Sheline, D. M. Barch, J. M. Donnelly, J. M. Ollinger, A. Z. Snyder and M. A. Mintun, *Biol. Psychiatry*, 2001, **50**, 651.
69. C. J. Harmer, C. E. Mackay, C. B. Reid, P. J. Cowen and G. M. Goodwin, *Biol. Psychiatry*, 2006, **59**, 816.
70. Y. Chen, H. I. Wan, J. P. O'Reardon, D. J. Wang, Z. Wang, M. Korczykowski and J. A. Detre, *Clin. Pharmacol. Ther.*, 2011, **89**, 251.
71. K. J. Black, J. M. Koller, M. C. Campbell, D. A. Gusnard and S. I. Bandak, *J. Neurosci.*, 2010, **30**, 16284.
72. W. Li, P. G. Antuono, C. Xie, G. Chen, J. L. Jones, B. D. Ward, M. B. Franczak, J. S. Goveas and S. J. Li, *Neuroimage*, 2012, **60**, 1083.
73. N. Khalili-Mahani, R. M. Zoethout, C. F. Beckmann, E. Baerends, M. L. de Kam, R. P. Soeter, A. Dahan, M. A. van Buchem, J. M. van Gerven and S. A. Rombouts, *Hum. Brain Mapp.*, 2012, **33**, 1003.
74. J. S. Goveas, C. Xie, B. D. Ward, Z. Wu, W. Li, M. Franczak, J. L. Jones, P. G. Antuono and S. J. Li, *J. Magn. Reson. Imaging*, 2011, **34**, 764.
75. J. T. Tauscher and A. J. Schwarz, in *Translational Medicine and Drug Discovery*, ed. B. H. Littman and R. Krishna, Cambridge University Press, New York, 2011, chap. 9, p. 222.
76. D. F. Wong, J. Tauscher and G. Grunder, *Neuropsychopharmacology*, 2009, **34**, 187.
77. E. A. Rabiner, J. Beaver, A. Makwana, G. Searle, C. Long, P. J. Nathan, R. D. Newbould, J. Howard, S. R. Miller, M. A. Bush, S. Hill, R. Reiley, J. Passchier, R. N. Gunn, P. M. Matthews and E. T. Bullmore, *Mol. Psychiatry*, 2011, **16**(785), 826–835.
78. P. J. Nathan, B. V. O'Neill, M. A. Bush, A. Koch, W. X. Tao, K. Maltby, A. Napolitano, A. C. Brooke, A. L. Skeggs, C. S. Herman, A. L. Larkin, D. M. Ignar, D. B. Richards, P. M. Williams and E. T. Bullmore, *J. Clin. Pharmacol.*, 2012, **52**, 464.
79. M. R. Yeomans and R. W. Gray, *Neurosci. Biobehav. R.*, 2002, **26**, 713.
80. M. R. Yeomans and P. Wright, *Appetite*, 1991, **16**, 249–259.
81. D. J. Wang, Y. Chen, M. A. Fernandez-Seara and J. A. Detre, *J. Pharmacol. Exp. Ther.*, 2011, **337**, 359.
82. A. Verma, R. Declercq, A. Coimbra and E. Achten, in *Imaging in CNS Drug Discovery and Development: Implications for Disease and Therapy*, ed. D. Borsook, L. Becerra, R. Hargreaves and E. Bullmore, Springer, New York, 2010, p. xviii, 396 pp.
83. C. R. Jack, Jr., M. A. Bernstein, B. J. Borowski, J. L. Gunter, N. C. Fox, P. M. Thompson, N. Schuff, G. Krueger, R. J. Killiany, C. S. Decarli, A. M. Dale, O. W. Carmichael, D. Tosun and M. W. Weiner, *Alzheimers Dement.*, 2010, **6**, 212.

84. M. Di Simplicio, R. Norbury and C. J. Harmer, *Mol. Psychiatry*, 2012, **17**, 503.

85. T. R. Franklin, Z. Wang, N. Sciortino, D. Harper, Y. Li, J. Hakun, S. Kildea, K. Kampman, R. Ehrman, J. A. Detre, C. P. O'Brien and A. R. Childres, *Drug Alcohol Depend.*, 2011, **117**, 176.

86. F. M. Filbey, E. Claus, A. R. Audette, M. Niculescu, M. T. Banich, J. Tanabe, Y. P. Du and K. E. Hutchison, *Neuropsychopharmacology*, 2008, **33**, 1391.

87. D. A. Kareken, E. D. Claus, M. Sabri, M. Dzemidzic, A. E. Kosobud, A. J. Radnovich, D. Hector, V. A. Ramchandani, S. J. O'Connor, M. Lowe and T. K. Li, *Alcohol. Clin. Exp. Res.*, 2004, **28**, 550–557.

88. F. B. Bolger, S. B. McHugh, R. Bennett, J. Li, K. Ishiwari, J. Francois, M. W. Conway, G. Gilmour, D. M. Bannerman, M. Fillenz, M. Tricklebank and J. P. Lowry, *J. Neurosci. Methods*, 2011, **195**, 135.

89. J. P. Lowry, K. Griffin, S. B. McHugh, A. S. Lowe, M. Tricklebank and N. R. Sibson, *Neuroimage*, 2010, **52**, 549.

90. J. Francois, M. W. Conway, J. P. Lowry, M. D. Tricklebank and G. Gilmour, *Neuroimage*, 2012, **60**, 2169.

91. A. H. Andersen, Z. Zhang, T. Barber, W. S. Rayens, J. Zhang, R. Grondin, P. Hardy, G. A. Gerhardt and D. M. Gash, *J. Neurosci. Methods*, 2002, **118**, 141.

92. G. Chen, F. Wang, B. C. Dillenburger, R. M. Friedman, L. M. Chen, J. C. Gore, M. J. Avison and A. W. Roe, *Magn. Reson. Imaging*, 2012, **30**, 36.

93. J. B. Goense, K. Whittingstall and N. K. Logothetis, *Methods*, 2010, **50**, 178.

94. A. J. Schwarz, L. Becerra, J. Upadhyay, J. Anderson, R. Baumgartner, A. Coimbra, J. Evelhoch, R. Hargreaves, B. Robertson, S. Iyengar, J. Tauscher, D. Bleakman and D. Borsook, *Drug Discov. Today*, 2011, **16**, 583.

95. A. J. Schwarz, L. Becerra, J. Upadhyay, J. Anderson, R. Baumgartner, A. Coimbra, J. Evelhoch, R. Hargreaves, B. Robertson, S. Iyengar, J. Tauscher, D. Bleakman and D. Borsook, *Drug Discov. Today*, 2011, **16**, 671.

96. K. J. Worsley and K. J. Friston, *Neuroimage*, 1995, **2**, 173–181.

97. T. E. Nichols and A. P. Holmes, *Hum. Brain Mapp.*, 2002, **15**, 1.

98. T. Nichols and S. Hayasaka, *Stat. Methods Med. Res.*, 2003, **12**, 419.

99. G. D. Mitsis, G. D. Iannetti, T. S. Smart, I. Tracey and R. G. Wise, *Neuroimage*, 2008, **40**, 121.

100. R. Buck, H. Singhal, J. Arora, H. Schlitt and R. T. Constable, *Neuroimage*, 2008, **40**, 1157.

101. J. Suckling, D. Ohlssen, C. Andrew, G. Johnson, S. C. Williams, M. Graves, C. H. Chen, D. Spiegelhalter and E. Bullmore, *Hum. Brain Mapp.*, 2008, **29**, 1111.

102. N. Kriegeskorte, W. K. Simmons, P. S. Bellgowan and C. I. Baker, *Nat. Neurosci.*, 2009, **12**, 535.

103. S. M. Smith, K. L. Miller, S. Moeller, J. Xu, E. J. Auerbach, M. W. Woolrich, C. F. Beckmann, M. Jenkinson, J. Andersson,

M. F. Glasser, D. C. van Essen, D. A. Feinberg, E. S. Yacoub and K. Ugurbil, *Proc. Natl Acad. Sci. USA*, 2012, **109**, 3131.

104. M. J. Brookes, M. Woolrich, H. Luckhoo, D. Price, J. R. Hale, M. C. Stephenson, G. R. Barnes, S. M. Smith and P. G. Morris, *Proc. Natl Acad. Sci. USA*, 2011, **108**, 16783.

105. V. Kiviniemi, T. Starck, J. Remes, X. Y. Long, J. Nikkinen, M. Haapea, J. Veijola, I. Moilanen, M. Isohanni, Y. F. Zang and O. Tervonen, *Hum. Brain Mapp.*, 2009, **30**, 3865.

106. W. R. Shirer, S. Ryali, E. Rykhlevskaia, V. Menon and M. D. Greicius, *Cereb. Cortex*, 2012, **22**, 158.

107. M. D. Greicius, G. Srivastava, A. L. Reiss and V. Menon, *Proc. Natl Acad. Sci. USA*, 2004, **101**, 4637.

108. A. Coimbra, R. Baumgartner, D. Feng, S. Wang, J. Upadhyay, A. Schwarz, J. Anderson, L. Nutile, G. Pendse, J. Bishop, E. George, S. Iyengar, D. Bleakman, R. Hargreaves, J. Evelhoch, D. Borsook and L. Becerra, International Society for Magnetic Resonance in Medicine, Stockholm, Sweeden, 2010.

109. N. Filippini, K. P. Ebmeier, B. J. MacIntosh, A. J. Trachtenberg, G. B. Frisoni, G. K. Wilcock, C. F. Beckmann, S. M. Smith, P. M. Matthews and C. E. Mackay, *Neuroimage*, 2011, **54**, 602.

110. D. J. Watts and S. H. Strogatz, *Nature*, 1998, **393**, 440.

111. M. Rubinov and O. Sporns, *Neuroimage*, 2010, **52**, 1059.

112. E. T. Bullmore and D. S. Bassett, *Annu. Rev. Clin. Psychol.*, 2011, **7**, 113.

113. M. Rubinov and O. Sporns, *Neuroimage*, 2011, **56**, 2068.

114. A. J. Schwarz and J. McGonigle, *Neuroimage*, 2011, **55**, 1132.

115. J. A. Mumford, S. Horvath, M. C. Oldham, P. Langfelder, D. H. Geschwind and R. A. Poldrack, *Neuroimage*, 2010, **52**, 1465.

116. C. Chang and G. H. Glover, *Neuroimage*, 2009, **47**, 1448.

117. K. Shmueli, P. van Gelderen, J. A. de Zwart, S. G. Horovitz, M. Fukunaga, J. M. Jansma and J. H. Duyn, *Neuroimage*, 2007, **38**, 306.

118. R. M. Birn, K. Murphy and P. A. Bandettini, *Hum. Brain Mapp.*, 2008, **29**, 740.

119. E. B. Beall and M. J. Lowe, *Neuroimage*, 2007, **37**, 1286.

120. R. H. Browne, *Stat. Med.*, 1995, **14**, 1933.

121. L. Friedman, H. Stern, G. G. Brown, D. H. Mathalon, J. Turner, G. H. Glover, R. L. Gollub, J. Lauriello, K. O. Lim, T. Cannon, D. N. Greve, H. J. Bockholt, A. Belger, B. Mueller, M. J. Doty, J. He, W. Wells, P. Smyth, S. Pieper, S. Kim, M. Kubicki, M. Vangel and S. G. Potkin, *Hum. Brain Mapp.*, 2008, **29**, 958.

122. G. H. Glover, B. A. Mueller, J. A. Turner, T. G. van Erp, T. T. Liu, D. N. Greve, J. T. Voyvodic, J. Rasmussen, G. G. Brown, D. B. Keator, V. D. Calhoun, H. J. Lee, J. M. Ford, D. H. Mathalon, M. Diaz, D. S. O'Leary, S. Gadde, A. Preda, K. O. Lim, C. G. Wible, H. S. Stern, A. Belger, G. McCarthy, B. Ozyurt and S. G. Potkin, *J. Magn. Reson. Imaging*, 2012, **36**, 39.

123. C. M. Bennett and M. B. Miller, *Ann. NY Acad. Sci.*, 2010, **1191**, 133.

CHAPTER 8

In Vivo Proton MR Spectroscopy: Animal and Human Applications at High Fields

IVAN TKÁČ* AND GÜLIN ÖZ

Center for Magnetic Resonance Research, University of Minnesota, Minneapolis, USA
*Email: ivan@cmr.umn.edu

8.1 Introduction

In vivo NMR spectroscopy, commonly referred to as MR spectroscopy (MRS), is a method that provides quantitative biochemical information from a selected volume of interest (VOI) inside the living body non-invasively. This unique feature of MRS presents great potential for many applications in drug discovery and development research. Similar to liquid NMR spectroscopy, high magnetic fields are beneficial for MRS due to increased sensitivity and chemical shift dispersion. However, MRS at high fields is challenged by a number of technical and methodological problems, such as B_0 shimming, homogeneity and strength of the excitation field B_1, and chemical shift displacement error, which need to be addressed in order to benefit fully from high fields. In this chapter we will focus on single voxel proton MRS of the brain at high magnetic fields, namely 3 T and above for human and 7 T and above for animal applications. First, we will review the most important methodological aspects of this

New Developments in NMR No. 2
New Applications of NMR in Drug Discovery and Development
Edited by Leoncio Garrido and Nicolau Beckmann
Published by the Royal Society of Chemistry, www.rsc.org

technique, including data acquisition, processing and quantification. Then, we will demonstrate the potential of high-field proton MRS for utilization in drug-discovery research by presenting selected examples of its applications for neurochemical profiling in the animal and human brain.

8.2 Methodology of Proton MRS at High Fields

8.2.1 Potentials for Neurochemical Profiling

The brain tissue contains thousands of chemical compounds, but approximately only 20 low-molecular-weight neurochemicals with sufficiently high concentration (in the millimolar range) contribute dominantly to the spectral pattern detectable by ^{1}H MRS. Fast relaxing signals of high-molecular-weight compounds, mostly proteins, superimpose into an underlying broad spectral pattern, designated as the macromolecule spectrum. All signals that originate from rigid membrane structures, mainly phospholipids, are extensively broadened by dipolar interactions and disappear in the baseline. A simulated high-resolution ^{1}H MR spectrum (Figure 8.1A) shows the complicated overlapping spectral patterns of metabolites that in theory can be quantified in the brain using MRS. However, such a spectral resolution cannot be achieved experimentally due to microscopic B_0 heterogeneities of the tissue, which increase the best achievable signal linewidth. In addition, this broadening is approximately proportional to the static magnetic field B_0 and consequently becomes the limiting factor for spectral resolution at high fields.[1,2]

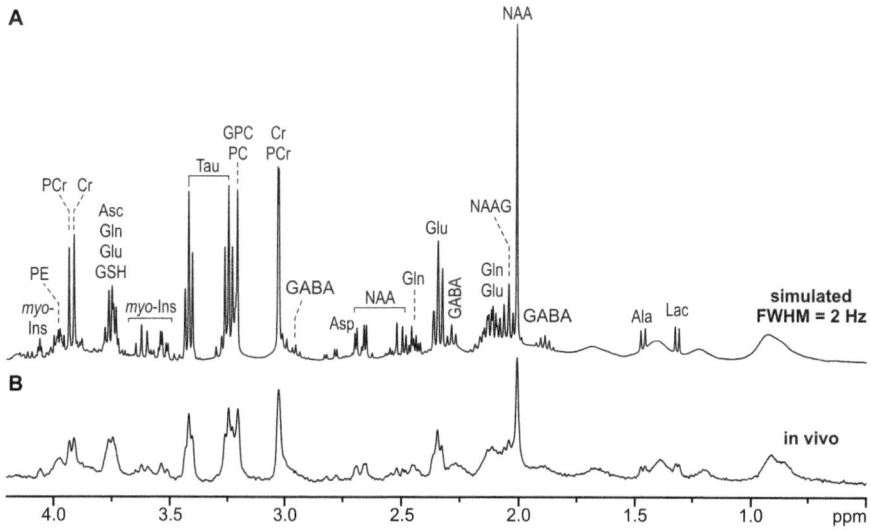

Figure 8.1   Simulated high-resolution (A) and experimentally measured *in vivo* ^{1}H NMR spectrum (B) of the rat brain at 9.4 T (FWHM – full width at half maximum).

Depending on the strength of magnetic field, the achieved spectral quality, the size of the VOI and the spectral processing and quantification methodology utilized, up to 20 neurochemicals can be quantified *in vivo* at high fields. These include primarily neuronal or glial neurochemicals, neurotransmitters, metabolites involved in energy production, antioxidants and many other metabolites that provide a wealth of information about the metabolic and functional status of the brain region under investigation, namely alanine (Ala), ascorbate (Asc), aspartate (Asp), creatine (Cr), phosphocreatine (PCr), γ-aminobutyric acid (GABA), glucose (Glc), glutamine (Gln), glutamate (Glu), glutathione (GSH), *myo*-inositol (*myo*-Ins), *scyllo*-inositol (*scyllo*-Ins), lactate (Lac), *N*-acetyl-aspartate (NAA), *N*-acetylaspartylglutamate (NAAG), phosphoethanolamine (PE), phosphocholine (PC), glycerophosphocholine (GPC), taurine (Tau) and macromolecules (MM).

An *in vivo* ^{1}H MR spectrum measured from the rat brain (Figure 8.1B) shows probably the highest spectral resolution experimentally achievable at 9.4 T. The high spectral overlap in combination with increased linewidths puts strict requirements on the quality of acquired spectra to preserve the reliability of neurochemical profiling.

8.2.2   Data Acquisition

The key difference between liquid and *in vivo* NMR spectroscopy is the spatial localization based on simultaneously applied radiofrequency (RF) pulses and magnetic field gradients. Typically the localization is achieved as a spin echo (point-resolved spectroscopy (PRESS) sequence[3]) or a stimulated echo (stimulated echo acquisition mode (STEAM) sequence[4]) in a selected volume generated by the intersection of three orthogonal slices. Sequences with short echo times (TEs) are favorable for high fields to minimize signal loss due to J-evolution of coupled spin systems and fast T_2 relaxation. Acquiring data with short TE and long repetition time (TR) is beneficial for "absolute" metabolite quantification, because in this approach relaxation effects can be neglected. An additional difference between liquid and *in vivo* NMR spectroscopy is the magnetic susceptibility heterogeneities of tissue, which originate from its cellular structure. These susceptibility heterogeneities translate into intrinsic microscopic B_0 heterogeneities, which are not correctable by shimming and increase the spectral linewidth. This line broadening increases with B_0 strength and limits the best achievable spectral resolution at high fields.[1,5] Luckily, this broadening is less important for resolving overlapped multiplets of coupled spin systems at high fields.[2] The quality of acquired spectra depends on multiple factors, for instance on the localization performance of the sequence, water suppression, adjustment of B_0 field homogeneity and performance of the RF coil. In the next sections we are going to describe these factors in more detail.

8.2.2.1   Shimming

The adjustment of B_0 homogeneity in the selected VOI, typically called B_0 shimming or simply shimming, is one of the key factors that influence spectral

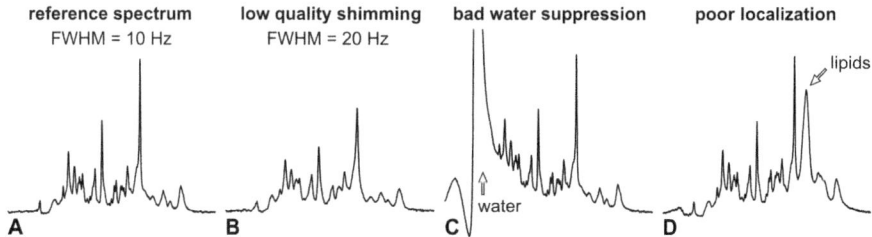

Figure 8.2   Key factors influencing the spectral quality. (A) Reference ^{1}H MR spectrum of the human brain at 7 T; (B) spectrum when only linear shims were adjusted; (C) spectrum with misadjusted water suppression; (D) spectrum with outer volume suppression turned off.

quality and the reliability of metabolite quantification. Local inhomogeneities in the B_0 field, mostly caused by susceptibility differences between tissue and air, are scaled with the B_0 field and become highly non-linear at ultrahigh magnetic fields. The elimination of this macroscopic B_0 inhomogeneity requires an efficient B_0 mapping method and a powerful higher-order shim system. Linear and strong second-order shims are typically sufficient to compensate B_0 field distortions in small VOIs selected for MRS. Different 3D B_0 field mapping methods[6] and methods using B_0 mapping along projections, *e.g.* FASTMAP (fast, automatic shimming technique by mapping along projections),[7,8] have been developed. Resolved resonances of creatine and phosphocreatine at ~3.9 ppm in the brain spectra of rodents (Figure 8.1B) are a suitable marker for successful shimming. The importance of second-order shimming is apparent from human brain ^{1}H MR spectra acquired at 7 T (Figure 8.2A and B). A two-fold increase in spectral linewidth and distorted lineshapes are typically observed when only linear shims are used.

8.2.2.2   Water Suppression

Brain tissue contains ~70–80% of water, which is ~80–90 mmol g^{-1} in water proton concentration, while the highest concentrations of brain metabolites, such as NAA, total creatine (tCr, the sum of Cr and PCr) and glutamate, are only around 10 µmol g^{-1}. This means that the biggest metabolite signals, namely the methyl resonances of NAA and tCr, are ~3000 times lower than the water signal. Insufficient water suppression causes baseline distortions in the spectrum (Figure 8.2C) and decreases the reliability of metabolite quantification, especially the assessment of weakly represented metabolites, such as Gln, GABA or GSH. In addition, satellites of the water resonance that originate from gradient coil vibrations may obscure the metabolite spectral pattern. A highly efficient multipulse water suppression scheme called VAPOR (variable pulse power and optimized relaxation delay) was developed for high-field MRS (Figure 8.3).[9,10] In this method, all parameters can be set fully automatically from a prior RF power calibration in the VOI and a 10 000-fold

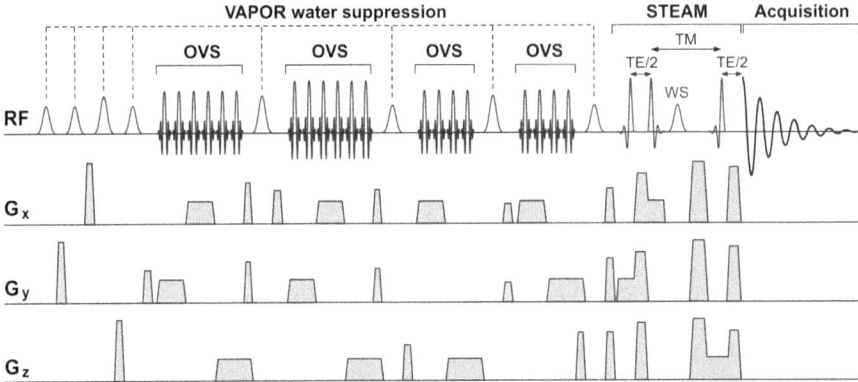

Figure 8.3 Schematic illustration of ultra-short echo time STEAM localization sequence for proton MRS of human brain at high magnetic fields. The VAPOR water suppression is interleaved with the outer volume suppression (OVS).
Modified with permission from Tkac *et al.*[9] © 2005 Springer.

water suppression efficiency is routinely achievable in the presence of inhomogeneous B_1 excitation fields.[11,12]

8.2.2.3 Localization Performance

The localization performance of the sequence can be defined as the ability to provide maximum signal from the selected VOI and to suppress all coherences that originate from outside of the VOI. The suppression of unwanted coherences is typically achieved by strong crusher gradients. Longer gradients improve the suppression efficiency, but the penalty is a longer TE. Contamination of spectra by signals that originate from outside of the VOI can also result from suboptimal excitation profiles of the RF pulses used for slice selection. This drawback can be improved by additional outer volume suppression (OVS) prior to the volume selection pulses (Figure 8.3). Unwanted coherences may also be eliminated by phase cycling, but this approach is not very robust due to small frequency fluctuations resulting from the respiration and cardiac cycle. The importance of OVS for improving the localization performance is demonstrated in Figure 8.2D, where turning off the OVS resulted in strong signal contributions from subcutaneous lipids.

8.2.2.4 Chemical Shift Displacement Error

The chemical shift displacement error (CSDE) is a general problem of all localization sequences used for MRS based on volume selection by slice selective pulses (band selective RF pulse simultaneously applied with magnetic field gradients). CSDE is the spatial displacement of volumes selected for off-resonance signals relative to the nominal VOI. This displacement becomes a significant problem at high magnetic fields due to increased chemical shift dispersion (in frequency units). Although this problem is not necessarily discernible in spectra, it may lead to misinterpretation of measured data. The CSDE is proportional to the ratio of the chemical shift range (in Hz) to the bandwidth of the slice selection RF pulse. Therefore, broadband RF pulses are required for high fields to minimize the CSDE. Sufficiently high transmit B_1 necessary for amplitude modulated broadband pulses is usually not a problem for small RF coils used for animal studies. However, this becomes a serious problem for human applications at 7 T. When the available peak B_1 is not sufficient, using adiabatic full passage (AFP) frequency modulated pulses in the localization sequence is an option.[13] However, this approach results in increased TE of the localization sequence.

8.2.2.5 Localization Pulse Sequences

An ultra-short echo time STEAM sequence has been developed for high field applications on animals[10] and humans.[2,9] This technique, based on stimulated echoes, provides only half of the available signal. However, this drawback is compensated by an outstanding localization performance due to additional OVS blocks interleaved with VAPOR water suppression and minimized CSDE resulting from broadband RF pulses (Figure 8.3). Due to the common problem of limited peak B_1 on human 7 T MR scanners, the STEAM sequence is preferential to the PRESS sequence because of reduced CSDE. In order to maintain the full intensity from the selected VOI, pulse sequences using pairs of AFP pulses, such as LASER (localization by adiabatic selective refocusing)[13] or semi-LASER[14,15] might be the methods of choice. The ultra-short TE SPECIAL (spin-echo full-intensity acquired localization) sequence was developed for animal studies at 9.4 T and 14.1 T.[16] The semi-adiabatic version of SPECIAL[17] is more appropriate for applications in humans. When using multichannel transmit RF coils, the efficiency of RF power transmission can be improved by B_1 shimming, which is essentially an optimization of phases and amplitudes of RF fields transmitted by individual coil elements to minimize destructive interferences.[18–20]

8.2.3 Data Processing

MRS techniques sample data from a living body. The respiratory and cardiac cycles may introduce frequency and phase fluctuations in the acquired data. In addition, a frequency drift may occur because the deuterium lock, which is standard for liquid NMR spectroscopy, is not used. Therefore, if the signal-to-noise ratio (SNR) is sufficient, single scans can be saved and corrected for frequency and phase fluctuations prior to summing the free induction decays (FIDs). Corrections of single scan data require FIDs free of unwanted coherences.[15] If the SNR of single-scan data is too low, data can be first summed in small blocks, *e.g.* 8 scans per block, and long-term frequency drifts can be corrected. In addition, data from compromised data blocks can be eliminated from final summation. Finally, effects of residual eddy currents affecting the signal line shapes can be removed from metabolite spectra using an unsuppressed water signal acquired from the same VOI.[9,21]

8.2.4 Metabolite Quantification

8.2.4.1 *Data Analysis and Prior Knowledge*

Spectra of individual metabolites are highly overlapped despite increased chemical shift dispersion at high magnetic fields (Figure 8.1). In addition, even when B_0 inhomogeneities are successfully corrected, intrinsic signal linewidths are substantially broader than is typical for liquid NMR spectroscopy. Therefore, meaningful quantification of MRS data requires extensive prior knowledge. In most cases, brain ^1H MR spectra consist of overlapped spectra of a well-defined group of metabolites. That is, typically the problem is not signal assignment of unknown compounds, but the quantification of a complex mixture of known metabolites. The most commonly used spectral fitting programs are jMRUI,[22] which performs fitting in the time domain, and LCModel,[23] which performs fitting in the frequency domain. Both programs use databases of metabolite spectra, which can be experimentally measured or simulated based on published metabolite chemical shifts and J-coupling constants.[24] Quantification errors are estimated by Cramér–Rao lower bounds (CRLBs). Note that these error estimates are based on the assumption that the model (metabolite spectra database) is correct and complete, which is obviously a simplification. The principle of LCModel analysis is shown in Figure 8.4, where the experimentally measured *in vivo* ^1H MR spectrum of the rat brain is modeled as a linear combination of metabolite spectra from the database. The model is calibrated in such a way that the coefficients of this linear combination correspond to metabolite concentrations. LCModel also quantifies the contribution of signals from fast relaxing macromolecules. The macromolecule spectra can be measured by the metabolite nulling technique using an inversion-recovery experiment.[25] Appropriate incorporation of MM signals in spectral analysis is crucial for accurate quantification of weakly represented metabolites, such as GABA and GSH.[26,27]

8.2.4.2 *Referencing*

Metabolite quantification requires appropriate referencing. The signal of tCr has been widely used as an internal reference in brain ^1H MRS, typically assuming a concentration of $8\,\mu mol\,g^{-1}$. Using the signal of total creatine as a reference is complicated by the fact that tCr concentration changes during brain development,[28] differs between brain regions[11,18,28,29] and may also change under pathological conditions.[30,31] Using the unsuppressed water signal from the same VOI as an internal reference is a useful approximation and was successfully used in multiple proton MRS applications.[11,12,28–30,32,33]

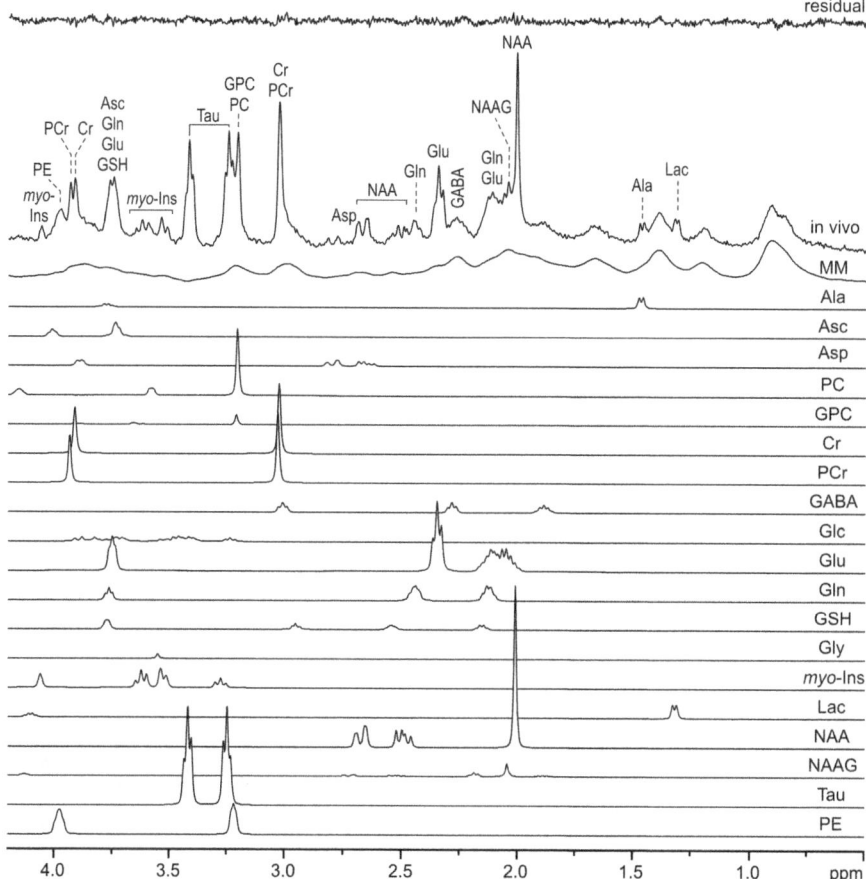

Figure 8.4 LCModel analysis of *in vivo* ¹H MR spectrum measured from the rat striatum at 9.4 T (STEAM, TE = 2 ms, TR = 5 s). *In vivo* spectrum can be modeled as a linear combination of brain metabolite spectra. The top trace shows the residual between the experimental and fitted spectrum.

8.2.4.3 Corrections for Relaxation

Setting the TR to a value close to the average T_1 of metabolites maximizes SNR, however it results in signal intensities that are T_1 weighted and therefore causes problems in absolute quantification. Increased duration of TE results in signal attenuation due to J-evolution of coupled spin systems and T_2 relaxation. While precise measurement of T_1 and T_2 values from ¹H MRS data is very difficult, these values can be estimated using some assumptions, *e.g.* assuming the same relaxation times for all CH_2 and CH protons within the same molecule.[17,34–36] The easiest approach for "absolute quantification" is to use ultra-short TE and long TR, under which conditions the relaxation effects can be neglected.

8.3   Applications of Proton MRS at High Fields

In the second part of this chapter we will present examples of using proton MRS for neurochemical profiling in animals and humans in order to demonstrate the potentials of this technique in drug discovery and development research. The strength of this technique is its non-invasiveness, which allows its use in longitudinal studies. Note that the following are selected examples of neurochemical profiling where high and ultrahigh field MRS was utilized. For a comprehensive review of applications of MRS to animal models of neuro-degenerative diseases, the reader is referred to reviews by Choi *et al.*[37] and Michaelis *et al.*[38] In addition, the current possibilities of high-field ^{1}H MRS for neurochemical profiling were recently reviewed by Duarte *et al.*[39]

8.3.1   Neurochemical Profiling in the Animal Brain

8.3.1.1   Developmental and Regional Changes

Major developmental neurochemical changes occur in the rodent brain during a four-week period after birth.[28,32,33] These changes are easily discernible from ^{1}H MR spectra (Figure 8.5A). Metabolite concentrations change not only with

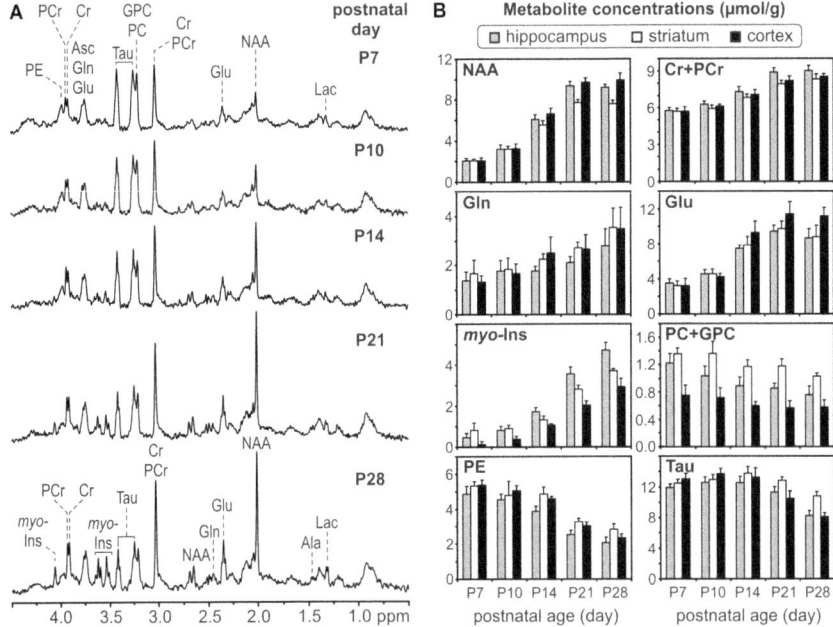

Figure 8.5   (A) *In vivo* [1]H MR spectra of the developing rat hippocampus measured during the postnatal period from P7 to P28. STEAM, TE = 2 ms, TR = 5 s, B_0 = 9.4 T. (B) Developmental changes in concentrations of selected brain metabolites measured in hippocampus, striatum and cerebral cortex of rat pups (postnatal days 7–28). Error bars indicate SD.
Modified with permission from Tkac *et al.*[28] © 2003 Wiley-Liss, Inc.

postnatal age but also with brain region[11,28,29] (Figure 8.5B). Developmental and regional variation of tCr clearly demonstrates that the signal of this compound cannot be used as an internal reference in these types of experiments. Using the unsuppressed water signal and the known age dependence of the brain water content appears to be more appropriate.[28,33,40] These developmental and regional changes in neurochemical profiles have to be taken into account when designing pre-clinical trials in animal models. MRS is highly advantageous for developmental studies because it allows longitudinal neurochemical profiling without the necessity of sacrificing animals at each time point.

8.3.1.2   Animal Physiology

Animals need to be immobilized using anesthesia during MRS experiments. General anesthesia does not appear to affect steady-state levels of MRS detectable metabolites except brain glucose and lactate.[41,42] No effects of recurrent anesthesia on neurochemical profiles were observed.[33] Proton MRS allows monitoring of physiological changes induced by interventions. Neurochemical changes in rat hippocampus during acute hypoglycemia are shown as an example[43] (Figure 8.6). Hypoglycemia (blood Glc $< 2.5\,\text{mmol}\,l^{-1}$) was induced by insulin injection and changes in metabolite levels were monitored for four hours. Major neurochemical changes resulting from anaplerotic processes in the starving brain started two hours after insulin administration. For example, a precipitous decrease of glutamine and glutamate indicated their use as alternative substrates entering the tricarboxylic acid cycle for energy production (Figure 8.6). Changes in the PCr/Cr ratio demonstrated insufficient adenosine triphosphate production under hypoglycemia. Changes in Cr, PCr and Glu were reversed by dextrose administration. The concentration changes in Asp, Gln, Glu, Cr and PCr can be easily visualized by the difference spectrum obtained by subtracting the spectrum acquired during the early phase (A) from that acquired during the late phase (B) of the hypoglycemia (Figure 8.6).

8.3.1.3   Iron Supplementation

Perinatal iron deficiency is common in humans and is associated with cognitive deficits that persist into adulthood. Proton MRS has been used in an animal model of iron deficiency to analyze its effects on brain development.[40,44] In addition, effects of different iron supplementation dosages on the neurochemical profile of the hippocampus were investigated in a rat model of iron deficiency[45] (Figure 8.7). Small but significant changes were observed in concentrations of NAA, NAAG, PE and the total choline (GPC + PC) between the iron-sufficient group and the two formerly iron-deficient groups that were supplemented with either the standard-dose ($40\,\text{mg}\,\text{kg}^{-1}$) or a high-dose ($400\,\text{mg}\,\text{kg}^{-1}$) iron in maternal diet from postnatal day 8 to 21. These metabolic changes indicate hypomyelination and abnormal phospholipid metabolism. Interestingly, high-dose iron supplementation (ID-400) normalized changes in

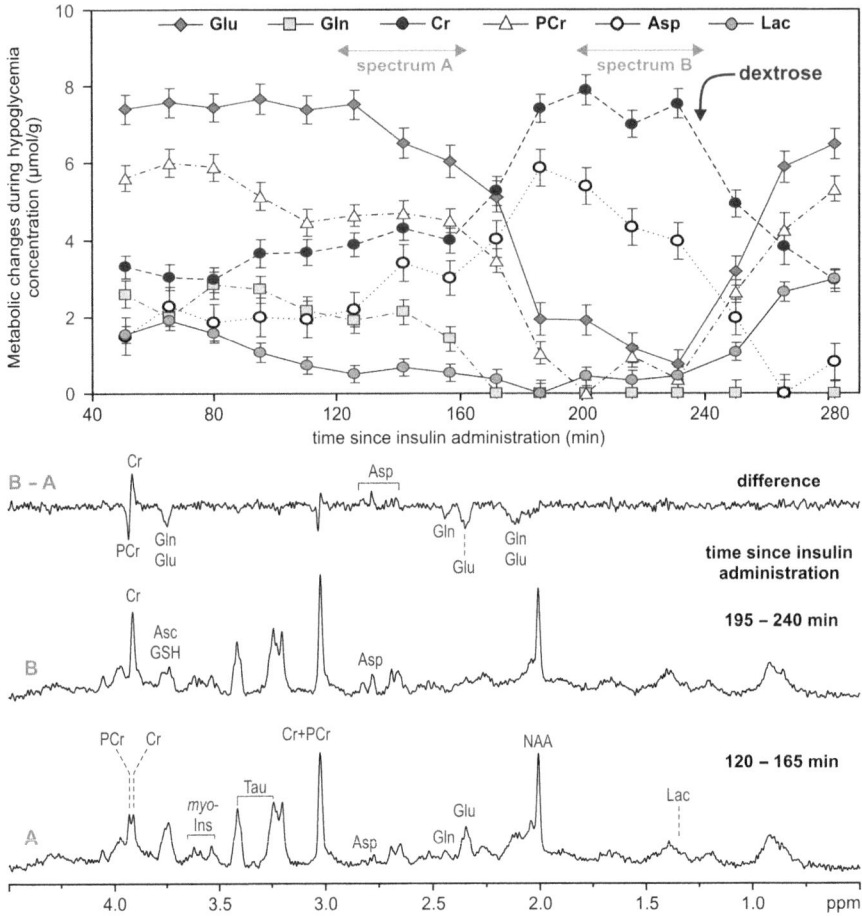

Figure 8.6 Top panel: Hippocampal neurochemical changes during acute hypo-glycemia and after dextrose administration (rat, P14). Error bars indicate CRLB. Bottom panel: *In vivo* [1]H MR spectrum acquired at early phase (A), late phase (B) and their difference (B – A). STEAM, TE = 2 ms, TR = 5 s, B_0 = 9.4 T.
Modified with permission from Rao *et al.*[43] © 2010 International Society for Neurochemistry.

NAA and NAAG that were observed with low-dose iron supplementation (ID-40) (Figure 8.7).

8.3.1.4 *Effect of Morphine on Developing Brain*

Opioids are frequently used in the neonatal intensive care unit for pain management and sedation of infants requiring mechanical ventilation. In addition, it was shown that adult rats exposed to recurrent administration of morphine have decreased neurogenesis and altered neurotransmission in the

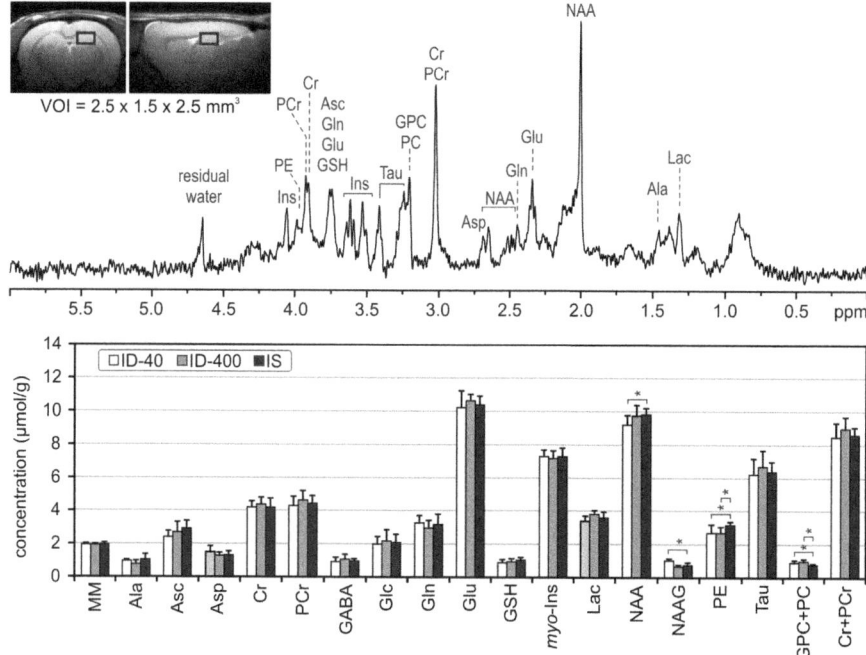

Figure 8.7 Top panel: *In vivo* [1]H MR spectrum of the rat hippocampus measured at 9.4 T (STEAM, TE = 2 ms, TR = 5 s). The position of VOI is shown on coronal and sagittal MRI. Bottom panel: Comparison of neurochemical profiles of formerly iron deficient (ID) rat pups supplemented by the standard-dose (ID-40, 40 mg kg^{-1}) or a high-dose (ID-400, 400 mg kg^{-1}) of iron in maternal diet (postnatal days 8–21) with the iron sufficient (IS) control group of rats. Error bars indicate SD, N = 7–8, *p < 0.05. Modified with permission from Rao *et al.*[45] © 2012 Nature Publishing Group.

hippocampus. Therefore, a rat model was utilized to study the effect of morphine on the developing brain using proton MRS.[46] Figure 8.8 shows *in vivo* [1]H MR spectra acquired on postnatal day 8 from the hippocampus of a rat exposed to morphine (2 mg kg^{-1} i.p., postnatal day 3 to 7) and a littermate control. Small but significant changes in the concentrations of GABA, GSH, *myo*-Ins, PE, Tau and GPC + PC were observed in the morphine group relative to control. Most of these metabolite changes can be visually recognized in the presented spectra (Figure 8.8). These neurochemical changes indicated effects of morphine on inhibitory neurotransmission (GABA, Tau), glial development (Gln), myelination (PE), osmoregulation (*myo*-Ins, Tau) and antioxidant status (GSH). The data presented in Figures 8.5, 8.6 and 8.8 demonstrate the feasibility of reliable neurochemical profiling in small brain regions of rat pups. In addition, they show that high-field proton MRS can provide much broader neurochemical information than was typical a decade ago, when the quantification was limited to three brain metabolites (NAA, total creatine and choline) in the majority of investigations.

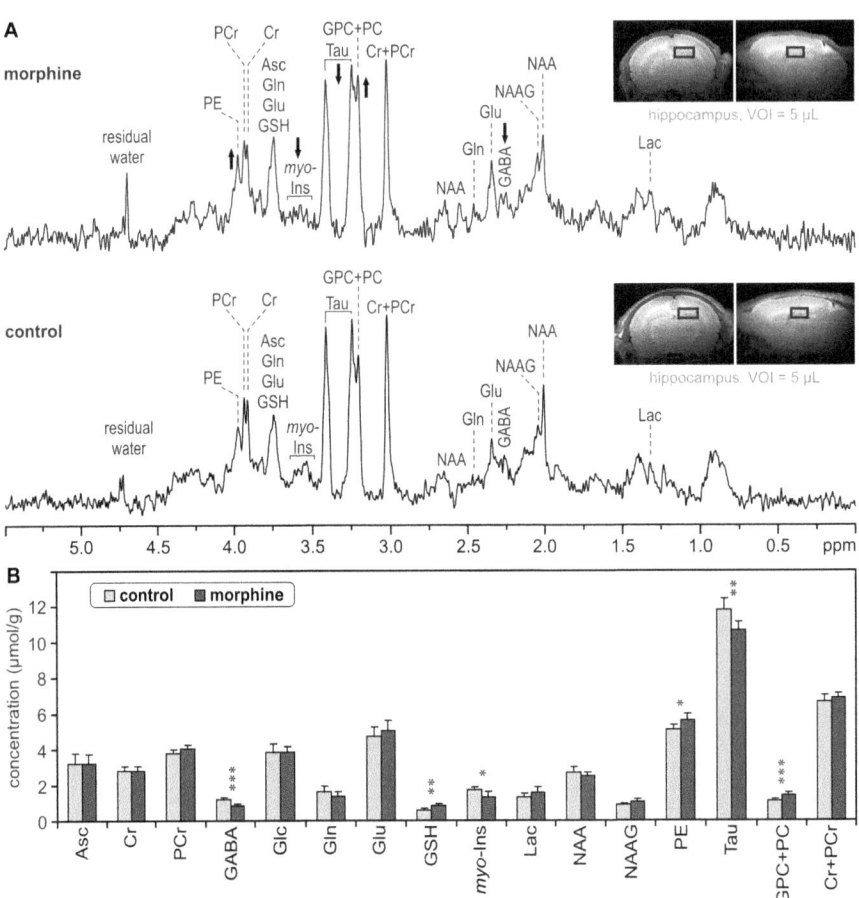

Figure 8.8 (A) *In vivo* ^{1}H MR spectra measured on postnatal day 8 from the hippocampus of a rat exposed to morphine (P3–P7) and a control rat (STEAM, TE = 2 ms, TR = 5 s, B_0 = 9.4 T). (B) Comparison of neurochemical profiles measured on P8 from rat pups exposed to morphine relative to littermate controls. Error bars indicate SD, N = 6–7, *p < 0.05, **p < 0.01, ***p < 0.001.
Modified with permission from Traudt *et al.*[46] © 2011 Wiley Periodicals, Inc.

8.3.1.5 Transgenic Mouse Models

Great progress has been made in the last decade in genetic engineering to generate transgenic and knock-in mouse models of human neurodegenerative diseases, such as Huntington's disease (HD), Alzheimer's disease (AD) and spinocerebellar ataxia (SCA). These mouse models require advanced diagnostic methods to characterize the disease progression and the effectiveness of therapies. MRS has become an invaluable method for non-invasive monitoring of brain neurochemistry in longitudinal studies using mouse models. For example, MRS was successfully applied in mouse models of AD.[47,48] Proton MRS of the mouse brain at high fields is rather challenging, but feasible.[11] Successful spectroscopy of mice requires a very powerful second-order shim system to compensate large B_0 inhomogeneities induced in the mouse brain. In order to demonstrate the power of *in vivo* neurochemical profiling in transgenic mouse models we chose examples of applications in HD and SCA models.

The R6/2 transgenic mouse model of HD is probably the most studied model of this disease. Proton MRS at 9.4 T can detect and quantify changes in multiple metabolites in striatal or cortical brain regions.[30,49] Representative spectra from the cerebral cortex of an R6/2 mouse and a wild-type control measured at 12 weeks of age are shown in Figure 8.9. The non-invasiveness of this technique enabled a longitudinal study design; the same mice were scanned

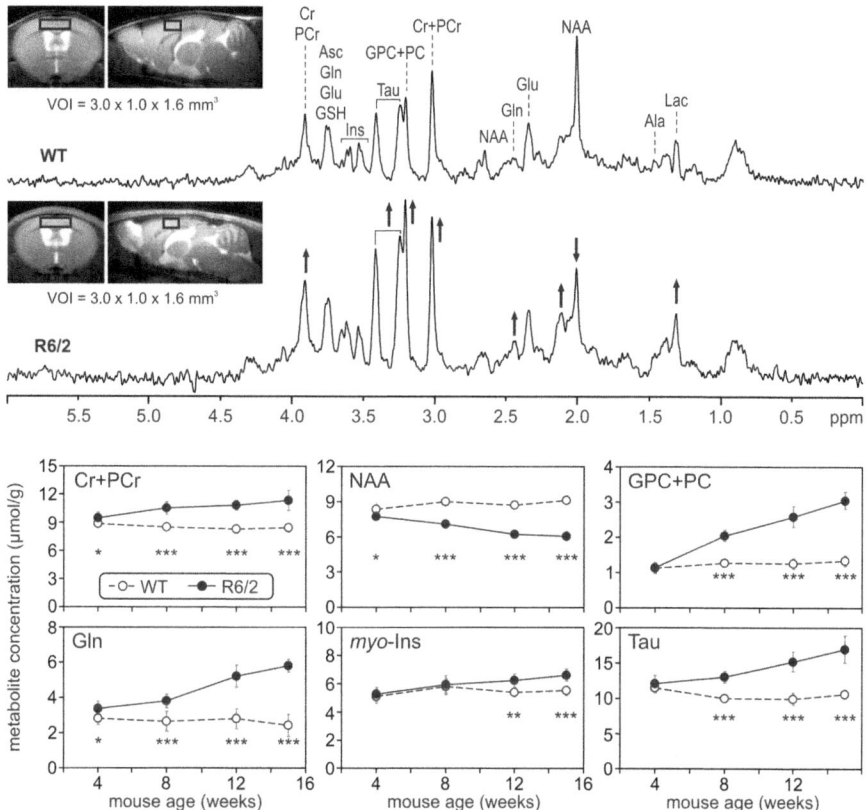

Figure 8.9 Top panel: *In vivo* [1]H MR spectra measured from the cerebral cortex of R6/2 transgenic mouse model of Huntington's disease and a wild-type (WT) control (STEAM, TE = 2 ms, TR = 5 s, B_0 = 9.4 T). Bottom panel: Age-dependent concentration changes of selected metabolites in the cerebral cortex of R6/2 mice and WT controls. Error bars indicate SD, N = 6–8, *p < 0.05, **p < 0.01, ***p < 0.001.
Modified with permission from Zacharoff *et al.*[30] © 2011 Nature Publishing Group.

every four weeks. Progressive changes in concentrations of selected metabolites are illustrated in Figure 8.9B. Changes in Cr + PCr, NAA, GPC + PC, Gln, *myo*-Ins and Tau indicated neurodegenerative processes resulting in neuronal loss and compromised neurotransmission, energy production and osmotic regulation. Progressive metabolite changes observed in the striatum were paralleled by those observed in the cortex,[30] indicating that this transgenic mouse model of HD does not closely mimic the phenotype observed in human HD, where the striatum is the region primarily affected by neurodegeneration. This example demonstrates that proton MRS may provide biomarkers for disease progression in HD, which may also potentially be used for evaluation of therapies.

Another example illustrating the strength of neurochemical profiling based on MRS comes from studies of mouse models of SCA. The progression of the neurodegeneration in the cerebellum of SCA type 1 (SCA1) mice is reflected in neurochemical levels measured by MRS.[50] Interestingly, the same neuro-chemicals that correlated significantly with semi-quantitative histological measures in this mouse model (NAA, *myo*-Ins, Glu) also correlated with clinical status in patients with SCA1,[51] indicating these metabolites as disease biomarkers. MRS was further applied to the study of a conditional mouse model of SCA1.[52] In this transgenic mouse model, the cerebellar pathology and ataxic phenotype can be reversed by administering doxycycline. In the example shown in Figure 8.10, mice were treated with doxycycline for 6 weeks to suppress transgene expression. In this study, blind monitoring of changes in the neurochemical levels of three metabolites (NAA, *myo*-Ins, Tau) simultaneously allowed the correct assignment of 17 out of 18 mice studied to the doxycycline treated *vs.* untreated groups, demonstrating the potential to monitor treatment effects in individual mice[52] (Figure 8.10). These data clearly demonstrate that biomarkers based on MRS measurements can accurately characterize the progression of neuropathology and its reversal by treatment. This implies a huge potential of this technique in drug-discovery research.

8.3.1.6   Vigabatrin and GABA

GABA is the major inhibitory neurotransmitter and its levels are altered in a number of neurological disorders. Anti-epileptic drugs such as vigabatrin, topiramat or gabapentin have been shown to increase brain GABA concen-trations by proton MRS.[53] The technique was also used in rat models to investigate cerebral GABA metabolism after administration of vigabatrin, an irreversible inhibitor of GABA transaminase, which is the enzyme responsible for GABA catabolism. The dynamics of GABA concentration after vigabatrin administration was successfully quantified using MRS.[54,55]

8.3.2   Neurochemical Profiling in the Human Brain

The methodology of MRS of the human brain is similar to animal applications, with only a few differences. First, the spatial resolution is

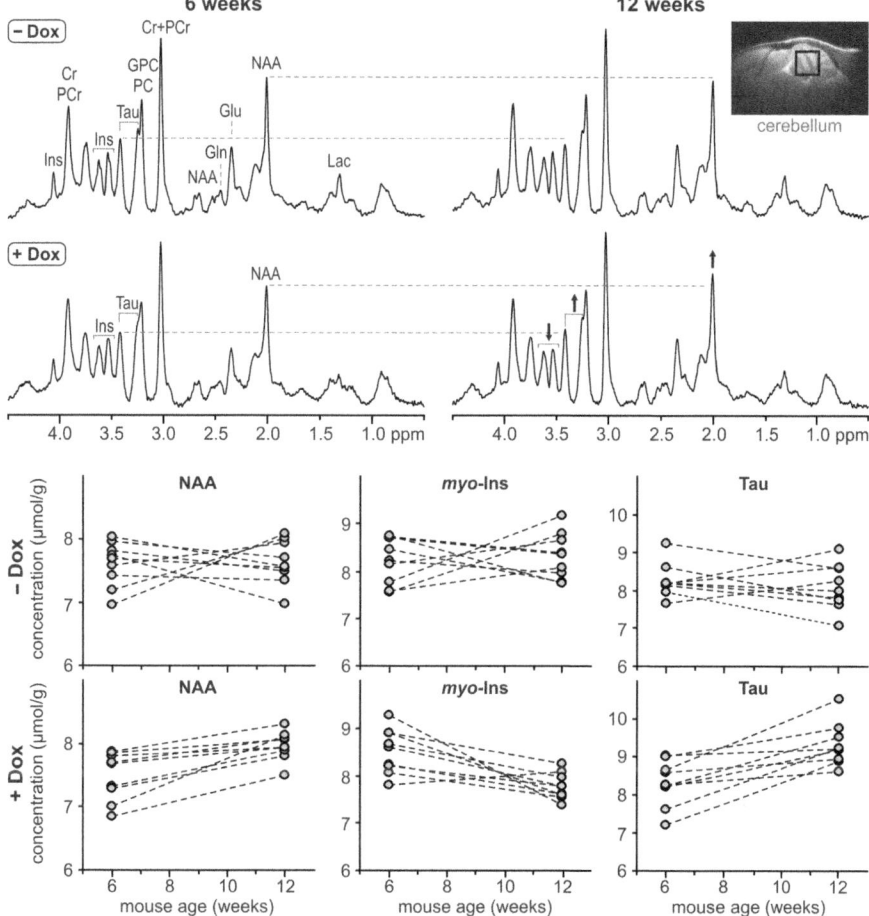

Figure 8.10   Top panel: *In vivo* ¹H MR spectra of a conditional SCA1 mouse treated with doxycycline (+Dox) and an untreated littermate control (−Dox). Data were acquired before (at 6 weeks) and after (at 12 weeks) treatment (LASER, TE = 15 ms, TR = 5 s, B_0 = 9.4 T). Bottom panel: Concentration changes in individual SCA1 mice treated with doxycycline (+Dox) and untreated littermate controls (−Dox). Eight of the nine treated mice displayed a normalization in all three metabolites (increase in NAA and taurine and decrease in *myo*-inositol), while none of the untreated mice showed a normalization in all three metabolites. Modified with permission from Öz *et al.*[52] © 2011 Elsevier.

different. The measured volume in human brain applications is typically in the milliliter range, which is approximately three orders of magnitude larger than in rodents (microliter range). The B_1 field management on 7 T human MR scanners is more complicated. Namely, multichannel transmit systems and B_1 shimming techniques help to meet the requirements for sufficiently high peak B_1.[18,20]

8.3.2.1 Current Possibilities at High Fields

Increased sensitivity and chemical shift dispersion at 7 T improve the reliability of metabolite quantification in the human brain.[2,9,12,18,56] To demonstrate the effect of increased B_0 field, the same group of healthy volunteers was scanned at 4 T and 7 T using very similar RF coil design, the same type of console and the same pulse sequence and data were post-processed and quantified using identical methods.[12] Representative 4 T and 7 T ^{1}H MR spectra measured from the same VOI location of the same subject are shown in Figure 8.11.

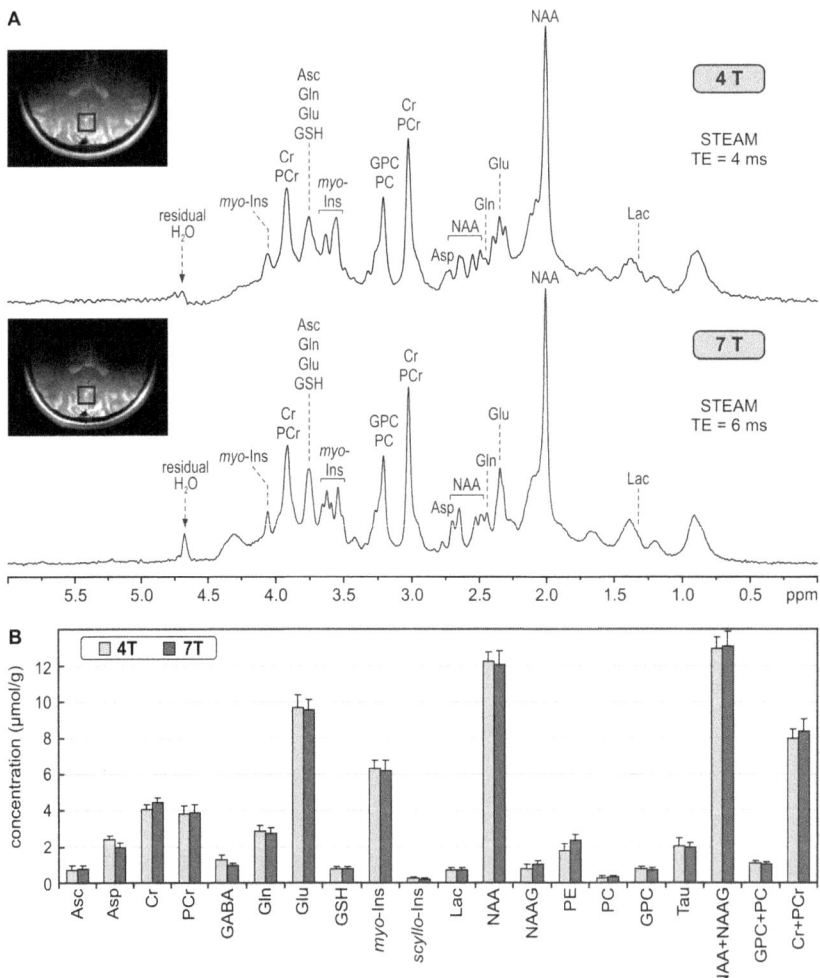

Figure 8.11   (A) *In vivo* ^{1}H MR spectra acquired at 4 T (TE = 4 ms) and 7 T (TE = 6 ms) from the brain of the same subject. (B) Comparison of neurochemical profiles determined at 4 T and 7 T from occipital cortex of 10 healthy volunteers (age = 24 ± 5 years). Error bars indicate SD. Modified with permission from Tkac *et al.*[12] © 2009 Wiley-Liss, Inc.

Neurochemical profiles quantified from the same subjects at 4 T and 7 T were nearly identical (Figure 8.11). The standard deviations of metabolite levels were similar between 4 T and 7 T, because the dominant contributions to metabolite level variations originated from intersubject differences. Consequently, positive correlations between metabolite concentrations quantified at 4 T and 7 T were observed.[12] This similarity in the intersubject variance between 4 T and 7 T does not mean that we do not gain at high field. The precision and accuracy of metabolite quantification in individual subjects was always higher at 7 T. For example, the estimated precision of metabolite quantification (CRLB) at 7 T using 8 scans was approximately the same as that achieved after 128 scans at 4 T.[12] In addition, the quantification precision for weakly represented metabolites at 7 T, such as Asc, GABA, GSH, NAAG, PE and Tau, appears to be nearly impossible to achieve at 4 T independently of the number of scans and achieved SNR. This high precision of metabolite quantification is critical for studies where time courses are monitored in individuals and small concentration changes are expected, for example in functional MRS aimed at measuring changes in metabolite levels during brain stimulation.[57] The neurochemical profiles quantified from human or rodent brain are extremely similar. The main difference is in taurine, the concentration of which in the human brain is approximately 5–10 times lower than in the mouse brain. In addition, *scyllo*-Ins can be quantified in the human brain, but is typically under the detection threshold in rodent brain spectra. Proton MRS of deep brain structures in the human brain highly benefits from multichannel RF coils and B_1 shimming at ultrahigh fields.[18]

8.3.2.2 Assessment of Adrenoleukodystrophy Lesions

Adrenoleukodystrophy (ALD) is a hereditary disease with a defect in beta-oxidation of very long chain fatty acids, which causes their accumulation and inflammatory demyelination of cerebral white matter in its most severe, childhood-onset phenotype. Hematopoietic stem cell transplantation (HSCT) is currently the only effective treatment of childhood-onset cerebral ALD and is most effective if performed at an early stage of the disease. Proton MRS at 4 T was used to establish biomarkers to plan for HSCT and to monitor the effects of treatment on abnormal neurochemical levels.[58] Figure 8.12A shows *in vivo* ^{1}H MR spectra of a patient with ALD before and one year after HSCT. Abnormally high lipid and lactate signals were substantially reduced after HSCT. The same trend was observed in other patients with ALD who underwent HSCT (Figure 8.12B). Therefore, proton MRS was capable of assessing the effects of HSCT in the brains of individual patients.

8.3.2.3 Assessment of Neurochemical Alterations in Movement Disorders

Proton MRS at 4 T was used to investigate patterns of neurochemical alterations in different types of hereditary and sporadic ataxias, which are rare

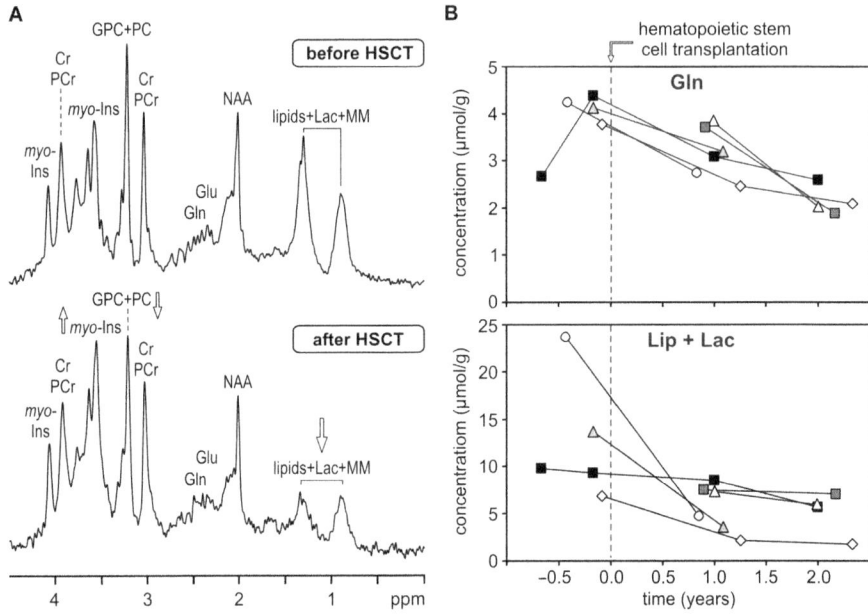

Figure 8.12   (A) *In vivo* [1]H MR spectra (STEAM, TE = 5 ms, TR = 4.5 s, B_0 = 4 T) of one patient with adrenoleukodystrophy (ALD) before and after hematopoietic stem cell transplantation (HSCT). (B) Decrease of glutamine and the sum of lipids and lactate in occipital white matter lesions of individual patients with ALD over time after HSCT.
Modified with permission from Öz *et al.*[58] © 2005 AAN Enterprises, Inc.

movement disorders that cause neurodegeneration in the cerebellum.[31,51,59] Neurochemical profiles of the vermis, pons and cerebellar hemispheres in patients with SCAs were different from those of healthy subjects and further displayed distinct differences between different SCA subtypes.[31] These findings demonstrated the potential of MRS to distinguish neurodegenerative diseases with similar pathology and clinical presentation.

High-field ^{1}H MRS was also used in more common movement disorders. For example, ultrahigh-field MRS revealed a previously undetected elevation in brainstem and striatal GABA levels in Parkinson's disease.[60] Future investigations into the effects of Parkinson's medications on these abnormal GABA levels are expected to provide valuable insights into the pathophysiology of this devastating disease.

8.4   Conclusions

This chapter aims at providing readers with a basic level of understanding of what is currently feasible to achieve using *in vivo* proton MRS at high fields. Note that the high performance *in vivo* MRS techniques presented in this chapter are not available as part of vendor-provided MRS packages on clinical scanners. Typically the standard pulse sequences for localized MR spectroscopy need substantial adjustments. However, the good news is that once the MRS methods are optimized, reliable quantification of more than 15 brain metabolites is feasible in animals and humans and these neurochemical profiles can be routinely utilized in pre-clinical and clinical applications. The non-invasiveness of proton MRS allows its broad applications in drug discovery and development efforts.

References

1. D. K. Deelchand, P. F. Van de Moortele, G. Adriany, I. Iltis, P. Andersen, J. P. Strupp, J. T. Vaughan, K. Ugurbil and P. G. Henry, *J. Magn. Reson.*, 2010, **206**, 74.
2. I. Tkac, P. Andersen, G. Adriany, H. Merkle, K. Ugurbil and R. Gruetter, *Magn. Reson. Med.*, 2001, **46**, 451.
3. P. A. Bottomley, *Ann. NY Acad. Sci.*, 1987, **508**, 333.
4. J. Frahm, K. D. Merboldt and W. Hanicke, *J. Magn. Reson.*, 1987, **72**, 502.
5. R. A. de Graaf, P. B. Brown, S. McIntyre, T. W. Nixon, K. L. Behar and D. L. Rothman, *Magn. Reson. Med.*, 2006, **56**, 386.
6. H. P. Hetherington, W. J. Chu, O. Gonen and J. W. Pan, *Magn. Reson. Med.*, 2006, **56**, 26.
7. R. Gruetter, *Magn. Reson. Med.*, 1993, **29**, 804.
8. R. Gruetter and I. Tkac, *Magn. Reson. Med.*, 2000, **43**, 319.
9. I. Tkac and R. Gruetter, *Appl. Magn. Reson.*, 2005, **29**, 139.
10. I. Tkac, Z. Starcuk, I. Y. Choi and R. Gruetter, *Magn. Reson. Med.*, 1999, **41**, 649.
11. I. Tkac, P. G. Henry, P. Andersen, C. D. Keene, W. C. Low and R. Gruetter, *Magn. Reson. Med.*, 2004, **52**, 478.
12. I. Tkac, G. Oz, G. Adriany, K. Ugurbil and R. Gruetter, *Magn. Reson. Med.*, 2009, **62**, 868.
13. M. Garwood and L. DelaBarre, *J. Magn. Reson.*, 2001, **153**, 155.
14. T. W. Scheenen, D. W. Klomp, J. P. Wijnen and A. Heerschap, *Magn. Reson. Med.*, 2008, **59**, 1.
15. G. Oz and I. Tkac, *Magn. Reson. Med.*, 2011, **65**, 901.
16. V. Mlynarik, G. Gambarota, H. Frenkel and R. Gruetter, *Magn. Reson. Med.*, 2006, **56**, 965.
17. L. Xin, B. Schaller, V. Mlynarik, H. Lu and R. Gruetter, *Magn. Reson. Med.*, 2013, **69**, 931.
18. U. E. Emir, E. J. Auerbach, P. F. Van De Moortele, M. Marjanska, K. Ugurbil, M. Terpstra, I. Tkac and G. Oz, *NMR Biomed.*, 2012, **25**, 152.
19. H. P. Hetherington, N. I. Avdievich, A. M. Kuznetsov and J. W. Pan, *Magn. Reson. Med.*, 2010, **63**, 9.
20. P. F. Van de Moortele, C. Akgun, G. Adriany, S. Moeller, J. Ritter, C. M. Collins, M. B. Smith, J. T. Vaughan and K. Ugurbil, *Magn. Reson. Med.*, 2005, **54**, 1503.
21. U. Klose, *Magn. Reson. Med.*, 1990, **14**, 26.

22. H. Ratiney, M. Sdika, Y. Coenradie, S. Cavassila, D. van Ormondt and D. Graveron-Demilly, *NMR Biomed.*, 2005, **18**, 1.
23. S. W. Provencher, *Magn. Reson. Med.*, 1993, **30**, 672.
24. V. Govindaraju, K. Young and A. A. Maudsley, *NMR Biomed.*, 2000, **13**, 129.
25. J. Pfeuffer, I. Tkac, S. W. Provencher and R. Gruetter, *J. Magn. Reson.*, 1999, **141**, 104.
26. N. Kunz, C. Cudalbu, V. Mlynarik, P. S. Huppi, S. V. Sizonenko and R. Gruetter, *Magn. Reson. Med.*, 2010, **64**, 939.
27. C. Cudalbu, V. Mlynarik and R. Gruetter, *J. Alzheimer's Dis.*, 2012, **31**(3), S101.
28. I. Tkac, R. Rao, M. K. Georgieff and R. Gruetter, *Magn. Reson. Med.*, 2003, **50**, 24.
29. L. Xin, G. Gambarota, J. M. Duarte, V. Mlynarik and R. Gruetter, *NMR Biomed.*, 2010, **23**, 1097.
30. L. Zacharoff, I. Tkac, Q. Song, C. Tang, P. J. Bolan, S. Mangia, P. G. Henry, T. Li and J. M. Dubinsky, *J. Cereb. Blood Flow Metab.*, 2012, **32**, 502.
31. G. Oz, I. Iltis, D. Hutter, W. Thomas, K. O. Bushara and C. M. Gomez, *Cerebellum*, 2011, **10**, 208.
32. J. M. das Neves Duarte, A. Kulak, M. M. Gholam-Razaee, M. Cuenod, R. Gruetter and K. Q. Do, *Biol. Psychiatry.*, 2012, **71**, 1006.
33. A. Kulak, J. M. Duarte, K. Q. Do and R. Gruetter, *J. Neurochem.*, 2010, **115**, 1466.
34. C. Cudalbu, V. Mlynarik, L. Xin and R. Gruetter, *Magn. Reson. Med.*, 2009, **62**, 862.
35. L. Xin, G. Gambarota, V. Mlynarik and R. Gruetter, *NMR Biomed.*, 2008, **21**, 396.
36. M. Marjanska, E. J. Auerbach, R. Valabregue, P. F. Van de Moortele, G. Adriany and M. Garwood, *NMR Biomed.*, 2012, **25**, 332.
37. I. Y. Choi, S. P. Lee, D. N. Guilfoyle and J. A. Helpern, *Neurochem. Res.*, 2003, **28**, 987.
38. T. Michaelis, S. Boretius and J. Frahm, *Prog. Nucl. Magn. Reson. Spectrosc.*, 2009, **55**, 1.
39. J. M. Duarte, H. Lei, V. Mlynarik and R. Gruetter, *NeuroImage*, 2012, **61**, 342.
40. R. Rao, I. Tkac, E. L. Townsend, R. Gruetter and M. K. Georgieff, *J. Nutr.*, 2003, **133**, 3215.
41. J. Valette, M. Guillermier, L. Besret, P. Hantraye, G. Bloch and V. Lebon, *J. Cereb. Blood Flow Metab.*, 2007, **27**, 588.
42. H. Lei, J. M. Duarte, V. Mlynarik, A. Python and R. Gruetter, *J. Neurosci. Res.*, 2010, **88**, 413.
43. R. Rao, K. Ennis, J. D. Long, K. Ugurbil, R. Gruetter and I. Tkac, *J. Neurochem.*, 2010, **114**, 728.
44. K. L. Ward, I. Tkac, Y. Jing, B. Felt, J. Beard, J. Connor, T. Schallert, M. K. Georgieff and R. Rao, *J. Nutr.*, 2007, **137**, 1043.

45. R. Rao, I. Tkac, E. L. Unger, K. Ennis, A. Hurst, T. Schallert, J. Connor, B. Felt and M. K. Georgieff, *Pediatr. Res.*, 2013, **73**, 31.
46. C. M. Traudt, I. Tkac, K. M. Ennis, L. M. Sutton, D. M. Mammel and R. Rao, *J. Neurosci. Res.*, 2012, **90**, 307.
47. M. Marjanska, G. L. Curran, T. M. Wengenack, P. G. Henry, R. L. Bliss, J. F. Poduslo, C. R. Jack, Jr., K. Ugurbil and M. Garwood, *Proc. Natl Acad. Sci. USA*, 2005, **102**, 11906.
48. V. Mlynarik, M. Cacquevel, L. Sun-Reimer, S. Janssens, C. Cudalbu, H. Lei, B. L. Schneider, P. Aebischer and R. Gruetter, *J. Alzheimer's Dis.*, 2012, **31**(3), S87.
49. I. Tkac, J. M. Dubinsky, C. D. Keene, R. Gruetter and W. C. Low, *J. Neurochem.*, 2007, **100**, 1397.
50. G. Oz, C. D. Nelson, D. M. Koski, P. G. Henry, M. Marjanska, D. K. Deelchand, R. Shanley, L. E. Eberly, H. T. Orr and H. B. Clark, *J. Neurosci.*, 2010, **30**, 3831.
51. G. Oz, D. Hutter, I. Tkac, H. B. Clark, M. D. Gross, H. Jiang, L. E. Eberly, K. O. Bushara and C. M. Gomez, *Mov. Disord.*, 2010, **25**, 1253.
52. G. Oz, M. L. Vollmers, C. D. Nelson, R. Shanley, L. E. Eberly, H. T. Orr and H. B. Clark, *Exp. Neurol.*, 2011, **232**, 290.
53. O. A. Petroff, F. Hyder, T. Collins, R. H. Mattson and D. L. Rothman, *Epilepsia*, 1999, **40**, 958.
54. R. A. de Graaf, A. B. Patel, D. L. Rothman and K. L. Behar, *Neurochem. Int.*, 2006, **48**, 508.
55. J. Yang and J. Shen, *J. Neural. Transm.*, 2009, **116**, 291.
56. R. Mekle, V. Mlynarik, G. Gambarota, M. Hergt, G. Krueger and R. Gruetter, *Magn. Reson. Med.*, 2009, **61**, 1279.
57. S. Mangia, I. Tkac, R. Gruetter, P. F. Van de Moortele, B. Maraviglia and K. Ugurbil, *J. Cereb. Blood Flow Metab.*, 2007, **27**, 1055.
58. G. Oz, I. Tkac, L. R. Charnas, I. Y. Choi, K. J. Bjoraker, E. G. Shapiro and R. Gruetter, *Neurology*, 2005, **64**, 434.
59. I. Iltis, D. Hutter, K. O. Bushara, H. B. Clark, M. Gross, L. E. Eberly, C. M. Gomez and G. Oz, *Brain Res.*, 2010, **1358**, 200.
60. U. E. Emir, P. J. Tuite and G. Oz, *PloS One*, 2012, **7**, e30918.

CHAPTER 9

Hyperpolarization: Concepts, Techniques and Applications

ARNAUD COMMENT

École Polytechnique Fédérale de Lausanne, Institute of Physics of Biological System, SB-IPSB GR-CO, Station 6, CH-1015, Lausanne, Switzerland
Email: arnaud.comment@epfl.ch

9.1 What is Hyperpolarization and Why Bother Using It?

Nuclear magnetic resonance (NMR) is arguably one of the most prevalent and versatile spectroscopic methods in chemistry and biochemistry. It is also widely used in biomedical and medical research, in particular in conjunction with magnetic resonance imaging (MRI), for pre-clinical studies and as a diagnostic tool in clinical environments. In the biomedical and medical fields, the high temporal and spatial resolution of MR makes it a powerful imaging modality allowing fast non-invasive detection of small morphological details even in rapidly moving organs like the heart (see Chapter 18). In fact, the modality is nowadays widespread and well accepted, and contrast agents are used extensively for improved contrast or perfusion examination (see also Chapter 6 for recent developments on contrast agents). MR is also a unique technique to obtain *in vivo* metabolic maps using the spectroscopic information that can be extracted from the time-domain acquisitions (see also Chapter 8). In particular, it is possible to monitor the biochemical transformations of specific substrates that are delivered to subjects. Because it gives access to the kinetics of the conversion of substrates into metabolites, MR spectroscopy (MRS) of the

New Developments in NMR No. 2
New Applications of NMR in Drug Discovery and Development
Edited by Leoncio Garrido and Nicolau Beckmann
© The Royal Society of Chemistry 2013
Published by the Royal Society of Chemistry, www.rsc.org

carbon nuclei (^{13}C) is one of the most powerful techniques to investigate intermediary metabolism. For instance, the investigation of cerebral and cardiac metabolism using ^{13}C MRS brought essential information on the use of energy substrate of the brain and of the heart. In particular, ^{13}C MRS results led to the demonstration that brain metabolism is compartmentalized between neurons and astrocytes, each of them having specific metabolic activities. ^{13}C MRS analysis of models of heart failure have built the concept that changes in metabolism modify the balance between energy provision and utilization, a fact that may lead to the progression towards decompensated heart failure. Besides glucose, the main cerebral fuel, other substrates such as ketone bodies, lactate and fatty acids can be oxidized by the brain[1] and the heart.[2] The central energy metabolic pathways and the ^{13}C label positions in the main metabolites are sketched in Figure 9.1.

Although powerful, conventional *in vivo* ^{13}C MRS only gives access to the most concentrated metabolites as a consequence of the low ^{13}C MR sensitivity and many metabolites cannot be detected. The poor sensitivity also limits the temporal resolution of the MRS measurements to several minutes when many biochemical transformations occur within seconds. These were two of the main motivations behind the recent development of hyperpolarization techniques. It is worth stating that although the sensitivity of ^{13}C and ^{15}N MRS is substantially lower than proton MRS, the contrast-to-noise ratio of ^{13}C and ^{15}N is considerably larger as a consequence of the low natural abundance of those isotopes leading to essentially null background signal.

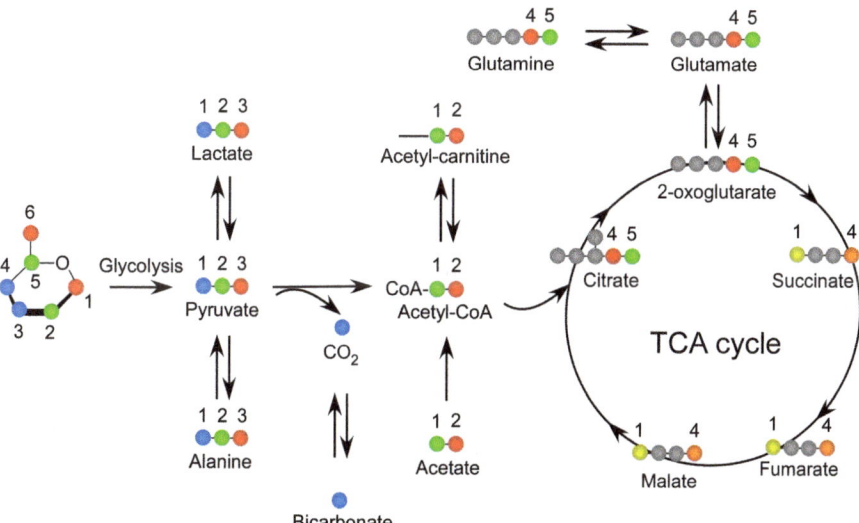

Figure 9.1   Central metabolic pathways: the color circles indicate the ^{13}C label positions in substrates and their corresponding metabolites that are of particular interest for hyperpolarized ^{13}C MR.

Although it is recognized for its multimodality, magnetic resonance (MR) is hampered by a relatively low sensitivity: the energy difference ΔE between the quantum states involved in the measurements, namely the spin up and spin down states,[*] is very small even in magnetic fields B_0 of several Tesla and many radiofrequency photons are required to build up a detectable signal. In addition, the thermal energy being much larger than ΔE for standard temperatures T, the population difference between the quantum states, referred to as polarization P in the context of MR, is only on the order of about 10^{-5} as given by Curie's law of paramagnetism:

$$P = C \frac{B_0}{T}, \qquad (9.1)$$

where C is the Curie constant. The net number of detectable photons following a radiofrequency excitation is thus very limited. In optical imaging, the transitions at the origin of the signal amplitude take place between molecular ground states and excited states, which are separated by a comparatively much larger energy difference. The energy of optical photons is thus much larger (by a factor of about 10^6) and the population difference between the involved states is essentially 1 even at room temperature. As a consequence optical methods are very sensitive and can provide cellular level resolution imaging, but they are restricted to external tissues due to the low penetration depth of optical photons.[3]

The development of higher-field superconducting magnets led to an important increase in the sensitivity of MR experiments, but further increasing the polarization by raising B_0 is becoming technologically very challenging and costly. Curie's law shows that a tremendous gain in polarization could be achieved if it was possible to lower the temperature of the spins. Although most biological and biomedical MR studies have to be performed around 37 °C, could it be possible to lower the temperature of the nuclear spins without lowering the temperature of the tissues in which they are located at the time of the MR measurements? This is precisely the challenge tackled by hyperpolarization techniques: establishing a nuclear polarization close to 1 in a room- or body-temperature biological sample.

The purpose of the present chapter is to introduce and describe three hyperpolarization techniques, namely parahydrogen induced polarization (PHIP), dissolution dynamic nuclear polarization (DNP) and the "brute force" method, and to present their established and potential applications for drug response analysis and drug development. Note that the discussion will be restricted to the techniques aiming at enhancing the liquid-state polarization and we will describe neither the optical pumping method to hyperpolarize noble gases (see Chapter 17) nor the solid-state DNP techniques combined with magic angle spinning (see Chapter 3).

---

[*]Although the same type of reasoning can be applied to larger spins, we will restrict the discussion to spin-1/2 particles in the present chapter.

9.1.1 Thermodynamic Considerations

As mentioned above, the aim of hyperpolarization techniques is to prepare molecules in which a specific nuclear spin state is much more populated than it would be at thermal equilibrium, *i.e.* a state with a nuclear polarization $P_{hyper} \gg P_{equ} = CB_0/T$. What we can control for this purpose are the temperature of the environment (also referred to as *bath* or *lattice*) and the external magnetic fields (static as well as alternating, *i.e.* radiofrequency or microwave). Thermodynamics is an appropriate formalism for describing the state of a system using macroscopic variables without needing any information on its internal microscopic configurations. The central concept on which a thermodynamic description of spin systems is based is that spins are weakly coupled to their environment. It thus takes a finite amount of time for the spins to be in equilibrium with the bath because the flow of energy that can be exchanged between the spin systems and their environment is limited. With the concept of spin temperature as introduced by Casimir and du Pré,[4] an analogy with classical thermodynamics is possible and the energy transfers between spin systems and lattice can be used to define the equilibrium state towards which they evolve.

Within this formalism, we can for instance describe an experiment in which the nuclear or electron spins of a sample are "saturated" by a strong radiofrequency or microwave field respectively. The experimental result is a disappearance of the NMR or electron spin resonance (ESR) signal as a consequence of the equilibration of the spin-up and spin-down populations, which leads to zero spin polarization. From Curie's law, we deduce that the spin temperature

$$T_S = CB_0/P \tag{9.2}$$

is infinite. In the usual case of a properly cooled sample and sufficiently low average radiofrequency or microwave power, the heat capacity of the bath is sufficient to cool the spins back to thermal equilibrium without substantially heating the sample (a non-negligible increase in temperature will be observed if proper care is not taken!) and the signal reappears on a time scale on the order of the spin-lattice relaxation time, which can be quite long. As in conventional thermodynamics, two systems in thermal contact evolve towards a common temperature.

The same thermodynamic considerations can be used to describe two spin systems that are coupled *via* energy-exchanging interactions and it shows that it is possible to design a sort of cooling machine to pump energy from one spin system to the other with the help of radiofrequency or microwave energy. The essential conceptual step to understand why two spin systems with very different energy level splitting ΔE (for instance ^{13}C and electron spins) can exchange energy was introduced by Alfred G. Redfield when he demonstrated in 1955 that spin thermodynamics is valid in a referential frame rotating at the frequency of the excitation radiofrequency or microwave field.[5] In this frame, the two spin systems have comparable ΔE and can thus exchange energy eventually to reach a common temperature. In addition, if one spin system is more isolated than the other from the lattice it will be dynamically cooled by the spin system that is in close contact with the lattice in which the energy is dumped. Note that this cooling machine is not equivalent to a classical refrigeration cycle since it cannot be modified to work in the reverse direction. Rather, it consists in two systems that are put in thermal contact and one of them is cooled by a thermostat (the lattice).

9.1.2 The Principles of Hyperpolarization and Dynamic Nuclear Polarization Methods

In situ and *ex situ* methods have been developed to enhance the polarization of nuclear spins. *In situ* techniques are designed to perform the polarization enhancement inside the MR probe that will be used to perform the targeted MR experiments and do not require the sample to be translated or any hardware modification between the polarization procedure and the MR measurements. The polarization enhancement is thus usually done in a high-field environment. *Ex situ* methods require a transfer of the sample from a dedicated preparation apparatus to the MR instrument (in some cases the same magnet can be used for both preparation and MR measurements). We propose to categorize the *in situ* and *ex situ* polarization methods that have been proposed so far in three groups: *low-temperature techniques, Overhauser techniques* and *parahydrogen induced polarization techniques.*

9.1.2.1 Low-Temperature Techniques

From Curie's law (see Eqn (9.1)), it appears that the most obvious way to enhance the nuclear spin polarization is to lower the temperature of the sample. Temperatures on the order of 10 mK can be achieved with appropriate commercially available cryogenic equipment. To avoid having to lower the temperature in the sub-Kelvin region, unpaired electron spins for which the Curie constant is much larger than for nuclear spins (~ 660 for ^{1}H, ~ 2400 for ^{13}C and ~ 6600 for ^{15}N) can be used to polarize the nuclear spins. It can be done efficiently at around 1 K by saturating the electron spin resonance and thus putting the nuclear spins in thermal contact with the electron spins to obtain a dynamic cooling as described in Section 9.1.1, the energy being exchanged *via* the dipolar coupling between electron and nuclear spins. This technique is called dynamic nuclear polarization (DNP),[6] or sometimes solid-state DNP, and in this context the unpaired electron spins can be referred to as *polarizing agents.*

Low-temperature techniques to enhance the polarization must be coupled with a rapid heating system since, as discussed in Section 9.1, the natural temperature for biological systems is around 37 °C. Because a temperature jump might damage the biological samples and because precisely controlling the temperature of a fast heating process is difficult, these techniques are inherently *ex situ* methods. The two methods based on these principles are the brute force method and dissolution DNP and will be discussed further in Sections 9.2.1 and 9.2.2, respectively. It is important to mention that Joo *et al.* performed *in situ* temperature jump DNP experiments using a laser to melt frozen solutions polarized at 90 K.[7] The method is attractive because a freezing-melting cycle can be repeated to perform data averaging and the sample is not diluted during the heating process (which also means that the concentration of polarizing agents is high in the liquid state). To the best of our knowledge, this method has not been further developed, most likely because it appears to be technically challenging and its applicability to biological samples seems to be limited.

9.1.2.2 Overhauser Techniques

Historically, Albert W. Overhauser was the first to predict that the polarization of nuclear spins coupled to electron spins can be enhanced by saturating the ESR and he considered the case of conduction electrons in a metal.[8] The experimental demonstration of what is now known as the *Overhauser effect* was performed by Thomas R. Carver and Charles P. Slichter in metallic lithium.[9] It was later shown that mobile paramagnetic centers (we shall also call them polarizing agents in analogy to solid-state DNP) saturated by an alternating field can induce the same effect in liquid solutions.[10,11] The essential difference between the Overhauser effect (sometimes also called DNP) and solid-state DNP is that, in the former, the polarizing agents are mobile and the nuclear polarization enhancement originates from diffusion-induced electron-nucleus cross-relaxation processes whereas, in the latter, the polarizing agents are fixed in the solid sample and the locally driven nuclear polarization enhancement propagates throughout the whole sample *via* nuclear spin diffusion.[12]

Because the electron-nucleus cross-relaxation rates decrease with increasing magnetic field (the energy level splitting ΔE of the electron spins becomes much larger than the ΔE of the nuclear spins and the two spin systems cannot efficiently exchange energy because they are not in good thermal contact), the Overhauser effect is much less efficient at high field. The obvious strategy thus consists in performing *ex situ* polarization enhancement at low field and transferring the solution to a high-field magnet for MR measurements. Several methods based on this concept have been proposed for *in vitro* measurements and they are sometimes collectively referred to as *shuttle DNP* methods.[13] In particular, Lingwood *et al.* showed that the polarization enhancement can be performed in the fringe field of a clinical MR scanner.[14] It was more recently demonstrated by Krummenacker *et al.* that an enhancement on the order of 100 can be obtained by directly polarizing inside a 1.5 T clinical scanner.[15] However, the amount of liquid that can be polarized per unit of time is rather small and adapting the method for the new standard MRI field of 3 T will certainly even lower the throughput as well as the enhancement factor. Note that Krummenacker *et al.* have also shown that it is possible to enhance the signal of metabolites in solution *in situ* at high-field but the maximum enhancement factor was only about 5.[16]

Finally, it is worth mentioning that the Overhauser effect was exploited to polarize liquids *in vivo*, a technique named Overhauser-enhanced MRI (OMRI).[17,18] The intrinsic concentration of paramagnetic species in cells and organisms is, however, too low to enhance significantly the nuclear spin polarization and it is necessary to inject a large concentration (typically on the order of 1 mmol kg^{-1}) of exogenous polarizing agents to obtain an enhancement of 1–2 orders of magnitude at very low field, which raises obvious toxicity issues.

The DNP techniques based on the Overhauser effect will not be further discussed in this chapter since the rather limited enhancement that can be obtained as compared to hyperpolarization techniques make them less generally applicable, in particular for *in vivo* applications, which we believe will be of the utmost importance in the context of drug research and development.

9.1.2.3   *Parahydrogen Induced Polarization Techniques*

The concepts behind the parahydrogen induced polarization (PHIP) methods are quite different from the ones involved in the techniques presented above: no paramagnetic centers are added, no saturating alternating field is required and there is no spin thermodynamics involved. This *ex situ* technique relies instead on the chemistry of hydrogenation of suitable molecules, starting from H_2 gas. Because H_2 is composed of only two protons and two electrons, the nuclear spins play a preponderant role in the selection of the allowed quantum mechanical wavefunctions of the molecule: H_2 has either a spherical shape, with a total nuclear spin of zero (parahydrogen), or a dumbbell shape, with a total nuclear spin of one (orthohydrogen), and the two configurations have different rotational angular momenta ($J = 0$ for the sphere and $J = 1$ for the dumbbell). The two J-states are split by about 170 K in zero magnetic field. Since the energy of the parahydrogen ground state is so much lower than the orthohydrogen ground state, it is possible to create pure parahydrogen samples by cooling it at a relatively convenient cryogenic temperature (~ 10 K). The H_2 molecules will stay in the para state upon rapid heating and, following a hydration reaction, it is possible to prepare molecules with highly polarized ^{13}C nuclear spins as will be discussed in Section 9.2.2.[19,20]

9.1.3   The Importance of Spin Relaxation

We have so far discussed neither the time scale of the preparation step of the hyperpolarized states nor the lifetime of these inherently out-of-equilibrium states. The lifetime considerations are essentially the same whichever technique is used to prepare the hyperpolarized states (see Section 9.1.3.2) and we will start by highlighting the crucial role of both nuclear and electron (if applicable) longitudinal spin relaxation times in the temporal considerations of the preparation step.

9.1.3.1 Preparation Step

At very low temperature, as in the case of the brute force method, the nuclear spin-lattice relaxation time is nearly infinite,[12,21] and it is necessary to add paramagnetic centers for the polarization to build up in a reasonable time. This issue was discussed by Krjukov *et al.* in the context of hyperpolarized noble gases.[22] For solid-state DNP, both the electron and the nuclear spin relaxation play a role and should be optimized:[†] on one hand, we wish to have a long nuclear spin relaxation time to avoid that the polarized nuclear spins relax while other nuclear spins are being polarized; on the other hand, we want the electron spin relaxation to be short enough so that the spins can be used to polarize as many nuclear spins as possible per unit of time. Therefore, in order to obtain the highest possible polarization, it is important to balance the choice of paramagnetic centers (type and concentration), temperature and magnetic field (it influences the relaxation times). In terms of spin thermodynamics, the aim is to isolate the nuclear spin system as much as possible from the lattice (there is always a so-called leakage through nuclear relaxation) and to tightly connect the electron spin system to the lattice to efficiently dump the energy extracted from the nuclear spin system. A possible trick to shorten the electron spin relaxation time without significantly influencing the nuclear spin relaxation times is to add a small quantity of Gd^{3+}-based contrast agent in the frozen solutions. Although the effect of those additional paramagnetic centers on the DNP mechanisms is not completely understood, their very large magnetic moment strongly influences the electron spin relaxation and the maximum achievable polarization can be considerably increased.[23,24]

For PHIP, the situation is somewhat analogous to the brute force method: the conversion rate from parahydrogen to orthohydrogen is prohibitively slow and requires the addition of paramagnetic centers to catalyze the transformation. From a spin thermodynamics point of view, the role of the catalyst is to absorb the energy resulting from the $J = 1$ to $J = 0$ state transition and to dump it into the lattice, *i.e.* the catalyst connects the hydrogen spin system to the lattice for efficient cooling.

9.1.3.2 Liquid-State Relaxation

Following the hyperpolarization process, the nuclear polarization will inevitably decrease back to its thermal equilibrium value. Let us assume that the polarization was enhanced by a factor ε, *i.e.* that at time $t = 0$ the polarization was $P_{\text{hyper}} = \varepsilon P_{\text{equ}}$. It will take a time on the order of $T_1 \ln(\varepsilon)$ for the enhanced polarization to decay to its equilibrium value. Thus, if the enhancement was for instance equal to 10^4, the polarization will reach its thermal equilibrium value after about $9 T_1$. The decay of the polarization will

---

[†]Strictly speaking the problem is similar in the brute force method since the T_1 of the paramagnetic centers also play a role and their concentration as well as the optimal temperature should be chosen such as to optimize the polarization.

additionally be accelerated by the application of the radiofrequency pulses that are required to perform the MR measurements. How long the signal will be detectable will depend on the sensitivity (noise level) and on the magnetization flip angle related to the radiofrequency pulses. The aim is obviously to conserve the enhanced nuclear polarization as long as possible and to make the best use of it. Several schemes have been proposed to perform multiecho acquisitions or to use spatially and spectrally selective radiofrequency excitations in order to increase the time domain over which the hyperpolarized nuclear spins can be measured by limiting the amount of enhanced longitudinal magnetization that is required for each acquisition.[25–28]

Three main parameters influence the relaxation times: the chemical environment of the nuclear spins, the applied magnetic field and the temperature. The chemical environment can be partially controlled by choosing an appropriate solvent (deuteration helps reduce the nuclear dipolar relaxation) and by reducing the concentration of paramagnetic centers dissolved in the hyperpolarized solutions.[29–31] Since all hyperpolarization techniques considered here are *ex situ* processes, the nuclear spins usually undergo drastic field and temperature variations from the preparation step to the MR measurements. As an example, typical temperature and field values experienced by the spins during a dissolution DNP experiment are sketched in Figure 9.2 (similar values would apply in a protocol based on the brute force method and, although the preparation step is done at much lower field and at room temperature, the spins hyperpolarized *via* PHIP will also have to travel through various non-optimal field values). The ^{13}C longitudinal relaxation varies from

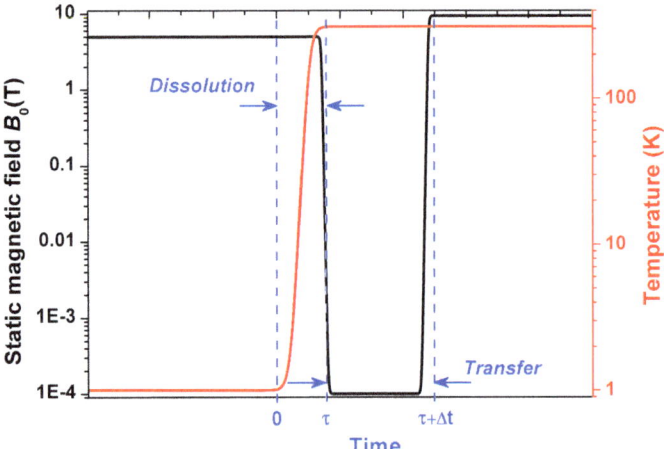

Figure 9.2   Typical temperatures and field values experienced by the spins during a dissolution DNP experiment. The dissolution step (from $t = 0$ to $t = \tau$) takes between a fraction of a second and a few seconds depending on the sample size. The transfer time Δt between the polarizer and the MR magnet is largely dependent on the local configuration of the experiment (it is on the order of 2 s in the setup presented in Figure 9.3).

several thousands of seconds at 1 K, inside the polarizer, to less than a minute in the measuring field at room temperature, and can in some cases be extremely short in the low-field region located between the two magnets.

Although the optimal magnetic field and temperature settings can be experimentally determined, they will be different for all types of nuclear spins and molecules and it will not always be possible to adjust those two parameters to reduce the losses due to relaxation processes (for instance, the temperature in *in vivo* experiments can obviously not be modified). It is also possible to highly enhance and detect nuclear spins with short relaxation times *via in vitro* or *in vivo* polarization transfer from long-T_1 hyperpolarized nuclear spins.[32–34] Such methods can potentially lead to increased sensitivity or increased spectral resolution even *in vivo*.[35] Finally, it has been shown that on specific molecules it is possible to store the enhanced polarization in long-lived singlet states, the relaxation times of which can be over 10 times longer than the T_1 of the nuclear spins forming the singlet states.[36,37]

9.1.4   Other Limitations

Besides the issues related to the finite relaxation time of the nuclear spins, which restrict the applications to specific spin-1/2 nuclei (note that ^6Li has a spin 1 but its quadrupolar constant is essentially negligible), hyperpolarization techniques as they are designed today have some practical limitations. First, if the solubility of the molecule of interest is low, the concentration of the hyperpolarized aqueous solution might be insufficient to obtain a satisfactorily large signal-to-noise ratio in the MR experiments. This is especially critical for the techniques based on a dissolution procedure (DNP and a potentially adapted brute force method) since the solutions are diluted by a factor of at least 10. Second, as some non-biocompatible substances are usually required to perform the hyperpolarization (radicals in DNP and possibly in brute force method and catalysts in PHIP), toxicity might become an issue especially for potential clinical applications. Filtering processes can be designed but they will have to be followed by quality control checks and the whole procedure will substantially increase the delay between the preparation step and the MR measurements. Finally, the sensitivity of MR being limited, the substrates concentration will be on the order of 1 mM even if the molecules are hyperpolarized since no averaging is possible and the experiments consist in single-shot acquisitions.[‡] For many molecules, such concentrations are clearly supraphysiological and can either lead to toxicity issues or alter the observed metabolism. For the above-mentioned reasons, it is clear that not all molecules will be interesting for hyperpolarized MR applications in biology and medicine.

[‡]Care has to be taken when comparing the sensitivity of thermal equilibrium and hyperpolarized MR measurements.

9.2   How to Hyperpolarize Nuclear Spins?

We will now focus on the practical considerations of the two techniques developed and currently used for applications in MRS and MRI, namely PHIP and Dissolution DNP. Although conceptually the simplest, the brute force method has never been implemented for hyperpolarizing spins in liquid-state solutions. We will nevertheless attempt to highlight the main requirements and issues that would need to be taken into consideration to develop the technique.

9.2.1   Brute Force Method

The most straightforward way to highly populate the low-energy spin state of a nuclear spin system is to reduce the thermal energy to a value that is on the order of or below the energy level (Zeeman) splitting ΔE. The amplitude of the splitting for ^{13}C or ^{15}N being around a few mK even in a field as high as 15 T, the use of a ^{3}He-^{4}He dilution cryostat is required. Once the spin temperature of the nuclear spin system has reached the lattice temperature, the sample could be rapidly extracted and warmed up to room temperature to obtain a hyper-polarized solution. Such a method has been proposed to hyperpolarize noble gases and is named brute force method.[22,38] Although the concept behind it is simple, its implementation is difficult for several reasons. First, the relaxation time of nuclear spins at 10 mK is quasi-infinite unless the sample is doped with a substantial amount of paramagnetic relaxing agents (see Section 9.1.3.1). Since these paramagnetic agents will also strongly accelerate the nuclear spin relaxation while the sample is warmed up, a fast heating procedure is required to avoid losing the enhanced nuclear polarization. Such a procedure has been proposed in conjunction with solid-state DNP techniques (the dissolution step described in Section 9.2.2) but it requires fast access to sample space to minimize the time the sample is at intermediate temperatures at which the relaxation induced by the paramagnetic agents is highly efficient. The fact that the ^{3}He-^{4}He mixture in the dilution cryostat is circulating in a closed cycle highly complicates sample access and it would be technically very challenging to implement such a procedure. In addition, although such cryostats are commercially available, the price and limited availability of ^{3}He also renders the setup impractical. For these reasons, no hyperpolarization setup based on this technique has been so far implemented.

9.2.2   Dissolution Dynamic Nuclear Polarization

As mentioned in Section 9.1.2.1, the efficiency of solid-state DNP is optimal at around 1 K, a temperature at which the electron polarization is essentially 1 and the relevant time constants for the preparation step are reasonable (preparation time on the order of 1 h). The original DNP hyperpolarizer was operating at 3.35 T, but it has recently been shown that larger maximum polarization can be obtained in a field of 5 T.[39,40] Polarization levels may dramatically vary depending on how the sample is prepared. Solvents and polarizing agents are the crucial components and must be carefully selected. Although nitroxyl radicals are the most widely available polarizing agents, trytil radicals and the stable free radical 1,3-bisdiphenylene-2-phenylallyl (BDPA) have proven to be the most efficient to date.[29,41] The sample preparation procedure consists in incorporating a relatively small quantity of polarizing agents (10–50 mM) in a solution containing a high concentration of molecules of interest (typically substrates isotopically enriched (^{13}C or ^{15}N) at long-T_1 positions). The solution is rapidly frozen to obtain a glassy sample (a visual control is usually sufficient to roughly assess the quality of the sample) and inserted inside the hyperpolarizer.

DNP is performed by irradiating the sample with microwaves at a frequency roughly corresponding to the difference between the ESR and the NMR frequencies. The microwave frequency is thus strongly dependent on the type of polarizing agent and in some cases (radicals with narrow ESR linewidth) it might vary depending on the type of targeted nuclei. Microwave power also has to be carefully adjusted since exceedingly large power will lead to sample heating, which will affect the polarization efficiency (the typical optimal power as measured at the microwave source output is on the order of 10–50 mW).

The dissolution step to obtain the liquid-state solution containing the hyperpolarized molecules is performed by projecting a hot solvent (typically superheated water) on the frozen solution to rapidly melt it (a volume ratio of 10–100 between the solvent and the frozen solution is required to completely melt the sample) and extract it prior to injecting it in an NMR tube, an animal or a human. Dissolution DNP was developed within Amersham Health Research and Development AB, a company that was later purchased by GE Healthcare.[41] The technique has been applied to hyperpolarize ^1H, ^6Li, ^{13}C, ^{15}N, ^{89}Y and both ^{107}Ag and ^{109}Ag,[34,41–47] and the liquid-state nuclear polarizations can be typically enhanced by 3–4 orders of magnitude compared to the room-temperature thermal polarizations obtained in standard fields (3–14 T).

Note that the distance between the hyperpolarizer and the MR instrument is usually non-negligible and a fast transfer procedure is of great importance as discussed in Section 9.1.3.2. An example of a full setup suitable for *in vivo* applications in rodents is shown in Figure 9.3.

9.2.3 Parahydrogen-Induced Polarization

Nearly pure parahydrogen can be prepared by flowing H_2 through a cryogenically cooled ($\sim$ 10 K), commercially available catalyst.[48] Parahydrogen gas can then be stored at room temperature in a pressurized container. The hydrogenation reaction required to hyperpolarize the selected substrates (typically ^{13}C-enriched) takes place in a warm reactor ($\sim$ 60 °C) containing a solvent (usually 5 mL of water) pressurized to about 10 bar with parahydrogen. A relatively high concentration ($\sim$ 0.1–1 M) of substrate along with a small quantity ($\sim$ 2.5 mM) of rhodium-based catalyst is introduced inside the reactor. The hydrogenation reaction takes only a few seconds to be completed. Two main methods have been developed to transfer the spin order from protons to

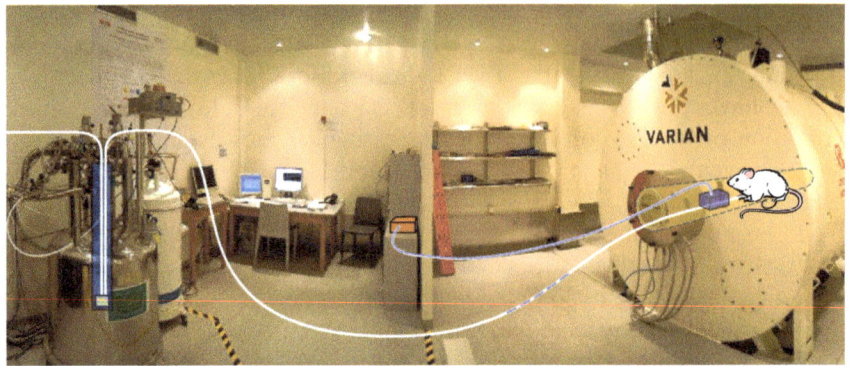

Figure 9.3   Dissolution DNP setup installed at the Center for Biomedical Imaging
(CIBM) of the École Polytechnique Fédérale de Lausanne (EPFL). A 5 T
polarizer is located about 4 m away from a 9.4 T rodent MR scanner.
A specific device minimizing the delay between the dissolution and the
infusion of DNP-enhanced molecules has been implemented for *in vivo*
applications.[67] The delay can be as short as 3 s.[31]

^{13}C in the hydrogenated substrates: a pulsed method and a field-cycling method
implemented by Golman *et al.*[48,49] Both are very low-field methods since the
idea is, in terms of spin thermodynamics, to put the ^1H and ^{13}C spin systems in
thermal contact as described in Section 9.1.1. A full description of the
equipment design and installation required to hyperpolarize molecules *via*
PHIP has been published by Hövener *et al.*[50]

The PHIP method can only be applied to molecules containing an alkene or
alkyne bond and so far the only compounds that have been used for *in vivo*
studies are [1-^{13}C]propionate,[49] which has a high toxicity and is not relevant for
metabolic studies, and [1-^{13}C]succinate.[51]

A derivation of the PHIP technique called SABRE (signal amplification by
reversible exchange) has been recently proposed to circumvent some of the
drawbacks of PHIP.[52,53] Instead of generating a hydrogenation reaction
between parahydrogen and the substrate, the two molecules are temporarily
associated through co-binding with a transition metal complex. The reaction,
which is reversible, is performed in a low magnetic field and the spin order is
transferred from parahydrogen to the substrate. In principle, there is no
restriction on the type of molecules that can be hyperpolarized using this
method although the achieved enhancements reported so far are not nearly as
large as what can be obtained *via* the original PHIP.

9.3   Applications of Hyperpolarized Magnetic Resonance in Drug Research

Although the inherently short lifetime of hyperpolarization states limits the
time scale of the experiments to, at most, a few minutes following the delivery
of the substrate, hyperpolarized MR has great potential for drug research and

development. The enhanced sensitivity along with the absence of background signal (most nuclei of interest have very low natural abundance) make the technique attractive for probing drug uptake and pharmacokinetics either *in vitro* or *in vivo*. By taking advantage of both the gain in sensitivity resulting from hyperpolarization and the high spectroscopic resolution of NMR, in particular for ^{13}C, it is also possible to follow metabolic transformation in real time non-invasively in cells, animals and even humans. No other technique is capable of providing such information and it is widely accepted as being the most promising application of hyperpolarized MR.

A large number of diseases are associated with impaired metabolism, which can be either the cause or a symptom. For instance, glycolytic rates are much higher in a wide variety of diseased cells. Fatty acid and protein synthesis are also substantially modified and reflect altered metabolism. The ability to probe metabolic pathways non-invasively in living organisms is obviously of great interest for developing and testing drugs. It is especially attractive for determining how the organism adjusts its metabolism in response to selected drugs.

To date, most biomedical applications of hyperpolarized MR were developed in the framework of heart disease and cancer research and, except for a couple of studies based on hyperpolarized fumarate, all studies involving drugs were performed using hyperpolarized pyruvate. The focus was on the Warburg effect,[54] which leads to a high production of lactate in cancer cells, and on the metabolic flux through pyruvate dehydrogenase (PDH) detected by measuring the ^{13}C labeling of bicarbonate from [1-^{13}C]pyruvate, as originally proposed by Merritt *et al.*[55] It certainly does not mean that the applications are restricted to only two substrates but, the technique being rather new, only the most straightforward applications have so far been explored. The following sections contain a non-exhaustive overview of the recent studies linked to drug research.

9.3.1 *In Vitro* Applications

By using hyperpolarized MR, the effect of a given drug on specific metabolic pathways can be probed in living cells in a controlled *in vitro* environment. The advantage of hyperpolarized MR over standard thermally polarized MR is that many intermediates as well as their transformation kinetics can be detected. In many cases, it is useful to determine how the drug affects cell metabolism prior to performing *in vivo* tests since the sensitivity is much higher and the metabolic pathways can be isolated. Venkatesh *et al.* demonstrated that hyperpolarized [1-^{13}C]pyruvate can be used to non-invasively assess the efficiency of anticancer drugs to inhibit the phosphatidylinositol 3-kinase/Akt/mammalian target of rapamycin (PI3K/Akt/mTOR) pathway, a pathway that is activated in more than 88% of glioblastomas.[56]

Hyperpolarized MR can also be used to evaluate the *in vitro* biodistribution and pharmacokinetics of specific therapeutic agents that cannot be directly detected by thermally polarized NMR. For instance, Lumata *et al.* demonstrated that silver complexes, which have been shown to have effective antimicrobial and anticancer properties, can be readily detected after having been hyperpolarized.[44] It is also possible to measure how long the drug stays in blood and how efficiently it is metabolized by the organism. Since this time period is in most cases much longer than the lifetime of the hyperpolarized state, it has been proposed to perform such measurements *in vitro* by hyperpolarizing blood samples.[57] It is also worth noting that Roth *et al.* attempted to enhance the ^{13}C spin polarization of barbituric acid using PHIP.[58] However, to the best of our knowledge, no other attempt to use PHIP in the framework of drug research has been reported.

9.3.2 *In Vivo* Applications

It was demonstrated that hyperpolarized MR allows following *in vivo* substrate uptake and performing metabolic imaging in real time.[59] It is thus possible to directly probe the effect of drugs on the metabolism of animal models and possibly even humans. The earliest demonstration of the usefulness of hyperpolarized MR for assessing drug response *in vivo* was the study performed by Day *et al.* in implanted tumors treated with etoposide, a chemotherapeutic drug.[60] The authors showed that the treatment induced a clear reduction of lactate production. More recently, Ward *et al.* tested, *in vivo* in mice, the potential of ^{13}C MRS following the injection of hyperpolarized pyruvate to detect early tumor metabolic responses to anticancer drugs targeting the phosphatidylinositol 3-kinase (PI3K) pathway.[61] Seth *et al.* showed that the effect of dichloroacetate, a drug that is currently undergoing clinical trials, also leads to a reduction of lactate production in subcutaneous tumors.[62]

In rodents implanted with lymphoma tumors, Gallagher *et al.* showed that the efficiency of the drug etoposide to target tumor cells can be monitored, starting as early as 24 h following treatment, by recording the production of malate from injected hyperpolarized ^{13}C-labeled fumarate (see Figure 9.4).[63] Hyperpolarized [1-^{13}C]pyruvate and [1,4-^{13}C$_2$]fumarate were also used to probe, *in vivo*, the effect of combretastatin-A4-phosphate, a vascular disrupting agent, on lymphoma tumors implanted in rodents.[64]

Hyperpolarized MR has also been used to evaluate the effect of drugs on cardiac metabolism. In particular, pre-clinical studies were performed *in vivo* in rat hearts using hyperpolarized [1-^{13}C]pyruvate to assess the influence of dichloroacetic acid on PDH flux by monitoring the kinetics of the pyruvate to bicarbonate metabolic transformation (see Figure 9.5).[65]

The biodistribution of specific drugs and its related kinetics can also be probed *in vivo* using hyperpolarized MR. Lithium salts, a class of drugs used for the treatment of manic-depressive (bipolar) and depressive disorders (see also Chapter 15), has for instance been detected in the rat brain.[47]

Although hyperpolarization methods have been around for more than ten years, only a few specific substrates have been extensively studied and tested *in vivo*. Despite spectacular signal enhancements, current hyperpolarization techniques only partially achieve the required sensitivity for allowing real-time metabolism to be measured *in vivo* at physiological concentrations of the probe molecules.

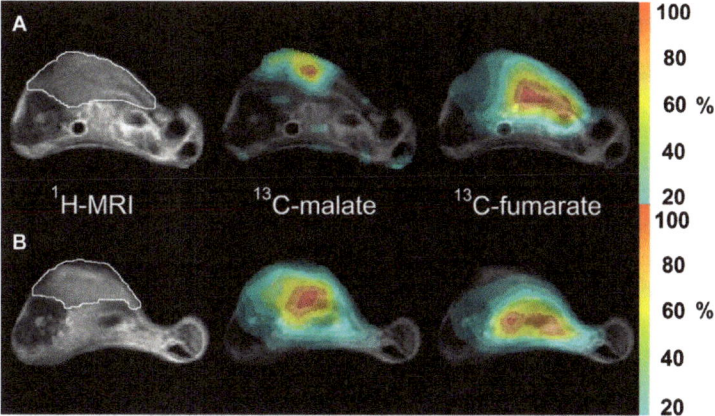

Figure 9.4   Representative transverse images from (A) untreated and (B) etoposide-treated mice with subcutaneously implanted lymphoma tumors. The proton image shows the anatomical location of the tumor, outlined in white. Adjacent to this proton image are false-color chemical shift images superimposed on the proton image, which demonstrate the spatial distribution of the total hyperpolarized ^{13}C-malate and ^{13}C-fumarate signals. The color scale indicates the relative signal intensity compared with the maximum intensity in each image. To reduce the effects of noise, voxels with signal intensities lower than 20% of the maximum signal intensity were removed from the image. Although the images have been scaled to their respective maxima, the maximum malate signal intensity in each experiment was 10–25% of the maximum fumarate signal. Reproduced with permission from Gallagher *et al.*[63] © 2009 National Academy of Sciences of the United States of America.

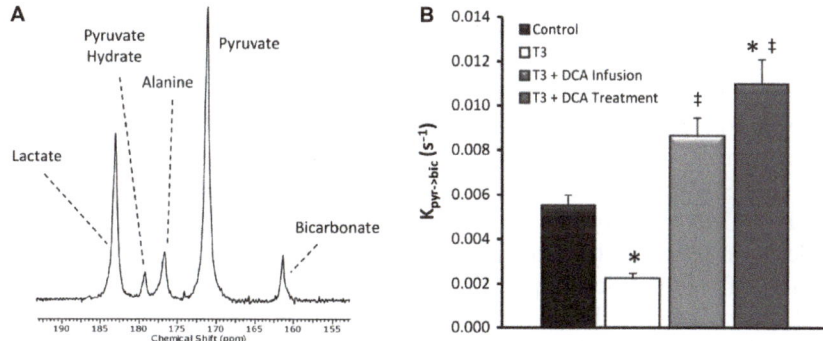

Figure 9.5   (A) Single representative magnetic resonance (MR) cardiac spectrum of a control rat at day 0 after infusion of [1-^{13}C]pyruvate recorded at $t = 10$ seconds. Pyruvate (and its equilibrium product, pyruvate hydrate), lactate, alanine and bicarbonate, metabolic products of pyruvate, are annotated. (B) The rate of exchange of the ^{13}C label from [1-^{13}C]pyruvate to [1-^{13}C]bicarbonate in each group assessed with hyperpolarized ^{13}C-MRS. Reproduced with permission from Atherton *et al.*[65] © 2011, American Heart Association, Inc.

9.3.3   Perspectives for Clinical Applications

One of the greatest promises of hyperpolarized MR is that it should be translatable to clinical applications. This technique is indeed not expected to have any adverse effects since it is based on endogenous substances and does not involve any ionizing radiation, as is the case with PET/CT techniques. It can also be combined with the multiple imaging capabilities that MRI offers. A white paper commissioned by the National Cancer Institute (NCI) of the National Institutes of Health (NIH) in the United States of America entitled *Analysis of Cancer Metabolism by Imaging Hyperpolarized Nuclei: Prospects for Translation to Clinical Research* was recently published,[66] and the first-in-human trial is in progress (expected to be completed by the end of 2012) at the University of California at San Francisco in patients diagnosed with prostate cancer. Beyond this proof of principles study, which mostly aims at testing how well the patients tolerate the injections of $[1-^{13}C]$pyruvate, several key points will need to be addressed for future clinical applications. It will be, for instance, important to reduce as much as possible the delay between the preparation step and the injection. This is not straightforward because of the necessity to filter out the polarizing agents or the catalyst (see Section 9.1.4). In addition, the preparation, transfer and injection of the hyperpolarized substrates have to be sufficiently reproducible to be useful for diagnosis purposes.

9.4   Conclusions

Hyperpolarization opened a wide new area of applications for MR and it is particularly useful to measure non-invasively fast metabolic processes. To date, it is the only technique allowing to probe metabolism *in vivo* in real time. Most diseases being associated to metabolic impairments, hyperpolarized MR could help in understanding the correlations between enzyme-catalyzed reactions and pathology in order to develop targeted therapy.

Dissolution DNP is remarkably versatile since it allows enhancing the polarization of several different types of nuclear spins in a wide variety of molecules. However, if the nuclei of interest can be hyperpolarized using PHIP, it would be advantageous to do so since it is much faster and the technique is simpler to implement. To date, most of the effort to develop hyperpolarization techniques for biomedical applications was focused on dissolution DNP, and it is still unclear whether PHIP (or possibly a brute force method) will be further developed for pre-clinical or even clinical applications. A tentative comparison between all three techniques is presented in Table 9.1.[§] Note that dissolution DNP is currently the only technique commercially available.

[§]The price for developing a brute force system was estimated by Dr. Giorgio Frossati (private communications). The estimation for a PHIP system was based on private communications with Dr. Eduard Y. Chekmenev. The price range for a dissolution DNP system includes the author's estimation based on the cost of the polarizers developed at the EPFL (see Figure 9.3) as well as the commercial prices suggested by the companies that can provide clinical or non-clinical systems.

Table 9.1 Comparison between all three hyperpolarization techniques presented in the present chapter.

	Preparation time	Costs[a]	Substrates selection	Availability	Suitable for use in humans
Brute force	>1 h[b]	0.4–0.9 M€	Unlimited	Not	Not currently
PHIP	~1 min.	0.05–0.15 M€	Limited	Limited	Not currently
DNP	~1 h	0.3–1 M€ (1.5–2.5 M€ for clinical systems)	Unlimited	Commercial	Yes

[a]These estimations include installation costs but not running costs.
[b]Based on estimation deduced from Eugeny V. Krjukov et al.[22]

It is generally accepted that one of the most important biomedical applications of hyperpolarized MR is to evaluate and monitor therapy. The technique might become an essential tool for testing and developing drugs targeting metabolic pathways and aiming at regulating metabolic dysfunctions. Pre-clinical studies performed in cells and animals could be potentially followed by clinical studies since hyperpolarized MR is a non-invasive, non-radioactive technique that is already being tested in humans.

Acknowledgements

The author wishes to thank Dr. Jacques van der Klink for his excellent suggestions, especially on spin thermodynamics, and for carefully reading this manuscript. The author is supported by the Swiss National Science Foundation (grant number PP00P2_133562).

References

1. C. Zwingmann and D. Leibfritz, *NMR Biomed.*, 2003, **16**, 370.
2. W. C. Stanley, F. A. Recchia and G. D. Lopaschuk, *Physiol. Rev.*, 2005, **85**, 1093.
3. E. M. Hillman, C. B. Amoozegar, T. Wang, A. F. H. McCaslin, M. B. Bouchard, J. Mansfield and R. M. Levenson, *Philos. Transact. A Math. Phys. Eng. Sci.*, 2011, **369**, 4620.
4. H. B. G. Casimir and F. K. du Pré, *Physica*, 1938, **5**, 507.
5. A. G. Redfield, *Phys. Rev.*, 1955, **98**, 1787.
6. A. Abragam, In *Proceedings of the International Conference on Polarized Targets and Ion Sources*, Centre d'Etudes Nucléaires de Saclay, Saclay, France, 1966.
7. C. G. Joo, K. N. Hu, J. A. Bryant and R. G. Griffin, *J. Am. Chem. Soc.*, 2006, **128**, 9428.
8. A. W. Overhauser, *Phys. Rev.*, 1953, **91**, 476.
9. T. R. Carver and C. P. Slichter, *Phys. Rev.*, 1953, **92**, 212.
10. K. H. Hausser, *J. Chim. Phys.*, 1964, **61**, 204.
11. K. H. Hausser and D. Stehlik, *Adv. Magn. Reson.*, 1968, **3**, 79.
12. N. Bloembergen, *Physica*, 1949, **15**, 386.

13. C. Griesinger, M. Bennati, H. M. Vieth, C. Luchinat, G. Parigi, P. Höfer, F. Engelke, S. J. Glaser, V. Denysenkov and T. F. Prisner, *Prog. Nucl. Magn. Reson. Spectrosc.*, 2012, **64**, 4.

14. M. D. Lingwood, T. A. Siaw, N. Sailasuta, B. D. Ross, P. Bhattacharya and S. G. Han, *J. Magn. Reson.*, 2010, **205**, 247.

15. J. G. Krummenacker, V. P. Denysenkov, M. Terekhov, L. M. Schreiber and T. F. Prisner, *J. Magn. Reson.*, 2012, **215**, 94.

16. J. Krummenacker, V. Denysenkov and T. Prisner, *Appl. Magn. Reson.*, 2012, **43**, 139.

17. K. Golman, I. Leunbach, J. S. Petersson, D. Holz and J. Overweg, *Acad. Radiol.*, 2002, **9**, S104.

18. D. J. Lurie, D. M. Bussell, L. H. Bell and J. R. Mallard, *J. Magn. Reson.*, 1988, **76**, 366.

19. C. R. Bowers and D. P. Weitekamp, *Phys. Rev. Lett.*, 1986, **57**, 2645.

20. C. R. Bowers and D. P. Weitekamp, *J. Am. Chem. Soc.*, 1987, **109**, 5541.

21. I. Waller, *Z. Phys.*, 1932, **79**, 370.

22. E. V. Krjukov, J. D. O'Neill and J. R. Owers-Bradley, *J. Low Temp. Phys.*, 2005, **140**, 397.

23. J. H. Ardenkjaer-Larsen, S. Macholl and H. Johannesson, *Appl. Magn. Reson.*, 2008, **34**, 509.

24. L. Lumata, M. E. Merritt, C. R. Malloy, A. D. Sherry and Z. Kovacs, *J. Phys. Chem. A*, 2012, **116**, 5129.

25. P. E. Larson, A. B. Kerr, A. P. Chen, M. S. Lustig, M. L. Zierhut, S. Hu, C. H. Cunningham, J. M. Pauly, J. Kurhanewicz and D. B. Vigneron, *J. Magn. Reson.*, 2008, **194**, 121.

26. A. Z. Lau, A. P. Chen, R. E. Hurd and C. H. Cunningham, *NMR Biomed.*, 2011, **24**, 988.

27. J. Leupold, S. Mansson, J. S. Petersson, J. Hennig and O. Wieben, *MAGMA*, 2009, **22**, 251.

28. C. von Morze, G. Reed, P. Shin, P. E. Z. Larson, S. Hu, R. Bok and D. B. Vigneron, *J. Magn. Reson.*, 2011, **211**, 109.

29. L. Lumata, S. J. Ratnakar, A. Jindal, M. Merritt, A. Comment, C. Malloy, A. D. Sherry and Z. Kovacs, *Chemistry*, 2011, **17**, 10825.

30. P. Miéville, P. Ahuja, R. Sarkar, S. Jannin, P. R. Vasos, S. Gerber-Lemaire, M. Mishkovsky, A. Comment, R. Gruetter, O. Ouari, P. Tordo and G. Bodenhausen, *Angew. Chem. Int. Ed. Engl.*, 2010, **49**, 6182.

31. T. Cheng, M. Mishkovsky, J. A. M. Bastiaansen, O. Ouari, P. Tordo, P. Hautle, B. Van den Brandt and A. Comment, *NMR Biomed.*, 2013, submitted.

32. E. Y. Chekmenev, V. A. Norton, D. P. Weitekamp and P. Bhattacharya, *J. Am. Chem. Soc.*, 2009, **131**, 3164.

33. M. Mishkovsky, T. Cheng, A. Comment and R. Gruetter, *Magn. Reson. Med.*, 2012, **68**, 349.

34. R. Sarkar, A. Comment, P. R. Vasos, S. Jannin, R. Gruetter, G. Bodenhausen, H. Hall, D. Kirik and V. P. Denisov, *J. Am. Chem. Soc.*, 2009, **131**, 16014.

35. M. Mishkovsky, A. Comment and R. Gruetter, *J. Cereb. Blood Flow Metab.*, 2012, **32**, 2108.
36. P. R. Vasos, A. Comment, R. Sarkar, P. Ahuja, S. Jannin, J. P. Ansermet, J. A. Konter, P. Hautle, B. van den Brandt and G. Bodenhausen, *Proc. Natl Acad. Sci. USA*, 2009, **106**, 18469.
37. W. S. Warren, E. Jenista, R. T. Branca and X. Chen, *Science*, 2009, **323**, 1711.
38. G. Frossati, *J. Low Temp. Phys.*, 1998, **111**, 521.
39. S. Jannin, A. Comment, F. Kurdzesau, J. A. Konter, P. Hautle, B. van den Brandt and J. J. van der Klink, *J. Chem. Phys.*, 2008, **128**, 241102.
40. H. Johanneson, S. Macholl and J. H. Ardenkjaer-Larsen, *J. Magn. Reson.*, 2009, **197**, 167.
41. J. H. Ardenkjaer-Larsen, B. Fridlund, A. Gram, G. Hansson, L. Hansson, M. H. Lerche, R. Servin, M. Thaning and K. Golman, *Proc. Natl Acad. Sci. USA*, 2003, **100**, 10158.
42. C. Gabellieri, S. Reynolds, A. Lavie, G. S. Payne, M. O. Leach and T. R. Eykyn, *J. Am. Chem. Soc.*, 2008, **130**, 4598.
43. F. A. Gallagher, M. I. Kettunen and K. M. Brindle, *Progr. Nucl. Magn. Reson. Spectr.*, 2009, **55**, 285.
44. L. Lumata, M. E. Merritt, Z. Hashami, S. J. Ratnakar and Z. Kovacs, *Angew. Chem. Int. Ed. Engl.*, 2012, **51**, 525.
45. M. E. Merritt, C. Harrison, Z. Kovacs, P. Kshirsagar, C. R. Malloy and A. D. Sherry, *J. Am. Chem. Soc.*, 2007, **129**, 12942.
46. M. Mishkovsky and L. Frydman, *ChemPhysChem*, 2008, **9**, 2340.
47. R. B. van Heeswijk, K. Uffmann, A. Comment, F. Kurdzesau, C. Perazzolo, C. Cudalbu, S. Jannin, J. A. Konter, P. Hautle, B. van den Brandt, G. Navon, J. J. van der Klink and R. Gruetter, *Magn. Reson. Med.*, 2009, **61**, 1489.
48. M. Goldman, H. Johannesson, O. Axelsson and M. Karlsson, *Magn. Reson. Imaging*, 2005, **23**, 153.
49. K. Golman, O. Axelsson, H. Johannesson, S. Mansson, C. Olofsson and J. S. Petersson, *Magn. Reson. Med.*, 2001, **46**, 1.
50. J. B. Hovener, E. Y. Chekmenev, K. C. Harris, W. H. Perman, L. W. Robertson, B. D. Ross and P. Bhattacharya, *MAGMA*, 2009, **22**, 111.
51. P. Bhattacharya, E. Y. Chekmenev, W. H. Perman, K. C. Harris, A. P. Lin, V. A. Norton, C. T. Tan, B. D. Ross and D. P. Weitekamp, *J. Magn. Reson.*, 2007, **186**, 150.
52. R. W. Adams, J. A. Aguilar, K. D. Atkinson, M. J. Cowley, P. I. Elliott, S. B. Duckett, G. G. Green, I. G. Khazal, J. López-Serrano and D. C. Williamson, *Science*, 2009, **323**, 1708.
53. K. D. Atkinson, M. J. Cowley, P. I. P. Elliott, S. B. Duckett, G. G. R. Green, J. Lopez-Serrano and A. C. Whitwood, *J. Am. Chem. Soc.*, 2009, **131**, 13362.
54. O. Warburg, *Science*, 1956, **123**, 309.
55. M. E. Merritt, C. Harrison, C. Storey, F. M. Jeffrey, A. D. Sherry and C. R. Malloy, *Proc. Natl Acad. Sci. USA*, 2007, **104**, 19773.

56. H. S. Venkatesh, M. M. Chaumeil, C. S. Ward, D. A. Haas-Kogan, C. D. James and S. M. Ronen, *Neuro. Oncol.*, 2012, **14**, 315.

57. M. H. Lerche, S. Meier, P. R. Jensen, S. O. Hustvedt, M. Karlsson, J. O. Duus and J. H. Ardenkjaer-Larsen, *NMR Biomed.*, 2011, **24**, 96.

58. M. Roth, J. Bargon, H. W. Spiess and A. Koch, *Magn. Reson. Chem.*, 2008, **46**, 713.

59. K. Golman, R. in't Zandt and M. Thaning, *Proc. Natl Acad. Sci. USA*, 2006, **103**, 11270.

60. S. E. Day, M. I. Kettunen, F. A. Gallagher, D. E. Hu, M. Lerche, J. Wolber, K. Golman, J. H. Ardenkjaer-Larsen and K. M. Brindle, *Nat. Med.*, 2007, **13**, 1382.

61. C. S. Ward, H. S. Venkatesh, M. M. Chaumeil, A. H. Brandes, M. Vancriekinge, H. Dafni, S. Sukumar, S. J. Nelson, D. B. Vigneron, J. Kurhanewicz, C. D. James, D. A. Haas-Kogan and S. M. Ronen, *Cancer Res.*, 2010, **70**, 1296.

62. P. Seth, A. Grant, J. Tang, E. Vinogradov, X. Wang, R. Lenkinski and V. P. Sukhatme, *Neoplasia*, 2011, **13**, 60.

63. F. A. Gallagher, M. I. Kettunen, D. E. Hu, P. R. Jensen, R. I. Zandt, M. Karlsson, A. Gisselsson, S. K. Nelson, T. H. Witney, S. E. Bohndiek, G. Hansson, T. Peitersen, M. H. Lerche and K. M. Brindle, *Proc. Natl Acad. Sci. USA*, 2009, **106**, 19801.

64. S. E. Bohndiek, M. I. Kettunen, D. E. Hu, T. H. Witney, B. W. Kennedy, F. A. Gallagher and K. M. Brindle, *Mol. Cancer Ther.*, 2010, **9**, 3278.

65. H. J. Atherton, M. S. Dodd, L. C. Heather, M. A. Schroeder, J. L. Griffin, G. K. Radda, K. Clarke and D. J. Tyler, *Circulation*, 2011, **123**, 2552.

66. J. Kurhanewicz, D. B. Vigneron, K. Brindle, E. Y. Chekmenev, A. Comment, C. H. Cunningham, R. J. Deberardinis, G. G. Green, M. O. Leach, S. S. Rajan, R. R. Rizi, B. D. Ross, W. S. Warren and C. R. Malloy, *Neoplasia*, 2011, **13**, 81.

67. A. Comment, B. van den Brandt, K. Uffmann, F. Kurdzesau, S. Jannin, J. A. Konter, P. Hautle, W. T. H. Wenckebach, R. Gruetter and J. J. van der Klink, *Concepts Magn. Reson.*, 2007, **31B**, 255.

CHAPTER 10

Combined PET/MRI for Improving Quantitative Imaging

TIM D. FRYER

Wolfson Brain Imaging Centre, Department of Clinical Neurosciences, Hills Rd, University of Cambridge, Cambridge, CB2 0QQ, UK
Email: tdf21@wbic.cam.ac.uk

10.1 Introduction

Positron emission tomography (PET) is an imaging technique that probes different aspects of *in vivo* physiology using radio-labelled tracers. Due to the wide variety of available tracers, high sensitivity and quantitative imaging ability, PET is the most powerful functional imaging technique. However, the anatomical information provided by PET is tracer dependent and limited at best, and the spatial resolution is lower than that provided by X-ray computerized tomography (CT) and magnetic resonance imaging (MRI). Consequently, PET images have long been combined with anatomical images from CT and MRI.

The most mature application of PET is brain imaging, where the ability to reliably co-register PET and MRI has led to the use of MRI-guided anatomical region-of-interest (ROI) definition and partial volume correction. Anatomically driven image reconstruction using MRI has also been the subject of research for brain PET, although it is not widely used for clinical studies.

Outside the brain, PET is usually combined with CT, with the data for both modalities being acquired with a combined PET/CT system,[1] which has been commercially available for just over a decade. A key driver for the development of combined PET/CT systems was the desire to overcome the problem of

New Developments in NMR No. 2
New Applications of NMR in Drug Discovery and Development
Edited by Leoncio Garrido and Nicolau Beckmann
© The Royal Society of Chemistry 2013
Published by the Royal Society of Chemistry, www.rsc.org

registering PET and CT images acquired on different scanners. Unlike the brain, where data from separate PET and MRI/CT sessions can be registered using rigid-body algorithms, other parts of the body are subject to non-rigid motion, which provides a far more difficult registration task.[2] Although the data are acquired sequentially, PET/CT scanners attempt to minimise the motion between the PET and CT data. An assumption is generally made that there is no motion between PET and CT data acquisition, and hence CT data are used for photon attenuation correction of PET, one of the most important corrections applied to achieve quantitative PET images. In addition, attenuation information is also used for the most commonly used scatter correction technique. However, motion does occur, with a particular issue being respiratory motion in the thorax.

Consequently, one of the most attractive features of simultaneous anatomical-functional imaging is minimization of motion between data sets from the different modalities, thereby providing the best possibility for data registration, especially outside the brain. Other advantages include the ability to image at the same physiological state, which could be especially attractive for studies conducted during drug treatment, and a reduction in overall scan time, which facilitates increased patient throughput.

As PET scans for pharmacokinetic analysis typically last 1–3 hours, with diagnostic scans usually taking approximately 30 minutes, simultaneous PET/CT is not a viable option due to the radiation dose from prolonged CT scanning. Combined PET/MRI provides a more attractive solution for simultaneous anatomical-functional imaging. Some combined PET/MRI systems, however, adopt the sequential imaging approach of PET/CT. Irrespective of the acquisition mode, the ultimate role of combined PET/MRI within the *in vivo* imaging portfolio has been the subject of much debate.[3–11]

This chapter will first consider combined PET/MRI system designs and then concentrate on areas where MR information can be used in the quantification of PET data, before concluding on clinical/pre-clinical applications of combined PET/MRI.

10.2   Combined PET/MRI Designs

Combined PET/MRI designs can be broadly classified into sequential and simultaneous imaging systems. The sequential approach obviates the need to change the PET detector technology from the traditional photo-multiplier tube (PMT) readout of scintillation light by ensuring that the PET detectors are located far enough away from the high magnetic field of the MR system (Figure 10.1).[12,13] As with PET/CT, the two systems are connected by a common patient bed. Sequential PET/MRI clinical systems also include a combination of PET/CT with MR (Trimodality, GE Healthcare, Waukesha, WI, USA), which, in addition to enhanced diagnostic ability, facilitates attenuation correction from CT data. A sequential PET/MRI system is also available for pre-clinical research (nanoScan, Mediso Medical Imaging Systems, Budapest, Hungary).

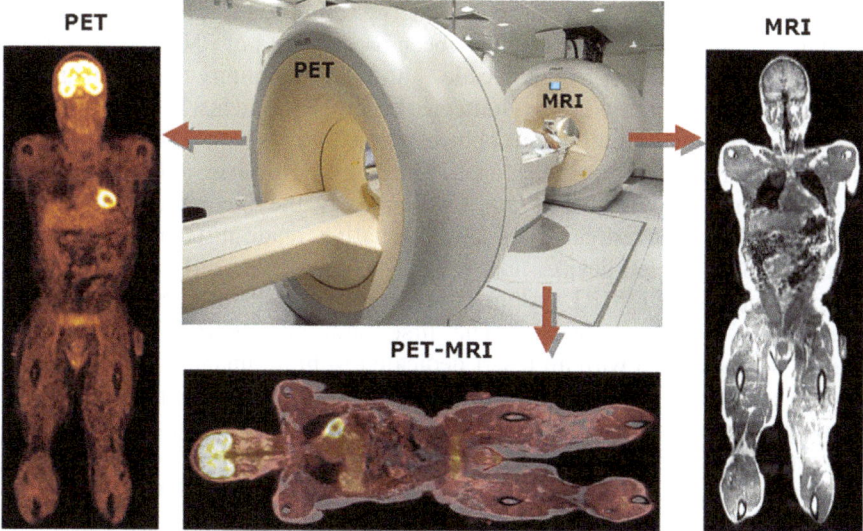

Figure 10.1 Sequential combined PET/MRI system (Philips Ingenuity, Philips Healthcare, Best, The Netherlands).
Image reproduced with permission from Zaidi and Del Guerra.[13] © 2011 American Association of Physicists in Medicine.

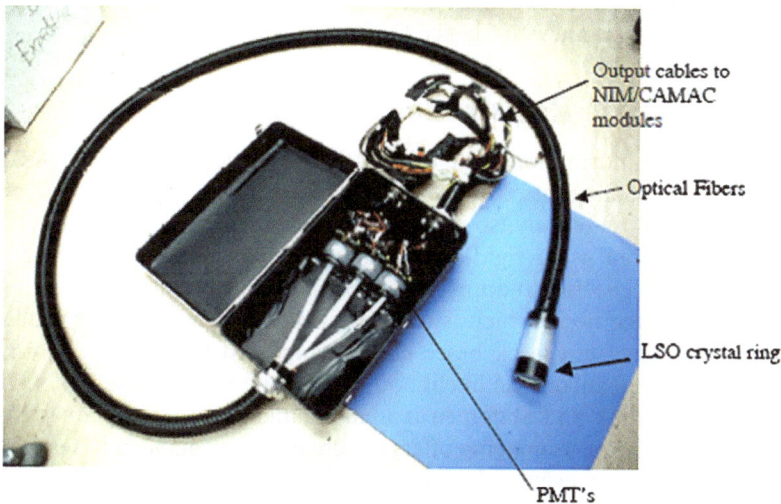

Figure 10.2 The first MR-compatible PET detector system (McPET).
Image reproduced with permission from Slates *et al.*[16] © 1999 IOP Publishing.

Simultaneous PET/MRI systems require a change to be made to the standard PET detector. The first prototype capable of simultaneous PET/MRI (McPET) used optical fibers to remove the PMTs from the MR environment (Figure 10.2).[14–16] The downside of this approach is the loss of light along the fibers, which results in degradation of energy, timing and spatial resolutions. Avalanche photo-diodes (APDs) have been used instead of PMTs in some pre-clinical PET scanners for over a decade,[17] and their immunity to high magnetic field has led to their use in simultaneous PET/MRI systems,[3,4,18,19] including the first commercial simultaneous clinical whole body system (Biograph mMR, Siemens Medical Solutions, Erlangen, Germany). However, APDs have low internal gain and slow signal output. Consequently, the silicon photo-multiplier (SiPM) is being investigated as an alternative readout technology[20] and prototype SiPM-based devices now exist.[21–23]

Simultaneous PET/MRI systems can be sub-divided into removable PET inserts and integrated devices. The first simultaneous PET/MRI clinical image[18] was acquired with an insert dedicated to brain imaging, which can be used in a standard MR system. More recent designs aimed at whole body imaging adopt an integrated approach, with the PET detectors incorporated into the construction of the MR hardware.[24]

A unique design has been used for pre-clinical imaging, whereby a small PET detector ring is fixed to the rat skull.[25] The RatCAP thereby inherently eliminates motion between the brain and the PET detectors for studies on conscious animals, which are of interest not only because any confounding influence of anesthesia is avoided but also because such studies open up the possibility of imaging responses to stimuli (see also Chapter 7, Section 7.4.1, for a discussion about fMRI studies in awake animals). The APD readout allows the system to be used in an MR scanner and hence fMRI is possible.

10.3   PET/MR Registration

In order for MR information to be used in the quantification of PET, the data from the two modalities need to be spatially registered. The skull and brain essentially form a rigid body, so images from separate PET and MRI systems can be co-registered using rigid-body registration. Typically brain PET and MRI can be co-registered to approximately 1 mm in each direction.[2]

In imaging of the neck, which is of interest not only for oncology but also cardiovascular disease, a neck brace can be employed to minimize non-rigid motion, thereby enhancing the likelihood of accurate registration.[26] Non-rigid motion in the thorax and abdomen is unavoidable and these are anatomical areas where simultaneous imaging offers the greatest advantage in terms of data registration. However, even with simultaneous acquisition the registration will be imperfect due to shim effects and frequency shifts affecting the localization of the MR signal.

10.4   Attenuation Correction

In order to obtain quantitatively accurate images, correction must be made for photon attenuation. The latter process is the loss of coincident photon pairs from their original path through scattering or absorption. PET-only systems use rotating radioactive rod or point sources to acquire two scans, one with the object being imaged in the field of view (transmission scan) and the other of an empty field of view (blank scan). The ratio of the blank-to-transmission scans along a line-of-response (LOR) gives the multiplicative factor needed to correct for attenuation. If the radioactive source used is a positron-emitter, the photon energy matches that of the radiotracer and the attenuation correction factors are quantitatively accurate (albeit contaminated with noise).

PET/CT scanners replace rotating radioactive sources with CT data; a CT image is a map of attenuation coefficients. However, the effective energy of the polyenergetic CT X-ray beam is much lower than that of the annihilation photons detected in PET. Consequently, as photon attenuation is energy dependent, the CT attenuation coefficients need to be scaled to estimate those for annihilation photons.[27] This process is inexact due to insufficient information about the atomic number and density of the attenuating materials; this can be improved using dual energy CT.[28,29]

Although there are some inaccuracies in CT-derived attenuation correction, obtaining accurate attenuation information from MR is potentially more prone to error as the MR signal is independent of photon attenuation. Furthermore, with standard MR sequences there is little or no signal obtained from bone and air. The subsequent difficulty in differentiating bone and air is problematic because photon attenuation from air is negligible, while bone has the highest attenuation coefficient of any tissue class.

The most advanced anatomical area for MR-based attenuation correction is the brain, where segmentation into tissue classes[18,30–35] and non-rigid registration of attenuation coefficient templates[36–43] have been used. To improve the segmentation of air and bone, which for brain imaging is especially problematic in the sinuses, ultra-short echo time (UTE) sequences have been employed (Figure 10.3).[33–35,44] A recent study demonstrated that reconstructed

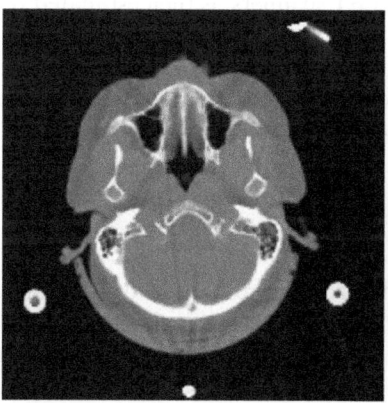 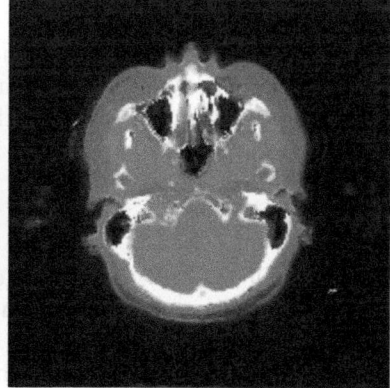

Figure 10.3 Transverse cranial CT (left) and pseudo-CT created from UTE MR (right). Note that the stereotactic frame and head holder on the CT scan was not modelled; the purpose was to illustrate the ability to model cranial CT from UTE MR.
Image reproduced with permission from Johansson *et al.*[44] © 2011 American Association of Physicists in Medicine.

radioactivity concentrations in brain tissue from a combined UTE/Dixon[45] four tissue class attenuation correction approach correlated tightly ($r^2 = 0.99$) with those produced using CT-based attenuation correction.[35]

More recently, work has been presented on MR-based attenuation correction for whole body imaging,[46–59] including the impact of surface coils and truncation artifacts from a limited MR transaxial field-of-view. Non-rigid registration of MR to CT in order to produce a pseudo-CT from MR for attenuation correction was not found to be as successful as the aforementioned non-rigid approaches for MR-based attenuation correction in the brain.[46] Recently, Dixon MR has been applied to differentiate air, lung, soft tissue and adipose tissue.[59] Refinement of this model, and ones similar to it that also differentiate bone, could be made using a joint estimation reconstruction algorithm, where both radioactivity concentrations and attenuation coefficients are updated to best fit the data.[55] The latter approach may be especially useful in the problematic lung region, which has a spectrum of attenuation coefficients due to its variable composition, and also bone, as a single attenuation co-efficient cannot be attributed to all types of bone.

Accurate attenuation information is not only important for attenuation correction but also for the most commonly used scatter correction technique – scatter modelling.[60] Scatter refers to photons that have scattered in the object being imaged, then been detected. Compared to unscattered photons, these events have lower spatial resolution, and they degrade image quantification. In whole body imaging, scatter can typically comprise 40% of the detected events, so correction is necessary. Scatter modelling analytically determines the single scatter distribution from an initial radioactivity concentration image and an attenuation coefficient map using the Klein–Nishina formula.[61] Multiple scatter, which constitutes a minor fraction of the total scatter for the narrow energy window typically used on modern PET scanners, is then estimated from the single scatter distribution.

In summary, accurate attenuation information is needed for both attenuation and scatter corrections. Encouraging results have recently been reported for brain imaging, but attenuation correction in the thorax is likely to be more problematic.

10.5   Motion Correction

Motion of the subject can significantly degrade the resolution of PET images and lead to quantification errors, with one issue being lack of registration between emission and attenuation data. The long duration of PET scans for pharmacokinetic analysis (up to 3–4 hours) makes these studies especially vulnerable to motion-induced errors; motion can significantly distort the tissue time-activity curve (TAC) data that are inputted into kinetic analysis algorithms. Similarly, motion also causes errors with image-based input function estimation by distorting the TAC extracted from a blood vessel or the left ventricular/atrial chamber. Consequently, motion correction methods have been investigated.

The simplest of these is registration of the PET images; for a dynamic scan intended for kinetic analysis up to approximately 70 images may be reconstructed to provide adequate temporal sampling. However, although this realignment reduces resolution loss, quantification errors due to emission/attenuation mismatch remain. For brain imaging, where the motion is well characterized as rigid motion, LOR rebinning has been applied prior to image reconstruction to address emission/attenuation mismatches, in addition to improving image resolution.[62–66] A number of studies have applied this technique using motion parameters determined with an infrared system, although this system requires a marker plate to be held firmly attached to the top of the skull.[64] As an alternative, LOR rebinning using MRI-derived motion information has recently been reported for a simultaneous brain PET/MRI system.[66] Non-rigid motion, however, cannot be corrected by LOR rebinning.

The main area where non-rigid motion correction has been applied is respiratory motion. For the latter, respiratory gating of the acquired data has been combined with either post-reconstruction registration (PRR) or motion-compensated image reconstruction (MCIR).[67–82] PRR reconstructs the data for each gate separately, ideally with matching attenuation information, then applies non-rigid registration to warp the gated images to one of the gates (reference gate) and averages the result to produce a high-resolution, low-noise image. MCIR directly reconstructs an image using all the gated data by embedding the motion information into the image reconstruction algorithm. Both techniques require accurate motion parameters, whose determination from PET data is hampered by noise and relatively poor anatomical detail, which is tracer dependent. Consequently, for PET/CT scanners, the deformation fields are most commonly determined from gated CT data, which ideally match the gated PET data. As the PET and CT data are acquired sequentially, variation in respiratory motion, which always occurs to a certain extent, leads to errors. There is also potential for non-respiratory motion of the patient between PET and CT. Ideally, the information used for motion correction is determined simultaneously with PET and, consequently, simultaneous PET/MRI systems have the potential to provide gold standard motion correction when non-rigid motion is involved.

A key issue with MR-derived motion correction is how to combine motion monitoring with diagnostic MRI and/or magnetic resonance spectroscopy (MRS), as continuously acquiring MR data for motion correction is incompatible with the desire to acquire diagnostic MR information during the PET scan. Furthermore, continuous monitoring will result in a large number of volumes, which need to be non-rigidly registered to a reference image; this will pose a significant computational burden. One solution to the latter two issues may be to acquire MR volumes over a number of respiratory cycles, thereby providing an opportunity to sample variations in respiration, and combine this with non-MR motion monitoring during PET to gate data into one of the MR volumes. The quality of fast MRI may also be an issue (Figure 10.4),[82] although tagged-MRI shows promise outside the lung.[83]

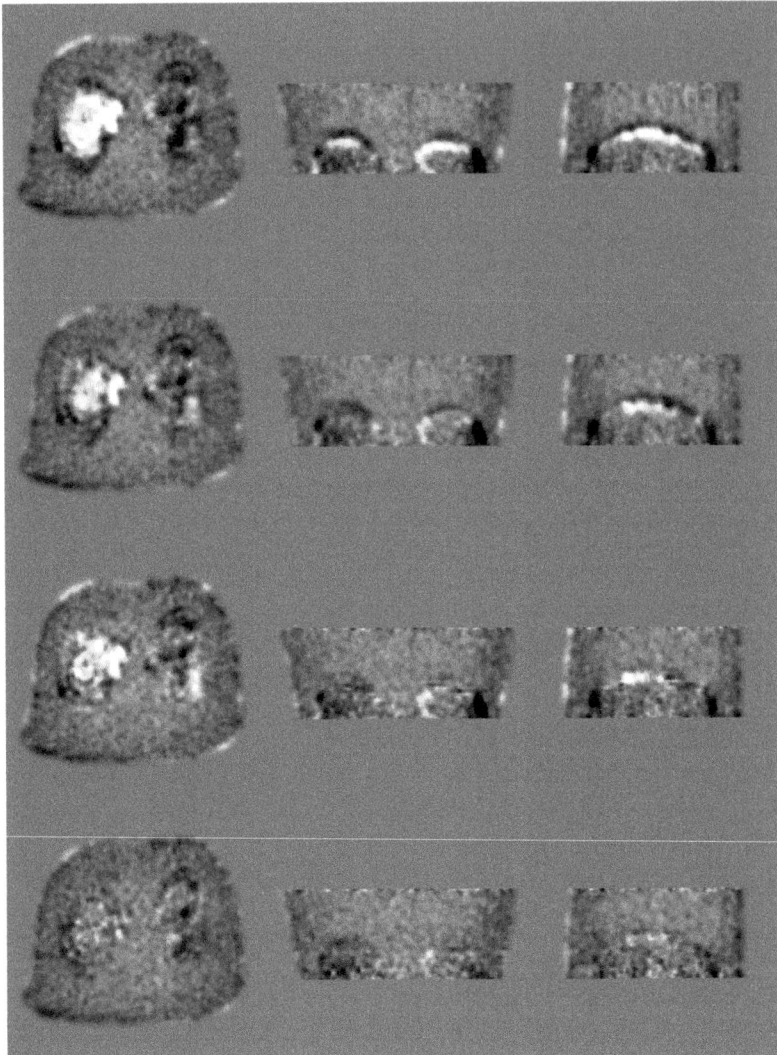

Figure 10.4 Transverse (left), coronal (centre) and sagittal (right) sections through
subtraction images between a reference gate PET image and PET images
produced from all respiratory gates with: (top row) no motion
correction; (second row) MR-based motion correction using PRR; (third
row) MR-based motion correction using MCIR; (bottom row) noise-free
PET-based motion correction using MCIR, which acts as a gold
standard. Improvements are seen with MR-based motion correction at
the lung/liver and lung/spleen boundaries, less so at soft tissue
boundaries of the liver and spleen.
Image reproduced with permission from Dikaios *et al.*[82] © 2012
Springer.

10.6 Anatomically Guided Image Reconstruction

Prior distributions are often utilized in iterative image reconstruction algorithms to encourage the reconstructed image to conform to *a priori* expectations about the local voxel intensity distribution, *i.e.* they act to regularize the solution. Most typically, a spatially invariant prior function is used but spatially variant priors can be used, with one class of these being anatomically guided priors, which are usually designed to penalize voxel intensity discontinuities within a tissue class but not at boundaries between tissue classes. These methods include resolution modelling to ensure that the edges encouraged by the prior do not conflict with the smooth boundaries in the data. The aim of these algorithms is to regularize noise in homogenous uptake areas, while also performing resolution recovery and hence reducing partial volume error. A study comparing the use of anatomical information during or post-reconstruction found fundamental advantages with the former.[84]

The concept of anatomically driven image reconstruction is not new[84–95] but one issue is that this approach requires accurate registration of PET and anatomical data. Hence, combined PET/MRI, especially simultaneous PET/MRI, is attractive in this regard, particularly for applications outside the brain. A major issue, however, is that the PET signal is not guaranteed to be homogeneous within a tissue class, which is especially true in disease. If PET heterogeneity is suspected *a priori* within a tissue class, the weighting of the prior and/or the form of the penalty function can be altered for that class; this has been applied for [^{18}F]fluoro-deoxyglucose (FDG) brain PET in epilepsy, where focal hypometabolism in grey matter is a signature.[96] To obviate the need for segmentation, priors have been applied that operate in local anatomical neighborhoods defined by similarity metrics rather than tissue segments.[97]

Little clinical verification of anatomically driven image reconstruction algorithms exists; however, one study indicated superiority over post-smoothed maximum likelihood reconstruction.[96]

10.7 Anatomically Driven Post-Reconstruction Resolution Recovery and Denoising

As described in the previous section, resolution modelling can be incorporated into the image reconstruction algorithm, which leads to a degree of resolution recovery and hence a reduction in partial volume error. Alternatively, partial volume correction can be applied post-reconstruction, both regionally[98–100] and at the voxel level.[101–106]

The most popular regional partial volume correction method is the geometric transfer matrix (GTM) approach.[98] The latter assumes that the image can be segmented into regions within which the radioactivity concentration is uniform. Using knowledge of the point spread function, the contribution made by each region to the signal in each region can be determined and estimates of true regional radioactivity concentration are obtained through matrix inversion.

For brain imaging, MRI is usually employed to define the regions, based on *a priori* assumptions of radioactivity homogeneity within certain anatomical domains. The method is sensitive to registration errors, so improvements in registration offered by combined PET/MRI will improve the accuracy of this approach.

The ill-conditioned nature of the GTM method, coupled with the need for matrix inversion, makes it unsuitable for partial volume correction at the voxel level. Anatomically driven voxel-wise approaches have been developed, which aim to provide superior results to deconvolution methods, such as Richardson–Lucy[107,108] and Van Cittert.[109] A notable class of algorithms are those that utilize wavelet decomposition to propagate high-frequency information from an anatomical image into a registered PET image (Figure 10.5).[103–105] A key issue with this approach is when to utilize the anatomical high-frequency information, due to mismatches between the

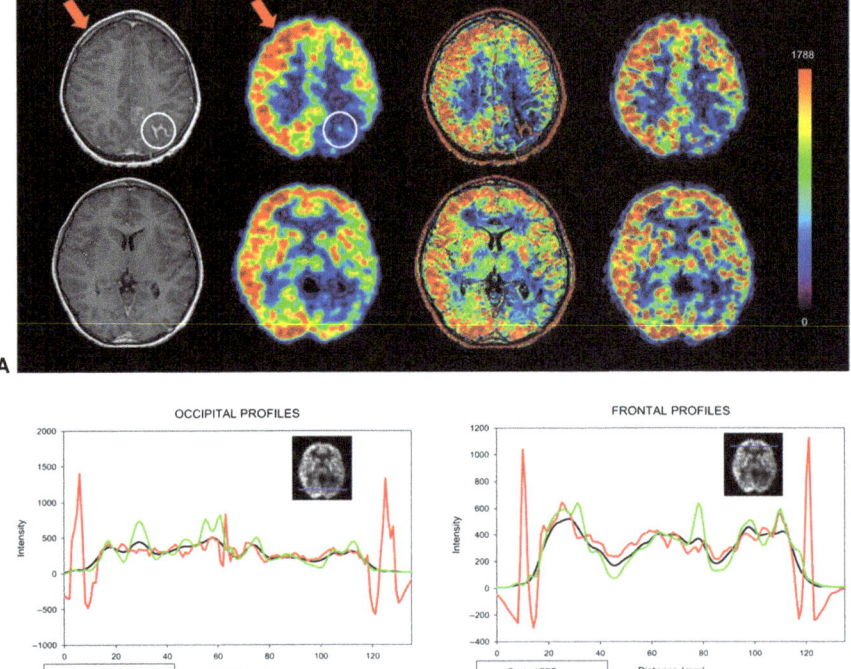

Figure 10.5   Comparison of raw FDG PET images (A, column 2) with images partial volume corrected using mutual multiresolution analysis with 2D global scaling (A, 3) and 3D local scaling (A, 4). Note that the global scaling for the 2D method propagates artifactual high-frequency detail (circled) from the gadolinium contrast MR (A, 1) into the PET image. Part B gives profiles through the three PET images shown in A.
Image reproduced with permission from Le Pogam *et al.*[105] © 2011 American Association of Physicists in Medicine.

radioactivity and anatomical distributions. Local scaling based on the ratio of PET and anatomical signals in the wavelet domain has been used, although further improvements have been demonstrated for brain imaging using PET image values in a set of probabilistic ROIs to create an anatomical map that better matches the PET distribution.[104] However, noise amplification still occurs, so it is recommended that the latter method be applied to parametric maps, rather than individual PET images from a dynamic series, even though there is an error in doing so as the partial volume error actually occurs in the individual PET images. Outside the brain, the aforementioned approach using probabilistic ROIs is (currently) not feasible, so recent developments for local scaling include replacing scaling factors determined voxel-wise with local median values that reduce the sensitivity to misregistration.[105]

It is worthy of note that the above wavelet approach can also be used for voxel-wise denoising of PET by first smoothing the registered anatomical image to match the resolution of PET.[110]

10.8   Image-Based Input Function Estimation

Gold standard analysis of radiotracer pharmacokinetics is provided by blood-based methods. Reference tissue algorithms for kinetic analysis exist but these require verification with blood-based approaches. The input to the tissue is provided by arterial blood, rather than venous, so blood sampling requires arterial puncture. The latter requires more specialist expertise than obtaining venous access, it carries a risk of complications and it is a deterrent to patient recruitment. For pre-clinical imaging, blood sampling is feasible in rats and larger species but very difficult in mice. Consequently, in situations where reference tissue analysis is not suitable and hence blood-based methods must be applied, alternative approaches for determining the blood input function are desirable.

One option that has been explored is determination of whole blood radioactivity concentration values from the PET images. There are, however, many issues with this approach, in addition to the aforementioned propensity to motion-induced error. First, even if a blood vessel can be accurately delineated, the signal within it is often heavily affected by partial volume error due to the small size of the vessel and the contrast in radioactivity concentration between blood and surrounding tissue. Use of the left ventricular or atrial chamber instead of a blood vessel should reduce the partial volume error but the heart will often not be present in the PET field-of-view. Secondly, the blood radioactivity concentrations estimated from the images will be significantly noisier than those obtained by counting blood samples with a gamma well counter. The situation will be even worse following partial volume correction, as this amplifies noise. Thirdly, although whole blood radioactivity is used for blood volume correction, the input function required for the tissue signal is the available radiotracer in plasma. Hence, image-based approaches will require multiplication of the whole blood radioactivity concentrations with population-based plasma-to-whole blood ratios. Furthermore, the majority of

PET radiotracers have radio-labelled metabolites that form part of the blood signal. So, in addition to multiplication by plasma-to-whole blood ratios, the image-derived blood signal will require multiplication by the parent radiotracer fraction in plasma over time. For some tracers, the use of population-based parent radiotracer fraction values has been shown to be sub-optimal compared to determining this information per subject. Finally, for some kinetic parameters, correction is also made for tracer binding to plasma proteins, which again necessitates blood sampling. So, in the vast majority of cases, image-based determination of whole blood radioactivity concentrations will have to be combined with information from blood samples, which in some cases needs to be determined per subject. The limitations of image-based input functions have recently been reported.[111–113]

Nevertheless, provided accurate registration is achieved, MR images acquired with a combined PET/MRI system could be used to delineate arterial ROIs on the PET images. In addition, the aforementioned MRI-guided resolution recovery techniques (*i.e.* during reconstruction or post-reconstruction) could be used to reduce partial volume error of the blood signal. However, for the image reconstruction approach one would need to be careful that the positivity constraint inherent in most likelihood functions does not result in positively biased blood values at late time points for tracers with rapid clearance.

Instead of using MRI to improve an input function determined from PET images, it has been postulated that an MR contrast agent could be co-injected with the radiotracer and an MR input function determined, which is then converted to a PET input function. However, the fact that the radiotracer and MR contrast agent will have different blood clearances, which are likely to vary independently on a per subject basis, means that producing a PET input function from an MR input function may not provide any advantage over the use of a population-based, blood sampling derived PET input function scaled for each subject using late venous samples.

10.9   Combined Functional Imaging

In addition to providing anatomical information to aid PET quantification and visual interpretation, combined PET/MRI systems will be capable of dual-modality functional imaging. The latter is possible with separate PET and MRI systems, particularly in brain applications, but combined PET/MRI systems may be the catalyst for an increase in such studies because of the enhanced image registration (especially outside the brain), the ability to image under the same physiological conditions and improved logistics.

For drug studies, there is likely to be interest in conducting fMRI and radioligand PET during the same drug exposure, especially if the occupancy of the drug is not sufficiently temporally invariant to allow unbiased sequential PET and MRI. Performing gold standard quantitative PET blood flow and/or oxygen metabolism scans simultaneously with fMRI may also be useful in interpreting fMRI findings.

Combined PET/MRI systems have also provoked interest in dual PET/MRI probes, although few currently exist. A number of those under development are based on nanoparticles,[114,115] with, for example, a superparamagnetic nano-particle providing the MR contrast agent, the nanoparticle shell incorporating a positron-emitting radionuclide, and a targeting moiety being attached to the nanoparticle. One issue with these dual-modality probes is the low sensitivity of MR (typically 1000 higher molar amounts must be administered compared to PET to achieve comparable sensitivity), so to provide enough MR signal the administered mass may violate the tracer principle for saturable systems (*e.g.* receptors), thereby providing biased information on the biological system under investigation. In addition, for certain compounds injecting such large masses will be prohibited due to the risk of side effects.

The ability of MRS to differentiate metabolites, which PET cannot, implies that acquiring MRS simultaneously with PET may provide sufficient information to allow compartmental modelling of PET tracers for which such analysis is otherwise intractable.[116] The potential of this approach is accen-tuated by the fact that positron-emitting and spectroscopic nuclides exist for both carbon (^{11}C and ^{13}C) and fluorine (^{18}F and ^{19}F), so pharmacologically similar compounds can be imaged with PET and MRS. The sensitivity of MRS, however, may limit the temporal resolution of the metabolite information.

Finally, functional PET and MRI could be combined in a parametric fusion approach. One recent study applied [^{11}C]choline PET and diffusion tensor MRI to the identification of prostate cancer.[117] Through comparison with histology, it was demonstrated that a fused parameter derived from the standardized uptake value (SUV) of [^{11}C]choline and the apparent diffusion coefficient (ADC) improved detection of high grade tumors compared to separate PET and MRI parameters.

10.10 Clinical/Pre-Clinical Applications

Although the first prototype simultaneous PET/MRI device was demonstrated 15 years ago, combined PET/MRI clinical systems have only been available since 2006 for brain imaging and 2010 for whole body imaging, with fully operational pre-clinical combined PET/MRI reported in 2008. Consequently, demonstration of the diagnostic ability of combined PET/MRI systems is still in its infancy. Studies reported so far include head and neck cancer,[118,119] prostate cancer,[117,120] brain tumours,[121,122] cardiac metabolism,[123] inflam-mation following myocardial infarction[124] and rheumatoid arthritis.[125] Commentaries also advocate use in infectious and inflammatory diseases,[10] and cardiovascular imaging.[11]

10.11 Conclusion

Combined PET/MRI systems are now commercially available. For brain imaging, data acquired on separate PET and MRI systems have been combined for many years, with, in particular, MRI used to delineate anatomical regions.

Dual-modality functional data have also been acquired using separate PET and MRI systems, primarily in the brain. So, "combined PET/MRI" is not new. The main advantage of combined PET/MRI systems, especially outside the brain with simultaneous data acquisition, is improved spatial registration of the PET and MR data, which will improve anatomical ROI definition, motion correction, resolution recovery and partial volume correction. Further improvements in PET quantification may be possible through use of MRS data in PET kinetic modelling. With regard to one of the weaknesses of these systems, namely determination of photon attenuation, recent results for MR-based attenuation correction of brain imaging are encouraging, with progress also being made for the more difficult task of whole body imaging. Finally, the reduction in radiation dose of PET/MRI compared to PET/CT, combined with possible enhancement of image-based input function determination, thereby obviating arterial blood sampling, will improve the feasibility of clinical drug trials involving longitudinal PET scanning.

Acknowledgements

The author would like to thank the authors who kindly allowed figures from their papers to appear in this chapter, together with the publishers of the journals from which the figures are reproduced.

References

1. T. Beyer, D. W. Townsend, T. Brun, P. E. Kinahan, M. Charron, R. Roddy, J. Jerin, J. Young, L. Byars and R. Nutt, *J. Nucl. Med.*, 2000, **41**, 1369.
2. P. J. Slomka and R. P. Baum, *Eur. J. Nucl. Med. Mol. Imaging*, 2009, **36**(1), 44.
3. C. Catana, D. Procissi, Y. Wu, M. S. Judenhofer, J. Qi, B. J. Pichler, R. E. Jacobs and S. R. Cherry, *Proc. Natl. Acad. Sci. USA*, 2008, **105**, 3705.
4. M. S. Judenhofer, H. F. Wehrl, D. F. Newport, C. Catana, S. B. Siegel, M. Becker, A. Thielscher, M. Kneilling, M. P. Lichy, M. Eichner, K. Klingel, G. Reischl, S. Widmaier, M. Röcken, R. E. Nutt, H.-J. Machulla, K. Uludag, S. R. Cherry, C. D. Claussen and B. J. Pichler, *Nat. Med.*, 2008, **14**, 459.
5. G. Antoch and A. Bockisch, *Eur. J. Nucl. Med. Mol. Imaging*, 2009, **36**, S113.
6. B. J. Pichler, A. Kolb, T. Nägele and H. P. Schlemmer, *J. Nucl. Med.*, 2010, **51**, 333.
7. T. Beyer, L. S. Freudenberg, J. Czernin and D. W. Townsend, *Insights Imaging*, 2011, **2**, 235.
8. J. A. Castelijns, *Eur. Radiol.*, 2011, **21**, 2425.
9. O. Ratib and T. Beyer, *Eur. J. Nucl. Med. Mol. Imaging*, 2011, **38**, 992.

10. A. W. Glaudemans, A. M. Quintero and A. Signore, *Eur. J. Nucl. Med. Mol. Imaging*, 2012, **39**, 745.
11. M. D. Majmudar and M. Nahrendorf, *J. Nucl. Med.*, 2012, **53**, 673.
12. H. Zaidi, N. Ojha, M. Morich, J. Griesmer, Z. Hu, P. Maniawski, O. Ratib, D. Izquierdo-Garcia, Z. A. Fayad and L. Shao, *Phys. Med. Biol.*, 2011, **56**, 3091.
13. H. Zaidi and A. Del Guerra, *Med. Phys.*, 2011, **38**, 5667.
14. P. B. Garlick, P. K. Marsden, A. C. Cave, H. G. Parkes, R. Slates, Y. Shao, R. W. Silverman and S. R. Cherry, *NMR Biomed.*, 1997, **10**, 138.
15. Y. Shao, S. R. Cherry, K. Farahani, K. Meadors, S. Siegel, R. W. Silverman and P. K. Marsden, *Phys. Med. Biol.*, 1997, **42**, 1965.
16. R. B. Slates, K. Farahani, Y. Shao, P. K. Marsden, J. Taylor, P. E. Summers, S. Williams, J. Beech and S. R. Cherry, *Phys. Med. Biol.*, 1999, **44**, 2015.
17. R. Lecomte, J. Cadorette, S. Rodrigue, D. Lapointe, D. Rouleau, M. Bentourkia, R. Yao and P. Msaki, *IEEE Trans. Nucl. Sci.*, 1996, **43**, 1952.
18. H. P. Schlemmer, B. J. Pichler, M. Schmand, Z. Burbar, C. Michel, R. Ladebeck, K. Jattke, D. Townsend, C. Nahmias, P. K. Jacob, W. D. Heiss and C. D. Claussen, *Radiology*, 2008, **248**, 1028.
19. G. Delso, S. Fürst, B. Jakoby, R. Ladebeck, C. Ganter, S. G. Nekolla, M. Schwaiger and S. I. Ziegler, *J. Nucl. Med.*, 2011, **52**, 1914.
20. E. Roncali and S. R. Cherry, *Ann. Biomedical Eng.*, 2011, **39**, 1358.
21. S. I. Kwon, J. S. Lee, H. S. Yoon, M. Ito, G. B. Ko, J. Y. Choi, S.-H. Lee, I. C. Song, J. M. Jeong, D. S. Lee and S. J. Hong, *J. Nucl. Med.*, 2011, **52**, 572.
22. S. Yamamoto, T. Watabe, H. Watabe, M. Aoki, E. Sugiyama, M. Imaizumi, Y. Kanai, E. Shimodegawa and J. Hatazawa, *Phys. Med. Biol.*, 2012, **57**, N1.
23. S. H. Yoon, G. B. Ko, S. I. Kwon, C. M. Lee, M. Ito, I. C. Song, D. S. Lee, S. J. Hong and J. Lee, *J. Nucl. Med.*, 2012, **53**, 608.
24. G. Delso and S. I. Ziegler, *Eur. J. Nucl. Med. Mol. Imaging*, 2009, **36**, S86.
25. S. H. Maramraju, S. D. Smith, S. S. Junnarkar, D. Schulz, S. Stoll, B. Ravindranath, M. L. Purschke, S. Rescia, S. Southekal, J. F. Pratte, P. Vaska, C. L. Woody and D. J. Schlyer, *Phys. Med. Biol.*, 2011, **56**, 2459.
26. J. H. F. Rudd, E. A. Warburton, T. D. Fryer, H. A. Jones, J. C. Clark, N. Antoun, P. Johnstrom, A. P. Davenport, P. J. Kirkpatrick, B. N. Arch, J. D. Pickard and P. L. Weissberg, *Circulation*, 2002, **105**, 2708.
27. P. E. Kinahan, D. W. Townsend, T. Beyer and D. Sashin, *Med. Phys.*, 1998, **25**, 2046.
28. M. J. Guy, I. A. Castellano-Smith, M. A. Flower, G. D. Flux, R. J. Ott and D. Visvikis, *IEEE Trans. Nucl. Sci.*, 1998, **45**, 1261.
29. N. S. Rehfeld, B. J. Heismann, J. Kupferschläger, P. Aschoff, G. Christ, A. C. Pfannenberg and B. J. Pichler, *Med. Phys.*, 2008, **35**, 1959.
30. H. Zaidi, M. L. Montandon and D. O. Slosman, *Med. Phys.*, 2003, **30**, 937.

31. E. R. Kops, G. Wagenknecht, J. Scheins, L. Tellmann and H. Herzog, *IEEE Nucl. Sci. Symp. Conf. Rec.*, 2009, 2530.

32. G. Wagenknecht, E. R. Kops, L. Tellmann and H. Herzog, *IEEE Nucl. Sci. Symp. Conf. Rec.*, 2009, 3338.

33. C. Catana, A. van der Kouwe, T. Benner, C. J. Michel, M. Hamm, M. Fenchel, B. Fischl, B. Rosen, M. Schmand and A. G. Sorensen, *J. Nucl. Med.*, 2010, **51**, 1431.

34. V. Keereman, Y. Fierens, T. Broux, Y. De Deene, M. Lonneux and S. Vandenberghe, *J. Nucl. Med.*, 2010, **51**, 812.

35. Y. Berker, J. Franke, A. Salomon, M. Palmowski, H. C. Donker, Y. Temur, F. M. Mottaghy, C. Kuhl, D. Izquierdo-Garcia, Z. A. Fayad, F. Kiessling and V. Schulz, *J. Nucl. Med.*, 2012, **53**, 796.

36. H. Zaidi, M. L. Montandon and D. O. Slosman, *Eur. J. Nucl. Med. Mol. Imaging*, 2004, **31**, 52.

37. M. L. Montandon and H. Zaidi, *NeuroImage*, 2005, **25**, 278.

38. E. R. Kops and H. Herzog, *IEEE Nucl. Sci. Symp. Conf. Rec.*, 2007, 4327.

39. M. L. Montandon and H. Zaidi, *Comput. Med. Imaging Graph.*, 2007, **31**, 28.

40. M. Hofmann, F. Steinke, V. Scheel, G. Charpiat, J. Farquhar, P. Aschoff, M. Brady, B. Schölkopf and B. J. Pichler, *J. Nucl. Med.*, 2008, **49**, 1875.

41. E. R. Kops and H. Herzog, *IEEE Nucl. Sci. Symp. Conf. Rec.*, 2008, 3786.

42. E. Schreibmann, J. A. Nye, D. M. Schuster, D. R. Martin, J. Votaw and T. Fox, *Med. Phys.*, 2010, **37**, 2101.

43. I. B. Malone, R. E. Ansorge, G. B. Williams, P. J. Nestor, T. A. Carpenter and T. D. Fryer, *J. Nucl. Med.*, 2011, **52**, 1142.

44. A. Johansson, M. Karlsson and T. Nyholm, *Med. Phys.*, 2011, **38**, 2708.

45. W. T. Dixon, *Radiology*, 1984, **153**, 189.

46. T. Beyer, M. Weigert, H. H. Quick, U. Piertrzyk, F. Vogt, C. Palm, G. Antoch, S. P. Müller and A. Bockisch, *Eur. J. Nucl. Med. Mol. Imaging*, 2008, **35**, 1142.

47. A. Martinez-Möller, M. Souvatzoglou, G. Delso, R. A. Bundschuh, C. Chefd'hotel, S. I. Ziegler, N. Navab, M. Schwaiger and S. G. Nekolla, *J. Nucl. Med.*, 2009, **50**, 520.

48. G. Delso, A. Martinez-Möller, R. A. Bundschuh, S. G. Nekolla and S. I. Ziegler, *Med. Phys.*, 2010, **37**, 2804.

49. G. Delso, A. Martinez-Möller, R. A. Bundschuh, R. Ladebeck, Y. Candidus, D. Faul and S. I. Ziegler, *Phys. Med. Biol.*, 2010, **55**, 4361.

50. J. Steinberg, G. Jia, S. Sammet, J. Zhang, N. Hall and M. V. Knopp, *Nucl. Med. Biol.*, 2010, **37**, 227.

51. M. Eiber, A. Martinez-Möller, M. Souvatzoglou, K. Holzapfel, A. Pickhard, D. Löffelbein, I. Santi, E. J. Rummeny, S. Ziegler, M. Schwaiger and S. G. Nekolla, *Eur. J. Nucl. Med. Mol. Imaging*, 2011, **38**, 1691.

52. M. Hofmann, I. Bezruhov, F. Mantlik, P. Aschoff, F. Steinke, T. Beyer, B. J. Pichler and B. Schölkopf, *J. Nucl. Med.*, 2011, **52**, 1392.

53. L. R. MacDonald, S. Kohlmyer, C. Liu, T. K. Lewellen and P. E. Kinahan, *Med. Phys.*, 2011, **38**, 2948.
54. F. Mantlik, M. Hofmann, M. K. Werner, A. Sauter, J. Kupferschläger, B. Schölkopf, B. J. Pichler and T. Beyer, *Eur. J. Nucl. Med. Mol. Imaging*, 2011, **38**, 920.
55. A. Salomon, A. Goedicke, B. Schweizer, T. Aach and V. Schulz, *IEEE Trans. Med. Imaging*, 2011, **30**, 804.
56. V. Schulz, I. Torres-Espallardo, S. Renisch, Z. Hu, N. Ojha, P. Börnert, M. Perkuhn, T. Niendorf, W. M. Schäfer, H. Brockmann, T. Krohn, A. Buhl, R. W. Günther, F. M. Mottaghy and G. A. Krombach, *Eur. J. Nucl. Med. Mol. Imaging*, 2011, **38**, 138.
57. L. Tellmann, H. H. Quick, A. Bockisch, H. Herzog and T. Beyer, *Med. Phys.*, 2011, **38**, 2795.
58. M. Eiber, M. Souvatzoglou, A. Pickhard, D. J. Loeffelbein, A. Knopf, K. Holzapfel, A. Martinez-Möller, S. G. Nekolla, E. Q. Scherer, M. Schwaiger, E. J. Rummeny and A. J. Beer, *Eur. J. Radiol.*, 2012, **81**, 2658.
59. A. Drzezga, M. Souvatzoglou, M. Eiber, A. J. Beer, S. Fürst, A. Martinez-Möller, S. G. Nekolla, S. Ziegler, C. Ganter, E. J. Rummeny and M. Schwaiger, *J. Nucl. Med.*, 2012, **53**, 845.
60. C. C. Watson, D. Newport, M. E. Casey, R. A. deKemp, R. S. Beanlands and M. Schmand, *IEEE Trans. Nucl. Sci.*, 1997, **44**, 90.
61. O. Klein and Y. Nishina, *Z. Phys.*, 1929, **52**, 853.
62. J. Qi and R. H. Huesman, *IEEE Int. Symp. Biol. Imaging*, 2002, 413.
63. S. Woo, H. Watabe, Y. Choi, K. Kim, C. Park and H. Iida, *IEEE Nucl. Sci. Symp. Conf. Rec.*, 2002, **2**, 830.
64. P. Bloomfield, T. Spinks, J. Reed, L. Schnorr, A. Westrip, L. Livieratos, R. Fulton and T. Jones, *Phys. Med. Biol.*, 2003, **48**, 959.
65. K. Thielemans, S. Mustafovic and L. Schnorr, *IEEE Nucl. Sci. Symp. Conf. Rec.*, 2003, **4**, 2401.
66. C. Catana, T. Benner, A. van der Kouwe, L. Byars, M. Hamm, D. B. Chonde, C. J. Michel, G. El Fakhri, M. Schmand and A. G. Sorensen, *J. Nucl. Med.*, 2011, **52**, 154.
67. Y. Picard and C. Thompson, *IEEE Trans. Med. Imaging*, 1997, **16**, 137.
68. M. Jacobson and J. Fessler, *IEEE Nucl. Sci. Symp. Conf. Rec.*, 2003, **5**, 3290.
69. A. Rahmim, P. Bloomfield, S. Houle, M. Lenox, C. Michel, K. R. Buckley, T. J. Ruth and V. Sossi, *IEEE Trans. Nucl. Sci.*, 2004, **51**, 2588.
70. M. Dawood, N. Lang, X. Jiang and K. P. Schäfers, *IEEE Trans. Med. Imaging*, 2006, **25**, 476.
71. T. Li, B. Thorndyke, E. Schreibmann, Y. Yang and L. Xing, *Med. Phys.*, 2006, **33**, 1288.
72. F. Qiao, T. Pan, J. W. Clark and O. R. Mawlawi, *Phys. Med. Biol.*, 2006, **51**, 3769.

73. F. Lamare, M. Ledesma Carbayo, T. Cresson, G. Kontaxakis, A. Santos, C. Cheze Le Rest, A. Reader and D. Visvikis, *Phys. Med. Biol.*, 2007, **52**, 5187.
74. M. Reyes, G. Malandain, P. M. Koulibaly, M. A. Gonzalez-Ballester and J. Darcourt, *Phys. Med. Biol.*, 2007, **52**, 3579.
75. M. Dawood, F. Büther, X. Jiang and K. P. Schäfers, *IEEE Trans. Med. Imaging*, 2008, **27**, 1164.
76. W. Bai and M. Brady, *Phys. Med. Biol.*, 2009, **54**, 2719.
77. J. Dey and M. A. King, *IEEE Trans. Nucl. Sci.*, 2009, **56**, 2739.
78. N. Dikaios and T. D. Fryer, *IEEE Nucl. Sci. Symp. Conf. Rec.*, 2009, 2806.
79. N. Dikaios and T. D. Fryer, *Med. Phys.*, 2011, **38**, 4958.
80. N. Dikaios and T. D. Fryer, *Phys. Med. Biol.*, 2011, **56**, 1695.
81. N. Dikaios and T. D. Fryer, *Med. Phys.*, 2012, **39**, 1253.
82. N. Dikaios, D. Izquierdo-Garcia, M. J. Graves, V. Mani, Z. A. Fayad and T. D. Fryer, *Eur. Radiol.*, 2012, **22**, 439.
83. B. Guerin, S. Cho, S. Y. Chun, X. Zhu, N. M. Alpert, G. El Fakhri, T. Reese and C. Catana, *Med. Phys.*, 2011, **38**, 3025.
84. J. Nuyts, K. Baete, D. Beque and P. Dupont, *IEEE Trans. Med. Imaging*, 2005, **24**, 667.
85. C. T. Chen, X. Ouyang, W. H. Wong, X. Hu, V. E. Johnson, C. E. Ordonez and C. E. Metz, *IEEE Trans. Nucl. Sci.*, 1991, **38**, 687.
86. V. E. Johnson, W. H. Wong, X. Hu and C. T. Chen, *IEEE Trans. Pattern Anal. Machine Intell.*, 1991, **13**, 413.
87. R. M. Leahy and X. Yan, in *Lecture Notes in Computer Science, Information Processing in Medical Imaging*, Spinger, Berlin, 1991, vol. 511, p. 105.
88. G. Gindi, M. Lee, A. Rangarajan and G. Zubal, *IEEE Trans. Med. Imaging*, 1993, **12**, 670.
89. X. Ouyang, W. H. Wong, V. E. Johnson, X. Hu and C. T. Chen, *IEEE Trans. Med. Imaging*, 1994, **13**, 627.
90. J. E. Bowsher, V. A. Johnson, T. G. Turkington, R. J. Jaszczak, C. R. Floyd and R. E. Coleman, *IEEE Trans. Med. Imaging*, 1996, **15**, 673.
91. B. Lipinski, H. Herzog, E. Rota Kops, W. Oberschelp and H. W. Müller-Gärtner, *IEEE Trans. Med. Imaging*, 1997, **16**, 129.
92. S. Sastry and R. E. Carson, *IEEE Trans. Med. Imaging*, 1997, **16**, 750.
93. C. Comtat, P. E. Kinahan, J. A. Fessler, T. Beyer, D. W. Townsend, M. Defrise and C. Michel, *Phys. Med. Biol.*, 2002, **47**, 1.
94. K. Baete, J. Nuyts, W. Van Paesschen, P. Suetens and P. Dupont, *IEEE Trans. Med. Imaging*, 2004, **23**, 510.
95. K. Baete, J. Nuyts, K. Van Laere, W. Van Paesschen, S. Ceyssens, L. De Ceuninck, O. Gheysens, A. Kelles, J. Van den Eynden, P. Suetens and P. Dupont, *NeuroImage*, 2004, **23**, 305.
96. K. Goffin, W. Van Paesschen, P. Dupont, K. Baete, A. Palmini, J. Nuyts and K. Van Laere, *Eur. J. Nucl. Med. Mol. Imaging*, 2010, **37**, 1148.

97. K. Vunckx, A. Atre, K. Baete, A. Reilhac, C. M. Deroose, K. Van Laere and J. Nuyts, *IEEE Trans. Med. Imaging*, 2012, **31**, 599.
98. O. G. Rousset, Y. Ma and A. C. Evans, *J. Nucl. Med.*, 1998, **39**, 904.
99. J. A. Aston, V. J. Cunningham, M. C. Asselin, A. Hammers, A. C. Evans and R. N. Gunn, *J. Cereb. Blood Flow Metab.*, 2002, **22**, 1019.
100. O. G. Rousset, D. L. Collins, A. Rahmim and D. F. Wong, *J. Nucl. Med.*, 2008, **49**, 1097.
101. H. W. Müller-Gärtner, J. M. Links, J. L. Prince, R. N. Bryan, E. McVeigh, J. P. Leal, C. Davatzikos and J. J. Frost, *J. Cereb. Blood Flow Metab.*, 1992, **12**, 571.
102. C. C. Meltzer, P. E. Kinahan, P. J. Greer, T. E. Nichols, C. Comtat, M. N. Cantwell, M. P. Lin and J. C. Price, *J. Nucl. Med.*, 1999, **40**, 2053.
103. N. Boussion, M. Hatt, F. Lamare, Y. Bizais, A. Turzo, C. Cheze-Le Rest and D. Visvikis, *Phys. Med. Biol.*, 2006, **51**, 1857.
104. M. Shidahara, C. Tsoumpas, A. Hammers, N. Boussion, D. Visvikis, T. Suhara, I. Kanno and F. E. Turkheimer, *NeuroImage*, 2009, **44**, 340.
105. A. Le Pogam, M. Hatt, P. Descourt, N. Boussion, C. Tsoumpas, F. E. Turkheimer, C. Prunier-Aesch, J.-L. Baulieu, D. Guilloteau and D. Visvikis, *Med. Phys.*, 2011, **38**, 4920.
106. H. Wang and B. Fei, *Med. Phys.*, 2012, **39**, 179.
107. W. H. Richardson, *J. Opt. Soc. Am.*, 1972, **62**, 55.
108. L. B. Lucy, *Astron. J.*, 1974, **79**, 745.
109. P. H. Van Cittert, *Z. Phys.*, 1931, **69**, 298.
110. F. E. Turkheimer, N. Boussion, A. N. Anderson, N. Pavese, P. Piccini and D. Visvikis, *J. Nucl. Med.*, 2008, **49**, 657.
111. P. Zanotti-Fregonara, E. M. Fadaili, R. Maroy, C. Comtat, A. Souloumiac, S. Jan, M. J. Ribeiro, V. Gaura, A. Bar-Hen and R. Trebossen, *J. Cereb. Blood Flow Metab.*, 2009, **29**, 1825.
112. P. Zanotti-Fregonara, K. Chen, J. S. Liow, M. Fujita and R. B. Innis, *J. Cereb. Blood Flow Metab.*, 2011, **31**, 1986.
113. P. Zanotti-Fregonara, R. Maroy, M. A. Peyronneau, R. Trebossen and M. Bottlaender, *Eur. J. Nucl. Med. Mol. Imaging*, 2012, **39**, 651.
114. C. Glaus, R. Rossin, M. J. Welch and G. Bao, *Bioconjug. Chem.*, 2010, **21**, 715.
115. D. Patel, A. Kell, B. Simard, B. Xiang, H. Y. Lin and G. Tian, *Biomaterials*, 2011, **32**, 1167.
116. W. Wolf, *Pharm. Res.*, 2011, **38**, 490.
117. H. Park, D. Wood, H. Hussain, C. R. Meyer, R. B. Shah, T. D. Johnson, T. Chenevert and M. Piert, *J. Nucl. Med.*, 2012, **53**, 546.
118. A. Boss, L. Stegger, S. Bisdas, A. Kolb, N. Schwenzer, M. Pfister, C. D. Claussen, B. J. Pichler and C. Pfannenberg, *Eur. Radiol.*, 2011, **21**, 1439.
119. B. Beuthien-Baumann, I. Platzek, I. Lauterbach, J. van den Hoff, G. Schramm, K. Zöphel, M. Laniado and J. Kotzerke, *Eur. J. Nucl. Med. Mol. Imaging*, 2012, **39**, 1087.

120. M. Lord, O. Ratib and J. P. Vallee, *Eur. J. Nucl. Med. Mol. Imaging*, 2011, **38**, 2288.

121. A. Boss, S. Bisdas, A. Kolb, M. Hofmann, U. Ernemann, C. D. Claussen, C. Pfannenberg, B. J. Pichler, M. Reimold and K. L. Stegger, *J. Nucl. Med.*, 2010, **51**, 1198.

122. D. Thorwarth, G. Henke, A. C. Müller, M. Reimold, T. Beyer, A. Boss, A. Kolb, B. Pichler and C. Pfannenberg, *Int. J. Radiation Oncology Biol. Phys.*, 2011, **81**, 277.

123. K. Büscher, M. S. Judenhofer, M. T. Kuhlmann, S. Hermann, H. F. Wehrl, K. P. Schäfers, M. Schäfers, B. J. Pichler and L. Stegger, *J. Nucl. Med.*, 2010, **51**, 1277.

124. W. W. Lee, B. Marinelli, A. M. van der Laan, B. Sena, R. Gorbatov, F. Leuschner, P. Dutta, Y. Iwamoto, T. Ueno, M. P. V. Begieneman, H. W. M. Niessen, J. J. Piek, C. Vinegoni, M. J. Pittet, F. K. Swirski, A. Tawakol, M. Di Carli, R. Weissleder and M. Nahrendorf, *J. Am. Coll. Cardiol.*, 2012, **59**, 153.

125. F. Miese, A. Scherer, B. Ostendorf, A. Heinzel, R. S. Lanzman, P. Kröpil, H. Hautzel, H.-J. Wittsack, M. Schneider, G. Antoch, H. Herzog and N. J. Shah, *Clin. Rheumatol.*, 2011, **30**, 1247.

CHAPTER 11

Magnetic Resonance-Based Cell Imaging Using Contrast Media and Reporter Genes

GREETJE VANDE VELDE AND UWE HIMMELREICH*

Biomedical MRI Unit and Molecular Small Animal Imaging Center, Department of Imaging and Pathology, O&N1, bus 505, Katholieke Universiteit Leuven, Herestraat 49, B-3000, Leuven, Belgium
*Email: uwe.himmelreich@med.kuleuven.be

11.1 Introduction

Cellular and molecular imaging comprise the non-invasive and repeated imaging of targeted tissue, cells and molecular processes in living organisms for the visualization of processes on a molecular level.[1] The purpose of visualizing molecular processes with close to cellular resolution is to better understand mechanisms of diseases and therapeutic approaches. Several imaging modalities are available that are able to perform non-invasive and repeated imaging of targeted cells in living organisms. Among these are single-photon emission computed tomography (SPECT), positron emission tomography (PET), optical imaging (fluorescence imaging and bioluminescence imaging (BLI)), magnetic resonance imaging (MRI), ultrasound and X-ray based methods. Molecular imaging was initially a domain for PET/SPECT and optical imaging.[1–6] Other imaging modalities were lacking either sensitivity and/or specificity for visualizing molecular events like detecting and quantifying the expression of certain enzymes. This has changed due to the

New Developments in NMR No. 2
New Applications of NMR in Drug Discovery and Development
Edited by Leoncio Garrido and Nicolau Beckmann
© The Royal Society of Chemistry 2013
Published by the Royal Society of Chemistry, www.rsc.org

development of contrast agents (see Chapter 6) linked to molecules that target particular enzymes and the utilization of reporter genes whose activity or products can be detected by the respective imaging technique.

Although some imaging techniques are superior to others for particular applications, all of them are imperfect in terms of resolution, sensitivity, specificity, target-to-background contrast or potential for clinical applications. Such deficits can be overcome by combining imaging techniques in so-called bi- and multimodal imaging approaches.[5] PET and MRI offer unlimited depth of interrogation and are capable of high-sensitivity (in the case of PET) or high-resolution (in the case of MRI) non-invasive imaging (see Chapter 10). When comparing both imaging modalities, MRI offers three-dimensional (3D) images with a routinely achieved isotropic resolution of down to 50 µm using hardware for small animal MRI.[7–9] Near-cellular resolution can be achieved when long acquisition times are permitted, high field strengths, sensitive, purpose-built coils (see Chapter 5) and/or highly sensitive contrast agents are used.[7,10–12] Because of this superb spatial resolution, MRI is ideally suited to provide information on the location and migration of (therapeutic) cells after transplantation or transfusion.[13,14] This approach requires labeling of the cells of interest with contrast agents to discriminate them from the surrounding tissue.

11.2   Magnetic Resonance Techniques

The most commonly used element for MR imaging applications is the hydrogen atom or proton (^{1}H) due to its high sensitivity that emerges from its magnetic properties and its high natural abundance in living organisms. Therefore, MR methods are often based on detecting proton signals of mainly water and lipids and the influence of the environment on these signals. Such "environmental" influences include the proton content, relaxation behavior (longitudinal relaxation – T_1, transverse relaxation – T_2, influences on magnetic field homogeneities – T_2*), tissue viscosity, flow, diffusivity and chemical composition among others. Exogenous contrast agents are frequently used to modulate T_1, T_2 and/or T_2* and thereby increase the contrast-to-noise ratio (CNR) for even better visualization of local anatomy with high resolution. In addition, conventional and more advanced experimental techniques also provide information on hemodynamic changes (blood flow, blood volume, tissue perfusion), metabolic changes (using MR spectroscopy, MRS, see also Chapter 8), functionality (using functional MRI, fMRI, see also Chapter 7) and cellular connectivity and tissue organization (using diffusion tensor imaging (DTI) and manganese enhanced MRI (MEMRI).[15–20] Also due to these diverse applications of MRI, it has developed into one of the most powerful tools for clinical diagnosis, the characterization of animal models and the evaluation of therapeutic strategies in biomedical research.

Although MRI contrast in endogenous tissues provides excellent sensitivity for detecting subtle changes in anatomy and function, MRI has partially poor specificity for attributing image contrast to pathologies. MRI also lacks the ability easily to acquire information on the viability of engrafted cells and their functional status like optical imaging methods can do. In contrast to PET, conventional MRI and MRS lack the sensitivity to provide metabolic/molecular information with high resolution. Apart from multimodal imaging approaches, recent methodological advances offer a broader applicability of MR techniques for molecular imaging. These concern mainly MR spectroscopic methods, including those using hyperpolarized compounds (see Chapter 9), the development of contrast agents (see Chapter 6) that target enzymes specifically expressed by certain cell types or transgenic cells and the genetic modification of cells to express MRI reporter genes.

11.2.1 Contrast Generation in MRI

11.2.1.1 Contrast Based on Differences in Relaxation Rates

In order to distinguish targeted tissue or cells from the background by using MRI, contrast needs to be generated either by utilizing intrinsic differences of MR related properties like spin density, T_1, T_2 or T_2^* relaxation, diffusivity, flow and chemical composition, or by using MR contrast agents (for reviews see references 15, 21–25). The T_1 relaxation time can be changed in three ways: by altering the rotational tumbling time of a magnetic dipole, altering the accessibility of water to a paramagnetic atom or altering the electron spin state of the dipole. T_2/T_2^* contrast can be influenced by altering the rotational tumbling time of a magnetic dipole (T_2) or by altering the local inhomogeneities in the local magnetic field (T_2^*). A variety of MRI contrast agents is available, based on interactions with the surrounding hydrogen atoms and hereby altering T_1, $T_2(^*)$ contrast or both to a greater or lesser extent. Superparamagnetic and paramagnetic contrast agents alter the intrinsic MR signal by changing the T_1 and T_2/T_2^* relaxation times.[22,26,27] Depending on the contrast agent used, one of the relaxation mechanisms may dominate. If the ratio of the respective relaxation rates ($R = 1/T$) R_2/R_1 is low, hyperintense (bright) contrast can be generated in T_1-weighted MRI.[26,28] Paramagnetic metals such as lanthanides (gadolinium, europium, *etc.*) or manganese have a strong effect on T_1 relaxation. Gadolinium (Gd), for example, has a large spin-dipole magnetic moment and seven unpaired electrons, and therefore has excellent characteristics to decrease T_1 relaxation time of the water molecules that are near the Gd metal ion.

Contrast agents with a high R_2/R_1 or R_2^*/R_1 ratio are often referred to as T_2/T_2^* contrast agents, generating hypointense (dark) contrast in T_2/T_2^*-weighted MR images, respectively.[22,27] Superparamagnetic iron oxide (SPIO) particles cause local perturbations in the magnetic field homogeneity and therefore loss of phase coherence and subsequent reduction of T_2 and T_2^* relaxation time. Therefore, such iron oxide particles are often referred to as T_2^* contrast agents. They are frequently used for highly sensitive cell labeling and visualization (for a review see references 11, 12, 22, 29, 30).

11.2.1.2 Contrast Based on Chemical Exchange

The exchange of protons between different environments (for example, the equilibrium between free water protons and protons in amide, amine or hydroxyl groups) has already been studied by magnetic resonance in the 1960s using magnetization transfer experiments.[31] Thereby, the chemical shift difference between those two (or more) populations of protons is used to saturate the magnetization of one proton pool and to detect the effect on the intensity of the other proton signal. Magnetization transfer MRI has already been used to generate contrast by selective saturation of signal contributions from protons with low T_2 values using off-resonance irradiation. Magnetization transfer from the saturated protons to mobile protons results in decreased signal intensity and produces substantial changes in contrast.[32] While many of the initial applications of magnetization transfer MRI focused on the utilization of OH and NH protons and their relatively small chemical shift difference to water protons (so called DIACEST agents), the introduction of paramagnetic agents improved the applicability of CEST based contrast significantly (see Chapter 6). Again, paramagnetic chelates have already been used in the early days of NMR spectroscopy as so-called paramagnetic shift reagents to simplify complex NMR spectra by changing the chemical shifts of signals of protons binding or exchanging with the respective agents.[33] The term chemical-exchange saturation transfer (CEST) for the generation of contrast in MR images by utilizing diamagnetic (DIACEST, DIAmagnetic Chemical Exchange Saturation Transfer) and paramagnetic (PARACEST, PARA-magnetic Chemical Exchange Saturation Transfer) agents was introduced by Ward *et al.*[34] The CEST efficiency depends on the rate of the chemical exchange and the chemical shift difference between the different proton populations.

The efficiency of CEST agents can be modified in two ways, either by changing the chemical exchange rate of the exchangeable hydrogen atoms with water, or by altering their chemical shift difference. On one hand, the exchange must be slow compared to the NMR time scale but still fast enough to avoid unwanted peak broadening, which makes selective radiofrequency (RF) irradiation difficult (Figure 11.1). On the other hand, larger frequency separation (for example, due to PARACEST agents) not only facilitates selective excitation but also allows shorter lifetimes of the different populations and thereby improves the sensitivity of detecting the CEST effect (for review, see reference 35). Applications range from the detection of pH,[36,37] pO_2,[38,39] lactate[40] or ion concentrations[41] to cell labeling and detection of several populations simultaneously.[42–44]

11.2.1.3   Contrast Based on Spin Density Differences Using Heteronuclear MRI

Contrast based on differences in spin density is difficult to achieve for cell tracking using ^1H MRI. However, if nuclei other than ^1H are monitored and their intrinsic MR detectable concentration in the organism is low (for example, the fluorine isotope ^{19}F), contrast agents based on those nuclei are also suitable for cell detection. As there are no MR-detectable intrinsic ^{19}F nuclei *in vivo*, MRI can specifically visualize and quantify accumulations of ^{19}F-based contrast agents with a sensitivity only marginally lower than for ^1H-based

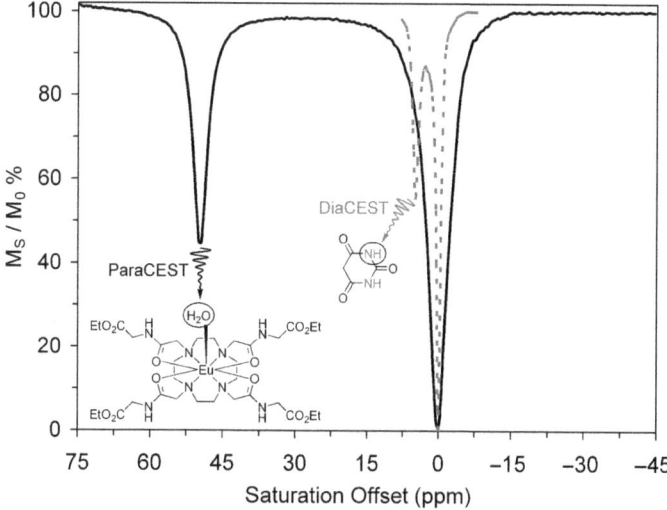

Figure 11.1   Changes in chemical shift differences between DIACEST and PARACEST agents. Due to the well-separated resonances of the water protons and protons coordinated to the europium ion, selective irradiation is feasible for *in vivo* applications using PARACEST agents. Adapted with permission from Hancu *et al.*[35] © 2010 The Foundation Acta Radiologica.

MRI. However, one has to keep in mind that local concentrations of fluorinated contrast agents (in the millimolar range) are much below the concentration of water usually detected by proton MRI. For anatomical information, overlay with ^{1}H MRI is necessary.[45–47] ^{19}F MRS can provide additional information on metabolic transformation of fluorinated compounds that can be used as molecular markers.[48] Additional technical aspects that require careful consideration for ^{19}F MRI are: 1) RF coils are required that are tunable to ^{1}H and ^{19}F, 2) the large frequency range of ^{19}F can lead to chemical-shift displacement errors for agents with more than one ^{19}F MR signal so that data acquisition from a narrow ^{19}F chemical shift range is favored by using either respective compounds or selective, adiabatic excitation pulses, and 3) relaxation times are often shorter for ^{19}F compared to similarly bound protons.

   Until a decade ago, generation of contrast based on other nuclei was rarely utilized in pre-clinical and clinical research. However, with the advancement of hyperpolarization methods, generation of contrast based on, for example, hyperpolarized ^{13}C is feasible and becoming increasingly popular.[49] The gains in sensitivity due to hyperpolarization techniques are due to the population difference between energy levels that are much greater than at thermal equilibrium.[50–52] Thereby, the enhancement of nuclear spin polarization (hyperpolarization) from Boltzmann distribution to non-thermal polarization has the potential to increase the number of spins contributing to the MR signal from 1–100 ppm by 4–5 orders of magnitude. Several methods are currently used for MR applications of hyperpolarized compounds ranging from optical pumping (laser polarization) of noble gases,[53] to polarization of ^{13}C and ^{15}N nuclei of biologically interesting compounds using dynamic nuclear polarization (DNP)[50,52] or parahydrogen induced polarization (PHIP).[51,54] For more details on theoretical and practical issues, we refer to a number of review articles (see also Chapter 9).[51,55,56] For tracing molecular events (biodistribution and metabolism), hyperpolarization of ^{13}C-labeled substrates by PHIP and DNP has resulted in 20–40% polarization. While DNP can in principle be applied to all molecules, PHIP is restricted to a few suitable molecules. However, DNP is more time consuming and expensive in terms of hardware and maintenance.[51,57]

Once a compound is hyperpolarized, its polarization will decay with T_1. Depending on the carbon atom in a molecule, T_1 is in the order of seconds to tens of seconds. This limits the carbon atoms suitable for hyperpolarization and the time available for monitoring metabolic processes (usually in the order of 1–3 minutes), but is also demanding on the kinetic modeling as substrates and metabolites will have different T_1 relaxation times. As the decay of magnetization towards thermal equilibrium is not recoverable, MRI/MRS methods for hyperpolarized substrates need to be rapid. In addition, one has to consider that excitation RF pulses irreversibly destroy a fraction of the longitudinal magnetization available within the excited slice. Therefore, low flip angles should be chosen and a compromise has to be made between the acquired matrix size and the required resolution. Similarly to conventional ^{13}C MRI/MRS, the large chemical shift frequency range may cause chemical shift displacement errors if a large frequency range needs to be covered. However, most hyperpolarized ^{13}C atoms used so far fall within a relatively narrow spectral width so that chemical shift displacement errors can be minimized. Mayer *et al.* have successfully tested spiral chemical shift imaging for metabolic mapping of hyperpolarized substrates.[58]

11.2.2 Contrast Generation in MR Spectroscopic Imaging

MRS is frequently used for detailed studies on metabolite composition and metabolic pathways. However, it lacks the sensitivity for acquiring this information with high resolution due to the low concentration of metabolites. Information provided by MRS is based on differences in chemical composition and the resulting resonance frequency (chemical shift) differences of protons from different metabolites. For localized MRS, the metabolite profile is acquired from a volume of interest (VOI), suppressing unwanted signals from outside the VOI. Metabolites of interest are present at best in millimolar concentrations and, therefore, their signals are several orders of magnitude lower in intensity than the water and lipid signals acquired by MRI. To be able to achieve detectable MRS signals, the resolution of localized *in vivo* MRS is in the order of cm^3 for clinical and mm^3 for experimental applications.[59,60] Broad applications of MRS mainly include the diagnosis and grading of neurological and neuro-oncological diseases.[61] In particular for metabolic diseases,[61] but to a certain extent also for brain abscesses,[62] MRS is the only non-invasive means for diagnosis.

Although ^{1}H MRS is the most sensitive, other MR active nuclei like ^{31}P and ^{13}C are increasingly used for *in vivo* applications. ^{31}P MRS is widely used for studying energy metabolism (like ATP, inorganic phosphate and phospho-creatine) but also in oncology.[63,64] The even less sensitive carbon is a useful marker for monitoring biochemical pathways (like glycolysis, TCA cycle, *etc.*) *in vivo*. Infusions of ^{13}C-labeled substrates are frequently used to monitor the conversion of these compounds (for example ^{13}C-labeled glucose, acetate or fatty acids) in metabolic pathways in humans and animal models.[65–67] Substrate uptake and metabolic flux can be studied non-invasively using metabolic models.[65,68] Although the resolution is still poor compared to nuclear methods, ^{13}C MRS has the advantage that not only can the location of those non-radioactive "tracers" be monitored but also their metabolism.

Although limited by the relatively fast relaxation (short "half-life") of the hyperpolarized compounds, dynamic *in vivo* studies on tissue perfusion and metabolism have been successful in humans and animal models using hyper-polarized ^{13}C-labeled substrates.[54,56] Using a clinical MRI scanner, spatial resolution of several mm and temporal resolution of 8 s can be achieved in the rat (Figure 11.2).[69,70]

11.3 MRI Contrast for Cellular and Molecular Imaging

11.3.1 Labeling Cells with Contrast Agents for MRI

For more than a decade, MRI contrast agents have been used for the visual-ization of intrinsic and engrafted cells.[8,71–74] These approaches have been further fine-tuned even to detect single cells,[10,75] enhance biocompatibility[76] and monitor cell location and migration *in vivo*.[9,14,70,77] Some of these approaches have also been applied in clinical studies.[47,78,79] These labeling approaches have been intensively reviewed. For more comprehensive reviews on MRI-based cell labeling and visualization techniques, we refer to references 11, 14, 22, 29, 30.

The most sensitive approach for the MRI-based cell visualization is the utilization of SPIOs as cell labeling agents (Figure 11.3). These SPIOs are prepared in a variety of size ranges that are generally classified as ultra-small (ultra-small superparamagnetic iron oxide, USPIO, <40 nm), small (SPIO, 40–100 nm) or micron-sized (microparticles of iron oxide, MPIO, 0.5–2 μm) iron oxide particles.[11,22,29] For better biocompatibility, those (U)SPIOs and MPIOs are usually coated with polymers (for example dextran), small molecules (for example citrate), lipids (for example magnetoliposomes) and others.[71,73,80–85] This coating can also contain fluorophores for later validation by fluorescence microscopy. (U)SPIOs have been widely used for targeted molecular imaging applications,[73,74,86] including *in vivo* tracking of stem cells[13,71,87] and tumor progression.[88] Limitations to the use of (U)SPIOs are the hypointense (negative) contrast that does not allow quantification, the size of the agents and the need to label cells *in vitro* with pre-synthesized particles prior

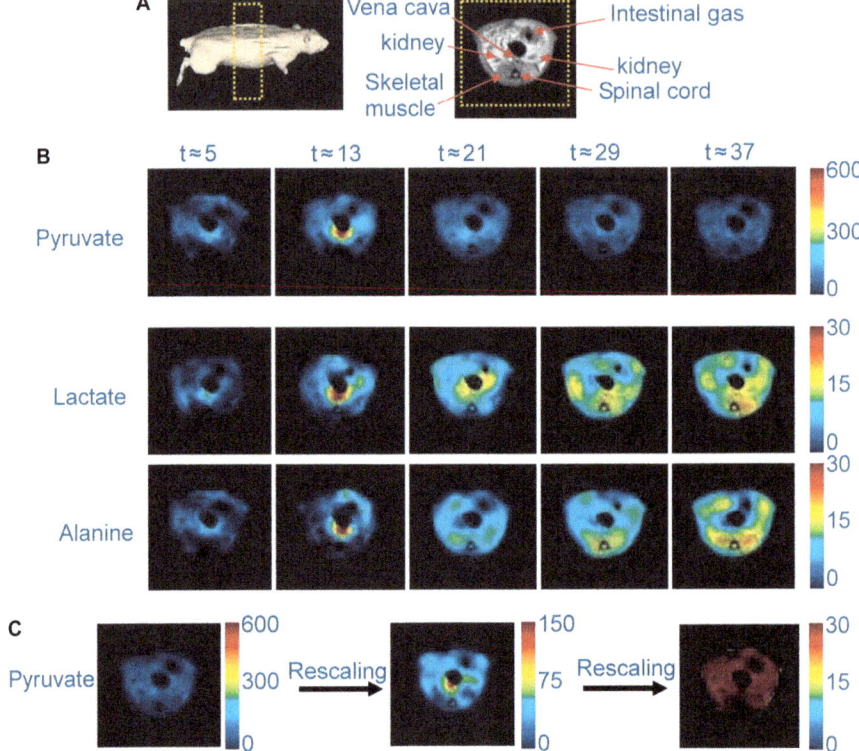

Figure 11.2 Real-time metabolic imaging with MRS. The time course of the build-up of pyruvate metabolites (lactate and alanine) in the skeletal muscle of a rat injected with hyperpolarized ^{13}C-enriched pyruvate. (A) Corresponding image slab and the transversal ^{1}H MRI of the rat. (B) Spectroscopic MR images (CSI) of the pyruvate, lactate and alanine signals were acquired simultaneously. The evolution of the metabolic maps as a function of time reveals the production of lactate and alanine in the skeletal muscle. To facilitate the interpretation, the ^{13}C-metabolic images have been super-imposed on the anatomical ^{1}H NMR image from A. The total amplitude of pyruvate varies with a factor of 5 between the bolus passage ($t = 5$ s) and the final image ($t = 37$ s). (C) To illustrate the presence of pyruvate in the whole image ($t = 37$ s), the pyruvate intensity was rescaled with factors of 1, 4 and 20. The image in C (right) shows the pyruvate amplitude being higher than the alanine and lactate amplitudes over the whole image. Reproduced with permission from Golman *et al.*[57] © 2006 National Academy of Sciences of the United States of America.

to cell transplantation, although recent efforts have resulted in protocols for *in vivo* labeling of neuroblasts in the rodent brain.[89,90]

Other approaches of cell labeling include the utilization of lanthanide chelates and other T_1 contrast agents like manganese-based chelates and particles but also the use of fluorinated molecules.[21,91–93]

Although excellent for providing high-resolution information on cell location, MR methods are limited in providing molecular information with the

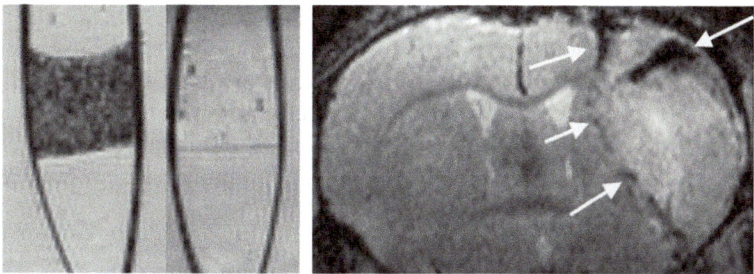

Figure 11.3 MRI of SPIO-labeled cells. (A) MR images of mesenchymal stem cells labeled with magnetoliposomes and suspended in agarose demonstrate single cell detection *in vitro*. The left image shows a suspension of 1000 cells μl^{-1} and the right image shows a suspension of 10 cells μl^{-1} (arrows pointing at individual cells). (B) Engraftment of those SPIO-labeled cells in a stroke mouse model shows distribution of cells from the site of engraftment towards the site of injury as highlighted by the arrows.

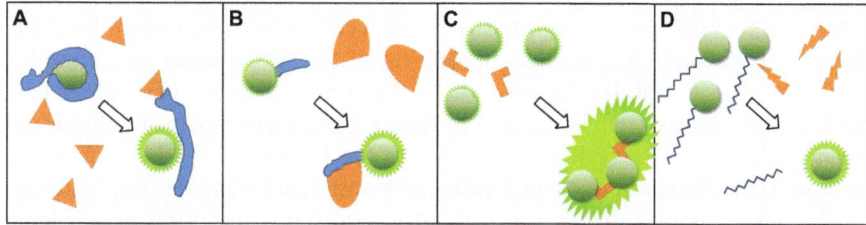

Figure 11.4 Mechanisms of action for smart contrast agents. Smart contrast agents can be designed to target particular cells or respond to physiological targets by several mechanisms. (A) A smart contrast agent (green) can be chemically modified by a cellular target molecule (orange), for example, by an enzyme that could modify the ligand of a lanthanide chelate to allow free coordination of water to the agent's metal ion (usually Gd), thereby changing its T_1 relaxivity. (B) Association of a lanthanide chelate (green) with a much larger binding partner (orange) results in an increase in rotational correlation time and thereby a decrease in T_1. (C) The presence of a target molecule (orange) induces aggregation of functionalized superparamagnetic nanoparticles (green), resulting in substantial changes in T_2 relaxation. (D) Insoluble and therefore inactive Gd-DTPA chelates can be activated by proteolytic cleavage of the long aliphatic side chains of the inactive compound. The contrast agent becomes soluble and is thereby activated in the presence of the expressed enzyme, causing a decrease in the T_1 relaxation time due to the subsequent soluble (free) lanthanide chelate.

same resolution. In order to make molecular events detectable by MR methods, different strategies involving chemically modified contrast agents can be used. Non-specific T_1 and T_2 contrast agents can be modified to report on their specific environment (Figure 11.4). A so-called "smart" contrast agent can report on physical properties like pH or ion strength, can be targeted to a specific cell by binding to particular receptors or transporters or can be modified by enzymes expressed by particular cell types. For all these mechanisms, their magnetic properties are altered due to interactions with the environment. Another strategy is to couple molecular events with reporter gene expression in transgenic cells by use of a specific promoter. Hereby, a gene product is responsible for contrast generation in combination with endogenously available molecules or an exogenously administered contrast agent (see Section 11.3.4, Genetic Reporter Systems).

11.3.2   MRI Contrast Agents Targeting Specific Cell Types

As the particle concentration within labeled cells decreases over time with cell proliferation, the use of non-specific contrast agents for long-term follow-up is limited. Another important limitation is that those agents cannot be linked directly to *in vivo* gene expression. To overcome these limitations, the contrast agent needs to be chemically modified to target particular receptors or transporters of specific cells. Thereby, a targeting molecule (for example, an antibody) can be covalently linked with an unspecific contrast agent. Upon addition of these magnetic labels to the targeted cells, specific uptake or binding follows. An example is the functionalization of (U)SPIOs using an antibody that will cause cellular uptake by transferrin-receptor (Tfr)-mediated endocytosis.[72] This system makes use of the endogenous presence of millions of copies of the Tfr on the cell's surface, which can be utilized for MR imaging. The anti-Tfr antibody mimics the binding of transferrin (Tf) to the Tfr, which is shuttling the complex inside the cell into endosomes, resulting in efficient labeling. The naturally very abundant endogenous presence of the Tfr[72] as well as over-expression of an engineered (to be constitutively over-expressed) Tfr[94] have been used to load cells with (U)SPIOs. Due to its abundance, no cell type-specific uptake *via* the Tfr was shown so far.

In order to achieve more specific labeling of particular cell types by systemic administration of a contrast agent, cell specific antibodies bound to (U)SPIOs or lanthanide chelates could accumulate at sites of high expression of the targeted receptors. The most common approach is to use trans-activating transcriptional activator (Tat) peptides to derivatize either coated nanoparticles or lanthanide chelates. Biotinylated antibodies can, for example, strongly bind several lanthanide chelates complexed with avidin. Biological applications of targeted contrast agents include inflammation,[95] angiogenesis,[96] apoptosis,[97] tumors[98] and atherosclerosis.[99] In another example for an antibody that enables a Gd-based contrast agent to bind to the adhesion molecule E-selectin, it was used to identify endothelial activation in the brain.[100] Similar to other applications of cell labeling, iron oxide particles show a higher sensitivity than lanthanide chelates.

A generic method to report on the expression of non-endogenous genes using (U)SPIOs was proposed, based on the use of an antigen-antibody reporting system and visualization by MRI.[101] In this method, the imaging marker gene encodes for a cell surface antigen that is non-endogenous to the cells studied. The antigen was recognized by a specific antibody that was conjugated to an

(U)SPIO particle. The gene encoding for the surface antigen can be introduced in the target cell population by inducible or specific viral vector systems, coupled to the expression of a transgene of interest. This idea was applied to follow up on the faith of embryonic stem cells (ESCs), transplanted in the injured mouse myocardium.[102] Before transplantation, the ESCs were genetically pre-labeled with a reporter gene expressing a transmembrane antigen. At different time points post-transplantation, SPIOs conjugated to an antibody recognizing this antigen were administered intravenously and allowed MRI-based localization, determination of viability and proliferation of these stem cells.

Although this method is still relying on the administration of a contrast agent (antibody-(U)SPIO conjugate) with potential drawbacks associated with delivery, biodistribution and clearance, it does not suffer from dilution of the contrast when compared to the traditional way of pre-labeling cells with (U)SPIOs. By administering the antibody conjugated SPIOs at every imaging time point, the method allows longitudinal tracking of the transplanted cell population. The use of this method is, however, limited to immunodeficient models due to possible immunogenic effects of the antigen, unless an existing extracellular target can be used like a known tumor cell surface antigen.

As gadolinium complexes are generally not permeable through the cell membrane, Meade and co-workers developed[103] and improved[104] an imaging strategy for tracking a specific cell signal transduction pathway by synthetizing steroid hormone gadolinium-conjugated contrast agents for MRI that may be internalized into the cells to bind the progesterone receptor. Where gadolinium chelates do not cross the cell membrane due to their hydrophilic nature, conjugating with steroids made the contrast agent more lipophilic and suitable for intracellular targeting. In addition, binding of the contrast agent to a macromolecule as the progesterone receptor resulted in intracellular accumulation, nuclear targeting, increased retention at the receptor site and increased rotational correlation time and hence contrast enhancement. The authors observed significant relaxivity increase[103,104] and showed the usefulness to increase MR contrast in uterus and breast cancer cells *in vivo*.[105]

Other approaches have also used either small molecules or particles, functionalized with small molecules that were direct substrates of cellular transporters. Malaisse *et al.* have used fluorinated mannoheptuloses that are specifically taken up by the glucose transporter-2 (GLUT2) of hepatocytes and pancreatic cells.[106,107] *In vitro* ^{19}F MRI experiments have indicated detectability limits in the order of thousands of cells per microliter. The *in vivo* applicability of this approach remains to be proven. Similarly, linking SPIOs with small molecules can be used to promote uptake in particular cell types. Soenen *et al.* have used magnetoliposomes that contained galactose-terminal phospholipids to promote specific uptake *via* asialoglycoprotein receptors.[108,109] These receptors are over-expressed on hepatocytes and hepatocyte-like cells so that systemic administration of those functionalized magnetoliposomes would allow the specific visualization of hepatocytes.[109]

11.3.3   Contrast Agents Responsive to Chemical Modification

Because exogenous MRI contrast agents are relatively insensitive and require a minimum detection threshold of typically 0.01–10 mM for *in vivo* applications, it is challenging to produce sufficient contrast for MRI when targeting a molecular event that is below this detection threshold. For example, nucleic acids are typically present at concentrations of 1–1000 pM, intracellular proteins at 1–1000 nM. Hence, it proves difficult to reach the threshold for *in vivo* detection with MRI when aiming at visualizing intracellular nucleic acids and proteins with the aid of MRI contrast agents. Extracellular targets (proteins) are more abundant (typical concentrations of 10 nM–10 mM) and therefore an easier target for contrast agents.

Changing the strategy by using a responsive agent that is irreversibly changed upon its interaction with the target, *e.g.* an enzyme, is an interesting alternative. Thereby, the contrast agent is produced at the imaging site and can accumulate to reach the detection threshold for MRI – provided there are no toxic effects or unwanted interactions with the molecular event of interest. In this way, the resulting contrast can report on even small concentrations of enzyme, taking advantage of the high specificity and turnover rate of the enzymatic reaction of interest.

Responsive contrast agents change contrast due to their changed chemical structure and subsequent changed magnetic properties (*e.g.* relaxation, chemical shift, rotational correlation time) in response to dynamic changes in enzymatic properties, but also in physiological and other metabolic properties, of the environment they report on.[11] Initially, contrast agents that are (chemically) modified by cellular changes have been developed for the reporting on the physiological status and metabolic activity of cells.[110]

11.3.3.1   Contrast Due to Modification of Coordination Sites of Lanthanide Chelates

One of the first examples of an MRI contrast agent that reported on gene expression was an enzyme-sensitive agent. This agent (Egad) was based on a macrocycle-chelated gadolinium (Gd) group with an additional galactose moiety that was attached through a bond, cleavable by β-galactosidase (β-gal), an enzyme encoded by the *LacZ* gene.[111] In its "inactive", native state, all coordination spheres were blocked and the effect on T_1 relaxation of its environment was minimal. In the presence of β-gal, the galactose group was enzymatically removed, opening access for water exchange with the gadolinium ion. This change in water coordination resulted in a change in relaxivity for the molecule after cleavage. This initial compound did not result in sufficient change of contrast in *in vivo* models. However, after chemical modification, the contrast agent EgadMe was able to detect expression of β-gal from the introduced *LacZ* gene in *Xenopus*.[112] Yet, the EgadMe molecule is a poor substrate to the enzyme as compared to the colorimetric biochemical β-gal substrate, *ortho*-nitrophenyl-β-galactoside (ONPG). Since the agent does not penetrate cells,[112] its major drawback is that it has to be introduced in the cells by microinjection. An alternative strategy for imaging of β-gal activity uses ^{19}F-labeled molecules as MR active substrates.[113] The same principle of enzyme-activatable contrast agents has been used in a recent study that reports on the first specific *in vivo* MR imaging of inhibitory GABAergic neurons based on a novel MRI contrast agent that is responsive to the neuronal enzyme glutamic acid decarboxylase (GAD) (Figure 11.5). Pre-labeled stem cells and GABAergic neurons engrafted in the mouse brain could be discriminated by *in vivo* MRI due to the intracellular changes in the T_1 relaxation time.[114]

11.3.3.2 Contrast Due to Modified Molecular Rotation of Contrast Agents

A different enzyme-based approach is a contrast agent based on blocking a functional group that binds with high affinity to a larger protein. The binding of an agent to a macromolecule substantially slows the molecular rotation of the gadolinium complex, thereby increasing the relaxivity and contrast. One compound based on this principle is a blood-pool contrast agent for MRI to assess blockage in arteries.[115] It is activated upon binding to human serum albumin in plasma when the groups blocking the high-affinity albumin binding sites are cleaved off enzymatically, which results in MRI detectable relaxivity enhancement.[116]

Another system for imaging specific enzyme activity was also based on changing the rotational correlation time. This so-called MR signal amplification (MRamp) strategy was founded on enzyme-mediated polymerization of paramagnetic substrates into oligomers of higher magnetic relaxivity. The substrates consisted of chelated gadolinium covalently bound to phenols, which served as electron donors during enzymatic hydrogen peroxide reduction by peroxidase. Subsequently, the oxidized monomers self-polymerized into paramagnetic oligomers, leading to a three-fold decrease in T_1 relaxation times. This paramagnetic polymerization probe allowed imaging of tissue peroxidase activity by MRI.[117] One advantage of the MRamp technique is that the same paramagnetic substrate has the potential to be used for identifying different molecular targets by attaching enzymes to antibodies or other target-seeking molecules. The highly reactive short-lived tyramide radicals formed in this reaction could bind to phenol moieties of tyrosine residues. Since the resulting modifications can alter protein function, this might hamper the use of the MRamp system for *in vivo* applications.[118]

11.3.3.3 Contrast Due to Modified Solubility of Contrast Agents

An enzyme-activated contrast agent that has been validated *in vivo*, is a protease-activated solubility-switchable MRI contrast agent, PCA2-switch.[119] This work was based on proof-of-concept studies demonstrating the feasibility of detecting matrix metalloproteinase (MMP) activity by MRI with contrast agents bearing a solubility switch.[120] The novel PCA2-switch contains a

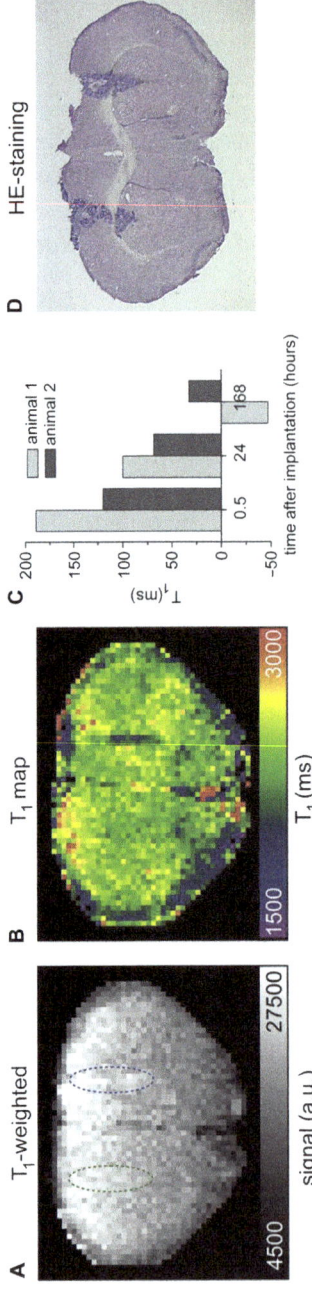

Figure 11.5 Discrimination of implanted native and differentiated cells based on T_1 values in the mouse brain. (A) Representative T_1 weighted image at 0.5 h after implantation of labeled differentiated cells (right striatum) and labeled undifferentiated cells (left striatum). (B) Corresponding T_1 map revealing distinct lower relaxation rate for the differentiated cell graft. (C) Quantitative evaluation of the difference in T_1 relaxivity (ΔR_1) between the embryonic stem cells and the differentiated sample from two animals monitored for up to 7 days revealed a decrease of contrast over time. (D) Histological proof of the localization of cell grafts after sacrificing the animal 7 days after implantation by hematoxylin-eosin staining. Reproduced with permission from Aswendt *et al.*[114] © 2012 Elsevier Inc.

paramagnetic gadolinium chelate (Gd-DOTA), attached *via* a hydrophobic chain to the N-terminus of a peptide sequence, cleavable by a specific enzyme. The presence of a polyethylene glycol (PEG) chain on the C-terminus of the peptide keeps the protease-modulated contrast agent in solution. Upon proteolytic cleavage of the peptide by a specific protease, demonstrated by using MMP-2 for its key role in tumor progression and metastasis, the PEG-chain is released from the protease-modulated contrast agent making it less water soluble so it is retained at the activation site within the tumor.[119] The feasibility of using this approach for detecting MMP-2 expression *in vivo* was shown in a mammary carcinoma mouse model. A significantly higher relative increase in the longitudinal relaxation rate was non-invasively observed with MRI in wild-type MMP-2-expressing tumors as compared to MMP-2 knockdown tumors.[121]

Himmelreich *et al.*[122] were following a similar approach. Insoluble and therefore inactive Gd-DTPA chelates can be activated by intracellular lipases after the contrast agent has been taken up by phagocytosis into dendritic cells. By cleavage of the long aliphatic side chains of the fatty acid moieties of the inactive compound, the contrast agent becomes soluble and is thereby activated in the presence of the expressed enzyme, causing an increase in R_1. This method allowed for *in vivo* observation of pre-labeled dendritic cells after implantation in rat brain.[122] One limitation of these insoluble contrast agents is the prerequisite that cells can take up the particles by phagocytosis.

11.3.3.4 Contrast Due to Aggregation of Superparamagnetic Nanoparticles

Strategies based on (U)SPIOs for sensing of gene expression levels with MRI, so-called nanosensors, have also been proposed. They are typically developed for *in vitro* MRI for probing enzyme activity. To detect changes in gene expression levels, a specifically designed oligonucleotide sequence can be incorporated in a probe that is conjugated to (U)SPIOs. When this probe hybridizes to a matching DNA sequence, single DNA-(U)SPIO particles form oligomeric complexes of larger size that reduce the T_2/T_2^* relaxation times.[123] It was also shown that nanosensors could be developed and applied for rapid screens of telomerase activity detection in biological samples with MRI.[124] The same principle can be applied when conjugating the nanosensor with a specific antibody for sensing the presence of specific protein targets.[123] Conversely, self-assembled particles can be prepared in such a way that they form substrates for specific enzymes, making them act as magnetic relaxation switches. The technique can be used to sense different types of reversible molecular interactions (DNA–DNA, protein–protein, protein–small molecule and enzyme reactions),[123] which makes this technology interesting for use in a high-throughput array. However, due to the large size of the non-activated clustered particles and the non-favorable bioavailability, they are not suitable for *in vivo* applications. Moreover, activation of the switch in the target tissue reduces the relaxivity, resulting in a reduction of the net contrast where a contrast increase would be desirable.

Much smaller are the protease-sensitive nanosensors developed by Schellenberger *et al.*[125,126] These are based on electrostatically stabilized, citrate-coated very small iron oxide particles (VSOPs). While sterically stabilized nanoparticles are larger (> 20 nm), electrostatic stabilization is independent of the coating thickness and allows the design of VSOPs with a diameter as small as 3 nm and a hydrodynamic diameter of 7.7 nm.[126] Smaller particles and targeted MRI probes developed with these particles should have improved the bioavailability for imaging of extravascular targets. Here, the VSOPs were targeted by coupling them to a cleavage motif of MMP-9,[125] an enzyme important in inflammation, atherosclerosis, tumor progression and many other diseases with alterations of the extracellular matrix. The resulting MMP-9-activatable protease-specific iron oxide particles were rapidly activated, resulting in aggregation and increased T_2*-relaxivity.[125] The same group reported on the use of VSOPs in the design of a targeted MR contrast agent for imaging of apoptosis.[126] These targeted particles have a hydrodynamic diameter of < 20 nm and should therefore be suitable for *in vivo* imaging by MRI.

11.3.3.5   Limitations and Challenges

Although the principle of the enzyme-based contrast agent activation for MRI-based molecular imaging is very appealing because of the seemingly endless applications for probing altered enzyme-states in diseases, few studies have proven their *in vivo* applicability. Among potential limitations are the relatively high contrast agent concentrations needed but also insufficient substrate access, penetration and delivery to the cells of interest, thereby limiting the potential for *in vivo* imaging for many of the developed enzyme-responsive contrast agents (see also Table 11.1). Due to the inherent low sensitivity, accumulation of enough contrast agent at the site of interest is challenging, *e.g.* CEST and PARACEST agents often require a high concentration of the probe (> 10 mM) and relatively intense irradiation pulses,[128] which can be difficult to translate to clinical practice.

The same hurdle, but to a lesser extent, has to be overcome by contrast agents responsive to physical or metabolic changes, like the production of metabolites, metal ions concentrations, temperature, pH and others. In addition, for a contrast agent to be able to report on physiological changes, it is crucial that the contrast depends only on the parameter that is to be measured and is entirely independent of the concentration of the contrast agent at the imaging site. Apart from direct monitoring by localized MR spectroscopy in the case of ^{19}F-based contrast agents, this requires some indirect measures. In the case of *in vivo* pH monitoring, the pH was measured in one study by subsequent injection of two different Gd(III)-based contrast agents, one responsive to the pH and one pH-irresponsive that was used as control to account for *in vivo* pharmacokinetics, which were assumed to be identical for the two agents (Figure 11.6).[129] Another more recent example was a single-injection approach. Two contrast agents were co-injected, a dysprosium-containing

Table 11.1 Contrast mechanisms responsive to intrinsically or transgenically expressed proteins. Examples of reporter systems that have a genetic component and are to be used in combination with an exogenously administered contrast agent are listed. These contrast agents are responsive to enzymes or proteins (*i.e.* transporters, receptors, *etc.*) that are intrinsically expressed or through genetic engineering of target cells.

Reporter	Contrast mechanism	Observed effect	Tested in	Reference
β-galactosidase	β-gal expressed from the *LacZ* gene cleaves a synthetic Gd^{3+} CA changing its relaxivity	Increase of 60% in T_1-weighted signal (12 T)	*Xenopus laevis* embryos transfected with *LacZ*	112
Transferrin receptor	SPIO-particles conjugated to transferrin bind to and are internalized by over-expressed Tfr	T_2-weighted MRI shows an 80% signal change (1.5 T)	Tfr-expressing gliosarcoma cells implanted in mouse	94
Tissue peroxidase	Mediates polymerization of paramagnetic substrates into oligomers of higher R_1 relaxivity	Measurable MR T_1-weighted signal intensity increase in supernatants of cells expressing E-selectin (1.5 T)	E-selectin expressing human endothelial cells	117
Matrix metallo-proteinase-2	Upon proteolytic cleavage by MMP-2, the PEG chain is detached from the Gd(III)-CA, which becomes less water soluble	Two-fold higher MMP-2 activity (R_1) in wt-carcinoma compared to MMP-2 knockdown tumor cells (7 T)	Mammary carcinoma mouse model	119
Lipase	Insoluble Gd(III)-chelate is activated by intracellular lipase, increasing R_1	20 000 cells per µl were detectable for up to 72 h after implantation on T_1-weighted MRI (7 T)	Pre-labeled dendritic cells implanted in rat brain	122
Progesterone receptor	Binds and internalizes steroid-hormone Gd(III)-conjugated CA	60% increase in R_1 compared to controls (9.4 T)	Breast cancer epithelial cell culture	104
Divalent metal transporter 1	Mediates intra-cellular uptake of systemically administered Mn^{2+}	Altered T_1 resulting in visible MEMRI contrast	Mouse brain glioma	127

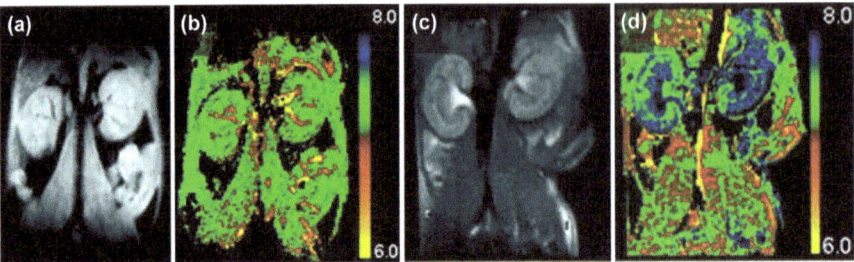

Figure 11.6 pH imaging by MRI. Left panel: pH image of a control female SCID
mouse. (a) Fat-suppressed, proton-density-weighted reference anatomic
image showing a coronal view of the kidneys and nearby tissues. (b)
Corresponding calculated pH image. A pH of 6.0 was measured in a
urine sample collected immediately following the imaging experiment.
Right panel: pH image of an acetazolamide-treated mouse. (c)
T_1-weighted, post-contrast reference anatomic image showing the
kidneys and nearby tissues. The cortex, medulla, calyx and proximal
segments of both ureters are visible. (d) Corresponding calculated pH
image. A pH of 8.2 was measured in a urine sample collected immediately
following the imaging experiment.
Images reproduced with permission from Raghunand et al.[129] © 2003
Wiley-Liss, Inc.

agent was used to produce a concentration map based on its T_2^* effect, the
other Gd-containing agent induced pH-dependent changes in T_1.[130] In a second
approach, a much better temporal resolution was reached by exploiting the
CEST effect of the two amide groups of iopamidol (a clinically approved X-ray
contrast agent). By using both exchangeable proton pools that display different
pH-dependent exchange rate constants, only the relative ratio of the saturation
transfer effect generated by selective irradiation of the pools and not the
absolute concentration of the contrast agent is important to report on the
pH.[131] Using the same ratiometric approach, Gallagher and co-workers
showed that tissue pH can be imaged in vivo from the ratio of the signal
intensities of hyperpolarized bicarbonate $H^{13}CO_3^-$ and $^{13}CO_2$ following
intravenous injection of hyperpolarized $H^{13}CO_3^-$.[132] Molecular imaging of pH
with MRI is a promising tool that would have important clinical applications in
ischemia, inflammation and cancer diagnosis and treatment, but all of these
methods still have serious hurdles to overcome on their way to the clinic.[133]

Despite the vast variety of novel contrast agents responsive to physiological
or enzymatic changes that have been developed over the past years (for more
extensive reviews and examples, see references 134, 135), only a few examples
report on successful in vivo application due to the limitations already
mentioned above.

11.3.4 Genetic Reporter Systems

Enzymes that are intrinsically expressed or proteins expressed after genetic
engineering can contribute to the generation of MR contrast or accumulate

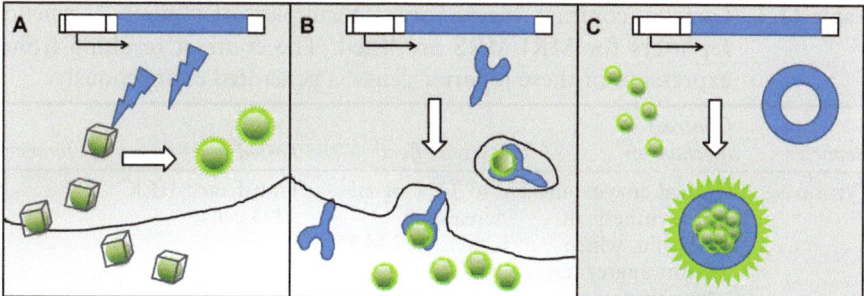

Figure 11.7 Mechanisms for (semi-)genetic control of MRI contrast. (A) A genetically expressed enzyme (blue) can alter an exogenously administrated contrast agent (green) that becomes active (green halo) upon enzyme processing. One enzyme molecule can process many contrast agent molecules, leading to signal amplification so that even low-level gene expression can (theoretically) yield a high level of activated contrast agent; see for example Louie *et al.*[112] (B) In an alternative semi-genetic system, gene expression leads to the synthesis of a cell surface protein (blue) that acts as a receptor for an exogenous contrast agent (green) and subsequently promotes internalization of the agent into vesicles (see for example Bluml *et al.*[60] and Lee *et al.*[103]). (C) MRI contrast can be purely genetically controlled, without the need for synthesis or delivery of exogenous agents, when a gene directs intracellular production of a metalloprotein (blue) that becomes a contrast agent in complex with endogenous metal ions (green).

MR contrast agents.[136] Thereby, the generated contrast can be used as a qualitative (on/off) or quantitative measure for protein expression and therefore as a genetic reporter. Different mechanisms for genetic reporter systems are illustrated in Figure 11.7.

11.3.4.1   Genetic Reporters for MRI

β-galactosidase, expressed from the *LacZ* gene, can be considered a "semi"-genetic reporter for MRI, since it is to be used in combination with the exogenous administration of a contrast agent that can be activated by the enzyme. This was first demonstrated for the visualization of gene expression in *Xenopus.*[111,112] Examples of "semi"-genetic and "purely" genetic reporter systems are given in Tables 11.1 and 11.2, respectively. The genes proposed as "pure" MRI reporters are typically genes encoding proteins that sequester endogenous paramagnetic ions such as iron (Fe^{2+}). The rationale is that if such a protein can be over-expressed in a genetically altered cell, the overall relaxivity of the cell might be changed in a sufficient way to distinguish it from the background of host tissue by using MRI. In contrast to *in vitro* labeled cells, it is the objective of reporter gene-based MRI to generate contrast that depends on cellular properties like enzyme expression. A first example of a proposed MRI reporter based on this principle is tyrosinase. It is the most critical enzyme in the synthesis of melanin, a high-molecular-mass pigment capable of binding

Table 11.2 Genetic contrast mechanisms. Examples of "purely" genetic reporters for MRI/MRS are listed. The contrast resulting from expression of these reporter genes is generated endogenously.

Reporter	Contrast mechanism	Observed effect	Tested in	Reference
Tyrosinase	Critical enzyme in the synthesis of melanin, which forms aggregates that bind intracellular iron	Up to 37% signal increase on T_1-weighted MRI (1.5 T)	Transfected HEK 293 cells	137
Ferritin	Sequesters iron from the intracellular labile iron pool	16.7 ± 1.6 T_2*-weighted CNR enhancement relative to untransduced tissue (for 24 voxels, 11.7 T)	Adenoviral vector transduction in mouse brain	141
MagA	A bacterial iron transporter involved in the production of magnetic iron oxide particles	Significant signal loss in T_2*-weighted MRI (9.4 T)	*MagA*-transduced cells transplanted into mouse brain	142
Artificial LRP reporter	A tandem of lysine-rich proteins (LRP) produces CEST contrast	Signal-intensity (SI) difference of ± 3.76 ppm and SI change of 57% (11.7 T)	Vector-transfected LRP glioma cells inoculated to mouse brain	143
Myoglobin	Heme iron protein and oxygen-sensor (T_2)	No statistically significant changes detected by ^{31}P NMR (4.7 T)	Myoglobin-over-expressing transgenic mice	144
Hemoglobin	Heme iron protein and oxygen-sensor (T_2)	± 50% changes in T_2 signal intensity when external O_2 levels are manipulated from 0% to 21% (14.1 T)	Hemoglobin-injected flies	145

iron, which results in modulation of T_1 and T_2/T_2* relaxation times.[137–139] Despite the fact that the melanin synthesis is regulated by different proteins and takes place in specialized membrane-bound organelles (melanosomes), it is possible to induce melanin synthesis by simply transfecting cells with the tyrosinase gene.[140] Over-expression of tyrosinase in cultured human cells increased metal uptake and changed MRI contrast as compared to controls.[137]

Apart from the readily detectable contrast evoked by the expression of melanin, another strategy involves the introduction of a paramagnetic contrast agent for the detection of tyrosinase activity. Interaction of the paramagnetic substrate with tyrosinase resulted in oligomers and conjugates of the contrast agent with albumin.[146] Both products resulted in a decrease in T_1 relaxation times. Though the exact mechanism of the observed signal enhancement is unknown, the study indicates that melanoma cells in a cancer mouse model can scavenge the contrast agent, which resulted in contrast enhancement on MRI. However, the necessity to administer a contrast agent limits its potential applicability.

There are some general constraints to the use of the tyrosinase gene as a reporter system for MRI. The cDNA of tyrosinase is large (1.8 kb), which makes the introduction of this gene in common (viral) vector systems difficult. Another disadvantage is the relatively high toxicity of the metabolites produced during melanogenesis and of melanin itself. Placing the tyrosinase expression under regulated control[139] can be part of a solution to this issue. Apart from toxicity concerns, enhancement of MRI sensitivity is needed before this imaging strategy can be used for *in vivo* studies. This is also illustrated by the lack of *in vivo* studies using melanin as a marker protein for MRI detection of gene expression.

Over-expression of the transferrin receptor (Tfr) and subsequent increased iron accumulation is another potential way to generate contrast using a genetic marker for MRI. Mouse fibroblasts were engineered to over-express the human Tfr, which led to detectable intracellular iron accumulation in ferritin and increased T_2* contrast on MRI.[147] However, the high endogenous Tfr levels require very high levels of Tfr over-expression to yield detectable contrast, limiting the applications of Tfr as a reporter gene. Over-expression of engineered Tfr in a mouse tumor model did not result in measurable contrast, showing that endogenous iron scavenging was insufficient to generate contrast. Nonetheless, this research has pointed to ferritin as a potential reporter gene. Using co-expression of the Tfr and ferritin (the heavy ferritin subunit, FerrH) as a way to induce cellular MRI contrast resulted in relaxivity enhancement in iron supplemented cells.[148] From this study, it cannot be judged how far Tfr contributed to the specific contrast, so that administration of exogenous transferrin coupled to magnetic particles is required for the Tfr approach.

Iron has a crucial role in the body, but excess of free iron can cause damage to cells because of the Fenton reaction, which generates reactive oxygen species.[149] The harmful excess of intracellular iron (Fe^{2+}) is therefore oxidized and stored in the biochemically "safe" (Fe^{3+}) (ferrihydrite) core of the ferritin protein. Ferritin is a heteropolymer that consists of 24 subunits of heavy (FerrH) and light (FerrL) chain subunits, forming a spherical shell 13 nm in diameter with an 8 nm inner cavity where iron is stored in a hydrated iron oxide mineral core that significantly affects T_2 relaxation.[150,151] The concept of using ferritin as a reporter gene for MRI is based on the natural iron sequestering function of this extensively studied protein. Over-expression of ferritin would shift the labile iron pool towards the ferritin bound storage form, which induces compensatory upregulation of the Tfr and increases iron uptake from the environment to restore iron homeostasis.[152] The *in vivo* Tfr upregulation and increased superparamagnetic iron content in FerrH transgenic grafts[153] provided further support to this model. To use ferritin over-expression as a reporter for MRI, it is engineered to be constitutively over-expressed. The first use of ferritin to track cells with MRI was carried out by stably transfecting C6 glioma cells with the murine FerrH and enhanced green fluorescent protein (eGFP) under tetracycline control.[154] Upon inoculation of these tumor cells in nude mice, significant decrease of T_2 relaxation times was observed without the addition of any contrast agent or iron supplementation. A significant difference in T_2 was also found between the tumor in a mouse flank formed by mouse embryonic stem cells that were transduced with a lentiviral vector encoding for FerrH and the wild-type tumor (Figure 11.8).[153] Transgenic mice expressing FerrH in a tissue-specific and tetracycline-dependent manner showed tissue-dependent changes in T_2 values that were MRI detectable.[155] The feasibility of

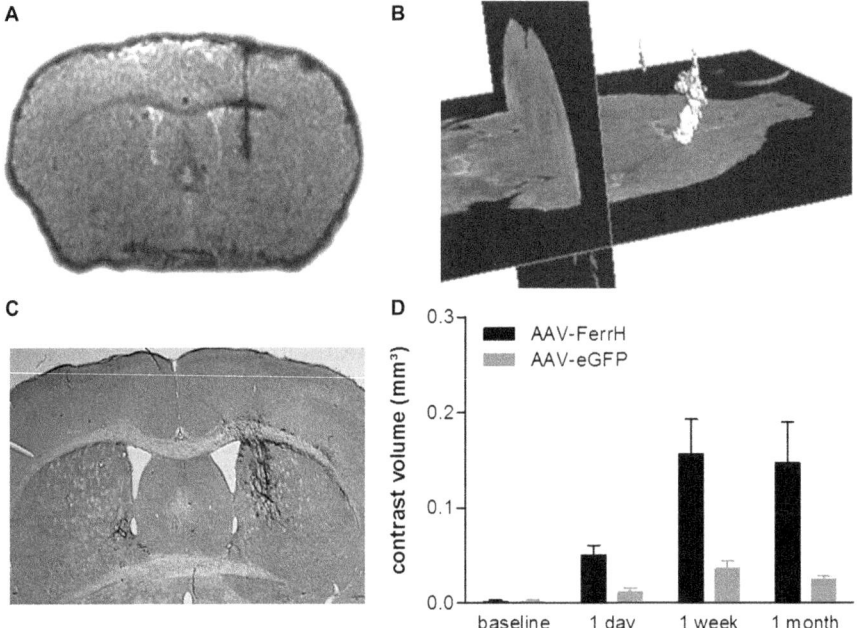

Figure 11.8 *In vivo* MRI detection of genetically induced MRI contrast in mouse brain. (A) 3D T_2*-weighted *in vivo* MR image of a mouse acquired 1 week after injection of AAV-FerrH-T2A-fLuc and AAV-eGFP-T2A-fLuc vectors in the right and left striatum, respectively. (B) 3D graphic representations of the contrast volume from the same mouse, delineated using image processing software on normalized 3D T_2*-weighted *in vivo* MRI data set acquired 1 month post-injection. (C) DAB-enhanced Prussian blue staining, showing the presence of iron at the site of AAV-FerrH injection (right striatum) in contrast to the contralateral control injection site (left striatum). (D) Quantification of the contrast volume using the normalized 3D T_2*-weighted *in vivo* MR images at different time points post-injection for AAV-FerrH-T2A-fLuc and AAV-eGFP-T2A-fLuc (for n = 5, error bars represent SEM).

using FerrH as an MRI reporter by *in situ* viral vector mediated delivery to the brain has been shown for different viral vector systems.[141,156,157] Upon injection of a FerrH over-expressing vector into the mouse brain, significant hypointense MRI contrast is produced at the site of vector injection.[141,156,157] Depending on the viral system used, vector-based delivery of MR reporter genes can elicit unspecific contrast due to the vector itself at the site of injection.[157,158] However, this confounding effect occurs at much smaller distances compared to the large susceptibility effect of SPIO-mediated labeling of the subventricular zone-cells that impedes tracking cells near the site of labeling.[89,90,159–162] Using lentiviral vectors expressing ferritin to label the endogenous neural stem cells, the stem cell progeny generated at the subventricular zone and migrating and integrating in the olfactory bulb could be detected and quantified with *ex vivo* MRI, but the system lacked sensitivity for *in vivo* tracking of the migrating stem cell progeny.[156] Ferritin is predominantly antiferromagnetic, meaning that most of its spins align in a regular pattern with neighboring spins aligned in opposite directions. Only the small fraction of unpaired spins on the surface of the ferritin crystal core contribute to a net magnetic moment,[163] which makes this approach several orders of magnitude less sensitive than (U)SPIO labeling.[164] Native ferritin is therefore a rather weak T_2 contrast agent. In an effort to improve the paramagnetic properties of ferritin, linking the light and heavy chains of ferritin resulted in a slightly larger protein cage and improved MR contrast *in vitro*.[165] Using an adenoviral vector encoding this chimeric ferritin reporter to label endogenous neural progenitor cells, migrating cells could be visualized along the rostral migratory stream with *in vivo* MRI at 11.7 T.[158] However, no contrast was seen in the olfactory bulb where these cells arrive and integrate, therefore improving the sensitivity of ferritin remains a challenge. Replacing its native iron core with a synthetic paramagnetic core[166] has proven to result in effective T_2/T_2^* contrast both *in vitro*[80] and *in vivo*.[167]

In the search for other MR reporter gene candidates, gene-mediated cellular production of magnetic iron-oxide nanoparticles using genes that are present in naturally occurring magnetotactic bacteria has been explored. These bacteria show motility that is thought to be directed by the earth's magnetic field.[168] They produce magnetosomes, which are naturally synthesized intracellular magnetic structures. These magnetosomes exist out of membrane-enveloped nanosized (< 100 nm) single crystalline magnetite (Fe_3O_4) units and vary in shape and size, depending on the species of bacteria. Unraveling the essential genes for magnetosome formation might be used to generate MR reporter genes for labeling of mammalian cells. The first magnetosome-associated gene studied as an MRI reporter candidate was *magA*, a bacterial gene involved with iron transport in magnetotactic bacteria. Upon implantation of *magA* trans-fected mammalian cells in mouse brain, magnetic iron-oxide nanoparticles were formed, which allowed subsequent visualization of the cells with *in vivo* MRI.[142] However, *magA* as a putative MR reporter gene is rather controversial as its implication in magnetosome formation has recently been contested.[169] A second magnetosome gene, *Mms6*, is being explored as a reporter for MRI contrast.[170] The Mms6 protein has a high affinity for iron proteins[171] and plays a critical role in the crystallization of magnetite within magnetosomes, most probably in controlling the magnetite crystal morphology.[172] Whether *Mms6* is the genetic MR reporter we are all looking for remains to be established.

So far, introduced MRI reporter genes were all based on the sequestration of available paramagnetic or superparamagnetic ions (*i.e.* iron). A completely different strategy was followed during the development of an artificial reporter gene system generating CEST contrast based on frequency-selective targeting of amide protons of expressed proteins.[143] Stable genetic alteration of glioma cells resulted in over-expression of lysine-rich protein (LRP) that allowed detection by CEST imaging based on the selective saturation of the amide proton frequency and observation of the reduced water frequency. Upon transplantation of these cells in the mouse brain, the genetically altered tumor cells were detectable using a saturation transfer sequence,[143] demonstrating that CEST systems such as this are applicable *in vivo*. One of the main advantages of CEST imaging is that the contrast can be turned on selectively by simply adding a selective saturation pulse to the MR imaging sequence.[173] Since amide proton exchange is base catalyzed and thus strongly pH dependent, the contrast is lowered by an order of magnitude during ischemia or cell death, giving LRP the potential for imaging cell viability. Labeling of multiple cell populations became a possibility with the introduction of additional frequency-specific reporters.[43] Since LRP is a highly charged molecule, this could lead to unwanted interactions, so the system might be engineered to reduce the charge. Because LRP contains 200 repeats, it is not compatible with reverse transcription and therefore not suitable for incorporation in a retroviral vector system, limiting the possibilities for gene transfer. As an alternative, the use of cytosine deaminase (CDase) was explored as a reporter gene for CEST MRI.[174] CDase catalyses the conversion of cytosine into uracil through a process of deamination, which could be detected by CEST MRI in bacteria and mammalian cells engineered to over-express CDase.[174] CDase also deaminates the prodrug 5-fluorocytosine (5-FC) to form the chemotherapeutic agent 5-fluorouracil (5-FU), whose detectability by CEST MRI was shown in transgenic cells.[174] This demonstrated the potential use of CDase as CEST reporter gene; however, the low sensitivity of CEST agents and contrast remains a hurdle for *in vivo* molecular MRI.

11.3.4.2   Genetic Reporters for MRS

The first reporter genes for MRS were genes encoding for enzymes that catalyze the conversion of ATP into ADP, which was monitored by ^{31}P MRS. One widely used example is the creatine kinase (CK) gene from mouse brain. If CK is specifically over-expressed in the liver, the synthesized phosphocreatine (PCr) could be directly detected by MRS due to the absence of endogenous PCr in the liver.[175] The CK reporter gene has been used in an adenoviral vector context to study gene therapy strategies in the mouse liver.[176] In addition, monitoring the PCr concentration with ^{31}P MRS indicated that CK can serve as a reporter for the vector-mediated gene delivery of the low-density lipoprotein- (LDL)-receptor to the liver of LDL-receptor knock-out mice.[177] It may also be usable as a marker gene in tissues that normally express CK, because the increment in CK activity can be determined by magnetization-transfer experiments as shown in mouse skeletal muscle.[178] The main limitations are the low sensitivity and subsequent low spatial resolution of MRS compared to MRI. Advantages are the absence of harmful side effects like an immunological response to the enzyme, since the end product is naturally occurring in muscle and brain.

The same strategy was employed when using the arginine kinase (AK) gene from *Drosophila melanogaster* in a viral vector for reporter gene transfer to the mouse muscle.[179] The product of the AK enzyme when converting ATP into ADP is phosphoarginine, a metabolite not present in vertebrate cells that could be detected by ^{31}P MRS after injection of an adenoviral vector encoding AK in the muscle. The expression and activity of AK could be followed *in vivo* for up to 8 months post-injection. Although AK produces a unique signal in mammalian muscle, it is a non-mammalian protein that is expected to induce an immune response.

Another MRS approach is based on the catalytic conversion of a prodrug by the reporter enzyme. This prodrug will be converted into an active drug in the targeted cells only upon internalization of the prodrug in the cells containing the reporter enzyme. An example is cytosine deaminase (CDase) that catalyzes the prodrug 5-fluorocytosine (5-FC) into 5-fluorouracil (5-FU), a chemo-therapeutic drug.[48,180,181] ^{19}F MRS can be used to monitor this reaction and subsequently treatments involving CDase for gene therapy of tumors.[48,180,181] The deamination of 5-FC to form 5-FU, which is then released to the tumor cells, can be followed by *in vivo* ^{19}F MRS for therapy assessment. Since 5-FC crosses the blood-brain barrier, this approach could be translated to brain studies as shown by visualization of intracerebral gliomas.[180] Similarly, bacteria expressing CDase were suggested to convert the non-toxic 5-FC into the cytotoxin 5-FU for more efficient tumor therapy.[48,182,183] *In vivo* ^{19}F MRS has demonstrated that this conversion can be monitored over time and be used for therapy assessment (Figure 11.9). Thereby, factors influencing the expression and efficacy of the therapeutic gene, for example due to the microenvironment and biological characteristics of the tumor, can be assessed.[48,184]

Based on the introduction of a fluorine atom to the colorimetric biochemical indicator, ONPG, 4-fluoro-2-nitrophenyl-β-D-galactopyranoside (PFONPG) was proposed to monitor the activity of β-galactosidase by MRS.[113] In cells transduced with a *LacZ*-encoding adenoviral vector, β-gal catalyzed the release of the aglycone (PFONP) from PFONPG that was accompanied by color formation and the appearance of the aglycone ^{19}F MRS signal. Since PFONP is pH-sensitive, it holds potential for direct pH determination at the site of enzyme activity. However, as PFONP is toxic and caused cell lysis in some cases,[113] PFONPG should be regarded only as an interesting prototype molecule for the use with MRS gene reporter molecules as β-gal. Nevertheless, it holds the potential to be used as a gene-activated chemotherapeutic as was proposed for CD. However, since PFONP is not trapped within the cell, the

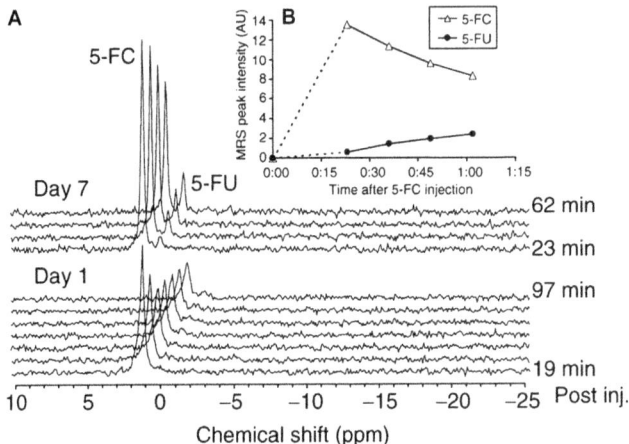

Figure 11.9   *In vivo* [19]F MR spectroscopy of a colon tumor mouse xenograft model
(A). The tumor was treated by intratumoral injection of *Salmonella
typhimurium* expressing cytosine deaminase. 5-fluorocytosine was
administered before ^{19}F MRS for the assessment of therapy after one
(lower series) and seven days after initiation of therapy by validating the
5-fluorocytosine and 5-fluorouracil signal. (B) Evolution of the MR
signals in the tumor at day 7.
Adapted with permission from Dresselaers *et al.*[48] ©2003 Cancer
Research UK.

information about β-gal activity would be difficult to localize to specific tissues.
A different reporter molecule (2-fluoro-4-nitrophenol-β-D-galactopyranoside
or OFPNPG), which appears to be less toxic, has been suggested for the
assessment by ^{19}F chemical shift imaging (CSI).[185]

There have been attempts to develop an imaging method based on the [1]H
NMR signals of myoglobin to measure tissue oxygenation *in vivo* (reference 186
and references therein). The authors reported that a chemical shift imaging
strategy using the deoxymyoglobin signal is feasible for monitoring both global
as well as localized tissue oxygen distribution in human gastrocnemius muscle.
The fast paramagnetic relaxation of the deoxymyoglobin signal and its
low cellular concentration pose a challenge to obtaining a high-resolution map
of the spatial distribution in the muscle as illustrated by the voxel sizes of
$100-300 \, cm^3$ reached in that study.

11.3.5   Direct Monitoring of Metabolic Pathways

Direct observation of metabolic transformation of substrates can be achieved
by using MR-active isotopes of low natural abundance. This is in particular the
case for ^{13}C MRS, where basically no background signal from the host tissue is
observed. The low natural abundance of ^{13}C requires the utilization of
[13]C-labeled compounds. A broad range of ^{13}C MRS applications followed the
metabolic fate of infused labeled compounds *in vivo*.[65-68] As shown in

Figure 11.2, even the localized monitoring of glucose utilization is possible in the mouse brain. In contrast to PET, ^{13}C MRS has the advantage to monitor the metabolic fate of the substrate. Its low sensitivity prevented its application for true imaging applications. This might change with the emerging field of hyperpolarized ^{13}C MRI/MRS (see also Chapter 9).[50,51,54,57,58,69,187] Many applications have already shown that the kinetic uptake and biodistribution of hyperpolarized ^{13}C-labelled compounds can be used for MRI-based perfusion and angiographic studies.[50,54] Due to the relatively long relaxation times of pyruvate and its metabolites, most MRS applications so far studied the conversion of pyruvate to alanine, lactate and HCO_3^-. Thereby, *in vivo* spatial resolutions of a few mm and a temporal resolution below 8 seconds is feasible using clinical MRI scanners.[57] Applications consisted in monitoring skeletal muscle (Figure 11.2)[57] or cardiac metabolism[188] in normal animals to assessing therapy response in tumors.[187] Similar to PET, contrast is generated based on metabolic differences in different tissue types within several half-lives of the hyperpolarized compound.

11.4 Limitations and Perspectives

The low sensitivity of MRI and MRS compared to other modalities used in molecular imaging applications will remain a major challenge. Many technical and methodological improvements like higher field strength, dedicated RF coils, superconducting RF coils and others have resulted in substantial improvements. However, only the applicability of hyperpolarized ^{13}C- and ^{15}N-labeled compounds has resulted in sensitivity improvements in the order of several orders of magnitude. This is still less sensitive than imaging with comparable PET tracers. However, in contrast to PET, MRI/MRS of hyperpolarized ^{13}C-labeled compounds not only provides information on the time profile of the biodistribution of the respective substrates but also on their metabolic fate. A disadvantage of MRI using hyperpolarized compounds is their relatively short relaxation time so that only rapid metabolic processes can be covered.

Although MRI provides excellent resolution that is frequently used in non-invasive cell imaging, it cannot easily provide direct information on enzyme activities. Contrast agents for MRI that can be changed due to direct inter-action with intrinsic enzymes are often based on lanthanide chelates that are even less sensitive than iron oxide particles. This restricts their applicability to situations where sufficiently high local concentrations of the contrast agents can be achieved. When using enzyme-substrate-based approaches, signal amplification can take place (one enzyme catalyzes more substrate conversions) or high relaxivity contrast agents like (U)SPIOs can be used. Similarly to other methods, systemic administration of contrast agents/tracers faces the challenge of delivery to the target site. The dependency of substrate administration has some inherent disadvantages, like the problem of sufficient contrast uptake and accumulation at the site of interest. There are methods in development to enhance uptake of nanoparticle-based agents, including attachment of membrane-translocating peptide sequences like the HIV-tat-peptide.[87] Non-specific MR contrast will also remain an issue of concern, in particular for applications targeting receptors/enzymes that are not exclusively present at or in a particular cell type.

"Pure" genetic reporters could be a solution to the problems involved with the administration of contrast agents as they rely on endogenously available molecules for contrast generation. In addition, genetic reporters can be incorporated in gene delivery systems like viral vectors or in expressing cassettes when generating transgenic animals, where their expression can be coupled to the expression of a (therapeutic) transgene of interest and to other reporter genes for multimodality imaging. In this case, dilution of contrast agents and subsequent signal loss upon cell division would not be a concern. Moreover, incorporation in an inducible expression system is possible, which holds the potential for dynamic sensing of gene expression under condition that the contrast can be cleared by normal cellular pathways to restore the signal to baseline when it needs to be in the "off" position. The latter is often not the case for MRI-based contrast. Also, the sensitivity of those systems is often relatively low.

The development of CEST contrast agents that can be coupled to the presence of certain enzymes and their first *in vivo* applications holds great promise in potential "multicolor" MRI applications. As CEST contrast is based on frequency differences and the applicability of labeling different cell populations with different CEST agents has been shown,[43] one can expect future applications where different molecular events can be visualized almost simultaneously. Again, the main bottleneck remains the low sensitivity.

The main advantages of MR techniques, namely superb spatial resolution, excellent soft tissue contrast and the ability to monitor metabolic processes non-invasively and longitudinally, will remain the driving forces for their application in molecular imaging. MRI has become a valuable tool in particular for following and guiding cell therapy studies. As other powerful modalities like nuclear and optical methods also have limitations, one potential solution to overcome them is to turn to multimodal imaging approaches.

Acknowledgements

We are grateful for support by the KU Leuven Program Financing "IMIR", the Flemish Science Foundation (FWO, G.0869.12N), the European FP7-NMP Vibrant (228933), the IWT SBO "iMAGiNe" (IWT 80017) and the EC-FP7 network "European Network for Cell Imaging and Tracking Expertise" (ENCITE, 2007-201842).

References

1. T. F. Massoud and S. S. Gambhir, *Genes Dev.*, 2003, **17**, 545.
2. M. Edinger, Y. A. Cao, Y. S. Hornig, D. E. Jenkins, M. R. Verneris, M. H. Bachmann, R. S. Negrin and C. H. Contag, *Eur. J. Cancer*, 2002, **16**, 2128.

3. R. Blasberg and J. Gelovani Tjuvajev, *Mol. Imaging*, 2002, **1**, 160.
4. O. Gheysens and F. M. Mottaghy, *Methods*, 2009, **48**, 139–145.
5. Y. Waerzeggers, P. Monfared, T. Viel, A. Winkeler, J. Voges and A. H. Jacobs, *Methods*, 2009, **48**, 139.
6. S. S. Gambhir, *Nat. Rev. Cancer*, 2002, **9**, 683.
7. J. J. Flint, C. H. Lee, B. Hansen, M. Fey, D. Schmidig, J. D. Bui, M. A. King, P. Vestergaard-Poulsen and S. J. Blackband, *Neuroimage*, 2009, **46**, 1037.
8. M. Hoehn, E. Küstermann, J. Blunk, D. Wiedermann, T. Trapp, S. Wecker, M. Föcking, H. Arnold, J. Hescheler, B. K. Fleischmann, W. Schwindt and C. Bührle, *Proc. Natl Acad. Sci. USA*, 2002, **99**, 16267.
9. C. Heyn, J. A. Ronald, S. S. Ramadan, J. A. Snir, A. M. Barry, L. T. MacKenzie, D. J. Mikulis, D. Palmieri, J. L. Bronder, P. S. Steeg, T. Yoneda, I. C. MacDonald, A. F. Chambers, B. K. Rutt and P. J. Foster, *Magn. Reson. Med.*, 2006, **56**, 1001.
10. E. M. Shapiro, K. Sharer, S. Skrtic and A. P. Koretsky, *Magn. Reson. Med.*, 2006, **55**, 242.
11. U. Himmelreich and T. Dresselaers, *Methods*, 2009, **48**, 112.
12. U. Himmelreich and M. Hoehn, *Minim. Invasive Ther. Allied Technol.*, 2008, **17**, 132.
13. M. Hoehn, E. Küstermann, J. Blunk, D. Wiedermann, T. Trapp, S. Wecker, M. Föcking, H. Arnold, J. Hescheler, B. K. Fleischmann, W. Schwindt and C. Bührle, *Proc. Natl Acad. Sci. USA*, 2002, **99**, 16267.
14. M. Hoehn, D. Wiedermann, C. Justicia, P. Ramos-Cabrer, K. Kruttwig, T. Farr and U. Himmelreich, *J. Physiol.*, 2007, **584**, 25.
15. E. T. Ahrens, P. T. Narasimhan, T. Nakada and R. E. Jacobs, *Prog. Nucl. Magn. Reson. Spectrosc.*, 2002, **40**, 275.
16. A. P. Koretsky, *Neuroimage*, 2012, **62**, 1208.
17. T. Barrett, M. Brechbiel, M. Bernardo and P. L. Choyke, *J. Magn. Reson. Imaging*, 2007, **26**, 235.
18. P. Mukherjee, S. W. Chung, J. I. Berman, C. P. Hess and R. G. Henry, *AJNR Am. J. Neuroradiol.*, 2008, **29**, 843.
19. P. Mukherjee, J. I. Berman, S. W. Chung, C. P. Hess and R. G. Henry, *AJNR Am. J. Neuroradiol.*, 2008, **29**, 632.
20. A. C. Silva and N. A. Bock, *Schizophr. Bull.*, 2008, **34**, 595.
21. D. Delli Castelli, E. Gianolio, S. Geninatti Crich, E. Terreno and S. Aime, *Coord. Chem. Rev.*, 2008, **252**, 2424.
22. J. W. M. Bulte and D. L. Kraitchman, *NMR Biomed.*, 2004, **17**, 484.
23. T. Michaelis, S. Boretius and J. Frahm, *Prog. Nucl. Magn. Reson. Spectrosc.*, 2009, **55**, 1.
24. I. R. Young, *Methods in Biomedical Magnetic Resonance Imaging and Spectroscopy*, John Wiley & Sons, New York, 2000.
25. M. A. Brown and R. C. Semelka, *MRI: Basic Principles and Applications*, Wiley-Liss, New York, 2003.
26. P. Caravan, J. J. Ellison, T. J. McMurry and R. B. Lauffer, *Chem. Rev.*, 1999, **99**, 2293.

27. C. F. Geraldes and S. Laurent, *Contrast Media Mol. Imaging*, 2009, **4**, 1.
28. S. Aime, C. Cabella, S. Colombatto, S. Geninatti Crich, E. Gianolio and F. Maggioni, *J. Magn. Reson. Imaging*, 2002, **16**, 394.
29. A. S. Arbab and J. A. Frank, *Regen. Med.*, 2008, **3**, 199.
30. W. Liu and J. A. Frank, *Eur. J. Radiol.*, 2009, **70**, 258.
31. S. Forsén and R. Hoffman, *J. Chem. Phys.*, 1963, **39**, 2892.
32. J. I. Tanttu, R. E. Sepponen, M. J. Lipton and T. Kuusela, *J. Comput. Assist. Tomogr.*, 1992, **16**, 19.
33. E. Gillies, W. A. Szarek and M. C. Baird, *Can. J. Chem.*, 1971, **49**, 211.
34. K. M. Ward, A. H. Aletras and R. S. Balaban, *J. Magn. Reson.*, 2000, **143**, 79.
35. I. Hancu, W. T. Dixon, M. Woods, E. Vinogradov, A. D. Sherry and R. E. Lenkinski, *Acta Radiol.*, 2010, **51**, 910.
36. M. P. Lowe, D. Parker, O. Reany, S. Aime, M. Botta, G. Castellano, E. Gianolio and R. Pagliarin, *J. Am. Chem. Soc.*, 2001, **123**, 7601.
37. K. M. Ward and R. S. Balaban, *Magn. Reson. Med.*, 2000, **44**, 799.
38. S. Aime, M. Botta, E. Gianolio and E. Terreno, *Angew. Chem. Int. Ed. Engl.*, 2000, **39**, 747.
39. L. Burai, R. Scopelliti and E. Toth, *Chem. Commun. (Camb.)*, 2002, **20**, 2366.
40. S. Aime, M. Botta, V. Mainero and E. Terreno, *Magn. Reson. Med.*, 2002, **47**, 10.
41. W. H. Li, G. Parigi, M. Fragai, C. Luchinat and T. J. Meade, *Inorg. Chem.*, 2002, **41**, 4018.
42. G. Ferrauto, D. D. Castelli, E. Terreno and S. Aime, *Magn. Reson. Med.*, 2012, in press.
43. S. Aime, C. Carrera, D. Delli Castelli, S. Geninatti Crich and E. Terreno, *Angew. Chem. Int. Ed. Engl.*, 2005, **44**, 1813.
44. M. T. McMahon, A. A. Gilad, M. A. DeLiso, S. M. Berman, J. W. Bulte and P. C. van Zijl, *Magn. Reson. Med.*, 2008, **60**, 803.
45. M. Srinivas, P. A. Morel, L. A. Ernst, D. H. Laidlaw and E. T. Ahrens, *Magn. Reson. Med.*, 2007, **58**, 725.
46. A. M. Morawski, P. M. Winter, X. Yu, R. W. Fuhrhop, M. J. Scott, F. Hockett, J. D. Robertson, P. J. Gaffney, G. M. Lanza and S. A. Wickline, *Magn. Reson. Med.*, 2004, **52**, 1255.
47. M. Srinivas, E. H. Aarntzen, J. W. Bulte, W. J. Oyen, A. Heerschap, I. J. de Vries and C. G. Figdor, *Adv. Drug. Deliv. Rev.*, 2010, **62**, 1080.
48. T. Dresselaers, J. Theys, S. Nuyts, B. Wouters, E. de Bruijn, J. Anne, P. Lambin, P. Van Hecke and W. Landuyt, *Br. J. Cancer*, 2003, **89**, 1796.
49. K. M. Brindle, S. E. Bohndiek, F. A. Gallagher and M. I. Kettunen, *Magn. Reson. Med.*, 2011, **66**, 505.
50. J. H. Ardenkjaer-Larsen, B. Fridlund, A. Gram, G. Hansson, L. Hansson, L. H. Lerche, R. Servin, M. Thaning and K. Golman, *Proc. Natl Acad. Sci. USA*, 2003, **100**, 10158.

51. J. B. Hövener, E. Y. Chekmenev, K. C. Harris, W. H. Perman, L. W. Robertson, B. D. Ross and P. Bhattacharya, *Magn. Reson. Mater. Phys. Biol. Med.*, 2009, **22**, 111.
52. A. Abragam and M. Goldman, *Rep. Prog. Phys.*, 1978, **41**, 395.
53. M. S. Albert, G. D. Cates, B. Driehuys, W. Happer, B. Saam, C. S. Springer and A. Wishnia, *Nature*, 1994, **370**, 199.
54. K. Golman, O. Axelsson, H. Johannesson, S. Mansson, C. Olofsson and J. S. Petersson, *Magn. Reson. Med.*, 2001, **46**, 1.
55. I. J. Rowland, E. T. Peterson, J. W. Gordon and S. B. Fain, *Curr. Pharm. Biotechnol.*, 2010, **11**, 709.
56. Y. F. Yen, K. Nagasawa and T. Nakada, *Magn. Reson. Med. Sci.*, 2011, **10**, 211.
57. K. Golman, R. in 't Zandt and M. Thaning, *Proc. Natl Acad. Sci. USA*, 2006, **103**, 11270.
58. D. Mayer, Y. S. Levin, R. E. Hurd, G. H. Glover and D. M. Spielman, *Magn. Reson. Med.*, 2006, **56**, 932.
59. I.-Y. Choi, A. Dedeoglu and B. G. Jenkins, *NMR Biomed.*, 2007, **20**, 216.
60. S. Bluml, A. Moreno, J. H. Hwang and B. D. Ross, *NMR Biomed.*, 2001, **14**, 19.
61. B. Ross and S. Bluml, *Anat. Rec.*, 2001, **265**, 54.
62. U. Himmelreich and R. K. Gupta, *Application of magnetic resonance for the diagnosis of infective brain lesions*, In: ed. G. A. Webb, Modern Magnetic Resonance. Springer, Dordrecht, 2006, vol. 2, pp. 991–999.
63. R. J. Gillies and D. L. Morse, *Annu. Rev. Biomed. Eng.*, 2005, **7**, 287.
64. G. J. Kemp, M. Meyerspeer and E. Moser, *NMR Biomed.*, 2007, **20**, 555.
65. P. Morris and H. Bachelard, *NMR Biomed.*, 2003, **16**, 303.
66. C. I. H. C. Nabuurs, D. W. J. Klomp, A. Veltien, H. E. Kan and A. Heerschap, *Magn. Reson. Med.*, 2008, **59**, 626.
67. P. G. Henry, I. Tkac and R. Gruetter, *Magn. Reson. Med.*, 2003, **50**, 684.
68. P. G. Henry, G. Adriany, D. Deelchand, R. Gruetter, M. Marjanska, G. Oz, E. R. Seaquist, A. Shestov and K. Ugurbil, *Magn. Reson. Imaging*, 2006, **24**, 527.
69. K. Golman and J. S. Petersson, *Acad. Radiol.*, 2006, **13**, 932.
70. B. J. Nieman, J. Bishop, J. Dazai, N. A. Bock, J. P. Lerch, A. Feintuch, X. J. Chen, J. G. Sled and R. M. Henkelman, *NMR Biomed.*, 2007, **20**, 291.
71. J. W. Bulte, T. Douglas, B. Witwer, S. C. Zhang, E. Strable, B. K. Lewis, H. Zywicke, B. Miller, P. van Gelderen, B. M. Moskowitz, I. D. Duncan and J. A. Frank, *Nat. Biotechnol.*, 2001, **19**, 1141.
72. J. W. M. Bulte, S.-C. Zhang, P. van Gelderen, V. Herynek, E. K. Jordan, I. D. Duncan and J. A. Frank, *Proc. Natl Acad. Sci. USA*, 1999, **96**, 15256.
73. R. Weissleder, *Magn. Reson. Med.*, 1991, **22**, 209; discussion 213.
74. L. G. Remsen, C. I. McCormick, S. Roman-Goldstein, G. Nilaver, R. Weissleder, A. Bogdanov, I. Hellstrom, R. A. Kroll and E. A. Neuwelt, *AJNR Am. J. Neuroradiol.*, 1996, **17**, 411.

75. E. Küstermann, U. Himmelreich, K. Kandal, T. Geelen, A. Ketkar, D. Wiedermann, K. Strecker, S. Esser, S. Arnold and M. Hoehn, *Contrast Media Mol. Imaging*, 2008, **3**, 27.

76. S. J. Soenen, M. De Cuyper, S. C. De Smedt and K. Braeckmans, *Methods Enzymol.*, 2012, **509**, 195.

77. C. Heyn, C. V. Bowen, B. K. Rutt and P. J. Foster, *Magn. Reson. Med.*, 2005, **53**, 312.

78. I. J. de Vries, W. J. Lesterhuis, J. O. Barentsz, P. Verdijk, J. H. van Krieken, O. C. Boerman, W. J. Oyen, J. J. Bonenkamp, J. B. Boezeman, G. J. Adema, J. W. Bulte, T. W. Scheenen, C. J. Punt, A. Heerschap and C. G. Figdor, *Nat. Biotechnol.*, 2005, **23**, 1407.

79. F. Saudek, D. Jirak, P. Girman, V. Herynek, M. Dezortova, J. Kriz, J. Peregrin, Z. Berkova, K. Zacharovova and M. Hajek, *Transplantation*, 2010, **90**, 1602.

80. J. W. M. Bulte, T. Douglas, S. Mann, R. B. Frankel, B. M. Moskowitz, R. A. Brooks, C. D. Baumgarner, J. Vymazal, M.-P. Strub and J. A. Frank, *J. Magn. Reson. Imaging*, 1994, **4**, 497.

81. C. Wilhelm, C. Billotey, J. Roger, J. N. Pons, J. C. Bacri and F. Gazeau, *Biomaterials*, 2003, **24**, 1001.

82. G. Fleige, F. Seeberger, D. Laux, M. Kresse, M. Taupitz, H. Pilgrimm and C. Zimmer, *Invest. Radiol.*, 2002, **37**, 482.

83. M. De Cuyper and M. Joniau, *Eur. Biophys. J.*, 1988, **15**, 311.

84. S. J. Soenen, U. Himmelreich, N. Nuytten, T. R. Pisanic, 2nd, A. Ferrari and M. De Cuyper, *Small*, 2010, **6**, 2136.

85. S. J. Soenen, U. Himmelreich, N. Nuytten and M. De Cuyper, *Biomaterials*, 2011, **32**, 195.

86. A. Moore, J. P. Basilion, E. A. Chiocca and R. Weissleder, *Biochim. Biophys. Acta*, 1998, **1402**, 239.

87. M. Lewin, N. Carlesso, C. H. Tung, X. W. Tang, D. Cory, D. T. Scadden and R. Weissleder, *Nat. Biotechnol.*, 2000, **18**, 410.

88. D. Artemov, N. Mori, B. Okollie and Z. M. Bhujwalla, *Magn. Reson. Med.*, 2003, **49**, 403.

89. E. M. Shapiro, O. Gonzalez-Perez, J. Manuel Garcia-Verdugo, A. Alvarez-Buylla and A. P. Koretsky, *Neuroimage*, 2006, **32**, 1150.

90. R. Vreys, G. Vande Velde, O. Krylychkina, M. Vellema, M. Verhoye, J. P. Timmermans, V. Baekelandt and A. Van der Linden, *Neuroimage*, 2010, **49**, 2094.

91. M. Srinivas, A. Heerschap, E. T. Ahrens, C. G. Figdor and I. J. de Vries, *Trends Biotechnol.*, 2010, **28**, 363.

92. E. Terreno, W. Dastru, D. Delli Castelli, E. Gianolio, S. Geninatti Crich, D. Longo and S. Aime, *Curr. Med. Chem.*, 2010, **17**, 3684.

93. E. M. Shapiro and A. P. Koretsky, *Magn. Reson. Med.*, 2008, **60**, 265.

94. R. Weissleder, A. Moore, U. Mahmood, R. Bhorade, H. Benveniste, E. A. Chiocca and J. P. Basilion, *Nat. Med.*, 2000, **6**, 351.

95. S. Laurent, L. Vander Elst, Y. Fu and R. N. Muller, *Bioconjug. Chem.*, 2003, **15**, 99.

96. P. M. Winter, A. M. Morawski, S. D. Caruthers, R. W. Fuhrhop, H. Zhang, T. A. Williams, J. S. Allen, E. K. Lacy, J. D. Robertson, G. M. Lanza and S. A. Wickline, *Circulation*, 2003, **108**, 2270.

97. M. Zhao, D. A. Beauregard, L. Loizou, B. Davletov and K. M. Brindle, *Nat. Med.*, 2001, **7**, 1241.

98. D. Artemov, N. Mori, R. Ravi and Z. M. Bhujwalla, *Cancer Res.*, 2003, **63**, 2723.

99. S. Flacke, S. Fischer, M. J. Scott, R. J. Fuhrhop, J. S. Allen, M. McLean, P. Winter, G. A. Sicard, P. J. Gaffney, S. A. Wickline and G. M. Lanza, *Circulation*, 2001, **104**, 1280.

100. N. R. Sibson, A. M. Blamire, M. Bernades-Silva, S. Laurent, S. Boutry, R. N. Muller, P. Styles and D. C. Anthony, *Magn. Reson. Med.*, 2004, **51**, 248.

101. P. W. So, S. Hotee, A. H. Herlihy and J. D. Bell, *Magn. Reson. Med.*, 2005, **54**, 218.

102. J. Chung, K. Kee, J. K. Barral, R. Dash, H. Kosuge, X. Wang, I. Weissman, R. C. Robbins, D. Nishimura, T. Quertermous, R. A. Reijo-Pera and P. C. Yang, *Magn. Reson. Med.*, 2011, **66**, 1374.

103. J. Lee, M. J. Zylka, D. J. Anderson, J. E. Burdette, T. K. Woodruff and T. J. Meade, *J. Am. Chem. Soc.*, 2005, **127**, 13164.

104. J. Lee, J. E. Burdette, K. W. MacRenaris, D. Mustafi, T. K. Woodruff and T. J. Meade, *Chem. Biol.*, 2007, **14**, 824.

105. P. A. Sukerkar, K. W. MacRenaris, T. J. Meade and J. E. Burdette, *Mol. Pharm.*, 2011, **8**, 1390.

106. W. J. Malaisse, Y. Zhang, K. Louchami, S. Sharma, T. Dresselaers, U. Himmelreich, G. W. Novotny, T. Mandrup-Poulsen, D. Waschke, Y. Leshch, J. Thimm, J. Thiem and A. Sener, *Arch. Biochem. Biophys.*, 2012, **517**, 138.

107. D. Waschke, Y. Leshch, J. Thimm, U. Himmelreich and J. Thiem, *Eur. J. Org. Chem.*, 2012, **2012**, 948.

108. S. J. Soenen, A. R. Brisson, E. Jonckheere, N. Nuytten, S. Tan, U. Himmelreich and M. De Cuyper, *Biomaterials*, 2011, **32**, 1748.

109. A. Atre, T. Struys, T. Notelaers, M. Hodenius, P. Roelandt, M. De Cuyper, C. Verfaillie and U. Himmelreich, *Magn. Reson. Mater. Phys.*, 2011, **24**(1), 38.

110. T. J. Meade, A. K. Taylor and S. R. Bull, *Curr. Opin. Neurobiol.*, 2003, **13**, 597.

111. R. A. Moats, S. E. Fraser and T. J. Meade, *Angew. Chem. Int. Ed. Engl.*, 1997, **36**, 726.

112. A. Y. Louie, M. M. Huber, E. T. Ahrens, U. Rothbacher, R. Moats, R. E. Jacobs, S. E. Fraser and T. J. Meade, *Nat. Biotechnol.*, 2000, **18**, 321.

113. W. Cui, P. Otten, Y. Li, K. S. Koeneman, J. Yu and R. P. Mason, *Magn. Reson. Med.*, 2004, **51**, 616.

114. M. Aswendt, E. Gianolio, G. Pariani, R. Napolitano, F. Fedeli, U. Himmelreich, S. Aime and M. Hoehn, *Neuroimage*, 2012, **62**, 1685.

115. P. Caravan, N. J. Cloutier, M. T. Greenfield, S. A. McDermid, S. U. Dunham, J. W. Bulte, J. C. Amedio, Jr., R. J. Looby, R. M. Supkowski, W. D. Horrocks, Jr., T. J. McMurry and R. B. Lauffer, *J. Am. Chem. Soc.*, 2002, **124**, 3152.

116. A. L. Nivorozhkin, A. F. Kolodziej, P. Caravan, M. T. Greenfield, R. B. Lauffer and T. J. McMurry, *Angew. Chem. Int. Ed.*, 2001, **40**, 2903.

117. A. Bogdanov, Jr., L. Matuszewski, C. Bremer, A. Petrovsky and R. Weissleder, *Mol. Imaging*, 2002, **1**, 16–23.

118. A. Louie, *Methods Mol. Med.*, 2006, **124**, 401.

119. B. Jastrzebska, R. Lebel, H. Therriault, J. O. McIntyre, E. Escher, B. Guerin, B. Paquette, W. A. Neugebauer and M. Lepage, *J. Med. Chem.*, 2009, **52**, 1576.

120. M. Lepage, W. C. Dow, M. Melchior, Y. You, B. Fingleton, C. C. Quarles, C. Pepin, J. C. Gore, L. M. Matrisian and J. O. McIntyre, *Mol. Imaging*, 2007, **6**, 393.

121. R. Lebel, B. Jastrzebska, H. Therriault, M. M. Cournoyer, J. O. McIntyre, E. Escher, W. Neugebauer, B. Paquette and M. Lepage, *Magn. Reson. Med.*, 2008, **60**, 1056.

122. U. Himmelreich, S. Aime, T. Hieronymus, C. Justicia, F. Uggeri, M. Zenke and M. Hoehn, *Neuroimage*, 2006, **32**, 1142.

123. J. M. Perez, L. Josephson, T. O'Loughlin, D. Hogemann and R. Weissleder, *Nat. Biotechnol.*, 2002, **20**, 816.

124. J. Grimm, J. M. Perez, L. Josephson and R. Weissleder, *Cancer Res.*, 2004, **64**, 639.

125. E. Schellenberger, F. Rudloff, C. Warmuth, M. Taupitz, B. Hamm and J. Schnorr, *Bioconjug. Chem.*, 2008, **19**, 2440.

126. E. Schellenberger, J. Schnorr, C. Reutelingsperger, L. Ungethum, W. Meyer, M. Taupitz and B. Hamm, *Small*, 2008, **4**, 225.

127. B. B. Bartelle, K. U. Szulc, G. A. Suero-Abreu, J. J. Rodriguez and D. H. Turnbull, *Magn. Reson. Med.*, 2012, in press.

128. X. Zhang, Y. Lin and R. J. Gillies, *J. Nucl. Med.*, 2010, **51**, 1167.

129. N. Raghunand, C. Howison, A. D. Sherry, S. Zhang and R. J. Gillies, *Magn. Reson. Med.*, 2003, **49**, 249.

130. G. V. Martinez, X. Zhang, M. L. Garcia-Martin, D. L. Morse, M. Woods, A. D. Sherry and R. J. Gillies, *NMR Biomed.*, 2011, **24**, 1380.

131. D. L. Longo, W. Dastru, G. Digilio, J. Keupp, S. Langereis, S. Lanzardo, S. Prestigio, O. Steinbach, E. Terreno, F. Uggeri and S. Aime, *Magn. Reson. Med.*, 2011, **65**, 202.

132. F. A. Gallagher, M. I. Kettunen, S. E. Day, D. E. Hu, J. H. Ardenkjaer-Larsen, R. Zandt, P. R. Jensen, M. Karlsson, K. Golman and M. H. Lerche, and K. M. Brindle, *Nature*, 2008, **453**, 940.

133. F. A. Gallagher, M. I. Kettunen and K. M. Brindle, *NMR Biomed.*, 2011, **24**, 1006.

134. B. Yoo and M. D. Pagel, *Front. Biosci.*, 2008, **13**, 1733.

135. M. Amanlou, S. D. Siadat, D. Norouzian, S. E. Ebrahimi, M. R. Aghasadeghi, M. Ghorbani, M. S. Alavidjeh, D. N. Inanlou, A. J. Arabzadeh and M. S. Ardestani, *Curr. Radiopharm.*, 2011, **4**, 31.
136. G. Vande Velde, V. Baekelandt, T. Dresselaers and U. Himmelreich, *Q. J. Nucl. Med. Mol. Imaging*, 2009, **53**, 565.
137. R. Weissleder, M. Simonova, A. Bogdanova, S. Bredow, W. S. Enochs and A. Bogdanov, Jr., *Radiology*, 1997, **204**, 425.
138. W. S. Enochs, P. Petherick, A. Bogdanova, U. Mohr and R. Weissleder, *Radiology*, 1997, **204**, 417.
139. H. Alfke, H. Stoppler, F. Nocken, J. T. Heverhagen, B. Kleb, F. Czubayko and K. J. Klose, *Radiology*, 2003, **228**, 488.
140. B. Bouchard, B. B. Fuller, S. Vijayasaradhi and A. N. Houghton, *J. Exp. Med.*, 1989, **169**, 2029.
141. G. Genove, U. DeMarco, H. Xu, W. F. Goins and E. T. Ahrens, *Nat. Med.*, 2005, **11**, 450.
142. O. Zurkiya, A. W. Chan and X. Hu, *Magn. Reson. Med.*, 2008, **59**, 1225.
143. A. A. Gilad, M. T. McMahon, P. Walczak, P. T. Winnard, Jr., V. Raman, H. W. van Laarhoven, C. M. Skoglund, J. W. Bulte and P. C. van Zijl, *Nat. Biotechnol.*, 2007, **25**, 217.
144. R. D. Shonat and A. P. Koretsky, *Adv. Exp. Med. Biol.*, 2003, **530**, 331.
145. P. Z. Sun, Z. B. Schoening and A. Jasanoff, *Magn. Reson. Med.*, 2003, **49**, 609.
146. M. Querol, D. G. Bennett, C. Sotak, H. W. Kang and A. Bogdanov, Jr., *ChemBioChem*, 2007, **8**, 1637.
147. A. P. Koretsky, Y. J. Lin, H. Schorle and R. Jaenisch, *Proc. Int. Soc. Magn. Reson. Med.*, 1996, **4**, 5471.
148. A. E. Deans, Y. Z. Wadghiri, L. M. Bernas, X. Yu, B. K. Rutt and D. H. Turnbull, *Magn. Reson. Med.*, 2006, **56**, 51.
149. M. T. D. Cronin, H. Morris and M. Valko, *Curr. Med. Chem.*, 2005, **12**, 1161.
150. Y. Gossuin, R. N. Muller and P. Gillis, *NMR Biomed.*, 2004, **17**, 427.
151. P. M. Harrison and P. Arosio, *Biochim. Biophys. Acta*, 1996, **1275**, 161.
152. A. Cozzi, B. Corsi, S. Levi, P. Santambrogio, A. Albertini and P. Arosio, *J. Biol. Chem.*, 2000, **275**, 25122.
153. J. Liu, E. C. Cheng, R. C. Long, Jr., S. H. Yang, L. Wang, P. H. Cheng, J. J. Yang, D. Wu, H. Mao and A. W. Chan, *Tissue Eng. Part C Methods*, 2009, **15**, 739.
154. B. Cohen, H. Dafni, G. Meir, A. Harmelin and M. Neeman, *Neoplasia*, 2005, **7**, 109.
155. B. Cohen, K. Ziv, V. Plaks, T. Israely, V. Kalchenko, A. Harmelin, L. E. Benjamin and M. Neeman, *Nat. Med.*, 2007, **13**, 498.
156. G. Vande Velde, J. Raman Rangarajan, R. Vreys, C. Guglielmetti, T. Dresselaers, M. Verhoye, A. Van der Linden, Z. Debyser, V. Baekelandt, F. Maes and U. Himmelreich, *Neuroimage*, 2012, **62**, 367.

157. G. Vande Velde, J. R. Rangarajan, J. Toelen, T. Dresselaers, A. Ibrahimi, O. Krylychkina, R. Vreys, A. Van der Linden, F. Maes, Z. Debyser, U. Himmelreich and V. Baekelandt, *Gene Ther.*, 2011, **18**, 594.

158. B. Iordanova and E. T. Ahrens, *Neuroimage*, 2012, **59**, 1004.

159. J. P. Sumner, E. M. Shapiro, D. Maric, R. Conroy and A. P. Koretsky, *Neuroimage*, 2009, **44**, 671.

160. B. J. Nieman, J. Y. Shyu, J. J. Rodriguez, A. D. Garcia, A. L. Joyner and D. H. Turnbull, *Neuroimage*, 2010, **50**, 456.

161. J. Yang, J. Liu, G. Niu, K. C. Chan, R. Wang, Y. Liu and E. X. Wu, *Neuroimage*, 2009, **48**, 319.

162. R. A. Panizzo, P. G. Kyrtatos, A. N. Price, D. G. Gadian, P. Ferretti and M. F. Lythgoe, *Neuroimage*, 2009, **44**, 1239.

163. R. A. Brooks, J. Vymazal, R. B. Goldfarb, J. W. M. Bulte and P. Aisen, *Magn. Reson. Med.*, 1998, **40**, 227.

164. A. A. Gilad, P. T. Winnard, Jr., P. C. van Zijl and J. W. Bulte, *NMR Biomed.*, 2007, **20**, 275.

165. B. Iordanova, C. S. Robison and E. T. Ahrens, *J. Biol. Inorg. Chem.*, 2010, **15**, 957.

166. J. W. M. Bulte, T. Douglas, S. Mann, R. B. Frankel, B. M. Moskowitz, R. A. Brooks, C. D. Baumbarner, J. Vymazal and J. A. Frank, *Invest. Radiol.*, 1994, **29**, S214.

167. J. W. M. Bulte, T. Douglas, S. Mann, J. Vymazal, P. G. Laughlin and J. A. Frank, *Acad. Radiol.*, 1995, **2**, 871.

168. C. Stephens, *Curr. Biol.*, 2006, **16**, R363.

169. R. Uebe, V. Henn and D. Schuler, *J. Bacteriol.*, 2012, **194**, 1018.

170. D. E. Goldhawk, R. Rohani, A. Sengupta, N. Gelman and F. S. Prato, *Wiley Interdiscip. Rev. Nanomed. Nanobiotechnol.*, 2012, **4**, 378.

171. A. Arakaki, J. Webb and T. Matsunaga, *J. Biol. Chem.*, 2003, **278**, 8745.

172. M. Tanaka, E. Mazuyama, A. Arakaki and T. Matsunaga, *J. Biol. Chem.*, 2011, **286**, 6386.

173. A. D. Sherry and M. Woods, *Annu. Rev. Biomed. Eng.*, 2008, **10**, 391.

174. G. Liu, Y. Liang, A. Bar-Shir, K. W. Y. Chan, C. S. Galpoththawela, S. M. Bernard, T. Tse, N. N. Yadav, P. Walczak, M. T. McMahon, J. W. M. Bulte, P. C. M. van Zijl and A. A. Gilad, *J. Am. Chem. Soc.*, 2011, **133**, 16326.

175. A. P. Koretsky, M. J. Brosnan, L. H. Chen, J. D. Chen and T. Van Dyke, *Proc. Natl Acad. Sci. USA*, 1990, **87**, 3112.

176. A. Auricchio, R. Zhou, J. M. Wilson and J. D. Glickson, *Proc. Natl Acad. Sci. USA*, 2001, **98**, 5205.

177. Z. Li, H. Qiao, C. Lebherz, S. R. Choi, X. Zhou, G. Gao, H. F. Kung, D. J. Rader, J. M. Wilson, J. D. Glickson and R. Zhou, *Hum. Gene Ther.*, 2005, **16**, 1429.

178. B. B. Roman, B. Wieringa and A. P. Koretsky, *J. Biol. Chem.*, 1997, **272**, 17790.

179. G. Walter, E. R. Barton and H. L. Sweeney, *Proc. Natl Acad. Sci. USA*, 2000, **97**, 5151.

180. D. A. Hamstra, K. C. Lee, J. M. Tychewicz, V. D. Schepkin, B. A. Moffat, M. Chen, K. J. Dornfeld, T. S. Lawrence, T. L. Chenevert, B. D. Ross, J. T. Gelovani and A. Rehemtulla, *Mol. Ther.*, 2004, **10**, 916.

181. L. D. Stegman, A. Rehemtulla, B. Beattie, E. Kievit, T. S. Lawrence, R. G. Blasberg, J. G. Tjuvajev and B. D. Ross, *Proc. Natl Acad. Sci. USA*, 1999, **96**, 9821.

182. J. Theys, W. Landuyt, S. Nuyts, L. Van Mellaert, A. van Oosterom, P. Lambin and J. Anne, *Cancer Gene Ther.*, 2001, **8**, 294.

183. K. B. Low, M. Ittensohn, T. Le, J. Platt, S. Sodi, M. Amoss, O. Ash, E. Carmichael, A. Chakraborty, J. Fischer, S. L. Lin, X. Luo, S. I. Miller, L. Zheng, I. King, J. M. Pawelek and D. Bermudes, *Nat. Biotechnol.*, 1999, **17**, 37.

184. L. Dubois, T. Dresselaers, W. Landuyt, K. Paesmans, A. Mengesha, B. G. Wouters, P. Van Hecke, J. Theys and P. Lambin, *Br. J. Cancer*, 2007, **96**, 758.

185. V. D. Kodibagkar, J. Yu, L. Liu, H. P. Hetherington and R. P. Mason, *Magn. Reson. Imaging*, 2006, **24**, 959.

186. N. S. Tuan-Khanh Tran, R. Hurd and T. Jue, *NMR Biomed.*, 1999, **12**, 26.

187. S. E. Day, M. I. Kettunen, F. A. Gallagher, D. E. Hu, M. Lerche, J. Wolber, K. Golman, J. H. Ardenkjaer-Larsen and K. M. Brindle, *Nat. Med.*, 2007, **13**, 1382.

188. K. Golman, J. S. Petersson, P. Magnusson, E. Johansson, P. Åkeson, C. M. Chai, G. Hansson and S. Månsson, *Magn. Reson. Med.*, 2008, **59**, 1005.

PART III
TRANSLATIONAL DRUG DISCOVERY: FROM BIOLOGICAL MODELS TO THE CLINICS

CHAPTER 12

Translational Magnetic Resonance Imaging and Spectroscopy: Opportunities and Challenges

JOHN C. WATERTON[a,b]

[a] Biomedical Imaging Institute, The University of Manchester, Manchester Academic Health Sciences Centre, Oxford Road, Manchester, M13 9PT, UK; [b] Personalised Healthcare & Biomarkers, AstraZeneca, Alderley Park, Macclesfield, Cheshire, SK10 4TG, UK
Email: john.waterton@manchester.ac.uk; john.waterton@astrazeneca.com

12.1 Translational Magnetic Resonance Imaging and Spectroscopy

The goal of drug development is to test the hypothesis that a specified molecule (the Candidate Drug), at specified dosage, in a specified patient population, will provide clinical benefit with tolerable risks and side effects. Clinical benefit is usually taken to mean "how a patient feels, functions, or survives".[1] Drug research, or discovery, in contradistinction, is the identification of that candidate drug molecule. If NMR (or MR; see Box 12.1) is to help drug development, then it will do so by providing tools to help test that hypothesis. Since each such hypothesis now costs over 10^9 to test,[2,3] and around 90% of hypotheses fail,[4] there is a definite need for improved tools.

New Developments in NMR No. 2
New Applications of NMR in Drug Discovery and Development
Edited by Leoncio Garrido and Nicolau Beckmann
© The Royal Society of Chemistry 2013
Published by the Royal Society of Chemistry, www.rsc.org

Box 12.1 NMR or MR?

While early studies in human and animal subjects described their technique as "NMR Imaging" or "NMR *in vivo*", the almost-universal convention since the 1980s has been to use "Magnetic Resonance Imaging (MRI)" or sometimes, in German-speaking countries, "Magnetic Resonance Tomography (MRT)". The reasons usually given for omitting the "Nuclear" are to avoid confusing patients, either with ionizing radiation, or with the hospital's Nuclear Medicine department, which conducts radio-isotope imaging using PET and SPECT. (Perversely, MRI studies are usually performed by the hospital's Radiology department, which administers far more ionizing radiation to patients than the Nuclear Medicine department.) This also has the minor consequence that non-nuclear magnetic resonance techniques are included within "MR", notably EPR and proton-electron double resonance. "MRS" (Magnetic Resonance Spectroscopy) is usually used to describe studies that yield an NMR spectrum of some tissue or organ *in vivo*. Such studies usually now include MRI to assist or confirm spatial localization. The term "NMR spectroscopy" on the other hand, in the medical context, is often limited to spectroscopic measurements *ex vivo* on biofluids such as blood or urine.

NMR is, of course, a rich source of measurement techniques, with several important attributes with potential to be useful in humans.

- Firstly, many biologically important molecules are readily detectable by NMR. This is probably the least important attribute for the drug developer in a translational setting: the vast majority of useful applications detect only ^1H_2O. Only occasionally are other resonances useful: a handful of other ^1H resonances,[5] notably the triglyceride C^1H_2 moiety, choline $N(C^1H_3)_3$ moieties (tCho) and the acetyl C^1H_3 in *N*-acetyl aspartate (NAA), $^{23}Na^+_{(aq)}$, some ^{31}P resonances (notably from phosphocreatine, ATP and $HPO_4^{2-}/H_2PO_4^-$) or occasionally ^{13}C at natural abundance (*e.g.* glycogen) or enriched (*e.g.* metabolites of 1-[^{13}C]-glucose). A very few fluorine-containing xenobiotics reach high enough concentrations in man to give useful ^{19}F signals[6,7] (as does the drug lithium, detectable *in vivo* with ^7Li MR;[8] see also Section 15.2 of Chapter 15). Hyperpolarization (of ^3He,[9,10] ^{129}Xe[10] or ^{13}C,[11] see also Chapter 9) slightly extends this short list of non-^1H_2O resonances for niche applications. Many other opportunities have been demonstrated, at least in animals, for this list to be extended to other substances and other resonances; notably ^{17}O, ^{15}N and ^2H may have potential. However, the challenges to make measurements in man that are actually useful for the drug developer are immense.
- Secondly, although relatively few endogenous molecules reach high enough *in vivo* concentrations in man to allow useful and practical MR

imaging measurements, the few that do (notably ^1H_2O, triglyceride C^1H_2, $^{23}Na^+_{(aq)}$ and hyperpolarized ^3He and ^{129}Xe) have interesting motion and relaxation properties, which can distinguish different normal tissues and pathologies, and respond to drug treatment. Contrast in MRI arises mainly from variations in longitudinal (T_1) and transverse (homogeneous, T_2, or inhomogeneous, T_2*) relaxation, from alteration of those relaxation times by contrast agents,[12] through modulation of the NMR signal by molecular diffusion, by flow, or by bulk physiologic motion and from differing concentrations of the substance whose NMR is detected.

- Thirdly, NMR does not expose the patient to ionizing radiation. This is an obvious benefit, since all ionizing radiation carries some risk (although, to be fair, there are many useful medical imaging tests that do employ ionizing radiation, but where the actual risks are extremely small). More importantly, repeated follow-up to monitor drug effects does not accumulate risk in the same way that ionizing-radiation-based modalities do. An additional, less appreciated, benefit is that iterative improvements to pulse sequences and measurement techniques are easy to undertake in healthy volunteers, without the ethical impediments encountered with ionizing-radiation-based modalities.

- Fourthly, of course, the signal can be spatially located. The insight is that, if the magnetic field is caused to vary across the body, the Larmor frequency will follow that variation, and be detectable by NMR.[13] The spatial resolution and accuracy of location are limited mainly by hardware (field strength, transceiver and gradient coil design, and signal-to-noise ratio), and compare very favorably with other imaging techniques using non-ionizing radiations such as ultrasonography,[14] optical tomographies,[15] electrical impedance tomography[16] or electrocardiographic imaging.[17]

- Fifthly, radiofrequency fields penetrate the adult human body very well, at least up to frequencies of 130 MHz (3 T for ^1H) or even 300 MHz (7 T). This is a major advantage over ultrasound- or near-infrared-based modalities, which are severely limited by depth penetration.

These advantages are all well known (anecdotally, soon after the very first NMR experiment at Stanford in 1946, Felix Bloch obtained a strong ^1H NMR signal from his own finger,[18] and arguably the first one-dimensional NMR image was created by Herman Carr as early as 1952[19]). Nevertheless it took more than three decades before these innovations were transformed into medically useful tools. It is now increasingly appreciated that the translation of a biophysical technique out of the lab and into a reliable tool that can be used to test biomedical hypotheses in patients is not simply a matter of performing further experiments, but requires a substantial cultural "gap" to be bridged. Moreover, a second "gap" is encountered when translating the tool out of the academic world of medical research, and into routine use in patient care throughout the healthcare system. Unless these two translational, or "Cooksey", gaps[20,21] are understood, it is unlikely that *in vivo* MR innovations will benefit patients.

12.2   NMR Measurements as Biomarkers

12.2.1   Biomarker Definition

If MR measurements are to cross the Cooksey gaps, and become useful in medical research or in healthcare, they need to be understood as "biomarkers" (Box 12.2). The term "biomarker", following the 1999 NIH Initiative on Biomarkers and Surrogate Endpoints,[1] admits a broad definition, encompassing almost any useful measurement (other than a clinical endpoint) made in a human. The distinction between biomarkers and clinical endpoints is reminiscent of the ancient medical distinction between signs (what the physician notices) and symptoms (what the patient complains of), except that the concept of a Biomarker incorporates a very strong sense of being "objectively quantified". By this inclusive definition, the category "biomarker" is not limited to metabolites measured in blood or urine *ex vivo* (such as glucose, creatinine or cholesterol) or *in vivo* (*e.g.* muscle phosphocreatine by ^{31}P MRS). Equally as important as "biomarkers" are genotypes; medical imaging measurements such as tumor size[22] or cartilage thickness in the knee joint;[23] physiologic measurements such as blood pressure or left ventricular ejection fraction[24] or even the results of certain standardized questionnaires.

Some biomarkers are so well trusted that they can be used by the drug developer as "surrogate endpoints", *i.e.* the regulatory authority may[25] approve and license the drug based on evidence that it improves the biomarker, in the absence of evidence of clinical benefit. For example, antihypertensives are prescribed *inter alia* because they reduce the patient's risk of stroke, thereby avoiding a potentially devastating impact on how a patient functions, or even survives (clinical endpoints); however, the drug approval may be on the basis not of the clinical endpoint, but of the surrogate endpoint "reduction of blood pressure" (which is a biomarker, because patients are

---

Box 12.2   Biomarkers and clinical endpoints

Clinical endpoint

A characteristic or variable that reflects how a patient feels, functions or survives.[1]

Biomarker (biological marker)

A characteristic that is objectively measured and evaluated as an indicator of normal biological processes, pathogenic processes or pharmacologic responses to a therapeutic intervention.[1] Intended to predict clinical outcome (benefit or lack of benefit or harm or lack of harm). A physical sign or laboratory measurement that occurs in association with a pathological process and that has putative diagnostic or prognostic utility.[87,88]

usually unaware of, and untroubled by, their elevated blood pressure before it is measured). Regulatory authorities are, however, extremely cautious in relying on surrogate endpoints for drug approvals, because of the public health risk that apparently beneficial effects on surrogate endpoints fail to translate to clinical benefit. As a notorious example, encainides, a class of antiarrhythmic drugs, which were approved on the basis of their ability to suppress arrhythmias (an electrophysiologic biomarker), were subsequently found not to decrease mortality (a clinical endpoint) but actually to increase mortality:[26] this led to the subsequent withdrawal of encainides, but only after the deaths had unfortunately already occurred.

Biomarkers, therefore, can only rarely, and with difficulty, be sufficiently validated to permit use as surrogate endpoints in regulatory approvals. Even very well-established biomarkers, such as tumor size measured by MRI or CT, may not be surrogate endpoints: many investigational anticancer drugs impressively shrink tumors without conferring a survival advantage. Tumor size was already used as a biomarker in the 1950s[27] (then by X-ray; now by CT or MRI). However, meta-analyses in, for example, metastatic breast cancer[28] (11 trials enrolling 3953 patients) and colorectal cancer[29] (25 trials enrolling 3791 patients) suggest that, although treatments that shrink tumors tend to improve survival, there are too many false-positives (drug shrinks tumor without survival benefit), as well as occasional false-negatives (drug improves survival without shrinking tumor) for the biomarker to serve as a surrogate endpoint. Statistical criteria established by Prentice[30] for validity of a surrogate seem so strict as to be unattainable. According to one recent analysis,[31] despite 150 000 papers on biomarkers, fewer than 100 biomarkers have become suitable for routine clinical practice. The implications appear dismal. Does every new MR biomarker need 50 years' evaluation, collecting a body of clinical trial data that is enormous even in comparison with Phase III drug development, and with no guarantee of success? Fortunately not! Biomarkers have many other uses in drug development, where lower confidence in the interpretation can be tolerated. They can be regarded as "qualified" for a specific, limited purpose, qualification being "a graded, fit-for purpose process dependent on the intended use".[32]

The goal of drug development is to test the hypothesis that the candidate drug will provide clinical benefit in a specific patient group, with tolerable risks and side effects (Figure 12.1a). Given the high costs, and risks of failure, drug developers have become increasingly interested in dividing the overarching hypothesis into small, interlocking, sub-hypotheses, each testable using biomarkers (Figure 12.1b). The idea is to test sequentially that the drug reaches the target, engages the target, impacts the pathway, changes the cell phenotype, modulates local physiology and causes microscopic and macroscopic structural change or amelioration. This has been described by Paul Workman as the pharmacological audit trail.[33] Biomarkers will be needed to test each of these sub-hypotheses, and each needs to be qualified for that purpose. If any sub-hypothesis fails, the drug developer can abandon the project, and avoid subsequent costs (as well as further ethical cost of futile exposure of sick

The Drug Development Hypothesis

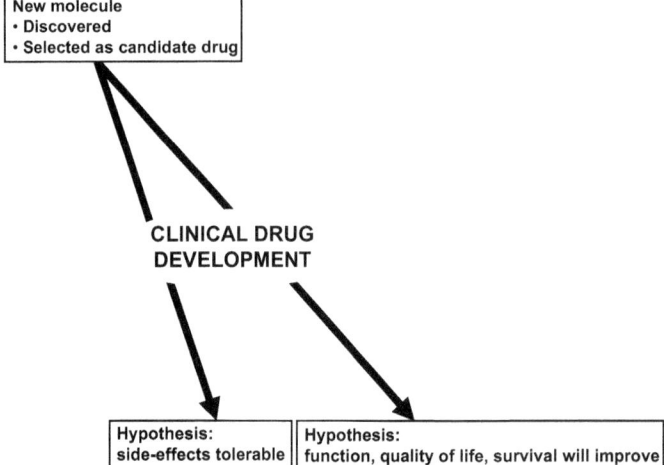

Figure 12.1a   The hypothesis that the Candidate Drug will provide clinical benefit in a specific patient group, with tolerable risks and side effects: this now costs over 10^9 to test, while around 90% of hypotheses fail.

patients to ineffective drugs). A biomarker categorization along similar lines has been proposed by Danhof *et al.*[34] and annotates Figure 12.1b. Similarly, the hypothesis that the drug does not carry intolerable risks or side effects can be tested to some extent with biomarkers ("safety biomarkers"[35]), and identification of the patient group most likely to benefit may be based on biomarkers (personalized healthcare or stratified medicine). Examples of MR biomarkers for each of these are given in Table 12.1.

12.2.2   Biomarker Qualification

The most difficult task for MR in translational drug development is to specify and conduct the correct portfolio of studies to qualify the biomarker. The greatest challenge is to minimize the risk or perception of false-negatives, as illustrated by the decision tree in Figure 12.2. By the time an investigational drug has reached clinical trials, a great deal of money, energy and scientific passion has already been invested. Negative results are always greeted with disappointment and dismay, and create the temptation to proceed down the dashed pathway in the decision tree. Did the MR biomarker truly measure what it purported to measure? Was the study design adequate? Were the time-points for observation incorrectly chosen? Have all the potential confounding factors that would invalidate the negative finding been identified and controlled? It is tempting to answer "No" to these questions, particularly if the questioner has a background in clinical medicine or molecular biology, and finds the language of magnetic resonance arcane and unintelligible. It then

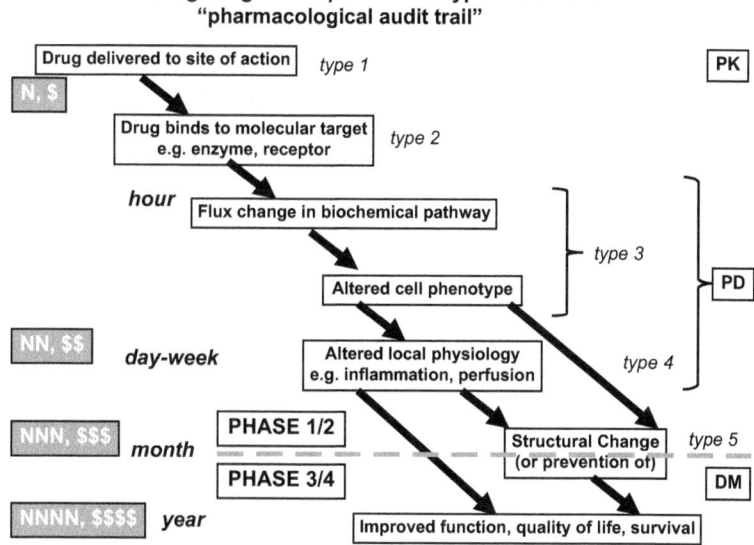

Figure 12.1b The overarching hypothesis can be divided into small, interlocking, sub-hypotheses, each testable using biomarkers. Successful test of one sub-hypothesis allows progression to the next. This is a general illustration for typical drug development, although it does not precisely describe development of every therapy in every disease. Early events (pharmacokinetics, PK) lead to the arrival of that drug at its site of action. PK trials are small, fast and inexpensive. Subsequent pharmacodynamic (PD) events include the engagement of the drug with its target and the immediate consequences. Typically MR PD biomarkers give a readout within the limits of the size and duration of a Phase II trial. However, in many diseases, structural remodeling or disease modification (DM), measured by MR morphologic biomarkers, needs larger and longer Phase II or Phase III trials to be detectable. An approximate mapping of the Danhof classification (type 1–5) is also shown: not included are Danhof type 0 (patient selection) and 6 (clinical outcome).

becomes difficult to justify the cost, complexity and burden to patients of the MR measurements, since both arms of the decision tree lead to the same place!

Qualification involves three distinct sets of activities: technical validation, biological validation and cost-effectiveness (Figure 12.3):

- "Technical validity" is simply the confidence that the MR biomarker can be measured reliably anywhere in the world or, if that is not achievable, an understanding of the circumstances in which the measurement is reliable. Technical validation does not address whether the measurement measures any useful biology, or makes any useful prediction about the patient's future health. Some degree of technical validity is obviously essential before any reliance is placed on the biomarker; however, it is much less challenging to establish technical validity of an MR biomarker to be used in a small clinical trial in a single expert center on a single MR scanner

Table 12.1 Some representative MR biomarkers illustrating the diversity of uses. The table is intended to be illustrative, not comprehensive.

MR biomarker	What biology does the biomarker measure?	Examples of use and standardization
Pharmacokinetics: active drug delivered to site of action;^{a} drug not delivered to site of toxicity. *Usually intensive, continuous variables, readily translated between animal models and man. Often ^{19}F MRS.*		
^{19}F MRS signal intensity from F-containing drug	Does the drug reach the tumor? Is it metabolized? Does the prodrug convert to the drug? Does the drug avoid potential sites of toxicity?	Few drugs have low enough potency to provide useful signals. Notable ^{19}F MRS exceptions can be seen in animal tumor studies of capecitabine[89] *in vivo*, and studies of 5-fluorouracil[7] uptake in human liver tumors[90]
Pharmacodynamics/toxicodynamics: drug engages target; modifies pathway; alters cell phenotype. *Usually intensive, continuous MRS or MRI variables, readily translated between animal models and man.*		
Brain NAA signal intensity by ^{1}H MRS (usually normalized *e.g.* to creatine)	NAA has been considered a marker of viable neurons	The correlation between NAA and neurodegeneration has been documented for a number of neurodegenerative conditions.[5] For example, in amytrophic lateral sclerosis, NAA/Cr is low in the motor cortex, and drugs that are clinically effective tend to improve NAA/Cr[91]
Rate of oxidative ATP synthesis estimated from the PCr resynthesis rate following exercise using ^{31}P MRS	Mitochondrial function	Quantification of ATP synthesis rates from ^{31}P MRS has been established.[92] ATP synthesis rates in skeletal muscle are reduced in mitochondrial myopathy[93]
Tumor $^{1}H_2O$ apparent diffusion coefficient (ADC), measurement (made avoiding confounds from perfusion)	Diffusional tortuosity has been considered a proxy for the non-viable fraction	Many studies in animal models and in patients show acute therapy-induced increases in ADC coincide with cell lysis and necrosis.[94,95] There is emerging consensus on standardization.[42,96]
Liver $^{1}H_2O$ R_2^*	R_2^* increases in liver iron overload because of sub-voxel heterogeneity in magnetic susceptibility	Several diseases are associated with liver iron overload, which can be successfully treated with iron chelation therapy. A relation between liver Fe and $^{1}H_2O$ R_2^* has been established,[97] which allows the efficacy of chelation therapy to be monitored[98]

Pharmacodynamics/toxicodynamics: drug modifies organ physiology.

Usually continuous MRI variables (intensive or extensive), which can be translated between animal and man if good models are available.

Lung ventilation volume, ventilation heterogeneity or ventilation defects measured from MRI of gas, or oxygen-enhanced ^1H MRI of lung tissue	Regional lung ventilation can be assessed by imaging an inhaled gas with a large enough signal for MRI, such as $C_3{}^{19}F_6$ gas; $S^{19}F_6$ gas; hyperpolarized ^3He gas[9] or hyperpolarized ^{129}Xe gas; or alternatively by inhaling a paramagnetic substance such as $^{99}O_2$, which reduces ^1H T_1 in lung tissue accessed by the gas	Lung ventilation volume and defect biomarkers have been used in animal models[100] and in humans, including normal volunteers, cigarette smokers, patients with emphysema and bronchitis and asthmatics, some following drug treatment.[101] However, because of the esoteric nature of many of the techniques, they are available only in a few specialized clinical centers
Tumor k^{trans} from DCEMRI using contrast agent such as gadopentetate	Measurement depends on tumor blood flow, vascular volume and endothelial permeability	Many studies in animal models and in patients[102] show acute decreases in k^{trans} following treatment with antiangiogenic agents or vascular disruptive agents. There is consensus on standardization[85]

Disease modification: drug modifies organ or lesion composition.

Usually extensive continuous MRI variables, which can be translated between animal and man if good models are available.

Carotid atheroma plaque composition	With high spatial resolution, and using different T_1- and T_2-weightings,[103] the proportions of plaque components such as wall, necrotic core and calcifications can be estimated[104]	Modulation of plaque composition with drug therapy has been measured in humans with carotid atherosclerosis.[105] The biomarker is difficult to translate into animals because of the very small size of plaques in well-known rodent models, and the difficulty in producing plaques in animals, which compositionally resemble human plaque
Number of multiple sclerosis lesions which enhance in ^{1}HMRI following contrast agent such as gadopentetate	Gadolinium-enhancing multiple sclerosis lesions represent active inflammation and demyelination[106]	Reduced number of gadolinium-enhancing lesions has been widely adopted to demonstrate proof of concept for agents that target the inflammatory component of the disease[107] and is supported by consensus documents.[106] MRI biomarkers have also been frequently used translationally in experimental allergic encephalomyelitis[108] in rodent and non-rodent models

Table 12.1 (*Continued*)

MR biomarker	What biology does the biomarker measure?	Examples of use and standardization
	Disease modification: drug modifies organ or lesion size.	
	Usually extensive continuous MRI variables, readily translated between animal models and man.	
Tumor size (^{1}H MRI ± other imaging modalities)	Tumor size is estimated from 1D, 2D or 3D measurements, with or without contrast media, considering whether the border seen in MRI represents the true pathological invading front of the tumor, and whether all the material inside the defined volume is indeed malignant	The Response Evaluation Criteria in Solid Tumors (RECIST),[22] based on a 1D estimate, are routinely used in clinical oncology trials. Since only very large changes, *e.g.* 20% increase in 1D equivalent to 73% isotropic increase in 3D, are considered clinically significant, very high precision is unnecessary. Tumor size is equally easily measured in rodent MRI (although for superficial tumors, caliper measurements can provide a simple alternative to MRI)
Cartilage thickness (^{1}H MRI)	Amount of articular hyaline cartilage and its loss from the knee and other involved joints in disease	Many studies have been performed in human subjects, mainly in knee osteoarthritis. There is some agreement on reproducibility,[23,109] nomenclature[110] and standardization[111] and there is evidence that cartilage thickness measurements can predict clinical outcome.[112] The biomarker has also been translated into animal models[54] and used for drug efficacy studies
Myocardial infarct size (^{1}H MRI post-contrast agent)	Late gadolinium enhancement (accumulation of extracellular contrast agent such as gadopentetate >10 minutes post i.v. injection) in infarcted regions of the myocardium represents a reduction in viable cell fraction	Cardiac infarct size has been used as a biomarker in clinical trials[24] and standardization guidelines have been proposed[113]

Disease modification: drug remodeling of organ or lesion size modifies function. *Usually extensive continuous 1H MRI variables, readily translated between animal models and man.*

Measure	Clinical context	Translational use
Volume of entire brain, or of brain subregions (*e.g.* hippocampus or white matter) (1H MRI)	Brain shrinkage, especially in neurodegenerative diseases such as Alzheimer's	There has been considerable progress towards standardization *inter alia* through the Alzheimer's Disease Neuroimaging Initiative[114] and brain volumetry has been used as a biomarker both prognostically and to assess response to therapy in patients as well as in rodent models
Derived cardiac physiological parameters such as left ventricular ejection fraction or myocardial strain (1H MRI)	It is well established that cardiac function is prognostic of survival, particularly in heart failure or following myocardial infarct	Morphology-derived functional cardiac MRI biomarkers have been extensively used in clinical trials[24] and have also been used translationally as biomarkers in rodent studies.[115,116] Standardization guidelines have been proposed[113]

Disease modification: drug modifies disease severity. *Equivalent scoring systems are seldom well developed for animal models.*

Usually discrete/categorical 1H MRI variables from clinical scoring schemes.

Measure	Clinical context	Translational use
Disease-specific scoring systems for 1H MRI in arthritis	Arthritides (*e.g.* rheumatoid arthritis,[117] osteoarthritis,[118–121] psoriatic arthritis[117] often involve multiple components of the musculoskeletal system. Composite MRI scores combine assessments of the severity of the disease in different joint compartments	The RAMRIS system for rheumatoid arthritis has been used as a biomarker of drug treatment efficacy in several clinical trials[122]
TNM stage in oncology (1H MRI ± other imaging modalities)	Most cancer types have a standard T(umour)-N(ode)-M(etastasis) staging scheme, which is largely or wholly determined by MRI and other imaging modalities. Change in TNM stage can be used to measure time-to-progression and progression-free-survival, which in some cases are accepted by regulatory authorities as valid surrogate endpoints	TNM systems are extensively described and standardized in the clinical cancer radiology literature[123] and used in diagnosis, drug development and patient management

Table 12.1 (*Continued*)

MR biomarker	What biology does the biomarker measure?	Examples of use and standardization
	Safety: drug does not cause harm.	
	Usually ¹H MRI variables, which may be intensive, extensive or discrete/categorical. Often characterized in animal toxicology. Many safety (harm) biomarkers are simply the converse of efficacy (benefit) biomarkers (e.g. adverse changes in cardiac function[124]), others are specific for harm/lack-of-harm.	
Vasogenic edema from T_1-weighted ¹H MRI	R_1 increases in vasogenic edema due to net influx of water to the brain	Vasogenic edema is a recognized risk with investigational immunotherapy for Alzheimer's disease[125] and MRI is commonly employed for safety monitoring[126] in clinical trials
	Personalized Healthcare (Stratified Medicine): MR biomarker identifies patients who will benefit from the drug, and patients who will not benefit and/or be harmed by the drug.	
	Usually discrete/categorical ¹H MRI variables from clinical scoring schemes. Equivalent scoring systems are seldom well developed for animal models.	
Absence of MR (or CT) evidence of cerebral hemorrhage coupled to treatment of acute stroke patients with alteplase	R_2^* increases in cerebral hemorrhage due to sub-voxel heterogeneity in magnetic susceptibility from erythrocytes packed with deoxyhemoglobin (high-spin Fe II)	Most strokes are due to blockage of a cerebral artery by a blood clot. Prompt treatment with alteplase can restore blood flow before irreversible brain damage has occurred.[127] Alteplase, however, can also cause fatal bleeding in the brain, so the regulatory approval and drug labeling for alteplase states that "treatment should only be initiated...after exclusion of intracranial hemorrhage by a...diagnostic imaging method sensitive for the presence of hemorrhage".[128]

---

[a]Although most drugs are too potent to permit direct detection by MRS, the observation of a pharmacodynamic effect logically also provides evidence of adequate pharmacokinetics (as well as target engagement and pathway activation).

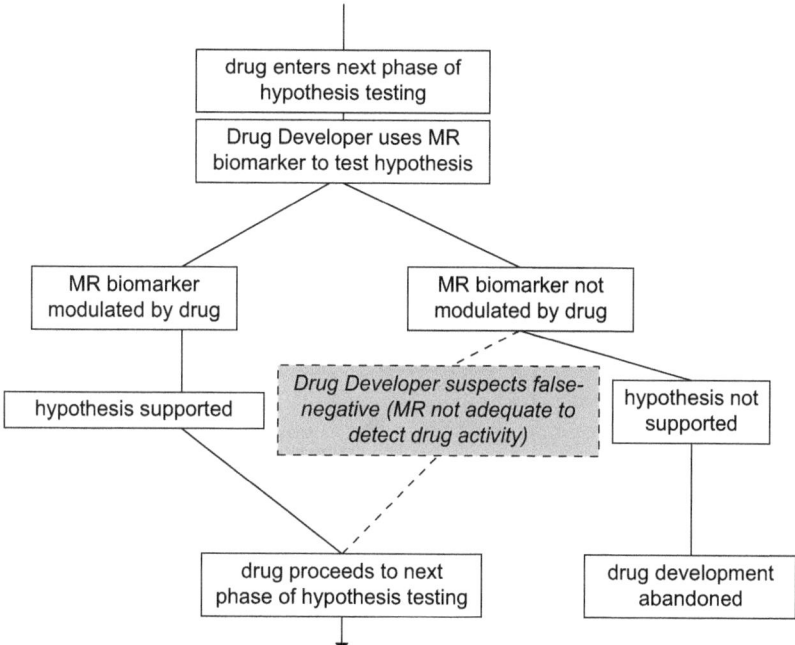

Figure 12.2   The false-negative trap. MR biomarkers only add value if the shaded box is eliminated from the decision tree.

than to establish technical validity of an MR biomarker for routine healthcare in hospitals around the world.

- "Biological validity" is the confidence that the biomarker correctly reports some underlying biology that is important to the patient's future clinical outcome, that changes in the biomarker correctly reflect important changes in the underlying biology and ultimately that the biomarker better forecasts future clinical outcome. Since the physician always has access to the patient's current clinical status and history, the biomarker is only useful if it provides a better forecast than these clinical data alone. (If the forecast is near-perfect, the biomarker might be considered a surrogate endpoint). However, it is entirely possible that a biomarker has high biological validity but only limited technical validity (*e.g.* only available in a single expert center). This is very often the case with high-field ($\geq 3\,T$) heteronuclear MR biomarkers, where coils for particular body parts specifically tuned to unusual frequencies may be rarely available.

- "Cost-effectiveness" is, of course, a third orthogonal consideration. MR biomarkers are always costly to measure. No matter how high the biological or technical validity of the MR biomarker, it will not be used in a drug development setting if there is a competing, easily measured, blood-borne biomarker. If the biomarker is intended to cross the second translational gap into the healthcare system (*e.g.* as a diagnostic), it will not be used, no matter how good, unless it is affordable.

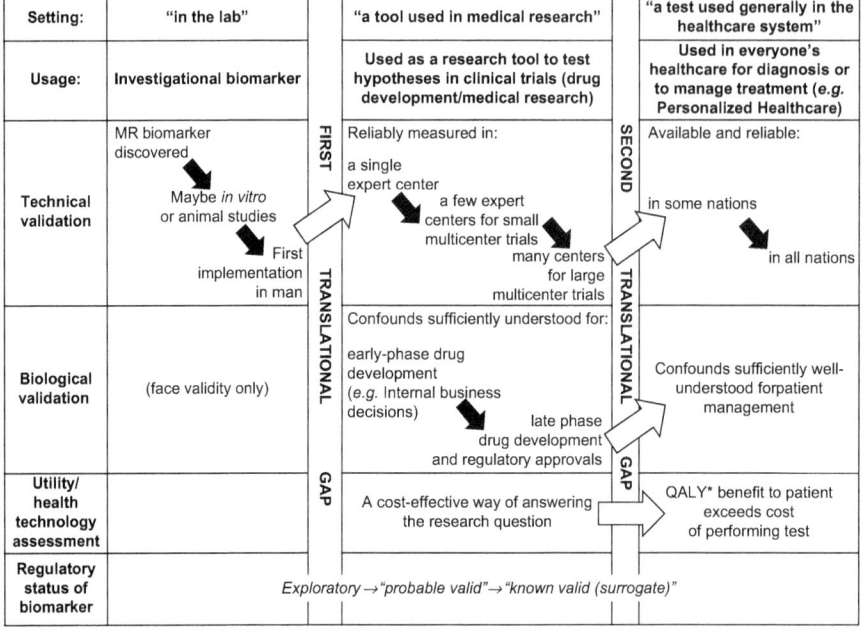

Setting:	"in the lab"	"a tool used in medical research"	"a test used generally in the healthcare system"
Usage:	Investigational biomarker	Used as a research tool to test hypotheses in clinical trials (drug development/medical research)	Used in everyone's healthcare for diagnosis or to manage treatment (*e.g.* Personalized Healthcare)
Technical validation	MR biomarker discovered Maybe *in vitro* or animal studies First implementation in man	**FIRST** Reliably measured in: a single expert center a few expert centers for small multicenter trials many centers for large multicenter trials **TRANSLATIONAL**	**SECOND** Available and reliable: in some nations in all nations **TRANSLATIONAL**
Biological validation	(face validity only)	Confounds sufficiently understood for: early-phase drug development (*e.g.* Internal business decisions) late phase drug development and regulatory approvals	Confounds sufficiently well-understood for patient management
Utility/ health technology assessment		**GAP** A cost-effective way of answering the research question	**GAP** QALY* benefit to patient exceeds cost of performing test
Regulatory status of biomarker		*Exploratory → "probable valid" → "known valid (surrogate)"*	

*QALY: Quality-Adjusted Life-Year

Figure 12.3 MR biomarker roadmap. After the MR biomarker has been discovered "in the lab" it must cross the first translational gap before it is "useful in medical research" and the second gap before it is generally used in healthcare. The time for this journey is measured in decades, not years.

12.2.3 Technical Validity

For technical validity we need to consider repeatability, reproducibility, accuracy and legal/ethical acceptability.

12.2.3.1 *Repeatability*

Repeatability and reproducibility are often used interchangeably in the literature, and for MR biomarkers the concepts overlap; however, it is helpful to create a distinction.

"Repeatability" is the concept that a test conducted twice on the same subject, in the same lab, using the same equipment, by the same operator, over a short period of time, should give the same result.[36] For a clinical trial of an investigational drug using a pharmacodynamic response biomarker, statistical power calculations and sample size calculations require an understanding of the repeatability of the test. As an (oversimplified) example, if the test-retest coefficient of variation for the MR biomarker in a given population over 7 days is 20%, and the drug is expected to reduce the biomarker by 20% over the same time interval, then a study in 13 patients will have 80% power to detect the effect of the drug with $P < 0.05$. If the repeatability is 25%, the number of patients required rises to 20. Of course, no biomarker is perfectly repeatable, and for MR biomarkers there are three main components of repeatability that must be understood: the patient, the equipment and the observer.

The patient: Consider that the patient is measured twice over an interval that is short in comparison with the progression of the disease. Even so, there are variations in normal physiology that affect the reproducibility. MR-determined cartilage thickness in the knee joint may vary according to the time of day[37] or according to patterns of use immediately preceding the scan.[38] For dynamic contrast-enhanced (DCE) MRI, recent caffeine consumption may affect the biomarker.[39] It is therefore prudent (although not always practical) to make the measurement at the same time of the day, with some control of prior exercise, eating, drinking and use of recreational drugs such as alcohol and tobacco.

The equipment: there are many small sources of scanner instability that can affect the repeatability of a biomarker. Clinical scanners usually undergo routine quality assurance scans and, while these may be perfect for the routine diagnostic radiology that provides most of the workload, they may be in-adequate for the most demanding applications. For example a 1% variation in all three gradients (small enough not to affect picture quality, and small enough to evade many quality control routines) produces >3% error in volume, which is disastrous for some musculoskeletal or neurology biomarkers where changes of <2% are of interest[23,40] (see also Chapter 16). Variations in scanner performance can be assessed using carefully designed phantoms. These need to be chemically and mechanically stable, and bear some relation to the *in vivo* situation. Many ^{1}H MRI biomarkers are measured in tissues with $T_1 \sim 1$ s and $T_2 < 0.1$ s; use of a large homogeneous water phantom with $T_1 \sim 3$ s and $T_2 \sim 2$ s may introduce esoteric and irrelevant artifacts due to echoes from persistent transverse magnetization, radiation damping or dielectric resonance. For biomarkers that are temperature-dependent (depending on diffusion or relaxation), temperature control is essential. Commonly used phantoms are the Eurospin phantom for relaxation,[41] and ice-water phantoms for diffusion biomarkers.[42]

The observer. While MR acquisitions are not as operator-dependent as image acquisitions for some other modalities such as ultrasound, the radiographer does play an important part in repeatability for some biomarkers. In rheumatoid- and osteoarthritis, where small changes are measured in the wrist or the knee, careful positioning is essential to ensure the patient is comfortable, and does not move. For measurements in the neck of atherosclerosis or cancer, the patient should be counseled not to cough or swallow. And for demented, psychotic or pediatric patients, special skill is needed by the radiographer to reassure the patient and minimize motion artifact. However, a much bigger source of observer error is in defining the region-of-interest (mark-up or segmentation). Most MR biomarkers (Table 12.1) are either *extensive* variables ("how much is there?" *e.g.* tumor volume, left ventricular wall thickness, hippocampal volume), or *intensive* variables ("what is the magnitude of some parameter such as flow or diffusion within the defined region?"); both are critically dependent on the ability of the segmenter to define accurately and consistently the structure of interest. This may be affected not only by obvious factors such as training, but also by the time between first and second segmentations[43] and even by lighting levels in the room where the images are displayed.[44]

12.2.3.2 Reproducibility

"Reproducibility" is the concept that a test conducted twice on comparable subjects, in different labs, using different equipment, by different operators, over different periods of time, should give comparable results.[36] For a small clinical trial of an investigational drug measuring changes in a pharmacodynamic biomarker over a short period of time using the same equipment and operator, reproducibility is less important than repeatability. Even then, if a scanner goes out of action during the trial, if key personnel leave or if new centers are included later, reproducibility still needs to be considered. In any case, grossly irreproducible biomarkers must always give concern, if only for the fundamental scientific reason that when a future investigator reads a published study, s/he ought to be able to repeat it in his/her own laboratory and obtain the same result. As is the case for repeatability, there are three main components of reproducibility that must be understood: the equipment, the observer and the patient.

The equipment: equipment variations make a major contribution to reproducibility of imaging biomarkers in ways that are not encountered with traditional blood or urine biomarkers. In the latter case, measurements can be made using a standardized *in vitro* diagnostic device, designed for task, with trained staff, in a central lab remote from the patient. However, for MR biomarkers, reproducibility is critically dependent on how the MR instrument performs while it is interfaced to the patient. There may be different MR equipment in each hospital. The MRI scanner was designed and marketed to provide pictures, and was not designed for quantitative biomarker tasks. The specific pulse sequences and hardware configuration may be unfamiliar to the radiographers in some of the clinical trial centers. If the drug developer intends to perform a trial with a specified number of patients in particular clinical centers in a specified time-frame, s/he may have little or no control over the MR equipment available. At the time of writing there are three main manufacturers supplying the European and North American markets, with other manufacturers also having significant market share particularly in Asia. Many use proprietary pulse sequences, which are not easily implemented on competitors' platforms. Different manufacturers' parallel imaging protocols create images with very different (and non-Rician) noise characteristics.[45] 1.5 T and 3 T scanners are both common; while 3 T scanners generally have better spatial resolution and sensitivity, the hardware is still evolving, and spatial inhomogeneity in the B_0, gradient and RF fields is more troublesome than at 1.5 T. It may be impractical to mandate the same specialized coils (*e.g.* head, neck, knee, wrist) on each scanner. Most clinical trials perform central analysis, but if locally written software is used to measure say T_2, ADC or K^{trans}, subtle differences in algorithm or implementation may affect reproducibility.

Two powerful approaches to assess reproducibility are the uniform-phantom and travelling-volunteer approaches. For example, ADC has been cross-validated across a number of scanners using an ice-water phantom,[42] while in knee osteoarthritis, cartilage thickness and cartilage T_2 were cross-validated by scanning volunteers with early osteoarthritis in 3 T scanners from three different manufacturers in three different hospitals.[46,47] There are, however, limitations to the travelling-volunteer approach. It is unrealistic to expect patients with advanced cancer and short life expectancy to volunteer for a study involving extensive scanning and travel to different cities. In contrast-enhanced protocols, variations in contrast agent between hospitals may affect reproducibility. Gadopentetate (dianion), gadoterate (monoanion) and gadodiamide (uncharged) are all commonly used. Charge certainly affects uptake at least in some settings.[48] However, it may be impractical to mandate the same contrast agent in each center (for example, the agents may have different safety profiles, and not all are available in all jurisdictions).

The Observer: in large studies it may be impractical to have all scans segmented by the same person simply because of the time required. It is reasonable to expect that between-segmenter variation will be larger than within-segmenter variation, although with careful training, cross-validation and quality control these differences can be minimized.[23]

The patient: a limitation of published studies of repeatability and reproducibility is that they do not always represent the trial populations of ultimate interest. In arthritis, studies in normal volunteers or early disease provide optimistic estimates of reproducibility that may not translate to advanced disease, where extensive joint damage makes segmentation difficult. In cancer, studies in newly presenting brain tumors will likely show better reproducibility than the common sites of metastasis typically encountered in Phase I trials such as lung and liver, which suffer motion artifact, and perhaps image complexity due to prior radiotherapy.

12.2.3.3 Accuracy

Accuracy is obviously highly desirable for any measurement, although for MRI/S biomarkers in practice it can be difficult or impossible to prove. Structural MRI measurements *in vivo* from humans or animals, such as hippocampal volume or cartilage thickness, can be compared with *post mortem* specimens. However, *post mortem* distortions and shrinkage during processing often lead to poor agreement. A more subtle confound is that the boundary between, say, a tumor and surrounding normal tissue may be quite diffuse and dependent on the measurement technique. More physiologic biomarkers, such as the transfer constant K^{trans} measured in DCEMRI, or BOLD activations in functional neuroimaging, have no real *post mortem* correlates at all. Even with MRS measurements, it can be difficult to derive and validate molar concentrations, *e.g.* due to non-Lorentzian lineshapes or B_1 inhomogeneity, and ratios (*e.g.* NAA:tCho) are commonly given instead. The unverified accuracy of MR biomarkers, therefore, is a common weakness in competition with more traditional biochemical biomarkers.

12.2.3.4 Legal and Ethical Acceptability

An MR biomarker that cannot legally be used in all the jurisdictions where it is needed, or cannot be ethically deployed in all the patients where it is needed, cannot be regarded as technically valid. The required MR devices (spectrometer, coil and perhaps analysis software) may not have regulatory approval in each jurisdiction where patients will be recruited. If a contrast agent is needed, that also needs marketing approval in each jurisdiction. Gadoterate (Dotarem), for example, has never been available in the United States, while gadofosveset (Ablavar), a unique protein-bound agent, is no longer available in Europe. It is almost impossible to conduct a trial using an investigational therapeutic drug and an investigational diagnostic (contrast agent) in the same trial.[12] There are unfortunately many contrast agents with unique MR properties shown in animal studies or clinical trials, without marketing approval. Gadomelitol (Vistarem) has a molecular weight an order of magnitude higher than other gadolinium-based contrast agents, thereby providing unique biomarkers of perfusion and permeability;[49] unfortunately it was never submitted by the manufacturer for regulatory approval and so is unavailable. Even agents that have received marketing approval may be withdrawn for commercial reasons. At the time of writing mangafodipir (Teslascan, the only manganese-based approved contrast agent), ferucarbotran (Resovist) iron oxide nanoparticles and (in Europe) gadofosveset (Vasovist) are all unavailable despite previous marketing approval.

Ethically one cannot, of course, expose patients to unnecessary risks. Many interesting pulse sequences are unavailable at medium (1.5 T) or high (3 T) field because of the tissue heating they would cause. Common contrast agents cannot be used in patients with poor renal function. But it is also ethically important to avoid unreasonable burdens. For a metastatic cancer patient with a short life expectancy, it is unreasonable to expect frequent attendance for short-term follow-up scans. Patients with painful conditions may not tolerate uncomfortable positions such as the so-called "Superman position" to bring the wrist to the magnet isocenter, or long scan duration for signal averaging. Endoscopic coils give better signal-to-noise but may not be tolerated. Even the noise and claustrophobia of MRI are significant issues for many patients.

12.2.4 Biological Validity

For MR biomarkers used in early drug development it may be impractical and unnecessary to meet the exacting standards of regulatory surrogacy. However, two important questions must be considered: the association between the biomarker and the underlying biology, and whether the biomarker improves forecasts of future clinical outcome. Direct correlation with the underlying biology may be difficult or impossible (few people will volunteer for a brain biopsy to help validate a neuroimaging biomarker), but well-designed animal studies can help provide a very strong evidence base. Indeed for many drug developers, this is the main or only rationale for performing *in vivo* MRI/S in animals.

Given the need to avoid the false negative trap (Figure 12.2), a single experiment rarely provides adequate biological validation. However, a good evidence base can be assembled using animal MR studies, complemented by limited cross-sectional and longitudinal studies in man, using the Bradford Hill criteria,[50,51] which were proposed many years ago in connection with predictivity in medical research. There follows a summary of different types of biological validation study, loosely aligned to Bradford Hill's criteria.

Plausibility (face validity). Probably the least important of the criteria, face validity is useful for communication. Most morphologic MRI biomarkers of disease modification (Table 12.1) have obvious face validity, which is helpful in explaining them to drug developers, physicians and patients. Brain volume and cartilage thickness "should not" decrease, while tumor volume, infarct volume, rheumatoid erosion count and multiple sclerosis lesion count "should" decrease. We have already seen, however, that even tumor volume can give false positives. In contrast to morphologic MRI biomarkers, pharmacodynamic MR biomarkers sometimes lack face validity. In cancer, ADC is not intuitively obvious as a biomarker of cell death, nor is ADC in the brain intuitively obvious as a biomarker of cytotoxic edema. It was not intuitively obvious that tumor nucleotide triphosphate should *increase* following effective cytotoxic chemotherapy,[52] nor is it intuitively obvious whether high tumor perfusion or low tumor perfusion should predict[53] worse outcome for cancer patients.

Coherence (1). Does the MR measurement agree with direct biochemical or pathologic correlates of the biomarker? Consider, for example, a drug program to prevent cartilage loss in osteoarthritis (considered to be a desirable objective). A strong correlation between disease severity, cartilage thickness measured by MRI and cartilage thickness measured pathologically in animals[54,55] and in humans[56] supports the biological validity of MRI-determined cartilage thickness as a biomarker.

Coherence (2). When drug intervention changes the MR biomarker, does the corresponding biochemical or pathologic measurement change likewise? For the drug developer, this is often a key animal experiment, particularly for novel pharmacologies where, by definition, there are no human data on response of the MR biomarker to the drug class at the outset. Well-conducted animal experiments are essential for the design and interpretation of subsequent clinical trials. Dynamic contrast-enhanced MRI biomarkers have been successfully employed in the clinical development of vascular destructive agents in cancer, in part because of the very clear correlation, in rodent tumors, between the response of the MR biomarker to the drug (contrast agent uptake abolished) and the modulation of the underlying pathology (intratumoral necrosis induced). In the early development of the vascular destructive agents, an extensive portfolio of animal DCEMRI studies with pathologic correlation[57-64] involving different doses, in different tumor models and species, with different times of observation, allowed robust design and interpretation of DCEMRI in the Phase I trials in patients with tumors in typical locations such as liver and pelvis.[58,65]

Biological gradient. When the drug induces a change on the MR biomarker does the clinical outcome improve (or in the case of a safety biomarker, worsen); does deterioration in the MR biomarker lead to a worse clinical outcome? Many studies, in animals and humans,[53] have shown antiangiogenic receptor tyrosine kinase inhibitors acutely reduce K^{trans}, a pharmacodynamic biomarker, in tumors. The biological validity of K^{trans} as a biomarker is strengthened by data in mice[66] and humans[67] showing that, as the magnitude of the change in K^{trans} increases, so does subsequent antitumor effect of the drug.

Temporality. Does the temporal evolution of the drug-induced changes in the MR biomarker follow the temporal evolution of corresponding biochemical or pathologic measurement and clinical outcome? In relapsing-remitting multiple sclerosis, contrast-enhancing lesions have been used as a biomarker for many years. Long studies are often conducted with monthly MRI measurements. The biological validity of the biomarker is supported through the temporal correlations[68] between change in the biomarker and disease progression, remission, relapse and response to therapy.

Specificity. If there is a well-characterized control group of non-responders, do these fail to elicit changes in the MR biomarker and the corresponding biochemical or pathologic measurement? For example, acute change in T_1 has been advocated[69] as a pharmacodynamic biomarker in cancer research. In support of this, 10 studies of different mouse tumors treated with different drugs were performed: in 8 cases where the drug was known to have antitumor efficacy, T_1 fell following treatment and in comparison to controls, whereas in the cases where the drug was known not to have antitumor efficacy, no change in T_1 occurred. The T_1 changes also generally correlated with pathologic changes.

Strength of association. How well do levels of the MR biomarker predict clinical outcome? Is there merely a weak, albeit statistically significant, correlation, or is there a strikingly high odds ratio linking the biomarker to the clinical outcome? As an example, in cardiac T_1-weighted MRI, small areas of late enhancement post-contrast medium show areas of scarring resulting from earlier, unrecognized myocardial infarcts. Evidence for the biological validity of this biomarker in patients without known prior myocardial infarct is provided by the very strong association[70] between late enhancement, and major adverse cardiac events or cardiac mortality (hazard ratios of 8.29 and 10.9, respectively).

Consistency: are the findings replicated by different investigators, in different labs, perhaps with different animal models or clinical cohorts? Replication of scientific findings has an obvious benefit, and similar results with somewhat different animal models in different labs are much more powerful than the repetition of identical experiments in the same lab. So returning to the example of development of vascular destructive agents in cancer, the fact that very similar findings have been reported[57–65] by different investigators, in different labs, using slightly different methods, in different tumor models in different species using structurally different vascular destructive agents, provides a much stronger evidence base than a single experiment, no matter how compelling, in a single lab. It should be noted, however, that while publication bias (*i.e.* apparently good, but spurious, biological validity because of failure to publish negative studies) is being eliminated from the clinical literature, although unfortunately it is still probably widespread in the animal literature,[71] so unless investigators publish negative findings,[72] consistency will be overestimated.

Analogy: are there similar findings with similar drugs and similar diseases? For example, successful use of DCEMRI biomarkers in oncology[53] has spurred interest in the use in other diseases such as benign lung disease,[73] atheroma[74] and the arthritides.[75–77]

12.3 MR Biomarkers in Practice

Despite these daunting challenges, drug developers do use MR biomarkers. Most current and recent clinical trials are listed in the clinicaltrials.gov registry,[78] although older trials from the 1980s, 1990s and early 2000s are missing. This registry lists 1276 trials with MRI or MRS biomarkers as outcome measures and with small-molecule or biological drug intervention, enrolling 186 047 patients or volunteers. These represent a variety of disease areas including oncology, neurology (*e.g.* stroke, dementia, multiple sclerosis, Parkinson's disease), psychiatry (*e.g.* bipolar disorder, depression, schizophrenia), musculoskeletal (*e.g.* osteoarthritis, rheumatoid arthritis, ankylosing spondylitis, *etc.*), cardiovascular (*e.g.* myocardial infarct, heart failure, atherosclerosis, aortic valve disease), diabetes, renal disease, sexual dysfunction, non-alcoholic fatty liver disease, respiratory (*e.g.* asthma, chronic obstructive pulmonary disease) and infections (*e.g.* HIV, fungal, neurocysticercosis). Representatives of different types of MR biomarkers are described in Table 12.1.

12.4 Unmet Needs and Opportunities

Several analyses[2,3] have recently been conducted of the challenges facing drug development, and of the tools, such as qualified biomarkers, that are now needed to improve the efficiency of drug development. Notably these include, in Europe, the Innovative Medicines Strategic Research Agenda,[79] and in the USA, the FDA Critical Path Opportunities List.[80] These describe unmet needs including imaging biomarkers of interest for MR in cardiovascular disease (atherosclerosis progression, measures of cardiac function in congestive heart failure), rheumatoid- and osteoarthritis (cartilage and joint soft tissue), chronic obstructive pulmonary disease progression, neurocognitive diseases (functional and morphologic measures in Parkinson's and Alzheimer's disease) and cancer.

Drug discovery faces some grand challenges that could be transformed by innovative MR biomarkers. In oncology, can we develop MR biomarkers that identify those tumors most likely to progress, metastasize and ultimately kill the patient? In neurology, can we develop MR biomarkers that identify which patients with mild cognitive impairment will progress to Alzheimer's disease? Following myocardial or cerebral infarct, can we predict which patients will tend to recover through remodeling of remaining healthy brain or heart tissue, and who will deteriorate clinically?

In response to these challenges, several major consortia have been established to standardize and evaluate MR (and other biomarkers), and/or correlate them with clinical outcome in large patient cohorts. These include ADNI (www.adni-info.org/; the Alzheimer's Disease Neuroimaging Initiative);[81] OAI (http://oai.epi-ucsf.org/; the US NIH Osteoarthritis Initiative);[82,83] QIBA (www.rsna.org/QIBA_.aspx; Quantitative Imaging Biomarkers Alliance of the Radiological Society of North America);[84] ECMC (www.ecmcnetwork.org.uk/network-groups/imaging-group/; the UK Experimental Cancer Medicine Center Network Imaging Group)[85]; QuIC-ConCePT (www.quic-concept.eu; the "Quantitative Imaging in Cancer Connecting Cellular Processes with Therapy" Consortium of the European Innovative Medicines Initiative)[86] and HRPI (www.hrpinitiative.com; the High Risk Plaque Initiative).

The MR literature is not short of ingenious measurements. Our greatest challenge is to bring them out of the MR lab, and across the Cooksey gaps, so they are useful in drug development, in other medical research and ultimately in healthcare.

References

1. Biomarkers Definitions Working Group, *Clin. Pharmacol. Ther.*, 2001, **69**, 89.
2. B. Munos, *Nat. Rev. Drug Discov.*, 2009, **8**, 959.
3. J. W. Scannell, A. Blanckley, H. Boldon and B. Warrington, *Nat. Rev. Drug Discov.*, 2012, **11**, 191.
4. S. M. Paul, D. S. Mytelka, C. T. Dunwiddie, C. C. Persinger, B. H. Munos, S. R. Lindborg and A. L. Schacht, *Nat. Rev. Drug Discov.*, 2010, **9**, 203.
5. J. M. N. Duarte, H. Lei, V. Mlynárik and R. Gruetter, *Neuroimage*, 2012, **61**, 342.
6. W. Wolf, C. A. Presant and V. Waluch, *Adv. Drug Deliv. Rev.*, 2000, **41**, 55.
7. H. W. van Laarhoven, C. J. Punt, Y. J. Kamm and A. Heerschap, *Crit. Rev. Oncol. Hematol.*, 2005, **56**, 321.
8. R. A. Komoroski, *NMR Biomed.*, 2005, **18**, 67.
9. S. Fain, M. L. Schiebler, D. G. McCormack and G. Parraga, *J. Magn. Reson. Imaging*, 2010, **32**, 1398.
10. S. Matsuoka, S. Patz, M. S. Albert, Y. Sun, R. R. Rizi, W. B. Gefter and H. Hatabu, *J. Thorac. Imaging*, 2009, **24**, 181.
11. K. Brindle, *Br. J. Radiol.*, 2012, **85**, 697.

12. J. C. Waterton, in *The Chemistry of Labels, Probes and Contrast Agents*, The Royal Society of Chemistry, Cambridge, UK, 2012, p. 1.

13. P. C. Lauterbur, *Nature*, 1973, **242**, 190.

14. J. C. Bamber, N. R. Miller and M. Tristam, in *Webb's Physics of Medical Imaging*, ed. M. A. Flower, CRC Press, Boca Raton FL, USA, 2nd edn, 2012, p. 351.

15. J. C. Hebden, in *Webb's Physics of Medical Imaging*, ed. M. A. Flower, CRC Press, Boca Raton FL, USA, 2nd edn, 2012, p. 665.

16. B. H. Brown and S. Webb, in *Webb's Physics of Medical Imaging*, ed. M. A. Flower, CRC Press, Boca Raton FL, USA, 2nd edn, 2012, p. 647.

17. C. Ramanathan, R. N. Ghanem, P. Jia, K. Ryu and Y. Rudy, *Nat. Med.*, 2004, **10**, 422.

18. E. D. Becker, C. L. Fisk and C. L. Kheptral, in *Encyclopedia of Magnetic Resonance*, ed. D. M. Grant and R. K. Harris, Wiley, Chichester, UK, 1996, vol. 1, p. 106.

19. H. Y. Carr, *Physics Today*, 2004, **57**, 83.

20. D. Cooksey, *A Review of UK Health Research Funding*, Her Majesty's Stationery Office, Norwich, UK, 2006.

21. S. H. Woolf, *JAMA*, 2008, **299**, 211.

22. E. A. Eisenhauer, P. Therasse, J. Bogaerts, L. H. Schwartz, D. Sargent, R. Ford, J. Dancey, S. Arbuck, S. Gwyther, M. Mooney, L. Rubinstein, L. Shankar, L. Dodd, R. Kaplan, D. Lacombe and J. Verweij, *Eur. J. Cancer*, 2009, **45**, 228.

23. F. Eckstein, F. Cicuttini, J.-P. Raynauld, J. C. Waterton and C. Peterfy, *Osteoarthritis Cartilage*, 2006, **14**, 46.

24. S. Desch, I. Eitel, S. de Waha, G. Fuernau, P. Lurz, M. Gutberlet, G. Schuler and H. Thiele, *Trials*, 2011, **12**, 204.

25. A. Breckenridge, M. Mello and B. M. Psaty, *Nat. Rev. Drug Discov.*, 2012, **11**, 501.

26. D. S. Echt, P. R. Liebson, L. B. Mitchell, R. W. Peters, D. Obias-Manno, A. H. Barker, D. Arensberg, A. Baker, L. Friedman, H. L. Greene, M. L. Huther and D. W. Richardson, *N. Engl. J. Med.*, 1991, **324**, 781.

27. C. O. Brindley, E. Markoff and M. A. Schneiderman, *Cancer*, 1959, **12**, 139.

28. T. Burzykowski, M. Buyse, M. J. Piccart-Gebhart, G. Sledge, J. Carmichael, H. J. Luck, J. R. Mackey, J. M. Nabholtz, R. Paridaens, L. Biganzoli, J. Jassem, M. Bontenbal, J. Bonneterre, S. Chan, G. A. Basaran and P. Therasse, *J. Clin. Oncol.*, 2008, **26**, 1987.

29. M. Buyse, P. Thirion, R. W. Carlson, T. Burzykowski, G. Molenberghs and P. Piedbois, *Lancet*, 2000, **356**, 373.

30. R. L. Prentice, *Stat. Med.*, 1989, **8**, 431.

31. G. Poste, *Nature*, 2011, **469**, 156.

32. J. A. Wagner, *Ann. Rev. Pharmacol. Toxicol.*, 2008, **48**, 631.

33. D. Sarker and P. Workman, *Adv. Cancer Res.*, 2007, **96**, 213.

34. M. Danhof, G. Alvan, S. G. Dahl, J. Kuhlmann and G. Paintaud, *Pharm. Res.*, 2005, **22**, 1432.

35. Institute of Medicine (US) Forum on Drug Discovery, Development, and Translation. Accelerating the Development of Biomarkers for Drug Safety: Workshop Summary. 2009, National Academies Press.

36. ISO, *Statistics – Vocabulary and Symbols – Part 2: Applied Statistics*, ISO 3534-2, 2006.

37. J. C. Waterton, S. Solloway, J. E. Foster, M. C. Keen, S. Gandy, B. J. Middleton, R. A. Maciewicz, I. Watt, P. A. Dieppe and C. J. Taylor, *Magn. Reson. Med.*, 2000, **43**, 126.

38. F. Eckstein, B. Lemberger, C. Gratzke, M. Hudelmaier, C. Glaser, K. H. Englmeier and M. Reiser, *Ann. Rheum. Dis.*, 2005, **64**, 291.

39. E. M. Haacke, C. L. Filleti, R. Gattu, C. Ciulla, A. Al-Bashir, K. Suryanarayanan, M. Li, Z. Latif, Z. DelProposto, V. Sehgal, T. Li, V. Torquato, R. Kanaparti, J. Jiang and J. Neelavalli, *Magn. Reson. Med.*, 2007, **58**, 463.

40. N. C. Fox, P. A. Freeborough and M. N. Rossor, *Lancet*, 1996, **348**, 94.

41. R. A. Lerski and J. D. de Certaines, *Magn. Reson. Imaging*, 1993, **11**, 817.

42. T. L. Chenevert, C. J. Galban, M. K. Ivancevic, S. E. Rohrer, F. J. Londy, T. C. Kwee, C. R. Meyer, T. D. Johnson, A. Rehemtulla and B. D. Ross, *J. Magn. Reson. Imaging*, 2011, **34**, 983.

43. F. Eckstein, L. Heudorfer, S. C. Faber, R. Burgkart, K. H. Englmeier and M. Reiser, *Osteoarthritis Cartilage*, 2002, **10**, 922.

44. P. C. Brennan, M. McEntee, M. Evanoff, P. Phillips, W. T. O'Connor and D. J. Manning, *Am. J. Roentgenol.*, 2007, **188**, W177.

45. O. Dietrich, J. G. Raya, S. B. Reeder, M. F. Reiser and S. O. Schoenberg, *J. Magn. Reson. Imaging*, 2007, **26**, 375.

46. S. Balamoody, T. G. Williams, J. C. Waterton, M. Bowes, R. Hodgson, C. J. Taylor and C. E. Hutchinson, *Arthrit. Res. Ther.*, 2010, **12**, R202.

47. S. Balamoody, T. Williams, C. Wolstenholme, J. Waterton, M. Bowes, R. Hodgson, S. Zhao, M. Scott, C. Taylor and C. Hutchinson, *Skeletal Radiol.*, in press.

48. W. Li, R. Scheidegger, Y. Wu, R. R. Edelman, M. Farley, N. Krishnan, D. Burstein and P. V. Prasad, *Magn. Reson. Med.*, 2010, **64**, 1267.

49. D. P. Bradley, J. L. Tessier, D. Checkley, H. Kuribayashi, J. C. Waterton, J. Kendrew and S. R. Wedge, *NMR Biomed.*, 2008, **21**, 302.

50. A. B. Hill, *Proc. R. Soc. Med.*, 1965, **58**, 295.

51. R. K. Chetty, J. S. Ozer, A. Lanevschi, I. Schuppe-Koistinen, D. McHale, J. S. Pears, J. Vonderscher, F. D. Sistare and F. Dieterle, *Clin. Pharmacol. Ther.*, 2010, **88**, 260.

52. R. G. Steen, *Cancer Res.*, 1989, **49**, 4075.

53. J. P. B. O'Connor and G. C. Jayson, *Clin. Cancer Res.*, 2012, **18**, 6588.

54. J. Bowyer and C. G. Heapy, J. K. Flannelly, J. C. Waterton and R. A. Maciewicz, *Int. J. Exp. Pathol.*, 2009, **90**, 174.

55. J. J. Tessier, J. Bowyer, N. J. Brownrigg, S. Peers, F. R. Westwood, J. C. Waterton and R. A. Maciewicz, *Osteoarthritis Cartilage*, 2003, **11**, 845.

56. F. Eckstein, M. Schnier, M. Haubner, J. Priebsch, C. Glaser, K. H. Englmeier and M. Reiser, *Clin. Orthop. Relat. Res.*, 1998, **352**, 137.
57. D. P. Bradley, J. J. Tessier, S. E. Ashton, J. C. Waterton, Z. Wilson, P. L. Worthington and A. J. Ryan, *Neoplasia*, 2007, **9**, 382.
58. J. L. Evelhoch, P. M. LoRusso, Z. He, Z. DelProposto, L. Polin, T. H. Corbett, P. Langmuir, C. Wheeler, A. Stone, J. Leadbetter, A. J. Ryan, D. C. Blakey and J. C. Waterton, *Clin. Cancer Res.*, 2004, **10**, 3650.
59. D. J. O. McIntyre, S. P. Robinson, F. A. Howe, J. R. Griffiths, A. J. Ryan, D. v. C. Blakey, I. S. Peers and J. C. Waterton, *Neoplasia*, 2004, **6**, 150.
60. S. P. Robinson, D. J. O. McIntyre, D. Checkley, J. J. Tessier, F. A. Howe, J. R. Griffiths, S. E. Ashton, A. J. Ryan, D. C. Blakey and J. C. Waterton, *Br. J. Cancer*, 2003, **88**, 1592.
61. D. A. Beauregard, S. A. Hill, D. J. Chaplin and K. M. Brindle, *Cancer Res.*, 2001, **61**, 6811.
62. D. A. Beauregard, R. B. Pedley, S. A. Hill and K. M. Brindle, *NMR Biomed.*, 2002, **15**, 99.
63. D. A. Beauregard, P. E. Thelwall, D. J. Chaplin, S. A. Hill, G. E. Adams and K. M. Brindle, *Br. J. Cancer*, 1998, **77**, 1761.
64. R. J. Maxwell, J. Wilson, V. E. Prise, B. Vojnovic, G. J. Rustin, M. A. Lodge and G. M. Tozer, *NMR Biomed.*, 2002, **15**, 89.
65. S. M. Galbraith, R. J. Maxwell, M. A. Lodge, G. M. Tozer, J. Wilson, N. J. Taylor, J. J. Stirling, L. Sena, A. R. Padhani and G. J. Rustin, *J. Clin. Oncol.*, 2003, **21**, 2831.
66. D. Checkley, J. J. Tessier, J. Kendrew, J. C. Waterton and S. R. Wedge, *Br. J. Cancer*, 2003, **89**, 1889.
67. C. de Bazelaire, D. C. Alsop, D. George, I. Pedrosa, Y. Wang, M. D. Michaelson and N. M. Rofsky, *Clin. Cancer Res.*, 2008, **14**, 5548.
68. D. H. Miller, *Clin. Neurol. Neurosurg.*, 2002, **104**, 236.
69. P. M. McSheehy, C. Weidensteiner, C. Cannet, S. Ferretti, D. Laurent, S. Ruetz, M. Stumm and P. R. Allegrini, *Clin. Cancer Res.*, 2010, **16**, 212.
70. R. Y. Kwong, A. K. Chan, K. A. Brown, C. W. Chan, H. G. Reynolds, S. Tsang and R. B. Davis, *Circulation*, 2006, **113**, 2733.
71. P. Perel, I. Roberts, E. Sena, P. Wheble, C. Briscoe, P. Sandercock, M. Macleod, L. E. Mignini, P. Jayaram and K. S. Khan, *BMJ*, 2007, **334**, 197.
72. J. K. Boult, Y. Jamin, V. Jacobs, L. D. Gilmour, S. Walker-Samuel, J. Halliday, P. Elvin, A. J. Ryan, J. C. Waterton and S. P. Robinson, *Br. J. Cancer*, 2012, **106**, 1960.
73. J. H. Naish, L. E. Kershaw, D. L. Buckley, A. Jackson, J. C. Waterton and G. J. M. Parker, *Magn. Reson. Med.*, 2009, **61**, 1507.
74. C. Calcagno, V. Mani, S. Ramachandran and Z. A. Fayad, *Angiogenesis*, 2010, **13**, 87.
75. R. Sanz, L. Marti-Bonmati, J. L. Rodrigo and D. Moratal, *J. Magn. Reson. Imaging*, 2008, **27**, 171.

76. R. Hodgson, A. Grainger, P. O'Connor, T. Barnes, S. Connolly and R. Moots, *Ann. Rheum. Dis.*, 2008, **67**, 270.
77. R. J. Hodgson, P. O'Connor and R. Moots, *Rheumatology*, 2008, **47**, 13.
78. www.clinicaltrials.gov.
79. Innovative Medicines Initiative. Scientific Research Agenda. Available at http://www.imi.europa.eu/content/research-agenda.
80. US Food and Drug Administration. Critical Path Opportunities List. 2006. Available at http://www.fda.gov/oc/initiatives/criticalpath/.
81. M. W. Weiner, D. P. Veitch, P. S. Aisen, L. A. Beckett, N. J. Cairns, R. C. Green, D. Harvey, C. R. Jack, W. Jagust, E. Liu, J. C. Morris, R. C. Petersen, A. J. Saykin, M. E. Schmidt, L. Shaw, J. A. Siuciak, H. Soares, A. W. Toga and J. Q. Trojanowski, *Alzheimers Dement.*, 2012, **8**, S1–68.
82. G. Lester, *J. Musculoskelet. Neuronal Interact.*, 2008, **8**, 313.
83. C. G. Peterfy, E. Schneider and M. Nevitt, *Osteoarthritis Cartilage*, 2008, **16**, 1433.
84. A. J. Buckler, L. Bresolin, N. R. Dunnick, D. C. Sullivan, H. J. Aerts, B. Bendriem, C. Bendtsen, R. Boellaard, J. M. Boone, P. E. Cole, J. J. Conklin, G. S. Dorfman, P. S. Douglas, W. Eidsaunet, C. Elsinger, R. A. Frank, C. Gatsonis, M. L. Giger, S. N. Gupta, D. Gustafson, O. S. Hoekstra, E. F. Jackson, L. Karam, G. J. Kelloff, P. E. Kinahan, G. McLennan, C. G. Miller, P. D. Mozley, K. E. Muller, R. Patt, D. Raunig, M. Rosen, H. Rupani, L. H. Schwartz, B. A. Siegel, A. G. Sorensen, R. L. Wahl, J. C. Waterton, W. Wolf, G. Zahlmann and B. Zimmerman, *Radiology*, 2011, **259**, 875.
85. M. O. Leach, B. Morgan, P. S. Tofts, D. L. Buckley, W. Huang, M. A. Horsfield, T. L. Chenevert, D. J. Collins, A. Jackson, D. Lomas, B. Whitcher, L. Clarke, R. Plummer, I. Judson, R. Jones, R. Alonzi, T. Brunner, D. M. Koh, P. Murphy, J. C. Waterton, G. Parker, M. J. Graves, T. W. J. Scheenen, T. W. Redpath, M. Orton, G. Karczmar, H. Huisman, J. Barentsz and A. Padhani, *Eur. Radiol.*, 2012, **22**, 1451.
86. J. C. Waterton and L. Pylkkanen, *Eur. J. Cancer*, 2012, **48**, 409.
87. M. Buyse, P. Thirion, R. W. Carlson, T. Burzykowski, G. Molenberghs and P. Piedbois, *Lancet*, 2000, **356**, 373.
88. L. J. Lesko and A. J. Atkinson, *Annu. Rev. Pharmacol. Toxicol.*, 2001, **41**, 347.
89. Y. L. Chung, H. Troy, I. R. Judson, R. Leek, M. O. Leach, M. Stubbs, A. L. Harris and J. R. Griffiths, *Clin. Cancer Res.*, 2004, **10**, 3863.
90. C. A. Presant, W. Wolf, V. Waluch, C. Wiseman, P. Kennedy, D. Blayney and R. R. Brechner, *Lancet*, 1994, **343**, 1184.
91. S. Kalra, P. Tai, A. Genge and D. L. Arnold, *J. Neurol.*, 2006, **253**, 1060.
92. G. J. Kemp, M. Meyerspeer and E. Moser, *NMR Biomed.*, 2007, **20**, 555.
93. D. J. Taylor, G. J. Kemp and G. K. Radda, *J. Neurol. Sci.*, 1994, **127**, 198.
94. R. Sinkus, B. E. Van Beers, V. Vilgrain, N. DeSouza and J. C. Waterton, *Eur. J. Cancer*, 2012, **48**, 425.

95. D. M. Patterson, A. R. Padhani and D. J. Collins, *Nat. Clin. Pract. Oncol.*, 2008, **5**, 220.

96. A. R. Padhani, G. Liu, D. M. Koh, T. L. Chenevert, H. C. Thoeny, T. Takahara, A. Dzik-Jurasz, B. D. Ross, M. Van Cauteren, D. Collins, D. A. Hammoud, G. J. Rustin, B. Taouli and P. L. Choyke, *Neoplasia*, 2009, **11**, 102.

97. J. C. Wood, C. Enriquez, N. Ghugre, J. M. Tyzka, S. Carson, M. D. Nelson and T. D. Coates, *Blood*, 2005, **106**, 1460.

98. C. B. Sirlin and S. B. Reeder, *Magn. Reson. Imaging Clin. N. Am.*, 2010, **18**, 359.

99. Y. Ohno and H. Hatabu, *Eur. J. Radiol.*, 2007, **64**, 320.

100. N. Beckmann, C. Cannet, H. Karmouty-Quintana, B. Tigani, S. Zurbruegg, F. X. Ble, Y. Cremillieux and A. Trifilieff, *Eur. J. Radiol.*, 2007, **64**, 381.

101. S. Samee, T. Altes, P. Powers, E. E. de Lange, J. Knight-Scott, G. Rakes, J. P. Mugler 3rd, J. M. Ciambotti, B. A. Alford, J. R. Brookeman and T. A. Platts-Mills, *J. Allergy Clin. Immun.*, 2003, **111**, 1205.

102. J. P. O'Connor, A. Jackson, G. J. Parker, C. Roberts and G. C. Jayson, *Nat. Rev. Clin. Oncol.*, 2012, **9**, 167.

103. N. Balu, J. Wang, L. Dong, F. Baluyot, H. Chen and C. Yuan, *Top. Magn. Reson. Imaging*, 2009, **20**, 203.

104. W. Kerwin, D. Xu, F. Liu, T. Saam, H. Underhill, N. Takaya, B. Chu, T. Hatsukami and C. Yuan, *Top. Magn. Reson. Imaging*, 2007, **18**, 371.

105. T. Hatsukami, X. Q. Zhao, L. W. Kraiss, D. L. Parker, J. Waterton, V. Cain, J. Raichlen and C. Yuan, *Am. Heart J.*, 2008, **155**, 584.e1.

106. D. H. Miller, P. S. Albert, F. Barkhof, G. Francis, J. A. Frank, S. Hodgkinson, F. D. Lublin, D. W. Paty, S. C. Reingold and J. Simon, *Ann. Neurol.*, 1996, **39**, 6.

107. F. Barkhof, P. A. Calabresi, D. H. Miller and S. C. Reingold, *Nat. Rev. Neurol.*, 2009, **5**, 256.

108. E. Aizman, A. Mor, J. Chapman, Y. Assaf and Y. Kloog, *J. Neuro-immunol.*, 2010, **229**, 192.

109. D. J. Hunter, W. Zhang, P. G. Conaghan, K. Hirko, L. Menashe, W. M. Reichmann and E. Losina, *Osteoarthritis Cartilage*, 2011, **19**, 589.

110. F. Eckstein, G. Ateshian, R. Burgkart, D. Burstein, F. Cicuttini, B. Dardzinski, M. Gray, T. M. Link, S. Majumdar, T. Mosher, C. Peterfy, S. Totterman, J. Waterton, C. S. Winalski and D. Felson, *Osteoarthritis Cartilage*, 2006, **14**, 974.

111. C. G. Peterfy, E. Schneider and M. Nevitt, *Osteoarthritis Cartilage*, 2008, **16**, 1433.

112. D. J. Hunter, W. Zhang, P. G. Conaghan, K. Hirko, L. Menashe, L. Li, W. M. Reichmann and E. Losina, *Osteoarthritis Cartilage*, 2011, **19**, 557.

113. C. M. Kramer, J. Barkhausen, S. D. Flamm, R. J. Kim and E. Nagel, Society for Cardiovascular Magnetic Resonance Board of Trustees Task Force on Standardized Protocols, *J. Cardiovasc. Magn. Reson.*, 2008, **10–35**, 1.

114. C. R. Jack, Jr, M. A. Bernstein, B. J. Borowski, J. L. Gunter, N. C. Fox, P. M. Thompson, N. Schuff, G. Krueger, R. J. Killiany, C. S. Decarli, A. M. Dale, O. W. Carmichael, D. Tosun and M. W. Weiner, *Alzheimers Dement.*, 2010, **6**, 212.

115. K. Umemura, W. Zierhut, M. Rudin, D. Novosel, E. Robertson, B. Pedersen and R. P. Hof, *J. Cardiovasc. Pharmacol.*, 1992, **19**, 375.

116. J. P. Vallee, M. K. Ivancevic, D. Nguyen, D. R. Morel and M. Jaconi, *MAGMA*, 2004, **17**, 149.

117. P. Boyesen, F. M. McQueen, F. Gandjbakhch, S. Lillegraven, L. Coates, C. Wiell, E. A. Haavardsholm, P. G. Conaghan, C. G. Peterfy, P. Bird and M. Ostergaard, *J. Rheumatol.*, 2011, **38**, 2034.

118. C. G. Peterfy, A. Guermazi, S. Zaim, P. F. Tirman, Y. Miaux, D. White, M. Kothari, Y. Lu, K. Fye, S. Zhao and H. K. Genant, *Osteoarthritis Cartilage*, 2004, **12**, 177.

119. P. R. Kornaat, R. Y. Ceulemans, H. M. Kroon, N. Riyazi, M. Kloppenburg, W. O. Carter, T. G. Woodworth and J. L. Bloem, *Skeletal Radiol.*, 2005, **34**, 95.

120. D. J. Hunter, G. H. Lo, D. Gale, A. J. Grainger, A. Guermazi and P. G. Conaghan, *Ann. Rheum. Dis.*, 2008, **67**, 206.

121. D. J. Hunter, A. Guermazi, G. H. Lo, A. J. Grainger, P. G. Conaghan, R. M. Boudreau and F. W. Roemer, *Osteoarthritis Cartilage*, 2011, **19**, 990.

122. M. A. Quinn, P. G. Conaghan, P. J. O'Connor, Z. Karim, A. Greenstein, A. Brown, C. Brown, A. Fraser, S. Jarret and P. Emery, *Arthrit. Rheum.*, 2005, **52**, 27.

123. S. B. Edge, D. R. Byrd, C. C. Compton, A. G. Fritz, F. L. Greene and A. Trotti (ed.), *AJCC Cancer Staging Manual*, Springer, Heidelberg, Germany, 7th edn, 2010.

124. J. B. Christian, J. K. Finkle, B. Ky, P. S. Douglas, D. E. Gutstein, P. D. Hockings, P. Lainee, D. J. Lenihan, J. W. Mason, P. T. Sager, T. G. Todaro, K. A. Hicks, R. C. Kane, H. Ko, J. Lindenfeld, E. L. Michelson, J. Milligan, J. Y. Munley, J. S. Raichlen, A. Shahlaee, C. Strnadova, B. Ye and J. R. Turner, *Am. Heart J.*, 2012, **164**, 846.

125. D. Morgan, *J. Intern. Med.*, 2011, **269**, 54.

126. S. Salloway, R. Sperling, S. Gilman, N. C. Fox, K. Blennow, M. Raskind, M. Sabbagh, L. S. Honig, R. Doody, C. H. van Dyck, R. Mulnard, J. Barakos, K. M. Gregg, E. Liu, I. Lieberburg, D. Schenk, R. Black and M. Grundman, *Neurology*, 2009, **73**, 2061.

127. J. M. Wardlaw, V. Murray, E. Berge and G. J. Del Zoppo, *Cochrane Database Syst. Rev.*, 2009, (4):CD000213.

128. Activase® (Alteplase) prescribing information, Genentech.

CHAPTER 13

In Vivo MRI/S for the Safety Evaluation of Pharmaceuticals

PAUL D. HOCKINGS*[a,b] AND HELEN POWELL[c]

[a] AstraZeneca R&D, PHB *In Vivo* Biomarkers, 43183 Mölndal, Sweden;
[b] MedTech West, Chalmers University of Technology, 41296 Gothenburg, Sweden; [c] AstraZeneca R&D, Global Safety Assessment, Macclesfield SK10 4TG, United Kingdom
*Email: paul.hockings@astrazeneca.com

13.1 Introduction

Despite the increasing investment in drug discovery and development, only one in nine new medicinal products that enter clinical studies actually reaches the registration phase, with approximately 30% of compounds failing due to toxicology findings, despite extensive safety testing.[1] In addition, a number of drugs that had been approved for use in humans have either been withdrawn or have been issued with black box warnings following more extensive clinical use.[2] In order to improve this attrition rate, there is a requirement for pharmaceutical companies to explore and employ increasingly sophisticated methods to assess the safety profile of compounds. The pharmaceutical industry uses medical imaging because it enhances the ability to quantify the impact of drugs on human health and can be used alongside ordinary or invasive clinical assessments. Medical imaging can decrease patient numbers and/or the duration of exposure in clinical trials and hence accelerate drug development. Magnetic resonance imaging (MRI) has been the most widely used medical imaging technique in the pharmaceutical industry because of its superb soft tissue contrast and capability of delivering quantitative 3D

New Developments in NMR No. 2
New Applications of NMR in Drug Discovery and Development
Edited by Leoncio Garrido and Nicolau Beckmann
© The Royal Society of Chemistry 2013
Published by the Royal Society of Chemistry, www.rsc.org

information on organ anatomy and function.[3] It is used in a variety of disease areas in both non-clinical and clinical drug efficacy studies; however, there are relatively few examples of the use of MRI in drug safety studies.[4-6] In 2007 one of the authors (PDH) conducted an informal survey of a number of non-clinical imaging groups in the pharmaceutical industry and showed that only approximately 5% of effort (range 0–20%) was devoted to safety imaging studies. This seems a disproportionally small effort considering that MRI is a powerful tool that could potentially be used to reduce attrition in the late pipeline.

The aim of non-clinical safety studies is to evaluate potential risk to humans of new medicinal products. By assessing both on- and off-target effects and defining the dose–response relationship of any observed adverse effects, safety margins can be calculated. Such studies therefore need to be conducted in species relevant to man in terms of pharmacology, exposure and metabolism, and at doses over and above those to be tested clinically. Current regulatory guidelines recommended by the International Conference on Harmonisation of Technical Requirements for Registration of Pharmaceuticals for Human Use (ICH; http://www.ich.org/products/guidelines/safety/article/safety-guidelines.html) require the evaluation of chronic toxicity, carcinogenicity, safety pharmacology and reproductive toxicity prior to regulatory approval. Chronic toxicity is assessed in one rodent and one non-rodent species, with studies of at least 6 or 9 months duration, respectively, with the examination of tissue morphology at the microscopic level and organ function by the measurement of clinical chemistry endpoints as readouts. The carcinogenic potential of pharmaceuticals is evaluated by the conduct of long-term carcinogenicity studies following the assessment of genotoxic potential both *in vitro* and *in vivo*. Reproductive toxicity studies investigate the effect of test compounds on male and female fertility, and embryonic, fetal and post-natal development. The safety pharmacology core battery of studies is used to investigate the effects of the test substance on the vital functions, including the cardiovascular, respiratory and central nervous systems with either follow-up or supplementary studies (*e.g.* of the renal/urinary system, autonomic nervous system or gastrointestinal system) as required. Other studies assessing, for example, immune or endocrine functions may also be conducted if warranted based on findings in standard studies. Despite such extensive testing, compounds deemed to be safe to administer to humans can be associated with unexpected toxicity in the clinic, identified either during early clinical studies or, worse still, following launch to the market. In particular, liver and cardiac toxicity have been associated with drugs that have either been withdrawn from use or have been issued with black box warnings limiting potential use.

Data from first time in man to registration during the period between 1991 and 2000 for the ten largest pharmaceutical companies indicate an average success rate of 11%, with 62% and 45% of all compounds entering Phase II and III, respectively, failing to progress to the next stage of clinical development.[1] Indeed, during this period 23% of compounds failed at the registration stage following completion of all the clinical trials and submission documentation. Whilst many factors may contribute to the attrition of drug candidates during non-clinical and clinical development, including efficacy, pharmacokinetic and commercial reasons, safety issues are estimated to account for 31% of all attrition, 65% of which is due to non-clinical toxicity and 35% of which is accounted for by human adverse events. With improvements in the estimation of pharmacokinetic parameters and bioavailability, toxicity is now the single most common cause of drug attrition.[1] Analysis of attrition rates from DuPont-Merck and Bristol-Myers Squibb during the period between 1993 and 2006 indicates that the two most common causes of attrition are cardiovascular and liver toxicity, accounting for approximately 27% and 15% of attrition of all advanced molecules, respectively.[2] The underlying mechanisms of toxicity are often complex; with on-target (mechanism-based) effects, hypersensitivity and immunological mechanisms, off-target effects, biotransformation and idiosyncratic mechanisms contributing to the overall rates of attrition.[7] Despite extensive monitoring prior to registration, adverse events have been demonstrated to be related to 6.5% of patient admissions to hospitals in the UK[8] and approximately one in seven in-patients experience an adverse drug reaction whilst in hospital, which is a significant cause of morbidity, increasing the length of stay of patients.[9] Such data indicate that in both drug development and clinical use, effective intervention strategies are urgently needed to reduce the risk of safety issues.

Whilst considerable effort and investment is being made to improve compound selection early in the drug discovery and development process by the incorporation of *in silico* models, better screening tools and *in vitro* and *in vivo* efficacy models, similar advances have not been made to the standard toxicology testing cascade. For example the pharmaceutical industry has invested heavily in the development of biomarkers to facilitate project progression[10] but relatively little in the development of safety biomarkers to indicate an adverse response to a test agent.[11] The term safety biomarker encompasses any measurement used to diagnose and monitor drug-induced toxicity, including traditional soluble markers found in biofluids such as plasma and urine, microscopic analysis of tissue biopsies, clinical tests such as electrocardiograms (ECG) and imaging (most commonly ultrasound). Biomarkers derived from medical imaging procedures have the advantage that they are comparatively non-invasive and good for following focal diseases such as cancer or atherosclerosis; however, they can be expensive to use and difficult to access.

An ideal clinical safety biomarker is a marker that informs about the lowest dose at which subtle, low-grade and reversible toxicities may appear, that will allow toxicities to be monitored in patients participating in studies and can be used to provide early information on benefit/risk-balance to inform project stop/go decisions. Non-clinical safety biomarkers can be used to a) distinguish between compounds in order to select the best possible pharmacological targets and molecules, taking account of the need to balance efficacy, kinetics, drug–drug interactions, safety, intellectual property and commercial considerations, b) enable molecules to progress efficiently from drug discovery into clinical development by providing a translational biomarker and c) predict susceptibility to an adverse response. New translational biomarkers have high potential impact and value and have often been the subject of consortia to align stakeholders and share costs as the resources required to validate/qualify them can be substantial. Non-clinical safety biomarkers are anchored by correlating the marker with histopathology, whereas clinical safety biomarkers generally cannot be. Therefore translation from non-clinical species to man is key for clinical safety biomarkers. While no differences exist in the technology used for efficacy and safety biomarkers, differences do exist in the context of use. There are two important aspects to consider when deciding whether a safety biomarker is fit for purpose. The first is validation, where the measurement performance characteristics of an assay and the range of conditions under which the assay will give reproducible and accurate data are assessed. The second is qualification, which is the fit-for-purpose evidentiary process of linking a biomarker with biological processes and clinical endpoints.[10] The expectation is that a safety biomarker will need a higher degree of validation and qualification than an efficacy biomarker to be considered fit for purpose.

Medical imaging is an important source of biomarkers for both clinical and non-clinical studies although these imaging biomarkers will not replace traditional endpoints in the short term. With increasing knowledge and experience this might partly change in the future, especially for

- species or animal models, which are costly, with limited availability and of high ethical/emotional value (*e.g.* non-human primates)
- assessment of lesion severity at the end of the dosing period and recovery after a drug holiday in animals allocated to recovery groups to ensure true reversal of effect
- assessment or evaluation of mechanisms of toxicity during investigative/ problem-solving studies
- longitudinal assessment of the development of a lesion in a single animal, replacing the need for necropsies at each individual timepoint, thereby reducing animal numbers.

Rather than provide a comprehensive literature review of the use of magnetic resonance imaging (MRI) and spectroscopy (MRS) in safety studies, this chapter will review the application of these techniques to study liver and cardiac toxicity as these result in the greatest number of drug projects failing during clinical development.

13.2 Hepatotoxicity

Drug-induced liver injury (DILI) is a recurrent cause of delayed progression and/or attrition of new drug candidates, failed drug licensing, drug withdrawal post-licensing and of serious illness in man. More drugs have been withdrawn from the market due to hepatotoxicity than for any other reason.[12,13] The consequences of these withdrawals for pharmaceutical companies have been enormous and there is much interest in preventing further post-approval attrition due to hepatotoxicity. One explanation for these withdrawals is that currently available clinical biomarkers for hepatotoxicity do not adequately predict which patients may suffer drug-induced hepatic injury. Specifically, DILI mechanisms vary depending on the drug and some types of liver injury such as hepatocellular damage and cholestasis are more important to be able to monitor than others.[14] As a result, there is intense interest in finding new non-clinical biomarkers that better predict hepatotoxicity in humans, and better clinical biomarkers that identify at-risk patients and provide earlier signals of hepatotoxicity. Currently available biomarkers of hepatocellular injury such as serum alanine aminotransferase (ALT) lack sensitivity and specificity and cannot distinguish between adaption *versus* progression to liver failure.[14]

Non-clinically, the emphasis is on finding and qualifying new biomarkers that help select compounds with no or markedly reduced potential for DILI and on translation of these new biomarkers from the non-clinical to the clinical arena. DILI occurs *via* complex mechanisms and can be initiated *via* multiple mechanisms such as reactive metabolites, mitochondrial injury, lysosomal injury, hepatobiliary transport inhibition and immune activation.[12] Downstream effects can include steatosis, cholestasis, necrosis and fibrosis. Biomarkers that assess individual mechanisms will improve mechanistic understanding and enable *in vivo* hazard identification and risk assessment. The primary value of DILI biomarkers in this regard is to enable

- selection of pharmacological targets that will not result in DILI
- selection of compounds having minimal possible potential to cause DILI
- monitoring of DILI in non-clinical species, using sensitive and specific biomarkers that ideally detect early stages of liver dysfunction prior to onset of irreversible liver changes and that translate to man.

13.2.1 Hepatic Steatosis

Hepatic steatosis is a reversible condition where vacuoles of lipid accumulate in hepatocytes. It is the most common liver disease in the Western world and its presence is associated with obesity and insulin resistance.[15] Some drugs such as the antiarrhythmic agent amiodarone, the antiviral nucleoside analogue fialuridine and the anti-estrogen agent tamoxifen have been reported to cause hepatic steatosis; however, there have been relatively few in-depth studies in humans or experimental animals.[16] This may be because hepatic steatosis has not historically been a reason for non-approval or withdrawal of drugs from the market. However, hepatic steatosis is known to increase the risk for progression to steatohepatitis and to make the liver more vulnerable to hepatocellular injury.[17] In some cases drug-induced hepatic steatosis patients can present with a rapid evolution of severe hepatic failure, lactic acidosis and ultimately death.[18]

Patients with hepatic steatosis are usually detected by elevated serum aminotransferase levels or ultrasonographic fatty liver; however, there is an absence of a predictable correlation between abnormalities in liver enzymes and histologic lesions.[19] Standard non-invasive tests such as the fatty liver index, SteatoTest and NashTest are based on a combination of age, body mass index (BMI) and sex combined with plasma biomarkers reflecting alterations in hepatic function such as ALT, aspartate aminotransferase and gamma-glutamyl transferase, which are not directly involved in the initiation and/or progression of liver disease.[20] Thus, liver biopsy remains the gold standard for diagnosis of hepatic steatosis. MRS and MRI are non-invasive techniques that have been shown to have high sensitivity in the detection of steatosis,[21] but have been described as being expensive and too complex to be used in the clinical setting.[22] Figure 13.1 shows a typical image and spectra used to evaluate hepatic steatosis in mouse liver. Localized spectroscopy sequences such as point-resolved spectroscopy (PRESS) or stimulated-echo acquisition mode (STEAM) are used and the spectra are analyzed by comparing the integrals of the water and lipid peaks. These techniques have been used in numerous non-clinical[23,24] and clinical[25,26] intervention studies where it was important to follow the relative changes in individuals over time and it may be in this setting

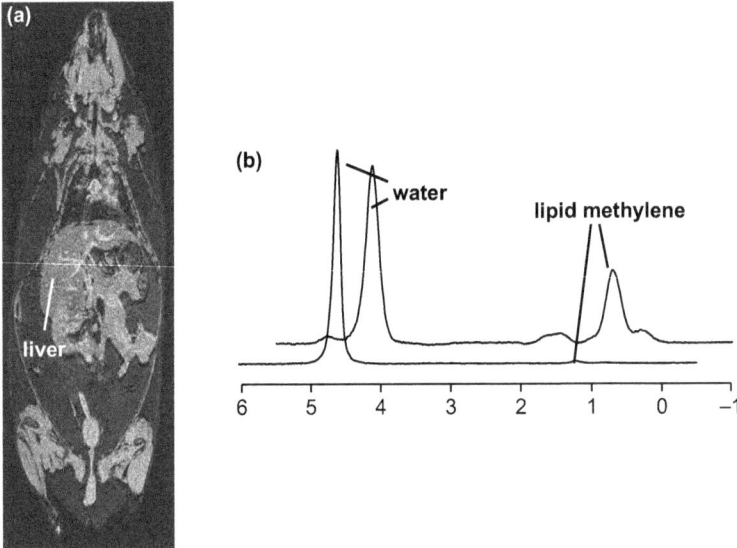

Figure 13.1   Assessment of hepatic steatosis in mice. (a) Coronal MRI slice through a mouse indicating the position of the liver. Image acquired with a 9.4 T/20 USR Bruker Biospec scanner using a high-resolution respiratory gated 3D FISP sequence with flip angle 4°, TR/TE 3.3 ms/1.7 ms, field of view $100\times45\times45$ mm and matrix size $428\times192\times192$. (b) Localized ^{1}H MRS from a $2\times2\times2$ mm^3 voxel positioned in the livers of two mice with a PRESS sequence, TR 3 s, TE 6.7 ms, SW 4006 Hz, 64 averages and 2048 data points. Bottom: control mouse with a normal fat : water ratio. Top: mouse on a high fat diet showing accumulation of lipid in the liver. Voxels are positioned well away from large blood vessels and fatty infiltrations.
Data provided by Abdel wahad Bidar, AstraZeneca, Sweden.

that they have their greatest application. There are also a small number of studies where MRI/S has been used to monitor hepatic steatosis as a safety biomarker. In 2004 Zhang *et al.* used a three-point Dixon method to quantitate fatty liver in rats induced by an experimental microsomal transfer protein (MTP) inhibitor in order to develop a safety biomarker that could be translated to human studies.[27] Cuchel *et al.* showed that inhibition of MTP by BMS-201038 resulted in a reduction of low density lipoprotein (LDL) cholesterol levels in patients with homozygous familial hypercholesterolemia, and that therapy was associated with elevated liver aminotransferase levels and hepatic fat accumulation as detected by MRS.[28] Visser *et al.* used MRS to monitor the potential accumulation of hepatic triglycerides on treatment with the apolipoprotein B-100 (ApoB) synthesis inhibitor mipomersen in familial hypercholesterolemia patients[29] and statin intolerant patients.[30] Given the difficulty of obtaining serial liver biopsies in clinical trials, MRI and MRS techniques are rapidly becoming the gold standard methodologies to assess hepatic steatosis.

13.2.2 Hepatobiliary Transporter Inhibition

DILI can arise as a result of multiple initiating factors, but a key mechanism is impairment of bile formation and flow resulting in cholestasis.[14] Impaired bile flow results in increased levels of bile acids and bilirubin in blood and symptoms such as pruritus, jaundice and eventually liver damage. Many pharmaceutical companies have developed *in vitro* transporter activity assays to identify candidate drugs that inhibit hepatobiliary transporters and therefore have the potential to induce cholestatic DILI in humans.[31] However, a current limitation of the *in vitro* transporter inhibition assays is that it is unclear how these data relate to transient or sustained elevations in plasma bile acids and other markers of impaired liver function, which in turn may or may not be indicative of cholestasis and/or hepatocellular damage. If imaging could be used to assess the functional consequence of transporter inhibition *in vivo* and thus determine the predictivity of the assays and their value in candidate selection, such approaches could be applied to risk assessment and problem solving studies in non-clinical species. The translational nature of these relatively non-invasive methods would also facilitate the assessment of hepatobiliary transporter activity in the clinic, in order to provide evidence to support future monitoring where indicated.

The MR contrast agent gadoxetate (Gd-EOB-DTPA, gadoxetic acid, Eovist or Primovist, Bayer HealthCare) is a clinically approved hepatobiliary-specific agent that is excreted by both liver and kidney. It is used to detect liver tumors as it is specifically taken up by hepatocytes and not tumor cells.[32] Gadoxetate is injected intravenously, transported from the extracellular space into the hepatocytes by the adenosine triphosphate (ATP) dependent organic anion transporting polypeptide 1 (OATP1/Oatp1 human/rat) and subsequently excreted into the biliary canaliculi by the multidrug resistance associated protein 2 (MRP2/Mrp2 human/rat).[33] Candidate drugs that inhibit OATP1 or MRP2 will inhibit the uptake or efflux of gadoxetate, respectively.

The MRI acquisition can be tuned so that image intensity is proportional to gadoxetate concentration. Ulloa *et al.*[34] acquired serial *in vivo* MR images of rat liver with a time resolution of one minute in order to determine the uptake of gadoxetate from plasma into the hepatocyte and excretion into the bile (Figure 13.2). The concentration of gadoxetate in the hepatocyte was calculated

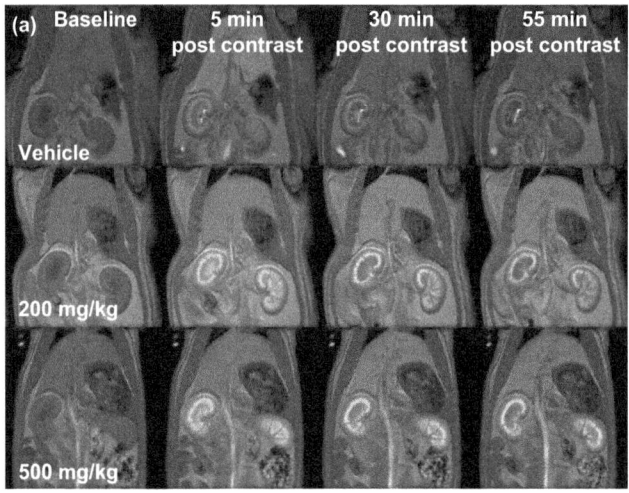

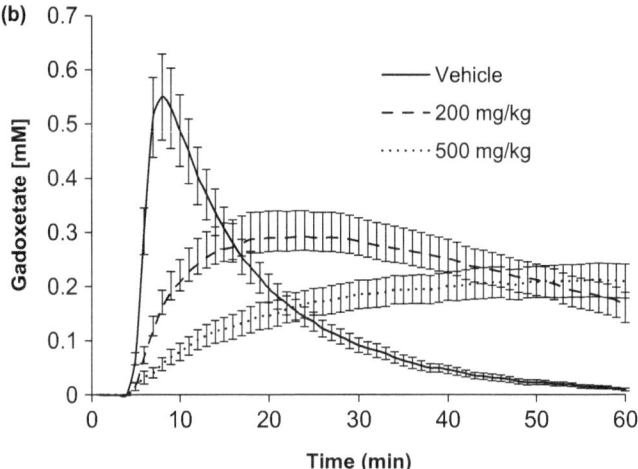

Figure 13.2 Uptake of gadoxetate from plasma into the hepatocyte and excretion into the bile in rats. (a) Examples of dynamic images for rats treated with vehicle (top), $200 \, \text{mg} \, \text{kg}^{-1}$ (middle) or $500 \, \text{mg} \, \text{kg}^{-1}$ (bottom) of a hepatobiliary transporter inhibitor at $t = 0$, 5, 30 and 55 min. after contrast injection. Note the enhancement of the small bowel lumen 30 min. after contrast injection in the vehicle treated animal. No enhancement was observed in the bowel of the animal treated with $500 \, \text{mg} \, \text{kg}^{-1}$. Images acquired with a $4.7 \, \text{T}/40$ Bruker Biospec and IntraGate FLASH TR/TE $60 \, \text{ms}/1.4 \, \text{ms}$, flip angle $30°$, field of view $60 \times 60 \, \text{mm}$ and matrix size 256×256. (b) Mean concentration of gadoxetate in hepatocytes. Error bars are SEM. 6 rats/group.

from the liver gadoxetate concentration corrected by the extracellular gadoxetate concentration and the volume fraction of extracellular space in the liver. The rate of change of gadoxetate concentration in the various compartments was used to model gadoxetate transport kinetics and therefore the function of the OATP1 and MRP2 hepatobiliary transporters. These researchers showed that dynamic contrast enhanced magnetic resonance imaging (DCE-MRI) could detect inhibition of gadoxetate uptake and efflux by an investigational chemokine agonist and that results correlated with clinical chemistry markers of DILI. This imaging technique has the potential to provide an *in vivo* biomarker of hepatobiliary transporter inhibition. It offers a relatively non-invasive alternative to bile duct cannulation studies for non-clinical assessment of effects of test compounds on bile flow *in vivo*, which enables reduction in animal numbers compared to the bile duct cannulation model, especially in longitudinal studies that otherwise can only be assessed using a sequential design. In addition this technique can be transferred to the clinical setting as gadoxetate is a clinically approved contrast agent.

13.3 Cardiotoxicity

Cardiotoxicity is another leading cause of drug attrition and is therefore a core subject in non-clinical and clinical safety testing of new drugs.[35,36] The mechanisms of cardiac toxicity are not completely understood but may result from interference with processes such as mitochondrial function, intracellular calcium homeostasis, selective membrane permeability and oxidant/antioxidant balance. Some drugs can cause irreversible myocardial damage *e.g.* doxorubicin, while others may cause reversible cardiotoxicity resulting in temporary left ventricular dysfunction. Cardiotoxicity as a result of cancer therapy has been widely studied as it is often the dose-limiting factor for patients. There are two important classes of cardiotoxic chemotherapeutic agents: anthracyclines and anti-HER2 (human epidermal growth factor receptor 2) directed therapies including trastuzumab, which induce left ventricular dysfunction and heart failure.[37] It is important to detect left ventricular dysfunction as early as possible in order to stop further treatment, reduce the chemotherapy dose and/or to commence standard medical treatment for heart failure. Cardiac imaging is recommended in all patients with symptoms of cardiotoxicity regardless of the cumulative dose, for asymptomatic patients on trastuzumab every 3 months while on therapy and for anthracycline treated patients at cumulative doses of 200, 300 and 400 $mg\,m^{-2}$, and every 50 $mg\,m^{-2}$ thereafter (doxorubicin equivalents).[37]

Cardiac imaging can be both a sensitive and specific methodology to measure small changes in cardiac function (see also Chapter 18).[38] Non-clinical imaging studies of potential cardiotoxicity can be used both to better inform risk-benefit evaluations before progressing into clinical development and to validate and qualify a safety imaging biomarker for clinical studies, by examining the reproducibility of the technique and the effect size of a compound challenge so that clinical studies can be appropriately powered. Imaging biomarkers can be coupled with non-imaging biomarkers such as plasma levels of cardiac troponins to improve the sensitivity and specificity for cardiac toxicity. No consensus regarding threshold values for these biomarkers has been established for evaluating drug-induced cardiotoxicity in non-clinical or clinical settings.

Cardiac MRI is used because it produces comprehensive structural and functional images with excellent soft tissue contrast without the need for ionizing radiation. This translates into superior interstudy reproducibility resulting in better reliability of observed changes and reduced patient numbers in clinical trials.[39] In contrast to 2D cardiac ultrasound, it does not rely on geometrical assumptions to calculate volumes and is therefore reliable even in remodeled hearts with complex geometry.[40] Three-dimensional echocardiography does not rely on geometrical assumptions either; however, there may be difficulties in full heart coverage for large hearts, and issues with image quality.[41] In addition, cardiac MRI can detect the presence of myocardial fibrosis and/or inflammation.

There are a number of disadvantages in cardiac MRI compared to alternative techniques such as echocardiography and radionuclide scans and these include a higher cost and the need to move patients to the imaging suite (compared to ultrasound). Contraindications to MRI include claustrophobia, metal implants such as pacemakers, defibrillators, insulin pumps, aneurysm clips, or any other foreign metallic body, and limitation in large body size (*e.g.* BMI $>38 \, \mathrm{kg \, m^{-2}}$). There is a requirement for a regular cardiac rhythm for ECG gated image acquisition. A specific problem for non-clinical safety studies is that the MRI scanners may not be co-localized with non-clinical safety facilities.

Hockings *et al.* showed that cardiac output measured non-invasively by MRI in the dog correlated well with invasive techniques and that cardiac MRI could detect changes in cardiac function induced by the plasma volume expander minoxidil.[42] Lightfoot *et al.* showed that doxorubicin-induced gadolinium late enhancement in the left ventricular myocardium is associated with a subsequent drop in left ventricular ejection fraction as well as histopathological evidence of intracellular vacuolization consistent with cardiotoxicity.[43] Similar doxorubicin induced impairment of cardiac function and increases in gadolinium late enhancement has also been observed by Woodhouse *et al.* (Figure 13.3 and Table 13.1).[44] Doxorubicin was also shown to induce decreases in murine cardiac energetics as measured by spatially localized ^{31}P MRS before left ventricular dysfunction became evident in the mouse.[45] The phosphocreatine-to-ATP ratio correlated with peak filling rate and ejection fraction, suggesting a relationship between cardiac energetics and both left ventricular systolic and diastolic dysfunction. The *MANTICORE 101 – Breast* clinical trial is designed to determine if conventional heart failure pharmacotherapy can prevent trastuzumab-mediated left ventricular remodeling among patients with HER2+ early breast cancer, as measured by the 12-month change in left ventricular end-diastolic volume using cardiac MRI.[46] In addition to these studies where imaging biomarkers of cardiac disease are used as endpoints in non-clinical and clinical studies, imaging may also be used to aid the

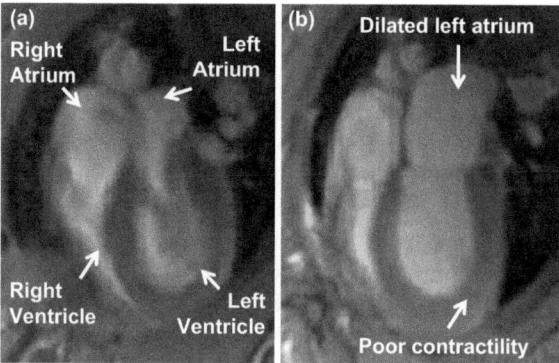

Figure 13.3 Doxorubicin-induced cardiac toxicity. Typical long axis images from (a) vehicle and (b) doxorubicin treated rats indicating dilated left atrium. Rats were imaged on a Bruker BioSpec 4.7 T system using a retrospectively gated IntraGate FLASH multislice cine sequence TR/TE 120 ms/1.4 ms, flip angle 30°, field of view 45×45 mm, matrix 192/256, slice thickness 1 mm, repetitions 150.

Table 13.1 Effect of 8 weeks doxorubicin or vehicle treatment on cardiac function and late gadolinium enhancement in 2 groups of rats (n = 6/group).

		LVM [g]	*EDV [mL]*	*ESV [mL]*	*SV [mL]*	*EF [%]*	*FS [%]*	*LGE [%]*
Vehicle	*Mean*	0.66	0.34	0.06	0.28	82	65	59
	95% C.I.	0.07	0.03	0.01	0.03	3	12	7
Doxorubicin	*Mean*	0.70	0.33	0.17	0.16	50	43	91
(1.25 mg kg⁻¹)	*95% C.I.*	0.04	0.04	0.05	0.03	13	13	7
	t-test	0.33	0.85	**0.003**	**0.0002**	**0.001**	**0.04**	**0.001**

LVM left ventricular mass; EDV end diastolic volume; ESV end systolic volume; SV stroke volume; EF ejection fraction; FS fractional shortening; LGE late gadolinium enhancement.

qualification of novel plasma biomarkers of cardiac damage. Such soluble biomarkers may be used either to predict susceptibility to cardiac toxicity or detect cardiac toxicity at an early stage in order to halt or modify the treatment regime.

13.4 Conclusion

Despite the emergence of new approaches to aid the selection of compounds with reduced toxicity liability and the harmonization of non-clinical safety studies to support human clinical trials, toxicity remains a major cause of attrition during drug development and post-marketing. Alongside further developments in *in silico* and *in vitro* screening, knock-out and humanized models, imaging has the potential to reduce attrition by providing a relatively non-invasive means of identifying and managing risk both non-clinically and clinically.

There are two types of safety biomarkers that need to be considered, namely biomarkers of harm and biomarkers of lack of harm. Biomarkers of harm can be used, for example, for internal decision making to close a project early or for patient monitoring. Biomarkers of lack of harm can be used to progress projects and to provide evidence to regulators that a compound is safe to dose into man *i.e.* if no safety signal is seen, or if the margins are sufficient, then the compound can be dosed into man with confidence. In both cases work is needed to validate and qualify the biomarker before it can be decision making. For obvious reasons a great deal more work is necessary in order to be confident that the absence of a safety signal will result in lack of harm. Good Laboratory Practice (GLP) and Good Clinical Practice (GCP) ensure that the data produced in safety studies are of high quality, reliable and valid. However, it is often difficult to implement non-clinical MRI studies to GLP as non-clinical MRI scanners are not currently equipped with software tools that guarantee consistent spectrometer operation or data transfer in compliance with GLP, and the burden of GLP documentation makes compliance for innovative imaging studies impractical. Regulatory agencies do accept investigatory studies that are not GLP compliant if the work is critical to a scientifically based risk assessment and has been conducted to an acceptable standard.

With all the potential benefits that imaging can bring to safety studies, it is essential to have a strategic view on which imaging biomarkers to prioritize both for incorporation in standard regulatory studies in the future, but also for use in investigative studies addressing mechanisms and contributing to risk assessment. Furthermore, since developing new biomarkers is a protracted process, being reactive (*i.e.* only promoting biomarker development for a specific project need) rather than proactive and deliberate increases the odds that, in the near term, the optimal set of biomarkers will not be validated and qualified for use in projects. In order to aid the development, validation and regulatory acceptance of imaging biomarkers, cross-company consortia have been established. One example is the Health and Environmental Sciences Institute (HESI), a non-profit institution that brings together scientists from academia, government and industry, which has recently established a project committee on the use of imaging in non-clinical safety assessment. Cardiac, brain and liver imaging have been prioritized as key areas of interest and work is ongoing to develop pre-competitive, cross-site validation studies in these areas. Importantly, this committee has input from the Food and Drug Administration (FDA), the body that regulates pharmaceutical approval and use in the USA, and thus it is hoped that by contributing to the development of imaging methods for risk assessment and management, such methods will be more readily accepted by regulatory agencies. The true measure of the importance of investment in developing imaging biomarkers lies in under-standing the potential cost of further unanticipated toxicity being observed in late stage projects or worse still once a compound has been marketed, balanced against the likelihood of investment in this area mitigating against project closure or drug withdrawal, or indeed providing project teams with the confidence to advance a compound.

In summary, MRI and MRS have the potential to provide relatively non-invasive, robust and reproducible biomarkers to predict and monitor safety that "plug the gaps" in our current approaches, and have the additional benefit of being applicable to both the non-clinical and the clinical settings. Investment in this area can not only underpin decisions to progress a compound either in the non-clinical or the clinical setting by providing a means of determining mechanisms of toxicity or monitoring for potential safety issues, but can also support the validation of *in vitro* assays to aid compound selection. Strategic planning and cross-company consortia are essential to exploit this technology and ensure that its potential in the safety assessment of pharmaceuticals is fulfilled.

References

1. I. Kola and J. Landis, *Nat. Rev. Drug Discovery*, 2004, **3**, 711.
2. F. P. Guengerich, *Drug Metab. Pharmacokinet.*, 2011, **26**, 3.
3. I. Rodríguez, S. Pérez-Rial, J. González-Jimenez, J. Pérez-Sánchez, F. Herranz, N. Beckmann and J. Ruíz-Cabello, *J. Pharm. Sci.*, 2008, **97**, 3637.
4. Y. J. Wang and S. Yan, *Lab. Anim.*, 2008, **42**, 433.
5. P. D. Hockings, in *Drug Discovery and Evaluation: Safety and Pharmacokinetic Assays*, ed. H. G. Vogel, F. J. Hock, J. Maas and D. Mayer, Springer, Heidelberg, 2006, p. 385.
6. M. W. Tengowski and J. J. Kotyk, in *Imaging in Drug Discovery and Early Clinical Trials*, ed. P. L. Herrling, A. Matter and M. Rudin, Birkhäuser Basel, Basel, 2005, p. 257.
7. D. C. Liebler and F. P. Guengerich, *Nat. Rev. Drug Discovery*, 2005, **4**, 410.
8. M. Pirmohamed, S. James, S. Meakin, C. Green, A. K. Scott, T. J. Walley, K. Farrar, B. K. Park and A. M. Breckenridge, *BMJ*, 2004, **329**, 15.
9. E. C. Davies, C. F. Green, S. Taylor, P. R. Williamson, D. R. Mottram and M. Pirmohamed, *PLoS One*, 2009, **4**, e4439.
10. J. A. Wagner, *Annu. Rev. Pharmacol. Toxicol.*, 2008, **48**, 631.
11. F. D. Sistare, F. Dieterle, S. Troth, D. J. Holder, D. Gerhold, D. Andrews-Cleavenger, W. Baer, G. Betton, D. Bounous and K. Carl, *Nat. Biotechnol.*, 2010, **28**, 446.
12. A. J. Pugh, A. J. Barve, K. Falkner, M. Patel and C. J. McClain, *Clinics in Liver Disease*, 2009, **13**, 277.
13. N. Chalasani, R. J. Fontana, H. L. Bonkovsky, P. B. Watkins, T. Davern, J. Serrano, H. Yang and J. Rochon, *Gastroenterology*, 2008, **135**, 1924.
14. W. C. Maddrey, *J. Clin. Gastroenterol.*, 2005, **39**, S83.
15. N. M. W. de Alwis and C. P. Day, *J. Hepatol.*, 2008, **48**(1), S104.
16. D. E. Amacher, *Toxicology*, 2011, **279**, 10.
17. C. P. Day and O. F. W. James, *Gastroenterology*, 1998, **114**, 842.
18. A. M. Diehl, *Semin. Liver Dis.*, 1999, **19**, 221.

19. J. M. Mato and S. C. Lu, *Hepatology*, 2011, **54**, 1115.
20. T. Poynard, V. Ratziu, S. Naveau, D. Thabut, F. Charlotte, D. Messous, D. Capron, A. Abella, J. Massard and Y. Ngo, *Comp. Hepatol.*, 2005, **4**, 10.
21. D. A. Raptis, M. A. Fischer, R. Graf, D. Nanz, A. Weber, W. Moritz, Y. Tian, C. E. Oberkofler and P. Clavien, *Gut*, 2012, **61**, 117.
22. X. Ma, N. Holalkere, R. A. Kambadakone, M. Mino-Kenudson, P. F. Hahn and D. V. Sahani, *Radiographics*, 2009, **29**, 1253.
23. P. D. Hockings, K. K. Changani, N. Saeed, D. G. Reid, J. Birmingham, P. O'Brien, J. Osborne, C. N. Toseland and R. E. Buckingham, *Diabetes, Obes. Metab.*, 2003, **5**, 234.
24. E. D. Berglund, D. G. Lustig, R. A. Baheza, C. M. Hasenour, R. S. Lee-Young, E. P. Donahue, S. E. Lynes, L. L. Swift, M. J. Charron, B. M. Damon and D. H. Wasserman, *Diabetes*, 2011, **60**, 2720.
25. D. G. Carey, G. J. Cowin, G. J. Galloway, N. P. Jones, J. C. Richards, N. Biswas and D. M. Doddrell, *Obes. Res.*, 2002, **10**, 1008.
26. T. Le, J. Chen, C. Changchien, M. R. Peterson, Y. Kono, H. Patton, B. L. Cohen, D. Brenner, C. Sirlin, R. Loomba and for the San Diego NAFLD Research Consortium (SINC), *Hepatology*, 2012, **56**, 922.
27. X. Zhang, M. Tengowski, L. Fasulo, S. Botts, S. A. Suddarth and G. A. Johnson, *Magn. Reson. Med.*, 2004, **51**, 697.
28. M. Cuchel, L. T. Bloedon, P. O. Szapary, D. M. Kolansky, M. L. Wolfe, A. Sarkis, J. S. Millar, K. Ikewaki, E. S. Siegelman, R. E. Gregg and D. J. Rader, *N. Engl. J. Med.*, 2007, **356**, 148.
29. M. E. Visser, F. Akdim, D. L. Tribble, A. J. Nederveen, T. J. Kwoh, J. J. P. Kastelein, M. D. Trip and E. S. G. Stroes, *J. Lipid Res.*, 2010, **51**, 1057.
30. M. E. Visser, G. Wagener, B. F. Baker, R. S. Geary, J. M. Donovan, U. H. W. Beuers, A. J. Nederveen, J. Verheij, M. D. Trip, D. C. G. Basart, J. J. P. Kastelein and E. S. G. Stroes, *Eur. Heart J.*, 2012, **33**, 1142.
31. S. Dawson, S. Stahl, N. Paul, J. Barber and J. G. Kenna, *Drug Metab. Dispos.*, 2012, **40**, 130.
32. K. I. Ringe, D. B. Husarik, C. B. Sirlin and E. M. Merkle, *Am. J. Roentgenol.*, 2010, **195**, 13.
33. N. Tsuda, K. Harada and O. Matsui, *J. Gastroenterol. Hepatol.*, 2011, **26**, 568.
34. J. L. Ulloa, S. Stahl, J. Yates, N. Woodhouse, J. G. Kenna, H. B. Jones, J. C. Waterton and P. D. Hockings, *NMR Biomed.*, 2013. doi: 10.1002/nbm.2946.
35. T. C. Stummann, M. Beilmann, G. Duker, B. Dumotier, J. M. Fredriksson, R. L. Jones, M. Hasiwa, Y. J. Kang, C. F. Mandenius and T. Meyer, *Cardiovasc. Toxicol.*, 2009, **9**, 107.
36. D. J. Leishman, T. W. Beck, N. Dybdal, D. J. Gallacher, B. D. Guth, M. Holbrook, B. Roche and R. M. Wallis, *J. Pharmacol. Toxicol. Methods*, 2012, **65**, 93.
37. R. M. Witteles, M. B. Fowler and M. L. Telli, *Heart Failure Clinics*, 2011, **7**, 333.

38. D. J. Pennell, *Circulation*, 2010, **121**, 692.
39. F. Grothues, G. C. Smith, J. C. C. Moon, N. G. Bellenger, P. Collins, H. U. Klein and D. J. Pennell, *Am. J. Cardiol.*, 2002, **90**, 29.
40. R. M. Lang, M. Bierig, R. B. Devereux, F. A. Flachskampf, E. Foster, P. A. Pellikka, M. H. Picard, M. J. Roman, J. Seward, J. S. Shanewise, S. D. Solomon, K. T. Spencer, M. St John Sutton and W. J. Stewart, *J. Am. Soc. Echocardiogr.*, 2005, **18**, 1440.
41. J. Walker, N. Bhullar, N. Fallah-Rad, M. Lytwyn, M. Golian, T. Fang, A. R. Summers, P. K. Singal, I. Barac, I. D. Kirkpatrick and D. S. Jassal, *J. Clin. Oncol.*, 2010, **28**, 3429.
42. P. D. Hockings, A. L. Busza, J. Byrne, B. Patel, S. C. Smart, D. G. Reid, H. L. Lloyd, A. White, K. Pointing, B. A. Farnfield, A. Criado-Gonzalez, G. A. Whelan, G. L. Taylor, J. M. Birmingham, M. R. Slaughter, J. A. Osborne, A. Krebs-Brown and D. Templeton, *Toxicol. Mech. Methods*, 2003, **13**, 39.
43. J. C. Lightfoot, R. B. D'Agostino, C. A. Hamilton, J. Jordan, F. M. Torti, N. D. Kock, J. Jordan, S. Workman and W. G. Hundley, *Circ. Cardiovasc. Imaging*, 2010, **3**, 550.
44. N. Woodhouse, H. R. Mellor, J. Ulloa, G. Healing, J. Kirk and P. D. Hockings, *Proc. Intl. Soc. Mag. Reson. Med.*, 2012, **20**, 1135.
45. M. Y. Maslov, V. P. Chacko, G. A. Hirsch, A. Akki, M. K. Leppo, C. Steenbergen and R. G. Weiss, *Am. J. Physiol.: Heart Circ. Physiol.*, 2010, **299**, H332.
46. E. Pituskin, M. Haykowsky, J. R. Mackey, R. B. Thompson, J. Ezekowitz, S. Koshman, G. Oudit, K. Chow, J. J. Pagano and I. Paterson, *BMC Cancer*, 2011, **11**, 318.

CHAPTER 14

Applications of MRI and MRS in Cartilage Therapeutics and Tissue Engineering

DAVID A. REITER*[a] AND RICHARD G. SPENCER[b]

[a] Clinical Research Branch, National Institutes of Health, National Institute on Aging, 3001 S. Hanover Street, 5th floor, Baltimore, MD 21225, USA;
[b] Magnetic Resonance Imaging and Spectroscopy Section, National Institutes of Health, National Institute on Aging, 3001 S. Hanover Street, 5th floor, Baltimore, MD 21225, USA
*Email: reiterda@mail.nih.gov

14.1 Introduction

Degenerative articular cartilage disease remains a significant cause of morbidity worldwide, with an estimated prevalence of symptomatic osteoarthritis (OA) of 10–20% in individuals over the age of 60.[1] A major contributor to the prevalence and severity of cartilage degeneration is the fact that turnover of cartilage matrix is exceedingly slow, even in response to injury. In addition, there are currently no disease-modifying interventions for OA in clinical use.

There has been extensive investigation of techniques to repair or replace damaged cartilage. Current approaches include microfracture,[2–5] autograft and allograft transplants[6–9] and autologous chondrocyte implantation.[10–14] However, long-term outcomes have been disappointing. Thus, development of functional engineered replacement tissues for cartilage remains an important area of research.

New Developments in NMR No. 2
New Applications of NMR in Drug Discovery and Development
Edited by Leoncio Garrido and Nicolau Beckmann
© The Royal Society of Chemistry 2013
Published by the Royal Society of Chemistry, www.rsc.org

Successful growth of an engineered replacement for cartilage requires appropriate stimuli to yield tissue with molecular, biochemical and biomechanical properties that approach those of native tissue. Engineered cartilage would ideally have the appropriate amounts of the primary matrix components – type II collagen and aggrecan, and water – and exhibit structural organization that promotes proper functional characteristics. Additional factors such as crosslinking of the collagen molecules[15–18] also influence tissue characteristics.

A multitude of methodological factors must be considered when engineering an optimal tissue for cartilage replacement, each demonstrating an impact on the outcome. Strategies that employ tissue engineering can involve the use of either differentiated cells, *i.e.* chondrocytes, or progenitor cells combined with growth promoting factors.[19–22] Scaffolds constructed from collagen or hydrogels are also often employed,[23–25] as are a variety of bioreactors.[26–32] Mechanical input has also been introduced as an additional stimulus for tissue growth[33–38] in order to recapitulate aspects of the *in vivo* environment. With the large number of possibilities, there remains no consensus as to the optimal approach.

Magnetic resonance (MR) imaging and spectroscopy have the potential to greatly improve the implementation of cartilage growth strategies and therapies through their capacity to evaluate the biophysical and biochemical properties of the tissue non-invasively. MR can potentially be used in tissue engineering studies to permit ongoing analysis of construct development and to guide interventions, as well as in the clinical setting to provide endpoints for tailoring ongoing therapy.

In the following, we present an overview of work focusing on the application of MR imaging and spectroscopy to cartilage tissue engineering, with the overall goal of improving therapy for articular cartilage disease. The first section (14.2) introduces work using MR-compatible bioreactors where basic relationships between biophysical and biochemical properties in engineered cartilage have been established. This section includes a variety of both MR and complementary non-MR modalities used to establish biophysical and biochemical correlates of tissue development. Section 14.3 discusses a variety of growth modulating factors such as anabolic and catabolic agents and mechanical stimuli. These studies not only provide insights into environmental influences on tissue development but also expand the dynamic range of matrix composition and tissue growth; this is particularly useful in establishing relationships between MR outcomes and tissue matrix constituents. Section 14.4 discusses the extension of MR methodologies to translational studies of cartilage constructs and integration into tissue. In the final section, 14.5, newly emerging analyses of MR tissue properties, multiexponential relaxation analysis and multivariate machine learning analysis are presented. These techniques are aimed at improving the sensitivity and specificity of MR outcomes to cartilage matrix components.

14.2 Hollow Fiber Bioreactor Studies

An important approach for advancing the development of tissue engineering therapies for cartilage repair has involved the development of a flexible and reliable MRI-compatible hollow fiber bioreactor (HFBR) system.[39] The use of MRI-compatible bioreactor systems for cellular studies is well established in the literature.[40–42] However, comparable studies of engineered tissue have been rather limited. Besides work in cartilage, the bioartificial liver system has been most extensively evaluated.[43–45]

A number of difficulties pertain to such MRI studies. The requirement of placement inside the MRI magnet imposes constraints on the size and construction of the system. It must be entirely non-magnetic, and even non-magnetic metallic construction elements must be a sufficient distance away from the imaging region in order to avoid imaging artifacts. The perfusion system must permit repeat placement and removal of the system in the magnet without compromising sterility. For experiments of long duration, such as heteronuclear spectroscopic studies, it is important to maintain incubator-like conditions throughout the data collection period.

The schematic of an MRI-compatible HFBR system is seen in Figure 14.1. This bioreactor system is constructed from glass tubing into which porous polypropylene hollow fibers are inserted. These fibers are arranged uniformly along the long axis of the bioreactor and held in place at either end using biomedical grade silicone rubber adhesive. Each bioreactor is designed with a side port covered with a rubber septum for cell inoculation. Perfusion medium is circulated through the hollow fibers by means of gas-permeable silicone tubing and a peristaltic pump. The entire bioreactor circuit is assembled and maintained under sterile conditions and kept within a tissue culture incubator or a homemade support chamber providing ideal temperature and oxygen and carbon dioxide levels.

With this system, conditions promoting the development of high-quality cartilage from cells can be studied intensively with full control over exposure of the developing neocartilage to growth factors, substrate composition, dissolved O_2 and CO_2 concentrations and temperature. The HFBR shares with other 3D culture systems the ability to support the hyaline cartilage phenotype.[46–48] As a tissue system, the HFBR permits true macroscopic growth.[49] Cell-matrix interactions and the effects of the matrix barrier to substrate delivery and metabolic product efflux are represented much more realistically than in monolayer systems. While *in situ* development of cartilage from cells in an organism will differ in important ways from the bioreactor conditions, *in vitro* studies will be able to point the way to appropriate conditions for development of functioning neocartilage from cells. Indeed, the quality of tissue produced in these bioreactors has demonstrated potential as a source of tissue for transplantation.

Although chondrocytes have been the primary cell line used for such studies, this system is amenable to other approaches including use of bone marrow stromal cells that can be directed to develop towards the chondrocyte phenotype. Exploration of such studies in realistic *in vitro* systems may represent an important step towards developing clinically viable protocols. Furthermore, as indicated above, the bioreactor system provides the ability to monitor tissue quality non-invasively during development. MRI is becoming increasingly accepted as a non-invasive tool for the measurement of cartilage

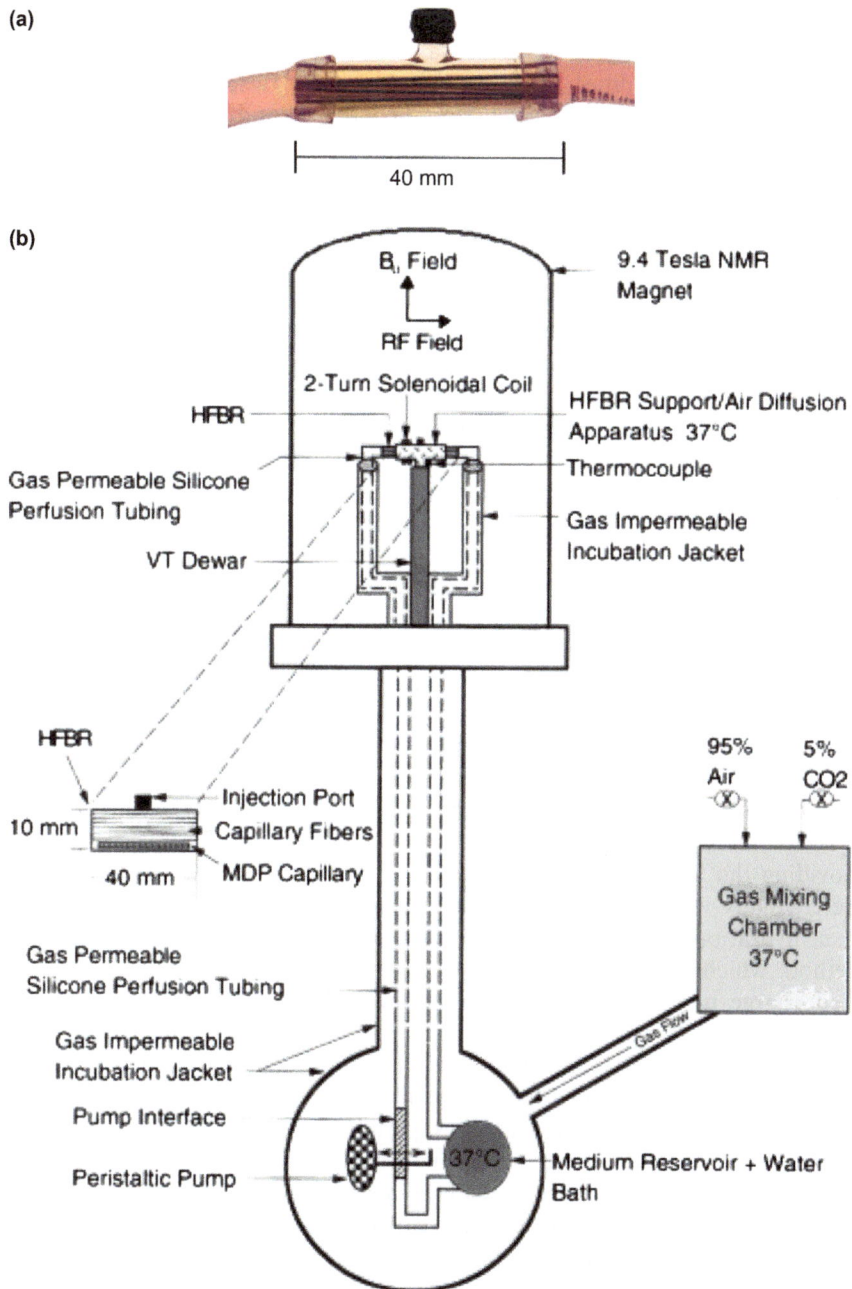

Figure 14.1 a) Photograph of a hollow fiber bioreactor (HFBR) inoculated with chondrocytes. b) Schematic drawing of an NMR compatible HFBR and incubation/perfusion apparatus.
Reprinted with permission from Petersen *et al.*[56] © 2000 Wiley-Liss, Inc.

thickness and volume, and identification of localized pathology. Establishment of the relationship between MRI measures of tissue quality *in vitro* with other matrix-specific modalities such as biochemistry, histology and Fourier transform infrared (FTIR) spectroscopy lays the groundwork for MRI-based clinical evaluation of constructs following implantation. Non-invasive evaluation with MR is clearly more desirable than existing minimally invasive methods such as arthroscopic biopsy.

14.2.1 ^{1}H MRI Studies of Hollow Fiber Bioreactor Cartilage

^{1}H MRI provides a useful means of morphological evaluation of HFBR tissue growth permitting longitudinal volumetric tracking of tissue deposition.[49] In addition to gross tissue morphology, MRI maps of parameters such as T_1, T_2, $T_{1\rho}$ and magnetization transfer (MT) ratio and MT rate (k_m) also show spatial contrast. This contrast is indicative of tissue heterogeneity and arises primarily from spatial variation in the mobility of matrix-associated water,[39] although other factors such as fibril orientation and water compartment sizes can also affect these parameters.

Quantitative evaluation of MRI parameters also provides a means of characterizing changes in tissue with maturation. For example, trends in these parameters have been observed with HFBR tissue maturation, with T_1, T_2 and apparent diffusion coefficient (ADC) decreasing with maturation while MT ratio increases.[49] Figure 14.2 shows more detailed comparisons in which trends are observed between MRI parameters and specific matrix macromolecules.[47,50,51] For example, both T_1 and T_2 correlate well with biochemically derived collagen and glycosaminoglycan (GAG) content, where GAG is the negatively charged primary subunit of aggrecan responsible for providing the tissue with its compressive stiffness.[51] The MT rate correlates preferentially with collagen concentration while ADC correlates with overall tissue hydration.[47,51] Additionally, through the use of the gadolinium exclusion method, which creates MRI contrast based on the concentration of negative fixed charge in the tissue (delayed gadolinium-enhanced MRI of cartilage, dGEMRIC), MRI-derived negative fixed charge density (FCD) correlates with biochemically derived GAG content in HFBR cartilage.[50] The dGEMRIC technique is based on the fact that, in cartilage, a lesser or greater concentration of GAG, with its fixed negative charges, permits a respectively greater or lesser degree of entry into the tissue of the negatively charged gadolinium chelate. Therefore, immature or degenerated cartilage, with a relatively lower concentration of GAG than mature or intact cartilage, exhibits enhanced relaxivity, that is, a shorter T_1 value in the presence of gadolinium.[52]

14.2.2   Hollow Fiber Bioreactor Studies Using Complementary Spectroscopic and Imaging Techniques

While ^{1}H MRI permits spatially resolved non-invasive characterization of tissue status throughout the entire extent of the tissue, outcomes are largely

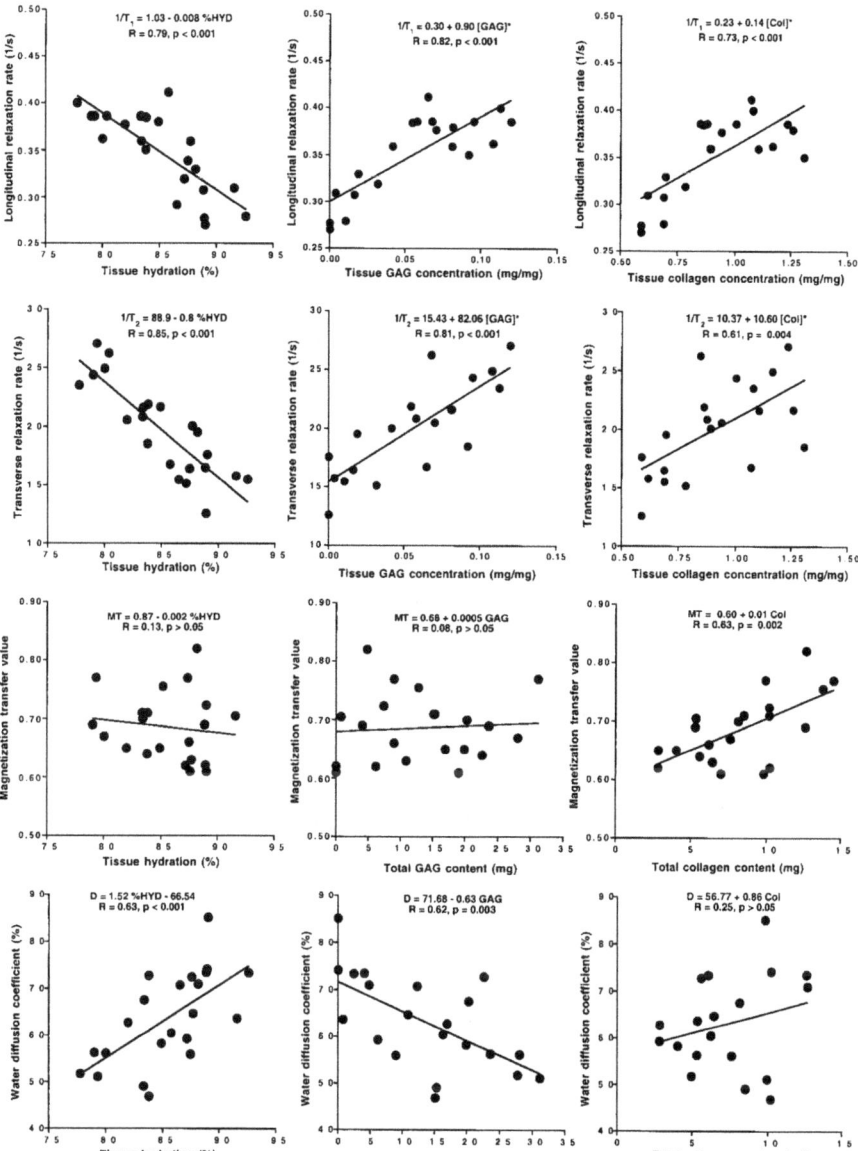

Figure 14.2 Graphs showing the relationship between tissue MR parameters R1, R2, MT ratio and water diffusion coefficient and biochemical tissue properties tissue hydration, total glycosaminoglycan (GAG) content and total collagen content.
Reprinted with permission from Potter *et al.*[51] © 2000 American College of Rheumatology.

indirect measures of central tissue characteristics, such as particular matrix macromolecules and metabolic substrates. Thus, an ongoing challenge is to introduce MR outcomes that are more sensitive and specific to particular cartilage matrix constituents. Towards this goal, a number of other spec-troscopic and imaging techniques have been found to be useful in providing complementary outcome measures as well as for examining the full potential of existing and future ^{1}H MRI outcome measures. The following sections highlight a variety of these complementary techniques applied to the investi-gation of HFBR cartilage tissue; these techniques include ^{31}P NMR spec-troscopy, electron paramagnetic resonance (EPR) and FTIR spectroscopy.

14.2.2.1 ^{31}P NMR Spectroscopy

^{31}P NMR spectroscopy permits direct and non-invasive measurement of several metabolites involved in high-energy phosphate metabolism such as inorganic phosphate (Pi), creatine phosphate (PCr) and adenosine tri-phosphate (ATP). Previous studies of native cartilage and isolated cells have provided a coherent view of chondrocyte bioenergetics, and indicated a strong dependence on the glycolytic pathway and less dependence on mitochondrial respiration.[53–55] This work also provided a view of changes in ^{31}P spectra due to a variety of stressors, demonstrating the sensitivity of these measurements to changes in cellular processes, such as the observation of increased free phospholipids ascribed to cell membrane degradation.[53] As a natural extension of work on native tissue, these measurements have been extended to HFBR cartilage, in which the high degree of cellularity permits ^{31}P spectra to be obtained much more readily than in native cartilage. Typical ^{31}P spectra of HFBR cartilage (Figure 14.3) show the presence of ATP, Pi and small amounts of PCr after 4 weeks of development. Observed ratios of ATP/Pi and PCr/Pi showed no changes in this system under standard incubation conditions over time, reflecting metabolic stability. An important feature of the HFBR system is the relatively easy access to the circulating cell culture media for assessment of circulating substrates with spectrophotometric analysis. Petersen *et al.* examined glucose utilization and lactate production in developing tissue under normal incubation conditions, and found results consistent with the predominantly glycolytic metabolism previously demon-strated in native cartilage.[39] This was also supported by measurement of tissue pH based on the chemical shift difference between PCr and Pi in the ^{31}P NMR spectra, which showed a decrease from 7.23 to 6.98 between 10 and 17 days of incubation.

In addition to the assessment of phosphorous-containing metabolite concentrations from spectral line amplitudes, relaxation measurements of individual metabolites provide information on the mobility of these molecules. Petersen *et al.* compared relaxation time measurements in cartilage developing in the HFBR from chondrocytes harvested from different anatomic sites of the embryonic chick sternum.[56] A significant increase in the T_1 of Pi over 4 weeks of development in tissue grown from proximal sternal chondrocytes was observed. In contrast, no change in the T_1 of Pi from tissue grown from distal sterna chondrocytes was observed over 4 weeks of development. These results are consistent with the fact that the proximal sternum of the chick undergoes

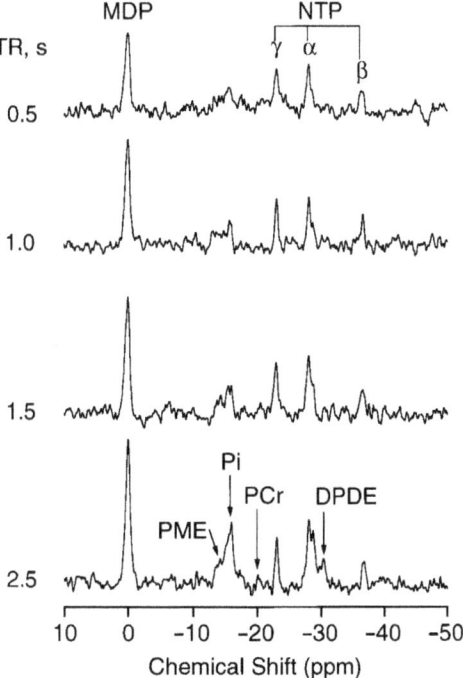

Figure 14.3 ^{31}P-NMR spectra of a HFBR 4 weeks after inoculation with 50×10^6 chondrocytes. Abbreviations include methylenediphosphonic acid (MDP), nucleotide phosphates (NTP), creatine phosphate (PCr), inorganic phosphate (Pi), phosphomonoesters (PME) and diphosphodiesters (DPDE). Reprinted with permission from Petersen *et al.*[56] © 2000 Wiley-Liss, Inc.

ossification, while the distal sternum remains cartilaginous in early development.

14.2.2.2 Electron Paramagnetic Resonance Imaging

While glycolysis is the primary pathway for energy provision in chondrocytes, adequate aerobic metabolism is thought to be important for cellular homeostasis and DNA synthesis. Furthermore, aerobic respiration may be coupled to matrix production.[57] This suggests the importance of assessing oxygen concentration in engineered cartilage. A novel method for this was developed using electron paramagnetic resonance (EPR) spectroscopic imaging.[58] This method takes advantage of the broadening effect of oxygen on the EPR linewidth. When properly calibrated (Figure 14.4), local oxygen concentration can therefore be determined from a linewidth map derived from the ratio of the local resonance amplitude measured as peak height to integrated spectral line intensity measured as area. Oxygen maps constructed according to this procedure were measured in the HFBR, and demonstrated sensitivity to decreased oxygen consumption in tissue exposed to cyanide, a potent inhibitor of oxidative phosphorylation.[59]

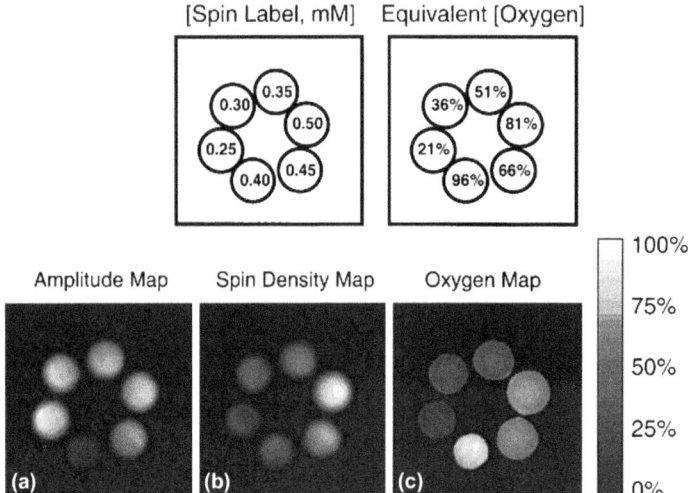

Figure 14.4   Phantom tube composition is illustrated above, showing the concentration of spin probe (^{15}N-PDT) and the level of oxygenation in each tube. a) Spatial map corresponding to maximum amplitude, which does not correspond to either [^{15}N-PDT] or the [O_2] levels. b) Spin density does correspond to [^{15}N-PDT], but not to [O_2]. c) In contrast, the map created by EPROM corresponds qualitatively and quantitatively to the distribution of [O_2] among the phantom tubes.
Reprinted with permission from Velan *et al.*[58] © 2000 Wiley-Liss, Inc.

14.2.2.3   *Infrared Spectroscopic Imaging*

As previously mentioned, while NMR and MRI studies are non-invasive and provide useful data reflecting tissue status, an ongoing challenge is to render the available outcomes more sensitive and specific to molecular-level matrix modifications occurring during development. In contrast to MR, Fourier transform infrared imaging spectroscopy (FT-IRIS) provides measurements that are highly specific to cartilage matrix components and degeneration, although it is invasive and typically destructive.[60–62] FT-IRIS measurements are based on absorption of radiation at specific infrared frequencies reflecting molecular resonances assigned to particular molecular vibration modes. FT-IRIS couples a Fourier transform infrared (FTIR) spectrometer to an optical microscope with an array detector. This permits spatially selective evaluation of matrix components and molecular characteristics of tissue sections, typically on the order of a few microns thick, with an in-plane spatial resolution of ~ 7 microns. The resulting data set consists of an infrared spectrum assigned to each spatial location across the sample; this spectrum reflects local molecular characteristics. As shown in Figure 14.5, the data can also be arranged into a spectroscopic image by mapping the amplitude of specific absorption bands. One use of this technique relies on the fact that collagen is the dominant contributor to the amide I band observed in FTIR spectra of cartilage. Thus, FT-IRIS can be used to map the variation of collagen concentration within a

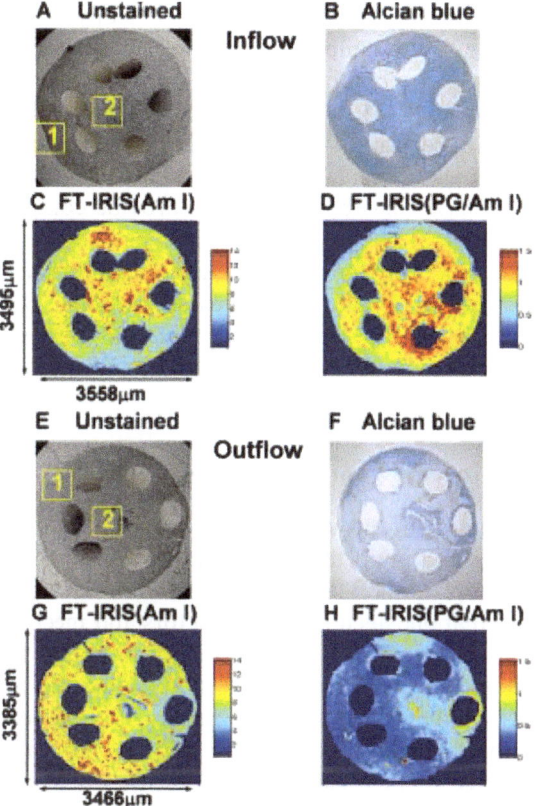

Figure 14.5 Cross-sections of inflow and outflow tissue from HFBR cartilage. Unstained sections on barium fluoride windows and histologic Alcian blue-stained sections from inflow (A,B) and outflow (E,F) tissue. FT-IRIS images of inflow (C,D) and outflow (G,H) tissues created based on amide I absorbance and PG/amide I absorbance, respectively. Regions labeled as "1" and "2" represent "surface" and "center" regions of tissue, respectively.
Reprinted with permission from Kim *et al.*[61] © 2005 Society of Photo-Optical Instrumentation Engineers.

tissue section. In contrast, bulk FTIR spectroscopy would indicate only the total collagen in the entire sample. Thus, FT-IRIS is a powerful technique for evaluating the quality of cartilage tissue at a molecular level with high spatial resolution.

Both near-IR and mid-IR wavelength regions have been used for FT-IR analysis of engineered cartilage matrix. Mid-FT-IR has been used to evaluate the quality of degenerated and replacement tissue at a molecular level with high spatial resolution.[63] However, radiation in the mid-IR band is fully absorbed by tissue within approximately 10 microns. This limits the applicability of mid-FT-IR to studies of tissue surfaces when used in reflectance mode, in which the detector is placed in the path of IR radiation reflected from the surface of intact tissue. Alternatively, using transmission spectroscopy in which infrared radiation passes through the sample before entering the detector, mid-FT-IR can be applied to histologically prepared thin tissue samples. In contrast, near-infrared spectroscopy (NIRS), which uses shorter wavelength radiation, penetrates tissue to a depth of several millimeters or even centimeters, permitting full-depth assessment of cartilage tissue. NIRS assessment of engineered cartilage matrix has been shown to provide comparable results to mid-IR analysis of cartilage.[64] These results form a basis for future applications of NIRS in arthroscopic evaluation of cartilage implantation procedures for repair and replacement of damaged or diseased cartilage. Such possibilities present exciting opportunities for *in vivo* molecule-specific evaluation of implanted cartilage for comparison to, or to be used in conjunction with, MRI indices.

14.3 Modulation of Neocartilage Growth

Modulation of tissue growth through the application of chemical and physical stimuli is an ongoing and important step towards the development of viable cartilage tissue for implantation. A variety of studies have demonstrated the effects of nutrients, anabolic agents, growth factors and mechanical stimuli on engineered cartilage growth.[49–51,65–67] Studies incorporating MR analyses have demonstrated both the effects of interventions on engineered cartilage and the sensitivity of MRI to these effects. Investigation of the effects of catabolic agents is also important for understanding the way in which implanted cells and tissue might be affected by their surroundings.

14.3.1 Nutrients, Anabolic Agents and Growth Factors

There is a complex relationship between growth-modifying compounds contained in or added to cell culture media, the local environment of the chondrocyte as defined, for example, by scaffold material and the quantity and quality of matrix produced. Modulation of growth conditions in an effort to produce viable implantable tissue is an area of active interest. The effects of subtle changes in growth conditions have been documented. Excised bovine cartilage chips cultured in Dulbecco's modified Eagle's medium with the addition of 20% fetal bovine serum (FBS) showed 3–4 times greater proteoglycan synthesis rates than tissue without serum.[65] In contrast, bovine chondrocytes seeded in a collagen I scaffold have been shown to produce more proteoglycan in low serum containing medium compared with tissue grown using 2% and 10% FBS.[67]

Non-invasive MRI studies may be of particular value in exploring these complex effects by permitting ongoing non-destructive analysis, in contrast to conventional biochemical and histologic methods. The effect of ascorbic acid, a required nutrient for efficient collagen synthesis, on the development of cartilage from bovine chondrocytes was evaluated by Nugent *et al.*[67] It was shown that incubation with media containing $50 \, \mu g \, mL^{-1}$ of ascorbate resulted in the production of significantly more collagen over 4 weeks as compared to tissue cultured without ascorbate. GAG production was also greatly augmented. The effects of ascorbate were accompanied by significantly greater MRI-derived negative FCD and k_m, parameters that correspond most closely to proteoglycan and collagen content, respectively.[68] A similar effect of ascorbate was found in HFBR tissue grown from chick embryo chondrocytes. Cells grown in incubation media receiving daily augmentation with $10 \, \mu g \, mL^{-1}$ ascorbic acid produced tissue with significantly greater collagen content than those without augmentation.[51] In addition increased k_m was observed in tissue treated with ascorbic acid, as well as significant decreases in T_1, T_2 and ADC.

MRI was also used for assessment of the effects of insulin-like growth factor 1 (IGF-1) and transformation growth factor beta (TGF-β) on engineered cartilage, both of which can promote the development of the chondrocyte phenotype from progenitor cells. Morphological MRI measurements in the HFBR system established an increase in tissue production by human secondary chondroprogenitor cells using media augmented with these growth factors.[49] In addition, MT ratio measurements showed comparable tissue quality between growth factor-treated tissue that developed from human chondroprogenitor cells, and tissue grown from chick embryo chondrocytes without growth factors. Quantitative MRI parameters were also used to demonstrate substantial augmentation of tissue growth from mesenchymal stem cells through growth factor augmentation. Li *et al.* reported a significant increase in k_m and decreases in T_1, T_2 and ADC in tissue treated with a combination of growth factors containing TGF-β1, insulin-transferrin-sodium selenite (ITS), ascorbate and dexamethasone.[66] These effects were accompanied by corresponding increases in GAG and type II collagen production.

14.3.2   Mechanical Stimulation

Mechanical stimulation has also been investigated as a means of improving the growth and quality of engineered cartilage for implantation. In particular, macroscopic cyclic mechanical stimulation increases matrix production under a wide variety of circumstances.[69] A more convenient form of mechanical stimulation is through application of an ultrasonic pressure wave, frequently based on the pulsed low-intensity ultrasound (PLIUS) used in clinical orthopedic practice for fracture healing. Ultrasound-induced increases in aggrecan gene expression and chondroitin sulfate synthesis have been found in many studies.[70–75] Augmentation of type II collagen production has also been found in several of these ultrasound-stimulated systems.[70,74–77] This is of particular interest due to the low collagen-to-proteoglycan ratio of typical engineered cartilage constructs.[78] Ultrasound has also been applied to several animal models of degenerative and damaged cartilage.[79–84] The Hartley guinea pig develops osteoarthritic pathology similar to that of humans.[85] Application of PLIUS to the stifle joints of Hartley guinea pigs over weeks to months showed beneficial effects in both prevention of OA and reversal of established disease. Outcome measures included uniformity of histological staining, surface fibrillation and deep fissuring.[86]

In terms of therapeutics, efficacy of PLIUS treatment has been demonstrated in cells, cartilage explants, developing engineered constructs and animal models of OA. With the expectation that human studies of PLIUS will follow this body of successful work, as well as in anticipation of other potential therapies, efforts have been made towards the development of non-invasive MRI endpoints sensitive to improvements in cartilage matrix quality. As previously mentioned, quantitative MRI parameters correlate with matrix development and maturation in the collagen I tissue engineering scaffold.[67] Additional work using this scaffold has demonstrated these MR parameters to be sensitive to PLIUS-induced changes in matrix development.[87] In particular, k_m was sensitive to differences in collagen content both longitudinally and with PLIUS treatment while T_2 showed a lack of specificity to a particular matrix component but rather was sensitive to differences in both collagen and proteoglycan content (Figure 14.6).

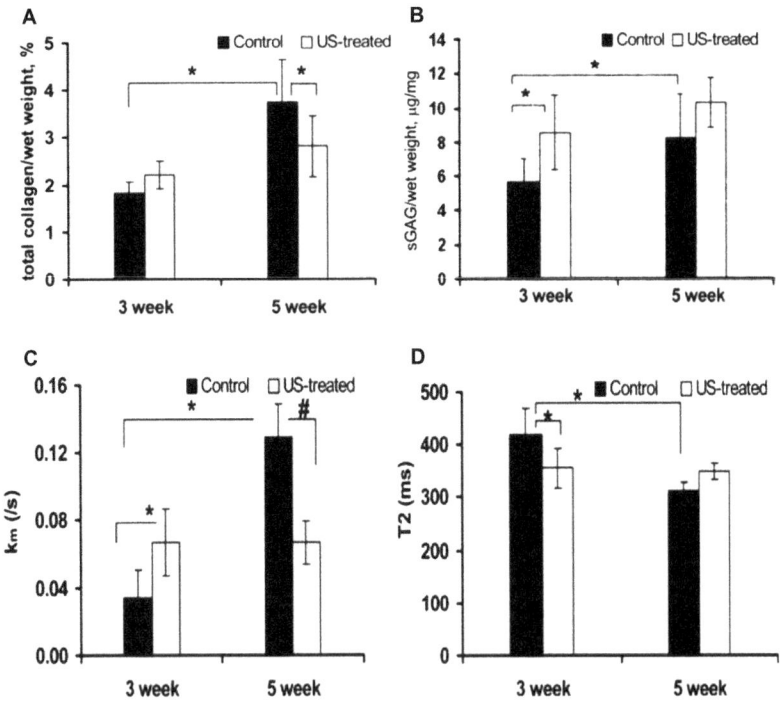

Figure 14.6   Effect of PLIUS treatment in chondrocyte-seeded collagen I scaffolds on A) total collagen content, B) sulfated glycosaminoglycan (sGAG) content, C) magnetization transfer rate (k_m) and D) T_2. Biochemical measurements (n = 6 per group) and MR measurements (n = 10) were made after 3 and 5 weeks of culture time.
Reprinted with permission from Irrechukwu *et al.*[87] © 2011 Mary Ann Liebert, Inc.

14.3.3 Catabolic Agents

The investigation of catabolic agents in engineered cartilage systems is also of substantial interest. These agents represent pathomimetic interventions for developing matrix-sensitive outcome measures and analyses. In addition, certain of these molecules are expected to be present or over-expressed in degenerative cartilage disease. An example of the strong effect of catabolic agents on tissue quality in engineered cartilage has been shown in the collagen I system seeded with bovine chondrocytes where, after 4 weeks of culture time, tissue treated for 1 week with the cytokines interleukin-1 beta (IL-1β) and tumor necrosis factor alpha (TNF-α) showed a reduction in proteoglycan content by a factor of more than two as compared to control tissue. This response to cytokines was accompanied by a corresponding loss of MRI-derived negative FCD.[67] Of importance, matrix production remained negligible over a 2-week growth period after withdrawal of cytokines. These results highlight the importance of potentially irreversible effects of cytokines on chondrocyte metabolism. Importantly, these results also demonstrate the potential use of MRI for non-invasive monitoring of these effects.

Catabolic agents, especially those that selectively target specific macro-molecules, are also useful for investigating relationships between macro-molecules and MRI indices. Such relationships are difficult to examine in normal tissue due to the high correlation between the amounts of individual cartilage matrix components. One example of this approach is described in the work of Potter *et al.*[51] where 3-week-old HFBR tissue was treated for 1 week with either retinoic acid or IL-1β. Tissue treated with retinoic acid showed greatly reduced proteoglycan content, greater water content and no significant difference in total collagen content. These biochemical changes were accompanied by increases in T_1, T_2 and ADC. IL-1β induced an increase in proteoglycan and collagen content on a per-cell basis compared with control tissue. These interventions, as well as provision of ascorbate as an anabolic agent, provided a wide dynamic range of tissue for development of correlations with MRI indices. T_1 and T_2 were found to be sensitive to both collagen and proteoglycan content, while k_m was, as expected, more sensitive to collagen content, and ADC to overall tissue hydration. A similar approach was used in the HFBR cartilage system through application of chondroitinase ABC. The resulting targeted degradation of proteoglycan permitted a correlation to be established between MRI-derived negative FCD and biochemically derived GAG content.[50]

14.4 Translational Studies: Tissue Integration and Assessment Using MRI

In addition to optimal growth of engineered cartilage, the overall success of tissue engineering-based therapy may depend upon successful integration of a construct into the surrounding tissue.

Injectable gel constructs have been investigated as an efficient means to place repair material into a defect through the use of minimally invasive arthroscopic procedures. This approach has the potential greatly to reduce complications associated with more traditional surgical implantation. A variety of gels have been created, such as polyethylene oxide (PEO)-based hydrogels, permitting control over the gelation process. Gelation in photopolymerizable hydrogels has been shown to influence cell distribution and scaffold structural properties.[88] PEO hydrogels have been demonstrated to support the development of tissue with favorable biochemical, mechanical and electromechanical properties.[88]

MRI has been used to evaluate PEO-based constructs implanted into cartilage plugs. Similar to other constructs, they exhibited a significant correlation between dGEMRIC-derived negative FCD and biochemically derived GAG content. This demonstrates the potential for non-invasive longitudinal evaluation of matrix production post-implantation.[89] In addition, cell tracking in these gels using MRI has been demonstrated using superparamagnetic iron oxide (SPIO) nanoparticles (see also Chapter 11). SPIO-labeled cells result in hypointense MRI contrast as compared with surrounding tissue due to the iron-induced increase in MRI relaxivity. Ramaswamy *et al.* demonstrated efficacy in labeling chondrocytes with Feridex IV, an FDA-approved SPIO contrast agent. No adverse effects of labeling were found with respect to cell phenotype or viability, or the production of major matrix constituents.[90] A close correspondence between MR image contrast and histologically determined cell distribution from corresponding sections of the construct was demonstrated.

Even following optimal *ex vivo* growth conditions and successful placement of a construct into a defect in native tissue, successful integration must be ensured. As an example, chondrocyte pellet cultures have been shown to demonstrate matrix composition, density and ultrastructure comparable to hyaline cartilage in culture.[91] However, following attempts at integration using centrifugation with cartilage explant tissue, interactions at the interface between constructs and native tissue resulted in chondrocyte death and abnormal matrix remodeling.[92] Such negative outcomes are expected to impact the long-term overall quality of cartilage repair.

Integration of injectable hydrogels has been studied *in vitro* and *in vivo* in order to investigate the effect of implant anchoring. MRI-derived indices have been compared with gold-standard histological techniques in these studies and have demonstrated the sensitivity of MRI to this aspect of implantation. For example, Ramaswamy *et al.* showed improvements in integration of PEO-based hydrogel cartilage implants *in situ* using tissue-initiated and chemical-initiated integration methods. In this work, the tissue-initiated integration involved the photo-oxidation of the native interface tissue while the chemically initiated method involved treatment of the native interface tissue with chondroitin sulfate-methacrylate-aldehyde.[93] Histologic assessment of these samples demonstrated a tissue void in the transition region of non-integrated samples, a sharp interface in the samples in which tissue-initiated integration was implemented and a more gradual metachromatic interface in samples using chemical initiation of integration (Figure 14.7). Consistent with these histologic results,

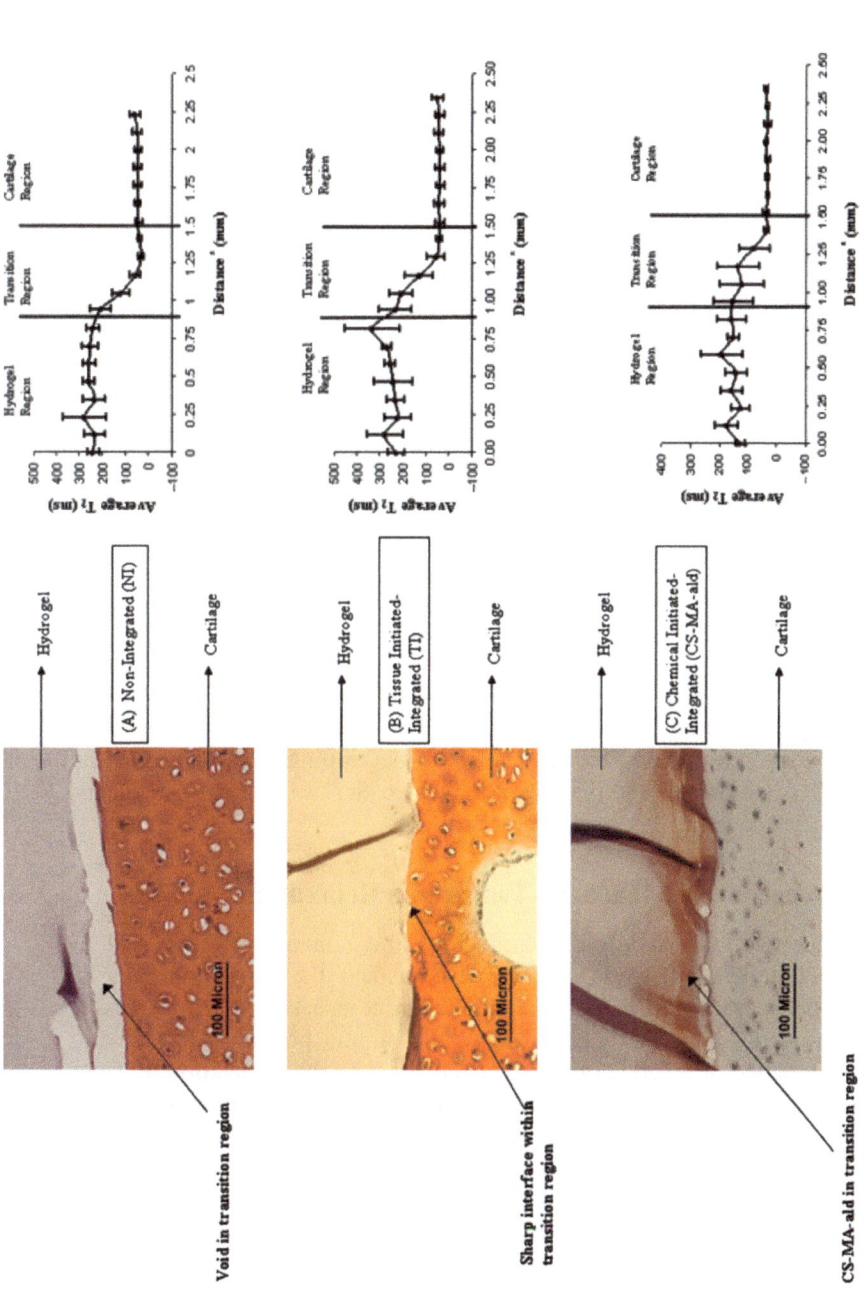

Figure 14.7 A) Histological sections using safranin-O staining showing transitions between native and implanted tissue (from top to bottom) non-integrated, tissue-integrated and chemically integrated methods. B) Spatially averaged T_2 distributions across the cartilage–implant interface in (from top to bottom) non-integrated, tissue-integrated and chemically integrated methods. Reprinted with permission from Ramaswamy *et al.*[93] © 2005 Wiley Periodicals, Inc.

MRI-derived T_2 maps showed a more gradual transition of T_2 values between implant and native cartilage in chemically initiated samples as compared with the other two sample groups. This demonstrates the sensitivity of T_2 to implant integration (Figure 14.7). Furthermore, the average T_2 value in the chemically initiated constructs was much shorter compared with the other two groups, indicating an overall greater amount of matrix elaboration.

T_2 measurements have also been applied *in vivo* for evaluating the degree of tissue fill in a subchondral defect rabbit model.[94] In this work, PEO-based hydrogel scaffolds were used to fill defects and tissue repair was quantified after 5 weeks. A linear correlation was established between T_2 and the histologically determined percent tissue fill within the defect. This suggests that T_2 may be used as a surrogate for the extent of implant fill of a tissue defect, again supporting the use of MRI as a means of evaluating tissue repair post-implantation. Such a non-invasive method would be much more desirable for longitudinal evaluations as compared with more invasive evaluation using arthroscopy.

14.5 Sensitivity and Specificity Improvements in Proton MRI

Numerous studies have clearly elucidated the general outline of the response of MR parameters to tissue status. Nevertheless, the sensitivity and specificity of MR indices to tissue constituents remains limited. Two recent approaches have been proposed to address this problem: i) alternative models for transverse (T_2) relaxation analysis and ii) multivariate machine learning approaches. This work is primarily motivated by the interest in developing more sensitive and specific MRI-based analyses that do not require the use of contrast agents, unusual pulse sequences or specialized hardware.

14.5.1 Alternative Models of Transverse Relaxation

While water proton relaxation is frequently and conveniently modeled as monoexponential in biological systems, it is evident that this represents an oversimplified description of complex tissue. In contrast to the single tissue water compartment implicitly assumed in the monoexponential framework, tissue is comprised of several compartments defined by the association of water with particular macromolecular components. Both native and engineered cartilage are expected to contain compartments defined by the association of water with collagen, the association of water with proteoglycans and relatively free, or mobile, water. Additional compartments may also be present. Multi-exponential analysis permits extraction of the concentrations and transverse relaxation rates of, and hence water mobility within, these compartments individually.

Multiexponential analyses of relaxation data from experiments using bovine nasal cartilage, a relatively homogenous and isotropic hyaline cartilage system that serves as a model for articular cartilage, were performed using a numerical inversion technique called non-negative least squares.[95] This technique makes no *a priori* assumptions about the number of underlying exponentials comprising the relaxation data. The analysis has clearly identified three exponential relaxation components.[96] Additional studies using bovine nasal cartilage have revealed a relationship between the weight fractions associated with specific T_2 components and biochemically derived proteoglycan content.[97] This suggests that these weight fractions may be used to quantify the amount of specific macromolecular components present in cartilage matrix in a variety of settings.

We note that while initial application of multiexponential analysis was to non-localized evaluation of entire individual cartilage samples in bulk, this method has now been extended to an imaging modality. This is achieved through performing the multiexponential analysis on a pixel-by-pixel basis in a multiecho imaging data set. The resulting component weights can therefore be assigned and mapped for each pixel.[98] In this way, this analysis permits the creation of matrix component-specific water fraction images using a standard MRI acquisition sequence that does not require specialized hardware or pulse sequences, or exogenous contrast agents.[99–101]

In addition to studies of native cartilage, multiexponential T_2 analysis has been extended to tissue engineered cartilage. In this setting as well, improved specificity to matrix components as compared to conventional monoexponential analysis was demonstrated. In engineered cartilage grown on a polyglycolic acid (PGA) scaffold, Reiter *et al.* identified a T_2 component with an intermediate relaxation time of $\sim 85\,\mathrm{ms}$, which was provisionally assigned to protons associated with proteoglycan (Figure 14.8). This intermediate-relaxing component was absent in empty hydrated PGA scaffolds and increased during development in concert with proteoglycan elaboration.[97] Similar studies were performed using a collagen I scaffold. Here, two intermediate components with relaxation times of $\sim 45\,\mathrm{ms}$ and $\sim 180\,\mathrm{ms}$ were found to correspond most closely to proteoglycan content. Again, this demonstrates a correlation between certain component fractions and biochemically derived sulfated glycosaminoglycan (sGAG) content.[102] The basis of the different number of proteoglycan-associated T_2 components in tissue developing on PGA *versus* collagen I scaffolds is unknown, but likely reflects, at least in part, differences in the molecular weight and hence mobility of the matrix components being elaborated by the cell. More detailed biochemical analysis is needed to support this hypothesis.

A distinct and potentially important new framework for describing transverse relaxation in cartilage has been introduced by Magin *et al.*[103] Fractional-order analyses, introduced in the context of anomalous diffusion in complex materials,[104] may represent a more appropriate framework to describe transverse relaxation in cartilage than either single exponential or multiexponential decay. Recent work has shown a correlation between a fractional order parameter, defined through the stretched exponential function and the stretched Mittag-Leffler function, and the macromolecular constituents of cartilage.[103] In gels, the fractional order parameter α was found to decrease, representing greater complexity, as macromolecular concentration increased. In cartilage, α increased, indicating less tissue structure, with pathomimetic

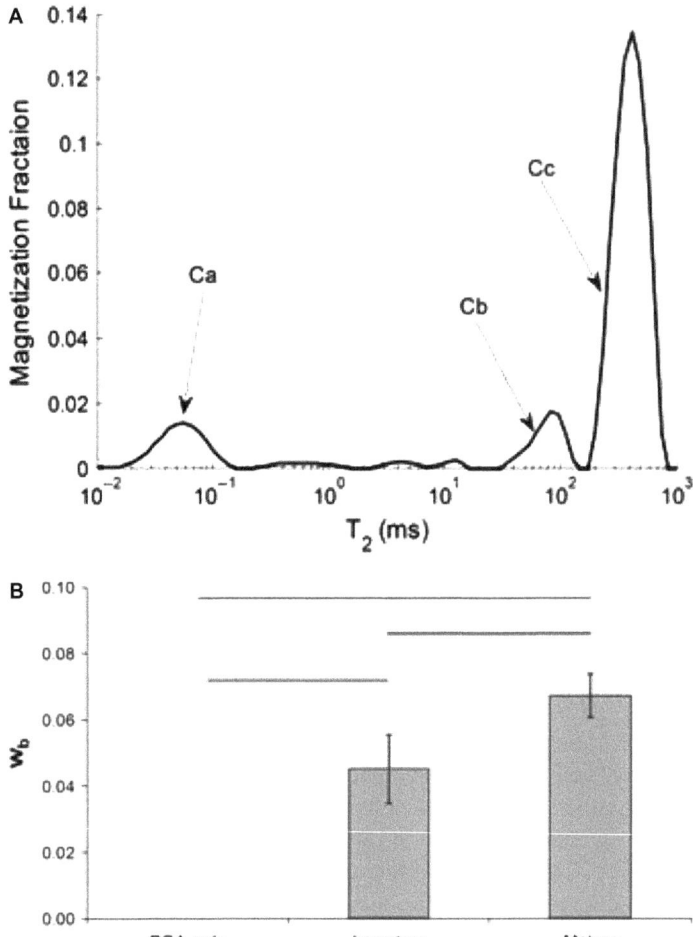

Figure 14.8   A) Typical T_2 distribution from a mature engineered cartilage sample showing the components Ca, Cb and Cc. B) Component weight fractions for empty polyglycolic acid (PGA) scaffolds and for immature and mature engineered cartilage ($p < 0.05$).
Reprinted with permission from Reiter *et al.*[111] © 2012 John Wiley & Sons, Ltd.

enzymatic degradation. Although this approach has not yet been investigated in engineered tissue, it represents a potentially important means of characterizing the matrix of developing constructs.

14.5.2   Multivariate Machine Learning Analysis of MRI Parameters

There is a substantial literature on the use of individual MRI parameters for characterization of cartilage.[105–109] Statistically significant changes in these parameters have been found to accompany tissue development and matrix elaboration in many circumstances. However, there is, in general, a substantial overlap in parameter values between different groups, in spite of group mean differences. This greatly limits the sensitivity and specificity of univariate classification of tissue, including native cartilage.[110] Similar findings were obtained in the univariate analysis of engineered cartilage.[111]

A number of multivariate approaches have been applied to MRI indices of cartilage matrix in an effort to define tissue status with greater sensitivity and specificity.[112,113] Similar to multiexponential T_2 analysis, this approach was initially applied to normal and enzymatically degraded cartilage. In the first of these studies, Lin *et al.* used a mild enzymatic degradation protocol to mimic relatively early OA. In contrast to the more severe degradation implemented in the previous work on univariate classification,[110] this mild enzymatic treatment resulted in small mean changes in MR parameter values, with significant differences seen only in T_1.[113] Univariate and multivariate classification into either the control or the degraded group was then performed. As expected, the accuracy of univariate classification based on parameter means was inferior to that seen under the more severe degradation performed previously.[110] With mild trypsin degradation, after establishing the classification cut-off in a training set, validation set sensitivity was 0.67 and the specificity was 0.62. The overall accuracy of classification, calculated as the average of sensitivity and specificity, was therefore only 0.65. Classification accuracy was dramatically improved using a multiparametric Gaussian mixture model established using the MCLUST algorithm.[114] The basis for this approach can be seen in Figure 14.9, which shows bivariate scatter plots of MR parameters from the control and trypsin-degraded groups with corresponding error ellipses derived from the Gaussian mixture model. For this particular set of cartilage samples, the best bivariate classifier was (T_1, k_m) with validation sensitivity of 0.81 and specificity of 0.78.

In spite of its success in the detection of cartilage degradation, there remain important limitations to multivariate Gaussian clustering for classification. The imposition of an elliptical data structure, while evidently a reasonable approximation in the present case, is restrictive. In addition, the problem of overfitting, in which performance in the validation set is substantially lower than performance in the training set, is clearly seen. Finally, the sensitivity and specificity for classification, while improved over univariate classification, still are not as good as one might desire. To address these limitations, the support vector machine (SVM) formalism has been applied to this classification problem. As shown in Figure 14.10, the SVM provides an extremely flexible framework for classification, making minimal assumptions regarding the distribution of data points in parameter space.[115] In the case of cartilage classification, this algorithm determines and orients a decision hyperplane in transformed MR parameter space, which is designed to discriminate between degraded and non-degraded samples.[112] In the work by Lin *et al.*,[112] the best univariate classifier, T_1, showed validation set classification accuracy of 0.6. SVM multivariate analysis demonstrated a large improvement, with the best bivariate classifier, (T_1, k_m), showing validation set classification accuracy of 0.92.

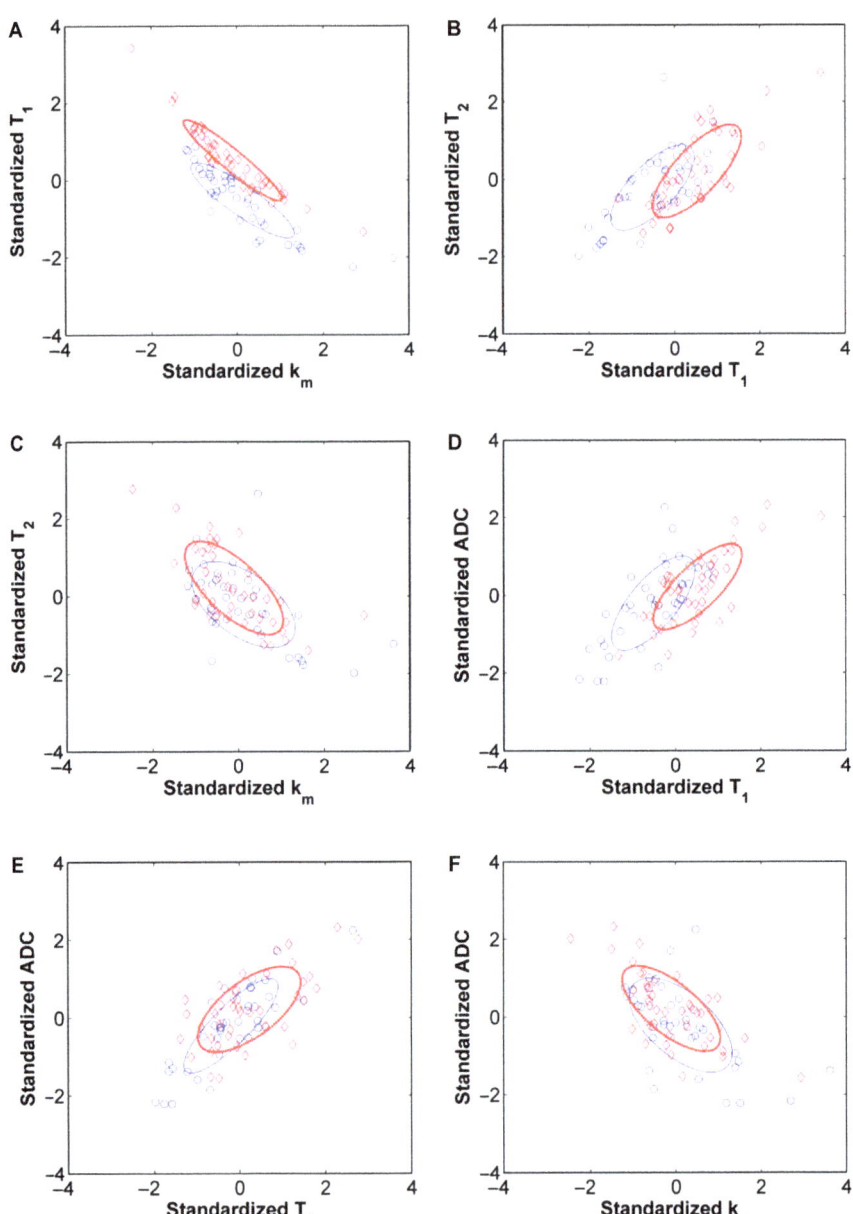

Figure 14.9 Bivariate scatter plots of data acquired from control and collagenase-digested bovine nasal cartilage samples. Covariance matrices for each pair of MRI parameters were calculated for control (blue; circle) and degraded (red; diamond) samples. The corresponding error ellipses illustrate the relationship between the parameter pairs indicated. A) (k_m, T_1), B) (T_1, T_2), C) (k_m, T_2), D) (T_1, ADC), E) (T_2, ADC) and F) (k_m, ADC) parameter pairs.
Reprinted with permission from Lin *et al.*[113] © 2009 Elsevier Inc.

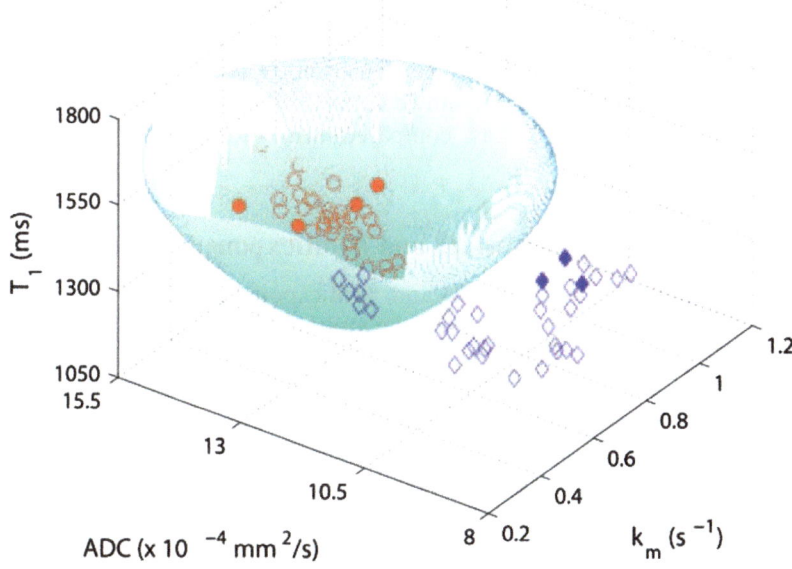

Figure 14.10 A scatter plot of MRI measurements collected on bovine nasal cartilage samples before and after 24-hour enzymatic digestion. A single trial of the 10-fold cross-validation is shown, with the division of the full set of data points into a training set (empty symbols) and a validation set (filled symbols) for classification. Each training or validation set data point was classified into the pre-degradation class (blue) or post-degradation class (red) through the support vector machine model established using the training set. The contour surface shown indicates the decision hypersurface on which the assignment probabilities for the pre-degradation and post-degradation classes are the same. Note that all data points from the pre-degradation group are located outside the bowl-shaped hypersurface, while all the points for the post-degradation class are inside the hypersurface; therefore, no classification errors occur for this particular trial.
Reprinted with permission from Lin *et al.*[112] © 2011 Wiley Periodicals, Inc.

Both of these multivariate approaches can be adapted to accommodate the graded, rather than binary, nature of cartilage degradation. In the Gaussian mixture model this was accomplished through assignment of cluster member-ship probabilities according to the multivariate Gaussian distribution.[113] For a given feature space defined by selection of classification parameters, this leads to a degradation probability for each individual sample. For SVM classifi-cation, degree of degradation can be defined according to a sigmoidal distance function from the decision hyperplane.[115] These metrics of graded degradation can easily be extended to tissue engineering applications in which the degree of matrix ellaboration, rather than degradation, is of interest.

The SVM approach has been extended to classification of maturation stage in engineered cartilage grown from bovine chondrocytes using a PGA scaffold.[111] The best univariate classifier for maturation, k_m, showed a validation set classification accuracy of 0.85. In contrast, the best bivariate classifier in these engineered cartilage constructs was (k_m, ADC), with an accuracy of 0.98. Thus, this work showed improved characterization of tissue maturation using the SVM as compared with conventional univariate classification.

14.5.3   Combined Multivariate and Multiexponential Analysis

We have described the multivariate and multiexponential techniques as two distinct analytic approaches. However, the multiexponential analysis, in effect, introduces new MR variables that can in turn be applied to tissue characterization through multivariate analysis. Initial efforts in this area made use of support vector regression (SVR), an extension of the SVM, to establish a model for prediction of biochemical properties using MR parameters. Engineered cartilage developing from bovine stifle joint chondrocytes within a collagen I scaffold was studied.[102] Multiexponential analysis was performed to determine component fraction sizes, which were then incorporated into an SVR model to predict tissue biochemical status. The resulting regression relations showed much larger correlation coefficients than did conventional regressions in which single MR parameters were used as biochemical correlates. For example, univariate linear regression analysis of the proteoglycan-associated water fraction resulted in a slope of 0.46 with an r^2 of 0.54 between predicted sGAG content and biochemically measured sGAG content. Using multiple component water fractions the slope between SVR predicted sGAG content and biochemically measured sGAG content was ~ 0.9 with an r^2 of 0.93. This work demonstrates the potential for SVR to produce high-quality estimates of tissue biochemistry from MR data.

14.6   Conclusions

There have been tremendous advances in the non-destructive assessment of tissue status and maturation in engineered constructs using MR, which provides a number of morphologic and compositional outcome measures. Advances in acquisition and analysis techniques have the potential to further improve the sensitivity and specificity of MR studies, and therefore to be of substantial importance in basic science studies, and potentially in the clinical research setting.

Acknowledgement

This work was supported by the Intramural Research Program of the NIH, National Institute on Aging.

References

1. A. Woolf and B. Pfleger, *B. World Health Organ.*, 2003, **81**, 646–656.
2. D. K. Bae, K. H. Yoon and S. J. Song, *Arthroscopy*, 2006, **22**, 367–374.
3. R. Gudas, E. Stankevicius, E. Monastyreckiene, D. Pranys and R. J. Kalesinskas, *Knee Surg. Sport. Tr. A.*, 2006, **14**, 834–842.
4. A. C. Kuo, J. J. Rodrigo, A. H. Reddi, S. Curtiss, E. Grotkopp and M. Chiu, *Osteoarthr. Cartilage*, 2006, **14**, 1126–1135.
5. K. Mithoefer, R. J. Williams, R. F. Warren, H. G. Potter, C. R. Spock, E. C. Jones, T. L. Wickiewicz and R. G. Marx, *J. Bone Joint Surg. Am.*, 2005, **87A**, 1911–1920.
6. R. E. Glenn, Jr., E. C. McCarty, H. G. Potter, S. F. Juliao, J. D. Gordon and K. P. Spindler, *Am. J. Sports Med.*, 2006, **34**, 1084–1093.
7. J. W. Alford and B. J. Cole, *Am. J. Sports Med.*, 2005, **33**, 443–460.
8. D. T. Ginat, S. Kenan and G. C. Steiner, *Acta Orthop.*, 2005, **76**, 934–938.
9. T. Malinin, H. T. Temple and B. E. Buck, *J. Bone Joint Surg. Am.*, 2006, **88**, 762–770.
10. A. A. Amin, W. Bartlett, C. R. Gooding, M. Sood, J. A. Skinner, R. W. Carrington, T. W. Briggs and G. Bentley, *Int. Orthop.*, 2006, **30**, 48–53.
11. C. R. Gooding, W. Bartlett, G. Bentley, J. A. Skinner, R. Carrington and A. Flanagan, *Knee*, 2006, **13**, 203–210.
12. W. Bartlett, J. A. Skinner, C. R. Gooding, R. W. Carrington, A. M. Flanagan, T. W. Briggs and G. Bentley, *J. Bone Joint Surg. Br.*, 2005, **87**, 640–645.
13. J. Wasiak, C. Clar and E. Villanueva, *Cochrane Database Syst. Rev.*, 2006, **3**, CD003323.
14. S. Roberts, A. P. Hollander, B. Caterson, J. Menage and J. B. Richardson, *Arthritis Rheum.*, 2001, **44**, 2586–2598.
15. J. Riesle, A. P. Hollander, R. Langer, L. E. Freed and G. Vunjak-Novakovic, *J. Cell Biochem.*, 1998, **71**, 313–327.
16. R. A. Bank, N. Verzijl, F. P. Lafeber and J. M. Tekoppele, *Osteoarthr. Cartilage*, 2002, **10**, 127–134.
17. N. Verzijl, J. DeGroot, Z. C. Ben, O. Brau-Benjamin, A. Maroudas, R. A. Bank, J. Mizrahi, C. G. Schalkwijk, S. R. Thorpe, J. W. Baynes, J. W. Bijlsma, F. P. Lafeber and J. M. TeKoppele, *Arthritis Rheum.*, 2002, **46**, 114–123.
18. S. C. Dickinson, T. J. Sims, L. Pittarello, C. Soranzo, A. Pavesio and A. P. Hollander, *Tissue Eng.*, 2005, **11**, 277–287.
19. I. Martin, R. Suetterlin, W. Baschong, M. Heberer, G. Vunjak-Novakovic and L. E. Freed, *J. Cell Biochem.*, 2001, **83**, 121–128.
20. C. K. Kuo, W. J. Li, R. L. Mauck and R. S. Tuan, *Curr. Opin. Rheumatol.*, 2006, **18**, 64–73.
21. M. Nawata, S. Wakitani, H. Nakaya, A. Tanigami, T. Seki, Y. Nakamura, N. Saito, K. Sano, E. Hidaka and K. Takaoka, *Arthritis Rheum.*, 2005, **52**, 155–163.

22. K. Potter, J. J. Butler, W. E. Horton and R. G. Spencer, *Arthritis Rheum.*, 2000, **43**, 1580–1590.
23. D. Nesic, R. Whiteside, M. Brittberg, D. Wendt, I. Martin and P. Mainil-Varlet, *Adv. Drug Deliv. Rev.*, 2006, **58**, 300–322.
24. P. J. Geutjes, W. F. Daamen, P. Buma, W. F. Feitz, K. A. Faraj and T. H. van Kuppevelt, *Adv. Exp. Med. Biol.*, 2006, **585**, 279–295.
25. J. Elisseeff, C. Puleo, F. Yang and B. Sharma, *Orthod. Craniofac. Res.*, 2005, **8**, 150–161.
26. C. T. Chen, K. W. Fishbein, P. A. Torzilli, A. Hilger, R. G. Spencer and W. E. Horton, Jr., *Arthritis Rheum.*, 2003, **48**, 1047–1056.
27. E. F. Petersen, K. W. Fishbein, E. W. McFarland and R. G. Spencer, *Magn. Reson. Med.*, 2000, **44**, 367–372.
28. K. Potter, J. J. Butler, C. Adams, K. W. Fishbein, E. W. McFarland, W. E. Horton and R. G. Spencer, *Matrix Biol.*, 1998, **17**, 513–523.
29. M. Pei, L. A. Solchaga, J. Seidel, L. Zeng, G. Vunjak-Novakovic, A. I. Caplan and L. E. Freed, *Faseb J.*, 2002, **16**, 1691–1694.
30. J. O. Seidel, M. Pei, M. L. Gray, R. Langer, L. E. Freed and G. Vunjak-Novakovic, *Biorheology*, 2004, **41**, 445–458.
31. G. Vunjak-Novakovic, B. Obradovic, I. Martin and L. E. Freed, *Biorheology*, 2002, **39**, 259–268.
32. D. Wendt, M. Jakob and I. Martin, *J. Biosci. Bioeng.*, 2005, **100**, 489–494.
33. C. J. Hunter, J. K. Mouw and M. E. Levenston, *Osteoarthr. Cartilage*, 2004, **12**, 117–130.
34. S. D. Waldman, C. G. Spiteri, M. D. Grynpas, R. M. Pilliar and R. A. Kandel, *Tissue Eng.*, 2004, **10**, 1323–1331.
35. S. D. Waldman, C. G. Spiteri, M. D. Grynpas, R. M. Pilliar and R. A. Kandel, *J Orthop. Res.*, 2003, **21**, 590–596.
36. S. D. Waldman, D. C. Couto, M. D. Grynpas, R. M. Pilliar and R. A. Kandel, *Osteoarthr. Cartilage*, 2006, **14**, 323–330.
37. M. J. Stoddart, L. Ettinger and H. J. Hauselmann, *Biotechnol. Bioeng.*, 2006, **95**, 1043–1051.
38. R. L. Mauck, S. B. Nicoll, S. L. Seyhan, G. A. Ateshian and C. T. Hung, *Tissue Eng.*, 2003, **9**, 597–611.
39. E. Petersen, K. Potter, J. Butler, K. W. Fishbein, W. Horton, R. G. S. Spencer and E. W. McFarland, *Int. J. Imag. Syst. Tech.*, 1997, **8**, 285–292.
40. J. L. Evelhoch, R. J. Gillies, G. S. Karczmar, J. A. Koutcher, R. J. Maxwell, O. Nalcioglu, N. Raghunand, S. M. Ronen, B. D. Ross and H. M. Swartz, *Neoplasia*, 2000, **2**, 152–165.
41. R. J. Gillies, J. P. Galons, K. A. McGovern, P. G. Scherer, Y. H. Lien, C. Job, R. Ratcliff, F. Chapa, S. Cerdan and B. E. Dale, *NMR Biomed.*, 1993, **6**, 95–104.
42. U. Pilatus, E. Ackerstaff, D. Artemov, N. Mori, R. J. Gillies and Z. M. Bhujwalla, *Neoplasia*, 2000, **2**, 273–279.
43. J. M. MacDonald, M. Grillo, O. Schmidlin, D. T. Tajiri and T. L. James, *NMR Biomed.*, 1998, **11**, 55–66.

44. J. M. MacDonald, S. P. Wolfe, I. Roy-Chowdhury, H. Kubota and L. M. Reid, *Ann. NY Acad. Sci.*, 2001, **944**, 334–343.
45. S. P. Wolfe, E. Hsu, L. M. Reid and J. M. MacDonald, *Biotechnol. Bioeng.*, 2002, **77**, 83–90.
46. N. J. Hickok, A. R. Haas and R. S. Tuan, *Microsc. Res. Techniq.*, 1998, **43**, 174–190.
47. K. Potter, J. J. Butler, C. Adams, K. W. Fishbein, E. W. McFarland, W. E. Horton and R. G. S. Spencer, *Matrix Biol.*, 1998, **17**, 513–523.
48. M. Schnabel, S. Marlovits, G. Eckhoff, I. Fichtel, L. Gotzen, V. Vecsei and J. Schlegel, *Osteoarthr. Cartilage*, 2002, **10**, 62–70.
49. K. Potter, K. W. Fishbein, W. E. Horton and R. G. Spencer, in *Spatially Resolved Magnetic Resonance*, ed. P. Blümler, B. Blümlich, R. Botto and E. Fukushima, Wiley-VCH Press, 1998, pp. 363–371.
50. C. T. Chen, K. W. Fishbein, P. A. Torzilli, A. Hilger, R. G. S. Spencer and W. E. Horton, *Arthritis Rheum.*, 2003, **48**, 1047–1056.
51. K. Potter, J. J. Butler, W. E. Horton and R. G. S. Spencer, *Arthritis Rheum.*, 2000, **43**, 1580–1590.
52. A. Bashir, M. L. Gray and D. Burstein, *Magn. Reson. Med.*, 1996, **36**, 665–673.
53. B. J. Kvam, E. Fragonas, A. Degrassi, C. Kvam, M. Matulova, P. Pollesello, F. Zanetti and F. Vittur, *Exp. Cell Res.*, 1995, **218**, 79–86.
54. F. Vittur, M. Grandolfo, E. Fragonas, C. Godeas, S. Paoletti, P. Pollesello, B. J. Kvam, F. Ruzzier, T. Starc, J. W. Mozrzymas, M. Martina and B. Debernard, *Exp. Cell Res.*, 1994, **210**, 130–136.
55. R. E. Wuthier, *J. Nutr.*, 1993, **123**, 301–309.
56. E. F. Petersen, K. W. Fishbein, E. W. McFarland and R. G. S. Spencer, *Magn. Reson. Med.*, 2000, **44**, 367–372.
57. K. Johnson, A. Jung, A. Murphy, A. Andreyev, J. Dykens and R. Terkeltaub, *Arthritis Rheum.*, 2000, **43**, 1560–1570.
58. S. S. Velan, R. G. S. Spencer, J. L. Zweier and P. Kuppusamy, *Magn. Reson. Med.*, 2000, **43**, 804–809.
59. S. J. Ellis, M. Velayutham, S. S. Velan, E. F. Petersen, J. L. Zweier, P. Kuppusamy and R. G. S. Spencer, *Magn. Reson. Med.*, 2001, **46**, 819–826.
60. K. Potter, L. H. Kidder, I. W. Levin, E. N. Lewis and R. G. Spencer, *Arthritis Rheum.*, 2001, **44**, 846–855.
61. M. Kim, X. Bi, W. E. Horton, R. G. Spencer and N. P. Camacho, *J. Biomed. Opt.*, 2005, **10**, 031105.
62. X. Bi, X. Yang, M. P. Bostrom, D. Bartusik, S. Ramaswamy, K. W. Fishbein, R. G. Spencer and N. P. Camacho, *Anal. Bioanal. Chem.*, 2007, **387**(5), 1601.
63. A. Boskey and N. Pleshko Camacho, *Biomaterials*, 2007, **28**, 2465–2478.
64. D. Baykal, O. Irrechukwu, P. C. Lin, K. Fritton, R. G. Spencer and N. Pleshko, *Appl. Spectrosc.*, 2010, **64**, 1160–1166.
65. V. C. Hascall, C. J. Handley, D. J. McQuillan, G. K. Hascall, H. C. Robinson and D. A. Lowther, *Arch. Biochem. Biophys.*, 1983, **224**, 206–223.

66. W. G. Li, L. Hong, L. P. Hu and R. L. Magin, *Tissue Eng. Pt. C-Meth.*, 2010, **16**, 1407–1415.
67. A. E. Nugent, D. A. Reiter, K. W. Fishbein, D. L. McBurney, T. Murray, D. Bartusik, S. Ramaswamy, R. G. Spencer and W. E. Horton, *Tissue Eng. Pt. A*, 2010, **16**, 2183–2196.
68. D. A. Reiter, K. W. Fishbein, S. Leen, A. E. Nugent, W. E. Horton and R. G. Spencer, Orthopaedic Research Society 54th Annual Meeting, San Francisco, CA, 2008.
69. A. J. Grodzinsky, M. E. Levenston, M. Jin and E. H. Frank, *Ann. Rev. Biomed. Eng.*, 2000, **2**, 691–713.
70. Z. J. Zhang, J. Huckle, C. A. Francomano and R. G. Spencer, *Ultrasound Med. Biol.*, 2003, **29**, 1645–1651.
71. J. Parvizi, C. C. Wu, D. G. Lewallen, J. F. Greenleaf and M. E. Bolander, *J. Orthop. Res.*, 1999, **17**, 488–494.
72. K. H. Yang, J. Parvizi, S. J. Wang, D. G. Lewallen, R. R. Kinnick, J. F. Greenleaf and M. E. Bolander, *J. Orthop. Res.*, 1996, **14**, 802–809.
73. T. Iwashina, J. Mochida, T. Miyazaki, T. Watanabe, S. Iwabuchi, K. Ando, T. Hotta and D. Sakai, *Biomaterials*, 2006, **27**, 354–361.
74. Z. J. Zhang, J. Huckle, C. A. Francomano and R. G. Spencer, *Ultrasound Med. Biol.*, 2002, **28**, 1547–1553.
75. B. H. Min, J. I. Woo, H. S. Cho, B. H. Choi, S. J. Park, M. J. Choi and S. R. Park, *Scand. J. Rheumatol.*, 2006, **35**, 305–311.
76. K. Miyamoto, H. S. An, R. L. Sah, K. Akeda, M. Okuma, L. Otten, E. J. Thonar and K. Masuda, *Spine*, 2005, **30**, 2398–2405.
77. B. H. Choi, J. I. Woo, B. H. Min and S. R. Park, *J. Biomed. Mater. Res. A*, 2006, **79**, 858–864.
78. K. J. Gooch, T. Blunk, D. L. Courter, A. L. Sieminski, P. M. Bursac, G. Vunjak-Novakovic and L. E. Freed, *Biochem. Biophys. Res. Commun.*, 2001, **286**, 909–915.
79. M. H. Huang, H. J. Ding, C. Y. Chai, Y. F. Huang and R. C. Yang, *J. Rheumatol.*, 1997, **24**, 1978–1984.
80. M. H. Huang, R. C. Yang, H. J. Ding and C. Y. Chai, *Arch. Phys. Med. Rehabil.*, 1999, **80**, 551–556.
81. R. Bhatia, V. K. Sobti and K. S. Roy, *Zentralbl. Veterinarmed [A] (J. Vet. Med. A)*, 1992, **39**, 168–173.
82. K. I. Singh, V. K. Sobti and K. S. Roy, *J. Equine Vet. Sci.*, 1997, **17**, 150–155.
83. S. D. Cook, S. L. Salkeld, L. S. Popich-Patron, J. P. Ryaby, D. G. Jones and R. L. Barrack, *Clin. Orthop.*, 2001, **391**, Suppl., S231.
84. S. R. Park, S. H. Park, K. W. Jang, H. S. Cho, J. H. Cui, H. J. An, M. J. Choi, S. I. Chung and B. H. Min, *Ultrasound Med. Biol.*, 2005, **31**, 1559–1566.
85. A. M. Bendele and J. F. Hulman, *Arthritis Rheum.*, 1988, **31**, 561–565.
86. I. Gurkan, A. Ranganathan, X. Yang, W. E. Horton, M. Todman, J. Huckle, N. Pleshko and R. G. Spencer, *Osteoarthr. Cartilage*, **18**, 724–733.

87. O. N. Irrechukwu, P. C. Lin, K. Fritton, S. Doty, N. Pleshko and R. G. Spencer, *Tissue Eng. Pt. A*, 2011, **17**, 407–415.
88. J. Elisseeff, W. McIntosh, K. Anseth, S. Riley, P. Ragan and R. Langer, *J. Biomed. Mater. Res.*, 2000, **51**, 164–171.
89. S. Ramaswamy, M. C. Uluer, S. Leen, P. Bajaj, K. W. Fishbein and R. G. Spencer, *Tissue Eng. Pt. C-Meth.*, 2008, **14**, 243–249.
90. S. Ramaswamy, J. B. Greco, M. C. Uluer, Z. J. Zhang, Z. L. Zhang, K. W. Fishbein and R. G. Spencer, *Tissue Eng. Pt. A*, 2009, **15**, 3899–3910.
91. Z. J. Zhang, J. M. McCaffery, R. G. S. Spencer and C. A. Francomano, *J. Anat.*, 2004, **205**, 229–237.
92. Z. J. Zhang, J. M. McCaffery, R. G. S. Spencer and C. A. Francomano, *J. Orthopaed. Res.*, 2005, **23**, 433–439.
93. S. Ramaswamy, D. A. Wang, K. W. Fishbein, J. H. Elisseeff and R. G. Spencer, *J. Biomed. Mater. Res. B*, 2006, **77B**, 144–148.
94. S. Ramaswamy, L. Gurkan, B. Sharma, B. Cascio, K. W. Fishbein and R. G. Spencer, *J. Biomed. Mater. Res. B*, 2008, **86B**, 375–380.
95. K. P. Whittall and A. L. Mackay, *J. Magn. Reson.*, 1989, **84**, 134–152.
96. D. A. Reiter, P. C. Lin, K. W. Fishbein and R. G. Spencer, *Magn. Reson. Med.*, 2009, **61**, 803–809.
97. D. A. Reiter, R. A. Roque, P. C. Lin, S. B. Doty, N. Pleshko and R. G. Spencer, *NMR Biomed.*, 2011, **24**, 1286–1294.
98. D. A. Reiter, R. A. Roque, P. C. Lin, O. Irrechukwu, S. Doty, D. L. Longo, N. Pleshko and R. G. Spencer, *Magn. Reson. Med.*, 2011, **65**, 377–384.
99. E. M. Shapiro, A. Borthakur, A. Gougoutas and R. Reddy, *Magn. Reson. Med.*, 2002, **47**, 284–291.
100. W. Ling, R. R. Regatte, G. Navon and A. Jerschow, *Proc. Natl Acad. Sci. USA*, 2008, **105**, 2266–2270.
101. L. M. Lesperance, M. L. Gray and D. Burstein, *J. Orthop. Res.*, 1992, **10**, 1–13.
102. O. N. Irrechukwu, D. A. Reiter, P. C. Lin, R. A. Roque, K. W. Fishbein and R. G. Spencer, *Tissue Eng. Pt. C-Meth.*, 2012, **18**, 433–443.
103. R. L. Magin, W. G. Li, M. P. Velasco, J. Trujillo, D. A. Reiter, A. Morgenstern and R. G. Spencer, *J. Magn. Reson.*, 2011, **210**, 184–191.
104. R. L. Magin, O. Abdullah, D. Baleanu and X. J. Zhou, *J. Magn. Reson.*, 2008, **190**, 255–270.
105. G. E. Gold and C. F. Beaulieu, *Semin. Musculoskel. R.*, 2001, **5**, 313–327.
106. D. Laurent, J. Wasvary, J. Y. Yin, M. Rudin, T. C. Pellas and E. O'Byrne, *Magn. Reson. Imag.*, 2001, **19**, 1279–1286.
107. N. M. Menezes, M. L. Gray, J. R. Hartke and D. Burstein, *Magn. Reson. Med.*, 2004, **51**, 503–509.
108. V. Mlynarik, I. Sulzbacher, M. Bittsansky, R. Fuiko and S. Trattnig, *J. Magn. Reson. Imag.*, 2003, **17**, 440–444.
109. T. J. Mosher and B. J. Dardzinski, *Semin. Musculoskel. R.*, 2004, **8**, 335–368.

110. P. C. Lin, D. A. Reiter and R. G. Spencer, *Magn. Reson. Med.*, 2009, **62**, 1311–1318.

111. D. A. Reiter, O. Irrechukwu, P. C. Lin, S. Moghadam, S. Von Thaer, N. Pleshko and R. G. Spencer, *NMR Biomed.*, 2012, **25**, 476–488.

112. P. C. Lin, O. Irrechukwu, R. Roque, B. Hancock, K. W. Fishbein and R. G. Spencer, *Magn. Reson. Med.*, 2012, **67**, 1815–1826.

113. P. C. Lin, D. A. Reiter and R. G. Spencer, *J. Magn. Reson.*, 2009, **201**, 61–71.

114. C. Fraley and A. Raftery, *MCLUST Version 3 for R: Normal Mixture Modeling and Model-based Clustering*, Department of Statistics, University of Washington, 2006.

115. T. Hastie, The Elements of Statistical Learning, Springer, 2001.

CHAPTER 15

Applications of Magnetic Resonance Spectroscopy to Psychiatric Disorders

RICHARD A. KOMOROSKI

Center for Imaging Research and Departments of Psychiatry & Behavioral Neuroscience and Biomedical Engineering, College of Medicine, University of Cincinnati, 231 Albert Sabin Way, Cincinnati, Ohio 45267-0583, USA
Email: richard.komoroski@uc.edu

15.1 Introduction

15.1.1 Scope

This chapter presents a brief overview on the use of magnetic resonance spectroscopy (MRS) to study psychiatric or mental disorders. These disorders are described categorically according to symptomatology in the *Diagnostic and Statistical Manual of Mental Disorders*, 4th edn, Text Revision (DSM-IV-TR).[1] The presentation given here reflects the current view of the author as to the status of MRS in psychiatric research, and is not intended to be a comprehensive review. Special reference is sometimes made to studies from the author's laboratory. Of necessity, the focus is primarily on *in vivo* studies in humans, although some consideration is given to animal models and *ex vivo* studies of brain tissue. In this regard, we use the term "MRS" to refer to *in vivo* studies, while retaining the term "NMR" for *ex vivo* and *in vitro* studies. Here we do not consider the related body of NMR work in peripheral cells such as erythrocytes from psychiatric patients.[2] The large literature on psychiatric

New Developments in NMR No. 2
New Applications of NMR in Drug Discovery and Development
Edited by Leoncio Garrido and Nicolau Beckmann
© The Royal Society of Chemistry 2013
Published by the Royal Society of Chemistry, www.rsc.org

applications of MRI-based techniques such as standard anatomic MRI, functional MRI (fMRI) and diffusion tensor imaging (DTI) is not addressed here, although all MRS applications *in vivo* utilize MRI spatial localization, and these techniques are often performed with, and may impact, MRS studies. A number of general reviews of psychiatric applications of MRS has appeared in recent years,[3–8] with two explicitly oriented toward drug development. Where appropriate, we make mention of considerations pertinent to pharmacologic treatment or drug analysis. Many disease-specific reviews are referenced later in the chapter.

15.1.2 *In Vivo* MRS in the Study of Psychiatric Disorders

Within biomedicine, psychiatric illnesses are in many ways unique. Although much has been learned over the last 60 years, the etiologies of psychiatric brain disorders, particularly at the molecular level, are not known. It is now universally accepted that psychiatric illnesses are brain disorders. But even broad characterizations, such as whether schizophrenia is primarily neurodevelopmental or neurodegenerative in origin, can be controversial. Because psychiatric illnesses largely involve mind, thought and cognition, they are uniquely human. That implies that much of our research in this area must be conducted and ultimately validated directly in the brain on humans, perhaps with less reliance on animal and *in vitro* approaches than for diseases such as cancer (see Section 15.1.3 below). The safe, non-invasive nature of NMR-based techniques, including MRS, makes them very attractive in this regard. Thus MRS studies of brain biochemistry, in comparison to normal subjects, have some potential to elucidate the molecular origins of psychiatric disorders. The non-invasive nature of MRS also makes it attractive for a variety of longitudinal and treatment studies. Because psychiatric illnesses are usually chronic and often follow a degenerating course, longitudinal MRS studies of the developing illness may shed light on its nature and trajectory. Longitudinal studies of the effects of treatment may provide a window into understanding drug response and efficacy. While the information content of MRS is certainly limited, much of what it provides is fundamental to basic brain biochemistry and neurotransmission (see Section 15.2 below).

Given the inherent difficulties in diagnosing mental disorders and the considerable overlap among many of their classifications in the DSM-IV-TR,[1] there has been much research directed toward identification of specific biomarkers, such as clinical chemistry measures, for individual illnesses.[9] However, no definitive biomarker has been found for any mental illness. Efforts have now moved to the realm of neuroimaging. Because MRS provides quantitative estimates of the concentrations of important metabolites in well-defined brain regions, it may yield biomarkers for psychiatric disorders and/or their time course. For example, in bipolar disorders (BPD) certain metabolite levels may be sensitive to whether the subject is in the manic, euthymic or depressed state. Importantly, longitudinal MRS may be a useful tool to follow treatment with drugs or other procedures such as electroconvulsive therapy and transcranial magnetic stimulation. Initial response to treatment can be slow, even if later adequate. Often several treatment courses are required to achieve acceptable response. Identifying reliable early markers of treatment response could dramatically improve treatment assignment and generate better response rates.

Several practical problems arise for psychiatric MRS studies. As is the case for all studies on human subjects, ethical considerations and clinical necessity limit the flexibility of treatment options and other interventions for study subjects. As might be expected, severely ill patients are often difficult to study by MR techniques, particularly at first appearance of the illness, or possibly at some later time when off medication. While medicated subjects are often studied, the variety of medications, doses, treatment durations and treatment responses can, unless closely controlled, provide significant confounds to the interpretation of results. As will be evident later and as is usually true for anatomic findings from MRI, group differences in MRS measures between psychiatric patients and normal subjects, when detected, are usually small. Coupled with the less-than-ideal precision of current *in vivo* MRS technology, this necessitates relatively large numbers of subjects for results to achieve statistical significance. Correlation of MRS results with clinical measures is further complicated by the inherent inadequacies of the psychiatric measures themselves.

15.1.3 Animal Models and *Ex Vivo* Studies

Because of the limitations involved with studying psychiatric subjects *in vivo*, whether by invasive or non-invasive techniques, much research has been directed at development of animal models. In general, animal models provide a powerful approach for the study of human illnesses. Wide varieties of animal models, typically in rodents, have been developed for psychiatric illnesses. However, because these disorders are uniquely human and their etiology is largely unknown, all animal models of these disorders are in some way problematic and limited.[10,11] Animal models are often designed to mimic one or a few behavioral aspects of an illness, but usually cannot reproduce the human symptoms (face validity) in sufficient detail. Animal models can be used in the development of psychoactive drugs. However, models are often validated on their response to known psychoactive drugs (predictive validity) in somewhat circular fashion. Without knowledge of the molecular etiology of a disorder (construct validity), animal models will remain of limited utility for psychiatric disorders for the foreseeable future. It is hoped that genetic approaches based on human findings will produce better animal models.[10]

Notwithstanding the above, animal studies retain a well-established place in psychiatric research. Within limitations imposed by the smaller brain size, *in vivo* MRS studies, which are very similar to those performed in humans, can be performed in animal brain, but with much more interventional flexibility. Some MR techniques are developed more readily in animals. Importantly, the ability to sample identical measures in parallel studies in humans and animals may serve as a useful tool for drug development. After *in vivo* sampling, the animal brains can be studied using the various detailed but invasive techniques of neuroscience. An example of MR technique development pertinent to psychoactive drug action and translatable to humans is our work on the environment of lithium (Li) in rat brain using ^{7}Li MRS (see Section 15.2.4 below).[12,13]

Studies that probe the etiology of psychiatric disorders often rely on *post mortem* tissue. Although the study of *post mortem* brain tissue has a long history in psychiatry, many challenges to its use remain.[14,15] MR techniques are seldom employed because most neurotransmitters, macromolecules and other metabolites of interest are not readily studied by high-resolution NMR *in situ* for reasons of either insufficient concentration or restricted molecular mobility. Even for MR-metabolites, other analytical techniques are faster and more inexpensive than NMR. Also, many interesting metabolites are visible *in vivo*, which is obviously the preferred MRS approach. However, the ability of NMR to detect many metabolites simultaneously and quantitatively occasionally gives it an advantage over chromatographic approaches. An example of a psychiatric application using *post mortem* tissue that is not possible *in vivo* is our ^{31}P NMR comparison of the phospholipid compositions of key anatomic regions of schizophrenic and normal brain.[16] Although phospholipid metabolites can be observed *in vivo* (see Section 15.2.2 below), the phospholipids themselves cannot, and must be studied after extraction from *post mortem* brain.

15.2   Information Content of *In Vivo* MRS of Brain

Here we do not review the many approaches to *in vivo* MRS of the brain. We refer the reader to the excellent text on the subject[17] and to Chapter 8 as well. Rather, we present the major features of the MR spectroscopy of the several nuclei pertinent to brain research and psychiatric illness.

15.2.1 ^{1}H MRS

In most cases precise spatial localization to the tissue region of interest is critical for informative *in vivo* MRS. The overwhelming majority of MRS studies in psychiatry employ ^{1}H. For ^{1}H MRS a wide variety of methods has been used to achieve either single-voxel or multi-voxel localization.[17] In the typical implementation of *in vivo* ^{1}H MRS, after acquisition of a localizer MRI, a rectangular voxel is defined in the brain region of interest. The spectrum is then acquired using one of several radiofrequency-pulse sequences to excite and detect MRS signals from only the voxel of interest, typically 1–8 cm^3. Common to ^{1}H MRS are single-voxel sequences such as point-resolved spectroscopy (PRESS) and stimulated-echo acquisition-mode (STEAM) spectroscopy.[17] The large signal from water is suppressed using variations of standard techniques. Figure 15.1 shows the ^{1}H MR spectrum from the left ventrolateral prefrontal cortex of a patient with BPD. The compounds that are most readily detected and measured are labeled. *N*-acetyl aspartate (NAA), the second-most

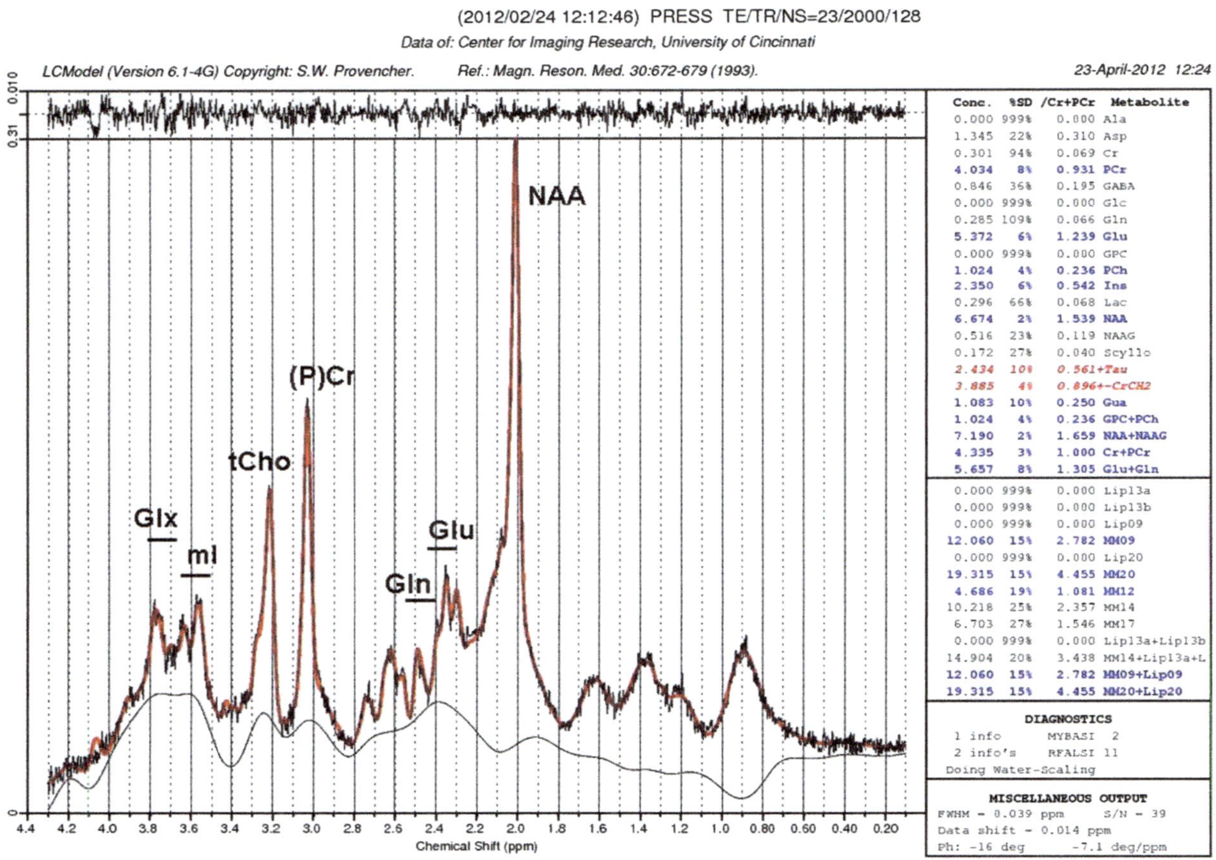

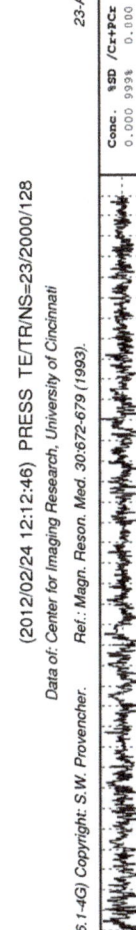

Figure 15.1 A PRESS-localized ^{1}H MR spectrum at 4 T from a subject with bipolar disorder, showing the typical analysis output from the computer program LCModel.[24] Major resonances of interest are labeled and described in the text. The region labeled Glx arises from both Glu and Gln. The red curve is the fit to the experimental spectrum; the difference between the fit and the experimental spectrum is shown at the top. Metabolite concentrations relative to tissue water with standard deviations (Cramer–Rao lower bounds) are given to the right.

abundant amino acid in the brain and typically the largest peak in the spectrum, has been proposed to function in many metabolic roles.[18] In the adult brain it occurs only in neurons, and is generally regarded as a measure of neuronal density and/or health.[19,20] For this reason NAA is the metabolite most studied in brain by *in vivo* MRS. The peak labeled "(P)Cr" arises primarily from creatine (Cr) and phosphocreatine (PCr), compounds intimately involved in cellular energy metabolism. It has often been used as an internal standard in quantitative studies, since the concentration of Cr + PCr was assumed to be relatively unchanging, an assumption that must be applied with caution.[21,22]

The peak labeled "tCho" arises from all molecularly mobile choline-containing compounds, such as choline (Cho) itself, phosphocholine (PC), glycerophosphocholine (GPC) and (as a very small percentage) the neuro-transmitter acetylcholine. Choline moieties in phospholipids such as phos-phatidylcholine in lipid bilayers do not contribute due to their restricted molecular mobility. Cho, PC and GPC are intimately involved in phospholipid metabolism, which is often thought to be aberrant in psychiatric illness.[16,23] The peaks labeled "mI" arise from *myo*-inositol, a compound involved in second-messenger metabolism, also thought to be involved with psychiatric illness.[23] Two important amino acids that appear in the 2.0–2.5-ppm region are glutamate (Glu), the primary excitatory neurotransmitter in the brain, and glutamine (Gln), an important metabolite synthesized from Glu in astrocytes and involved in Glu-Gln cycling between neurons and astrocytes. At fields lower than about 4 T, these latter two compounds are not resolved in the 2.0–2.5-ppm region, and are quantified together and referred to as "Glx". While there are several approaches to quantifying metabolites in an *in vivo* ^{1}H MR spectrum, a standard approach is to fit the spectrum to the sum of the most common metabolites known to contribute, using a computer program such as LCModel.[24] A typical LCModel analysis is shown in Figure 15.1.

The dominant metabolites in the spectrum typically occur in the 4–10 mM range. Several interesting metabolites occurring at about the 1 mM level are obscured in the spectrum. Using spectral editing techniques to remove inter-ference from the dominant resonances,[17] γ-aminobutyric acid (GABA), the primary inhibitory neurotransmitter in the brain, and lactate (Lac), a product of anaerobic metabolism, can be measured. Both of these compounds are of keen interest for psychiatric studies.[23,25,26]

15.2.2 ^{31}P MRS

A recent general review of *in vivo* ^{31}P MRS with some coverage of psychiatric applications has appeared.[27] Phosphorus-31 MRS is an excellent technique for the study of *in vivo* brain metabolism.[17] It provides information on a different group of compounds from ^{1}H MRS. Although not as sensitive as ^{1}H MRS, ^{31}P MRS does not require water suppression and has a substantially larger chemical-shift range. Because of the short T_2s and large spectral width relative to ^{1}H MRS, the choices are fewer for localization of ^{31}P MRS in practice.

Image-selected *in vivo* spectroscopy (ISIS), often combined with outer-volume suppression, remains the single-voxel technique of choice for ^{31}P MRS.[17] Pulse-acquire 4D MRSI (MR spectroscopic imaging) is well suited for multivoxel acquisition of ^{31}P MR spectra of brain. High magnetic fields have allowed substantially reduced voxel sizes relative to early, low-field studies. Actual voxel sizes of about 15 cm^3 can be expected at 4 T,[28] whereas 6–8 cm^3 can be achieved at 7 T.[29]

The technique provides information on two classes of compounds – high-energy-phosphorus metabolites and phospholipid metabolites. The key high-energy phosphorus metabolites adenosine triphosphate (ATP) and PCr are readily measured. These are undoubtedly of fundamental interest for studies of brain energy metabolism in psychiatric disorders, as mitochondrial dysfunction has been proposed to underlie schizophrenia and BPD.[23,30] The phosphomono-esters (PMEs) PC and phosphoethanolamine (PEth), which are precursors to the derivative phospholipids, are observable either separately (at high field or with ^{1}H decoupling) or in combination, as are the phosphodiester (PDE) degradation products of phospholipid metabolism, GPC and glycerophospho-ethanolamine (GPEth). There is considerable evidence that phospholipid metabolism is abnormal in some psychiatric disorders.[16,23,31] Inorganic phosphate (P$_i$), a degradation product of all biological phosphorus compounds, is also observed in the ^{31}P spectrum. Estimates of intracellular pH and intra-cellular Mg^{2+} concentration can be made from the chemical shift separations of P$_i$ and PCr, and β-ATP and PCr, respectively.[32,33] Figure 15.2 shows a ^{31}P spectrum at 4 T of an MRSI voxel from the brain of a schizophrenic subject.[34]

15.2.3 ^{13}C MRS

Localized ^{13}C MRS is seeing increasing use in biomedicine and in the brain *in vivo*.[35–37] Applications to humans are increasing, but are virtually non-existent for psychiatric illness. Although natural-abundance ^{13}C spectra can be acquired for brain, the power of the technique is only realized with adminis-tration of compounds isotopically enriched in the ^{13}C isotope or possibly with sensitivity enhancement *via* hyperpolarization techniques (see also Chapter 9).[38] Administration of 1-^{13}C-glucose makes it possible to follow the fate of the isotopic label as it enters the tricarboxylic-acid (TCA) cycle through pyruvate and eventually is incorporated into MRS-visible metabolites such as Glu, Gln, aspartate and NAA. The major advantage of ^{13}C MRS *in vivo* is the ability to measure net synthesis rates of key metabolites. These rates may be more sensitive to disease processes than steady-state metabolite concentrations. In addition to the need for multinuclear capability on the MRI scanner, the disadvantages of ^{13}C MRS are the need for intravenous infusion of large quantities (5–20 g) of the labeled compound, which is very expensive.

The approach has been applied preliminarily to schizophrenia by the Ross group.[39,40] They found decreased incorporation of ^{13}C label into Glu, as well as a slightly increased synthesis rate for NAA in schizophrenia relative to normal controls.[39,40]

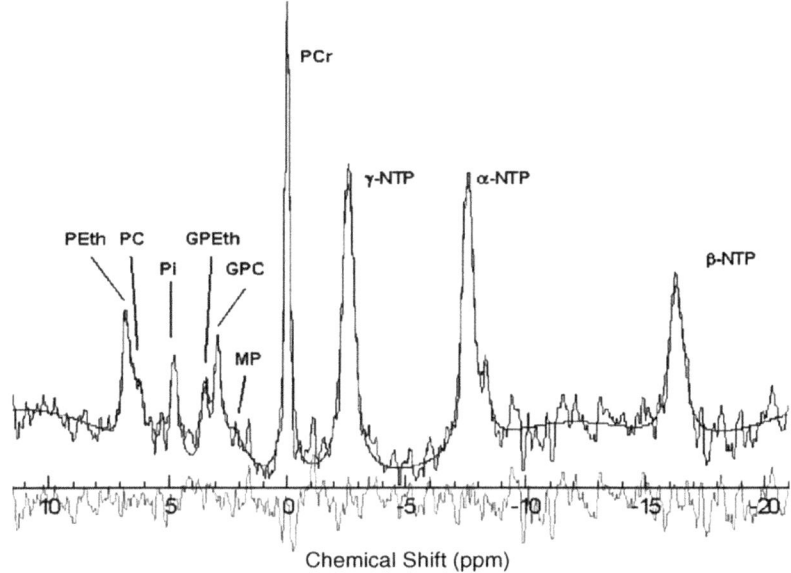

Chemical Shift (ppm)

Figure 15.2 A ^{31}P MRSI spectrum from a 15-cm^3 voxel in the brain of a schizophrenic
subject, acquired at 4 T. PEth, phosphoethanolamine; PC, phosphocholine;
Pi, inorganic phosphate; GPEth, glycerophosphoethanolamine;
GPC, glycerophosphocholine; MP, mobile phospholipids; PCr, phosphocreatine; α-,β-,γ-NTP, α-,β-,γ-phosphorus of nucleoside triphos-
phates, respectively.
Reprinted with permission from Jensen et al.[34] © 2006 Elsevier Ltd.

15.2.4 Direct Detection of Psychoactive Drugs by ^{19}F and ^{7}Li MRS

A possible alternative to studying endogenous metabolites by MRS is to
monitor directly the drugs used in the treatment of psychiatric illnesses.[41] The
overwhelming majority of psychotropic drugs are at too low a concentration in
the brain during normal therapy for direct detection by ^{1}H MRS *in vivo* against
the background signal of endogenous metabolites (Figure 15.1).[42]

However, many drugs, including some used in psychiatry, contain fluorine,[43]
an element with a sensitive spin-1/2 MR isotope (^{19}F, with 83.8% of the
sensitivity of ^{1}H) and no endogenous background signal. Fluorine-19 MRS has
been used in a variety of applications in biomedicine.[44] Unfortunately, even
though drugs like fluoxetine (Prozac®, used to treat major depression and other
conditions) accumulate to *relatively* high levels in the brain,[43] the ^{19}F signal
strength is still usually too low for usable, spatially localized MRS at 1.5 T.[45]
Currently the only potential application of ^{19}F MRS in psychiatry is the estimation of whole-brain drug concentration during treatment. The utilization of
high-field magnets and phased-array coils may provide sufficient signal-to-noise
ratio (SNR) for localization to major brain regions.

The unique psychoactive drug that achieves a relatively high concentration (~ 0.5 to 1 mM) in the brain is the Li cation, administered orally as Li_2CO_3, which is used to treat acute mania and for prophylaxis in BPD.[46] Because ^7Li ($I = 3/2$) has an MR sensitivity that is 27.2% of ^1H, relatively narrow lines for a quadrupolar nucleus and no detectable background signal, it is straightforward to detect in brain *in vivo* under normal treatment. Applications and technique developments have been reviewed.[46] Most published studies have focused on technique development or correlation of brain and serum Li levels in both humans and animals.[46,12,13] The major application so far has been the measurement of whole-brain Li. Very little work has dealt with correlation between brain Li levels and treatment measures. However, as an indicator of the technique's potential at high field, Lee *et al.*[47] recently demonstrated spatial variations of Li concentration in brain for bipolar patients using 4D ^7Li MRSI at 4 T. Figure 15.3 shows slices, presented as overlays on a ^1H anatomic MRI, from the ^7Li MRSI of a bipolar patient.

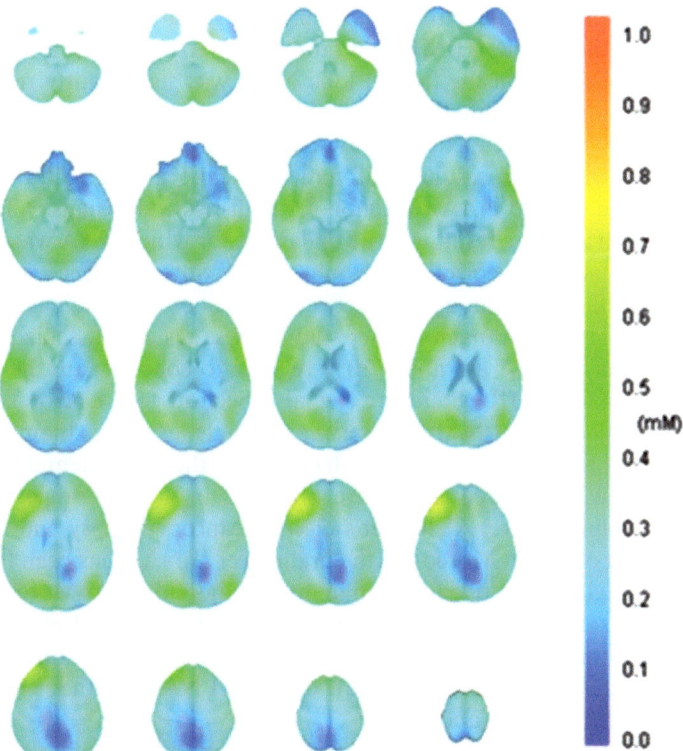

Figure 15.3   A 4D ^7Li MR spectroscopic image at 4 T of a bipolar patient, presented as an interpolated overlay on an average ^1H T_1 anatomical MRI from the International Consortium for Brain Mapping 452 T_1 Atlas. Lithium concentration scale is at right. Average brain Li concentration, 0.30 mM; maximum brain Li concentration, 0.74 mM; serum Li concentration, 0.80 mM. See reference 47 for data-processing details.

15.3   Applications to Specific Disorders

Space does not permit in-depth coverage of MRS applications for the many psychiatric disorders. Recent focused reviews for individual disorders are referenced in each subsection below. These reviews should be consulted for the many details that cannot be covered here. Rather, we very briefly present the current status of MRS studies of the three major psychiatric illnesses and mention others that have garnered significant attention with MRS.

15.3.1   Schizophrenia

Several reviews on the application of ^{1}H MRS techniques specifically to schizophrenia have recently appeared.[18,26,48–50] These reviews have largely or exclusively focused on NAA, the metabolite of interest in the overwhelming majority of all ^{1}H MRS studies of brain.[48–50] Although inconsistencies abound concerning the MRS technique and data-processing procedures among studies, as well as the ranges of patient and brain-region selection, two meta-analyses[48,50] of the usable ^{1}H-MRS literature on NAA revealed a number of important findings.[48,50] In the frontal lobe NAA was significantly reduced in both chronic (long-medicated) and first-episode schizophrenics relative to controls. No significant reduction in NAA was found in high-risk individuals (*e.g.* siblings).[50] A similar result was found for temporal lobe, although here there was a trend toward NAA reduction for high-risk individuals.[50] The only other region with a strong result was the thalamus, for which NAA was reduced in both chronic and first-episode schizophrenics.[50] Some evidence was found for medication effects,[50] although such effects, as well as longitudinal studies of disease progression, merit much further study. These results for schizophrenia, where reductions in NAA are about 10% relative to controls, are the clearest within the psychiatric MRS literature (see also Section 15.4.1 below). Unfortunately, it is unlikely that NAA can serve as a single, exclusive biomarker for schizophrenia or any other mental disorder. Beyond the other mental disorders considered below, NAA reductions are observed in a wide range of brain disorders, including stroke, Alzheimer's disease, cancer, epilepsy and multiple sclerosis, among others.[18] Moreover, the brain level of NAA has been found to correlate with higher cognitive function in normal individuals.[51] More research is necessary to sort out the many potential roles of NAA and elucidate the relationship of steady-state NAA levels to brain and neuronal function. We note again the use of ^{13}C MRS to follow abnormal NAA synthesis in schizophrenia (see Section 15.2.3 above).[39,40]

Much less attention has been paid to Cho-containing metabolites and (P)Cr than to NAA in schizophrenia. Both early[52] and more recent[53,54] results have been mixed. This can be attributed to both the smaller effect sizes and the reduced reproducibility seen for the tCho and (P)Cr resonances relative to NAA in ^{1}H MRS.[55]

Much interest and research in recent years have been centered on dysfunctional glutamatergic neurotransmission and the *N*-methyl-D-aspartate

(NMDA) receptor in schizophrenia. The focus of ^{1}H MRS has recently been shifting to the measurement of brain Glu and Gln, either separately or together (Glx), in schizophrenia.[26,49,56] A recent meta-analysis[56] found that medial-frontal Glu was decreased and Gln increased in schizophrenics relative to controls. No differences were seen for other brain regions, primarily because there was an insufficient number of studies for meta-analysis.[56] Along with Glu and Gln, GABA, the primary inhibitory neurotransmitter in the brain, has recently become a focus of ^{1}H-MRS studies of schizophrenia. To date the few reported studies have yielded mixed results.[26,57] A very recent study[57] found elevated GABA and Glx in medial prefrontal cortex in unmedicated, but not medicated, patients relative to controls.

Beginning with the pioneering work of Pettegrew *et al.*,[58] *in vivo* ^{31}P MRS has also been applied to schizophrenia. One early review[59] and two later partial reviews[27,60] have appeared. One hypothesis concerning the pathophysiology of schizophrenia postulates that the disorder involves abnormalities in cell membrane phospholipid metabolism and composition.[16,31] In this regard, several *in vivo* studies have found reduction in PMEs in both drug-naïve and chronic schizophrenics, while results for PDEs have been mixed.[27,59,60] Our *in vitro* results on *post mortem* brain provided no significant support for the phospholipid membrane hypothesis of schizophrenia.[16,31] *In vivo* work in this area continues, as demonstrated by a very recent longitudinal study in which spectral changes suggest continuing membrane breakdown in schizophrenia.[61]

Early results for high-energy phosphates were somewhat contradictory.[59] More recently, several interesting studies have appeared.[27] Of particular note is the report by Jayakumar *et al.*[62] on the effects of antipsychotic treatment. They found that untreated schizophrenic patients had lower PCr/ATP ratios than healthy controls at baseline. These ratios were normalized at follow-up after an average of one year of treatment with a variety of antipsychotics. They also found a positive correlation between PCr/ATP and the degree of improvement on the Positive and Negative Syndrome Scale, a common measure of treatment efficacy.[62]

15.3.2 Bipolar Disorder (BPD)

As mood disorders, BPD and major depressive disorder (MDD) are often compared and contrasted. State-trait considerations and possible medication effects make BPD somewhat more complex to study and describe as patients can be in the manic, euthymic or depressed states, as well as medicated or not, when studied. Within these contexts, the application of ^{1}H-MRS techniques to BPD has been reviewed in several places.[63–65] One of these[64] focused on MR-technical as well as clinical issues. NAA was reduced in frontal lobe and hippocampus in euthymic BPD subjects.[63,64] Glx appears to be elevated in several brain regions in BPD.[65] Results for other metabolites appear unchanging between groups or generally inconsistent, due perhaps in part to technical variations among studies.[64]

The effect of Li on MRS measures in BPD has been of considerable interest. Although results have been somewhat inconsistent,[64] the latest work[66] suggests that Li restores reduced NAA levels in BPD to the levels found for normal subjects. This is consistent with the demonstrated neuroprotective effects of Li in cellular studies.[67] However, it appears that Li (and the alternative drug valproate) does not reduce mI levels as might be expected based on the "inositol-depletion hypothesis" of Li efficacy in BPD.[68]

Phosphorus-31 MRS has been applied to BPD for over 20 years. Both an early[69] and a later[27] review have appeared. One finding is that PME values depend on mood state, being lower in euthymic BPD subjects relative to depressed and manic subjects and normal controls.[27,69] No differences were detected for PDE levels.[69] Both the intracellular pH and the PCr concentration were found to be decreased significantly in BPD.[27] These observations, coupled with the findings for NAA mentioned above, strongly suggest that mitochondrial dysfunction may play a role in the etiology of BPD.

15.3.3   Major Depressive Disorder (MDD)

In addition to the two general reviews on mood disorders previously cited,[64,65] there have been several reviews specifically aimed at MDD, or depression.[70–72] Unlike for other psychiatric disorders, there appear to be no consistent changes in NAA. Perhaps the major finding is the decrease in Glx in areas of the frontal cortex in untreated adult and pediatric depression.[70–72] The ratio Cho/Cr was elevated in adult MDD in the basal ganglia. GABA was reduced in occipital lobe in MDD and was normalized with treatment.[64]

There is some evidence that mood disorders may involve changes in cerebral energy metabolism. Thus ^{31}P MRS may prove to be an important technique to study MDD and its treatment. Here we cite two recent studies[73,74] that employ ^{31}P MRS to follow changes in cerebral high-energy-phosphate metabolism with treatment. Both studies suggest that ^{31}P MRS may be valuable in the development of pharmacologic treatments for MDD.

15.3.4   Other Psychiatric Disorders

Status reports on the applications of MRS to several other psychiatric disorders have been given in general reviews referenced earlier.[4,5,7] Without individual discussion, we here reference recent applications to or reviews on schizoaffective disorder,[75] attention-deficit/hyperactivity disorder,[27,76] post-traumatic stress disorder,[77] obsessive-compulsive disorder,[78] generalized anxiety disorder,[79] borderline personality disorder[80] and severe mood dysregulation.[81] Common features of these studies are 1) metabolite-concentration differences between patient and control groups are small or non-existent and 2) like for the major psychiatric disorders considered earlier, NAA, if changed, will always be lower in the patient group than in healthy subjects. While this lack of specificity for NAA does not necessarily rule it out as a biomarker for treatment efficacy in a particular disorder, future applications of MRS to these disorders will likely focus on other metabolites such as Glu and GABA.

15.4 Future Directions

15.4.1 Technically Oriented Directions

Applications of multinuclear MRS to biomedicine in general and psychiatric illnesses in particular will continue to advance as the technology becomes more powerful and widespread. Of course, MRS will not be the primary motivation for technological advances in MR, but rather will follow the increasing power and utility of MRI-based techniques such as fMRI, DTI and MR angiography. Magnetic fields of 3 T to 7 T are becoming common and have improved SNR and hence spatial resolution and/or imaging time in anatomic MRI and related techniques.[82] The situation is similar for MRS, but with the added benefit of better spectral resolution of chemically shifted species in both ^{1}H and ^{31}P MRS at higher field. Further increases in SNR from hardware improvements such as phased-array coils promise the smaller voxel sizes needed to avoid partial-volume effects for key brain regions like hippocampus and amygdala. Signals from nuclei such as ^{19}F and ^{7}Li may then be sufficient for useful, localized MRS or MRSI. Better shimming capabilities such as higher-order shim sets and dynamic shimming will improve spectral resolution, particularly at high magnetic fields and for difficult-to-shim brain regions. All of these improvements will benefit precise quantitation, which is a necessity given the small group differences usually seen for metabolite levels between psychiatric patients and normal subjects. Development of robust spectral editing ^{1}H-MRS pulse sequences for hard-to-detect metabolites such as GABA, Gln, Lac and glutathione will boost the available information content of *in vivo* ^{1}H MRS, making it more attractive for clinical studies. Finally, the ability to measure metabolic rates using the ^{13}C isotope could add a new dimension to *in vivo* MRS studies of psychiatric disorders.[35–37]

15.4.2 Clinically Oriented Directions

Recruitment of suitable psychiatric subjects, particularly unmedicated subjects at illness onset, can be a slow process. To quickly probe the applicability of this novel technique, many early MRS applications in psychiatry paid less attention to patient selection and numbers, detailed diagnosis, correlation with clinical and neuropsychiatric measures and medication effects than might otherwise have been the case. The design of psychiatric studies using MRS has seen continuing improvement, with larger subject samples, better delineation of clinical subgroups, and more attention to medication effects. Because psychiatric illnesses are usually chronic conditions demanding lifetime management, both short- and long-term longitudinal studies of disease course and treatment efficacy, though difficult and expensive, are likely to become increasingly common. Some studies will need to address identification of neuroimaging biomarkers indicative of an underlying disease *trait vs.* those associated with the current disease *state* in the study subject. The non-invasive nature of MRS and other MRI-based techniques (see Chapter 16) suggests that they may eventually play a role in early diagnosis of psychiatric illness, identification of individuals who are at-risk for either genetic or environmental reasons and early assessment of treatment efficacy. As better, genetically based animal models are developed for psychiatric disorders,[10] MRS can function as a non-invasive bridge technique to follow metabolic biomarkers and treatment effects in parallel in both model and patients. While this chapter has focused on the primary importance of *in vivo* studies, *in vitro* analysis of body fluids and peripheral and *post mortem* tissue by NMR-based metabonomics may also play a significant role in the future (see also Chapter 3).[83,84]

Acknowledgements

I thank Dr. Stephen Strakowski, Dr. Jing-Huei Lee and Matthew Norris of the Center for Imaging Research of the University of Cincinnati, for providing Figures 15.1 and 15.3. Some of the work described in this chapter was supported by grants from the National Institutes of Health of the USA, including P50MH077138 to S. M. Strakowski, and R21MH081000 and R21MH083139 to R. A. Komoroski.

References

1. American Psychiatric Association, *Diagnostic and Statistical Manual of Mental Disorders, 4th edn, Text Revision*, American Psychiatric Association, Washington, DC, 2000.
2. D. Mota de Freitas, J. Silberberg, M. T. Espanol, E. Dorus, A. Abraha, W. Dorus, E. Elenz and W. Whang, *Biol. Psychiatry*, 1990, **28**, 415.
3. G. F. Mason and J. H. Krystal, *NMR Biomed.*, 2006, **19**, 690.
4. S. R. Dager, N. M. Oskin, T. L. Richards and S. Posse, *Topics Magn. Reson. Imaging*, 2008, **19**, 81.
5. N. Agarwal, J. D. Port, M. Bazzocchi and P. F. Renshaw, *Radiology*, 2010, **255**, 23.
6. N. Agarwal and P. F. Renshaw, *AJNR Am. J. Neuroradiol.*, 2012, **33**, 595.
7. R. J. Maddock and M. H. Buonocore, *Curr. Top. Behav. Neurosci.*, 2012, **11**, 199.
8. D. Borsook and L. Becerra, *Curr. Opin. Investig. Drugs*, 2010, **11**, 771.
9. D. J. Kupfer, M. B. First and D. A. Regier (ed.), *A Research Agenda for DSM-V*, American Psychiatric Publishing, Washington, DC, 2002.
10. E. J. Nestler and S. E. Hyman, *Nature Neurosci.*, 2010, **13**, 1161.
11. J. Flint and S. Shifman, *Curr. Opinion Genet. Development*, 2008, **18**, 235.
12. J. M. Pearce, M. Lyon and R. A. Komoroski, *Magn. Reson. Med.*, 2004, **52**, 1087.
13. R. A. Komoroski and J. M. Pearce, *Magn. Reson. Med.*, 2008, **60**, 21.
14. D.A. Lewis, *Neuropsychopharmacology*, 2002, **26**, 143.

15. R. E. McCullumsmith and J. H. Meador-Woodruff, *Biol. Psychiatry*, 2011, **69**, 127.
16. J. M. Pearce, R. A. Komoroski and R. E. Mrak, *Magn. Reson. Med.*, 2009, **61**, 28.
17. R. A. de Graaf, *In Vivo NMR Spectroscopy, Principles and Techniques*, John Wiley & Sons, Ltd, Chichester, 2007.
18. J. R. Moffett, B. Ross, P. Arun, C. N. Madhavarao and A. M. A. Namboodiri, *Prog. Neurobiol.*, 2007, **81**, 89.
19. N. DeStefano, P. M. Matthews and D. L. Arnold, *Magn. Reson. Med.*, 1995, **34**, 721.
20. A. R. Guimaraes, P. Schwartz, M. R. Prakash, C. A. Carr, U. V. Berger, B. G. Jenkins, J. T. Coyle and R. G. Gonzalez, *Neuroscience*, 1995, **69**, 1095.
21. J. M. Hakumäki, M. Ala-Korpela and R. A. Kauppinen, *Curr. Top. Neurochem.*, 1997, **1**, 59.
22. B. S. Y. Li, H. Wang and O. Gonen, *Magn. Reson. Imaging*, 2003, **21**, 923.
23. C. Stork and P. F. Renshaw, *Mol. Psychiatry*, 2005, **10**, 900.
24. S. W. Provencher, *Magn. Reson. Med.*, 1993, **30**, 672.
25. L. Chang, C. C. Cloak and T. Ernst, *J. Clin. Psychiatry*, 2003, **64**(3), 7.
26. J. D. Port and N. Agarwal, *J. Magn. Reson. Imaging*, 2011, **34**, 1251.
27. J.-H. Lee, R. A. Komoroski, W.-J. Chu and J. A. Dudley, *Ann. Rep. NMR Spectrosc.*, 2012, **75**, 115.
28. J. E. Jensen, D. J. Drost, R. S. Menon and P. C. Williamson, *NMR Biomed.*, 2002, **15**, 338.
29. M. Lagemaat, T. Kobus, S. Orzada, A. Bitz, A. Heerschap and T. Scheenen, *Proc. Intl Soc. Mag. Reson. Med.*, 2011, **19**, 1055.
30. H. B. Clay, S. Sillivan and C. Konradi, *Int. J. Dev. Neurosci.*, 2011, **29**, 311.
31. R. A. Komoroski, J. M. Pearce and R. E. Mrak, *Magn. Reson. Med.*, 2008, **59**, 469.
32. O. A. C. Petroff, J. W. Prichard, K. L. Behar, J. R. Alger, J. A. den Hollander and R. G. Shulman, *Neurology*, 1985, **35**, 781.
33. H. R. Halvorson, A. M. Q. Vande Linde, J. A. Helpern and K. M. A. Welch, *NMR Biomed.*, 1992, **5**, 53.
34. J. E. Jensen, J. Miller, P. C. Williamson, R. W. J. Neufeld, R. S. Menon, A. Malla, R. Manchanda, B. Schaefer, M. Densmore and D. J. Drost, *Psychiatry Res.: Neuroimaging*, 2006, **146**, 127.
35. P.-G. Henry, G. Adriany, D. Deelchand, R. Gruetter, M. Marjanska, G. Oz, E. R. Seaquist, A. Shestov and K. Ugurbil, *Magn. Reson. Imaging*, 2006, **24**, 527.
36. R. A. de Graaf, D. L. Rothman and K. L. Behar, *NMR Biomed.*, 2011, **24**, 958.
37. D. L. Rothman, H. M. DeFeyter, R. A. de Graaf, G. F. Mason and K. L. Behar, *NMR Biomed.*, 2011, **24**, 943.
38. F. A. Gallagher, M. I. Kettunen and K. M. Brindle, *Prog. NMR Spectrosc.*, 2009, **55**, 285.
39. K. Harris, P. Bhattacharya, A. P. Lin, B. C. Schweinburg, I. Grant and B. D. Ross, *Proc. Intl Soc. Mag. Reson. Med.*, 2004, **11**, 299.

40. K. Harris, A. Lin, P. Bhattacharya, T. Tran, W. Wong and B. Ross, *Adv. Exp. Med. Biol.*, 2006, **576**, 263.
41. R. A. Komoroski, *Anal. Chem.*, 1994, **66**, 1024A.
42. J. Leib, J. Braun, A. Schilling, C. Klingner, S. Seyfert, W. Vollmann, E. Gedat and J. Bernarding, *Neuroradiology*, 2004, **46**, 363.
43. R. A. Komoroski, in *Recent Advances in MR Imaging and Spectroscopy*, ed. N. R. Jagannathan, Jaypee Bros. Medical Publishers, New Delhi, 2005, ch. 18, p. 428.
44. J. Ruiz-Cabello, B. P. Barnett, P. A. Bottomley and J. W. M. Bulte, *NMR Biomed.*, 2011, **24**, 114.
45. R. A. Komoroski, J. E. O. Newton, D. Cardwell, J. Sprigg and C. N. Karson, *Magn. Reson. Med.*, 1994, **31**, 204.
46. R. A. Komoroski, *NMR Biomed.*, 2005, **18**, 67.
47. J. H. Lee, C. Adler, M. Norris, W.-J. Chu, E. M. Fugate, S. M. Strakowski and R. A. Komoroski, *Magn. Reson. Med.*, 2012, **68**, 363.
48. R. G. Steen, R. M. Hamer and J. A. Lieberman, *Neuropsychopharmacology*, 2005, **30**, 1949.
49. C. Abbott and J. Bustillo, *Curr. Opin. Psychiatry*, 2006, **19**, 135.
50. S. Brugger, J. M. Davis, S. Leucht and J. M. Stone, *Biol. Psychiatry*, 2011, **69**, 495.
51. A. J. Ross and P. S. Sachdev, *Brain Res. Rev.*, 2004, **44**, 83.
52. M. S. Keshavan, J. A. Stanley and J. W. Pettegrew, *Biol. Psychiatry*, 2000, **48**, 369.
53. P. Ohrmann, A. Siegmund, T. Suslow, A. Pedersen, K. Spitzberg, A. Kersting, M. Rothermundt, V. Arolt, W. Heindel and B. Pfleiderer, *J. Psychiatric Res.*, 2007, **41**, 625.
54. E. S. Lutkenhoff, T. G. van Erp, M. A. Thomas, S. Therman, M. Manninen, M. O. Huttunen, J. Kaprio, J. Lönnqvist, J. O'Neill and T. D. Cannon, *Mol. Psychiatry*, 2010, **15**, 308.
55. P. G. Mullins, L. Rowland, J. Bustillo, E. J. Bedrick, J. Lauriello and W. M. Brooks, *Magn. Reson. Med.*, 2003, **50**, 704.
56. A. Marsman, M. P. van den Heuvel, D. W. J. Klomp, R. S. Kahn, P. R. Luijten and H. E. H. Pol, *Schizophrenia Bull.*, 2013, **39**, 120.
57. L. S. Kegeles, X. Mao, A. D. Stanford, R. Girgis, N. Ojeil, X. Xu, R. Gil, M. Slifstein, A. Abi-Dargham, S. H. Lisanby and D. C. Shungu, *Arch. Gen. Psychiatry*, 2012, **69**, 449.
58. J. W. Pettegrew, M. S. Keshavan, K. Panchalingam, S. Strychor, D. B. Kaplan, M. G. Tretta and M. Allen, *Arch. Gen. Psychiatry*, 1991, **48**, 563.
59. H. Fukuzako, *World J. Biol. Psychiatry*, 2001, **2**, 70.
60. R. Reddy and M. S. Keshavan, *Prostaglan. Leukotri. Essen. Fatty Acids*, 2003, **69**, 401.
61. J. Miller, D. J. Drost, E. Jensen, R. Manchuda, S. Northcott, R. W. J. Neufeld, R. Menon, N. Rajakumar, W. Pavlosky, M. Densmore, B. Schaefer and P. Williamson, *Psychiatry Res.: Neuroimaging*, 2012, **201**, 25.

62. P. N. Jayakumar, B. N. Gangadhar, G. Venkatasubramanian, S. Desai, L. Velayudhan, D. Subbakrishna and M. S. Keshavan, *Psychiatry Res.: Neuroimaging*, 2010, **181**, 237.
63. A. Yildiz-Yesiloglu and D. P. Ankerst, *Prog. Neuro-Psychopharmacology Biol. Psychiatry*, 2006, **30**, 969.
64. A. A. Capizzano, R. E. Jorge, L. C. Acion and R. G. Robinson, *J. Magn. Reson. Imaging*, 2007, **26**, 1378.
65. C. Yüksel and D. Öngür, *Biol. Psychiatry*, 2010, **68**, 785.
66. T. Hajek, M. Bauer, A. Pfennig, J. Cullis, J. Ploch, C. O'Donovan, G. Bohner, R. Klingebiel, L. T. Young, G. M. MacQueen and M. Alda, *J. Psychiatry Neurosci.*, 2012, **37**, 185.
67. D.-M. Chuang and H. K. Manji, *Biol. Psychiatry*, 2007, **62**, 4.
68. P. H. Silverstone and B. M. McGrath, *Int. Rev. Psychiatry*, 2009, **21**, 414.
69. A. Yildiz, G. S. Sachs, D. J. Dorer and P. F. Renshaw, *Psychiatry Res.: Neuroimaging*, 2001, **106**, 181.
70. A. Yildiz-Yesiloglu and D. P. Ankerst, *Psychiatry Res.: Neuroimaging*, 2006, **147**, 1.
71. G. Ende, T. Demirakca and H. Tost, *Prog. Brain Res.*, 2006, **156**, 481.
72. L. A. Hulvershorn, K. Cullen and A. Anand, *Brain Imaging Behav.*, 2011, **5**, 307.
73. B. P. Forrester, D. G. Harper, J. E. Jensen, C. Ravichandran, B. Jordan, P. F. Renshaw and B. M. Cohen, *Int. J. Geriatr. Psychiatry*, 2009, **24**, 788.
74. D. G. Kondo, Y.-H. Sung, T. L. Hellem, K. K. Fiedler, X. Shi, E.-K. Jeong and P. F. Renshaw, *J. Affective Disorders*, 2011, **135**, 354.
75. D. Kalayci, O. Özdel, G. Sözeri-Varma, Y. Kiroğlu and S. Tümkaya, *Prog. Neuro-Psychopharmacology Biol. Psychiatry*, 2012, **37**, 176.
76. E. Perlov, A. Philipsen, S. Matthies, T. Drieling, S. Maier, E. Bubl, B. Hesslinger, M. Buechert, J. Henning, D. Ebert and L. T. Van Elst, *World J. Biol. Psychiatry*, 2009, **10**, 355.
77. A. Karl and A. Werner, *Neurosci. Biobehav. Rev.*, 2010, **34**, 7.
78. M.-J. Bedard and S. Chantal, *Psychiatry Res.: Neuroimaging*, 2011, **192**, 45.
79. S. J. Mathew, R. B. Price, X. Mao, E. L. P. Smith, J. D. Coplan, D. S. Charney and D. C. Shungu, *Biol. Psychiatry*, 2008, **63**, 891.
80. M. Hoerst, W. Weber-Fahr, N. Tunc-Skarka, M. Ruf, M. Bohus, C. Schmahl and G. Ende, *Arch. Gen. Psychiatry*, 2010, **67**, 946.
81. D. P. Dickstein, J. W. van der Veen, L. Knopf, K. E. Towbin, D. S. Pine and E. Leibenluft, *Psychiatry Res.: Neuroimaging*, 2008, **163**, 30.
82. E. Moser, F. Stahlberg, M. E. Ladd and S. Trattnig, *NMR Biomed.*, 2011, **25**, 695.
83. E. Holmes, T. M. Tsang, J. T. J. Huang, F. M. Leweke, D. Koethe, C. W. Gerth, B. M. Holden, S. Gross, D. Schreiber, J. K. Nicholson and S. Bahn, *PLoS Med.*, 2006, **3**, 1420.
84. M. J. Lan, G. A. McLoughlin, J. L. Griffin, T. M. Tsang, J. T. J. Huang, P. Yuan, H. Manji, E. Holmes and S. Bahn, *Mol. Psychiatry*, 2009, **14**, 269.

CHAPTER 16

Structural Magnetic Resonance Imaging Biomarkers in Neurodegenerative Disease

LINDA K. McEVOY,*[a] DOMINIC HOLLAND[b] AND
ANDERS M. DALE[a,b]

[a] Department of Radiology, University of California, San Diego, 9500
Gilman Drive, La Jolla, CA 92093, USA; [b] Department of Neurosciences,
University of California, San Diego, 9500 Gilman Drive, La Jolla, CA
92093, USA
*Email: lkmcevoy@ucsd.edu

16.1 Introduction

Structural magnetic resonance imaging (MRI) of the brain provides a
non-invasive, detailed view of brain structures. When appropriate acquisition
protocols are used, efficient and reliable automated image analysis algorithms
can be applied to sensitively quantify thickness or volume of individual brain
structures, or of the whole brain. With precise co-registration of serial MRIs,
subtle change over time in these structures can be quantified. These properties
make structural MRI very well suited for detecting and tracking progressive
regional brain atrophy associated with neurodegenerative disorders such as
Alzheimer's disease (AD). Unlike clinical or cognitive measures that can reflect
symptomatic as well as disease-modifying treatment effects, structural MRI
can supply evidence for treatment-related slowing of neurodegeneration,
providing the basis for use as an outcome measure in clinical trials to establish

New Developments in NMR No. 2
New Applications of NMR in Drug Discovery and Development
Edited by Leoncio Garrido and Nicolau Beckmann
© The Royal Society of Chemistry 2013
Published by the Royal Society of Chemistry, www.rsc.org

disease-modifying effects of a therapy. The sensitivity of MRI to brain changes in the earliest clinically detectable stages of AD, as well as in pre-symptomatic stages, suggests that it may prove useful as a surrogate outcome measure in trials aimed at the early stages of the disease, where traditional clinical outcome variables lack sensitivity. This increased sensitivity over clinical measures increases the power to detect subtle effects and enables smaller sample sizes. The sensitivity to brain changes in pre-clinical and prodromal stages also can aid in the identification of trial participants most likely to decline over the period of a trial, enabling further reductions in sample size. However, to realize the potential of structural MRI as a biomarker for clinical trials, several technical and theoretical issues must be adequately addressed in study design.

This chapter describes the technical issues related to MRI data acquisition and analyses in the context of clinical trials, and discusses how structural MRI biomarkers can be incorporated in clinical trial design. We focus on AD since it is the most common cause of dementia in the elderly, and a large number of therapeutic agents are currently being investigated in clinical trials for their potential to slow or halt this devastating and ultimately fatal disorder. Despite the focus here on AD, similar principles will apply to the use of structural MRI as a biomarker for any neurodegenerative disorder.

16.2 Technical Considerations

In traditional clinical neuroimaging, visual assessment of structural MRI is often sufficient to inform diagnosis. However, for use as a biomarker in clinical trials, visual assessment is neither sensitive nor reliable enough to track changes associated with disease progression, or with potentially subtle slowing of progression with effective treatments. Precise, efficient, operator-independent quantification of brain structure and changes in brain structure over time, as required in clinical trials, presents a number of technical challenges that must be overcome for structural MRI to be useful as a clinical biomarker. This includes the need for standardized high-resolution acquisition protocols suitable for use across multiple clinical sites and MRI platforms; procedures to remove or minimize the influence of artifacts in images; automated algorithms to quantify regional atrophy in individual patients; precise and symmetrical image co-registration techniques to align serial MRIs; and sensitive automated algorithms to quantify subtle changes over time in small brain structures. We will discuss each of these issues in turn, before addressing clinical trial design issues relevant to the incorporation of MRI biomarkers.

16.2.1 Standardized Data Acquisition Protocols

Due to the large number of subjects required to demonstrate clinical efficacy, and the need to obtain representative study samples, most clinical trials collect data across multiple sites. Pooling data across sites requires the use of standardized data acquisition protocols that are robust to differences in scanner platforms. To enable automated quantification of brain structures, acquisition protocols must provide high grey matter/white matter contrast yet be of short duration to minimize participant burden. These issues were addressed by the Alzheimer's Disease Neuroimaging Initiative (ADNI), which developed standardized acquisition parameters for structural MRI data that were successfully applied across the 80+ sites involved in ADNI.[1,2] The ADNI included a preparatory phase in which images acquired with different acquisition protocols were evaluated and compared.[1] The outcome of these efforts was the selection of a sagittal inversion-prepared three-dimensional T_1-weighted gradient-echo sequence (magnetization-prepared rapid acquisition gradient echo, MP-RAGE, or equivalent) that provides high grey/white contrast and with an acquisition time of less than 10 minutes.[1] This acquisition protocol has become the standard for multisite trials incorporating structural neuroimaging, and is being used in similar large-scale multisite trials of aging and AD around the world.[3–5] The use of standardized image acquisition protocols facilitates pooling of data across these different studies, and will enable generation of large normative databases, which will facilitate detection of pathological atrophy.

16.2.2   Artifact Minimization

16.2.2.1   *Prospective Motion Correction*

Artifacts caused by head or body motion can induce image distortions that significantly impair the ability to detect between-group differences or to detect subtle changes in brain structure over time within individual patients. Although motion artifacts can be problematic in any subject population, they may be more pronounced in studies of dementia patients who may have difficulty remembering instructions to remain still throughout the scan. Not only can motion artifact result in greater variance in estimates of atrophy (translating into smaller effect sizes), but differences in degree of motion may differ with disease severity (and thus potentially with treatment), confounding group or treatment comparisons.

   Patient motion during acquisition of three-dimensional pulse sequences, such as those standardized by the ADNI, is particularly problematic since such within-scan motion produces k-space data inconsistencies, which result in ringing, ghosting and blurring of boundaries between structures.[6] These artifacts can cause automated segmentation algorithms to fail. Post-acquisition mitigation of within-volume artifacts requires precise k-space interpolation and grid-readjustment to reconstruct the images accurately,[7–10] but still results in some degree of image blurring or distortion due to interpolation that can affect the ability to detect subtle differences.[11] Furthermore, these methods cannot correct for motion that occurs orthogonal to the plane of image acquisition (through-plane motion).[11] Prospective motion correction algorithms that modify the pulse-sequence during image acquisition to keep the coordinate system fixed with respect to the patient's head throughout the scan do not share these drawbacks,[12–15] but can substantially increase scan time. Tracking head motion using an external optical tracking device enables more rapid real-time motion correction without the need to modify pulse sequences, but requires additional expensive hardware.[16,17]

A recently developed image-based approach for prospective motion correction in MRI (PROMO) allows for rapid real-time rigid-body motion tracking and correction through the use of spiral navigator scans interspersed within the dead time of standard image acquisition, keeping the coordinate system fixed with respect to the position of the brain.[6] The method employs non-iterative recursive filters that allow the technique to combine current position information with motion trajectory data to predict brain position in the upcoming acquisition. The method automatically rescans images acquired during intervals where significant motion was detected, with an adjustable rescan threshold to balance acquisition of motion-free data with minimizing time in scanner. The ability of this method to mitigate motion artifact was demonstrated in another subject group particularly prone to motion artifact – children. In these subjects, PROMO resulted in improved image clarity, decreased between-scan variability, higher image quality ratings and enabled successful use of automated image segmentation algorithms in participants with high levels of motion, which otherwise would have failed.[11,18] An example of the ability of real-time motion correction to improve MRI signal quality is shown in Figure 16.1. Motion correction is not yet routinely incorporated into clinical trials. However, it has strong potential to improve accuracy of volumetric assessment, and will be critical in future if trials are to incorporate other types of MRI that may be even more affected by motion, such as diffusion-weighted imaging or functional MRI.

16.2.2.2 Pre-Processing Image Corrections

Gradient field non-linearities induce geometrical distortions and intensity inaccuracies in structural MR images. These distortions vary across MRI manufacturers and across scanner versions built by the same manufacturer.[19] Even on the same scanner distortions can differ between scans of the same patient due to differences in head position. These artifacts substantially degrade the ability to detect differences between patient groups and to detect change over time within patients. The heightened variability in morphometric measures induced by these distortions substantially reduces effect sizes.

A gradient field distortion-correction method has been developed that transforms the distorted image into a corrected image by displacing each voxel into its estimated correct location (Figure 16.2). This is calculated from the non-linear terms, supplied by the vendor, that characterize the non-uniformity of the magnetic field generated by each gradient coil. Each voxel's intensity is then scaled to account for voxel size distortions.[19] The accuracy of this method has been demonstrated using phantom data, and its ability to improve image reproducibility has been shown through application to imaging data acquired from the same subject on different scanner platforms.[19]

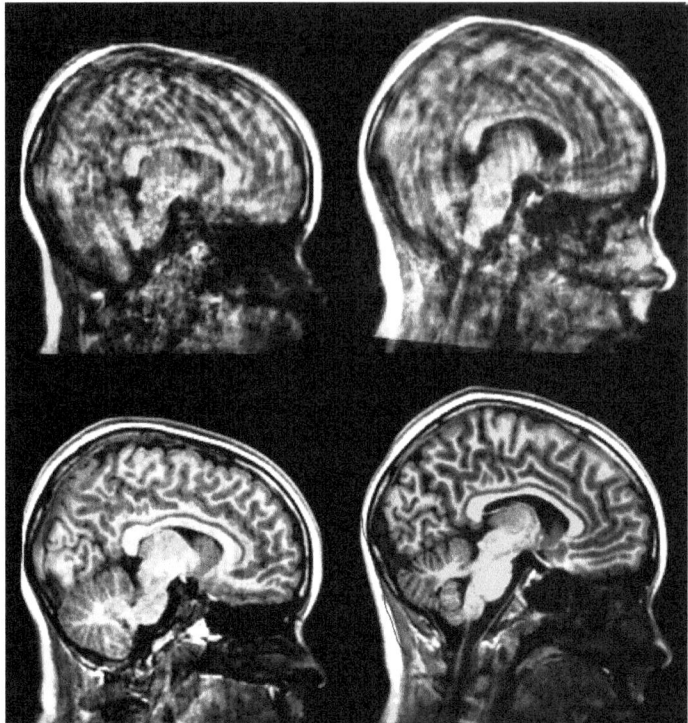

Figure 16.1 Top. MRI data collected without motion correction. Bottom: Real time
motion correction applied during acquisition while participant engaged
in same degree of motion. Examples from two participants are shown.

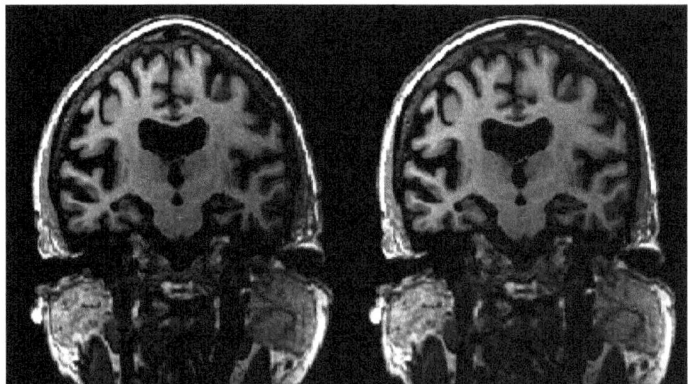

Figure 16.2 Left: Uncorrected sagittal MRI. Right: Same MRI after correction for
gradient field non-linearity.

Image intensity inhomogeneity arising from non-uniformity of the radiofrequency coil is characterized by smoothly varying intensity across the image.
Such intensity variations, while not interfering with visual assessment of

images, can degrade performance of automated segmentation methods that assume intensity homogeneity within tissue classes. A non-parametric, non-uniform intensity normalization method (N3) has been developed that models the inhomogeneity as a smooth multiplicative field.[20] This automated method has been shown to correct intensity non-uniformity in three-dimensional MR volumes irrespective of the pulse sequence.[20] Corrections for gradient field non-linearities and intensity normalization must be incorporated as standard pre-processing steps prior to analysis of MRI data in clinical trials to optimize performance of automated image analysis algorithms.

16.2.3 Automated Image Segmentation and Quantification

Since a Phase III clinical trial for a potential disease-modifying treatment could easily involve several hundred patients, each of whom would undergo repeated scans to assess the treatment's effect on atrophy rate, fully automated, high-throughput image analysis methods are essential. Such methods must produce reliable, operator-independent volumetric quantification that is robust to gross individual differences in brain morphology. Additionally, since a treatment's effect may be minimal in brain areas not affected by disease pathology, and maximal in regions most vulnerable to the disease, methods that provide regional quantification are likely to provide a more sensitive biomarker than methods that provide global measures of brain atrophy such as ventricular or whole brain volumes.[21–23]

A number of automated methods have been developed to quantify atrophy on MRIs.[21,24–36] Generally these methods strip the skull from the brain, segment the brain image into tissue types (grey matter, white matter, cerebrospinal fluid, CSF) based on intensity differences and spatially normalize images so that they are in a common reference system, enabling comparison of structures across subjects. Some methods do this through registering the image with a canonical brain (*e.g.* Talairach–Tournoux or Montreal Neurological Institute template) or a customized group template. Significant errors can occur at this step if the individual's brain topography differs substantially from that of the template, as can occur when individuals with varying degrees of atrophy are included in a study.[37–40] Methods that analyze data in native space are not prone to these registration errors, and are highly suited for studies in which brain morphometry may differ substantially between subjects, as is the case in neurodegenerative disorders. Such methods are of even greater value when atrophy may be prominent in small but critical brain structures. For example, in AD, pathology first develops in a small medial temporal lobe structure, the entorhinal cortex, then spreads to hippocampus and other limbic regions, before affecting wider cortical areas.[41] Thus detection of neuropathology associated with the earliest stage of AD requires the ability precisely to quantify atrophy in this small structure that varies considerably across individuals, and across disease stage.[42]

A large number of studies, using a variety of automated image analysis techniques, have demonstrated the sensitivity of structural MRI for detecting atrophy as a function of AD severity (including differences in pre-clinical and prodromal stages) in cross-sectional studies, as recently reviewed.[43–45] The degree of atrophy correlates with symptom severity[46–49] and predicts subsequent clinical decline.[50–61] The ability to predict the risk of clinical decline based on a single screening MRI offers an important opportunity for MRI to serve as an enrichment strategy for clinical trials, as will be discussed in greater detail in section 16.3.1.

16.2.4 Longitudinal Image Analysis

For use as a biomarker for detecting change in disease progression, MRI measures must show high sensitivity to change within subjects over time. For longitudinal image analysis, key issues concern co-registration of serial images and determination of change across images in a bias-free manner. Bias in image registration is a well-recognized problem[62,63] that may arise from several sources, including asymmetries in image smoothing or interpolation and asymmetry in the image matching or regularization term in the cost function used in image registration.[64,65] Several methods have been developed to deal with bias in longitudinal image registration.[30,64,66–70] Although most longitudinal analysis methods incorporate procedures to minimize bias, a recent comparative analysis of different longitudinal image analysis methods applied to ADNI data revealed a substantial contribution of bias to change estimates from two methodologies.[71] In the worst case, bias accounted for more than 55% of the observed six-month change in AD subjects, and more than 65% of the change in subjects with mild cognitive impairment (MCI). This artifactual inflation of the effect size (*i.e.* the magnitude of estimated disease-related atrophy) results in substantial sample size underestimation: we showed that when these change measures were corrected for bias, sample size estimates increased four-fold.[71] Unfortunately, even after an attempt was made to improve the method,[72] significant bias remained, accounting for 48% of the change observed in participants with MCI over six months, leading to an underestimation of sample sizes by more than 50%.[71] Unrealistically low sample size estimates resulting from use of bias-afflicted analysis methods have been reported in numerous publications.[2,45,73,74]

As we[71,75] and others[65,76] have noted, several steps can be taken to ensure the fidelity of measurement of longitudinal change from serial MRIs. The registration procedure should be fully symmetric. That is, the degree of change obtained when image 1 is registered to image 2 should be the same as that obtained when image 2 is registered to image 1. A simple method for ensuring such inverse consistency is to measure the change in both directions independently, and then average the results.[71,75] It is also important to ensure the transitivity of the results, if more than one follow-up image is obtained. That is, the difference between image 1 and image 3 should equal the sum of the differences between image 1 and image 2, and between image 2 and image 3. Methods can also be validated by assessing change in images obtained over a very short time span, when no biological change is expected. As a minimal

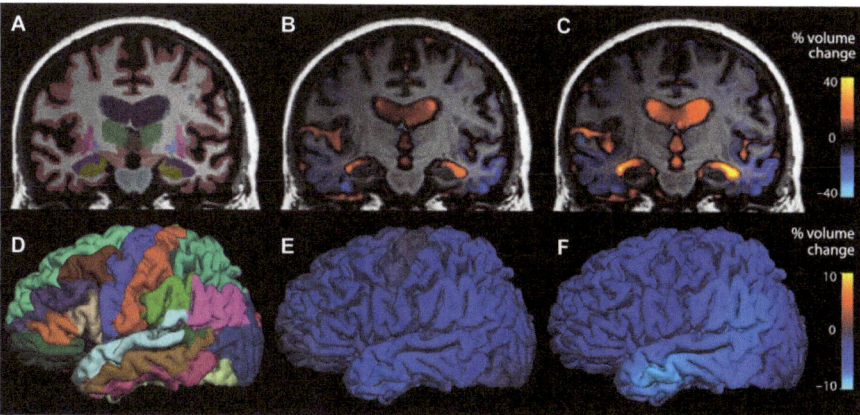

Figure 16.3 Tissue segmentation, with 6- and 12-month volume change fields for a patient with MCI from the Alzheiemer's Disease Neuroimaging Initiative. (A) Segmentation of the baseline MRI scan, with different brain structures represented in different colors. The hippocampus is shown in gold. (B) Corresponding coronal slice overlain with a heat map showing change in volume at 6 months and (C) 12 months. Blue indicates areas undergoing shrinkage, red/yellow indicates expansion. (D) Left hemisphere cortical parcellation of the baseline MRI scan. (E) Cortical surface overlain with a heat map representation of the estimates of cortical volumetric change at 6 months and (F) 12 months. Even in this prodromal phase of the disorder, atrophy is widespread across the cortex, and maximal in temporal regions.
Reproduced with permission from Holland *et al.*[22] © 2009 National Academy of Sciences of the United States of America.

check for bias-free registration, there should be no difference when an image is registered to itself.

Despite these methodological challenges, numerous studies have demonstrated that serial MRI can be used to detect neurodegeneration associated with AD and with normal aging.[77–81] Atrophy rate has been shown to vary with disease stage,[81,82] to increase prior to the development of dementia in individuals with familial AD[83] and to be useful for predicting risk of developing dementia in individual patients with MCI.[59] An example of the degree of atrophy experienced by a patient with MCI over a 6- and 12-month period is shown in Figure 16.3. In addition to its sensitivity and predictive ability, atrophy rate is far less variable between patients with MCI or AD than are clinical measures,[22,78,82] making it highly suited for use as outcome measures in clinical trials, as will be discussed in greater detail in a subsequent section.

16.3 Clinical Trial Design

In addition to consideration of technical issues for obtaining sensitive measures of brain atrophy and atrophy rate over time, there are important clinical trial design issues to consider when incorporating biomarkers into clinical trials.

Biomarkers may serve several roles in a clinical trial: they may be used as a selection criterion to enrich the study sample; as a covariate in the analyses to improve sensitivity for detecting effects; or as outcome measures to demonstrate efficacy of a treatment for slowing neurodegeneration. When employing structural MRI in any of these roles, several issues need to be considered to ensure appropriate trial design.

16.3.1   MRI for Clinical Trial Enrichment

In neurodegenerative disorders there is strong interest in applying treatments to slow or halt disease progression as early as possible, so that patients can retain maximum cognitive ability. There is also growing concern that by the time a patient has become demented, treatments may be too late to interrupt the degenerative cascade.[84–86] For example, in AD, amyloid-β (Aβ) pathology is believed by many to trigger a neurodegenerative process that, at some point, becomes independent of amyloid.[84] This would imply that anti-amyloid agents, the most common type of potential treatment currently under investigation for AD, may be maximally (or solely) effective in the early stages of the disorder.

Demonstrating the efficacy of a disease-modifying effect of a treatment in asymptomatic individuals or those with very mild symptoms is challenging for a number of reasons. It is difficult to identify accurately individuals who are in a prodromal – let alone pre-clinical – phase of a neurodegenerative disorder since there are many causes of cognitive impairment in the elderly that could lead to a similar presentation of MCI. Decline over time in individuals in early stages of the disorder is also likely to be small and difficult to detect using standard clinical outcome measures.

Biomarkers for disease detection and progression will thus be critically important for enabling clinical trials in early disease stages. Biomarkers of Aβ pathology will be vital to identify candidates for anti-amyloid therapeutic trials since individuals without Aβ pathology should not be exposed to risks associated with these treatments. However, because Aβ pathology may occur a decade or more prior to the onset of dementia, and because it is not yet clear whether all individuals with Aβ pathology will eventually develop dementia, trials of asymptomatic Aβ positive individuals, or those with only mild impairment, may need to be very long and very large to demonstrate a significant clinical effect. In these cases, structural MRIs may be useful for identifying those amyloid-positive individuals who are likely to experience measureable decline over the period of a typical clinical trial (often 18 months). This can substantially increase the power of the trial, enabling smaller sample sizes.

We illustrated the potential of an atrophy-based enrichment strategy using data from the ADNI.[87] We found that relative to a study that selected participants based on standard criteria for defining MCI,[88] a study that further restricted enrollment to those who showed a pattern of regional atrophy characteristic of mild AD[58] on a screening MRI could achieve the same power for detecting 25% slowing on the rate of decline on the Clinical

Dementia Rating Scale, sum of boxes score CDR-SB, with a 56% smaller sample size.[87]

When considering enrichment strategies, it is important to consider the potential limitations in clinical usage that may be imposed due to clinical trial enrollment criteria. Regulatory agencies may require that labeled use be restricted to patients meeting the same criteria.[89] This would require that the procedures used to establish trial eligibility be widely available and approved for clinical use. Thus the enrichment strategy cited above, though demonstrating the potential of MRI for selecting at-risk subjects, is unlikely to be used in a clinical trial, given that the methods used for detecting patterns of regional atrophy are unlikely to become widely clinically available. We therefore used an automated method for quantifying hippocampal atrophy that has been cleared for clinical use by the Food and Drug Administration (FDA) and is currently in use in a number of clinical centers across the USA.[35,90] An enrichment strategy based on selectively enrolling MCI patients who show a similar degree of hippocampal atrophy on a screening MRI as mild AD patients[60] would enable a 45% reduction in sample size, from 565 (95% confidence interval, C.I.: 428–781) patients per arm to 309 (95% C.I.: 234–425) patients per arm. These estimates were obtained assuming a 24-month trial with a 6-month assessment interval. Power calculations were performed with the requirement that the trial have 80% to detect a 25% reduction in rate of decline on the CDR-SB, with a 2-sided significance of 0.05.

An alternative to selectively enrolling patients with atrophy on a screening MRI to increase study power is to include the degree of baseline atrophy as a covariate in the analysis. This can increase power by reducing between-subject variability.[91] Atrophy at baseline can also be used as a stratification factor to enable examination of differential treatment effects as a function of disease severity. This can increase sensitivity for finding effects without imposing limits on subsequent clinical use of the agent.

16.3.2 MRI as an Outcome Variable

In addition to offering potential as an enrichment strategy to better identify patients suffering from a neurodegenerative disorder, MRI biomarkers may also serve as outcome measures. As noted in a recent comprehensive discussion of biomarkers for AD clinical trials, regulatory agencies are unlikely to accept any biomarker as a primary outcome measure when clinical outcome measures can reasonably be used to detect treatment effects.[92] Thus for trials aimed at dementia patients, clinical measures, which have been developed and validated for detecting clinical change in AD, will continue to serve as primary outcome measures. To facilitate establishing disease-modifying effects of therapies, however, MRI biomarkers are now routinely included in clinical trials as secondary outcome measures. Since MRI measures are less variable than clinical measures[78,93] far fewer patients are needed to detect significant effects. For example, we found that estimated sample sizes for detecting slowing in rate of hippocampal atrophy in AD patients can be less than one-third of the sample size needed to detect slowing in the CDR-SB.[22] Thus it would be possible to obtain evidence for a therapy's ability to slow neurodegeneration by collecting structural MRI data on a subset of participants only, thereby reducing costs of the trial.

As mentioned above, there is growing concern that attempts to treat AD at such advanced stages are unlikely to be successful.[84–86] Clinical trials are increasingly being aimed at patients with less advanced disease, such as those with MCI. Consideration is also being given to testing treatments in asymptomatic patients who test positive for Aβ pathology.[86] Since these patients are not yet exhibiting clinical impairment, and may not do so for years, surrogate biomarkers will be necessary to provide evidence of potential clinical effects of a therapy. No biomarkers, however, have yet been validated as surrogate outcome measures for AD, and many challenges must be overcome before such validation occurs.[94] Structural MRI offers strong potential as a surrogate outcome measure due to its face validity as a measure of a treatment's effect on rate of brain atrophy, its sensitivity to brain changes in early stages of the disease[59,83,95] and the ability of atrophy rate measures to predict development of dementia.[23,59,96]

In these cases, when a clinical trial may be powered to detect a change in atrophy rate, it is vital to ensure that the potential effect size is calculated in a meaningful fashion. Unlike AD clinical measures, which are, by design, insensitive to normal age-related changes, normal aging is accompanied by measureable brain atrophy.[80,97–100] There is little reason to believe that a therapeutic agent that targets AD pathology will affect rate of change that is not associated with that pathology. Thus modeling the potential treatable effect size as the magnitude of the full change observed over time in the patient group, without first subtracting the degree of change that occurs in normal aging, will overestimate the magnitude of potential change that could be affected by the therapy, leading to an underestimation of sample size required to detect slowing in disease-related change. This problem is particularly acute for trials focused on early disease stages, where a much larger portion of the observed change over time can be attributable to normal aging, not to disease.[64] The majority of papers that report or compare sample sizes when atrophy rates are used as outcome measures fail to account for effects of normal aging[72–74,101–106] (although there are exceptions[22,36,65,78,87,91,107]). Most studies do not provide a rationale for including change due to normal aging as part of the potentially treatable effect sizes. An implicit assumption may be that most, if not all, change with age is due to pre-clinical AD. Current evidence suggests that Aβ begins to accumulate a decade or more prior to symptom onset, and thus many older, apparently healthy individuals are likely to be in a pre-clinical stage of Alzheimer's disease; atrophy in these individuals could thus be due to AD. However, findings that healthy older individuals who show no evidence of Aβ pathology experience similar levels of atrophy as those who test positive argue against this view (see Figure 16.4).[64] Although attractively smaller samples sizes are obtained when changes observed in normal aging are not taken into account, an under-powered clinical trial could jeopardize the likelihood of

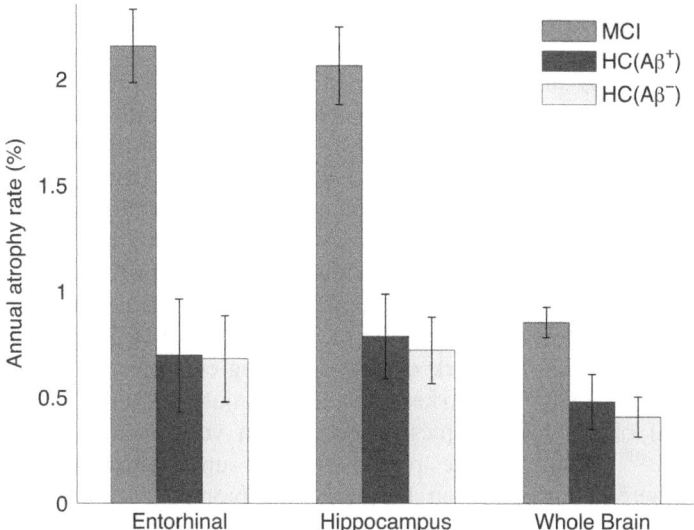

Figure 16.4   Annual rate of volume change in entorhinal cortex (ERC), hippocampus (Hipp) and whole brain, with 95% confidence intervals, for patients with mild cognitive impairment (MCI) and for healthy controls (HC) who tested positive or negative on a cerebrospinal fluid biomarker of brain amyloid-β (Aβ) pathology. Atrophy rates are calculated using an analysis method robust to bias, Quarc.[64,71] Atrophy rates are higher, and differ more between MCI patients and controls for temporal lobe structures than for whole brain. There is little difference in atrophy rate between healthy controls who test positive or negative for amyloid pathology, indicating that all atrophy in aging is not due to incipient Alzheimer's disease. Using the absolute atrophy rate as the estimate of the maximum effect size will overestimate the power to detect a beneficial effect of a treatment that slows disease-related atrophy to the rate experienced by healthy controls.

detecting a beneficial effect of a treatment that actually slowed disease-related neurodegeneration. The failure to detect such a potentially beneficial treatment effect, leading to abandonment of a possibly effective therapy, would be much more costly to a pharmaceutical company, and to society as a whole, than the cost of a larger, but suitably powered, trial.

16.4   Caveats and Conclusions

Structural MRI is a sensitive biomarker for tracking progression of neuro-degenerative disease. Numerous cross-sectional and longitudinal studies have demonstrated its sensitivity to AD across varying stages of the disorder, and have shown that degree and rate of atrophy correlate well with clinical symptoms and predict progressive decline. This provides a strong basis for the use of structural MRI as an outcome measure for detecting beneficial effects of disease-slowing treatments. However, until such treatments are discovered, it will not be possible to validate structural MRI as an outcome variable. Since potential treatments may have a variety of physiological effects that are unlikely to be fully understood during initial testing and use, contradictory effects on MRI may be observed. In the immunotherapy with human Aβ 1–42 peptide (AN1792) clinical trial, antibody responders showed a paradoxical increase in whole brain atrophy rate and ventricular expansion, despite showing reduced cognitive decline and reduced levels of tau protein in the cerebrospinal fluid, a marker of neuronal injury.[108] There are many potential explanations for such an unexpected finding, as discussed by the study authors. Treatment, which induced meningoencephalitis in some patients (causing the trial to be discontinued), could have accelerated neuronal loss. This was deemed unlikely due to the lack of corresponding cognitive decline, and the reduction in tau, a CSF biomarker of neuronal injury. Other possibilities are that removal of amyloid plaques reduced brain volume, and altered CSF reasborption.[108] Whatever the reason for the unexpected findings, they illustrate inherent limitations in biomarkers – biomarkers provide indirect measures of disease progression and they may be influenced by a variety of factors, other than disease progression. Results from ongoing clinical trials that have included structural MRI measures can be expected to help establish their validity as biomarkers for detecting treatment effects.

References

1. C. R. Jack, Jr., M. A. Bernstein, N. C. Fox, P. Thompson, G. Alexander, D. Harvey, B. Borowski, P. J. Britson, J. L. Whitwell, C. Ward, A. M. Dale, J. P. Felmlee, J. L. Gunter, D. L. Hill, R. Killiany, N. Schuff, S. Fox-Bosetti, C. Lin, C. Studholme, C. S. Decarli, H. A. Ward, G. J. Metzger, K. T. Scott, R. Mallozzi, D. Blezek, J. Levy, J. P. Debbins, A. S. Fleisher, M. Albert, R. Green, G. Bartzokis, G. Glover, J. Mugler and M. W. Weiner, *J. Magn. Reson. Imaging*, 2008, **27**, 685.
2. C. R. Jack, Jr., M. A. Bernstein, B. J. Borowski, J. L. Gunter, N. C. Fox, P. M. Thompson, N. Schuff, G. Krueger, R. J. Killiany, C. S. Decarli, A. M. Dale, O. W. Carmichael, D. Tosun and M. W. Weiner, *Alzheimers Dement.*, 2010, **6**, 212.
3. G. B. Frisoni, W. J. Henneman, M. W. Weiner, P. Scheltens, B. Vellas, E. Reynish, J. Hudecova, H. Hampel, K. Burger, K. Blennow, G. Waldemar, P. Johannsen, L. O. Wahlund, G. Zito, P. M. Rossini, B. Winblad and F. Barkhof, *Alzheimers Dement.*, 2008, **4**, 255.
4. G. B. Frisoni, *Alzheimers Dement.*, 2010, **6**, 280.
5. T. Iwatsubo, *Alzheimers Dement.*, 2010, **6**, 297.
6. N. White, C. Roddey, A. Shankaranarayanan, E. Han, D. Rettmann, J. Santos, J. Kuperman and A. Dale, *Magn. Reson. Med.*, 2010, **63**, 91.
7. J. G. Pipe, *Magn. Reson. Med.*, 1999, **42**, 963.
8. A. Manduca, K. P. McGee, E. B. Welch, J. P. Felmlee, R. C. Grimm and R. L. Ehman, *Radiology*, 2000, **215**, 904.

9. P. Kochunov, J. L. Lancaster, D. C. Glahn, D. Purdy, A. R. Laird, F. Gao and P. Fox, *Hum. Brain Mapp.*, 2006, **27**, 957.
10. C. Liu, R. Bammer, D. H. Kim and M. E. Moseley, *Magn. Reson. Med.*, 2004, **52**, 1388.
11. T. T. Brown, J. M. Kuperman, M. Erhart, N. S. White, J. C. Roddey, A. Shankaranarayanan, E. T. Han, D. Rettmann and A. M. Dale, *Neuroimage*, 2010, **53**, 139.
12. H. A. Ward, S. J. Riederer, R. C. Grimm, R. L. Ehman, J. P. Felmlee and C. R. Jack, Jr., *Magn. Reson. Med.*, 2000, **43**, 459.
13. D. G. Norris and W. Driesel, *Magn. Reson. Med.*, 2001, **45**, 729.
14. E. B. Welch, A. Manduca, R. C. Grimm, H. A. Ward and C. R. Jack, Jr., *Magn. Reson. Med.*, 2002, **47**, 32.
15. A. J. van der Kouwe, T. Benner and A. M. Dale, *Magn. Reson. Med.*, 2006, **56**, 1019.
16. C. Dold, M. Zaitsev, O. Speck, E. A. Firle, J. Hennig and G. Sakas, *Med. Image Comput. Comput. Assist. Interv.*, 2005, **8**, 482.
17. M. Zaitsev, C. Dold, G. Sakas, J. Hennig and O. Speck, *Neuroimage*, 2006, **31**, 1038.
18. J. M. Kuperman, T. T. Brown, M. E. Ahmadi, M. J. Erhart, N. S. White, J. C. Roddey, A. Shankaranarayanan, E. T. Han, D. Rettmann and A. M. Dale, *Pediatr. Radiol.*, 2011, **41**, 1578.
19. J. Jovicich, S. Czanner, D. Greve, E. Haley, A. van der Kouwe, R. Gollub, D. Kennedy, F. Schmitt, G. Brown, J. Macfall, B. Fischl and A. Dale, *Neuroimage*, 2006, **30**, 436.
20. J. G. Sled, A. P. Zijdenbos and A. C. Evans, *IEEE Trans. Med. Imaging*, 1998, **17**, 87.
21. J. Ashburner, J. G. Csernansky, C. Davatzikos, N. C. Fox, G. B. Frisoni and P. M. Thompson, *Lancet Neurol.*, 2003, **2**, 79.
22. D. Holland, J. B. Brewer, D. J. Hagler, C. Fennema-Notestine and A. M. Dale, *Proc. Natl Acad. Sci. USA*, 2009, **106**, 20954.
23. W. J. Henneman, J. D. Sluimer, J. Barnes, W. M. van der Flier, I. C. Sluimer, N. C. Fox, P. Scheltens, H. Vrenken and F. Barkhof, *Neurology*, 2009, **72**, 999.
24. C. Davatzikos and N. Bryan, *IEEE Trans. Med. Imaging*, 1996, **15**, 785.
25. A. M. Dale, B. Fischl and M. I. Sereno, *Neuroimage*, 1999, **9**, 179.
26. B. Fischl and A. M. Dale, *Proc. Natl Acad. Sci. USA*, 2000, **97**, 11050.
27. B. Fischl, D. H. Salat, E. Busa, M. Albert, M. Dieterich, C. Haselgrove, A. van der Kouwe, R. Killiany, D. Kennedy, S. Klaveness, A. Montillo, N. Makris, B. Rosen and A. M. Dale, *Neuron*, 2002, **33**, 341.
28. P. M. Thompson, M. S. Mega, R. P. Woods, C. I. Zoumalan, C. J. Lindshield, R. E. Blanton, J. Moussai, C. J. Holmes, J. L. Cummings and A. W. Toga, *Cereb. Cortex*, 2001, **11**, 1.
29. D. W. Shattuck and R. M. Leahy, *Med. Image Anal.*, 2002, **6**, 129.
30. S. M. Smith, Y. Zhang, M. Jenkinson, J. Chen, P. M. Matthews, A. Federico and N. De Stefano, *Neuroimage*, 2002, **17**, 479.

31. X. Han, D. L. Pham, D. Tosun, M. E. Rettmann, C. Xu and J. L. Prince, *Neuroimage*, 2004, **23**, 997.

32. J. S. Kim, V. Singh, J. K. Lee, J. Lerch, Y. Ad-Dab'bagh, D. MacDonald, J. M. Lee, S. I. Kim and A. C. Evans, *Neuroimage*, 2005, **27**, 210.

33. O. Colliot, G. Chetelat, M. Chupin, B. Desgranges, B. Magnin, H. Benali, B. Dubois, L. Garnero, F. Eustache and S. Lehericy, *Radiology*, 2008, **248**, 194.

34. J. H. Morra, Z. Tu, L. G. Apostolova, A. E. Green, C. Avedissian, S. K. Madsen, N. Parikshak, X. Hua, A. W. Toga, C. R. Jack, Jr., M. W. Weiner and P. M. Thompson, *Neuroimage*, 2008, **43**, 59.

35. J. B. Brewer, S. Magda, C. Airriess and M. E. Smith, *AJNR Am. J. Neuroradiol.*, 2009, **30**, 578.

36. K. K. Leung, J. Barnes, G. R. Ridgway, J. W. Bartlett, M. J. Clarkson, K. Macdonald, N. Schuff, N. C. Fox and S. Ourselin, *Neuroimage*, 2010, **51**, 1345.

37. J. C. Baron, G. Chetelat, B. Desgranges, G. Perchey, B. Landeau, V. de la Sayette and F. Eustache, *Neuroimage*, 2001, **14**, 298.

38. J. L. Whitwell, *J. Neurosci.*, 2009, **29**, 9661.

39. T. M. Seibert, D. S. Majid, A. R. Aron, J. Corey-Bloom and J. B. Brewer, *Neuroimage*, 2012, **59**, 2452.

40. T. M. Seibert and J. B. Brewer, *J. Neurosci. Methods*, 2011, **198**, 301.

41. H. Braak and E. Braak, *Acta Neuropathol. (Berl.)*, 1991, **82**, 239.

42. C. Fennema-Notestine, D. J. Hagler, Jr., L. K. McEvoy, A. S. Fleisher, E. H. Wu, D. S. Karow and A. M. Dale, *Hum. Brain Mapp.*, 2009, **30**, 3238.

43. M. Atiya, B. T. Hyman, M. S. Albert and R. Killiany, *Alzheimer Dis. Assoc. Disord.*, 2003, **17**, 177.

44. L. K. McEvoy and J. B. Brewer, *Expert Rev. Neurother.*, 2010, **10**, 1675.

45. M. W. Weiner, D. P. Veitch, P. S. Aisen, L. A. Beckett, N. J. Cairns, R. C. Green, D. Harvey, C. R. Jack, W. Jagust, E. Liu, J. C. Morris, R. C. Petersen, A. J. Saykin, M. E. Schmidt, L. Shaw, J. A. Siuciak, H. Soares, A. W. Toga and J. Q. Trojanowski, *Alzheimers Dement.*, 2012, **8**(1 Suppl), S1–68.

46. K. B. Walhovd, A. M. Fjell, A. M. Dale, L. K. McEvoy, J. Brewer, D. S. Karow, D. P. Salmon and C. Fennema-Notestine, *Neurobiol. Aging*, 2010, **31**, 1107.

47. B. C. Dickerson, A. Bakkour, D. H. Salat, E. Feczko, J. Pacheco, D. N. Greve, F. Grodstein, C. I. Wright, D. Blacker, H. D. Rosas, R. A. Sperling, A. Atri, J. H. Growdon, B. T. Hyman, J. C. Morris, B. Fischl and R. L. Buckner, *Cereb. Cortex*, 2009, **19**, 497.

48. P. Vemuri, H. J. Wiste, S. D. Weigand, L. M. Shaw, J. Q. Trojanowski, M. W. Weiner, D. S. Knopman, R. C. Petersen and C. R. Jack, Jr., *Neurology*, 2009, **73**, 287.

49. Y. L. Chang, M. W. Jacobson, C. Fennema-Notestine, D. J. Hagler, Jr., R. G. Jennings, A. M. Dale and L. K. McEvoy, *Cereb. Cortex*, 2010, **20**, 1305.

50. C. R. Jack, Jr., R. C. Petersen, Y. C. Xu, P. C. O'Brien, G. E. Smith, R. J. Ivnik, B. F. Boeve, S. C. Waring, E. G. Tangalos and E. Kokmen, *Neurology*, 1999, **52**, 1397.

51. M. Grundman, D. Sencakova, C. R. Jack, Jr., R. C. Petersen, H. T. Kim, A. Schultz, M. F. Weiner, C. DeCarli, S. T. DeKosky, C. van Dyck, R. G. Thomas and L. J. Thal, *J. Mol. Neurosci.*, 2002, **19**, 23.

52. L. deToledo-Morrell, T. R. Stoub, M. Bulgakova, R. S. Wilson, D. A. Bennett, S. Leurgans, J. Wuu and D. A. Turner, *Neurobiol. Aging*, 2004, **25**, 1197.

53. L. G. Apostolova, R. A. Dutton, I. D. Dinov, K. M. Hayashi, A. W. Toga, J. L. Cummings and P. M. Thompson, *Arch. Neurol.*, 2006, **63**, 693.

54. D. P. Devanand, G. Pradhaban, X. Liu, A. Khandji, S. De Santi, S. Segal, H. Rusinek, G. H. Pelton, L. S. Honig, R. Mayeux, Y. Stern, M. H. Tabert and M. J. de Leon, *Neurology*, 2007, **68**, 828.

55. Y. Fan, N. Batmanghelich, C. M. Clark and C. Davatzikos, *Neuroimage*, 2008, **39**, 1731.

56. R. S. Desikan, H. J. Cabral, F. Settecase, C. P. Hess, W. P. Dillon, C. M. Glastonbury, M. W. Weiner, N. J. Schmansky, D. H. Salat and B. Fischl, *Neurobiol. Aging*, 2010, **31**, 1364.

57. M. Ewers, C. Walsh, J. Q. Trojanowski, L. M. Shaw, R. C. Petersen, C. R. Jack, Jr., H. H. Feldman, A. L. Bokde, G. E. Alexander, P. Scheltens, B. Vellas, B. Dubois, M. Weiner and H. Hampel, *Neurobiol. Aging*, 2012, **33**, 1203.

58. L. K. McEvoy, C. Fennema-Notestine, J. C. Roddey, D. J. Hagler, Jr., D. Holland, D. S. Karow, C. J. Pung, J. B. Brewer and A. M. Dale, *Radiology*, 2009, **251**, 195.

59. L. K. McEvoy, D. Holland, D. J. Hagler, Jr., C. Fennema-Notestine, J. B. Brewer and A. M. Dale, *Radiology*, 2011, **259**, 834.

60. D. Heister, J. B. Brewer, S. Magda, K. Blennow and L. K. McEvoy, *Neurology*, 2011, **77**, 1619.

61. I. A. van Rossum, P. J. Visser, D. L. Knol, W. M. van der Flier, C. E. Teunissen, F. Barkhof, M. A. Blankenstein and P. Scheltens, *J. Alzheimers Dis.*, 2012, **29**, 319.

62. G. E. Christensen and H. J. Johnson, *IEEE Trans. Med. Imaging*, 2001, **20**, 568.

63. J. Ashburner and K. J. Friston, *Neuroimage*, 2000, **11**, 805.

64. D. Holland and A. M. Dale, *Med. Image Anal.*, 2011, **15**, 489.

65. N. C. Fox, G. R. Ridgway and J. M. Schott, *Neuroimage*, 2011, **57**, 15.

66. A. D. Leow, I. Yanovsky, M. C. Chiang, A. D. Lee, A. D. Klunder, A. Lu, J. T. Becker, S. W. Davis, A. W. Toga and P. M. Thompson, *IEEE Trans. Med. Imaging*, 2007, **26**, 822.

67. I. Yanovsky, A. D. Leow, S. Lee, S. J. Osher and P. M. Thompson, *Med. Image Anal.*, 2009, **13**, 679.

68. P. A. Yushkevich, B. B. Avants, S. R. Das, J. Pluta, M. Altinay and C. Craige, *Neuroimage*, 2010, **50**, 434.

69. M. Reuter, H. D. Rosas and B. Fischl, *Neuroimage*, 2010, **53**, 1181.

70. K. K. Leung, G. R. Ridgway, S. Ourselin and N. C. Fox, *Neuroimage*, 2011, **59**, 3995.

71. D. Holland, L. K. McEvoy and A. M. Dale, *Hum. Brain Mapp.*, 2012, **33**, 2586.

72. X. Hua, B. Gutman, C. P. Boyle, P. Rajagopalan, A. D. Leow, I. Yanovsky, A. R. Kumar, A. W. Toga, C. R. Jack, Jr., N. Schuff, G. E. Alexander, K. Chen, E. M. Reiman, M. W. Weiner and P. M. Thompson, *Neuroimage*, 2011, **57**, 5.

73. L. A. Beckett, D. J. Harvey, A. Gamst, M. Donohue, J. Kornak, H. Zhang and J. H. Kuo, *Alzheimers Dement.*, 2010, **6**, 257.

74. J. L. Cummings, *Neurobiol. Aging*, 2010, **31**, 1481.

75. W. K. Thompson and D. Holland, *Neuroimage*, 2011, **57**, 1.

76. A. Klein, J. Andersson, B. A. Ardekani, J. Ashburner, B. Avants, M. C. Chiang, G. E. Christensen, D. L. Collins, J. Gee, P. Hellier, J. H. Song, M. Jenkinson, C. Lepage, D. Rueckert, P. Thompson, T. Vercauteren, R. P. Woods, J. J. Mann and R. V. Parsey, *Neuroimage*, 2009, **46**, 786.

77. C. R. Jack, Jr., R. C. Petersen, Y. Xu, P. C. O'Brien, G. E. Smith, R. J. Ivnik, E. G. Tangalos and E. Kokmen, *Neurology*, 1998, **51**, 993.

78. N. C. Fox, S. Cousens, R. Scahill, R. J. Harvey and M. N. Rossor, *Arch Neurol.*, 2000, **57**, 339.

79. C. R. Jack, Jr., V. J. Lowe, S. D. Weigand, H. J. Wiste, M. L. Senjem, D. S. Knopman, M. M. Shiung, J. L. Gunter, B. F. Boeve, B. J. Kemp, M. Weiner and R. C. Petersen, *Brain*, 2009, **132**, 1355.

80. A. M. Fjell, K. B. Walhovd, C. Fennema-Notestine, L. K. McEvoy, D. J. Hagler, D. Holland, J. B. Brewer and A. M. Dale, *J Neurosci.*, 2009, **29**, 15223.

81. C. R. McDonald, L. K. McEvoy, L. Gharapetian, C. Fennema-Notestine, D. J. Hagler, Jr., D. Holland, A. Koyama, J. B. Brewer and A. M. Dale, *Neurology*, 2009, **73**, 457.

82. C. R. Jack, Jr., M. M. Shiung, J. L. Gunter, P. C. O'Brien, S. D. Weigand, D. S. Knopman, B. F. Boeve, R. J. Ivnik, G. E. Smith, R. H. Cha, E. G. Tangalos and R. C. Petersen, *Neurology*, 2004, **62**, 591.

83. B. H. Ridha, J. Barnes, J. W. Bartlett, A. Godbolt, T. Pepple, M. N. Rossor and N. C. Fox, *Lancet Neurol.*, 2006, **5**, 828.

84. T. E. Golde, L. S. Schneider and E. H. Koo, *Neuron*, 2011, **69**, 203.

85. D. J. Selkoe, *Nat. Med.*, 2011, **17**, 1060.

86. R. A. Sperling, C. R. Jack, Jr. and P. S. Aisen, *Sci. Transl. Med.*, 2011, **3**, 111cm33.

87. L. K. McEvoy, S. D. Edland, D. Holland, D. J. Hagler, J. C. Roddey, C. Fennema-Notestine, D. P. Salmon, A. K. Koyama, P. S. Aisen, J. B. Brewer and A. M. Dale, *Alzheimer Dis. Assoc. Disord.*, 2010, **24**, 269.

88. R. C. Petersen, P. S. Aisen, L. A. Beckett, M. C. Donohue, A. C. Gamst, D. J. Harvey, C. R. Jack, Jr., W. J. Jagust, L. M. Shaw, A. W. Toga, J. Q. Trojanowski and M. W. Weiner, *Neurology*, 2010, **74**, 201.

89. C. R. Jack, Jr., F. Barkhof, M. A. Bernstein, M. Cantillon, P. E. Cole, C. Decarli, B. Dubois, S. Duchesne, N. C. Fox, G. B. Frisoni, H. Hampel, D. L. Hill, K. Johnson, J. F. Mangin, P. Scheltens, A. J. Schwarz, R. Sperling, J. Suhy, P. M. Thompson, M. Weiner and N. L. Foster, *Alzheimers Dement.*, 2011, **7**, 474.
90. L. K. McEvoy and J. B. Brewer, *Imaging in Medicine*, 2012, **4**, 343.
91. J. M. Schott, J. W. Bartlett, J. Barnes, K. K. Leung, S. Ourselin and N. C. Fox, *Neurobiol. Aging*, 2010, **31**, 1452.
92. H. Hampel, G. Wilcock, S. Andrieu, P. Aisen, K. Blennow, K. Broich, M. Carrillo, N. C. Fox, G. B. Frisoni, M. Isaac, S. Lovestone, A. Nordberg, D. Prvulovic, C. Sampaio, P. Scheltens, M. Weiner, B. Winblad, N. Coley and B. Vellas, *Prog. Neurobiol.*, 2011, **95**, 579.
93. C. R. Jack, Jr., M. Slomkowski, S. Gracon, T. M. Hoover, J. P. Felmlee, K. Stewart, Y. Xu, M. Shiung, P. C. O'Brien, R. Cha, D. Knopman and R. C. Petersen, *Neurology*, 2003, **60**, 253.
94. R. Katz, *NeuroRx*, 2004, **1**, 189.
95. J. L. Whitwell, M. M. Shiung, S. A. Przybelski, S. D. Weigand, D. S. Knopman, B. F. Boeve, R. C. Petersen and C. R. Jack, Jr., *Neurology*, 2008, **70**, 512.
96. C. R. Jack, Jr., M. M. Shiung, S. D. Weigand, P. C. O'Brien, J. L. Gunter, B. F. Boeve, D. S. Knopman, G. E. Smith, R. J. Ivnik, E. G. Tangalos and R. C. Petersen, *Neurology*, 2005, **65**, 1227.
97. P. A. Freeborough and N. C. Fox, *IEEE Trans. Med. Imaging*, 1997, **16**, 623.
98. J. Barnes, J. W. Bartlett, L. A. van de Pol, C. T. Loy, R. I. Scahill, C. Frost, P. Thompson and N. C. Fox, *Neurobiol. Aging*, 2009, **30**, 1711.
99. A. M. Fjell, L. T. Westlye, I. Amlien, T. Espeseth, I. Reinvang, N. Raz, I. Agartz, D. H. Salat, D. N. Greve, B. Fischl, A. M. Dale and K. B. Walhovd, *Cereb. Cortex*, 2009, **19**, 2001.
100. K. B. Walhovd, L. T. Westlye, I. Amlien, T. Espeseth, I. Reinvang, N. Raz, I. Agartz, D. H. Salat, D. N. Greve, B. Fischl, A. M. Dale and A. M. Fjell, *Neurobiol. Aging*, 2009, **32**, 916.
101. X. Hua, S. Lee, I. Yanovsky, A. D. Leow, Y. Y. Chou, A. J. Ho, B. Gutman, A. W. Toga, C. R. Jack, Jr., M. A. Bernstein, E. M. Reiman, D. J. Harvey, J. Kornak, N. Schuff, G. E. Alexander, M. W. Weiner and P. M. Thompson, *Neuroimage*, 2009, **48**, 668.
102. N. Schuff, N. Woerner, L. Boreta, T. Kornfield, L. M. Shaw, J. Q. Trojanowski, P. M. Thompson, C. R. Jack, Jr. and M. W. Weiner, *Brain*, 2009, **132**, 1067.
103. X. Hua, S. Lee, D. P. Hibar, I. Yanovsky, A. D. Leow, A. W. Toga, C. R. Jack, Jr., M. A. Bernstein, E. M. Reiman, D. J. Harvey, J. Kornak, N. Schuff, G. E. Alexander, M. W. Weiner and P. M. Thompson, *Neuroimage*, 2010, **51**, 63.
104. M. Lorenzi, M. Donohue, D. Paternico, C. Scarpazza, S. Ostrowitzki, O. Blin, E. Irving and G. B. Frisoni, *Neurobiol. Aging*, 2010, **31**, 1443.

105. S. L. Risacher, L. Shen, J. D. West, S. Kim, B. C. McDonald, L. A. Beckett, D. J. Harvey, C. R. Jack, Jr., M. W. Weiner and A. J. Saykin, *Neurobiol. Aging*, 2010, **31**, 1401.
106. P. Vemuri, H. J. Wiste, S. D. Weigand, D. S. Knopman, J. Q. Trojanowski, L. M. Shaw, M. A. Bernstein, P. S. Aisen, M. Weiner, R. C. Petersen and C. R. Jack, Jr., *Neurology*, 2010, **75**, 143.
107. J. M. Schott, S. L. Price, C. Frost, J. L. Whitwell, M. N. Rossor and N. C. Fox, *Neurology*, 2005, **65**, 119.
108. N. C. Fox, R. S. Black, S. Gilman, M. N. Rossor, S. G. Griffith, L. Jenkins and M. Koller, *Neurology*, 2005, **64**, 1563.

Magnetic Resonance Imaging in Respiratory Diseases: From Diagnosis to Pharmaceutical Research and Development

NICOLAU BECKMANN,*[a] ALEXANDRE TRIFILIEFF,[b] CHRISTINE EGGER[a,c] AND YANNICK CRÉMILLIEUX[d]

[a] Novartis Institutes for BioMedical Research, Global Imaging Group, CH-4056 Basel, Switzerland; [b] Novartis Institutes for BioMedical Research, Respiratory Diseases Department, CH-4056 Basel, Switzerland; [c] Biocenter of the University of Basel, CH-4056 Basel, Switzerland; [d] Université Bordeaux 2, F-33076 Bordeaux, France
*Email: nicolau.beckmann@novartis.com

17.1 Introduction

Diseases of the airways such as asthma and chronic obstructive pulmonary disease (COPD) involve a complex interplay of many inflammatory and structural cell types, all of which can release inflammatory mediators including cytokines, chemokines, growth factors, and adhesion molecules. Activated eosinophils are considered particularly important in asthma, contributing to epithelial cell damage, bronchial hyperresponsiveness, plasma exudation and edema of the airway mucosa, as well as smooth muscle hypertrophy and mucus plugging, through the release of enzymes and proteins.[1,2] In COPD, inflammation of the small airways and lung parenchyma with the involvement of

New Developments in NMR No. 2
New Applications of NMR in Drug Discovery and Development
Edited by Leoncio Garrido and Nicolau Beckmann
© The Royal Society of Chemistry 2013
Published by the Royal Society of Chemistry, www.rsc.org

neutrophils, macrophages and T-lymphocytes results in chronic obstructive bronchitis, destruction of the lung parenchyma by proteolytic enzymes (emphysema) and mucus hypersecretion leading to severe airflow limitation.[1,2] Pulmonary fibrosis is a progressive and lethal lung disease involving an over-exuberant repair process, characterized by accumulation of inflammatory cells, excessive fibroblast proliferation, increase in collagen content and deposition of extracellular matrix in the lungs.[3,4]

Spirometry measuring the volume and flow of inhalation and exhalation is the most common approach for diagnosing lung function abnormalities in humans. This method lacks regional information and can have a large range of variation based on patient effort and cooperation. Moreover, subtle changes due to *e.g.* treatment may be masked by the global assessment of lung function since non-diseased lung tissue may compensate for the functional impairment of diseased lung tissue.

Laboratory animals provide models for airways diseases in humans to help developing novel therapies. Terminal procedures such as broncho-alveolar lavage (BAL) fluid analysis, histology and weighing of lungs are commonly used to analyze such models. Pulmonary function is assessed either non-invasively in conscious, unrestrained animals (plethysmography), or invasively requiring intubation or tracheotomy and artificial ventilation.[5] The main concern with whole body plethysmography is that it provides respiratory measures that are so tenuously linked to respiratory mechanics that it is debatable if they can be considered as meaningful indicators of lung function.[5] Having access to non-invasive, spatially resolved readouts is therefore highly desirable for both ethical and scientific reasons.

A potentially improved diagnosis capability is the motivation for the introduction of magnetic resonance imaging (MRI) in the context of development of new therapies for respiratory diseases. The present contribution addresses the use of MRI both in animal models and in humans. The main focus is on the applications to derive information on several aspects of pulmonary diseases, ranging from inflammation to fibrosis, with the ultimate objective to support and facilitate the drug discovery and development process in this medical area.

17.2   Lung Imaging: Basic Considerations

Table 17.1 provides an overview of current imaging modalities of interest for imaging the lungs. The interested reader is referred to other reviews for detailed descriptions of each method for small animal[6–8] or human applications.[9–12] In many respects the imaging techniques are complementary; there is no "all-in-one" imaging modality providing optimal sensitivity, specificity and temporo-spatial resolution. For instance, due to its relatively low sensitivity, MRI is of limited value for detecting molecular processes *in vivo*; nevertheless, its high spatial resolution provides a good anatomical reference for molecular data obtained with high-sensitivity, low-resolution imaging modalities. Co-registration may be achieved by post-processing of data obtained in different

Table 17.1 Current imaging modalities of interest in pulmonary drug research and discovery.

| Technique | Spatial resolution and time scale | | Application | Main characteristics |
	Small rodents	Humans		
Computerized tomography (CT)	50–100 μm; s to min.	∼600 μm; <1 min.	Anatomical, functional	Ionizing radiation; poor soft tissue contrast
Scintigraphy	No	∼20 mm	Functional	Planar information
Single photon emission tomography (SPECT) (low-energy γ-rays)	≤1 mm; ∼20 min.	∼10 mm; 10–30 min.	Functional	Ionizing radiation; radioisotopes have longer half-lives than those used in PET; sensitivity 10 to 100 times smaller than PET
Positron emission tomography (PET) (high-energy γ-rays)	1–2 mm; ∼20 min.	∼4 mm; ∼20 s	Metabolic, functional, molecular	Ionizing radiation; high sensitivity (picomolar concentrations); cyclotron needed
Magnetic resonance imaging (MRI)	80–140 μm; s to h	∼1.5 mm; s to min.	Anatomical, functional, molecular	High spatial resolution and soft tissue contrast; low sensitivity
Near-infrared fluorescence (NIRF) optical imaging	1–3 mm; s to min.	No	Molecular	High sensitivity (nanomolar concentrations); max. penetration depth <10 cm

imaging sessions or by simultaneous multimodality imaging such as PET-MRI (see also Chapter 10),[13] PET-CT[14] and SPECT-CT.[15]

A challenge in lung imaging is that cardiac and respiratory motion can cause marked image artifacts. In humans, image acquisition may be performed during breathhold, or by gating it by an electrocardiogram. Problems are more evident in small rodents, because of their higher cardiac and respiratory rates. To address this issue, measurements are often performed in artificially ventilated animals to maintain a constant breathing rate and/or image acquisition is triggered by the electrocardiogram. However, for compound testing *in vivo* in animal models of airways diseases, it is important to keep acquisition conditions as simple as possible so that repeated measurements interfering minimally with the physiology and the well-being of the animals can be performed longitudinally. For instance, one needs to carefully consider possible interferences between the pathophysiology of the disease models and lung injury complications that might potentially be caused by mechanical ventilation,[16] especially if this is applied repeatedly. Indeed, it has been reported that mechanical ventilation of healthy rats can cause an increase of neutrophils in broncho-alveolar lavage (BAL) fluid, pulmonary edema and even hypoxemia that may lead to progressive circulatory failure and death. Consequently, mechanical ventilation should be avoided whenever possible for the longitudinal investigation of lung disease models with expected inflammatory responses. Signal averaging allows the acquisition of lung images from spontaneously breathing rats and mice, without any gating.[17–19]

For ethical reasons, animals are kept anesthetized during imaging. One needs to consider what influence anesthesia may have on functional readouts and whether it may interfere with the development of the disease model, especially under repeated inductions. It is recommended to keep the anesthesia time to ≤ 30 min. for each session.

Besides cardiac and respiratory motion, lung MRI faces two main issues, namely (i) the rapid dephasing of transverse magnetization (transverse relaxation times measured in lung tissue are typically ranging between 0.5 and 2 ms, depending on the strength of the static magnetic field) and (ii) the low spin density (related to the presence of alveolar space) as compared to other organs of interest such as the heart or brain. Dedicated proton or non-proton nuclei MR techniques are available and can help overcome these difficulties.

Proton MRI sequences referred as UTE (ultra-short echo time) techniques are associated with echo times below 1 ms that allow the acquisition of the MRI signal from lung parenchyma before decay of the magnetization.[20] When performed with a radial scanning of k-space, UTE sequences are less sensitive to respiratory and cardiac motion as compared to Cartesian acquisition.[19,21]

Alternatively, MRI of nuclei located on gaseous molecules or atoms can be performed to image the airway compartment in the lung. MRI of fluorinated gases associated with a large number of ^{19}F nuclei (SF_6, C_2F_6, CF4 or C_3F_8) has been demonstrated in animal and human studies.[22] Hyperpolarization techniques based on optical pumping approaches can be used to further improve the detection sensitivity of a gas (see references 23 and 24 for reviews on HP techniques). Following hyperpolarization, ^3He and ^{129}Xe NMR signals can be increased by five orders of magnitude. MRI of hyperpolarized (HP) gases has been applied in animal models of airways diseases and in emphysema, cystic fibrosis (CF) or asthma patients (see next section).[24] Most small rodent studies have been performed in artificially ventilated animals in order to better control gas delivery; however, measurements can be performed in spontaneously respiring animals as well.[25–27]

17.3 MRI in Respiratory Diseases: From Animal Models to Patients

In the next subsections, the use of imaging to characterize non-invasively several aspects of lung diseases will be discussed. Whenever possible, references are made to pharmacological studies involving imaging, either in animal models or in patients. For many applications, however, studies involving compounds have not been reported. Nonetheless, the applications are addressed because of their potential in becoming useful tools for drug discovery in the near future.

17.3.1 Airway Inflammation

A characteristic feature of lung inflammation is edema in the airways due to an increase in the permeability of the microvasculature. Proton MRI has been used to quantify edema in the lungs of spontaneously breathing mice[18,28] or rats[17,29,30] actively sensitized to and challenged with ovalbumin (OVA). The MRI signals following OVA challenge correlated significantly with a variety of inflammatory parameters determined in the BAL fluid recovered from the same animals. Importantly, the fluid signals detected by MRI correlated significantly with the perivascular edema assessed by histology.[18,31]

When assessing the effects of anti-inflammatory drugs administered prior to disease induction in these models, a dose-related reduction of the MRI signals has been shown for compounds such as the glucocorticosteroids, budesonide[17,31,32] and mometasone,[33] and a mitogen-activated protein kinase inhibitor.[33] Moreover, the pharmacology of sphingosine-1 receptors has been studied *in vivo* using MRI (Figure 17.1).[32] Imaging data correlated with changes in the parameters of inflammation assessed in the BAL fluid. MRI was also applied to address the effects of compounds on established allergic inflammation. Treatment with budesonide, mometasone or with a phosphodiesterase-4 inhibitor at 24 h after OVA challenge reduced MRI signals already at 3 h after drug administration. The decline in MRI signals correlated significantly with a reduction in perivascular edema quantified by histology.[17,31,33] No changes in BAL parameters were observed at this early timepoint. These observations indicate that proton MRI is more suitable to detect early effects of compounds on established inflammation than the traditional BAL fluid analysis. Effects of corticosteroid treatment on airway inflammation, mechanics and hyperpolarized ^{3}He MRI in an allergic mouse model have been reported recently.[34]

Compared to single dosing, repeated OVA challenge in actively sensitized rats induced an attenuation of the inflammatory response as evidenced by proton MRI and BAL fluid analysis.[35] Moreover, vascular remodeling has been detected by MRI in the model: the decrease in vascular permeability assessed by dynamic contrast enhanced (DCE)-MRI was consistent with the thickening of the vascular wall for vessels of diameter up to 300 μm as revealed histologically.[35] Clinically, the assessment of lung edema with MRI has not been widely applied yet.[10]

While MRI can be efficiently used to image the exudative component of acute inflammation in the lungs, very few imaging techniques are available to estimate the leukocyte activity in acute or chronic lung inflammation. It has been reported that detection of lung inflammation in COPD was feasible using ^{18}fluoro-deoxyglucose (^{18}FDG) and positron emission tomography (PET).

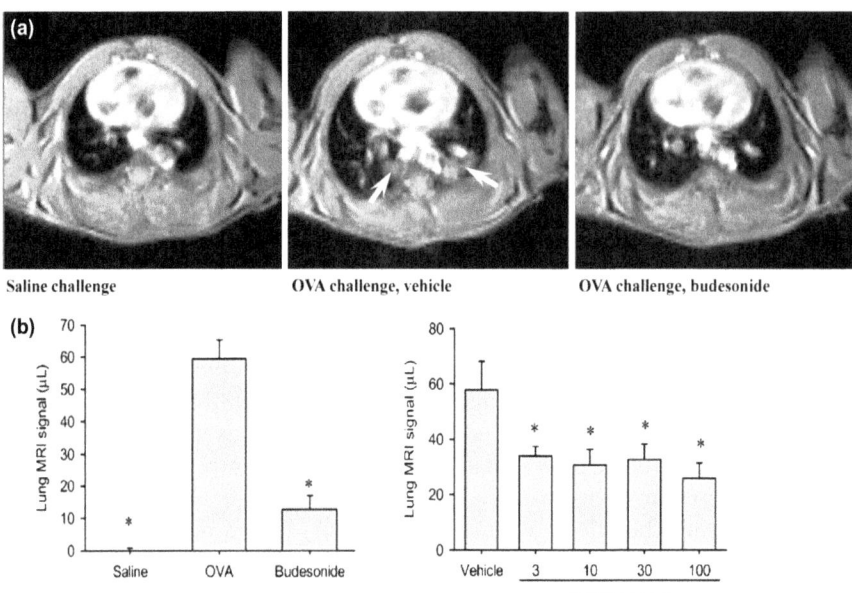

Figure 17.1 Proton MRI in a murine model of lung inflammation. (a) Images from three actively sensitized mice, intranasally challenged with saline or OVA. The OVA-challenged animals received either vehicle or budesonide ($3 \, mg \, kg^{-1}$) treatment, *via* the intranasal route. Arrows indicate MRI lung fluid signals. Measurements were performed in spontaneously breathing mice. (b) Summary of MRI fluid signal volumes for saline- or OVA-challenged, actively sensitized mice. Animals receiving OVA were treated with either budesonide ($3 \, mg \, kg^{-1}$) or the sphingosine-1-phosphate (S1P) receptor agonist, FTY720. The significant reduction of signal in the lungs of budesonide- and FTY720-treated mice reflect the anti-inflammatory effects of the compounds. For details, see Blé *et al.*[32]

Since inflammatory cells utilize glucose as a source of energy during their activation, it was suggested that [18]FDG uptake by inflammatory cells in the lung could be used as an *in vivo* measurement of regional lung inflammation.[36–38] More recently, Ebner *et al.*[39] investigated the lipopolysaccharide (LPS)-induced inflammation process by means of emulsified perfluorocarbons (PFC). Intravenous application of PFC particles in mice resulted in their accumulation in inflammatory regions of the lungs as detected with ^{19}F MRI (Figure 17.2). The authors showed that PFC particles were transported to the site of inflammation *via* circulating monocytes/macrophages. Furthermore, the regions where PFC infiltration was detected showed the presence of edema on proton MR images at the later timepoints, indicating that this approach was able to selectively localize the cellular components of inflammation in the lungs. Additionally, the authors demonstrated the ability of the technique to monitor the effect of anti-inflammatory therapies in this animal model of lung inflammation.

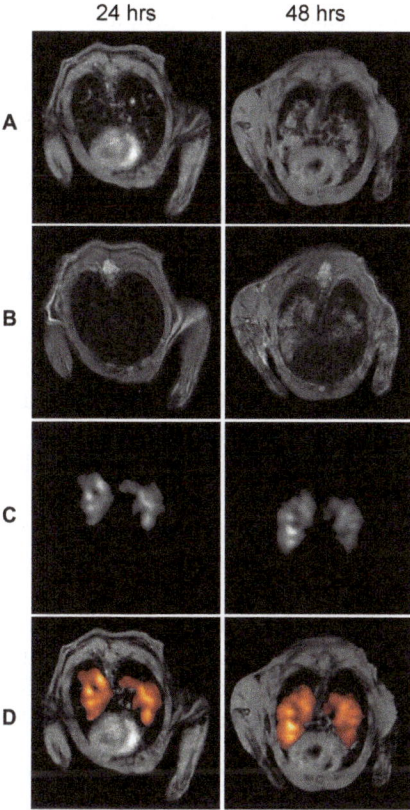

Figure 17.2 ^{19}F signals precede structural changes in LPS-induced inflammation. Representative respiratory-gated proton (A, gradient-echo; B, fast spin echo) and [19]F (C) MRI of the mouse thorax 24 and 48 h after intratracheal administration of LPS ($0.3 \, \mu g \, g^{-1}$ body weight). For merging of data sets (D), a hot iron color table was applied to the ^{19}F images. On day 1, ^{19}F MRI clearly detected the accumulation of perfluorocarbons in both lung lobes, whereas no hyperintense signals were observed in proton images. Follow-up proton MRI investigation on day 2 revealed bilateral infiltrates in the same anatomic region where ^{19}F signals had been observed at day 1.

Reproduced with permission from Ebner *et al.*[39] Copyright © 2010 American Heart Association.

17.3.2 Mucus Secretion and Clearance

Chronic mucus hypersecretion and dysfunctions in mucociliary clearance are associated with the accelerated loss of lung function in several respiratory diseases. Measurements of mucus clearance in the mouse lungs have been performed using scintigraphy in combination with the administration of ^{99m}Tc-labeled sulfur colloid.[40] The same approach has been applied to monitor the retention and clearance of radiolabeled human serum albumin and sulfur colloids in dogs.[41] Also clinically, radiolabeled, inert particles have been widely used to assess mucociliary clearance. From the ratio of particle deposition in central *versus* peripheral regions of the lungs it could for instance be clearly demonstrated that treatment of CF patients with hypertonic saline significantly increased mucociliary clearance.[42]

Exposure of Brown Norway rats to LPS induces mucus release.[43] MRI can detect secreted mucus in the lungs of spontaneously respiring rats up to 8 days following a single LPS challenge.[29,44] Bilabeled amino dextran-based probes binding specifically to mucus have been synthetized to extract information on mucus dynamics in this model.[45]

An upregulation of sensory-efferent neural pathways is implicated in asthma and COPD. The acute effects of sensory nerve stimulation by capsaicin in the rat lung have been studied by MRI.[46] Capsaicin-induced MRI signals reflected the release of mucus following activation of sensory nerves. The transient receptor potential vanilloid-1 antagonist, capsazepine, the dual neurokinin-1,2 receptor antagonist, DNK333, and the mast cell stabilizer, di-sodium cromoglycate, blocked the effects of capsaicin in the airways.

The epithelial sodium channel (ENaC) regulates airway mucosal hydration and mucus clearance. The lack of such regulation in CF patients leads to desiccation of the airway lumen, resulting in mucostasis that establishes the environment for infections.[47] Osmotic agents and negative ENaC regulators can be used to restore mucosal hydration. Proton MRI has been shown to provide a target-related readout to study modulators of lung fluid hydration in spontaneously breathing rats.[48]

Puderbach *et al.*[49] explored the clinical suitability of MRI to assess mucus plugging by comparison with high-resolution computerized tomography (HRCT) and found a median lobe-related concordance of 77% for mucus plugging. HRCT provided superior image quality, but could not discriminate bronchial wall thickening from mucus. By using contrast agents, which distribute into the bronchial wall but not into intrabronchial secretions, MRI is able to discriminate both elements.[50]

^{3}He MRI in CF patients was first reported in 1999[51] and since then HP ^{3}He was used to depict ventilation defects and to estimate CF disease severity in studies in adults[52] and in children.[53,54] Mentore *et al.*[52] studied the effect of aerosol treatment (albuterol and DNAse) and chest physiotherapy on ventilation function in a population of young CF patients with a large range of lung dysfunction (mean FEV_1 of 67% $\pm$ 27). They quantified the number of ventilation defects per slice and observed a significant decrease of ventilation defects after albuterol and a trend towards increasing ventilation defects following DNAse treatment and chest physiotherapy. A poor correlation between FEV_1 values and ^{3}He readouts was reported.

In pediatric CF patients it was reported that, although global lung ventilation volume stayed constant, lung ventilation varied regionally following chest physiotherapy.[55] Such changes were attributed to mucus plug movement in response to treatment. Bannier *et al.*[56] observed ventilation changes (in both number of defects and size) after chest physiotherapy in young CF individuals

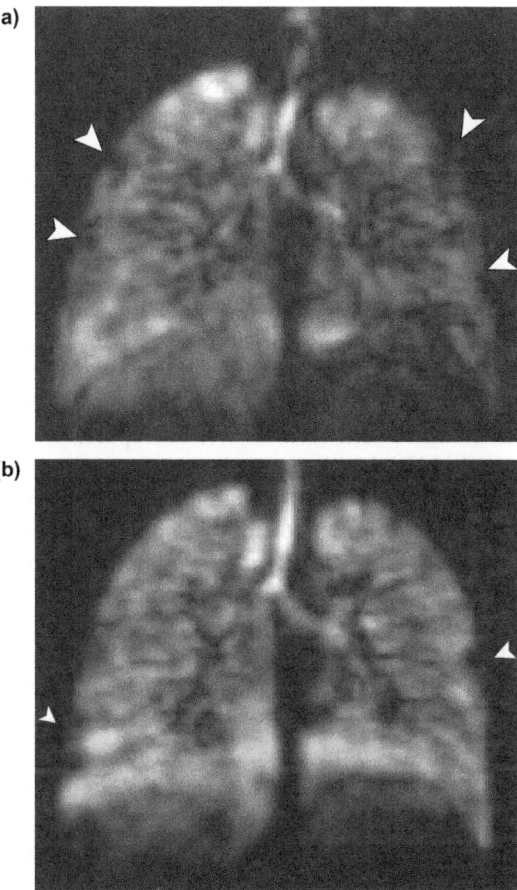

Figure 17.3 ^{3}He-MR images from a CF patient acquired (a) before and (b) after a single chest physiotherapy session. Ventilation defects are reduced in size and number, and the overall gas distribution looks more homogeneous after chest physiotherapy. Arrowheads = subsegmental ventilation defects in apical, anterior and medial segments.
Reproduced with permission from Bannier *et al.*[56] © 2010 American Radiological Society.

with normal spirometric results (mean FEV_1 of 122% ± 14.5). These results illustrate the high sensitivity of ^{3}He MRI in depicting ventilation defects (Figure 17.3).

17.3.3 Emphysema

HRCT is still the method of choice to assess emphysematous pathology in patients and has been shown to correlate well with both morphometric analysis and functional assessment.[10] Following standardization, the emphysematous destruction of lung parenchyma in patients can be well quantified with HRCT.[57] So far, clinical MRI could not be successfully applied for this task. For inflammation (see Section 17.3.1), hypoxic vasoconstriction (Section 17.3.4.2) and pulmonary arterial hypertension (Section 17.3.6) that accompany emphysema, MRI shows more clinical promise.

In contrast, MRI has been applied successfully in animal studies. Using a single-point imaging technique to achieve short echo times, Olsson et al.[58] determined the relaxation time T_2^* to detect emphysematous changes in the lungs of tight-skin mice, which spontaneously develop emphysema-like alveolar enlargement. Tight-skin mice displayed significantly shorter T_2^* values than control, age-matched mice, because their larger alveoli result in an increased air/tissue ratio and hence an increase in the internal susceptibility gradients. The T_2^* of the lung parenchyma at 4.7 T was approximately 0.46 ms for control mice, being in excellent agreement with the 0.48 ms reported earlier.[59] Also ultra-short echo time (UTE) MRI has been used to compare normal and emphyse-matous lungs in mutant mice.[60] Using a similar MRI sequence, Zurek et al.[61] investigated the correlations between lung histomorphological parameters (mean chord length in lung tissue) and MR parameters (signal intensity and T_2^* values) in an elastase-instilled mouse model of emphysema (Figure 17.4). The results showed an excellent agreement between MR findings and histological morphometry and indicated that proton MRI allows structural changes at alveolar level to be monitored longitudinally. In the clinics, T_2^* assessments with UTE sequences have been demonstrated to be potentially as useful as thin-section multidetector-row computerized tomography (CT) for pulmonary function loss assessment and clinical stage classification of COPD in smokers.[62]

Gradient-echo proton MRI has been used to detect elastase-induced changes in the lungs of spontaneously breathing rats.[63] Reductions in MRI signal intensity of the lung parenchyma detected from 2 to 8 weeks following the insult correlated significantly with the loss of alveolar structure assessed by histology, suggesting that the MRI signal reflected elastase-induced alveolar destruction.[7] Treatment with retinoic acid did not elicit a reversal of lung damage as measured by MRI and histology.

Emphysema has also been studied in small rodents and clinically using HP ^{3}He-diffusion MRI. The apparent diffusion coefficient (ADC) is a sensitive measure for the airspace size. Significantly increased ADC values have been obtained in elastase-induced emphysema in rats and mice[64–66] and in patients. Good correlations between ADC values and histology have been reported in humans.[67] The initial enthusiasm for this approach has been slightly dampened by two longitudinal studies. In a one-year study on COPD patients, disease progression was apparent from an increase in functional residual capacity, which was not matched by an increase in ^{3}He ADC values, but neither by a significant change in FEV_1 or diffusion of CO. However, lacunarity, a measure of the inhomogeneity in distribution of ADC values, did correlate with disease progression.[68] In a two-year study in α1-antitrypsin deficient patients, significant changes were reported for FEV_1, diffusion of CO, tissue density and emphysema index from CT, but the increase in ^{3}He ADC values just missed significance.[69]

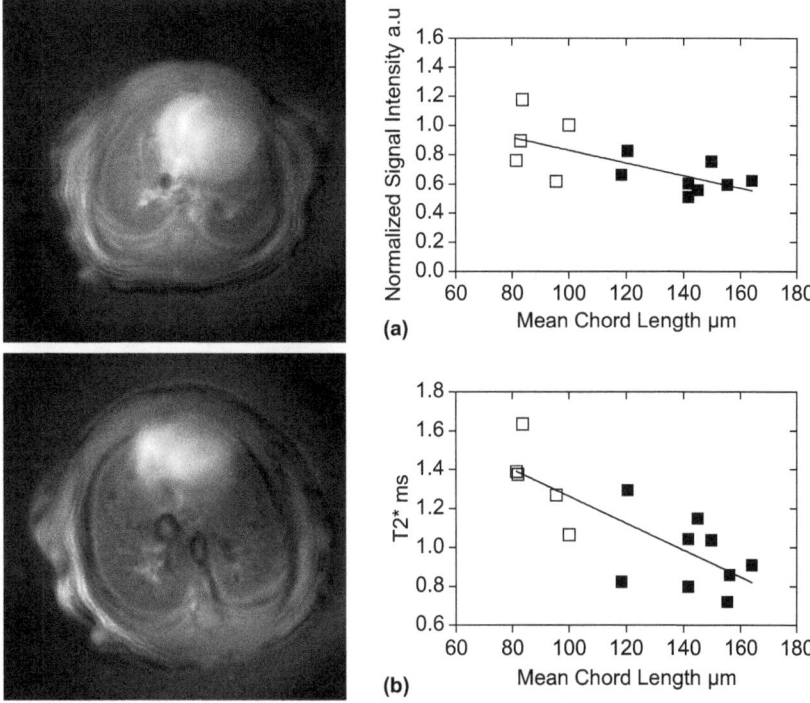

Figure 17.4 Left: Representative axial proton MR images of the mouse thorax acquired at 4.7 T using a UTE sequence (echo time 0.55 ms, slice thickness 1.7 mm). The top image was obtained in a control animal instilled with saline while the bottom image corresponds to an animal instilled with elastase. Both images were acquired 8 weeks following the instillation. Right: Correlation between the mean chord length measured histologically and MR signal intensity (a) and mean T_2^* values (b). Correlation includes the control (open squares) and elastase-instilled animals (full squares).
Reproduced with permission from Zurek *et al.*[61] © 2011 Wiley Periodicals, Inc.

Given its greater availability and lower cost, ^{129}Xe ADC MRI offers an alternative to ^{3}He ADC MRI. Kaushik *et al.*[70] demonstrated recently the feasibility of HP ^{129}Xe ADC MRI on COPD subjects (Figure 17.5). The mean parenchymal ADC was 0.036 ± 0.003 cm^2 s^{-1} for healthy volunteers, 0.043 ± 0.006 cm^2 s^{-1} for age-matched healthy controls and 0.056 ± 0.008 cm^2 s^{-1} for COPD subjects with emphysema. In healthy individuals, but not the COPD group, ADC decreased significantly in the anterior-posterior direction by $\sim 22\%$, likely because of gravity-induced tissue compression. The COPD group exhibited a significantly larger superior-inferior ADC reduction ($\sim 28\%$) than the healthy groups ($\sim 24\%$), consistent with smoking-related tissue destruction in the superior lung. Superior-inferior gradients in healthy subjects may result from regional differences in xenon concentration. ADC was

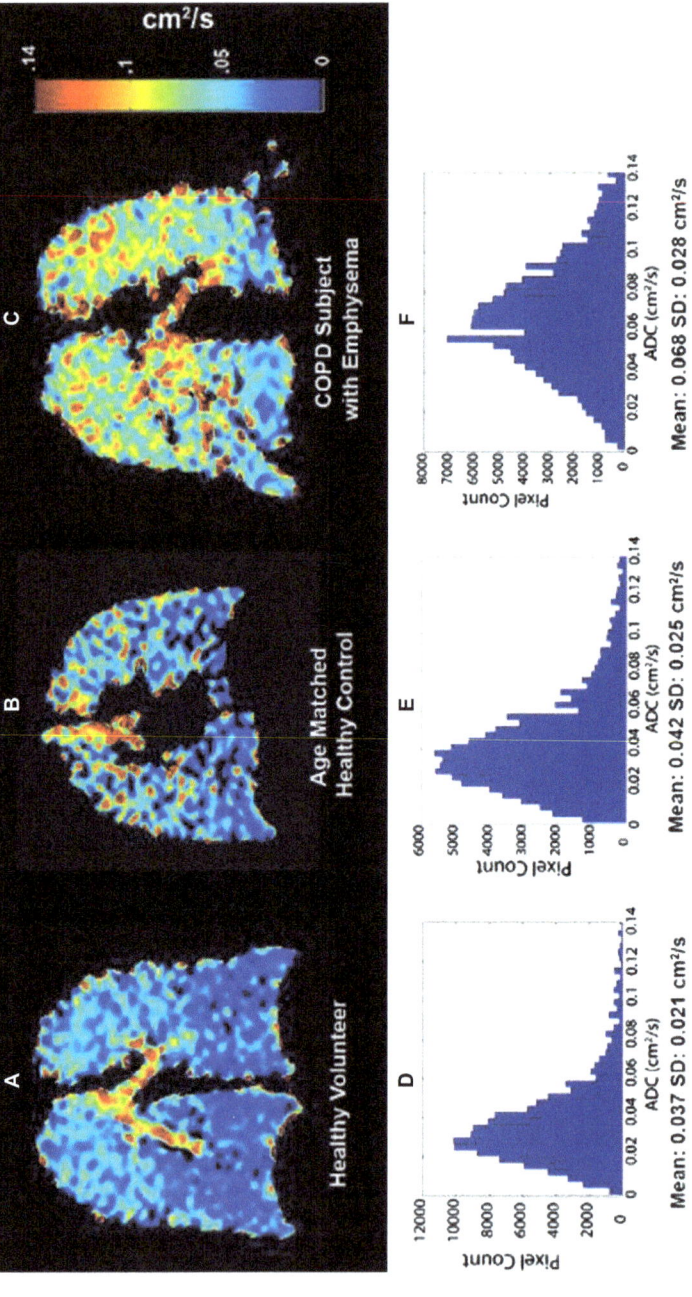

Figure 17.5   Representative slices from ^{129}Xe ADC maps and corresponding whole-lung ADC histograms. A: Healthy volunteer (age = 28 years) with a low mean ADC of $0.037 \pm 0.021 \ cm^2 \ s^{-1}$ indicating normal alveolar microstructure. The ADC values in the airways are higher ($0.083 \pm 0.029 \ cm^2 \ s^{-1}$) and reflect nearly free diffusion. B: Age-matched healthy control displaying similarly low parenchymal ADC values ($0.042 \pm 0.025 \ cm^2 \ s^{-1}$). C: COPD subject with emphysema shows high ADC values ($0.068 \pm 0.028 \ cm^2 \ s^{-1}$) in the parenchyma, indicating alveolar destruction. D: Whole-lung histogram for the healthy volunteer in panel A showing narrow ADC distribution. E: Whole-lung histogram for the age-matched control in panel B, exhibits a similarly homogenous distribution. F: Whole-lung histogram corresponding to the COPD subject with emphysema in panel C, exhibiting a moderately broader distribution.
Reproduced with permission from Kaushik *et al.*[70] © 2010 Wiley-Liss, Inc.

significantly correlated with pulmonary function tests. In healthy groups, ADC increased with age by $0.0002\,cm^2\,s^{-1}$ year^{-1}.[70] These data indicate that ^{129}Xe ADC MRI is sufficiently sensitive to distinguish healthy volunteers from subjects with emphysema, and detects age- and posture-dependent changes.

17.3.4 Lung Ventilation and Perfusion

Ventilation and perfusion distribution in the lung form the foundation of pulmonary physiology and remain cornerstones in pathology. Efficient gas exchange in the lungs can only occur through intimate matching of regional ventilation and perfusion. The non-uniform distribution of regional lung blood flow and ventilation were first demonstrated utilizing radioactive tracers and external scintillation detectors that registered the distribution of radioactivity within the lung. After intravenous injection, particles larger than red blood cells (RBCs) are trapped in the first capillary bed that they encounter.

17.3.4.1 Ventilation

Ventilation imaging has been shown to be sensitive to a variety of lung disease models, including asthma in mice,[71] emphysema in rats[72] and pulmonary embolism in sheep.[73] Ventilation can be quantified from the dynamic change in image signal following application of an inhaled contrast agent. For instance, ventilation imaging has been demonstrated in animals with ^{3}He MRI.[71]

Hyperpolarized ^{3}He-MRI has been used to measure the fractional ventilation in artificially ventilated small rodents.[74,75] The approach detected early changes of lung function and structure in a rat model of elastase-induced emphysema at mild and moderate severities.[75] The fractional ventilation declined primarily in the first 5 weeks, while enlargement of alveolar diameters appeared primarily between the 5th and 10th week post-elastase. Further HP ^{3}He-MRI studies focused on airway constriction induced chemically. High-resolution ^{3}He-MRI was used to depict regional ventilation changes and airway narrowing in artificially ventilated mice[76] or rats[77] challenged with methacholine. Mosbah *et al.*[26] demonstrated in a spatially resolved manner the effects of serotonin-induced bronchoconstriction on lung ventilation in spontaneously breathing rats. Dynamic ventilation ^{3}He MR images spanning a respiratory cycle were obtained using a retrospective CINE image reconstruction procedure.[25] Clinical applications of HP noble gas MRI have also slowly matured. In particular, ^{3}He-MRI has been used to characterize ventilation heterogeneity and defects in asthma,[78] CF[52,54] and COPD patients.[79] Individuals with asthma demonstrated ventilation defects both before and after a challenge with methacholine or exercise[80,81] and a significant overlap of ^{3}He defects and hyperlucency on HRCT was found.[78] In COPD patients, mostly ^{3}He diffusion MRI has been applied to examine alveolar air space size as mentioned in Section 17.3.3. However, this parameter can only be assessed in reasonably well-ventilated lung tissue. To date, very few studies employing ^{3}He MRI have reported on therapeutic interventions: preliminary results of bronchodilator

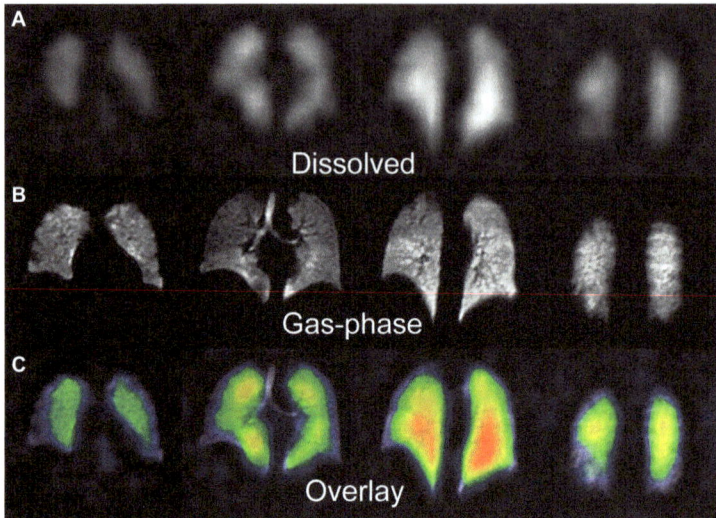

Figure 17.6 HP ^{129}Xe MR imaging. Panels are arranged with the more anterior portions of the lungs shown to the left and posterior portions to the right. (A) 15-mm-thick sections from a dissolved-phase HP ^{129}Xe image (12.5×12.5 mm^2 in-plane resolution) of a healthy human volunteer. (B) Corresponding 15-mm-thick slices from a gas-phase HP ^{129}Xe image of the same subject (3.2×3.2 mm^2 in-plane resolution). (C) Dissolved ^{129}Xe image from (A) displayed in color and overlaid on the grayscale ventilation image from (B).
Reproduced with permission from Cleveland *et al.*[83] © 2010 Cleveland *et al.*

therapy in asthmatics[80] and immediate effects of standard chest physiotherapy on regional ventilation with spirometry unchanged in CF patients.[55] Due to a potential shortage of ^{3}He, improved polarization technology for ^{129}Xe and the solubility of Xe in tissue, enabling simultaneous ventilation and dissolved phase images, have renewed interest in clinical ^{129}Xe MRI (Figure 17.6).[82,83] However, the anesthetic properties of Xe are of some concern.

Other techniques not relying on the administration of gases have been used to derive information on ventilation. Proton MRI detected, in spontaneously breathing rats, the effects of broncho-modulating agents or of inflammation-induced airway remodeling and hyporesponsiveness in OVA- or LPS-challenged animals, respectively.[84,85] The approach consists in detecting modulations of lung parenchymal proton signals induced by changes in oxygenation levels.[86] Oxygen-enhanced MRI has been further explored clinically in smoking-induced COPD patients[87,88] and in asthmatics.[89] Ohno *et al.*[88] showed recently in a large cohort of individuals that dynamic O_2-enhanced MRI is as efficacious as thin-section multidetector-row CT to quantify COPD-related changes in patients (Figure 17.7). It needs to be emphasized that the oxygen-enhanced MRI maps show not only correlation with ventilation limitations in the lungs, but also with oxygen transport capacity.

17.3.4.2 Perfusion

Clinically, regional lung perfusion is mostly assessed with radionuclide scintigraphy, which, however, suffers from limited spatial and temporal resolution. Three-dimensional (3D) MRI has provided the required high spatial and temporal resolution to analyze the first passage through the lungs of a bolus of MRI contrast agent. MRI lung perfusion has been used to demonstrate perfusion abnormalities in patients with CF[90] and emphysema.[91]

Fourier decomposition (FD) MRI was recently introduced, enabling lung perfusion and ventilation imaging without the administration of contrast agents.[92] Information about regional perfusion- and ventilation-related information (parenchyma density) is obtained during a single study using spectral analysis of MRI data. Rapid image acquisition with *e.g.* the TrueFISP sequence enables the observation of signal changes caused by respiratory motion and pulsation of the pulmonary blood, and a Fourier decomposition approach is used to separate blood signal and lung parenchyma signal. The technique has been compared to HP ^{3}He and DCE-MRI in a controlled experiment in pigs.[93] Baseline FD, ^{3}He and DCE MRI in healthy animals showed homogeneous ventilation and perfusion. Pulmonary embolism was then artificially produced, and FD and DCE MRI perfusion measurements were repeated. Subsequently, atelectasis and air trapping were induced, followed by FD MRI and ^{3}He MRI ventilation measurements. Functional defects were detected by all MRI techniques at identical anatomical locations (Figure 17.8). Signal intensity in ventilation- and perfusion-weighted FD images was significantly lower in pathological than in healthy lung parenchyma. The study has demonstrated FD MRI to be an alternative, non-invasive and easily implementable technique for the assessment of acute changes in lung function.

The combined challenges of high temporal and spatial resolution have rendered routine quantitative perfusion imaging difficult in small rodents. MRI perfusion assessments in animals have been primarily accomplished using contrast-enhanced techniques comprising the dynamic acquisition of images in combination with the administration of a paramagnetic contrast agent.[94] Such an approach has been used to analyze a rabbit model of pulmonary embolism[95] and a newborn piglet model of pulmonary hypertension.[96] Mistry *et al.*[97] developed a CINE technique based on the acquisition of radial images during repeated contrast agent injection matched to the physiology of the animal using a microinjector, enabling perfusion imaging at high spatial and temporal resolution in artificially ventilated small rodents. The feasibility of lung perfusion with arterial spin labeling precluding the administration of contrast material has been demonstrated in rabbit models of pulmonary embolism,[98] and during repeated balloon occlusion of a segmental pulmonary artery as well as during pharmacological stimulation in pigs.[99]

Driehuys *et al.*[100] showed in artificially ventilated rats that regional evaluation of pulmonary perfusion and gas exchange can be obtained by intravenous injection of saline saturated with HP ^{129}Xe and subsequent MRI of the gas phase in the alveolar airspaces. After a single injection, the emerging

^{129}Xe gas could be detected separately from ^{129}Xe remaining in the blood because of chemical shift differences. The features observed in dissolved-phase ^{129}Xe MR images are consistent with gravity-dependent lung deformation, which produces increased ventilation, reduced alveolar size (*i.e.* higher surface-to-volume ratios), higher tissue densities and increased perfusion in the dependent portions of the lungs. These results suggest that dissolved HP ^{129}Xe imaging reports on pulmonary function at a fundamental level.

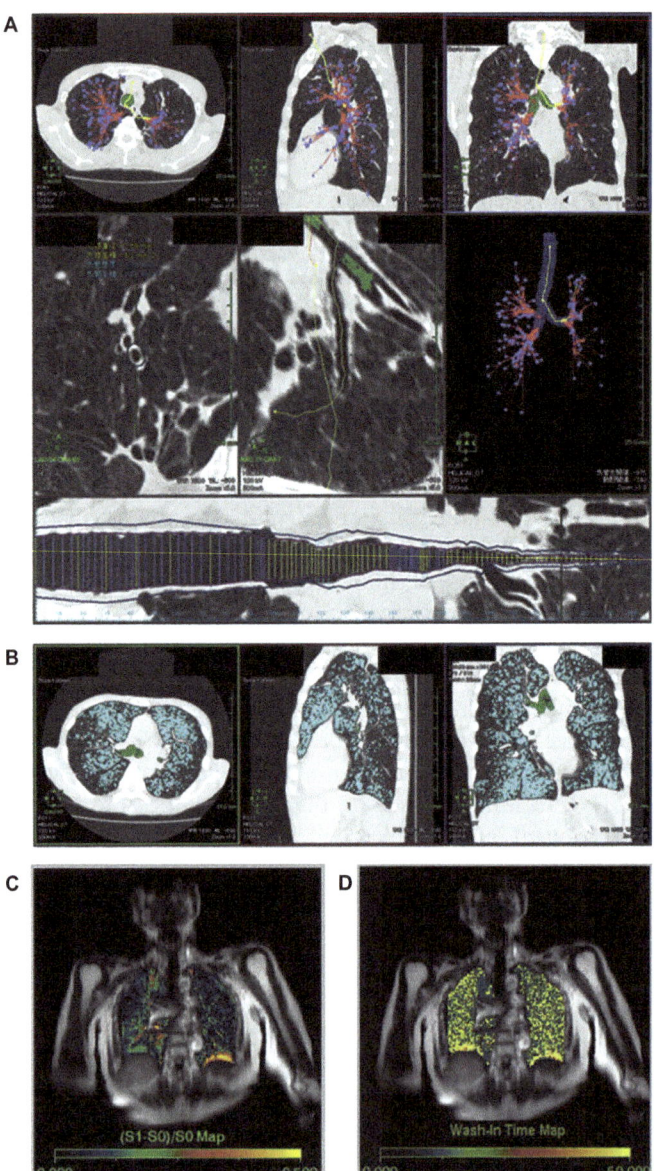

17.3.5 Lung Fibrosis

Currently no efficacious treatment exists for lung fibrosis. Animal models are important to investigate pathological mechanisms and for pre-clinical evaluation of novel therapies. The best characterized and most commonly adopted model in small rodents is of bleomycin-elicited injury. Bleomycin, a derivative of *Streptomyces verticillus*, is an anti-neoplastic antibiotic that has been used clinically for over 30 years.[101] It has the well-known side effect of producing inflammation and fibrosis specific to the lung.[102] Local instillation of bleomycin in rodents is often used to model lung fibrosis.[103,104]

Proton MRI has been used to follow bleomycin-induced injury in rats.[105–107] The initial response in rats, in the first two weeks post-insult, characterized predominantly by diffuse MRI signals, was primarily inflammation-related. At later time points, up to 70 days post-bleomycin, increased MRI signals reflected tissue remodeling involved in fibrosis development, as suggested by histology revealing prominent collagen deposition in the same areas where MRI signals had been detected *in vivo* (Figure 17.9). Topical administration of budesonide showed an effect on inflammation but not on fibrosis.

Proton MRI has been adopted as well to follow the course of bleomycin-induced lung injury in mice and to investigate two knock-out mouse lines with the aim of providing potential therapeutic targets.[108] A long-lasting response following repeated administration of bleomycin has been detected in the lungs of male C57BL/6 mice. Histology showed that, from day 14–70 after bleomycin, fibrosis was the predominant component of the injury. Female C57BL/6 mice displayed a smaller response than males, an observation that is consistent with the fact that estrogen plays a protective role against airway fibrosis in female mice.[109] Bleomycin-induced injury was significantly more pronounced in C57BL/6 than in Balb/C mice. Early studies have shown that Balb/C are more resistant than C57BL/6 mice to the development of bleomycin-induced pulmonary fibrosis.[110,111] This likely reflects strain-dependent differences

---

Figure 17.7   Imaging from a 76-year-old male smoker with moderate COPD. (A) Apical bronchus in the left upper lobe is shown as an example of quantitative measurement of six-generation bronchus. WA% was 80%. (B) For assessment of CT-based FLV in the lungs, pixels showing less than -960 HU are shown as sky blue. CT-based FLV was 62.0%. (C) Relative enhancement map from dynamic O_2-enhanced MR data demonstrates heterogeneous and markedly reduced oxygen enhancement within the lung. Calculated MRER is 0.10. The relative enhancement in each pixel is expressed as a color-coded map showing pixels with 0–50% enhancement progressing from dark green to yellow. (D) Wash-in time map from dynamic O_2-enhanced MR data demonstrates heterogeneous and markedly prolonged wash-in time within the lung; calculated MWT is 43 s. The wash-in time in each pixel is expressed as a color-coded map showing pixels with 0–50 s enhancement progressing from dark green to yellow.
Reproduced with permission from Ohno *et al.*[88] © 2011 Elsevier Ireland Ltd.

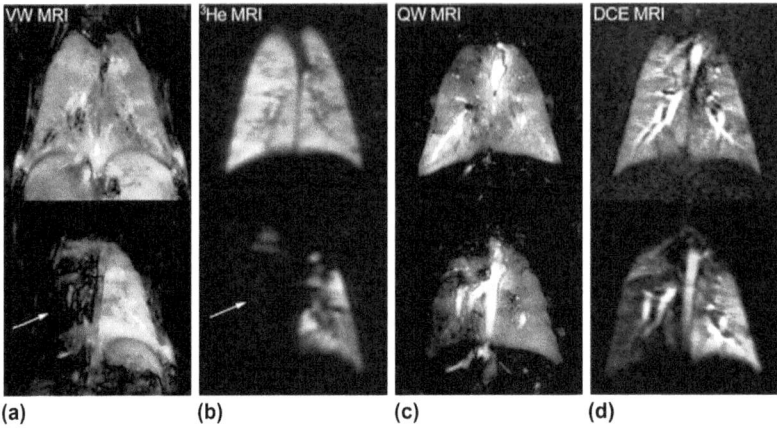

Figure 17.8   Images acquired from a pig during the baseline measurement (top row) and after bronchial obstruction (bottom row) by using ventilation-weighted (VW) FD MRI (a), ^{3}He MRI (b), perfusion-weighted (QW) FD MRI (c) and DCE MRI (d). Data obtained before bronchial obstruction show homogeneous signal intensity distribution across the lung. Blockage of the lobar bronchus and injection of 100 mL nitrogen led to air trapping detected on VW FD MRI and ^{3}He MRI (arrows) as well as redistribution of pulmonary blood flow visible in QW FD MRI and DCE MRI (bottom row).
Reproduced with permission from Bauman *et al.*[93] © 2012 Wiley Periodicals, Inc.

in the expression of the inactivating enzyme, bleomycin hydrolase,[112] or may be related to differential expression of genes involved in apoptosis regulation and oxidative stress.[113] MRI and histology demonstrated a protection against bleomycin insult in female heterozygous and male homozygous cancer Osaka thyroid (COT) kinase knock-out animals. In contrast, no protection was seen in cadherin-11 knock-out animals. The rationale for the analysis of COT kinase deficient mice is the fact that COT regulates the production of tumor necrosis factor-α and interleukin-1β,[114] which have been shown to be implicated in fibrosis[115] and wound healing.[116] Cadherin-11 knock-out mice were investigated in the bleomycin model because it has been shown that cadherin-11 is over-expressed in human renal tubular cells during epithelial to mesenchymal transition (EMT),[117] a process that can adversely cause organ fibrosis.[118] EMT has also been reported to occur in bleomycin-induced pulmonary fibrosis.[119] With the ability for repetitive measurements in the same animal, the technique is attractive for *in vivo* target analysis and compound profiling in this murine model.

Effective pulmonary gas exchange relies on the free diffusion of gases across the thin tissue barrier separating airspace from the capillary RBCs. An increased blood–gas barrier thickness is present in pathologies as inflammation, fibrosis and edema. Driehuys *et al.*[120] demonstrated in a rat model of unilateral bleomycin lung injury the feasibility of detecting such impairment using

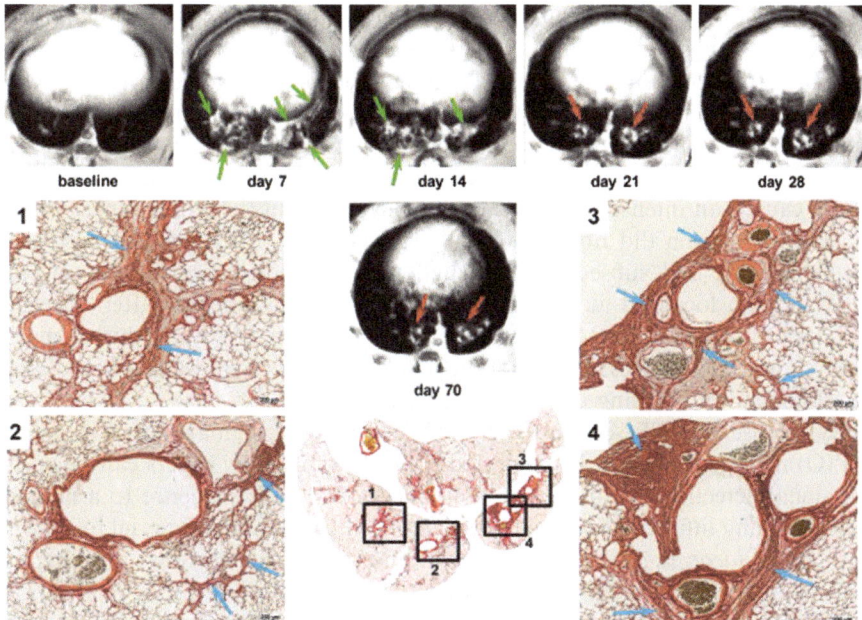

Figure 17.9   Effects of bleomycin in a rat lung fibrosis model. Transverse MRI scans at approximately the same anatomical location show the lungs of a naïve rat at baseline and at different time points after intratracheal bleomycin $(4\,mg\,kg^{-1})$. Signals in the initial phase had a diffuse nature (green arrows) and reflected primarily inflammation, whereas signals at later time points (red arrows) were concentrated in distal regions. At day 70, histological analysis with picrosirius staining revealed areas of dense staining for collagen (blue arrows), at zones corresponding to increased MRI signals. The alignment of the histology section was akin to that of the MR image.
Reproduced with permission from Babin *et al.*[107] © 2011 Wiley-Liss, Inc.

^{129}Xe-MRI and by exploiting the fact that ^{129}Xe resonates at three distinct frequencies in the airspace, tissue barrier and RBC compartments. Based on a simple diffusion model, the authors estimated that this MRI method for measuring ^{129}Xe alveolar-capillary transfer is sensitive to changes in blood–gas barrier thickness of $\approx 5\,\mu m$.

17.3.6   Infections

Research on the pathogenesis and therapy of influenza and other emerging respiratory viral infections would be aided by methods that directly visualize pathophysiologic processes. MRI has the potential to advance *in vivo* studies of respiratory viral infections both in animals and in patients.

   MRI can be used to track the development of pulmonary lesions and characterize inflammatory responses in murine models of bacterial infection.

Marzola *et al.*[121] reported inflammatory lesions localized mainly in the apical part of the lungs, in medial and peribronchial regions, 48 h after intranasal administration of about 10^6 colony-forming units of *Streptococcus pneumoniae*. The anatomical localization of the lesions was confirmed by histology. Tournebize *et al.*[122] followed the development and regression of inflammatory lesions caused by infection by *Klebsiella pneumoniae* in mouse lungs. A virulent strain caused an intense inflammation within 2 days in the whole lungs, while an avirulent strain did not show significant changes. Mice infected with the virulent strain and subsequently treated with antibiotics presented a severe inflammation localized mainly in the left lung that disappeared after a week. The lesions observed by MRI correlated with the damage seen by histological analysis and a 3D representation of the tissue allowed better visualization of the development and healing of inflammatory lesions.

By the use of UTE sequences, T_2^* effects of superparamagnetic iron oxide (SPIO) nanoparticles can be neglected while T_1 shortening effects can be used for signal detection. Strobel *et al.*[123] applied a 3D UTE sequence to a mouse model of lung inflammation, which was induced by systemic bacterial infection with *Staphylococcus aureus*. The bacteria cause a systemic infection, resulting in inflammation of the lung.[124] Following the systemic application of SPIO, a significant signal increase in the lung of infected animals was detected already at 24 h post-infection, compared to control mice (17%, $P < 0.001$). Iron accumulation in the lung parenchyma as consequence of the host immune response was histologically confirmed. The fact that by conventional T_2^*- and T_2-weighted imaging neither structural changes nor formation of substantial edema were observed indicates the potential of UTE in combination with the administration of SPIO to increase the sensitivity to detect inflammatory processes.

This publication suggests that the use of UTE-like sequences in combination with MRI-compatible cell labeling[125] (see Chapter 11 for the application of cell labelling to studies in the brain and other organs) may provide the unique opportunity to study pulmonary influenza or other respiratory viral infections as well as potential therapies. For example, inflammatory cells loaded *ex vivo* with antiviral drugs or nanoparticles have been demonstrated to localize to sites of infection. The effectiveness of anti-retroviral therapies (ART) depends on its ultimate ability to clear reservoirs of continuous human immunodeficiency (HIV) virus infection. Dou *et al.*[126,127] demonstrated that, being a principal vehicle for viral dissemination, the mononuclear phagocytes could also serve as an ART transporter and as such improve therapeutic indices. A nanoparticle-indinavir (NP-IDV) formulation was made and taken up into and released from vacuoles of human monocyte-derived macrophages (MDM). Following a single NP-IDV dose, drug levels within and outside MDM remained constant for 6 days without cytotoxicity. Administration of NP-IDV when compared to equal drug levels of free soluble indinavir significantly blocked induction of multinucleated giant cells, production of reverse transcriptase activity in culture fluids and cell-associated HIV-1p24 antigens after HIV-1 infection. After a single administration, single-photon emission computed tomography (SPECT), histology and reverse-phase-high-performance liquid chromatography demonstrated robust bone marrow-derived macrophages and drug distribution in the lung, liver and spleen. These data provide proof of concept for the use of macrophage-based nanoparticle delivery systems for human HIV-1 infections. Labeling the bone marrow macrophages with *e.g.* SPIO instead of with [111]indium oxyquinoline for SPECT would provide the opportunity to perform measurements at higher spatial resolution using UTE-MRI.

17.3.7 Pulmonary Arterial Hypertension

Right heart catheterization (RHC) is the gold standard for determination of pulmonary arterial pressure (PAP) and vascular resistance, but is invasive and not attractive for monitoring drug therapy. Echocardiography is the most widely used non-invasive modality to measure elevated PAP and pulmonary vascular resistance and to monitor progression of ensuing right heart failure. However, ~50% of patients cannot be analyzed because of failure to trace the entire right ventricle,[128] and PAP can only be assessed by Doppler echocardiography when significant tricuspid regurgitation is present. In contrast, MRI is less patient-dependent and has better reproducibility and lower intraobserver variability than echocardiography.[129] Sanz *et al.*[130] found strong correlations of RHC with mean PAP, systolic PAP and pulmonary vascular resistance index derived from average velocity in the pulmonary artery obtained with phase-contrast MRI. Using related methods, Reiter *et al.*[131] demonstrated that the duration of a flow vortex in the pulmonary artery had even better correlation with RHC-derived mean PAP. Although only possible in the presence of tricuspid regurgitation, Nogami *et al.*[132] found MRI-assessed regurgitation velocity and PAP to be better correlated with RHC than echocardiography-derived PAP. Also the distensibility of the proximal pulmonary artery can be assessed with MRI and its relative area change was the only baseline parameter that correlated with the 6-minute walk test after one year of treatment of chronic thromboembolic pulmonary hypertension patients with sildenafil.[133]

Comparable technology has been introduced to study a rat model of monocrotaline-induced pulmonary hypertension. Flow-derived PAP and right ventricular mass and function determined with MRI correlated better with RHC than echocardiography-assessed parameters.[134] In the same model, treatment with anti-oxidant reduced the three-fold increase in end-systolic right ventricle volume by 42%.[135]

Recently, Ohno *et al.*[136] compared the therapeutic effect assessment capability of multidetector-row CT, MR angiography (MRA) and dynamic perfusion MRI for chronic thromboembolic pulmonary hypertension (CTEPH) patients. Individuals treated with conventional therapy underwent pre- and post-therapeutic multidetector-row CT, MRA, dynamic perfusion MRI, 6-minute walk distance, cardiac ultrasound and right heart catheterization. The results showed that dynamic perfusion MRI has better capability for assessment of therapeutic effect on CTEPH patients than does multidetector-row CT.

17.3.8   Side Effects from Inhaled Pollutants or Particles

As a non-invasive and non-ionizing technique, MRI has been proposed as an alternative and a complementary diagnostic approach to CT or PET for assessing side effects and diseases in lungs induced by inhaled gaseous pollutants or particles. The quantitative or qualitative evaluation of side effects can rely on the anatomical, functional or metabolic changes induced by the exogenous agent. Additionally, the intensity changes in MR images generated by the intrinsic relaxivity of the inhaled particles can be used to localize depositions of particles in the lungs.

The effects of subchronic ozone (O_3) exposure on rat lung ventilation were investigated using HP ^{3}He MRI *in vivo*.[137] Ozone, a powerful oxidant and a major gaseous pollutant, was shown to induce heterogeneous obstructive patterns in this experimental animal model of ozone exposure (2 and 6 days alternate/continuous exposure at 0.5 ppm ozone level). Ventilation defects, appearing as delayed lung filling regions and heterogeneous lung filling, were observed in the dynamic lung ventilation image series. The percentage of animals with ventilation defects in the control, two-day and six-day exposed groups were equal to 20%, 43% and 75%, respectively. In a subgroup of animals exposed six days for 12 hours per day, the percentage of animals exhibiting ventilation defects was equal to 85%. These findings were consistent with the increase of airway resistance and the narrowing of the peripheral small airways, a well-known consequence of ozone exposure.

Regarding exposure to asbestos, MRI could be regarded as suitable for potentially repeated examinations following initial screening by CT.[138,139] In an experimental model of malignant mesothelioma, Hasegawa *et al.*[140] reported the use of manganese-enhanced MRI (MEMRI) for selective mesothelioma imaging *in vivo*. The authors made use of the high-level expression of manganese-superoxide dismutase (Mn-SOD) observed in malignant mesotheliomas to induce accumulation of manganese (Mn^{2+}) ions in tumors, resulting in significant T_1 signal enhancement *in vitro* and *in vivo* in xenografted pleural tumors (Figure 17.10). Moreover, in a more clinically relevant setting, H226 xenografted pleural tumor was markedly enhanced and readily detected by MEMRI using manganese dipyridoxyl diphosphate (MnDPDP), a clinically used contrast agent, as well as $MnCl_2$. Thus, MEMRI can be a potentially powerful method for non-invasive detection of malignant mesothelioma, with high spatial resolution and marked signal enhancement, by targeting Mn-SOD.

MRI is able to assess the biodistribution and clearance of instilled or inhaled magnetically labeled nanoparticles. The follow-up of the biodistribution of instilled USPIO (ultra-small superparamagnetic iron oxide) in small animals using ^{3}He and proton MRI was reported by Al Faraj *et al.*[141] In a 2-week longitudinal imaging study, performed on rats instilled with a 0.5 mg magnetite solution, hypointense and void signal regions associated with intrapulmonary USPIO were observed throughout the study in ^{3}He ventilation images, whereas no USPIO-related proton signal intensity changes were found in other organs (liver, spleen or kidneys).

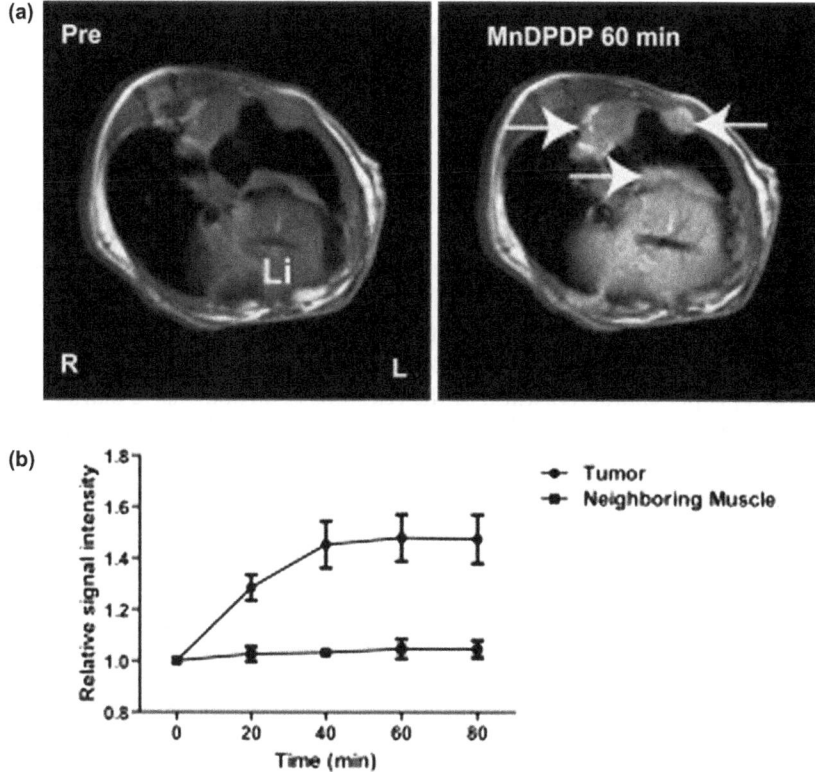

Figure 17.10 Detection of intrapleural H226 xenografted tumors by MnDPDP. (a) Typical T_1-weighted images before (left) and 60 min. after (right) MnDPDP administration are shown. White arrows indicate the tumors. Note that liver was also enhanced because MnDPDP is a contrast agent for liver MRI. Li: liver; R: right; L: left. (b) Relative change of signal intensities of the tumors and neighboring muscle after systemic $MnCl_2$ administration (n = 4).
Reproduced with permission from Hasegawa *et al.*[140] © 2010 Union for International Cancer Control.

The same authors imaged the biodistribution and the anatomical changes induced by an instilled solution of single wall carbon nanotubes (SWCNT) in rodents using ^{3}He and proton MRI.[142,143] The intrinsic relaxivity of non-purified SWCNT associated with iron impurities was used to localize the presence of carbon nanotubes in the lungs based on signal intensity changes on ^{3}He images. In a 3-month follow-up study, proton MRI was used to evaluate tissue inflammation by monitoring local changes in signal intensity. One month after SWCNT exposition, proton MRI revealed the presence of hyper-intense pixels, which could be related to small detectable inflammatory nodules in animals instilled with 0.5 and 1 mg of SWCNTs, and to the same degree in the 0.5 mg SWCNT instilled group after 3 months. The MRI findings were confirmed using *ex vivo* microscopy techniques showing the presence of granulomatous and inflammatory reactions produced in a time- and dose-dependent manner by instilled raw SWCNTs.

17.4   Final Remarks

In the past few years, lung MRI methods have been considerably refined, allowing examinations of humans and small rodents at the anatomical, functional and even molecular or target levels at high spatial and temporal resolution. The main challenge in using such techniques within the framework of drug discovery and development is two-fold: (i) like any biomarker, imaging signatures need to be properly validated and qualified. In animal models, this involves a proper characterization of the readouts against invasive markers like cell influx into BAL fluid and histology. Then, the ability of modulating a given imaging signature by reference to pharmacological agents of known activity needs to be demonstrated, in order to verify the sensitivity of the potential biomarker; (ii) it needs to be shown that the increased technical capabilities of imaging translate into improved disease diagnosis as compared to standard approaches. The hope is that with imaging, earlier phases of the disease can be diagnosed, thus enhancing the chance of pharmacological interventions. Moreover, imaging could facilitate patient stratification in clinical trials.

It cannot be expected that all imaging readouts satisfy these requirements. It may happen that, for a certain indication, imaging can be used at the experimental level both in small rodents and in humans, thus facilitating the examination of translational aspects. The other situation may also happen, *i.e.* imaging being used exclusively in animals or in humans, or not being adopted at all. In general terms, pre-clinical and clinical activities should be mutually supportive. For instance, small animal imaging may help to improve the characterization of clinical readouts and, conversely, imaging in humans may support the improvement of animal models. For both regulatory and cost reasons, imaging will be primarily used in proof-of-concept studies involving a small patient population, rather than in large Phase III clinical trials.

Given the plethora of imaging techniques (Table 17.1), advantages and limitations of each approach have to be understood for every application. Also, different modalities might be used in small animals and in humans. For instance, a certain molecular probe may be labeled for fluorescence optical imaging in mice and rats, and then obtain a radioactive label for nuclear imaging in patients. Also, particular care needs to be taken about radiation exposure when using nuclear imaging in pharmacological studies. Therefore, although HRCT is currently the clinical imaging of choice for the diagnosis of several pulmonary disease conditions in patients, there is room for improvement of MRI approaches.

Generally speaking, efforts in developing imaging biomarkers for respiratory diseases are going to benefit both drug development and diagnosis. Patients will certainly profit from advancements in this field of research.

References

1. P. J. Barnes, *Nat. Rev. Immunol.*, 2008, **8**, 183.
2. P. J. Barnes, *J. Clin. Invest.*, 2008, **118**, 3546.
3. R. M. Strieter and B. Mehrad, *Chest*, 2009, **136**, 1364.
4. T. E. King, Jr., A. Pardo and M. Selman, *Lancet*, 2011, **378**, 1949.
5. H. G. Hoymann, *J. Pharmacol. Toxicol. Methods*, 2007, **55**, 16.
6. N. Beckmann, R. Kneuer, H. U. Gremlich, H. Karmouty-Quintana, F. X. Blé and M. Müller, *NMR Biomed.*, 2007, **20**, 154.
7. N. Beckmann, C. Cannet, H. Karmouty-Quintana, B. Tigani, S. Zurbruegg, F. X. Blé, Y. Crémillieux and A. Trifilieff, *Eur. J. Radiol.*, 2007, **64**, 381.
8. R. A. De Kemp, F. H. Epstein, C. Catana, B. M. Tsui and E. L. Ritman, *J. Nucl. Med.*, 2010, **51**(1), 18S.
9. R. S. Harris and D. P. Schuster, *J. Appl. Physiol.*, 2007, **102**, 448.
10. J. Ley-Zaporozhan, S. Ley and H. U. Kauczor, *Eur. Radiol.*, 2008, **18**, 510.
11. J. Petersson, A. Sánchez-Crespo, S. A. Larsson and M. Mure, *J. Appl. Physiol.*, 2007, **102**, 468.
12. B. Sundaram, A. R. Chughtai and E. A. Kazerooni, *J. Thorac. Imaging*, 2010, **25**, 125.
13. H. P. Schlemmer, B. J. Pichler, M. Schmand, Z. Burbar, C. Michel, R. Ladebeck, K. Jattke, D. Townsend, C. Nahmias, P. K. Jacob, W. D. Heiss and C. D. Claussen, *Radiology*, 2008, **248**, 1028.
14. T. D. Poeppel, B. J. Krause, T. A. Heusner, C. Boy, A. Bockisch and G. Antoch, *Eur. J. Radiol.*, 2009, **70**, 382.
15. G. Mariani, L. Bruselli, T. Kuwert, E. E. Kim, A. Flotats, O. Israel, M. Dondi and N. Watanabe, *Eur. J. Nucl. Med. Mol. Imaging*, 2010, **37**, 1959.
16. B. Walder, E. Fontao, M. Totsch and D. R. Morel, *Eur. J. Anaesthesiol.*, 2005, **22**, 785.
17. N. Beckmann, B. Tigani, D. Ekatodramis, R. Borer, L. Mazzoni and J. R. Fozard, *Magn. Reson. Med.*, 2001, **45**, 88.
18. F. X. Blé, C. Cannet, S. Zurbruegg, H. Karmouty-Quintana, N. Frossard, A. Trifilieff and N. Beckmann, *Radiology*, 2008, **248**, 834.
19. M. Zurek, A. Bessaad, K. Cieslar and Y. Crémillieux, *Magn. Reson. Med.*, 2010, **64**, 401.
20. G. H. Glover and J. M. Pauly, *Magn. Reson. Med.*, 1992, **28**, 275.
21. C. J. Bergin, D. C. Noll, J. M. Pauly, G. H. Glover and A. Macovski, *Radiology*, 1992, **183**, 673.
22. K. Mosbah, J. Ruiz-Cabello, Y. Berthezène and Y. Crémillieux, *Contrast Media Mol. Imaging*, 2008, **3**, 173.
23. B. M. Goodson, *J. Magn. Reson.*, 2002, **155**, 157.
24. H. E. Möller, X. J. Chen, B. Saam, K. D. Hagspiel, G. A. Johnson, T. A. Altes, E. E. de Lange and H. U. Kauczor, *Magn. Reson. Med.*, 2002, **47**, 1029.

25. V. Stupar, E. Canet-Soulas, S. Gaillard, H. Alsaid, N. Beckmann and Y. Crémillieux, *NMR Biomed.*, 2007, **20**, 104.

26. K. Mosbah, V. Stupar, Y. Berthezène, N. Beckmann and Y. Crémillieux, *Magn. Reson. Med.*, 2010, **63**, 1669.

27. H. Imai, A. Kimura, Y. Hori, S. Iguchi, T. Kitao, E. Okubo, T. Ito, T. Matsuzaki and H. Fujiwara, *NMR Biomed.*, 2011, **24**, 1343.

28. G. Conti, S. Tambalo, G. Villetti, S. Catinella, C. Carnini, F. Bassani, N. Sonato, A. Sbarbati and P. Marzola, *MAGMA*, 2010, **23**, 93.

29. B. Tigani, E. Schaeublin, R. Sugar, A. D. Jackson, J. R. Fozard and N. Beckmann, *Biochem. Biophys. Res. Commun.*, 2002, **292**, 216.

30. H. K. Quintana, C. Cannet, E. Schaeublin, S. Zurbruegg, R. Sugar, L. Mazzoni, C. P. Page, J. R. Fozard and N. Beckmann, *Am. J. Physiol. Lung Cell. Mol. Physiol.*, 2006, **291**, L651.

31. B. Tigani, C. Cannet, S. Zurbruegg, E. Schaeublin, L. Mazzoni, J. R. Fozard and N. Beckmann, *Br. J. Pharmacol.*, 2003, **140**, 239.

32. F. X. Blé, C. Cannet, S. Zurbruegg, C. Gérard, N. Frossard, N. Beckmann and A. Trifilieff, *Br. J. Pharmacol.*, 2009, **158**, 1295.

33. B. Tigani, F. Di Padova, S. Zurbruegg, E. Schaeublin, L. Revesz, J. R. Fozard and N. Beckmann, *Eur. J. Pharmacol.*, 2003, **482**, 319.

34. A. C. Thomas, S. S. Kaushik, J. Nouls, E. N. Potts, D. M. Slipetz, W. M. Foster and B. Driehuys, *J. Appl. Physiol.*, 2012, **112**, 1437.

35. B. Tigani, C. Cannet, H. K. Quintana, F. X. Blé, S. Zurbruegg, E. Schaeublin, J. R. Fozard and N. Beckmann, *Am. J. Physiol. Lung Cell. Mol. Physiol.*, 2007, **292**, L644.

36. N. R. Labiris, C. Nahmias, A. P. Freitag, M. L. Thompson and M. B. Dolovich, *Eur. Respir. J.*, 2003, **21**, 848.

37. D. L. Chen and D. P. Schuster, *Am. J. Physiol. Lung Cell. Mol. Physiol.*, 2004, **286**, L834.

38. D. L. Chen, T. J. Bedient, J. Kozlowski, D. B. Rosenbluth, W. Isakow, T. W. Ferkol, B. Thomas, M. A. Mintun, D. P. Schuster and M. J. Walter, *Am. J. Respir. Crit. Care Med.*, 2009, **180**, 533.

39. B. Ebner, P. Behm, C. Jacoby, S. Burghoff, B. A. French, J. Schrader and U. Flögel, *Circ. Cardiovasc. Imaging*, 2010, **3**, 202.

40. W. M. Foster, D. M. Walters, M. Longphre, K. Macri and L. M. Miller, *J. Appl. Physiol.*, 2001, **90**, 1111.

41. J. C. Lay, M. R. Stang, P. E. Fisher, J. R. Yankaskas and W. D. Bennett, *J. Aerosol Med.*, 2003, **16**, 153.

42. S. H. Donaldson, W. D. Bennett, K. L. Zeman, M. R. Knowles, R. Tarran and R. C. Boucher, *N. Engl. J. Med.*, 2006, **354**, 241.

43. Y. Tesfaigzi, M. J. Fischer, A. J. Martin and J. Seagrave, *Am. J. Physiol. Lung Cell. Mol. Physiol.*, 2000, **279**, L1210.

44. N. Beckmann, B. Tigani, R. Sugar, A. D. Jackson, G. Jones, L. Mazzoni and J. R. Fozard, *Am. J. Physiol. Lung Cell. Mol. Physiol.*, 2002, **283**, L22.

45. F. X. Blé, P. Schmidt, C. Cannet, R. Kneuer, H. Karmouty-Quintana, R. Bergmann, K. Coote, H. Danahay, S. Zurbruegg, H. U. Gremlich and N. Beckmann, *Magn. Reson. Med.*, 2009, **62**, 1164.

46. H. Karmouty-Quintana, C. Cannet, R. Sugar, J. R. Fozard, C. P. Page and N. Beckmann, *Br. J. Pharmacol.*, 2007, **150**, 1022.
47. R. C. Boucher, *Annu. Rev. Med.*, 2007, **58**, 157.
48. F. X. Blé, C. Cannet, S. Collingwood, H. Danahay and N. Beckmann, *Br. J. Pharmacol.*, 2010, **160**, 1008.
49. M. Puderbach, M. Eichinger, J. Gahr, S. Ley, S. Tuengerthal, A. Schmähl, C. Fink, C. Plathow, M. Wiebel, F. M. Müller and H. U. Kauczor, *Eur. Radiol.*, 2007, **17**, 716.
50. J. Ley-Zaporozhan, M. Puderbach and H. U. Kauczor, *Magn. Reson. Imaging Clin. N. Am.*, 2008, **16**, 291.
51. L. F. Donnelly, J. R. MacFall, H. P. McAdams, J. M. Majure, J. Smith, D. P. Frush, P. Bogonad, H. C. Charles and C. E. Ravin, *Radiology*, 1999, **212**, 885.
52. K. Mentore, D. K. Froh, E. E. de Lange, J. R. Brookeman, A. O. Paget-Brown and T. A. Altes, *Acad. Radiol.*, 2005, **12**, 1423.
53. P. Koumellis, E. J. van Beek, N. Woodhouse, S. Fichele, A. J. Swift, M. N. Paley, C. Hill, C. J. Taylor and J. M. Wild, *J. Magn. Reson. Imaging*, 2005, **22**, 420.
54. E. J. van Beek, C. Hill, N. Woodhouse, S. Fichele, S. Fleming, B. Howe, S. Bott, J. M. Wild and C. J. Taylor, *Eur. Radiol.*, 2007, **17**, 1018.
55. N. Woodhouse, J. M. Wild, E. J. van Beek, N. Hoggard, N. Barker and C. J. Taylor, *J. Magn. Reson. Imaging*, 2009, **30**, 981.
56. E. Bannier, K. Cieslar, K. Mosbah, F. Aubert, F. Duboeuf, Z. Salhi, S. Gaillard, Y. Berthezène, Y. Crémillieux and P. Reix, *Radiology*, 2010, **255**, 225.
57. B. C. Stoel and J. Stolk, *Invest. Radiol.*, 2004, **39**, 681.
58. L. E. Olsson, M. Lindahl, P. O. Onnervik, L. B. Johansson, M. Palmér, M. K. Reimer, L. Hultin and P. D. Hockings, *J. Magn. Reson. Imaging*, 2007, **25**, 488.
59. N. Beckmann, B. Tigani, L. Mazzoni and J. R. Fozard, *NMR Biomed.*, 2001, **14**, 297.
60. M. Takahashi, O. Togao, M. Obara, M. van Cauteren, Y. Ohno, S. Do, M. Kuro-o, C. Malloy, C. C. Hsia and I. Dimitrov, *J. Magn. Reson. Imaging*, 2010, **32**, 326.
61. M. Zurek, L. Boyer, P. Caramelle, J. Boczkowski and Y. Crémillieux, *Magn. Reson. Med.*, 2012, **68**, 898.
62. Y. Ohno, H. Koyama, T. Yoshikawa, K. Matsumoto, M. Takahashi, M. van Cauteren and K. Sugimura, *A.J.R. Am. J. Roentgenol.*, 2011, **197**, W279.
63. H. K. Quintana, C. Cannet, S. Zurbruegg, F. X. Blé, J. R. Fozard, C. P. Page and N. Beckmann, *Magn. Reson. Med.*, 2006, **56**, 1242.
64. X. J. Chen, L. W. Hedlund, H. E. Moller, M. S. Chawla, R. R. Maronpot and G. A. Johnson, *Proc. Natl Acad. Sci. USA*, 2000, **97**, 11478.
65. J. P. Dugas, J. R. Garbow, D. K. Kobayashi and M. S. Conradi, *Magn. Reson. Med.*, 2004, **52**, 1310.

66. G. Peces-Barba, J. Ruiz-Cabello, Y. Crémillieux, I. Rodriguez, D. Dupuich, V. Callot, M. Ortega, M. L. Rubio Arbo, M. Cortijo and N. Gonzalez-Mangado, *Eur. Respir. J.*, 2003, **22**, 14.

67. J. C. Woods, C. K. Choong, D. A. Yablonskiy, J. Bentley, J. Wong, J. A. Pierce, J. D. Cooper, P. T. Macklem, M. S. Conradi and J. C. Hogg, *Magn. Reson. Med.*, 2006, **56**, 1293.

68. S. Diaz, I. Casselbrant, E. Piitulainen, P. Magnusson, B. Peterson, E. Pickering, T. Tuthill, O. Ekberg and P. Akeson, *Acad. Radiol.*, 2009, **16**, 700.

69. T. Stavngaard, L. V. Sogaard, M. Batz, L. M. Schreiber and A. Dirksen, *Acta Radiol.*, 2009, **50**, 1019.

70. S. S. Kaushik, Z. I. Cleveland, G. P. Cofer, G. Metz, D. Beaver, J. Nouls, M. Kraft, W. Auffermann, J. Wolber, H. P. McAdams and B. Driehuys, *Magn. Reson. Med.*, 2011, **65**, 1154.

71. A. Haczku, K. Emami, M. C. Fischer, S. Kadlecek, M. Ishii, R. A. Panettieri and R. R. Rizi, *Acad. Radiol.*, 2005, **12**, 1362.

72. Z. Z. Spector, K. Emami, M. C. Fischer, J. Zhu, M. Ishii, V. Vahdat, J. Yu, S. Kadlecek, B. Driehuys, D. A. Lipson, W. Gefter, J. Shrager and R. R. Rizi, *Magn. Reson. Med.*, 2005, **53**, 1341.

73. T. J. Wellman, T. Winkler, E. L. Costa, G. Musch, R. S. Harris, J. G. Venegas and M. F. Melo, *J. Nucl. Med.*, 2010, **51**, 646.

74. A. J. Deninger, S. Mansson, J. S. Petersson, G. Pettersson, P. Magnusson, J. Svensson, B. Fridlund, G. Hansson, I. Erjefeldt, P. Wollmer and K. Golman, *Magn. Reson. Med.*, 2002, **48**, 223.

75. K. Emami, R. V. Cadman, J. M. Woodburn, M. C. Fischer, S. J. Kadlecek, J. Zhu, S. Pickup, R. A. Guyer, M. Law, V. Vahdat, M. E. Friscia, M. Ishii, J. Yu, W. B. Gefter, J. B. Shrager and R. R. Rizi, *J. Appl. Physiol.*, 2008, **104**, 773.

76. N. N. Mistry, A. Thomas, S. S. Kaushik, G. A. Johnson and B. Driehuys, *Magn. Reson. Med.*, 2010, **63**, 658.

77. B. T. Chen and G. A. Johnson, *Magn. Reson. Med.*, 2004, **52**, 1080.

78. S. B. Fain, G. Gonzalez-Fernandez, E. T. Peterson, M. D. Evans, R. L. Sorkness, N. N. Jarjour, W. W. Busse and J. E. Kuhlman, *Acad. Radiol.*, 2008, **15**, 753.

79. E. J. R. van Beek, J. M. Wild, W. Schreiber, H. U. Kauczor, J. Mugler III and E. E. de Lange, *J. Magn. Reson. Imaging*, 2004, **20**, 540.

80. S. Samee, T. A. Altes, P. Powers, E. E. de Lange, J. Knight-Scott, G. Rakes, J. P. Mugler III, J. M. Ciambotti, B. A. Alford, J. R. Brookeman and T. A. E. Platts-Mills, *J. Allergy Clin. Immunol.*, 2003, **111**, 1205.

81. Y. S. Tzeng, K. Lutchen and M. Albert, *J. Appl. Physiol.*, 2009, **106**, 813.

82. S. Patz, I. Muradian, M. I. Hrovat, I. C. Ruset, G. Topulos, S. D. Covrig, E. Frederick, H. Hatabu, F. W. Hersman and J. P. Butler, *Acad. Radiol.*, 2008, **15**, 713.

83. Z. I. Cleveland, G. P. Cofer, G. Metz, D. Beaver, J. Nouls, S. S. Kaushik, M. Kraft, J. Wolber, K. T. Kelly, H. P. McAdams and B. Driehuys, *PLoS One*, 2010, **5**, e12192.

84. N. Beckmann, Y. Crémillieux, B. Tigani, H. K. Quintana, F. X. Blé, J. R. Fozard, in *In Vivo MR Techniques in Drug Discovery and Development*, ed. N. Beckmann, Informa, New York, 2006, p. 351.
85. N. Beckmann, C. Cannet, S. Zurbruegg, M. Rudin and B. Tigani, *Magn. Reson. Med.*, 2004, **52**, 258.
86. R. R. Edelman, H. Hatabu, E. Tadamura, W. Li and P. V. Prasad, *Nat. Med.*, 1996, **2**, 1236.
87. Y. Ohno, T. Iwasawa, J. B. Seo, H. Koyama, H. Takahashi, Y. M. Oh, Y. Nishimura and K. Sugimura, *Am. J. Respir. Crit. Care Med.*, 2008, **177**, 1095.
88. Y. Ohno, H. Koyama, T. Yoshikawa, K. Matsumoto, N. Aoyama, Y. Onishi, D. Takenaka, S. Matsumoto, Y. Nishimura and K. Sugimura, *Eur. J. Radiol.*, 2012, **81**, 1068.
89. Y. Ohno, H. Koyama, K. Matsumoto, Y. Onishi, M. Nogami, D. Takenaka, S. Matsumoto and K. Sugimura, *Eur. J. Radiol.*, 2011, **77**, 85.
90. M. Eichinger, M. Puderbach, C. Fink, J. Gahr, S. Ley, C. Plathow, S. Tuengerthal, I. Zuna, F. M. Muller and H. U. Kauczor, *Eur. Radiol.*, 2006, **16**, 2147.
91. J. Ley-Zaporozhan, S. Ley, R. Eberhardt, O. Weinheimer, C. Fink, M. Puderbach, M. Eichinger, F. Herth and H. U. Kauczor, *Eur. J. Radiol.*, 2007, **63**, 76.
92. G. Bauman, M. Puderbach, M. Deimling, V. Jellus, C. Chefd'hotel, J. Dinkel, C. Hintze, H. U. Kauczor and L. R. Schad, *Magn. Reson. Med.*, 2009, **62**, 656.
93. G. Bauman, A. Scholz, J. Rivoire, M. Terekhov, J. Friedrich, A. de Oliveira, W. Semmler, L. M. Schreiber and M. Puderbach, *Magn. Reson. Med.*, 2013, **69**, 229.
94. D. Neeb, R. P. Kunz, S. Ley, G. Szábo, L. G. Strauss, H. U. Kauczor, K. F. Kreitner and L. M. Schreiber, *Magn. Reson. Med.*, 2009, **62**, 476.
95. S. D. Keilholz, U. Bozlar, N. Fujiwara, J. F. Mata, S. S. Berr, C. Corot and K. D. Hagspiel, *Korean J. Radiol.*, 2009, **10**, 447.
96. P. Ryhammer, M. Pedersen, S. Ringgaard and H. Ravn, *J. Magn. Reson. Imaging*, 2007, **26**, 296.
97. N. N. Mistry, J. Pollaro, J. Y. Song, M. D. Lin and G. A. Johnson, *Magn. Reson. Med.*, 2008, **59**, 289.
98. T. A. Altes, V. M. Mai, T. M. Munger, J. R. Brookeman and K. D. Hagspiel, *J. Vasc. Interv. Radiol.*, 2005, **16**, 999.
99. D. A. Roberts, R. R. Rizi, D. A. Lipson, M. A. Ferrante, L. Bearn, L. Rolf, J. Baumgardner, A. Yamomoto, H. Hatabu, J. Hansen-Flaschen, W. B. Gefter and M. D. Schnall, *J. Magn. Reson. Imaging*, 2001, **14**, 175.
100. B. Driehuys, H. E. Möller, Z. I. Cleveland, J. Pollaro and L. W. Hedlund, *Radiology*, 2009, **252**, 386.
101. J. S. Lazo, S. M. Sebti and J. H. Schellens, *Cancer Chemother. Biol. Response Modif.*, 1996, **16**, 39.
102. J. M. O'Sullivan, R. A. Huddart, A. R. Norman, J. Nicholls, D. P. Dearnaley and A. Horwich, *Ann. Oncol.*, 2003, **14**, 91.

103. B. B. Moore and C. M. Hogaboam, *Am. J. Physiol. Lung Cell. Mol. Physiol.*, 2008, **294**, 152.

104. F. Chua, J. Gauldie and G. J. Laurent, *Am. J. Respir. Cell Mol. Biol.*, 2005, **33**, 9.

105. H. Karmouty-Quintana, C. Cannet, S. Zurbruegg, F. X. Blé, J. R. Fozard, C. P. Page and N. Beckmann, *J. Magn. Reson. Imaging*, 2007, **26**, 941.

106. R. E. Jacob, B. G. Amidan, J. Soelberg and K. R. Minard, *J. Magn. Reson. Imaging*, 2010, **31**, 1091.

107. A. L. Babin, C. Cannet, C. Gérard, D. Wyss, C. P. Page and N. Beckmann, *J. Magn. Reson. Imaging*, 2011, **33**, 603.

108. A. L. Babin, C. Cannet, C. Gérard, P. Saint-Mezard, C. P. Page, H. Sparrer, T. Matsuguchi and N. Beckmann, *Magn. Reson. Med.*, 2012, **67**, 499.

109. E. D. Lekgabe, S. G. Royce, T. D. Hewitson, M. L. Tang, C. Zhao, X. L. Moore, G. W. Tregear, R. A. Bathgate, X. J. Du and C. S. Samuel, *Endocrinology*, 2006, **147**, 5575.

110. D. J. Schrier, R. G. Kunkel and S. H. Phan, *Am. Rev. Respir. Dis.*, 1983, **127**, 63.

111. G. A. Rossi, S. Szapiel, V. J. Ferrans and R. G. Crystal, *Am. Rev. Respir. Dis.*, 1987, **135**, 448.

112. S. M. Sebti, J. E. Mignano, J. P. Jani, S. Srimatkandada and J. S. Lazo, *Biochemistry*, 1989, **28**, 6544.

113. C. K. Haston, T. G. Tomko, N. Godin, L. Kerckhoff and M. T. Hallett, *J. Med. Genet.*, 2005, **42**, 464.

114. J. Miyoshi, T. Higashi, H. Mukai, T. Ohuchi and T. Kakunaga, *Mol. Cell. Biol.*, 1991, **11**, 4088.

115. D. E. Sullivan, M. Ferris, D. Pociask and A. R. Brody, *Am. J. Respir. Cell. Mol. Biol.*, 2005, **32**, 342.

116. A. A. Thomay, J. M. Daley, E. Sabo, P. J. Worth, L. J. Shelton, M. W. Harty, J. S. Reichner and J. E. Albina, *Am. J. Pathol.*, 2009, **174**, 2129.

117. M. Forino, R. Torregrossa, M. Ceol, L. Murer, M. Della Vella, D. Del Prete, A. D'Angelo and F. Anglani, *Int. J. Exp. Pathol.*, 2006, **87**, 197.

118. J. P. Thiery, H. Acloque, R. Y. Huang and M. A. Nieto, *Cell*, 2009, **139**, 871.

119. N. Hashimoto, S. H. Phan, K. Imaizumi, M. Matsuo, H. Nakashima, T. Kawabe, K. Shimokata and Y. Hasegawa, *Am. J. Respir. Cell. Mol. Biol.*, 2010, **43**, 161.

120. B. Driehuys, G. P. Cofer, J. Pollaro, J. B. Mackel, L. W. Hedlund and G. A. Johnson, *Proc. Natl Acad. Sci. USA*, 2006, **103**, 18278.

121. P. Marzola, A. Lanzoni, E. Nicolato, V. Di Modugno, P. Cristofori, F. Osculati and A. Sbarbati, *J. Magn. Reson. Imaging*, 2005, **22**, 170.

122. R. Tournebize, B. T. Doan, M. A. Dillies, S. Maurin, J. C. Beloeil and P. J. Sansonetti, *Cell. Microbiol.*, 2006, **8**, 33.

123. K. Strobel, V. Hoerr, F. Schmid, L. Wachsmuth, B. Löffler and C. Faber, *Magn. Reson. Med.*, 2012, **68**, 1924.

124. C. Yi, Y. Cao, S. H. Mao, H. Liu, L. L. Ji, S. Y. Xu, M. Zhang and Y. Huang, *Inflamm. Res.*, 2009, **58**, 855.

125. U. Himmelreich and T. Dresselaers, *Methods*, 2009, **48**, 112.
126. H. Dou, C. J. Destache, J. R. Morehead, R. L. Mosley, M. D. Boska, J. Kingsley, S. Gorantla, L. Poluektova, J. A. Nelson, M. Chaubal, J. Werling, J. Kipp, B. E. Rabinow and H. E. Gendelman, *Blood*, 2006, **108**, 2827.
127. H. Dou, J. Morehead, C. J. Destache, J. D. Kingsley, L. Shlyakhtenko, Y. Zhou, M. Chaubal, J. Werling, J. Kipp, B. E. Rabinow and H. E. Gendelman, *Virology*, 2007, **358**, 148.
128. G. B. Bleeker, P. Steendijk, E. R. Holman, C. M. Yu, O. A. Breithardt, T. A. Kaandorp, M. J. Schalij, E. E. van der Wall, P. Nihoyannopoulos and J. J. Bax, *Heart*, 2006, **92**(1), i19.
129. W. M. Bradlow, M. L. Hughes, N. G. Keenan, C. Bucciarelli-Ducci, R. Assomull, J. S. Gibbs and R. H. Mohiaddin, *J. Magn. Reson. Imaging*, 2010, **31**, 117.
130. J. Sanz, P. Kuschnir, T. Rius, R. Salguero, R. Sulica, A. J. Einstein, S. Dellegrottaglie, V. Fuster, S. Rajagopalan and M. Poon, *Radiology*, 2007, **243**, 70.
131. G. Reiter, U. Reiter, G. Kovacs, B. Kainz, K. Schmidt, R. Maier, H. Olschewski and R. Rienmueller, *Circ. Cardiovasc. Imaging*, 2008, **1**, 23.
132. M. Nogami, Y. Ohno, H. Koyama, A. Kono, D. Takenaka, T. Kataoka, H. Kawai, H. Kawamitsu, Y. Onishi, K. Matsumoto, S. Matsumoto and K. Sugimura, *J. Magn. Reson. Imaging*, 2009, **30**, 973.
133. M. R. Toshner, D. Gopalan, J. Suntharalingam, C. Treacy, E. Soon, K. K. Sheares, N. W. Morrell, N. Screaton and J. Pepke-Zaba, *J. Heart Lung Transplant.*, 2010, **29**, 610.
134. D. Urboniene, I. Haber, Y. H. Fang, T. Thenappan and S. L. Archer, *Am. J. Physiol. Lung Cell. Mol. Physiol.*, 2010, **299**, L401.
135. E. M. Redout, A. van der Toorn, M. J. Zuidwijk, C. W. van de Kolk, C. J. van Echteld, R. J. Musters, C. van Hardeveld, W. J. Paulus and W. S. Simonides, *Am. J. Physiol. Heart Circ. Physiol.*, 2010, **298**, H1038.
136. Y. Ohno, H. Koyama, T. Yoshikawa, M. Nishio, S. Matsumoto, K. Matsumoto, N. Aoyama, M. Nogami, K. Murase and K. Sugimura, *J. Magn. Reson. Imaging*, 2012, **36**, 612.
137. Y. Crémillieux, S. Servais, Y. Berthezene, D. Dupuich, A. Boussouar, V. Stupar and J. M. Pequignot, *J. Magn. Reson. Imaging*, 2008, **27**, 771.
138. J. Podobnik, I. Kocijancic, V. Kovac and I. Sersa, *Rad. Onc.*, 2010, **44**, 92.
139. M. F. Carette, *Rev. Mal. Resp.*, 2012, **29**, 529.
140. S. Hasegawa, M. Koshikawa-Yano, S. Saito, Y. Morokoshi, T. Furukawa, I. Aoki and T. Saga, *Int. J. Cancer*, 2011, **128**, 2138.
141. A. Al Faraj, G. Lacroix, H. Alsaid, D. Elgrabi, V. Stupar, F. Robidel, S. Gaillard, E. Canet-Soulas and Y. Crémillieux, *Magn. Reson. Med.*, 2008, **59**, 1298.
142. A. Al Faraj, K. Cieslar, G. Lacroix, S. Gaillard, E. Canet-Soulas and Y. Crémillieux, *Nanoletters*, 2009, **9**, 1023.
143. A. Al Faraj, A. Bessaad, K. Cieslar, G. Lacroix, E. Canet-Soulas and Y. Crémillieux, *Nanotechnology*, 2010, **21**, 175103.

CHAPTER 18

Cardiovascular Magnetic Resonance

ALEX PITCHER, THEODOROS D. KARAMITSOS,
JÜRGEN E. SCHNEIDER AND STEFAN NEUBAUER*

Oxford Centre for Clinical Magnetic Resonance Research, Division of
Cardiovascular Medicine, Radcliffe Department of Medicine, John Radcliffe
Hospital, University of Oxford, Headley Way, Oxford, UK
*Email: stefan.neubauer@cardiov.ox.ac.uk

18.1 Scope of the Problem: the Global Burden of Cardiovascular Disease

Cardiovascular disease is the leading cause of death in the West, and is a growing cause of disability and death in many parts of the rest of the world.[1] Complications of atherosclerotic disease, including myocardial infarction, stroke and peripheral vascular disease, hypertension and its complications and congestive cardiac failure arising as a consequence of these or as a consequence of valvular or intrinsic cardiomyopathies account for much of this burden.[1]

18.2 Definition of CMR and an Overview of the Clinical Role of CMR

Cardiovascular magnetic resonance (CMR) is the term used for a range of powerful imaging and spectroscopic techniques based on the NMR principle, with important developments to allow optimal evaluation of the heart and blood vessels.[2,3] The clinical application of CMR imaging is now widespread

New Developments in NMR No. 2
New Applications of NMR in Drug Discovery and Development
Edited by Leoncio Garrido and Nicolau Beckmann
© The Royal Society of Chemistry 2013
Published by the Royal Society of Chemistry, www.rsc.org

(though unevenly distributed),[4] and CMR is frequently regarded as the gold standard method for evaluating a number of aspects of cardiovascular structure and function. MR imaging and spectroscopy are both widely and increasingly used in research contexts aiming to understand mechanisms of disease, to allow detailed and often quantitative phenotyping of disease, to allow precise quantification of disease severity and to develop risk stratification approaches to identify patients at particular hazard from their disease or who are particularly likely (or unlikely) to experience net benefit from therapy. In the evaluation of candidate therapies for drug development, CMR can be applied to characterize and monitor *in vivo* responses to candidate therapies in animal models of disease. In the clinical phase, it can be used to identify suitable participants, to ensure baseline comparability of treatment arms and to generate markers of disease presence, severity or activity for use as outcome measures in clinical trials.[5]

18.3 Overview of the Current Status of CMR (Imaging and Spectroscopy) and the Rationale for the Use of CMR in Drug Evaluation

CMR has a number of features that make it particularly suitable for the evaluation of novel therapies. It is non-invasive and well tolerated by the vast majority of subjects. The technique itself carries no known risks (as long as safety precautions are followed), and specifically does not require the use of ionizing radiation (unlike CT and nuclear methods). It is therefore particularly suitable for studies making repeated measures of the same individual, *e.g.* before and after treatment phases. In the context of clinical trials, this may lead to potentially important reductions in sample size (and hence cost) for a given power, statistical significance threshold and predicted treatment effect size (Figure 18.1).[5,6]

CMR is extremely versatile, and this versatility arises from the very wide range of pulse sequences available.[7,8] A limited number of sequences are typically required to answer specific questions and these can often be completed in a single study lasting ~ 30–45 minutes. CMR provides a quantitative assessment of many aspects of cardiovascular structure and function, such as left and right ventricular function (Figure 18.2),[9,10] valve function,[11] vascular function[12] and blood flow.[13] It can determine the presence and extent of myocardial scar,[14] fibrosis[15] and edema,[16] and can quantitatively evaluate myocardial perfusion (Figure 18.3),[17] metabolism[18] and myocardial tissue characteristics.[19,20]

CMR allows assessment of parameters that cannot be evaluated in any other way, and therefore provides a unique insight into some aspects of cardiovascular disease processes: examples include T_2^* imaging for the assessment of myocardial iron overload,[21,22] and T_1 mapping for the evaluation of myocardial edema and diffuse fibrosis.[23]

There is now a robust evidence base for the validation and reproducibility of many CMR techniques and an emerging pool of data regarding the prognostic implications of CMR surrogate endpoints.[24] Each CMR technique differs in

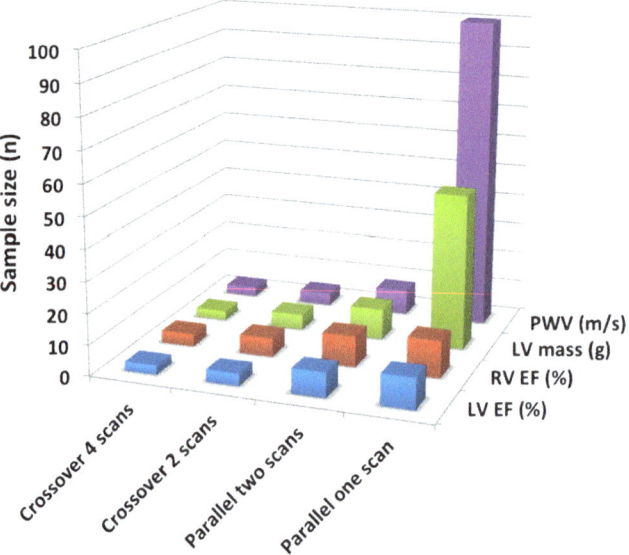

Figure 18.1   Effect of study design on sample size, assuming a given level of power and statistical significance, to discriminate a clinically important difference for four cardiovascular magnetic resonance surrogate markers. EF, ejection fraction; LV, left ventricle; PWV, pulse wave velocity; RV, right ventricle. Reproduced with permission from Pitcher et al.[5] © 2011 BMJ Publishing Group.

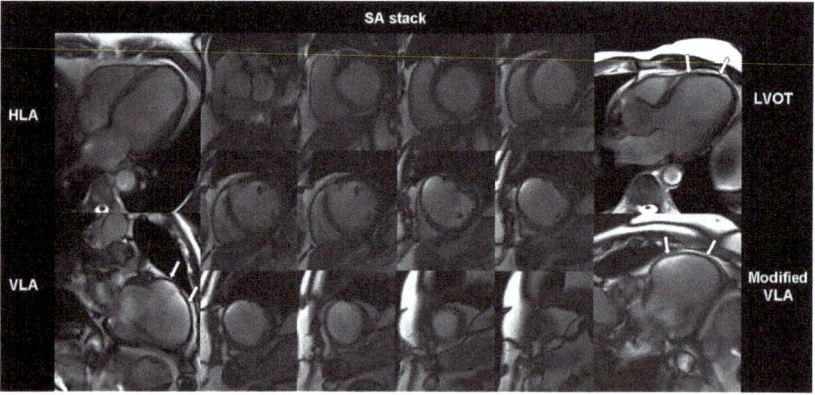

Figure 18.2   A 76-year-old man with history of ventricular aneurysm post-anterior myocardial infarction. End-diastolic images of cine CMR obtained using a steady-state free precession sequence at 3 T. Note the extensive aneurysmal area (arrows) in the anterior wall and the anterior septum. Cardiovascular magnetic resonance has the ability to measure volumes and ejection fraction in grossly remodeled hearts without the need of geometric assumptions. HLA indicates horizontal long axis; LVOT, LV outflow track; SA, short axis; VLA, vertical long axis.

Reproduced with permission from T. D. Karamitsos, E. Dall'Armellina, R. P. Choudhury and S. Neubauer, *Am. Heart J.*, 2011, **162**, 16. © 2011 Mosby, Inc.

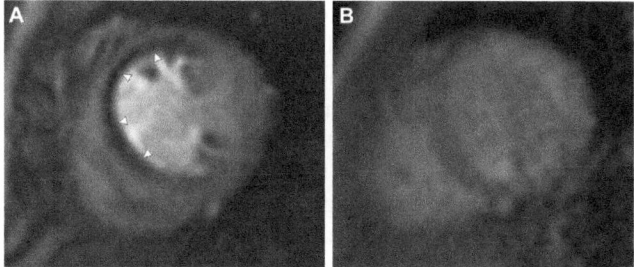

Figure 18.3 Two representative examples of CMR perfusion scans during adenosine stress. A. Patient with a significant stenosis of the left anterior descending coronary artery. Note the area of hypoenhancement in the anteroseptum and anterior wall (arrowheads), representing a perfusion defect. B. Healthy individual with normal coronary arteries. Note the homogeneous enhancement in all myocardial regions during the infusion of adenosine.
Reproduced with permission from T. D. Karamitsos, E. Dall'Armellina, R. P. Choudhury and S. Neubauer, *Am. Heart J.*, 2011, **162**, 16. © 2011 Mosby, Inc.

terms of its accuracy, precision, reproducibility, sensitivity, link with clinical endpoints and ease of use in terms of both acquisition and analysis. These strengths have led to a rapid increase in the use of CMR in clinical trials of new therapies, a trend that is likely to continue.

18.4 CMR in Pre-Clinical Drug Development

Small animals such as mice and rats have been a mainstay of basic cardiovascular research for several decades as they are anatomically and genetically close to humans, and provide obvious logistical advantages (*i.e.* housing costs, reproduction cycle *etc.*) compared to large animal models. Importantly, the genomes of mice, rats and humans have now been fully analyzed, which paves the way to systematically investigate the function of specific genes by overexpression, deletion or mutation of these genes and their products in rodents (*e.g.* reference 25 for review). These recent advances in gene manipulation provide a powerful platform to investigate the role of genes in certain diseases, and to develop novel treatments.

Furthermore, dedicated surgical techniques, such as transverse aortic constriction or ligation of the left coronary artery, have been developed in rodents in order to model cardiac disease conditions that resemble those found in patients with heart disease. Specifically, aortic constriction in rodents causes pressure overload in the heart, resulting in progressive hypertrophy of the left ventricle as a compensatory mechanism, eventually leading to heart failure.[26,27] Transient or permanent ligation of the left coronary artery causes myocardial ischemia and subsequently acute infarction, leading to ventricular dilatation, depression of cardiac function and remodeling of the non-infarcted residual myocardium in the chronic phase.

CMR techniques in mice and rats allow – analogously to clinical CMR – the assessment of systolic and diastolic function by measuring left ventricular volumes and mass with high accuracy and low inter- and intra-observer variability.[28,29] Regional cardiac function and transmural wall motion can be quantified using phase-contrast,[30,31] tagging[32–35] and displacement encoding[36–38] techniques. The latter two CMR methods enable the direct calculation of circumferential and longitudinal strain. Myocardial blood flow can be determined using arterial spin-labeling[39,40] and first-pass perfusion.[41,42] Combined with infarct size[43–45] and area-at-risk[46] measurements, these techniques provide sensitive tools for evaluating pharmacological interventions aiming to reduce (irreversible) ischemia-reperfusion damage in acute myocardial infarction. CMR techniques (*i.e.* ^{1}H-/^{13}C-/^{23}Na- and ^{31}P-MRS) characterizing the metabolic state of healthy and diseased myocardium have been reported in mice and rats *in vivo*,[47–49] but are less commonly applied due to the significant methodological and technical challenges.

The potential of CMR in rodents for drug development is demonstrated in a study by Naumova *et al.*, who used ^{1}H-MRI and ^{31}P-MRS to investigate the metabolic and contractile effects of xanthine oxidase inhibitors (XOIs) in failing mouse hearts post myocardial infarction.[50] This study found that in failing hearts XOI treatment reduced left ventricular (LV) dilatation, improved ejection fraction and normalized cardiac energetics compared to the placebo group.[51] Following this study, improved cardiac energetics was subsequently confirmed in a clinical trial (NCT00181155) in the failing human heart, which demonstrated that the cardiac high-energy phosphate metabolism can also be pharmacologically augmented in humans.[52]

Different metabolic and pharmacokinetic rates and increased/decreased sensitivity to active compounds, when compared to humans, place obvious limitations on the use of small animals for developing and testing novel drugs. Nevertheless, the various models, combined with a non-invasive imaging modality such as CMR, represent a powerful basic science tool in the fight against human heart disease.

18.5 CMR in Clinical Drug Development: Current Status of CMR for Interrogating Aspects of Cardiovascular Tissue and Function

18.5.1 Vascular Imaging

Atherosclerosis is the leading cause of death in developed countries, and emerging therapies will continue to focus on disrupting the pathways that lead to the progression and destabilization of atherosclerotic plaque, and so prevent ischemia and infarction of distal vascular territories. Circulating biomarkers of atherosclerosis (*e.g.* LDL, HDL, HBA1c) have been used as surrogate outcome measures in several recent clinical trials but have limitations. The particular strength of imaging surrogate endpoints for clinical trials of new antiatherogenic drugs is that they assess the biological process further downstream than circulating biomarkers, and so capture the end product of the mechanistic pathways that lead to atheroma. This means that they can potentially capture the activation of both known and unknown pro-atherogenic pathways. The failure of circulating biomarkers to do so may have contributed to the high-profile failure of recent biomarker studies to predict adverse cardiovascular events in subsequent clinical studies.[53]

Atherosclerosis imaging by CMR can measure a number of aspects of large-artery (aorta and carotid) arterial wall structure and function.[54] The wealth of information available at these sites is remarkable, and allows serial imaging of the same region of vasculature, and sometimes even the same plaque, over time. Several measures have now been standardized and can be routinely measured by experienced core-labs: wall area and wall volume both measure plaque volume, whereas minimal lumen area and plaque eccentricity measure the degree of intrusion of the plaque into the lumen and disposition of the plaque, respectively. These measures are highly reproducible[55–57] and agree closely with histological and *ex vivo* MR measures.[58,59]

As a result of progress in atherosclerosis imaging, there is a change of emphasis, from measures of plaque extent to attempts to evaluate the risk posed by a plaque, using techniques that assess plaque composition and biological activity, including inflammation or markers of vascular elastic function. CMR can assess the fibrous cap, the lipid rich/necrotic core, the presence, extent and age of intraplaque hemorrhage and the relative contribution of loose and dense fibrous tissue to the plaque. American Heart Association (AHA) lesion classification, soft plaque identification and plaque risk assessment may be achieved.[60] Current research efforts in CMR athero-sclerosis imaging, for example, dynamic contrast-enhanced CMR, are investigating putative markers of the biological activity of the plaque, particularly with respect to inflammation. CMR can evaluate the functional consequences of atheroma burden on the elastic function of arteries, and can measure both regional and global aortic stiffness.

Saam *et al.* have provided reproducibility calculations for several CMR atherosclerosis measures based on serial MR data from the placebo group of a randomized controlled trial.[56,57,61] Several multicenter trials have used CMR atherosclerosis imaging as an endpoint. Corti *et al.* used CMR to demonstrate reductions in vessel wall area and vessel thickness in both the carotid and thoracic aorta in a randomized controlled trial using simvastatin for 24 months in subjects with known atherosclerotic plaque.[55] Lee *et al.* used CMR to demonstrate reductions in atherosclerotic burden in statin-treated patients using high-dose modified-release nicotinic acid.[62]

At present, high-resolution arterial wall imaging and plaque characterization is available only for large, relatively immobile arteries, and the walls of the coronary arteries in particular cannot be assessed in this way, owing to their small size, mobility and tortuosity. While carotid and abdominal aortic atheroma is of clinical relevance, and probably broadly reflects overall atheroma burden, extrapolation of atheroma regression at one site to other vascular beds may not always be justified. For instance, in a study by Yonemura *et al.* the reduction in plaque volume seen in the carotid artery with high-dose atorvastatin was not seen in the abdominal aorta.[63] Finally, while it has been shown that there is correlation between Framingham risk and CMR markers of atherosclerosis,[64] the predictive value of such imaging approaches beyond standard risk prediction scores has not yet been proven.[53]

18.5.2 Ischemic Heart Disease

The development of new therapies for the treatment of ongoing or completed myocardial infarction is an important goal in cardiovascular drug and device development. The improvements in mortality and morbidity following myocardial infarction in recent years mean that demonstrating a further reduction requires large sample sizes. CMR affords a highly accurate means of quantifying the size and transmural extent of myocardial infarction, and this, combined with accurate measures of left ventricular volumes, systolic function, wall thickness, wall thickening and area at risk (Figure 18.4), allows the accurate characterization of the "severity" of myocardial infarction with much greater sensitivity than clinical endpoints such as death or need for revascularization, which are relatively uncommon over the short duration of many clinical trials nowadays. CMR can use a range of sequences before and after contrast administration to provide a comprehensive assessment of myocardial size, function, presence and extent of scar, viability and perfusion, as well as area at risk, and this information can be acquired in around 45 minutes, using a time-optimized protocol (Figure 18.5). Infarct size is a powerful predictor of subsequent clinical outcome,[65] and recent studies have reported that late

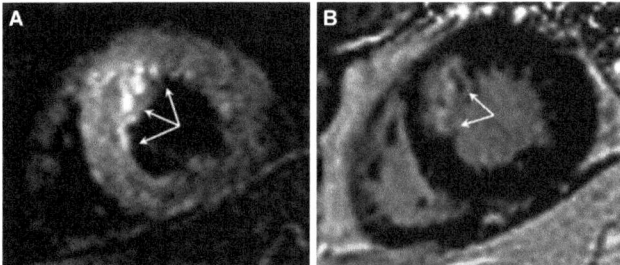

Figure 18.4   Patient with acute anteroseptal myocardial infarction. A. A short tau inversion-recovery T_2-weighed image that shows an area of transmurally increased signal intensity involving the septum (arrows). B. The corresponding LGE image. Note that the spatial extent of myocardial injury in the edema-sensitive T_2 image is larger than that of the necrosis-sensitive late enhancement (arrows).
Reproduced with permission from T. D. Karamitsos, E. Dall'Armellina, R. P. Choudhury and S. Neubauer, *Am. Heart J.*, 2011, **162**, 16. © 2011 Mosby, Inc.

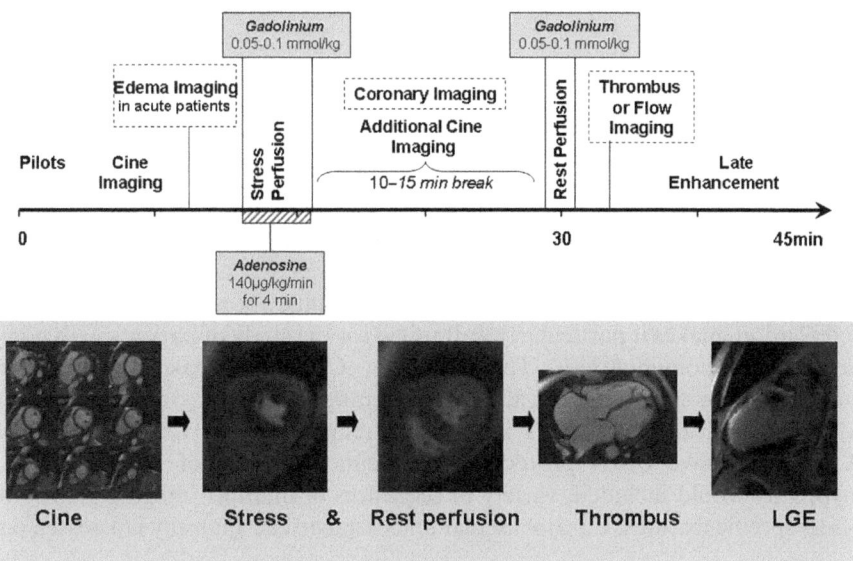

Figure 18.5   Fully time-optimized protocol for a multiparametric CMR study in ischemic heart disease. The major components of this 45-minute protocol are cine imaging, first-pass perfusion at stress and rest, and LGE. During the break between stress and rest perfusion, cine imaging (*e.g.* short-axis stack) can be performed. Coronary imaging is an optional addition and can also be performed during this break. The second short break after rest perfusion can be used for early post-contrast images (thrombus imaging) or flow imaging if a patient has concomitant valvular disease. In patients with suspected acute coronary syndrome, T_2-weighted edema imaging can be added to this protocol (before contrast administration), extending its duration by 5 to 10 minutes.

Reproduced with permission from T. D. Karamitsos, E. Dall'Armellina, R. P. Choudhury and S. Neubauer, *Am. Heart J.*, 2011, **162**, 16. © 2011 Mosby, Inc.

gadolinium enhancement-CMR (LGE-CMR)-measured infarct size is a stronger predictor of clinical outcome than LV ejection fraction or LV volumes.[66-68] Myocardial edema imaging allows an estimation of the area at risk, and a number of methods are currently in use to estimate the "salvaged" myocardium after reperfusion of acute myocardial infarction. These methods are sensitive to the timing of imaging, and there is a need to agree on optimal imaging methods among those available. Kim *et al.* have recently reviewed the role of CMR in both the clinical management of myocardial infarction and in infarct size measurement in clinical trials of myocardial infarction.[69] They give detailed advice regarding the correct technique for measurement of infarct size, and make several important points relevant to the use of this technique. In particular, they emphasize that infarct size evolves over the days and weeks following a myocardial infarction and this means that careful decisions regarding timing of imaging studies and transparency in reporting time after infarct are essential.

18.5.3 Cardiomyopathy

Imaging plays a crucial role in the diagnosis, management and prognosis assessment of patients with non-ischemic cardiomyopathies (Figures 18.6 and 18.7). Over the past decade, the role of CMR imaging in clinical practice has been rapidly expanding. The technique's unsurpassed accuracy in defining cardiac morphology and function, and its ability to provide tissue characterization, makes it particularly well suited for the study of patients with non-ischemic cardiomyopathies. The ability of CMR to measure ventricular volumes, mass and function accurately, reproducibly and on serial studies without significant patient risk is a major strength of CMR for clinical trials. A comprehensive CMR protocol for patients suspected of having cardiomyopathy would include a variety of sequences in multiple imaging planes to assess specific features. Cardiovascular and extracardiac anatomy is assessed on

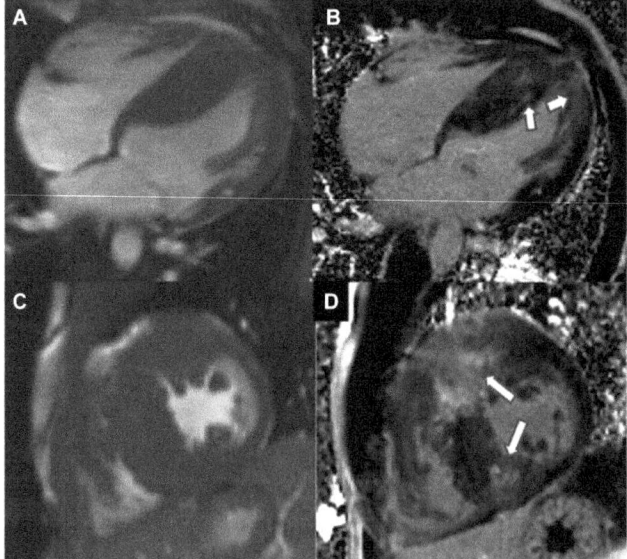

Figure 18.6 Cardiovascular magnetic resonance scan of a patient with hypertrophic cardiomyopathy. A. Still frame from horizontal long-axis cine showing marked septal hypertrophy. B. Same view with LGE imaging showing patchy areas of fibrosis in the hypertrophied septum and in the apical lateral wall (arrows). C. Still frame from short-axis cine of the same patient illustrating the asymmetric septal hypertrophy. D. Same short-axis view with LGE imaging showing the marked diffuse fibrosis within the thickened septum involving both LV and RV junctions (arrows). Reproduced with permission from Karamitsos *et al.*[15] © 2011 Elsevier Inc.

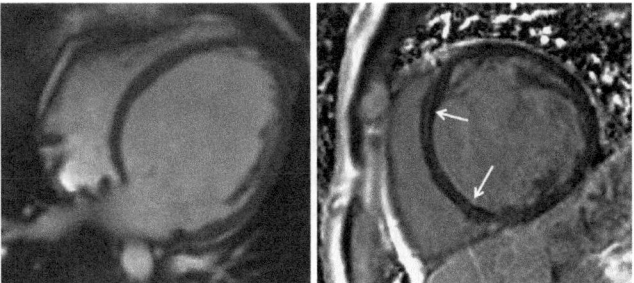

Figure 18.7 Cardiovascular magnetic resonance scan of a patient with dilated cardiomyopathy. Left panel is the end-diastolic frame from horizontal long-axis cine showing a dilated left ventricle with low-normal wall thickness (6–8 mm). Right panel is a short-axis view with LGE showing mid-wall septal fibrosis (arrows).
Reproduced with permission from Karamitsos *et al.*[15] © 2011 Elsevier Inc.

black and/or bright blood images in standard orthogonal (*e.g.* transverse, coronal, sagittal) and sometimes in individualized double-angulated (*e.g.* oblique sagittal) planes. The evaluation of cardiac volumes, function and mass using steady-state free precession (SSFP) sequences is a major component of a CMR protocol for cardiomyopathies. These sequences provide excellent endocardial border definition and generate cine images with high signal and contrast ratio between blood, myocardium and surrounding structures. This has profound implications for the accuracy and reproducibility of CMR measurements of cardiac chambers and volumes, which is excellent for both ventricles.[52,23] Unlike standard two-dimensional echocardiography, the measurement of left ventricular (LV) and right ventricular (RV) volumes is obtained without the need for geometrical assumptions, and is valid even in grossly remodeled hearts where ellipsoidal shape has been lost. Substantial sample size reductions have been predicted (of between 81% and 97%) if CMR measures of volumes, mass and function were serially measured compared to echocardiography.[6] The sample sizes required for LV mass in particular were small (n = 9 for each group to detect a 10 g change in LV mass with a power of 90% and p value of <0.05), based on the high reproducibility of this measure by CMR. This is of relevance to trials in which measurement of ventricular performance and their changes are important, and perhaps as part of broader cardiovascular safety studies, for both cardiovascular and non-cardiovascular drugs.

A unique feature of CMR is its ability to perform myocardial tissue characterization based on proton density, T_1 and T_2 relaxation times (two distinct magnetic resonance relaxation processes that affect the net magnetization and may vary substantially for different tissues and disease states) and by modulating specific parameters of magnetic resonance sequences. For example, in T_1-weighted images, myocardial tissue is relatively dark, whereas fat is bright. By applying special pre-pulses, the signal from fat can be specifically suppressed (saturated). On the other hand, T_2-weighted images highlight unbound water in the myocardium and are used to show myocardial edema caused by inflammation or acute ischemia. Recently, techniques that allow fully quantitative analysis of myocardial T_1 and T_2 values have been developed (relaxation times "mapping"), in an effort to better characterize and quantify global and regional myocardial tissue changes.[19,20] T_2^* is a parameter related to but distinct from T_2, which decreases when regional magnetic field homogeneity is reduced in the vicinity of iron. Consequently, T_2^* imaging is particularly useful to identify cardiac or liver iron overload.[70] Lastly, by using intravenous gadolinium-chelated contrast agents, additional tissue characterization is possible. Areas of fibrosis or scarring increase the volume of gadolinium distribution and thereby produce T_1 shortening, which manifests as hyper-enhancement (high signal) in areas of abnormality on so-called inversion-recovery CMR images.[71] By "nulling" the normal myocardial signal, the area of myocardial injury appears extremely bright (late gadolinium enhancement, LGE), with high signal contrast relative to the black normal myocardium. The LGE technique has widespread application for imaging of infarcted areas, which can be depicted with unprecedented resolution. Areas of myocardial infarction follow a coronary distribution, always involve the subendocardial layer and may extend transmurally depending on the size of the infarct.[72,73] Different LGE patterns (mid-wall, sub-epicardial and circumferential subendocardial) have been described for non-ischemic cardiomyopathies, providing valuable diagnostic and pathophysiological insights.[74] Novel techniques, such as pre- and post-contrast T_1-mapping and equilibrium contrast CMR,[75-77] have been used with promising results in the assessment of diffuse fibrosis in patients with non-ischemic cardiomyopathies, but large-scale clinical data are lacking.

18.6   Limitations of CMR

Several limitations must be recognized in determining whether a CMR evaluation is safe and appropriate for a particular patient, and these must be kept in mind when determining whether CMR will be safe and appropriate in a clinical trial. Implanted pacemakers, defibrillators and resynchronization devices generally preclude CMR imaging on safety grounds, although MR compatible versions are in development,[78] and, importantly, nearly all prosthetic cardiac valves, coronary and vascular stents, and orthopedic implants are safe in a 3 T (or less) MR environment, although current MR safety information should always be consulted before embarking on a study, and for the assessment of participants prior to entering the magnetic field. Gadolinium-containing contrast agents have recently been associated with the development of a rare systemic disorder called nephrogenic systemic fibrosis. The patients at risk for developing this disease are those with acute or chronic severe renal insufficiency (glomerular filtration rate $<30\,\text{mL}\,\text{min}^{-1}\,1.73\,\text{m}^{-2}$), or acute renal dysfunction of any severity due to the hepato-renal syndrome or in the peri-operative liver transplantation period.[79] To date, there is no evidence that other patient groups are at risk. Therefore, gadolinium-based contrast media should be avoided in high-risk patients. Individuals who are obese or claustrophobic can be evaluated in experienced units but, when severe, these may preclude or limit CMR.

CMR mandates a period of relative inaccessibility of the participant while data are acquired, and so cannot be performed at certain times in a patient's clinical care. This can limit its usefulness in some types of study. An example is patients sustaining ST-segment elevation myocardial infarction (STEMI) undergoing primary angioplasty in whom it is clearly not feasible to perform CMR in the phase of admission prior to urgent reperfusion therapy. This limits the use of before-and-after study designs for the evaluation of new therapeutic adjuncts to primary angioplasty in STEMI. It should be noted, however, that safe imaging of patients with recent myocardial infarction is highly feasible in experienced specialist units with careful safety planning.[80]

Optimal time point selection is an important aspect of the application of CMR to clinical trials. Several CMR surrogate endpoints of disease visualize and quantify a highly dynamic process, such as the evolving extent of markers of reversible and irreversible myocardial damage, as scar and ventricle remodel in the weeks following an infarction.

A barrier to CMR use in clinical trials is logistics. CMR systems require major investment in hardware, software, infrastructure and maintenance, and access is limited in many parts of the world. The technique is complex, and substantial training and experience are required to allow accurate measurements. There is also limited transferability of sequences across vendors and field strengths, which currently constrains multicenter studies. Standardized protocols for data acquisition and analysis, often with core-lab data analysis, are essential in the clinical trial setting.

18.7 Future Perspectives: Emerging CMR Techniques Likely to Aid Evaluation of Emerging Therapies (Pre-Clinical and Clinical Sections)

A number of developments in the near future are likely to impact on the use of CMR in drug development and discovery. CMR scanning will be faster and more automated, with the ultimate goal of whole-heart imaging during a single breath-holding. The use of accelerated acquisition techniques heralds improved spatial and temporal resolution for CMR perfusion imaging,[81] with 3D whole heart perfusion CMR imaging comparing favorably with invasive fractional flow reserve assessment of significant coronary artery disease.[82] Moreover, oxygenation imaging as a more direct measure of tissue ischemia, using blood oxygen level dependent CMR may become a useful adjunct and, in some cases, a non-contrast alternative to perfusion imaging for the detection of myocardial ischemia.[83] High-resolution CMR with its ability to visualize atherosclerotic plaque and to discriminate its components holds promise to become the preferred non-invasive modality to study atherothrombotic disease in several vessels such as the aorta, the carotid arteries and eventually the coronary arteries.[84] The great accuracy and reproducibility of CMR for the assessment of myocardial function, perfusion and viability, coupled with molecular imaging techniques such as superparamagnetic particles of iron oxide, make it an attractive imaging modality to monitor the short-term distribution and long-term survival of regenerative cardiac stem cells therapies[85,86] while molecular imaging approaches may identify arterial inflammation.[87] Imaging at higher field strengths (≥ 7 T) will further improve spatial and temporal resolution, providing significant benefits for techniques such as MR spectroscopy, which currently has limited clinical applicability at 1.5 T.[88] This, coupled with new techniques such as ^{13}C-hyperpolarization imaging (see also Chapter 9), should allow dynamic changes in a wide range of metabolite levels to be measured, and therefore provide a reliable method to investigate cardiac energetics during stress-induced myocardial ischemia and for monitoring response to medical and interventional therapies.

Acknowledgements

Parts of the text and the figures of this chapter have been reproduced with permission from the following sources: Cardiovascular MRI in clinical trials: expanded applications through novel surrogate endpoints. A. Pitcher, D. Ashby, P. Elliott and S. E. Petersen, *Heart*, 2011, **97**, 1286. © 2011 BMJ Publishing Group (Text and Figure 18.1); the current and emerging role of cardiovascular magnetic resonance in the diagnosis of non-ischemic cardiomyopathies. T. D. Karamitsos, J. M. Francis and S. Neubauer, *Prog. Cardiovasc. Dis.*, 2011, **54**, 253. © 2011 Elsevier Inc. (Text and Figures 18.6 and 18.7); ischemic heart disease: comprehensive evaluation by cardiovascular magnetic resonance. T. D. Karamitsos, E. Dall'Armellina, R. P. Choudhury and S. Neubauer, *Am. Heart J.*, 2011, **162**, 16. © 2011 Elsevier Inc. (Text and Figures 18.2–18.5). Stefan Neubauer recognizes support from the Oxford NIHR Biomedical Research Centre, and the Oxford BHF Centre of Research Excellence.

References

1. R. O. Bonow, D. L. Mann, D. P. Zipes and P. Libby, *Braunwald's Heart Disease: A Textbook of Cardiovascular Medicine*, Elsevier, Philadelphia, 9th edn, 2012.
2. D. J. Pennell, *Circulation*, 2010, **121**, 692.
3. T. D. Karamitsos and S. Neubauer, *Prog. Cardiovasc. Dis.*, 2011, **54**, 179.
4. R. Antony, M. Daghem, G. P. McCann, S. Daghem, J. Moon, D. J. Pennell, S. Neubauer, H. J. Dargie, C. Berry, J. Payne, M. C. Petrie and N. M. Hawkins, *J. Cardiovasc. Magn. Reson.*, 2011, **13**, 57.
5. A. Pitcher, D. Ashby, P. Elliott and S. E. Petersen, *Heart*, 2011, **97**, 1286.
6. N. G. Bellenger, L. C. Davies, J. M. Francis, A. J. Coats and D. J. Pennell, *J. Cardiovasc. Magn. Reson.*, 2000, **2**, 271.

7. M. A. Bernstein, K. F. King and Z. J. Zhou, *Handbook of MRI Pulse Sequences*, Academic Press, Boston, 2004.
8. C. T. Rodgers and M. D. Robson, *Prog. Cardiovasc. Dis.*, 2011, **54**, 181.
9. F. Grothues, G. C. Smith, J. C. Moon, N. G. Bellenger, P. Collins, H. U. Klein and D. J. Pennell, *Am. J. Cardiol.*, 2002, **90**, 29.
10. F. Grothues, J. C. Moon, N. G. Bellenger, G. S. Smith, H. U. Klein and D. J. Pennell, *Am. Heart J.*, 2004, **147**, 218.
11. T. D. Karamitsos and S. G. Myerson, *Prog. Cardiovasc. Dis.*, 2011, **54**, 276.
12. C. P. Leeson, M. Robinson, J. M. Francis, M. D. Robson, K. M. Channon, S. Neubauer and F. Wiesmann, *J. Cardiovasc. Magn. Reson.*, 2006, **8**, 381.
13. J. Lotz, C. Meier, A. Leppert and M. Galanski, *Radiographics*, 2002, **22**, 651.
14. R. J. Kim, E. Wu, A. Rafael, E. L. Chen, M. A. Parker, O. Simonetti, F. J. Klocke, R. O. Bonow and R. M. Judd, *N. Engl. J. Med.*, 2000, **16**, 343.
15. T. D. Karamitsos, J. M. Francis and S. Neubauer, *Prog. Cardiovasc. Dis.*, 2011, **54**, 253.
16. I. Carbone and M. G. Friedrich, *Curr. Cardiol. Rep.*, 2012, **14**, 1.
17. C. J. Holloway, J. Suttie, S. Dass and S. Neubauer, *Prog. Cardiovasc. Dis.*, 2011, **54**, 320.
18. B. L. Gerber, S. V. Raman, K. Nayak, F. H. Epstein, P. Ferreira, L. Axel and D. L. Kraitchman, *J. Cardiovasc. Magn. Reson.*, 2008, **28**, 10.
19. S. K. Piechnik, V. M. Ferreira, E. Dall'Armellina, L. E. Cochlin, A. Greiser, S. Neubauer and M. D. Robson, *J. Cardiovasc. Magn. Reson.*, 2010, **12**, 69.
20. S. Giri, Y. C. Chung, A. Merchant, G. Mihai, S. Rajagopalan, S. V. Raman and O. P. Simonetti, *J. Cardiovasc. Magn. Reson.*, 2009, **11**, 56.
21. L. J. Anderson, S. Holden, B. Davis, E. Prescott, C. C. Charrier, N. H. Bunce, D. N. Firmin, B. Wonke, J. Porter, J. M. Walker and D. J. Pennell, *Eur. Heart J.*, 2001, **22**, 2171.
22. M. A. Tanner, R. Galanello, C. Dessi, G. C. Smith, M. A. Westwood, A. Agus, M. Roughton, R. Assomull, S. V. Nair, J. M. Walker and D. J. Pennell, *Circulation*, 2007, **115**, 1876.
23. O. Catalano, S. Antonaci, C. Opasich, G. Moro, M. Mussida, M. Perotti, G. Calsamiglia, M. Frascaroli, M. Baldi and F. Cobelli, *J. Cardiovasc. Med. (Hagerstown)*, 2007, **8**, 807.
24. V. Hombach, N. Merkle, P. Bernhard, V. Rasche and W. Rottbauer, *Cardiol. J.*, 2010, **17**, 549.
25. W. M. Franz, O. J. Mueller, R. Hartong, N. Frey and H. A. Katus, *J. Mol. Med.*, 1997, **75**, 115.
26. H. A. Rockman, R. S. Ross, A. N. Harris, K. U. Knowlton, M. E. Steinhelper, L. J. Field, J. Ross Jr. and K. R. Chien, *Proc. Natl Acad. Sci. USA*, 1991, **88**, 8277.
27. Y. Liao, F. Ishikura, S. Beppu, M. Asakura, S. Takashima, H. Asanuma, S. Sanada, J. Kim, H. Ogita, T. Kuzuya, K. Node, M. Kitakaze and M. Hori, *Am. J. Physiol. Heart Circ. Physiol.*, 2002, **282**, H1703.

28. J. Ruff, F. Wiesmann, K. H. Hiller, S. Voll, M. von Kienlin, W. R. Bauer, E. Rommel, S. Neubauer and A. Haase, *Magn. Reson. Med.*, 1998, **40**, 43.

29. J. E. Schneider, P. J. Cassidy, C. Lygate, D. J. Tyler, F. Wiesmann, S. M. Grieve, K. Hulbert, K. Clarke and S. Neubauer, *J. Magn. Reson. Imaging*, 2003, **18**, 691.

30. J. U. Streif, V. Herold, M. Szimtenings, T. E. Lanz, M. Nahrendorf, F. Wiesmann, E. Rommel and A. Haase, *Magn. Reson. Med.*, 2003, **49**, 315.

31. E. Dall'Armellina, B. A. Jung, C. A. Lygate, S. Neubauer, M. Markl and J. E. Schneider, *Magn. Reson. Med.*, 2012, **67**, 541.

32. R. Zhou, S. Pickup, J. D. Glickson, C. H. Scott and V. A. Ferrari, *Magn. Reson. Med.*, 2003, **49**, 760.

33. A. A. Young, B. A. French, Z. Yang, B. R. Cowan, W. D. Gilson, S. S. Berr, C. M. Kramer and F. H. Epstein, *J. Cardiovasc. Magn. Reson.*, 2006, **8**, 685.

34. J. S. Chuang, A. Zemljic-Harpf, R. S. Ross, L. R. Frank, A. D. McCulloch and J. H. Omens, *Magn. Reson. Med.*, 2010, **64**, 1281.

35. F. H. Epstein, Z. Yang, W. D. Gilson, S. S. Berr, C. M. Kramer and B. A. French, *Magn. Reson. Med.*, 2002, **48**, 399.

36. W. D. Gilson, Z. Yang, B. A. French and F. H. Epstein, *Magn. Reson. Med.*, 2004, **51**, 744.

37. W. D. Gilson, Z. Yang, B. A. French and F. H. Epstein, *Am. J. Physiol. Heart Circ. Physiol.*, 2005, **288**, H1491.

38. W. D. Gilson, Z. Yang, F. C. Sureau, B. A. French and F. H. Epstein, *Proc. Intl Soc. Mag. Reson. Med.*, 2004, **11**, 1789.

39. F. Kober, I. Iltis, M. Izquierdo, M. Desrois, D. Ibarrola, P. J. Cozzone and M. Bernard, *Magn. Reson. Med.*, 2004, **51**, 62.

40. J. U. Streif, M. Nahrendorf, K. H. Hiller, C. Waller, F. Wiesmann, E. Rommel, A. Haase and W. R. Bauer, *Magn. Reson. Med.*, 2005, **53**, 584.

41. B. F. Coolen, R. P. Moonen, L. E. Paulis, T. Geelen, K. Nicolay and G. J. Strijkers, *Magn. Reson. Med.*, 2010, **64**, 1658.

42. M. Makowski, C. Jansen, I. Webb, A. Chiribiri, E. Nagel, R. Botnar, S. Kozerke and S. Plein, *Magn. Reson. Med.*, 2010, **64**, 1592.

43. J. E. Schneider, F. Wiesmann, C. A. Lygate and S. Neubauer, *J. Cardiovasc. Magn. Reson.*, 2006, **8**, 693.

44. Z. Yang, S. S. Berr, W. D. Gilson, M. C. Toufektsian and B. A. French, *Circulation*, 2004, **109**, 1161.

45. S. Bohl, C. A. Lygate, H. Barnes, D. Medway, L. A. Stork, J. Schulz-Menger, S. Neubauer and J. E. Schneider, *Am. J. Physiol. Heart Circ. Physiol.*, 2009, **296**, H1200.

46. R. J. Beyers, R. S. Smith, Y. Xu, B. A. Piras, M. Salerno, S. S. Berr, C. H. Meyer, C. M. Kramer, B. A. French and F. H. Epstein, *Magn. Reson. Med.*, 2012, **67**, 201.

47. J. A. Bittl, J. A. Balschi and J. S. Ingwall, *J. Clin. Invest.*, 1987, **79**, 1852.

48. J. E. Schneider, D. J. Tyler, M. ten Hove, A. E. Sang, P. J. Cassidy, A. Fischer, J. Wallis, L. M. Sebag-Montefiore, H. Watkins, D. Isbrandt, K. Clarke and S. Neubauer, *Magn. Reson. Med.*, 2004, **52**, 1029.

49. A. Gupta, A. Akki, Y. Wang, M. K. Leppo, V. P. Chacko, D. B. Foster, V. Caceres, S. Shi, J. A. Kirk, J. Su, S. Lai, N. Paolocci, C. Steenbergen, G. Gerstenblith and R. G. Weiss, *J. Clin. Invest.*, 2012, **122**, 291.

50. A. V. Naumova, V. P. Chacko, R. Ouwerkerk, L. Stull, E. Marban and R. G. Weiss, *Am. J. Physiol. Heart Circ. Physiol.*, 2006, **290**, H837.

51. G. A. Hirsch, P. A. Bottomley, G. Gerstenblith and R. G. Weiss, *J. Am. Coll. Cardiol.*, 2012, **59**, 802.

52. T. D. Karamitsos, L. E. Hudsmith, J. B. Selvanayagam, S. Neubauer and J. M. Francis, *J. Cardiovasc. Magn. Reson.*, 2007, **9**, 777.

53. R. Duivenvoorden, E. de Groot, E. S. Stroes and J. J. Kastelein, *Atherosclerosis*, 2009, **206**, 8.

54. H. R. Underhill, T. S. Hatsukami, Z. A. Fayad, V. Fuster and C. Yuan, *Nat. Rev. Cardiol.*, 2010, **7**, 165.

55. C. Yuan, W. S. Kerwin, V. L. Yarnykh, J. Cai, T. Saam, B. Chu, N. Takaya, M. S. Ferguson, H. Underhill, D. Xu, F. Liu and T. S. Hatsukami, *NMR Biomed.*, 2006, **19**, 636.

56. T. Saam, W. S. Kerwin, B. Chu, J. Cai, A. Kampschulte, T. S. Hatsukami, X. Q. Zhao, N. L. Polissar, B. Neradilek, V. L. Yarnykh, K. Flemming, J. Huston 3rd, W. Insull Jr., J. D. Morrisett, S. D. Rand, K. J. DeMarco and C. Yuan, *J. Cardiovasc. Magn. Reson.*, 2005, **7**, 799.

57. R. Corti, V. Fuster, Z. A. Fayad, S. G. Worthley, G. Helft, D. Smith, J. Weinberger, J. Wentzel, G. Mizsei, M. Mercuri and J. J. Badimon, *Circulation*, 2002, **106**, 2884.

58. L. M. Mitsumori, T. S. Hatsukami, M. S. Ferguson, W. S. Kerwin, J. Cai and C. Yuan, *J. Magn. Reson. Imaging*, 2003, **17**, 410.

59. X. Kang, N. L. Polissar, C. Han, E. Lin and C. Yuan, *Magn. Reson. Med.*, 2000, **44**, 968.

60. C. Yuan, L. M. Mitsumori, M. S. Ferguson, N. L. Polissar, D. Echelard, G. Ortiz, R. Small, J. W. Davies, W. S. Kerwin and T. S. Hatsukami, *Circulation*, 2001, **104**, 2051.

61. R. Corti, Z. A. Fayad, V. Fuster, S. G. Worthley, G. Helft, J. Chesebro, M. Mercuri and J. J. Badimon, *Circulation*, 2001, **104**, 249.

62. J. M. Lee, M. D. Robson, L. M. Yu, C. C. Shirodaria, C. Cunnington, I. Kylintireas, J. E. Digby, T. Bannister, A. Handa, F. Wiesmann, P. N. Durrington, K. M. Channon, S. Neubauer and R. P. Choudhury, *J. Am. Coll. Cardiol.*, 2009, **54**, 1787.

63. A. Yonemura, Y. Momiyama, Z. A. Fayad, M. Ayaori, R. Ohmori, K. Higashi, T. Kihara, S. Sawada, N. Iwamoto, M. Ogura, H. Taniguchi, M. Kusuhara, M. Nagata, H. Nakamura, S. Tamai and F. Ohsuzu, *J. Am. Coll. Cardiol.*, 2005, **45**, 733.

64. F. A. Jaffer, C. J. O'Donnell, M. G. Larson, S. K. Chan, K. V. Kissinger, M. J. Kupka, C. Salton, R. M. Botnar, D. Levy and W. J. Manning, *Arterioscler. Thromb. Vasc. Biol.*, 2002, **22**, 849.

65. R. J. Burns, R. J. Gibbons, Q. Yi, R. S. Roberts, T. D. Miller, G. L. Schaer, J. L. Anderson and S. Yusuf, CORE Study Investigators, *J. Am. Coll. Cardiol.*, 2002, **39**, 30.

66. S. Kelle, S. D. Roes, C. Klein, T. Kokocinski, A. de Roos, E. Fleck, J. J. Bax and E. Nagel, *J. Am. Coll. Cardiol.*, 2009, **54**, 1770.

67. S. D. Roes, S. Kelle, T. A. Kaandorp, T. Kokocinski, D. Poldermans, H. J. Lamb, E. Boersma, E. E. van der Wall, E. Fleck, A. de Roos, E. Nagel and J. J. Bax, *Am. J. Cardiol.*, 2007, **100**, 930.

68. E. Wu, J. T. Ortiz, P. Tejedor, D. C. Lee, C. Bucciarelli-Ducci, P. Kansal, J. C. Carr, T. A. Holly, D. Lloyd-Jones, F. J. Klocke and R. O. Bonow, *Heart*, 2008, **94**, 730.

69. H. W. Kim, A. Farzaneh-Far and R. J. Kim, *J. Am. Coll. Cardiol.*, 2009, **55**, 1.

70. L. J. Anderson, S. Holden, B. Davis, E. Prescott, C. C. Charrier, N. H. Bunce, D. N. Firmin, B. Wonke, J. Porter, J. M. Walker and D. J. Pennell, *Eur. Heart J.*, 2001, **22**, 2171.

71. T. D. Karamitsos and S. Neubauer, *Curr. Cardiol. Rep.*, 2011, **13**, 210.

72. R. J. Kim, D. S. Fieno, T. B. Parrish, K. Harris, E. L. Chen, O. Simonetti, J. Bundy, J. P. Finn, F. J. Klocke and R. M. Judd, *Circulation*, 1999, **100**, 1992.

73. R. J. Kim, E. Wu, A. Rafael, E. L. Chen, M. A. Parker, O. Simonetti, F. J. Klocke, R. O. Bonow and R. M. Judd, *N. Engl. J. Med.*, 2000, **343**, 1445.

74. T. D. Karamitsos, J. M. Francis, S. Myerson, J. B. Selvanayagam and S. Neubauer, *J. Am. Coll. Cardiol.*, 2009, **54**, 1407.

75. L. Iles, H. Pfluger, A. Phrommintikul, J. Cherayath, P. Aksit, S. N. Gupta, D. M. Kaye and A. J. Taylor, *J. Am. Coll. Cardiol.*, 2008, **52**, 1574.

76. M. Jerosch-Herold, D. C. Sheridan, J. D. Kushner, D. Nauman, D. Burgess, D. Dutton, R. Alharethi, D. Li and R. E. Hershberger, *Am. J. Physiol. Heart Circ. Physiol.*, 2008, **295**, H1234.

77. A. S. Flett, M. P. Hayward, M. T. Ashworth, M. S. Hansen, A. M. Taylor, P. M. Elliott, C. McGregor and J. C. Moon, *Circulation*, 2010, **122**, 138.

78. B. L. Wilkoff, D. Bello, M. Taborsky, J. Vymazal, E. Kanal, H. Heuer, K. Hecking, W. B. Johnson, W. Young, B. Ramza, N. Akhtar, B. Kuepper, P. Hunold, R. Luechinger, H. Puererfellner, F. Duru, M. J. Gotte, R. Sutton and T. Sommer, EnRhythm MRI SureScan Pacing System Study Investigators, *Heart Rhythm*, 2011, **8**, 65.

79. A. Kribben, O. Witzke, U. Hillen, J. Barkhausen, A. E. Daul and R. Erbel, *J. Am. Coll. Cardiol.*, 2009, **53**, 1621.

80. E. Dall'Armellina and R. P. Choudhury, *Prog. Cardiovasc. Dis.*, 2011, **54**, 230.

81. S. Plein, J. Schwitter, D. Suerder, J. P. Greenwood, P. Boesiger and S. Kozerke, *Radiology*, 2008, **249**, 493.

82. R. Jogiya, S. Kozerke, G. Morton, K. De Silva, S. Redwood, D. Perera, E. Nagel and S. Plein, *J. Am. Coll. Cardiol.*, 2012, **60**, 756.

83. T. D. Karamitsos, L. Leccisotti, J. R. Arnold, A. Recio-Mayoral, P. Bhamra-Ariza, R. K. Howells, N. Searle, M. D. Robson, O. E. Rimoldi,

P. G. Camici, S. Neubauer and J. B. Selvanayagam, *Circ. Cardiovasc. Imaging*, 2010, **3**, 32.

84. A. C. Lindsay and R. P. Choudhury, *Nat. Rev. Drug Discov.*, 2008, **7**, 517.
85. S. L. Beeres, F. M. Bengel, J. Bartunek, D. E. Atsma, J. M. Hill, M. Vanderheyden, M. Penicka, M. J. Schalij, W. Wijns and J. J. Bax, *J. Am. Coll. Cardiol.*, 2007, **49**, 1137.
86. M. A. McAteer, J. E. Schneider, Z. A. Ali, N. Warrick, C. A. Bursill, C. von zur Muhlen, D. R. Greaves, S. Neubauer, K. M. Channon and R. P. Choudhury, *Arterioscler. Thromb. Vasc. Biol.*, 2008, **28**, 77.
87. R. P. Choudhury and E. A. Fisher, *Arterioscler. Thromb. Vasc. Biol.*, 2009, **29**, 983.
88. L. E. Hudsmith and S. Neubauer, *JACC*, 2009, **2**, 87.

CHAPTER 19

Magnetic Resonance Imaging Techniques in Cancer

D. M. MORRIS,[a] J. P. B. O'CONNOR[a] AND A. JACKSON*[b]

[a] The University of Manchester, Centre for Imaging Sciences, Stopford Building, Oxford Road, Manchester, M13 9PT, UK; [b] The University of Manchester, Wolfson Molecular Imaging Institute, 27 Palatine Road, Manchester, M20 3LJ, UK
*Email: Alan.Jackson@manchester.ac.uk

19.1 Overview

MRI has become an integral and essential part of cancer management, providing non-invasive, radiation-free visualization of the presence, extent and spread of tumors within the body. Clinical and pre-clinical oncology researchers have proven to be early adopters of advanced MR technology and many groups focus on the development and validation of novel imaging approaches. The clinical uptake of more advanced MR imaging techniques such as diffusion weighted imaging (DWI) and dynamic contrast enhanced MRI (DCE-MRI) is slower and has occurred only in recent years. This relatively late adoption largely reflects the difficulties in translating complex acquisition or analysis techniques into the clinical setting but also the delay inherent in validating novel imaging techniques for clinical use.

Increasingly, there is a requirement for quantitative metrics, which reflect physiological or disease processes (e.g. quantified images of tumor blood flow). These "imaging biomarkers" and their identification, validation and clinical translation have become the paradigm for the development of novel techniques in oncological MRI.[1] Typically, the development of a novel biomarker begins

New Developments in NMR No. 2
New Applications of NMR in Drug Discovery and Development
Edited by Leoncio Garrido and Nicolau Beckmann
© The Royal Society of Chemistry 2013
Published by the Royal Society of Chemistry, www.rsc.org

with the identification of a biological process of interest (*e.g.* hypoxia, apoptosis or angiogenesis), followed by the development of candidate biomarkers, usually in a pre-clinical setting. The development of such a biomarker will commonly involve the design of novel acquisition protocols and matching analysis techniques. Candidate biomarkers must be validated for methodological stability, reproducibility and biological relevance in both the pre-clinical and subsequently clinical settings (see also Chapter 12). Many novel biomarkers are developed in parallel with, and for the study of, novel therapeutic agents that target a specific biological process. Validation of imaging biomarkers must therefore provide sufficient understanding of their relationship to the underlying biological process to allow the user to understand whether observing, or failing to observe, a predicted change in a specific biomarker provides sufficient information to support go/no-go decisions in drug development.[2]

In this chapter we will provide an overview of some of the more topical areas of advanced MR imaging and their importance and relevance in oncology.

19.2 Relaxation Times and Proton Density

In MRI, a radiofrequency pulse (denoted B_1) is applied to material in the presence of an external magnetic field (termed B_0). The pulse is limited to a few milliseconds after which it ceases and the energy imparted to the tissue is lost by one of two relaxation processes. Longitudinal relaxation time (T_1; spin-lattice relaxation time) describes how energy is dissipated from spinning protons to the surrounding tissue lattice composed of macromolecular structures whereas transverse relaxation time (T_2; spin-spin relaxation time) describes coherence loss as spinning protons interact with one another in a classical spin echo experiment. If a gradient echo sequence is used, the true T_2 is not measured due to inhomogeneity in the magnetic field and the effective T_2 (denoted T_2* and always shorter than the corresponding T_2) is measured instead.

All normal and pathological tissues can be characterized by their native T_1, T_2, T_2* relaxation properties and by their proton density. Clinical MRI sequences are programmed to use different echo and repetition times so that they can simultaneously maximize and minimize the differential contributions of the relaxation times and proton density towards the collected signal. This approach gives rise to sequences that are dominated by one process, such as a T_1-weighted, T_2-weighted or proton density-weighted sequence. These three sequences, particularly the first two, form the basis of diagnostic MRI since different tissues have subtle differences in their T_1, T_2 and proton density and can therefore be distinguished from one another when signals are spatially encoded to form images. In practice, tissues such as fat, liver or grey matter have a range of T_1 and T_2 times, rather than one exact value.[3] However, this is clinically useful, for example in the brain, where grey matter and white matter have distinct and predictable differences in their T_1 and T_2 times that help determine anatomy.

19.2.1   Clinical Use of "Weighted" Sequences

There has been interest since the early days of clinical MRI in whether or not the relaxation properties of pathological tissue such as tumors were sufficiently different from normal tissue to allow characterization of disease – for example that a specific tissue T_1 value would accurately describe malignant cancer. Although some initial studies suggested that the relaxation properties of tumors might be distinctly different from normal tissues,[4] considerable investigation over the last four decades has shown that this is not consistently true in clinical practice. Although differences exist between mean relaxation parameters for tumor and normal tissues – that allow classical MRI sequences to detect tumors – the overlap in parameter range has resulted in the failure of attempts to employ individual parameters (proton density, T_1 or T_2) as a quantitative diagnostic marker or to define individual tissues by their native relaxation properties or proton density.[5–8]

This difference does, however, allow tumor detection on MRI by visual inspection of the difference in T_1, T_2, T_2* or proton density from surrounding normal tissue, a difference that may be variably marked or subtle. In Figure 19.1, images are shown from a patient with an aggressive primary brain

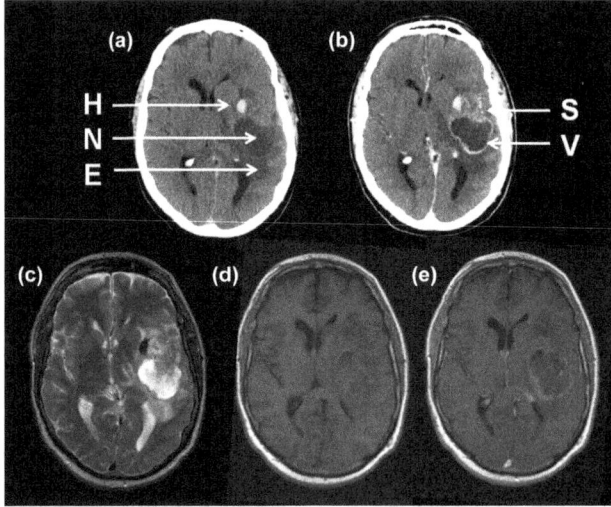

Figure 19.1   Computerized tomography and MR images of a left temporal lobe glioblastoma multiforme in a 47-year-old man. (a, b) Pre-contrast and post-contrast CT images demonstrate a spatially heterogeneous tumor with central necrosis (N; focal low density on both images), hemorrhage (H; high density on pre-contrast images), soft tissue components (S) and vascular ring enhancement (V; increased density on post-contrast images compared to pre-contrast). Surrounding edema is present (E; ill defined, low density). (c) T_2-weighted MRI shows corresponding areas of necrosis (focal high signal), hemorrhage (very low signal) and edema (ill defined, high signal). (d, e) T_1-weighted MRI shows clear vascular contrast enhancement (peripheral high signal). Other key radiological features are tumor size, position and relationship to other structures and degree of mass effect.

tumor, glioblastoma multiforme. Normal grey matter and white matter have been replaced by pathological tissues such as perfused tumour, necrosis, cystic change, hemorrhage and edema, which can be characterized by their appearances on T_1- and T_2-weighted images.

The degree of contrast between tissues and, therefore, diagnostic usefulness depends on physical properties of the MRI sequence (including scanner static magnetic field B_0, local field inhomogeneity, which relaxation process is weighting the sequence and signal-to-noise ratio), the underlying tumor biology (including how cellular, vascular, hemorrhagic, cystic and edematous the tumor is) and variation in native tissue relaxation properties. For example patients whose livers have infiltration with fat or fibrous tissue will have different native T_1 to patients with healthy hepatocytes that will alter the contrast between background liver and a tumor.

In clinical practice, many tumors demonstrate hypointense signal on T_1-weighted images and hyperintense signal on T_2-weighted images, allowing reliable clinical identification of tumor in solid organs. This largely reflects the increased permeability of solid tumors, which causes edema within the tumor and resultant increase in proton density, T_1 and T_2 relative to surrounding tissues.[9,10] These endogenous MRI contrast mechanisms have been employed successfully since the early 1980s to detect tumors.[11]

Although T_1-weighted and T_2-weighted images are the basis of clinical MRI, other sequences are also routinely used in oncology. A commonly used approach is to derive endogenous contrast by applying an initial inversion pulse before the RF pulse designed to attenuate signal from a specific tissue. In theory, any particular tissue with a relatively well-defined T_1 range can be suppressed if an inversion pulse is applied and the time between the inversion pulse and radiofrequency (RF) pulse set to approximately 70% (natural logarithm of 2) of the tissue T_1 to null signal from that tissue. One common clinical example is fat signal suppression (short T_1), where the inversion time is set to around 140–150 ms (since fat has a T_1 of around 220 ms at 1.5 T). This enables fat tissue to have near zero longitudinal magnetization and therefore produce negligible signal in voxels whose constituents are mainly fat. In distinction, all other tissues have longer T_1 times and hence have variable, but available, magnetization to be influenced by an RF pulse applied after the initial inversion pulse. A similar approach, termed Fluid Attenuated Inversion Recovery (FLAIR),[12] can be used to suppress free water signal. Since water has a much longer T_1, an inversion time of around 1800–2500 ms is selected depending on magnet performance.

These approaches help to detect pathology due to fat-containing lesions (for example, focal fatty infiltration in the liver, lipoma or dermoid cyst) and lesions that contain clear fluid. An alternative approach is diffusion-weighted imaging, where the random Brownian motion of free water molecules in the presence of local magnetic gradients is quantified as a measure of tissue cellularity, which is frequently different in normal tissue and in tumors.[13]

Finally, exogenous contrast agents have been used in many oncology applications for several decades.[14] Most commonly, the T_1-weighted sequences are performed not only as described above, but also in the presence of an intravenously administered paramagnetic contrast agent (typically a gadolinium chelate) that acts to markedly shorten the longitudinal relaxation times. Since most solid tumors have elevated tumor vascular permeability, increased flow and vascular pooling, this further aids identification of malignant tumors (Figure 19.1b and e), as well as helping to identify biologically aggressive foci of tumor within a lesion to target for biopsy.[15]

19.2.2 Quantifying T_1 and T_2*

There are several research applications that require quantification of tissue relaxation times. Most common in oncology are DCE-MRI studies that seek to quantify the passage of gadolinium-based tracer through abnormal tumor microvessels[16] and require measurement of either T_1 or T_2*. Other applications include T_1-weighted oxygen-enhanced MRI in normal tissues, diseased lungs and tumors[17–19] and Blood Oxygen Level Dependent (BOLD) imaging of renal disease and cancer[20,21] (see below). Recently, the native T_1 of tumors has been investigated as a biomarker of response to therapy and several studies have shown reductions in tumor T_1 in response to various cytotoxic and anti-vascular endothelial growth factor therapies.[22,23]

Measurement of T_1 *in vivo* is a non-trivial problem and accurate T_1 measurements can take a long time to obtain (15–20 minutes for a single slice of MRI data). In general, the methods for T_1 measurement fall into two main categories, namely inversion or saturation recovery prepared techniques and variable saturation/variable flip angle techniques, using various imaging sequences (echo planar imaging EPI, spin echo or gradient echo). Quantification using the inversion recovery method requires inversion of magnetization with a 180° RF pulse and variable delay (inversion times) prior to excitation with B_1. This results in very long acquisition times so that, while this remains the most accurate method, it is usually impractical for clinical imaging applications. Saturation recovery with a varied repetition time (TR) provides a simple alternative that can be acquired more quickly and is sometimes used for clinical imaging. In practice, the variable flip angle method is more practical and, if the TR and flip angle used are very small, volume coverage can be obtained in just a few seconds using gradient echo sequences.

In all cases, measurement errors can occur, due to incorrect spoiling of T_2 effects, inhomogeneities in the magnetic field, inaccuracy in the applied RF pulses (particularly a problem in variable flip angle acquisitions) and the presence of significant measurement noise. Inaccurate measurements will, of course, seriously hinder generation of biologically meaningful data that rely on T_1 measurement and, for this reason, considerable research time has been invested on measurement accuracy as discussed elsewhere.[24]

In contrast, measurement of T_2* is more straightforward using either single or multiple gradient echo techniques. These methods are relatively insensitive to variation in RF pulse, but are highly sensitive to spatial variation in the field within the magnetic bore. This is most easily minimized using shimming gradients to keep inhomogeneity at an acceptable level.

19.3 Dynamic Contrast Enhanced MRI (DCE-MRI)

As contrast enhanced MRI (CE-MRI) using contrast agents based on gadolinium chelates became a standard for the diagnosis of human cancer it soon became apparent that additional biological information could be obtained by observing the time course of the enhancement process. In breast and lung tumors significant additional diagnostic information was provided by the rate of tumor enhancement over a period of several minutes to half an hour after intravenous contrast agent administration. The use of multiple delayed acquisitions to document enhancement pattern became routine practice with radiologists applying subjective classification criteria based on enhancement rate and pattern.[25]

With the advent of faster image acquisition sequences, the ability to collect dynamic contrast enhanced data following the intravenous injection of contrast material as a bolus improved dramatically, allowing both higher temporal and spatial resolution. The use of dynamic contrast enhanced MRI (DCE-MRI) data combined with appropriate image analysis can now provide powerful and specific biomarkers of the tumor vascular microstructure and microenvironment. Both acquisition sequence and data analysis must be matched to the clinical research question and, consequently, a wide range of DCE-MRI techniques has become established.[22]

19.3.1 Dynamic Susceptibility Contrast Enhanced-MRI (DSCE-MRI)

The earliest applications of DCE-MRI used T_2^* weighted images to study the transit of contrast media through the brain.[26] The contrast in these images results largely from susceptibility effects so that signal change is produced from water molecules at some distance from the contrast molecule producing the effect. In vascular beds where the proportion of vascular tissue is low and the spacing between individual vessels is relatively high, this susceptibility contrast mechanism provides good signal-to-noise ratio by inducing signal change in non-vascular tissue.[27]

Dynamic susceptibility contrast enhanced MRI (DSCE-MRI) has a number of other significant advantages. High temporal resolution collections can be performed using simple gradient echo-based acquisitions together with a rapid data sampling technique such as EPI. Even on clinical scanners these combinations allow acceptable temporal and spatial resolution typically providing coverage of the whole brain with a spatial resolution of approximately $1.5 \times 1.5 \times 3$ mm and temporal resolution in the region of 2 seconds.

In normal brain, small-molecular-weight contrast agents remain entirely intravascular allowing the application of simple flow-based models to extract physiological parameters of interest. Typically, contrast is injected into an antecubital vein as a bolus over a period of 4–6 seconds. This is followed by a flush of normal saline to ensure that the bolus remains coherent as it enters into the central circulation. As a result, examination of the signal time course data from an area of normal brain demonstrates a clear short-lived signal decrease as this initial bolus passes through the brain. As the bolus passes around the body it will broaden by diffusion within the vascular space and it will also leak into peripheral tissues around the body. Thus, the first pass peak is followed by a recirculation phase, which demonstrates a much lower overall concentration and a gradual decay due to distribution into peripheral tissue spaces combined with renal excretion (Figure 19.2).[28]

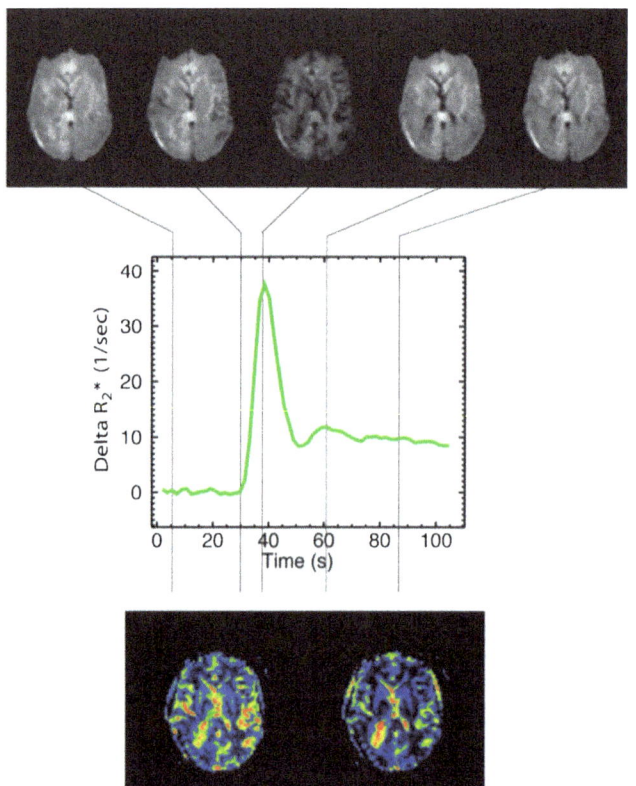

Figure 19.2 DSCE-MRI in a patient with a right-sided (left side of the image) occipital glioblastoma. (Top) Dynamic time course series through the brain using a T_2*-weighted acquisition during passage of bolus of contrast agent. There is a decrease in signal intensity within the brain as the contrast agent bolus traverses the grey and white matter. (Middle) Plot of contrast concentration changes in the tumor. (Bottom) Calculated images of blood flow (F; right) and regional blood volume (CBV; left) showing increased flow and blood volume within the tumor.

These dynamic data typically have good signal-to-noise characteristics and can be analyzed using a number of different approaches.[28–31] A number of important biomarkers can be derived from the first-pass bolus by fitting a gamma variant curve to the first pass data (Figure 19.2). The area under this curve will reflect the relative cerebral blood volume (rCBV) within the voxel and absolute measurements can be calculated by assuming that major vessels correspond to a value of 100%. The time of arrival (T_0) and time to peak (TTP) contrast concentrations can be easily extracted and are of considerable value in cerebral vascular disease. It is also, in theory, possible to calculate maps of cerebral blood flow (CBF) using these data. Correction for variations in the shape and timing of the bolus arriving within the brain must be made if this is to be achieved. Identification of a major input vessel at the base of the brain is therefore performed in order to identify an arterial input function (AIF). Individual tissue residue functions can then be deconvolved using the measured AIF to produce parametric maps of blood flow.[29]

The use of DSCE-MRI in tumors has been basically limited to the brain where it is now relatively routine in clinic. Use of the technique in peripheral tissues is limited both by the presence of extensive susceptibility artifacts within the body and, more importantly, by the presence of contrast leakage. In tumors, where contrast leakage occurs, significant modification of acquisition and or analysis techniques is required. The presence of static contrast in the extravascular extracellular space leads to significant shortening of T_1 so that, where the imaging sequence is sensitive to both T_2^* and T_1 effects, these will produce competing signal loss and signal enhancement respectively. This "T_1 shine-through" effect leads to highly inaccurate estimations of the first pass signal change, with consequent underestimation of CBV.[30] The T_1 effects can be mitigated in one of three ways. Firstly, the T_1 sensitivity of the basic sequence can be reduced, typically by reduction of the flip angle. This will result in a pure intravascular signal but will also significantly reduce signal-to-noise ratio (SNR). Secondly, a bolus of contrast agent can be administered several minutes prior to the dynamic acquisition. This has the effect of saturating T_1 effects whilst allowing the use of a larger flip angle with better SNR. This method is increasingly becoming routine in clinical practice.[30] The third approach is to incorporate the mixed signal into an explicit model that attempts to separate the effects. This is relatively complex, becomes unstable at high rates of leakage and is seldom used.[27]

The use of DSCE-MRI in cerebral tumors has proven to have significant clinical benefit.[32] In particular, CBV measurements show clear differences between glial cell tumors and potential tumor mimics such as abscess, radiation necrosis, atypical demyelination, pseudo-progression and lymphoma. DSCE-MRI measurement of CBV has consequently become a mainstream component of diagnostic imaging. In addition, CBV measurements show a close correlation with tumor grade and survival in glioma and early malignant dedifferentiation in low-grade glioma can be detected up to 6 months earlier by measurement of CBV compared to simple contrast enhanced imaging.[33]

19.3.2 Dynamic Relaxivity Contrast Enhanced MRI (DRCE-MRI)

DCE-MRI using T_1-weighted sequences presents both advantages and disadvantages compared to DSCE-MRI. With T_1-weighted images both intravascular and extravasated contrast will cause increases in signal intensity (Figure 19.3). This makes separation of the intravascular and extravascular components of the time course curves more straightforward than with DSCE-MRI. This has led to the development of a wide range of semiquantitative metrics, which typically described the degree of enhancement as either an absolute or a relative change; the rate of this change or the integral of

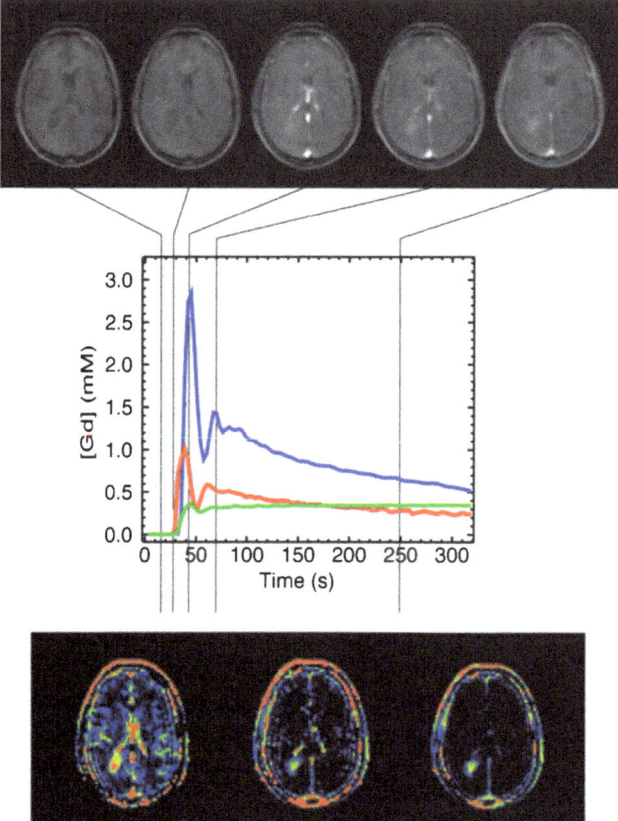

Figure 19.3 DRCE-MRI from the patient shown in Figure 19.1. (Top) Dynamic time course series through the brain using a T_1-weighted acquisition during the passage of a bolus of contrast agent. There is increase in signal intensity within the brain as the contrast agent bolus traverses the grey and white matter. Early persistent enhancement is seen in the tumor. (Middle) Concentration–time course curves from the middle cerebral artery (red), superior sagittal sinus (blue) and tumor (green). (Bottom) Parametric maps of v_p (left), K^{trans} (center) and v_e (right).

contrast induced signal changes over a set time period.[28,34] These semi-quantitative metrics are very simple to calculate and have proven useful in the investigation and diagnosis of some tumor types, particularly breast and liver. Unfortunately, many of these semi-quantitative metrics can be adversely affected by variations in sequence design or between scanners so that reproducibility, particularly across multiple centers, is typically poor and comparison of results from multiple centers is often not possible.

Many groups have modeled changes in T_1-weighted signal using pharmacokinetic models combined with curve-fitting techniques to calculate specific physiological parameters. In order to use this approach, it is essential to transform signal intensity changes into contrast concentration measurements. Unfortunately, this requires measurement of baseline T_1 values, which adds complexity to both acquisition and analysis. Assuming that baseline T_1 measurements can be acquired, then transformation of signal change into contrast concentration time course curves for each voxel in the imaging volume can be performed. An appropriate AIF can be identified within the volume to support pharmacokinetic analysis of individual tissue residue functions using one of a number of available models.[35] Where an AIF cannot be measured then a number of population-based standardized AIF functions are available in the literature.[36]

Pharmacokinetic models range from simple two-compartment approaches to extremely complex models, designed to estimate separately a wide range of microvascular parameters.[37,38] The simplest models measure a contrast transfer coefficient (K^{trans}) and the fractional volume of the extravascular, extracellular space into which contrast is distributed (v_e). Unfortunately, this model clearly lacks specificity and local blood volume; blood flow and the permeability and surface area of the capillary endothelial membrane will affect estimates of K^{trans}. These simple models are inappropriate in most tumor types and the most common approach is to explicitly include the contribution of intravascular contrast by direct estimation of the fractional volume of the vascular space (v_p) (Figure 19.4). In this case estimates of K^{trans} will be affected by blood flow and by the permeability surface area product of the capillary endothelium. Although it is almost always desirable to separate out these contributions, the addition of further complexity to the analysis by the use of an appropriate model introduces significant instability into the curve-fitting procedure with resultant reductions in accuracy and precision.[38] In all pharmacokinetic analyses the choice of model must be based on the specific biological change predicted in response to the intervention being studied, the specificity and therefore complexity of the pharmacokinetic model required to detect it and the ability of the scanning system to deliver an acquisition sequence with sufficient temporal resolution, spatial resolution and SNR to support the analytical model.

Despite these complexities, DRCE-MRI is widely used, particularly in early phase drug trials of novel anticancer agents. The technique has shown considerable utility in detecting therapy-induced changes at an early stage with a range of novel therapeutic agents, particularly those that target angiogenesis or tumor vasculature.[16,22]

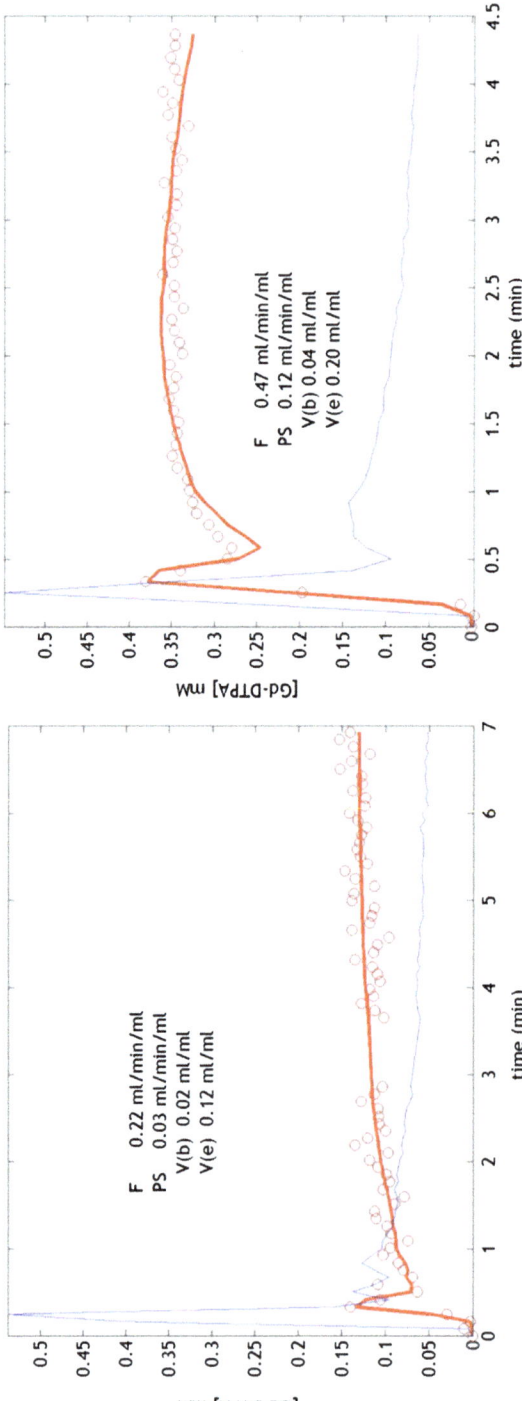

Figure 19.4 Contrast time course curves from a low-grade (left) and high-grade (right) glioma fitted with the adiabatic tissue homogeneity model showing calculated values for flow (F), endothelial permeability surface area product (PS), fractional blood volume (v_b) and fractional extravascular extracellular space volume (v_e). Individual points represent measured data averaged for the entire tumor in order to increase signal-to-noise ratio and the dark line the resulting fit. The pale line shows the measured AIF.

19.4 Diffusion Weighted Imaging

Diffusion weighted imaging (DWI) has a long history in nuclear magnetic resonance after the discovery by Stejskal and Tanner, Le Bihan, Chenevert[39–41] and others that it was possible to use tissue water as an endogenous contrast agent. Typically, diffusional motion of protons is detected by use of matched de-phasing and re-phasing pulses (Stejskal–Tanner sequence). Where protons have physically moved between the de-phasing and re-phasing pulses a decrease in returned signal amplitude will occur. Changing the strength of the de-phasing pulses varies the sensitivity of the pulse sequence to movement. This can be achieved by changing the duration or amplitude of the gradient; the strength of the pulses is quantified by the area under the pulse, commonly referred to as the b value. Changes in the diffusion measured by these techniques represent variations in the microstructure of the tumor, affecting the way the water diffuses and this can act as a biomarker for the identification and characterization of cancer.

19.4.1 Apparent Diffusion Coefficient

The parameter that quantifies diffusion is the diffusion coefficient. However, diffusion observed in MR imaging is not true diffusion as it is restricted, being lower than the free diffusion of water. Consequently, it is quantified by the apparent diffusion coefficient (ADC). This will be affected by the presence, or lack, of membranes or other cellular architecture and the density of cellular packing. The typical diffusion weighted (DW) MRI imaging experiment acquires data less than 100 ms after the tissue has been sensitized to the movement of molecules otherwise signal decay makes measurement impossible. Within this time scale thermal Brownian motion will have allowed the water to interact with cells and their hydrophilic membranes. This means that, despite the physical resolution of DW MRI being a few mm, it is sensitive to motion occurring on the cellular scale. It must be noted that beside the cellular architecture other processes may affect ADC that do not provide the same information. Molecular transport, flow and perfusion will all affect the dynamic of water within the tissue whereas cardiac or respiratory motion may result in apparent motion of the water when there is none present.

ADC may be derived directly from the signal decreases observed in the imaging experiment and is ideally a mono-exponential fit with increasing diffusion weighting (b). This is an idealized representation. With several factors contributing to the DWI signal, a multiexponential decay with increasing values of b is often observed. There are difficulties in relating these observations to the underlying biophysics and the construction of tissue/tumor mimicking phantoms is an active area of research.[42,43] This is clearly observed in the case of vascular tissue where there is significant small vessel perfusion, which may influence ADC measurements. In this case there will be both fast, characterizing perfusion, and slow diffusion indicative of the cellular architecture of interest. Typically, high b values ($>100\,\mathrm{s\,mm^{-2}}$) are used in order to nullify the effects of perfusion on the identified signal decrease.

Malignant tumors typically display lower ADC values than benign masses. This is believed to be related to the increased cellularity of aggressively proliferating tumors[44] but has still to be determined with certainty. Regardless of mechanism, the definition of ADC threshold is dependent not only on the tumor type, its differentiation or the amount of necrosis, but also on the acquisition strategy employed. It may be that the changes in ADC are driven by necrosis within the tumor that is typically associated with a significantly worse outcome. While an increase in ADC following treatment appears to indicate response, this may also be accompanied by a transient decrease. The time scales of these responses are different for various tumor types.

19.4.2   Clinical Applications

ADC has been used to characterize tissue and while the reasons for the lower ADC in malignant tumors are complex, possibly resulting from increased cellularity, disruption of tissue architecture and/or increased extra- *vs.* intracellular space resulting in more restricted diffusion, this correlation has been noted in a number of studies.[44,45] Some biological processes such as inflammation, infection and cyst formation, which are commonly seen in tumors, will increase the ADC. These effects may produce paradoxical changes in ADC and lead to false-negative findings. Indeed, some studies[46] using ADC estimation in lymph nodes have reported apparent increases in ADC in nodes with malignant infiltration. The picture is the same when assessing treatment response where an increased ADC, towards normal tissue values, has been observed as a positive predictor in multiple cancers.[47-49] However, other effects in responding tissue such as dehydration or fibrosis will cause the ADC to fall. This is very difficult to interpret as cellular swelling early in treatment generally causes an initial fall in ADC before the prognostic change is observed.

The true additional benefit of ADC is its ability to differentiate between a re-occurrence of the tumor and any residual cancer, which should be treated differently and may well be confounded by the response of the tumor to treatment. In the case of head and neck tumors, the higher the ADC the more likely it is to be a therapeutic reaction rather than residual disease. The observation that higher ADC is predictive of a positive outcome can also be extended to pre-treatment assessment for a variety of treatments across a number of different tumor types. In the complex picture presented by ADC this is not true for all tumor types and better definition of the histopathology underlying these results is required.

19.4.3   Drug Development

Interest in ADC is largely driven by its potential to act as a biomarker of clinical response, occurring before changes in conventional measurements of tumor size.[50,51] This has been demonstrated in Phase I and II clinical trials[47,52] and opens up the possibility for a personalized form of medicine. Here the effectiveness of a drug can be determined early in the treatment before any therapeutic response, allowing for the selection of a more appropriate therapeutic regime at an early stage. This kind of personalized therapy is desirable to optimize treatment response and to avoid the expense and potential side effects of many novel targeted therapies.

Such early identification of response has important implications for drug development as a means of triaging the large number of possible compounds that are available before beginning an expensive clinical trial. Early stop/go decisions are possible using ADC as it is a measure of the fundamental tumor structure that a variety of different drug pathways may be able to affect. Therefore, it is most likely to be useful in treatments that induce apoptosis in the tumor.

While ADC has begun to be used routinely in clinical practice for individual patients at individual centers, a standardization of techniques and analysis is required before it can be used in multicenters to realize its potential as a biomarker in drug development and personalized medicine, particularly in relation to the interpretation of results.[13]

19.5 Imaging Oxygenation Using MRI

The use of MRI for the determination of tissue oxygenation is a relatively new field but the ability to determine the oxygenation state of tissue non-invasively is of particular relevance in cancer.[53] Cellular hypoxia, reduction in intracellular oxygen concentration, is a common feature in cancer. As tumors outgrow their blood supply hypoxaemia provides a significant stimulus for the production of pro-angiogenic cytokines that induce new blood vessel formation. In many cases tumor cell growth continues to outstrip vascular development leaving areas of the tumor chronically hypoxic. Hypoxia significantly reduces the sensitivity of tumor cells to both radiotherapy and chemotherapy. In addition, hypoxia stimulates localized genetic mutation, which can produce further treatment resistant tumor cell clones. Identification of tumor hypoxia allows improved radiotherapy planning and treatment selection. In principle, hypoxia imaging could also improve the selection of patients who have tumors appropriate for cytoreductive agents, which target hypoxic tissue.

While the current gold standard for hypoxia imaging in cancer is considered to be positron emission tomography (PET) using nitromidazole-based tracers such as [18F] fluoromisonidazole (FMISO) or [11Cu] diacetyl-bis(N4-methyl-thiosemicarbazone) (Cu-ATSM), the expense and logistic problems associated with the production of radio pharmaceuticals has limited the applicability of such techniques. MRI, available in most general hospitals, has demonstrated how the oxygen environment affects the relaxation time of tissues and recently more systematic quantification of the oxygen environment in tumors has been investigated.

19.5.1 Quantification of Oxygen Status with MRI

There are a number of important factors relevant to the determination of oxygenation status *in vivo* using MRI. The oxygen in tissue is a function of the blood volume present and the oxygen in the blood will be in one of two compartments: the first bound to hemoglobin, normally expressed as the oxygen saturation (SO_2; %), the second dissolved in the blood plasma, expressed as a concentration in terms of the partial pressure of oxygen (PO_2; mmHg). The oxygen in these two compartments will have different effects on the relaxation times of the tissue. Both T_1 (sensitive to plasma partial pressure of oxygen) and T_2^* (affected by the concentration of deoxyhemoglobin) relaxation times can be measured to characterize the oxygen environment. The difference in response to the compartment of the oxygen is related to the size and mobility of the oxygen atoms. When oxygen is bound to hemoglobin this creates a large molecule that has difficulty interacting with the molecular lattice, thus affecting T_1. A hemoglobin molecule is paramagnetic when it does not have four oxygen molecules bound to it (deoxyhemoglobin) while the oxyhemoglobin, found in saturated blood, is diamagnetic. The diamagnetic blood aligns with the magnetic field reducing inhomogeneities in the magnetic field, which contribute to the rate of T_2^* decay. The more saturated the blood is the longer the T_2^* decay constant of the tissue will be. The oxygen molecules dissolved in blood plasma are paramagnetic. This has the effect of dissipating the energy more rapidly with the lattice and reducing the T_1 of the tissue as the concentration of molecular oxygen increases.

   The difficulty with these techniques is that they do not give a direct measurement of SO_2 or PO_2 for the tissue of interest. This could be possible *via* a calibration procedure to determine the T_1 and T_2^* of the blood for a given SO_2 or PO_2 and an estimate of the blood fraction of the tissue, but physiological variability would mean that the calibration constant might need to be calculated on an individual basis rendering this approach impractical. In MRI the way this has been compensated for is by use of a physiological oxygen challenge; that is, changing the amount of oxygen administered to the patient and observing the relative changes in the relaxation parameters.[20,54–56] The rates of change can be determined experimentally for blood and allow for absolute quantification, although these calibration factors have yet to be validated. The main application will be to use the oxygen disassociation curve relating PO_2 and SaO_2. On administration of oxygen, if the blood is at low saturation, we would expect the T_2^* to increase as the saturation is increased but changes in the T_1 to remain constant as the oxygen impulse is bound to hemoglobin preferentially rather than dissolving in blood. In the well-oxygenated case the blood would be fully saturated and there would be no change in oxygen saturation, or T_2^*, but the T_1 would decrease as the amount of paramagnetic oxygen present in the blood plasma increases. There are possible confounds to this simple paradigm in that we are assuming that the supply of blood to the tissue is a constant and if this is variable, or in fact hypoxia is as a result of impaired blood flow, differentiation of hypoxic and normal tissue is compromised.

19.5.2   Pre-Clinical and Clinical Research

The true potential of the oxygenation techniques is in the qualification of the oxygen environment present in the tumor. The quantification has been investigated in only a few studies to date. The practicality of using dual T_1 and T_2^* has been demonstrated in normal tissue and in a small cohort of solid tumors.[19] A significant reduction in T_1 was observed in tumors in the presence of maintained perfusion whilst, conversely, those tumors with poorly perfused cores demonstrated a reduced T_1 change consistent with hypoxia caused by a reduction in flow. Carbogen (98% oxygen with 2% CO_2) has also been used as a gas challenge in prostate tumors, the small amount of CO_2 being intended to overcome oxygen induced vasoconstriction.[57] These studies showed a significant increase in T_2^* in response to oxygen challenge. This suggests that the oxygen status of the tumor was improved by the administration of carbogen indicating it may be used as a sensitizing agent for radiotherapy. This result was also validated in several murine tumor models[58] and previous studies[59] have shown that the oxygen dynamics of tumors in mice are very different depending on whether the tumors are hypoxic or not.

While the use of MRI to determine oxygen is in its infancy, the importance of oxygen status in the treatment of cancer means that applications to oncology are at the forefront of this field.

19.6   Magnetization Transfer Imaging

Magnetization transfer (MT) imaging allows the interactions of the water protons to be probed in more detail and also the states in which they are located. While MT imaging has been around for nearly 20 years, the applications to cancer imaging have been limited by the time taken to acquire the images, and currently the technique is limited to the brain where it is able to answer specific questions relating to cerebral diseases of which cancer is only one. It is also the case that it is in competition with the more developed and technically easier to perform diffusion imaging, which provides a biomarker sensitive to the same physiological processes.

19.6.1   Magnetization Transfer Contrast

Magnetization transfer contrast (MTC) imaging is based on the difference in T_2 relaxation time between free (unbound) protons and protons bound to macromolecules in the same environment.[60] The unbound protons are associated with tissue water with a longer T_2. The very short T_2 of the bound protons is because they are associated with macromolecules or are in the hydration layers. The effect of these bound protons cannot be directly imaged but can be seen in the signal of the free protons if there is a transfer of the magnetization between the two pools. To remove the contribution from the bound protons these spins need to be saturated such that, when their magnetization is transferred into the free proton pool, it results in a reduction of the signal observed. This saturation is normally achieved by an off-resonance RF pulse. The contrast of the new image will be affected by the relative water content of different types of tissue or pools. For example, in the brain white matter has a greater water content than grey matter meaning the contrast between the two signals will be increased in an MTC image.

While the difference in contrast may make the identification of masses or tumors easier, a more quantitative measurement would allow one to quantify the amount of protons in either of these two pools to describe the amount of water present relative to the structure of the tissue. This is simply done by taking the signal with saturation (M_s) and the signal without saturation (M_0) and building the ratio

$$MTR = (M_0 - M_s)/M_0 \qquad (19.1)$$

to give the magnetization transfer ratio (MTR). This takes twice as long since the second unsaturated image is required. The whole process is also extended because the repeated use of radiofrequency pulses, as in most rapid imaging experiments, will saturate the bound protons as well, meaning a longer acquisition is required both with and without saturation. While there are methods using multiple saturations and differing imaging parameters to try and probe deeper into the reasons for this signal loss,[61,62] none of these more complex models has been applied to cancer.

19.6.2 Cancer Studies Using MTR

It has been shown that in the brain the MTR can be used as a biomarker to distinguish between a wide variety of malignancies and tumors, both cystic and solid.[63] The MTR values observed were significantly lower than those observed in areas of infarcted or infected tissue, all of which were lower than normal tissue. However, there has been no significant MTR difference shown between tumor grades. The use of the MTR images has improved the sensitivity of identification of tumors and allows for the differentiation of types of brain tumors based on their physical characteristics.[64]

Since MT imaging is giving us some insight into the extracellular membranes, it is hoped this can be used as a method for monitoring tumor response.[65] At the most basic level thresholding of MTR maps allows for accurate identification of cancerous tissue, increasing the efficacy of any volume assessment of tumor size changes in response to treatment. In Figure 19.5 T_1 contrast enhanced (c–d), T_2 (a) and MTR (b) images are shown along with volume of tumor defined. Whilst the T_2 volume (e) is bigger than the T_1 volume (d) neither define the tumor to the extent of the MTR (f). Beyond this, studies of breast tumors have shown that the MTR is different between tissue that has been irradiated and that which has not.[66] The MTR was higher in the treatment case suggesting that MTR can be used to monitor the late effects of treatment. It has also been shown in breast that the MT ratios between cancerous and benign tumors are significantly different[67] but this requires further validation.

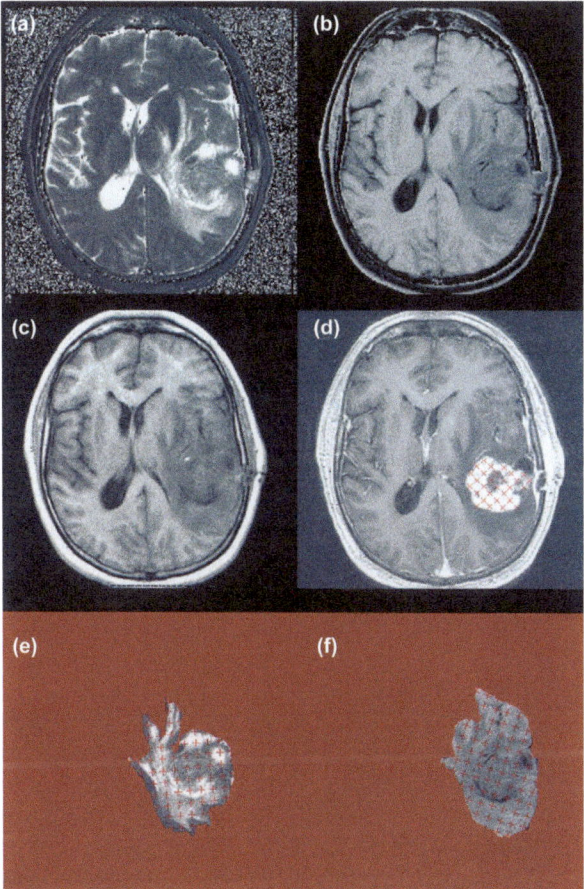

Figure 19.5 Sample images from a 61-year-old patient having a grade III glioma. (a) T_2 map; (b) magnetization transfer ratio (MTR) map; (c) pre-contrast T_1-weighted image; (d) post-contrast T_1-weighted image. The T_2 map (e) and MTR map (f) are thresholded so that the total abnormality is seen within the red shading. Note that stereological test systems for point counting are overlaid on the enhancing abnormality (d), T_2 total enhancing abnormality (e) and MTR total abnormality (f) with a uniform random position to obtain unbiased volume estimation. Each red cross (+) signifies one test point. Details of the calculation of tumor volume and precision are given in the text.
Reproduced with permission from Gong *et al.*[65] © 2004 The British Institute of Radiology.

19.7 Spectroscopy

Magnetic resonance spectroscopy (MRS) is distinct from other MRI applications that may be applied to cancer because rather than forming an image based upon the distribution and relaxation properties of the nucleus (normally

508   *Chapter 19*

^{1}H) excited in water, MRS looks at the resonance spectra of metabolites other than water. While the interpretation and quantification of these spectra is considerably more complicated than traditional MRI, the ability to gain an insight into the chemical composition of tissue in cancer is very important. *In vivo* investigations of neoplasm can allow for identification of type and grade.[68] Figure 19.6 shows spectra of extracts from a ductal breast tumor and from normal control tissue. In the spectrum from the tumor extract, glycerophosphocholine (GPC) is reduced and phosphocholine (PC) is increased (Figure 19.6, upper section), while glucose is reduced (Figure 19.6, lower section). These signatures are indicative of malignant processes.

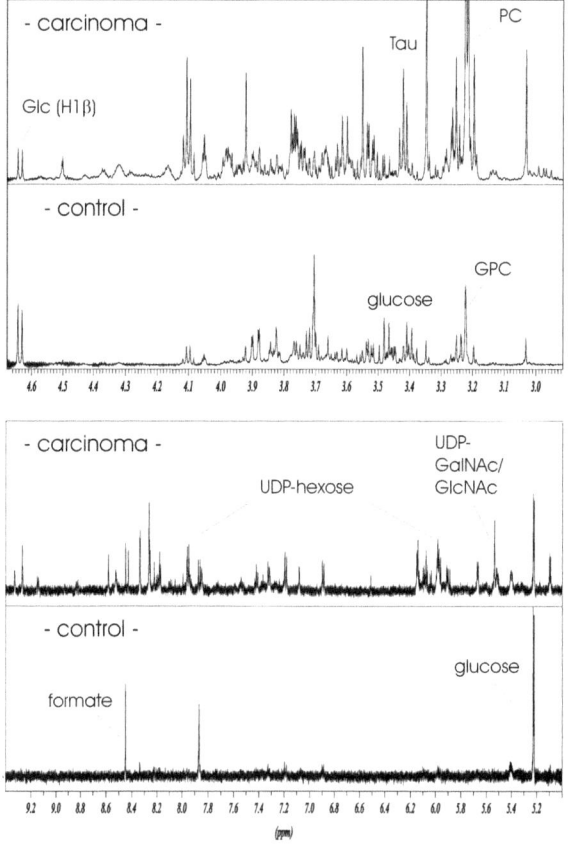

Figure 19.6 Sections of NMR spectra of extracts of healthy and malignant tissues (water-soluble compounds). The upper traces in the dual plots correspond to the spectrum of a ductal carcinoma of grade G3. The lower traces correspond to the spectrum of a control-tissue sample from outside the tumor margin.
Reproduced with permission from Beckonert *et al.*[68] © 2003 John Wiley & Sons, Ltd.

19.7.1 Technical Aspects of Spectroscopy

Technically MRS is challenging because water is the most abundant source of hydrogen nuclei and this signal must be suppressed as it will dwarf the other resonances observed. The paucity of the other nuclei also means that the SNR of these other resonances is very low. This is typically improved by the use of a large (>20) number of signal averages, increasing the duration of the scan and making it less suitable for regions where motion may be significant. One approach that has been employed to overcome this lack of sensitivity is the use of large voxels to extract a sufficient signal for the rarer metabolites at the expense of information about the heterogeneity of the tissue under investigation. Two techniques are used to generate these spectra, Point RESolved Spectroscopy (PRESS) and STimulated Echo Acquisition Mode (STEAM). While these two sequences accomplish the same thing, they are suited to different types of metabolites. The simulated echoes of the STEAM sequence are ideal for the detection of metabolites with very short relaxation times, which are lost in the longer echoes of the PRESS sequence. The longer echoes have made the PRESS sequence the more commonly used as it is less susceptible to motion and diffusion effects during the acquisition than STEAM, making the longer relaxation time metabolites more commonly investigated.

The logical extension of these single voxel techniques is Chemical Shift Imaging (CSI) where the spectrum is phase encoded, allowing for the localization of different spectra within the volume of interest. This is important in making MRS of greater clinical utility given the inhomogeneous nature of tumors. This spatial encoding results in a further lengthening of scanning time. CSI is normally restricted to 2D imaging experiments *in vivo* where it is combined with both PRESS and STEAM.

Quantification of these spectra is an unresolved issue. At the present time quantification of metabolite peaks is carried out by the relative ratio of a peak to another (reference) peak, which is assumed to remain constant and be unaffected by the condition under investigation or any treatment applied. This has the potential to introduce errors into the analysis of such spectra and has led to a drive towards absolute quantification.

19.7.2 Metabolite Images

The ^{1}H spectra observed allow for the quantification of a number of metabolites that are important when investigating different types of cancer. Increased levels of choline are likely to be indicative of a cancerous tumor in the brain as the glial cells, involved in protective and restorative functions in response to cancer, produce choline as they proliferate.[69] Single metabolites are rarely examined in isolation.[70] It has also been observed that in brain tumors an increase in creatine is a marker of cancer as is an increase in *myo*-inositol.[71] In Figure 19.7 spectra acquired at short and long echo times, emphasizing different metabolites, are shown for the brain of a patient with a gliomatosis

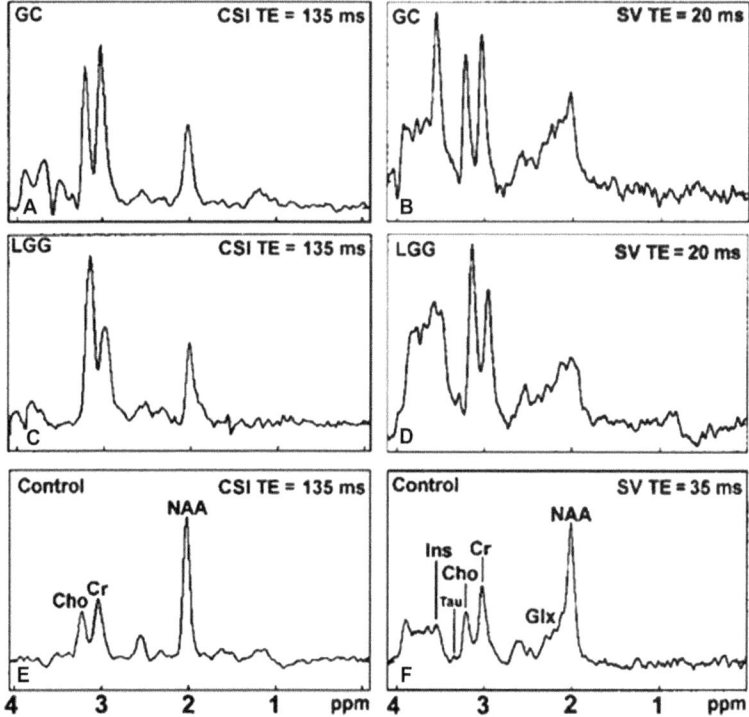

Figure 19.7   Spectra recorded from white matter at long (135 ms) and short (20 ms) echo times in a patient with gliomatosis cerebri (A and B), a patient with a low grade glioma (C and D) and a healthy volunteer (E and F). All spectra recorded at long and short echo times are displayed at the same scale. The tumor process in the two patients harboring neoplasms is characterized by markedly elevated Ins and reduced NAA levels. The Cho level is more elevated and the NAA level more reduced in patients with LGG than in those with GC and healthy volunteers.
Reproduced with permission from Galanaud *et al.*[71] © 2003 JNS Publishing Group.

cerebri (A,B), a patient with a low grade glioma (C,D) and a normal volunteer (E,F). The malignant tissue is characterized by an increase in inositol and a decrease in *N*-acetyl aspartate (NAA). The creatine levels are also increased in the higher grade glioma but are not significantly increased in the lower grade. While these metabolites allow for the differentiation of cancer from normal tissue, other metabolites may provide information on treatment, being related to tumor stage and progression. Increased lactate levels are a sign that the tumor is metabolically active and proliferating; since the tumor can only utilize anaerobic glycolysis this will lead to an increase in lactate.

While the proton or ^{1}H nucleus is the most abundant in the human body and the easiest to investigate using clinical MR systems since it requires no additional hardware, other nuclei are of interest as well when investigating spectra *in vivo*. These include the sodium (^{23}Na) nuclei, which are actively pumped from cells into the extracellular extravascular space, and phosphorous (^{31}P) nuclei, the monitoring of which allows insights into the ATP cycle.[72,73]

19.7.3 Clinical Applications

The majority of cancer applications have focused on the brain where the long acquisition times required are not adversely affected by physiological motion.

In the prostate where the conventional treatment is typically conservative and relies on histopathology to determine whether the tumor is idle or aggressive, MRS has allowed for the improved differentiation of tumor types based on their metabolite profiles and given improved confidence to the choice of treatment.[74] Currently meta studies seek to confirm this.[75] Similarly in the esophagus the choline-to-creatine ratio is different between cancerous and normal tissue, but more importantly the same differences are seen in the precursor condition to cancerous growth, showing that these metabolic changes precede tumor growth making MRS a valuable screening tool.[76]

Changes in lipid concentrations, measurable by MRS, are also very important in cancer diagnosis. Pancreatic cancer and chronic focal pancreatitis are difficult to distinguish and by the time symptoms are present the cancer may be beyond treatment.[77] Differences in MRS spectra, namely less lipid in the non-malignant growths, allow for the investigation of the phospholipid membrane metabolism, an increase in phosphatidylcholine using ^{31}P MRS being observed in cancer.[78] In breast cancer the appearance of a phospho-choline peak is indicative of tumor with levels considerably greater than normal tissue.[79]

While care must be taken in the interpretation of MRS results, the ability to assess the relative metabolite concentrations makes this a powerful tool in the investigation of cancer.

19.8 Non-Proton MRI

Nuclei that contain an odd mass number have net angular momentum and as such are MRI visible. These MRI active nuclei can align their axis of rotation to an applied magnetic field and can, in theory, be used to derive spectra or images. In practice non-proton imaging is limited by the abundance of the magnetic species in the human body and by the size of the magnetic moment. These techniques can be performed on conventional clinical and pre-clinical systems but require tuning a dedicated RF coil. Despite their potential benefits most of these techniques remain in the research arena and have not yet impacted on clinical practice.

Phosphorus spectroscopy is a long-established technique that has been used to investigate tumor biology for several decades. Since ^{31}P has a lower Larmor frequency than hydrogen, higher field strengths are required to achieve acceptable spatial resolution. In addition, phosphorus is much less abundant than hydrogen in the body (by a factor of over 1000-fold), so that SNR is poor. This has largely limited ^{31}P investigations to spectroscopy and often at higher field strength (4 T or higher) in pre-clinical systems. One application of ^{31}P in oncology is to measure tumor adenosine triphosphate (ATP) metabolism, where cellular energetics can be followed in animal models of tumors following chemotherapy.[80]

Sodium (^{23}Na) is MRI visible but is at very low concentrations within the human body. A handful of studies have evaluated the role of ^{23}Na in measuring the extravascular, extracellular space (EES) of tumors, based on the fact that the concentration of Na^+ ions in the EES is regulated to be around 135–145 mM, compared to approximately 5 mM in cells. Although the Na^+ ion concentration in the vasculature is also between 135 and 145 mM (as it is in equilibrium with the EES) the microvasculature is a relatively small compartment in most tumor voxels (around 2–10%) and so, in simple terms, the concentration of Na^+ ions can be used to monitor the change in the size of the EES. This method has been used to monitor the temporal evolution of necrosis and other cell death processes, following various chemotherapy regimens in animal models.[81–84]

There is considerable current interest in imaging carbon with NMR techniques, since carbon is an abundant atom that is a vital component of organic substrates and metabolites. Nearly all carbon exists in the ^{12}C isotope with only a fraction in an MRI visible form. Only 1.1% of carbon exists as ^{13}C, but a technique called dynamic nuclear polarization (DNP) has been described that enables molecules containing ^{13}C to be hyperpolarized (see also Chapter 9). This process results in an increase in sensitivity of more than 10 000-fold, allowing the spatial distribution of an injected labeled molecule to be imaged. Although this process was described nearly 60 years ago,[85] it has only been recently used practically.

In DNP, the injected molecule can be distinguished from ^{13}C-containing metabolites since they resonate at different frequencies. DNP has been applied to endogenous molecules such as CO_2 and bicarbonate ions and from this the pH of a tumor can be mapped (from the Henderson–Hasselbalch equation)[86] (Figure 19.8). Other applications include measuring amino acid metabolism as a biomarker of tumor proliferation[87] and assessing tumor necrosis and monitoring response to targeted therapy.[88] The major limitation of this technique is the short half-life of the hyperpolarized state, which is often less than a minute and therefore requires rapid image acquisition.

Clinical systems based on DNP require production of adequate volumes of sterile hyperpolarized molecules in a robust manner with appropriate quality control and automation.[89] Hyperpolarized ^{13}C imaging is currently undergoing clinical investigation, with preliminary results reported in a proof-of-concept Phase I study in patients with prostate cancer.[90] The widespread translation of this technique into the clinic faces many challenges and is at the moment some way off being ready for widespread clinical use, but DNP has potential to aid diagnosis, to identify disease heterogeneity, to target tumor biopsy, to determine response to therapy and to predict outcome.

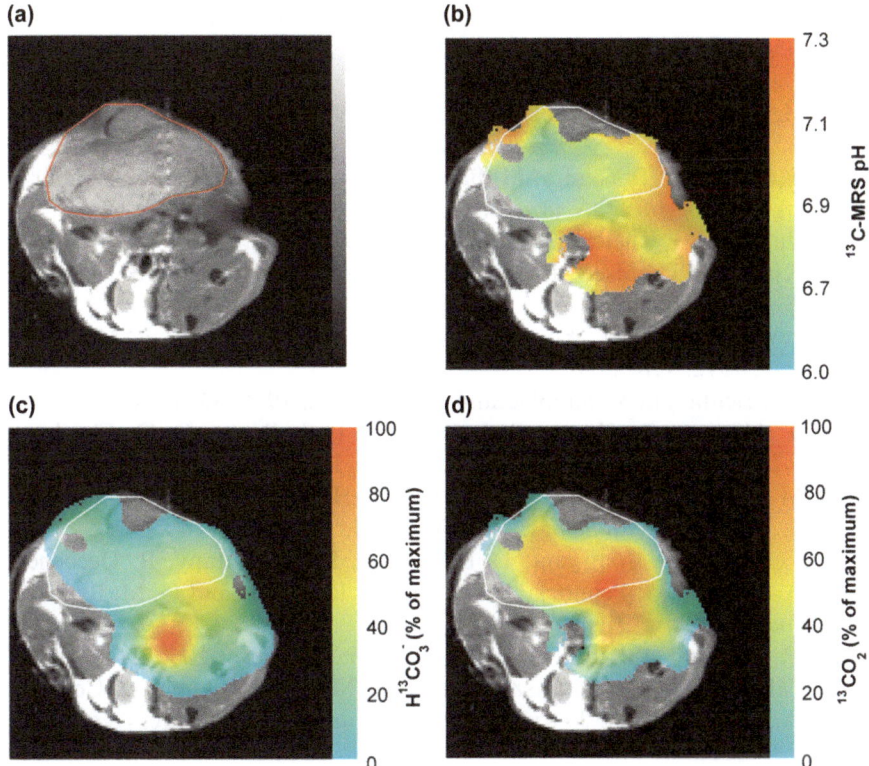

Figure 19.8   Hyperpolarized MRI images of tumor pH following the intravenous injection of hyperpolarized carbon-13 labeled sodium bicarbonate. Axial slices through a mouse with an implanted subcutaneous lymphoma tumor, outlined in red. (a) Grayscale proton MR image. (b) pH map superimposed over the proton image as a false-color map, with an acidic tumor (green). The pH has been derived from the spatial distribution of bicarbonate (c) and carbon dioxide (d) using the Henderson–Hasselbalch equation.
Reproduced with permission from Gallagher *et al.*[86] © 2008 Nature Publishing Group.

Acknowledgements

David Morris is funded by the Manchester Cancer Research Centre.
James O'Connor is a CRUK funded clinical scientist.

References

1. J. C. Waterton and L. Pylkkanen, *Eur. J. Cancer*, 2012, **48**, 409.
2. Biomarkers Definition Working Group, *Clin. Pharmacol. Ther.*, 2001, **69**, 89.

3. C. M. de Bazelaire, G. D. Duhamel, N. M. Rofsky and D. C. Alsop, *Radiology*, 2004, **230**, 652.
4. R. Damadian, *Science*, 1971, **171**, 1151.
5. C. M. Mills, L. E. Crooks, L. Kaufman and M. Brantzawadzki, *Radiology*, 1984, **150**, 87.
6. M. Komiyama, H. Yagura, M. Baba, T. Yasui, A. Hakuba, S. Nishimura and Y. Inoue, *Am. J. Neuroradiol.*, 1987, **8**, 65.
7. M. Just and M. Thelen, *Radiology*, 1988, **169**, 779.
8. L. Kjaer, C. Thomsen, F. Gjerris, B. Mosdal and O. Henriksen, *Acta Radiol.*, 1991, **32**, 498.
9. I. D. Weisman, M. W. Woods, D. Burk, L. H. Bennett and L. R. Maxwell, *Science*, 1972, **178**, 1288.
10. I. C. Kiricuta and V. Simplaceanu, *Cancer Res.*, 1975, **35**, 1164.
11. R. J. Alfidi, J. R. Haaga, S. J. Elyousef, P. J. Bryan, B. D. Fletcher, J. P. Lipuma, S. C. Morrison, B. Kaufman, J. B. Richey, W. S. Hinshaw, D. M. Kramer, H. N. Yeung, A. M. Cohen, H. E. Butler, A. E. Ament and J. M. Lieberman, *Radiology*, 1982, **143**, 175.
12. J. V. Hajnal, D. J. Bryant, L. Kasuboski, P. M. Pattany, B. Decoene, P. D. Lewis, J. M. Pennock, A. Oatridge, I. R. Young and G. M. Bydder, *J. Comput. Assist. Tomogr.*, 1992, **16**, 841.
13. A. R. Padhani, G. Liu, D. M. Koh, T. L. Chenevert, H. C. Thoeny, T. Takahara, A. Dzik-Jurasz, B. D. Ross, M. Van Cauteren, D. Collins, D. A. Hammoud, G. J. Rustin, B. Taouli and P. L. Choyke, *Neoplasia*, 2009, **11**, 102.
14. D. H. Carr, J. Brown, G. M. Bydder, H. J. Weinmann, U. Speck, D. J. Thomas and I. R. Young, *Lancet*, 1984, **1**, 484.
15. E. A. Knopp, S. Cha, G. Johnson, A. Mazumdar, J. G. Golfinos, D. Zagzag, D. C. Miller, P. J. Kelly and I. I. Kricheff, *Radiology*, 1999, **211**, 791.
16. J. P. O'Connor, A. Jackson, G. J. Parker, C. Roberts and G. C. Jayson, *Nat. Rev. Clin. Oncol.*, 2012, **9**, 167.
17. R. R. Edelman, H. Hatabu, E. Tadamura, W. Li and P. V. Prasad, *Nat. Med.*, 1996, **2**, 1236.
18. Y. Ohno, H. Koyama, M. Nogami, D. Takenaka, S. Matsumoto, M. Obara and K. Sugimura, *AJR Am. J. Roentgenol.*, 2008, **190**, W93.
19. J. P. O'Connor, J. H. Naish, G. J. Parker, J. C. Waterton, Y. Watson, G. C. Jayson, G. A. Buonaccorsi, S. Cheung, D. L. Buckley, D. M. McGrath, C. M. West, S. E. Davidson, C. Roberts, S. J. Mills, C. L. Mitchell, L. Hope, N. C. Ton and A. Jackson, *Int. J. Radiat. Oncol. Biol. Phys.*, 2009, **75**, 1209.
20. P. V. Prasad, R. R. Edelman and F. H. Epstein, *Circulation*, 1996, **94**, 3271.
21. J. R. Griffiths, N. J. Taylor, F. A. Howe, M. I. Saunders, S. P. Robinson, P. J. Hoskin, M. E. Powell, M. Thoumine, L. A. Caine and H. Baddeley, *Int. J. Radiat. Oncol. Biol. Phys.*, 1997, **39**, 697.
22. J. P. O'Connor, R. A. Carano, A. R. Clamp, J. Ross, C. C. Ho, A. Jackson, G. J. Parker, C. J. Rose, F. V. Peale, M. Friesenhahn, C. L. Mitchell,

Y. Watson, C. Roberts, L. Hope, S. Cheung, H. B. Reslan, M. A. Go, G. J. Pacheco, X. M. Wu, T. C. Cao, S. Ross, G. A. Buonaccorsi, K. Davies, J. Hasan, P. Thornton, O. del Puerto, N. Ferrara, N. van Bruggen and G. C. Jayson, *Clin. Cancer Res.*, 2009, **15**, 6674.

23. P. M. McSheehy, C. Weidensteiner, C. Cannet, S. Ferretti, D. Laurent, S. Ruetz, M. Stumm and P. R. Allegrini, *Clin. Cancer Res.*, 2010, **16**, 212.
24. R. D. Dortch and M. D. Does, in *Quantitative MRI in Cancer*, ed. T. E. Yankeelov, D. R. Pickens and R. R. Price, Taylor & Francis, New York, 2012, pp. 53–66.
25. A. Jackson and D. Nicholson, in *Dynamic Contrast-enhanced Magnetic Resonance Imaging in Oncology*, ed. A. Jackson, D. L. Buckley and G. J. M. Parker, Springer Verlag, 2005, pp. 239–261.
26. A. Jackson, *Br. J. Radiol.*, 2003, **76**(2), S159.
27. J. L. Boxerman, L. M. Hamberg, B. R. Rosen and R. M. Weisskoff, *Magn. Reson. Med.*, 1995, **34**, 555.
28. A. Jackson, *Br. J. Radiol.*, 2004, **77**(2), S154.
29. L. Ostergaard, D. Chesler, R. Weisskoff, A. Sorensen and B. Rosen, *J. Cereb. Blood Flow Metab.*, 1999, **19**, 690.
30. A. Kassner, D. J. Annesley, X. P. Zhu, K. L. Li, I. D. Kamaly-Asl, Y. Watson and A. Jackson, *J. Magn. Reson. Imaging*, 2000, **11**, 103.
31. N. A. Thacker, M. L. Scott and A. Jackson, *J. Magn. Reson. Imaging*, 2003, **17**, 241.
32. G. Thompson, S. J. Mills, S. M. Stivaros and A. Jackson, *Neuroimaging Clin. N. Am.*, 2010, **20**, 337.
33. G. B. Caseiras, S. Chheang, J. Babb, J. H. Rees, N. Pecerrelli, D. J. Tozer, C. Benton, D. Zagzag, G. Johnson, A. D. Waldman, H. R. Jager and M. Law, *Eur. J. Radiol.*, 2010, **73**, 215.
34. J. P. O'Connor, P. S. Tofts, K. A. Miles, L. M. Parkes, G. Thompson and A. Jackson, *Brit. J. Radiol.*, 2011, **84**, S112.
35. G. J. Parker and P. S. Tofts, *Top. Magn. Reson. Imaging*, 1999, **10**, 130.
36. G. J. Parker, C. Roberts, A. Macdonald, G. A. Buonaccorsi, S. Cheung, D. L. Buckley, A. Jackson, Y. Watson, K. Davies and G. C. Jayson, *Magn. Reson. Med.*, 2006, **56**, 993.
37. J. H. Naish, L. E. Kershaw, D. L. Buckley, A. Jackson, J. C. Waterton and G. J. Parker, *Magn. Reson. Med.*, 2009, **61**, 1507.
38. S. B. Donaldson, C. M. West, S. E. Davidson, B. M. Carrington, G. Hutchison, A. P. Jones, S. P. Sourbron and D. L. Buckley, *Magn. Reson. Med.*, 2010, **63**, 691.
39. E. O. Stejskal and J. E. Tanner, *J. Chem. Phys.*, 1965, **42**, 288.
40. D. Lebihan, E. Breton, D. Lallemand, M. L. Aubin, J. Vignaud and M. Lavaljeantet, *Radiology*, 1988, **168**, 497.
41. T. L. Chenevert, J. A. Brunberg and J. G. Pipe, *Radiology*, 1990, **177**, 401.
42. P. S. Tofts, D. Lloyd, C. A. Clark, G. J. Barker, G. J. M. Parker, P. McConville, C. Baldock and J. M. Pope, *Magn. Reson. Med.*, 2000, **43**, 368.

43. P. L. Hubbard and G. J. M. Parker, in *Diffusion MRI: From Quantitative Measurement to in-vivo Neuroanatomy*, ed. H. Johansen-Berg and T. E. J. Behrens, Elsevier, London, 2009, pp. 354–372.

44. B. D. Ross, B. A. Moffat, T. S. Lawrence, S. K. Mukherji, S. S. Gebarski, D. J. Quint, T. D. Johnson, L. Junck, P. L. Robertson, K. M. Muraszko, Q. Dong, C. R. Meyer, P. H. Bland, P. McConville, H. R. Geng, A. Rehemtulla and T. L. Chenevert, *Mol. Cancer Ther.*, 2003, **2**, 581.

45. A. Dzik-Jurasz, C. Domenig, M. George, J. Wolber, A. Padhani, G. Brown and S. Doran, *Lancet*, 2002, **360**, 307.

46. M. Sumi, N. Sakihama, T. Sumi, M. Morikawa, M. Uetani, H. Kabasawa, K. Shigeno, K. Hayashi, H. Takahashi and T. Nakamura, *Am. J. Neuroradiol.*, 2003, **24**, 1627.

47. Y. Mardor, R. Pfeffer, R. Spiegelmann, Y. Roth, S. E. Maier, O. Nissim, R. Berger, A. Glicksman, J. Baram, A. Orenstein, J. S. Cohen and T. Tichler, *J. Clin. Oncol.*, 2003, **21**, 1094.

48. R. J. Theilmann, R. Borders, T. P. Trouard, G. W. Xia, E. Outwater, J. Ranger-Moore, R. J. Gillies and A. Stopeck, *Neoplasia*, 2004, **6**, 831.

49. M. D. Pickles, P. Gibbs, M. Lowry and L. W. Turnbull, *Magn. Reson. Imaging*, 2006, **24**, 843.

50. T. L. Chenevert, P. E. McKeever and B. D. Ross, *Clin. Cancer Res.*, 1997, **3**, 1457.

51. J.-P. Galons, M. I. Altbach, G. D. Paine-Murrieta, C. W. Taylor and R. J. Gillies, *Neoplasia*, 1999, **1**, 113.

52. Y. Mardor, Y. Roth, Z. Lidar, T. Jonas, R. Pfeffer, S. E. Maier, M. Faibel, D. Nass, M. Hadani, A. Orenstein, J. S. Cohen and Z. Ram, *Cancer Res.*, 2001, **61**, 4971.

53. A. R. Padhani, *Cancer Imaging*, 2005, **5**, 128.

54. G. A. Wright, B. S. Hu and A. Macovski, *J. Magn. Reson. Imaging*, 1991, **1**, 275.

55. R. A. Jones, M. Ries, C. T. Moonen and N. Grenier, *Magn. Reson. Med.*, 2002, **47**, 728.

56. D. M. McGrath, J. H. Naish, J. P. B. O'Connor, C. E. Hutchinson, J. C. Waterton, C. J. Taylor and G. J. M. Parker, *Magn. Reson. Imaging*, 2008, **26**, 221.

57. R. Alonzi, A. R. Padhani, R. J. Maxwell, N. J. Taylor, J. J. Stirling, J. I. Wilson, J. A. d'Arcy, D. J. Collins, M. I. Saunders and P. J. Hoskin, *Br. J. Cancer*, 2009, **100**, 644.

58. D. Zhao, S. Ran, A. Constantinescu, E. W. Hahn and R. P. Mason, *Neoplasia*, 2003, **5**, 308.

59. D. W. Zhao, A. Constantinescu, E. W. Hahn and R. P. Mason, *Int. J. Radiat. Oncol. Biol. Phys.*, 2002, **53**, 744.

60. R. M. Henkelman, G. J. Stanisz and S. J. Graham, *NMR Biomed.*, 2001, **14**, 57.

61. A. Ramani, C. Dalton, D. H. Miller, P. S. Tofts and G. J. Barker, *Magn. Reson. Imaging*, 2002, **20**, 721.

62. M. Cercignani and G. J. Barker, *J. Magn. Reson.*, 2008, **191**, 171.

63. M. H. Pui, *J. Magn. Reson. Imaging*, 2000, **12**, 395.
64. A. Okumura, K. Takenaka, Y. Nishimura, Y. Asano, N. Sakai, K. Kuwata and S. Era, *Neurol. Res.*, 1999, **21**, 250.
65. Q. Y. Gong, P. R. Eldridge, A. R. Brodbelt, M. Garcia-Finana, A. Zaman, B. Jones and N. Roberts, *Br. J. Radiol.*, 2004, **77**, 405.
66. Y. Ito, S. Matsushima, Y. Kinosada, N. Fuwa and Y. Kikuchi, *Nippon Acta Radiol.*, 1997, **57**, 67.
67. R. H. Bonini, D. Zeotti, L. A. Saraiva, C. S. Trad, J. M. Filho, H. H. Carrara, J. M. de Andrade, A. C. Santos and V. F. Muglia, *Magn. Reson. Med.*, 2008, **59**, 1030.
68. O. Beckonert, K. Monnerjahn, U. Bonk and D. Leibfritz, *NMR Biomed.*, 2003, **16**, 1.
69. G. Tedeschi, N. Lundbom, R. Raman, S. Bonavita, J. H. Duyn, J. R. Alger and G. DiChiro, *J. Neurosurg.*, 1997, **87**, 516.
70. N. Shah, A. Sattar, M. Benanti, S. Hollander and L. Cheuck, *J. Am. Osteopath. Assoc.*, 2006, **106**, 23.
71. D. Galanaud, O. Chinot, F. Nicoli, S. Confort-Gouny, Y. Le Fur, M. Barrie-Attarian, J. P. Ranjeva, S. Fuentes, P. Viout, D. Figarella-Branger and P. J. Cozzone, *J. Neurosurg.*, 2003, **98**, 269.
72. F. Arias-Mendoza, G. S. Payne, K. L. Zakian, A. J. Schwarz, M. Stubbs, R. Stoyanova, D. Ballon, F. A. Howe, J. A. Koutcher, M. O. Leach, J. R. Griffiths, A. Heerschap, J. D. Glickson, S. J. Nelson, J. L. Evelhoch, H. C. Charles, T. R. Brown and M. R. S. Cooperative Grp, *NMR Biomed.*, 2006, **19**, 504.
73. R. Ouwerkerk, K. B. Bleich, J. S. Gillen, M. G. Pomper and P. A. Bottomley, *Radiology*, 2003, **227**, 529.
74. P. Swindle, S. McCredie, P. Russell, U. Himmelreich, M. Khadra, C. Lean and C. Mountford, *Radiology*, 2003, **228**, 144.
75. M. Umbehr, L. M. Bachmann, U. Held, T. M. Kessler, T. Sulser, D. Weishaupt, J. Kurhanewicz and J. Steurer, *Eur. Urol.*, 2009, **55**, 575.
76. S. T. Doran, G. L. Falk, R. L. Somorjai, C. L. Lean, U. Himmelreich, J. Philips, P. Russell, B. Dolenko, A. E. Nikulin and C. E. Mountford, *Am. J. Surg.*, 2003, **185**, 232.
77. S. G. Cho, D. H. Lee, K. Y. Lee, H. Ji, K. H. Lee, P. R. Ros and C. H. Suh, *J. Comput. Assist. Tomogr.*, 2005, **29**, 163.
78. J. Solivera, S. Cerdan, J. Maria Pascual, L. Barrios and J. Maria Roda, *NMR Biomed.*, 2009, **22**, 663.
79. R. Katz-Brull, P. T. Lavin and R. E. Lenkinski, *J. Natl Cancer Inst.*, 2002, **94**, 1197.
80. P. M. Winter, H. Poptani and N. Bansal, *Cancer Res.*, 2001, **61**, 2002.
81. V. D. Schepkin, K. C. Lee, K. Kuszpit, M. Muthuswami, T. D. Johnson, T. L. Chenevert, A. Rehemtulla and B. D. Ross, *NMR Biomed.*, 2006, **19**, 1035.
82. V. D. Schepkin, B. D. Ross, T. L. Chenevert, A. Rehemtulla, S. Sharma, M. Kumar and J. Stojanovska, *Magn. Reson. Med.*, 2005, **53**, 85.

83. A. M. Babsky, S. K. Hekmatyar, H. Zhang, J. L. Solomon and N. Bansal, *Neoplasia*, 2005, **7**, 658.
84. R. Sharma, R. P. Kline, E. X. Wu and J. K. Katz, *Cancer Cell Internat.*, 2005, **5**, 26.
85. A. W. Overhauser, *Phys. Rev.*, 1953, **91**, 476.
86. F. A. Gallagher, M. I. Kettunen, S. E. Day, D. E. Hu, J. H. Ardenkjaer-Larsen, R. Zandt, P. R. Jensen, M. Karlsson, K. Golman, M. H. Lerche and K. M. Brindle, *Nature*, 2008, **453**, 940.
87. F. A. Gallagher, M. I. Kettunen, S. E. Day, M. Lerche and K. M. Brindle, *Magn. Reson. Med.*, 2008, **60**, 253.
88. S. E. Bohndiek, M. I. Kettunen, D.-e. Hu, T. H. Witney, B. W. C. Kennedy, F. A. Gallagher and K. M. Brindle, *Mol. Cancer Ther.*, 2010, **9**, 3278.
89. J. H. Ardenkjaer-Larsen, A. M. Leach, N. Clarke, J. Urbahn, D. Anderson and T. W. Skloss, *NMR Biomed.*, 2011, **24**, 927.
90. S. J. Nelson, J. Kurhanewicz, D. B. Vigneron, P. Larson, A. Harzstarck, M. Ferrone, M. van Criekinge, J. Chang, R. Bok, I. Park, G. Reed, L. Carvajal, J. Crane, J. H. Ardenkjaer-Larsen, A. Chen, R. Hurd, L. I. Odegardstuen and J. Tropp, *Proceedings ISMRM*, 2012, **20**, 274.

Abbreviations

^{18}FDG	^{18}fluoro-deoxyglucose
2D	two-dimensional
3D	three-dimensional
3D HCF	three-dimensional high constant flow
5-FC	5-fluorocytosine
5-FU	5-fluorouracil
αSN	α-synuclein
β-gal	β-galactosidase
δ_{iso}	isotropic chemical shift
σ	chemical shielding
σ_{iso}	isotropic chemical shielding
γ	gyromagnetic ratio
τ_c	correlation time
τ_R	molecular reorientational time
ω	Larmor frequency
Aβ	amyloid-β
AD	Alzheimer's disease
ADC	apparent diffusion coefficient
ADF	Amsterdam Density Functional
ADME	absorption, distribution, metabolism and elimination
ADNI	Alzheimer's Disease Neuroimaging Initiative
AFP	adiabatic full passage
AHA	American Heart Association
AIF	arterial input function
AK	arginine kinase
AKI	acute kidney injury
Ala	alanine
ALD	adrenoleukodystrophy

New Developments in NMR No. 2
New Applications of NMR in Drug Discovery and Development
Edited by Leoncio Garrido and Nicolau Beckmann
© The Royal Society of Chemistry 2013 2013
Published by the Royal Society of Chemistry, www.rsc.org

ALS	amyotrophic lateral sclerosis
ALT	alanine aminotransferase
ANOVA-PCA	ANOVA principal component analysis
APD	avalanche photo-diode
API	active pharmaceutical ingredient
ApoB	apolipoprotein B
ApoE	apolipoprotein E
APT	attached proton test
ART	anti-retroviral therapy
Asc	ascorbate
ASL	arterial spin labeling
Asp	aspartate
ATP	adenosine triphosphate
AUIM	ataxin-3 ubiquitin interacting motif
B3LYP	Becke, three-parameter, Lee–Yang–Parr
BAL	broncho-alveolar lavage
Bcl-xL	B-cell lymphoma-extra large
BDPA	1,3-bisdiphenylene-2-phenylallyl
BLI	bioluminescence imaging
BMI	body mass index
BOLD	blood oxygen-level dependent
BPD	bipolar disorder
bpPFGLED	bipolar pulsed field gradient longitudinal eddy-current delay
CA	contrast agent
CB1-R	cannabinoid-1-receptor
CBF	cerebral blood flow
CBIM	Center for Biomedical Imaging
CBV	cerebral blood volume
CCR5	C-C chemokine receptor type 5
CDase	cytosine deaminase
CEST	chemical-exchange saturation transfer
CE-μMRA	contrast-enhanced MR microangiography
CE-MRI	contrast-enhanced MRI
CF	cystic fibrosis
CFA	complete Freund adjuvant
CHB	1-cyano-2-hydroxy-3-butene
Cho	choline
CI2	chymotrypsin inhibitor-2
CK	creatine kinase
CLL	chronic lymphocytic leukemia
CLOUDS	Classification of Unknowns by Density Superposition
CMR	cardiovascular magnetic resonance
CNR	contrast-to-noise ratio
CNS	central nervous system
COPD	chronic obstructive pulmonary disease
COSY	correlation spectroscopy

COT	cancer Osaka thyroid
CP	cross-polarization
CP-HETCOR	cross-polarization heteronuclear correlation
CP-MAS	cross-polarization magic angle spinning
CPMG	Carr–Purcell–Meiboom–Gill
CPP	cell-penetrating peptide
CP-TOSS	cross-polarization total sideband suppression
C_Q	quadrupolar coupling constant
Cr	creatine
CRAMPS	combined rotation and multiple pulse sequence
CRIPT	cross-relaxation induced polarization transfer
CRLB	Cramér–Rao lower bound
CRP	cryogenic probe
CSA	chemical shift anisotropy
CSDE	chemical shift displacement error
CSF	cerebrospinal fluid
CSI	chemical shift imaging
CST	chemical shift tensor
CT	computerized tomography
CTEPH	chronic thromboembolic pulmonary hypertension
CTUC-COSY	constant-time, uniform-sign cross-peak correlation spectroscopy
Cu-ATSM	[^{11}Cu] diacetyl-bis(N4-methylthiosemicarbazone)
CW	continuous wave
CYP3A4	cytochrome P450 3A4
DARR	dipolar assisted rotational resonance
DCE CMR	dynamic contrast-enhanced cardiovascular magnetic resonance
DCE-MRI	dynamic contrast-enhanced magnetic resonance imaging
DFT	density functional theory
dGEMRIC	delayed gadolinium-enhanced MRI of cartilage
DIACEST	diamagnetic chemical exchange saturation transfer
DILI	drug-induced liver injury
DIRECTION	difference of inversion recovery rate with and without target irradiation
DMPK	drug metabolism and pharmacokinetics
DN	diabetic nephropathy
DNP	dynamic nuclear polarization
DOSY	diffusion ordered spectroscopy
Dox	doxycycline
DP-MAS	direct polarization magic-angle spinning
DQ	double-quantum
DQ-BABA	double-quantum back-to-back
DQ-CRAMPS	double-quantum-combined rotation and multiple pulse sequence
DRCE-MRI	dynamic relaxivity contrast-enhanced MRI

DSC	differential scanning calorimetry
DSCE-MRI	dynamic susceptibility contrast-enhanced magnetic resonance imaging
DSM-IV-TR	Diagnostic and Statistical Manual of Mental Disorders (4th edn, text revision)
DTI	diffusion tensor imaging
DTT	dithiothreitol
DUB	deubiquitylase
DUMBO	decoupling using mind boggling optimization
DW	diffusion weighted
DWI	diffusion weighted imaging
E. coli	*Escherichia coli*
EAE	experimental autoimmune/allergic encephalomyelitis
ECG	electrocardiogram
EDTA	ethylenediaminetetraacetic acid
EES	extravascular, extracellular space
EFG	electric field gradient
Egad	(4,7,10-tri(acetic acid)-l-(2-~-galactopyranosylethoxy)-1,4,7,10-tetraaza-cyc1ododecane)gadolinium
EgadMe	1-[2-(β-galactopyranosyloxy)propyl]-4,7,10-tris(carboxy-methyl)-1,4,7,10-tetraazacyclododec-ane) gadolinium(III)
eGFP	enhanced green fluorescent protein
EMT	epithelial to mesenchymal transition
ENaC	epithelium sodium channel
EPFL	École Polytechnique Fédérale de Lausanne
EPI	echo-planar imaging
EpoA	epothilone A
EPR	electron paramagnetic resonance
ESC	embryonic stem cell
ESR	electron spin resonance
FASTMAP	fast, automatic shimming technique by mapping along projections
FBDD	fragment-based drug discovery
fBIRN	functional Bioinformatics Research Network (fBIRN)
FBS	fetal bovine serum
FCD	fixed charge density
FD	Fourier decomposition
FDA	Food and Drug Administration
FDG	fluoro-deoxyglucose
FerrH	heavy subunit of ferritin
FerrL	light subunit of ferritin
FEV1	forced expiratory volume in 1 second
FID	free induction decay
FIREMAT	five π replicated magic angle turning
FITC	fluorescein isothiocyanate
FKBP	FK506 binding protein

FLAIR	fluid attenuated inversion recovery
FLT	[^{18}F]fluorothymidine
FMISO	[^{18}F]fluoromisonidazole
fMRI	functional magnetic resonance imaging
fp-RFDR	finite-pulse RF-driven dipolar recoupling
FRB	fkbp12-rapamycin binding
FSLG	frequency-switched Lee–Goldburg
FS-REDOR	frequency-selective transferred-echo double resonance (REDOR)
FTIR	Fourier transform infrared
FT-IRIS	Fourier transform infrared imaging spectroscopy
FWHM	Full width at half maximum
GABA	γ-aminobutyric acid
GAD	glutamic acid decarboxylase
GAG	glycosaminoglycan
GalP	galactose-H+ symport protein
GB1	B1 domain of streptococcal protein G
GCP	good clinical practice
Gd-EOB-DTPA	gadolinium ethoxybenzyl diethylenetriamine pentaacetic acid (gadoxetate)
GE	gradient echo
GE-EPI	gradient echo-echo planar imaging
GFT	G-matrix Fourier Transform
GGA	generalized gradient approximation
GIP	good imaging practice
GIPAW	gauge-including projector augmented wave
Glc	glucose
GlcNac	*N*-acetylglucosamine
GLM	general linear model
Gln	glutamine
GLP	good laboratory practice
Glu	glutamate
GLUT2	glucose transporter-2
GOF	goodness-of-fit
GPC	glycerophosphocholine
GPCR	G-protein coupled receptor
GPEth	glycerophosphoethanolamine
GRF	Gaussian random fields
GSH	glutathione
GTM	geometric transfer matrix chronic lymphocytic leukemia
h	Planck's constant
ℏ	reduced Planck constant
HAS	human serum albumin
HBA1c	hemoglobin A1c
HD	Huntington's disease

HDL	high-density lipoprotein
HER2	human epidermal growth factor receptor 2
HESI	Health and Environmental Sciences Institute
HETCOR	heteronuclear correlation
HFBR	hollow fiber bioreactor
HIV	human immunodeficiency virus
HIV-1	human immunodeficiency virus-1
HMBC	heteronuclear multiple-bond correlation spectroscopy
HMQC	heteronuclear multiple-quantum coherence
HMQC-J-MAS	heteronuclear multiple-quantum coherence J-coupling magic-angle spinning
HOESY	heteronuclear Overhauser effect spectroscopy
HPLC	high-performance liquid chromatography
HPMC-AS	hydroxypropylmethylcellulose acetate succinate
HRCT	high-resolution computerized tomography
HR-MAS	high-resolution magic-angle spinning
HRPI	High Risk Plaque Initiative
Hrs	hepatocyte growth factor-regulated tyrosine kinase substrate
HS	hypertonic saline
HSA	human serum albumin
HSCT	hematopoietic stem cell transplantation
hSOD1	human copper, zinc superoxide dismutase 1
HSQC	heteronuclear single quantum correlation
HSQC-NOESY	heteronuclear single quantum correlation nuclear Overhauser enhancement spectroscopy
HTS	high-throughput screening
IBD	inflammatory bowel disease
IC	independent component
ICA	independent component analysis
ICH	International Conference on Harmonisation of Technical Requirements for Registration of Pharmaceuticals for Human Use
ID	iron deficient
IDL	intermediate density lipoprotein
IGF-1	insulin-like growth factor-1
IL-1β	interleukin-1β
IMCL	intramyocellular lipid
INADEQUATE	incredible natural abundance double quantum transfer
INEPT	insensitive nuclei enhanced by polarization transfer
INOE	inter-ligand nuclear Overhauser effect
INPHARMA	inter-ligand nuclear Overhauser effect for pharmacophore mapping
IS	iron sufficient
ISIS	image-selected *in vivo* spectroscopy
ITAM	immunoreceptor tyrosine-based activation motif

ITS	insulin-transferrin-sodium selenite
Jak1	Janus kinase-1
JRES	J-resolved
K_D	dissociation constant
k_m	magnetization transfer rate
K^{trans}	contrast transfer coefficient
Lac	lactate
LASER	localization by adiabatic selective refocusing
LC	liquid chromatography
LC-MS	liquid chromatography-mass spectrometry
LDA	local density approximation
LDL	low-density lipoprotein
LGCP	Lee–Goldburg cross-polarization
LGE	late gadolinium enhancement
LOR	line-of-response
LPS	lipopolysaccharide
LRP	lysine-rich protein
LTBS	Liver Toxicity Biomarker Study
LV	left ventricular
MACS	magic angle coil spinning
MANTICORE	Multidisciplinary Approach to Novel Therapies in Cardiology and Oncology Research
MAS	magic angle spinning
MBP	myelin basic protein
MCA	middle cerebral artery
MCI	mild cognition impairment
MCIR	motion-compensated image reconstruction
MDD	major depressive disorder
MELODRAMA	melding of spin-locking and dipolar recovery at the magic angle
MEMRI	manganese-enhanced magnetic resonance imaging
MISSISSIPPI	multiple intense solvent suppression intended for sensitive spectroscopic investigation of protonated proteins instantly
MM	macromolecules
MMP	matrix metalloproteinase
MMP-12	matrix metalloproteinase-12
MMP-2	matrix-metalloproteinase-2
MMP-9	matrix-metalloproteinase-9
MnDPDP	manganese dipyridoxyl diphosphate
Mn-SOD	manganese-superoxide dismutase
MPIO	microparticles of iron oxide
MP-RAGE	magnetization-prepared rapid acquisition gradient echo
MQMAS	multiple-quantum magic angle spinning
MR	magnetic resonance
MRA	magnetic resonance angiography
MRamp	MR signal amplification

MRI	magnetic resonance imaging
MRP2	multidrug resistance associated protein 2
MRS	magnetic resonance spectroscopy
MRSI	MR spectroscopic imaging
MT	magnetization transfer
MTC	magnetization transfer contrast
MTD	maximum tolerated dose
mTOR	mammalian target of rapamycin
MTP	microsomal transfer protein
MTR	magnetization transfer ratio
myo-Ins	*myo*-inositol (*myo*-Ins)
N3	non-parametric, non-uniform intensity normalization method
NAA	*N*-acetyl aspartate
NAAG	*N*-acetylaspartylglutamate
NAFLD	non-alcoholic fatty liver disease
NCAM	neural cell adhesion molecule
NCI	National Cancer Institute
NCTR	National Center for Toxicological Research
NHP	non-human primate
NIH	National Institutes of Health
NIRS	near-infrared spectroscopy
NK	neurokinin
NMDA	*N*-methyl-D-aspartate
NMR	nuclear magnetic resonance
NMR-DOC	nuclear magnetic resonance docking of compounds
NMR-SOLVE	nuclear magnetic resonance structurally oriented library valency engineering
NOE	nuclear Overhauser effect
NOESY	nuclear Overhauser effect spectroscopy
NOI	networks of interest
NSAID	non-steroidal anti-inflammatory drug
OA	osteoarthritis
OAI	Osteoarthritis Initiative
OATP	organic anion transporting polypeptide
OATP1	organic anion transporting polypeptide 1
OFPNPG	2-fluoro-4-nitrophenol-β-D-galactopyranoside
OMRI	Overhauser-enhanced MRI
ONIOM	Our own N-layered Integrated molecular Orbital and Molecular Mechanics
ONPG	*ortho*-nitrophenyl-β-galactoside
O-PLS-DA	orthogonal partial least squares discriminant analysis
OVA	ovalbumin
OVS	outer volume suppression
PAGE	polyacrylamide gel electrophoresis
PAIN	proton-assisted insensitive nuclei

PAP	pulmonary arterial pressure
PAPS	3-phosphoadenosine-5-phosphosulfate
PARACEST	paramagnetic chemical exchange saturation transfer
PASS	phase-adjusted spinning sideband
PBE	Perdew–Burke–Ernzerhof
PC	phosphocholine
PCA	principal components analysis
PCr	phosphocreatine
PCS	pseudocontact shift
PD	pharmacodynamics
PDE	phosphodiester
PDH	pyruvate dehydrogenase
PDSD	proton-driven spin diffusion
PE	phosphoethanolamine
PEG	polyethylene glycol
PEO	polyethylene oxide
PET	positron emission tomography
PEth	phosphoethanolamine
PFC	perfluorocarbon
PFG-SE	pulsed field gradient spin echo
PFG-STE	pulsed field gradient stimulated echo
PFONPG	4-fluoro-2-nitrophenyl-β-D-galactopyranoside
PGA	polyglycolic acid
PHIP	parahydrogen induced polarization
phMRI	pharmacological magnetic resonance imaging
PI	propidium iodide
PI3K	phosphatidylinositol 3-kinase
PK	pharmacokinetic
PK/PD	pharmacokinetic/pharmacodynamic
PLIUS	pulsed low-intensity ultrasound
PLS-DA	partial least squares discriminant analysis
PME	phosphomonoester
PMLG	phase-modulated Lee–Goldburg
PMT	photo-multiplier tube
PRE	paramagnetic relaxation enhancement
PRESS	point-resolved spectroscopy
PROMO	prospective motion correction
PRR	post-reconstruction registration
PSC	percent signal change
PTM	post-translational modification
PUFA	polyunsaturated fatty acid
PVP	polyvinylpyrrolidone
PW91	Perdew–Wang-91
PXRD	powder X-ray diffraction
Q	radiofrequency coil quality factor
QC	quality control

QCPMG	quadrupolar Carr–Purcell–Meiboom–Gill
QIBA	Quantitative Imaging Biomarkers Alliance
QuIC-ConCePT	Quantitative Imaging in Cancer Connecting Cellular Processes with Therapy
R_1	longitudinal relaxation rate
R_2	transverse relaxation rate
R^2	coefficient of correlation (R-square)
RAMRIS	Rheumatoid Arthritis MRI Scoring System
RBC	red blood cell
rCBV	relative cerebral blood volume
RECIST	Response Evaluation Criteria in Solid Tumors
REDOR	rotational-echo double resonance
RF	radiofrequency
RFDR	radiofrequency-driven dipolar recoupling
RFT	random field theory
RHC	right heart catheterization
rmsd	root mean square deviation
RO	receptor occupancy
ROI	region of interest
RSN	resting state network
S. cerevisiae	*Saccharomyces cerevisiae*
SABRE	signal amplification by reversible exchange
SAH	S-adenosylhomocysteine
SALMON	solvent accessibility, ligand binding and mapping of ligand orientation by NMR spectroscopy
SAMP	senescence-accelerated mouse prone
SAP	square anti-prismatic
SAR	structure-activity relationship
SCA	spinocerebellar ataxia
SCA1	spinocerebellar ataxia type 1
SCXRD	single crystal X-ray diffraction
scyllo-Ins	*scyllo*-inositol
SD	standard deviation
SDS-PAGE	sodium dodecyl sulfate polyacrylamide gel electrophoresis
SE	spin echo
SEA-TROSY	solvent exposed amides with TROSY
sGAG	sulfated glycosaminoglycan
sICA	spatial independent component analysis
SIMPSON	simulation program for solid-state NMR spectroscopy
SiPM	silicon photomultiplier
SLAPSTIC	spin labels attached to protein side chains as tool to identify interacting compounds
SLO	streptolysin O
SMILI	screening of small molecule interactor library
SMO	smoothened
SNR	signal-to-noise ratio

SOFAST-HMQC	selective optimized flip-angle short-transient heteronuclear multiple quantum coherence
SPECIAL	spin-echo full-intensity acquired localized
SPECT	single-photon emission computed tomography
SPINAL	small phase incremental alteration
SPIO	superparamagnetic iron oxide
SSFP	steady-state free precession
SSNMR	solid-state nuclear magnetic resonance
ST%	percentile saturation transfer
STAM	signal-transducing adapter molecule
STD	saturation transfer difference
STDD	saturation transfer double difference
STEAM	stimulated-echo acquisition mode
STEMI	ST-segment elevation myocardial infarction
STINT-NMR	structural interactions NMR
STMAS	satellite transition magic angle spinning
STOCSY	statistical total correlation spectroscopy
STOCSYS	statistical total correlation spectroscopy scaling
SUV	standardized uptake value
SVM	support vector machine
SVR	support vector regression
SWCNT	single wall carbon nanotubes
T. thermophilus	*Thermus thermophilus*
T_1	longitudinal relaxation or spin-lattice relaxation time
$T_{1\rho}$	longitudinal relaxation or spin-lattice relaxation time in the rotating frame
T_2	transverse relaxation or spin-spin relaxation time
T_2^*	effective or inhomogenous spin-spin relaxation time
TAC	time-activity curve
Tat	trans-activating transcriptional activator
Tau	taurine
TCA	tricarboxylic acid
TCEP	*tris*(2-carboxyethyl)phosphine
tCr	total creatine
TE	echo time
TEDOR	transferred-echo double resonance
TET	2,2,2-trifluoroethanethiol
Tf	transferrin
Tfr	transferrin-receptor
Tg	glass transition temperature
TGF-β	transformation growth factor-β
tICA	temporal independent component analysis
TINS	target immobilized NMR screening
TMS	tetramethylsilane
TNF-α	tumor necrosis factor-α
TNM	tumor-node-metastasis

TOBSY	total through bond correlation spectroscopy
TOCSY	total correlation spectroscopy
TOF-MRA	time-of-flight magnetic resonance angiography
TOSS	total sideband suppression
TPPM	two-pulse phase modulation
TR	repetition time
tr-NOE	transferred nuclear Overhauser effect
TR-NOESY	transferred nuclear Overhauser effect spectroscopy
TROSY	transverse relaxation optimized spectroscopy
TSAP	twisted square anti-prismatic
tSNR	time-series signal-to-noise ratio
Tβ4	thymosin β4
UCH	ubiquitin C-terminal hydrolase
UDP-GlcNAc	uridine diphospho-N-acetylglucosamine
UIM	ubiquitin-interacting motif
UIP	ubiquitin-interacting protein
USPIO	ultra-small superparamagnetic iron oxide
UTE	ultra-short echo time
VAPOR	variable pulse power and optimized relaxation delay
VOI	volume of interest
VSOP	very small iron oxide particle
WaterLOGSY	water-ligand optimized gradient spectroscopy
WT	wild-type
X. laevis	*Xenopus laevis*
XAFS	X-ray absorption fine structure
XOI	xanthine oxidase inhibitor
ZF-TEDOR	z-filtered transferred-echo double resonance

Subject Index